"Scientific research on the connections between the brain, biology, and social behavior is among the most cutting-edge inquiry taking place in the communication discipline today. Floyd and Weber have gathered the most influential scholars in that field and highlighted their contributions in the *Handbook of Communication Science and Biology*. This significant volume pushes the boundaries of neuroscience and communication science and offers guidance and resources for anyone seeking to understand how the mind and body intersect with human communication behavior. This contemporary text is sure to be valuable to scholars and students alike."

Michael Gazzaniga, *Distinguished Professor of Psychology and Director of the SAGE Center for the Study of the Mind, University of California, Santa Barbara*

"Communication science and biology is one of the latest and most prominent extensions of the communication discipline, founded by a group of young, brilliant, and interdisciplinary-minded scholars who have successfully established a new interest group within the International Communication Association. Floyd and Weber now provide what I am sure will soon become the key volume of this area."

Peter Vorderer, *Professor of Media and Communication Studies; Fellow and former President of the International Communication Association*

THE HANDBOOK OF COMMUNICATION SCIENCE AND BIOLOGY

The Handbook of Communication Science and Biology charts the state of the art in the field, describing relevant areas of communication studies where a biological approach has been successfully applied. The book synthesizes theoretical and empirical development in this area thus far and proposes a roadmap for future research.

As the biological approach to understanding communication has grown, one challenge has been the separate evolution of research focused on media use and effects and research focused on interpersonal and organizational communication, often with little intellectual conversation between the two areas. *The Handbook of Communication Science and Biology* is the only book to bridge the gap between media studies and human communication, spurring new work in both areas of focus.

With contributions from the field's foremost scholars around the globe, this unique book serves as a seminal resource for the training of the current and next generation of communication scientists, and will be of particular interest to media and psychology scholars as well.

Kory Floyd is Professor of Communication and Professor of Psychology at the University of Arizona.

René Weber is Professor of Communication at the University of California Santa Barbara and the Director of UCSB's Media Neuroscience Lab (http://medianeuroscience.org).

The editors' names appear alphabetically to denote equal contributions to this volume.

INTERNATIONAL COMMUNICATION ASSOCIATION (ICA) HANDBOOK SERIES

Robert T. Craig, Series Editor

Selected titles include:

Edited by Jesper Strömbäck and Lynda Lee Kaid—*The Handbook of Election News Coverage Around the World*

Edited by George Cheney, Steve May, and Debashish Munshi—*The Handbook of Communication Ethics*

Edited by Frank Esser and Thomas Hanitzsch—*The Handbook of Comparative Communication Research*

Edited by Howard Giles—*The Handbook of Intergroup Communication*

Edited by Peter Simonson, Janice Peck, Robert T. Craig, and John Jackson—*The Handbook of Communication History*

Edited by Donal Carbaugh—*The Handbook of Communication in Cross-cultural Perspective*

Edited by Bryan C. Taylor and Hamilton Bean—*The Handbook of Communication and Security*

Edited by Karin Wahl-Jorgensen and Thomas Hanitzsch—*The Handbook of Journalism Studies, 2nd edition*

Edited by Kory Floyd and René Weber—*The Handbook of Communication Science and Biology*

For more information on this series, please visit:
www.routledge.com/ICA-Handbook-Series/book-series/ICAHAND.

THE HANDBOOK OF COMMUNICATION SCIENCE AND BIOLOGY

Edited by
Kory Floyd and
René Weber

NEW YORK AND LONDON

First published 2020
by Routledge
52 Vanderbilt Avenue, New York, NY 10017

and by Routledge
2 Park Square, Milton Park, Abingdon, Oxon, OX14 4RN

Routledge is an imprint of the Taylor & Francis Group, an informa business

Library of Congress Cataloging-in-Publication Data
Names: Floyd, Kory, editor. | Weber, René, Dr., editor.
Title: The handbook of communication science and biology / edited by Kory Floyd, René Weber.
Description: New York, NY : Routledge, 2020. | Includes bibliographical references and index.
Identifiers: LCCN 2019054079 (print) | LCCN 2019054080 (ebook) | ISBN 9780815376712 (hardback) | ISBN 9780815376736 (paperback) | ISBN 9781351235587 (ebook)
Subjects: LCSH: Human information processing–Physiological aspects. | Communication–Physiological aspects. | Communication–Psychological aspects. | Mass media–Psychological aspects. | Neurophysiology. | Cognitive neuroscience. | Communication–Research.
Classification: LCC QP396 .H36 2020 (print) | LCC QP396 (ebook) | DDC 612.8–dc23
LC record available at https://lccn.loc.gov/2019054079
LC ebook record available at https://lccn.loc.gov/2019054080

ISBN: 978-0-8153-7671-2 (hbk)
ISBN: 978-0-8153-7673-6 (pbk)
ISBN: 978-1-351-23558-7 (ebk)

Typeset in Times New Roman and Helvetica
by Wearset Ltd, Boldon, Tyne and Wear

To all our fellow scholars ...
... who study the complexity of human communication systems and processes
... who value and uphold the scientific mission in communication research
... who strive for rigorous, replicable, reproducible, and open research
... who support diversity and collaboration across scientific backgrounds and perspectives
... who reject ideology, authority, and status in science, and are foremost committed to finding conditional scientific truth, and
... who are proud to be recognized as communication scientists and represent the scientific mission with knowledge, courage, creativity, innovation, rigor, openness, and integrity

Contents

PART III: COMMUNICATION AND MEDIA NEUROSCIENCE

PART IV: INTERPERSONAL COMMUNICATION

PART V: INTEGRATED PERSPECTIVES

Figures

Tables

Series Editor's Foreword

Robert T. Craig

The Handbook of Communication Science and Biology, edited by Kory Floyd and René Weber, is a comprehensive reference covering the most current theories, methods, and research findings on the biological dimension of human social behavior. The volume's 34 chapters have been contributed by leading researchers in media psychology, interpersonal communication, cognitive neuroscience, evolutionary psychology, psychophysiology, and related fields. By bringing together lines of work in media and interpersonal areas of communication science that have tended to develop separately, this *Handbook* is intended to stimulate cross-fertilization and integrative studies. It will be an indispensable resource on the biology of communication for researchers and advanced students.

While it may be most directly useful to specialists in communication science and biology, the *Handbook* is generally readable by non-specialists and has much to offer those like myself who are broadly interested in ideas about communication or who enjoy the serendipitous insights they encounter when venturing into less familiar fields. For example, scholars in persuasion, rhetoric, and discourse studies may be interested to learn how rhetorically powerful speeches produce synchronized brain responses among audience members (Chapter 8). Others may benefit from learning about the relations between brain function and socially relevant issues like multitasking (Chapter 13), video game addiction (Chapter 14), and media violence (Chapter 15), or about the mutual influence between interpersonal relationships and psychophysiological stress (Chapters 24 and 26). And who would have thought that theoretical formalisms from quantum physics could be applied in communication studies to illuminate phenomena such as cognitive uncertainty (Chapter 33) or processes in which an interpersonal couple simultaneously comes together and comes apart while interacting (Chapter 25)?

The *Handbook* demonstrates throughout that the community of researchers in communication and biology is strongly committed to a scientific approach. Scholars interested in scientific epistemology and methodology in communication research will find numerous exemplars in this, a subfield that seems especially well poised for scientific success. Consider, for example, the interesting and notably good-natured debate between Lang (Chapter 29) and Fisher et al. (Chapter 30), which has a sort of man-bites-dog news value for scientific epistemologists. Whereas scientists are often dogged defenders of their own theories, Annie Lang's engaging chapter explains why she has abandoned her Limited Capacity Model of Motivated Mediated Message Processing (LC4MP) and no longer considers it useful, while her critics respond in the following chapter that the LC4MP has stronger empirical support than ever and continues to be important.

As a communication theorist, I am especially interested in the implications of a biological conception of communication. The case for a distinct biological tradition in communication theory is raised by the editors (Chapter 1) and argued forcefully by Joseph N. Cappella, a pioneer in the field as one of the first communication scholars to study the biological foundations of human interaction (Chapter 2). Chapter 3, by the distinguished evolutionary psychologists John Tooby and Leda Cosmides, implicitly advances the case by developing a fundamental conception of communication based on principles of physics and natural selection. Especially for those of us not already well steeped in evolutionary theory, this extensive and carefully argued chapter will reward close study.

The ICA Handbook Series has published several volumes devoted to themes that cut across established subfields of communication research and constitute what are arguably fundamental dimensions of communication study, such as ethics, history, and culture. Whether or not biology should be counted as a distinct tradition of communication theory (this is not the place for me to engage in that debate), I am convinced that biology is one of those fundamental dimensions that are potentially relevant throughout the field.

About the Editors*

Kory Floyd (Ph.D., University of Arizona) is Professor of Communication and Professor of Psychology at the University of Arizona. His research focuses on the communication of affection in close relationships and on the intersection between interpersonal behavior and health. He has written 16 books and over 100 journal articles and chapters on interpersonal and family communication, nonverbal behavior, and psychophysiology. He is a past editor of *Communication Monographs* and the *Journal of Family Communication*, past chair of the National Communication Association Family Communication Division, and a Fellow of the International Communication Association. His work has been recognized with both the Charles H. Woolbert Award and the Bernard J. Brommel Award from the National Communication Association, the Gerald R. Miller Early Career Achievement Award from the International Association for Relationship Research, and the Distinguished Scholar Award and the B. Aubrey Fisher Award from the Western States Communication Association.

René Weber (M.D., RWTH University Aachen; Ph.D., Berlin University of Technology, Germany) is Professor of Communication at the University of California, Santa Barbara, and the Director of the Media Neuroscience Lab (http://medianeuroscience.org). As a cognitive neuroscientist and communication scholar, his research focuses on cognitive responses to content in traditional and new technology media. He develops social scientific, statistical, and neuroscientific methodology for the rigorous test of media-related theories. He has published more than 120 journal articles and book chapters and has written three books. He currently serves as Associate Editor of *Computational Communication Research*, as Area Editor of the *Proceedings of the National Academy of Sciences*, and has edited special issues in the area of communication science and biology in the *Journal of Media Psychology* and in *Communication Methods and Measures*. He has received numerous awards for his work, including the Distinguished Article Award from the National Communication Association in 2018, the Paul Lazarsfeld Professorship from the University of Vienna in 2017, and the Best Article of the Year Awards from the Advertising Research Foundation in 2016 and from the Association for Education in Journalism and Mass Communication in 2012 and 2018. He also has received more than $2.3 million in funding for communication research. He is a Fellow of the International Communication Association, past chair of ICA's Mass Communication Division, and founder and past chair of ICA's Communication Science and Biology Interest Group.

*The editors' names appear alphabetically to denote equal contributions to this volume.

Contributors

Tamara D. Afifi is Professor in the Department of Communication at the University of California, Santa Barbara. Her research focuses on: (1) how family members communicate when they are stressed and its impact on personal/relational health and biological stress responses; and (2) information regulation. She has published over 130 articles and has received numerous research awards, including the Young Scholar Award from the International Communication Association and the Bernard J. Brommel Award from the National Communication Association. She is a Fellow of the International Communication Association and former editor of *Communication Monographs*.

Bradly Alicea (Ph.D., Michigan State University) is the founder and currently Head Scientist of Orthogonal Research (orthogonal-research.weebly.com) and a Senior Contributor at the OpenWorm Foundation (openworm.org). His interests span the biological, computational, and social sciences. He has published in multiple academic venues, including *Nature Reviews Neuroscience*, *Stem Cells and Development*, *Biosystems*, and *Proceedings of Artificial Life*. Please see Bradly's research website (bradly-alicea.weebly.com) or blog (syntheticdaisies.blogspot.com) for more information.

Grace L. Anderson earned her doctoral degree from the Department of Communication at the University of California, Santa Barbara. She is currently the Director of Institutional Research and Effectiveness at Lewis-Clark State College, Lewiston, ID.

Elisa C. Baek (Ph.D., University of Pennsylvania) is a postdoctoral fellow in the Department of Psychology and Department of Mathematics (joint) at the University of California, Los Angeles. Her research focuses on the neural and psychological drivers behind information sharing, and on the variables that are associated with whether such attempts lead to successful information propagation. Specifically, she is interested in the neural, psychological, and social network variables that distinguish successful communication. Her work is supported by an NSF SBE Postdoctoral Research Fellowship.

Steven Bellman (Ph.D., University of New South Wales) is Professor of Marketing in the Ehrenberg-Bass Institute at the University of South Australia. His research on advertising and the changing media landscape has been published in the *Journal of Marketing*, *Management Science*, and the *Journal of Communication*. He sits on several editorial boards, including the *Journal of Advertising* and *Communication Methods and Measures*.

Margaret Bennett (M.A., University of Missouri, St. Louis) is a doctoral student in the Department of Communication at the University of Connecticut. Her research interests focus on interpersonal communication, with an emphasis on sexual communication as it relates to sexual desire and arousal.

Gary Bente (Ph.D., University of Trier) is Professor in the Department of Communication at Michigan State University and the founder and Scientific Director of the Center for Avatar Research and Immersive Social Media Applications (CARISMA). He has published over 250 articles in major peer-reviewed journals in the fields of psychology, communication science, cognitive science, and neuroscience. His research focuses on human movement and nonverbal communication in real-life encounters and shared virtual environments. He specializes in the use of character animation as a communication research tool and works on the integration of behavioral and neurophysiological measurements to study human experience within Virtual and Augmented Realities.

Paul D. Bolls (Ph.D., Indiana University) is Professor and Associate Director of the Center for Communication Research in the College of Media & Communication at Texas Tech University. His research focuses on applying psychophysiological measures to study how individuals process and are affected by media content and technologies. He is a neuromarketing consultant who has worked with large industry clients and set up labs in both market research companies and academic institutions. His research has been published in top tier journals, such as *Media Psychology*, the *Journal of Media Psychology*, *Health Communication*, and the *Journal of Advertising.*

Justin P. Boren (Ph.D., Arizona State University) is Associate Professor in the Department of Communication at Santa Clara University in Santa Clara, CA. His research focuses on the way that social support networks influence the experience of co-worker stress and burnout. He also studies the work/life balance, co-rumination, organizational culture, psychological and physiological stress, and employee relationships. His work has appeared in top-tier communication and management journals.

Lorraine Borghetti is a Ph.D. candidate in the School of Communication at the Ohio State University and started her National Academies of Sciences, Engineering, and Medicine postdoc position at the Air Force Research Laboratory in late 2019. She is interested in understanding the mechanisms of information processing, inference, and biased decision-making. She uses psychophysical methods in her research.

Michael Brill (Ph.D., University of Würzburg) is a researcher and the Chair of Media Psychology at the University of Würzburg, Germany. His research interests include evolutionary and emotional aspects of media use, with a focus on media technology of movies and video games. He is also interested in observational methods and in the analysis of behavioral structure and time series data.

Gregory A. Bryant (Ph.D., University of California, Santa Cruz) is Professor of Communication at the University of California, Los Angeles. His research focuses primarily on vocal communication from an evolutionary perspective, including laughter, infant-directed speech, vocal emotion, and perceptions of physical strength and body size. Much of his work involves cross-cultural experiments, including small-scale societies. Additionally, he has worked in other areas such as the evolution and perception of music and experimental pragmatics.

Erik P. Bucy (Ph.D., University of Maryland, College Park) holds the Marshall and Sharleen Formby Regents Chair of Strategic Communication in the College of Media and Communication at Texas Tech University. He is the author of *Image Bite Politics: News and the Visual Framing of Elections* (with Maria Elizabeth Grabe, OUP, 2009) and is editor of the *Sourcebook for Political Communication Research: Methods, Measures, and Analytical Techniques* (with R. Lance Holbert, Routledge, 2013). His research interests include visual and nonverbal analysis of political news, normative theories of media and democracy, digital disinformation, and public opinion about the press.

Jerome Busemeyer is Distinguished Professor in Psychological and Brain Sciences, Cognitive Science, and Statistics at Indiana University-Bloomington. He won the Warren Medal in 2015 for outstanding achievement in Experimental Psychology. He is a Fellow of the Society of Experimental Psychologists and the American Academy of Arts and Sciences. His research includes mathematical models of learning and decision-making, and he has formulated a dynamic theory of human decision-making called the decision field theory. More recently, he has been working on a new theory applying quantum probability to model human judgment and decision-making.

Joseph N. Cappella (Ph.D., Michigan State University) is the Gerald R. Miller Professor of Communication at the Annenberg School for Communication, University of Pennsylvania. He has authored more than 175 articles and chapters and four books in the areas of health and political communication, social interaction, nonverbal behavior, media effects, and statistical methods. He is a Fellow of the International Communication Association and its past president, a distinguished scholar of the National Communication Association, and recipient of the B. Aubrey Fisher Mentorship Award.

Jason C. Coronel (Ph.D., University of Illinois, Urbana-Champaign) is Assistant Professor in the School of Communication at the Ohio State University. His research examines how the information environment, in combination with psychological processes, influences political decision-making. His approach is interdisciplinary in nature, bringing together concepts and data from behavioral, psychological, and neurobiological levels of analysis. He uses a combination of techniques including event-related potentials, eye movement monitoring, fMRI, and tDCS to examine the cognitive processes that underlie political decision-making.

Leda Cosmides (Ph.D., Harvard University) is Distinguished Professor of Psychological & Brain Sciences at the University of California, Santa Barbara, where she co-directs the Center for Evolutionary Psychology with John Tooby. Her research helped launch the field of evolutionary psychology. Cosmides has received the AAAS Prize for Behavioral Science Research, the American Psychological Association's Distinguished Scientific Award for an Early Career Contribution to Psychology, and a Lifetime Career Award from the Human Behavior and Evolution Society.

Benjamin E. Custer (M.A., Stanford University) is a doctoral student in communication at the University of Arizona. He has presented his research at the National Communication Association convention and the Western States Communication Association convention, where he received the 2018 Top Debut Paper in the Media Studies Interest Group. His research examines the effects of communication technologies on personal and relational outcomes.

Maria DelGreco (M.A., University of Hawai'i at Mānoa) is a Ph.D. student and Graduate Teaching Assistant in the Department of Communication at the University of Connecticut. Her area of specialization is interpersonal communication, with emphases in gender communication, health communication, and the use of communication to understand and reduce instances of gender-based harassment and assault. Her research has been presented at national and international communication conferences and published in top communication journals.

Amanda Denes (Ph.D., University of California, Santa Barbara) is Associate Professor in the Department of Communication at the University of Connecticut. Her primary area of specialization is interpersonal communication, with emphases in biosocial models of communication, sexual communication, and communication processes related to maintaining successful relationships. Her research has been published in top communication and interdisciplinary journals and she is the co-editor of the *Oxford Handbook of the Physiology of Interpersonal Communication*.

Emily B. Falk (Ph.D., UCLA) is Professor of Communication, Psychology, and Marketing at the University of Pennsylvania, the Director of Penn's Communication Neuroscience Lab, and a Distinguished Fellow of the Annenberg Public Policy Center. Her research examines what makes messages persuasive, why and how ideas spread, and what makes people effective communicators. Falk has received numerous awards, including the ICA Young Scholar Award, SPSP Attitudes and Social Influence Division Early Career Award, a Fulbright grant, the NIH Director's New Innovator Award, and DARPA Young Faculty Award.

Jacob T. Fisher (M.A., Texas Tech University) is a Ph.D. candidate in the Department of Communication at the University of California, Santa Barbara, a researcher in the Media Neuroscience Lab, and a trainee at the National Science Foundation IGERT in Network Science and Big Data. He conducts research at the intersection of neuroscience, media psychology, and big data. His current work leverages functional neuroimaging and behavioral measures to investigate the relationships between various digital technology habits and neurodevelopmental changes in attention networks in the brain.

Douglas A. Gentile (Ph.D., University of Minnesota) is Professor of Psychology at Iowa State University. His research focuses on the positive and negative effects of media on children and adolescents, and he was a pioneer in studying gaming disorder in the United States. Author of three books and over 125 peer-reviewed articles, his work was instrumental in informing the inclusion of Internet Gaming Disorder in the DSM-5 and Gaming Disorder in ICD-11.

Rebecca J. Gilbertson (Ph.D., University of Kentucky) is Assistant Professor in the Department of Psychology at the University of Minnesota-Duluth. Her research interests are directed toward understanding health-related factors that influence neurocognitive function, including substance dependence (particularly alcohol). Behavioral addiction is an emerging area of interest. She has presented her work nationally and internationally and has published in journals including *Alcoholism: Clinical and Experimental Research* and *The Journal of Studies on Alcohol and Drugs*.

Clare Grall (M.A., Michigan State University) is a Ph.D. student of communication science in the Department of Communication, Michigan State University. She studies communication neuroscience with a general focus on how the brain processes narratives as dynamic, complex stimuli. Specifically, she uses fMRI to study the affective processing of narratives.

Kathryn Harrison received her doctorate from the Department of Communication at the University of California, Santa Barbara, in 2019. Her research centers on two primary areas: (1) how people communicate when they are stressed and its effect on personal and relational health; and (2) computer-mediated technology. Her dissertation examined whether virtual reality can be used as a relationship maintenance tool in residential care communities to build communal orientation among residents, subsequently increasing vitality and reducing negative mental health indices (e.g., depressive symptoms, loneliness).

Colin Hesse (Ph.D., Arizona State University) is Associate Professor of Speech Communication at Oregon State University. His research interests include the communication of emotion, health communication, and family communication. Specific topics include alexithymia, affection, and family communication patterns. He has conducted several studies dealing with wet psychophysiology, examining markers such as cortisol, testosterone, oxytocin, and cholesterol.

James M. Honeycutt (Ph.D., University of Illinois) is retired Distinguished Professor of Communication Studies at Louisiana State University and Adjunct Professor at the University of

Texas at Dallas. He has written and edited 10 books and over 100 articles on interpersonal and intercultural communication dealing with cognition, imagined interactions, physiology, personality, marital/family relationships, nonverbal signals, and relational violence.

Frederic R. Hopp (M.A., University of California, Santa Barbara) is a Ph.D. student in the Department of Communication at the University of California, Santa Barbara, a graduate student researcher in UCSB's Media Neuroscience Lab, and an affiliate of the NSF IGERT Network Science and Big Data program. His research integrates computational and neuroscientific approaches to explore large-scale media processes and the effects of morally laden media contents.

Richard Huskey (Ph.D., University of California, Santa Barbara) is Assistant Professor in the Department of Communication at the University of California Davis, principal investigator of the Cognitive Communication Science Lab (https://cogcommscience.com/), and an officer of the Communication Science and Biology Interest Group of the International Communication Association. He specializes in media psychology and cognitive neuroscience. His research investigates how motivation shapes attitudes and behavior.

Lauren Keblusek received a B.A. in philosophy-neuroscience-psychology from Washington University in St. Louis and an M.A. and Ph.D. in communication from the University of California, Santa Barbara. She is currently working as a market research project manager in Seattle, Washington. Her research examines human communication processes in evolutionary context, with a focus on indirect or social forms of aggression, including gossip.

Justin Robert Keene (Ph.D., Indiana University) is Assistant Professor of Creative Media Industries and the Director of the Cognition & Emotion Lab in the College of Media & Communication at Texas Tech University. His research interests are in motivated cognition, dynamic systems, drug prevention messaging, and the effect of individual differences on cognition.

Martin Klasen (Ph.D.) is a psychologist and postdoctoral researcher at Rheinisch-Westfälische Technische Hochschule, Aachen, Germany. His research focuses on the neural basis of aggression and violence, as well as on the neurobiology of psychiatric disorders. He investigates violent media with modern brain imaging techniques, psychological methods, and genetic markers. He is the author of various publications on the effects of violent video games on the brain and on the neurogenetics of aggression.

Annie Lang (Ph.D., University of Wisconsin-Madison) is Distinguished Professor of Communication Science and Cognitive Science at Indiana University. Her research focuses on understanding human communication as an evolved, embedded, dynamic system. She is a past editor of *Media Psychology*, a Fellow of the International Communication Association, and a winner of the AEJMC's Krieghbaum Under-40 Award and ICA's Chaffee Career Productivity Award.

Chelsea A. Lonergan (M.A., University of California, Santa Barbara) is a Ph.D. student in the Department of Communication at the University of California, Santa Barbara, and a researcher in the Media Neuroscience Lab. Her research interests include media selection and effects, sex-based hormonal and biological variables, and cognitive communication science. Her research centers primarily on cognitive sex differences in media selection and preferences. She is also interested in the integration and application of evolutionary theories to the field of communication.

Klaus Mathiak (M.D., Ph.D.) is a mathematician, neuroscientist, and psychiatrist at Rheinisch-Westfälische Technische Hochschule, Aachen, Germany. His research focuses on the neural basis of psychiatric disorders and social communication using innovative imaging methods encompassing natural behavior paradigms and neurofeedback. He has authored more than 150 peer-reviewed publications.

Dar Meshi (Ph.D., Columbia University) is Assistant Professor in the College of Communication Arts and Sciences at Michigan State University. His research focuses on the neural processing of socially communicated information, with a special interest in understanding motivation to use online social networking sites.

Alan C. Mikkelson (Ph.D., Arizona State University) is Professor of Communication Studies at Whitworth University, Spokane. His research interests include relational communication, health communication, and workplace communication. Specific topics include relational decision-making, relational messages, and affectionate communication. He has conducted several studies dealing with physiology and affectionate communication, examining markers such as cortisol and cholesterol.

Eric Novotny (M.A., University at Buffalo, The State University of New York) is a Ph.D. candidate in the Department of Communication at Michigan State University and the current lab manager for the Center for Avatar Research and Immersive Social Media Applications (CAR-ISMA). His primary area of interest is basic nonverbal communication processes with new media, with a focus on interpersonal synchrony. As the lab manager, Eric has integrated these interests with the virtual reality and motion capture systems to examine coordinated movement's capacity to promote rapport as well as reduce outgroup bias.

Ceylan Özdem-Mertens (Ph.D., Vrije Universiteit Brussel) is a postdoctoral research associate in the College of Communication Arts and Sciences at Michigan State University. Her research focuses on social cognition, the representation of the self-perspective on online social networking sites, and virtual interaction. She broadcasts scientific news on her social media accounts to transfer knowledge to society.

Nicholas A. Palomares (Ph.D., University of California, Santa Barbara) is Professor of Communication at the University of California, Davis. His research aims to shed light on the goals people pursue, how they pursue them, and how others gain an understanding of those goals. He also has expertise in gender-based communication. His work focuses on the fundamental processes of communication that transcend contexts, means, and modes of communication.

Robert F. Potter (Ph.D., Indiana University) is Professor of Communication Science in the Media School at Indiana University, Bloomington. His research focuses on the effects of auditory elements on information processing of media and psychophysiological measures as indicators of cognitive and emotional responses to media. His work has been published in *Media Psychology*, *Communication Research*, *Human Communication Research*, *Communication Monographs*, and the *Journal of Advertising*, among others.

Jennifer Priem (Ph.D., Penn State University, 2009) is Associate Professor in the Department of Communication at Wake Forest University, Winston-Salem, NC. Her work focuses on how emotional support messages influence physiological stress and how romantic couples influence each other to co-create supportive interactions in ways that have health consequences. Her work has received many awards from the National Communication Association, including the outstanding dissertation award, as well as an article award and top paper awards.

Ryan D. Rasner (M.A., University of Wyoming) is a doctoral student at Louisiana State University. His research focuses on quantum modeling of relationships, evolutionary communication, cognition, and physiology.

Scott A. Reid (Ph.D., University of Queensland) is Professor of Communication at the University of California, Santa Barbara. His research interests are in group processes and intergroup relations, which he studies using social identity and evolutionary perspectives.

Chris R. Sawyer (Ph.D., University of North Texas) is Professor of Communication Studies in the Bob Schieffer College of Communication at Texas Christian University. His research interests include speech anxiety, instructional communication, and the physiology of human stress during social interaction. Along with James M. Honeycutt and Shaughan A. Keaton, he was an editor and contributor to volume 7 in the Peter Lang Health Communication series, *The Influence of Communication on Physiology and Health*.

Ralf Schmälzle (Ph.D., University of Konstanz) is Assistant Professor of Communication Science in the Department of Communication, Michigan State University. He directs the Neuroscience of Messages (NOM) Lab (nom.cas.msu.edu), which focuses on motivational responses to messages, particularly in the domains of health and risk communication, entertainment, and speeches.

Christin Scholz (Ph.D., University of Pennsylvania) is Assistant Professor of Persuasive Communication at the University of Amsterdam, the Netherlands. She uses neuroimaging to study neural and psychological mechanisms that explain daily health behaviors as a function of diverse information sources such as social influence from peers and professional media interventions. Her work is supported by a Marie Curie Fellowship from the European Community.

Frank Schwab (Ph.D., University of Würzburg) is Professor and Chair of Media Psychology at the University of Würzburg, Germany. His research focuses on emotional and evolutionary aspects of mass and individual media. He is also interested in entertaining content and health communication. He is an expert on the analysis of facial behavior (FACS), time series analysis, emotions in the media, and emotions triggered by media.

Ron Tamborini is Professor of Communication at Michigan State University, where he teaches courses in media psychology and methods of communication inquiry. His research examines processes that govern the selection, valuation, and production of media content. More recent work focuses on the salience of intuitive motivations represented in human cognition and media exemplars, and the reciprocal relationship that shapes these representations.

John Tooby (Ph.D., Harvard University) is Distinguished Professor of Anthropology at the University of California, Santa Barbara, where he co-directs the Center for Evolutionary Psychology with Leda Cosmides. He is best known for his work on the foundations of evolutionary psychology. Tooby received the National Science Foundation's Presidential Young Investigator Award, a J.S. Guggenheim Fellowship, and a Lifetime Career Award from the Human Behavior and Evolution Society.

Benjamin O. Turner (Ph.D., University of California, Santa Barbara) is Assistant Professor of Communication in the Wee Kim Wee School of Communication and Information at Nanyang Technological University, Singapore. His graduate and postgraduate work focused primarily on using functional magnetic resonance imaging to understand processes of human learning and memory, and how those processes differ between individuals, whereas his ongoing research applies a similar approach to study media and communication.

Duane Varan (Ph.D., University of Texas, Austin) is CEO of MediaScience, a leading provider of lab-based audience research. Previously, he was Professor of Audience Research at Murdoch University, Australia. Varan is the recipient of numerous awards, including the Australian Prime Minister's University Teacher of the Year Award.

Alice E. Veksler (Ph.D., Arizona State University) is Associate Professor of Communication and the Director of the Health Communication Research Laboratory at Christopher Newport University in Newport News, VA. Her research focuses on ways that interpersonal interactions affect human health and physiology. She is particularly interested in stress, conflict, and affectionate communication as related to various health processes.

Zheng (Joyce) Wang is Professor in the School of Communication, Translational Data Analytics Institute, Center for Cognitive and Brain Sciences, and Decision Science at the Ohio State University. She directs the Communication and Psychophysiology (CAP) Lab. One of her research foci is to study how people process and use media. Another research focus is to understand contextual influences on decision, cognition, and communication by building new probabilistic and dynamic systems based upon quantum rather than classical probability theory.

Nathan T. Woo (M.A., San Diego State University) is a doctoral student in communication at the University of Arizona. His research focuses on the cognitive and behavioral outcomes of affection and loneliness in interpersonal contexts. He has presented his research at the Western States Communication Association, National Communication Association, and International Communication Association conventions, and he is a reviewer for the *Western Journal of Communication*.

Chance York (Ph.D., Louisiana State University) is Assistant Professor of Mass Communication at Kent State University, OH. His research explores how genes and parental socialization shape communication behavior and media use throughout the life course. His research investigating these topics has been published in journals such as *Computers in Human Behavior*, *Journalism & Mass Communication Quarterly*, and *Communication Monographs*.

Nicole Zamanzadeh received her doctorate from the Department of Communication at the University of California, Santa Barbara in 2019. Her research interests lie at the intersection of technology, interpersonal communication and stress, resilience, and thriving. Her current work investigates the effects of technology use habits such as media multitasking on the stress or resilience of adolescents and family units.

Jinguang Zhang is Assistant Professor in the Department of Communicology, University of Hawai'i at Mānoa. Zhang studies human voice, censorship attitudes, and trust from an evolutionary perspective, and his work has appeared in influential journals such as *Evolution and Human Behavior*, *Evolutionary Psychology*, and *Political Psychology*.

Acknowledgments

A significant volume such as this one comes together only through the investment of many individuals, and it is our pleasure to acknowledge those whose contributions have made this book possible. First and foremost, we are indebted to the fine scholars whose scientific works populate these pages. Our vision for this handbook was to represent a broad range of questions and methods related to the intersection of communication, science, and biology, and we are delighted at the efforts of all our contributors toward helping us realize that goal.

Equally valuable has been the support of ICA Handbooks Series editor Robert Craig, Routledge Communication Studies publisher Felisa Salvago-Keyes, and the entire editorial, production, and marketing staff at Routledge/Taylor & Francis. We have greatly appreciated their diligence throughout the process of assembling this handbook.

Kory's individual acknowledgment: I am grateful for the support of many important colleagues and friends. First, my colleagues at the University of Arizona are a joy to work with and to know, and I appreciate their support every day. I am thankful for many current and former graduate students who have collaborated with me in the study of communication and biology over the last two decades, particularly Justin Boren, Lou Clark, Ben Custer, Dana Dinsmore, Jen Eden, Lisa Farinelli, Mark Generous, Colin Hesse, Bree McEwan, Alan Mikkelson, Perry Pauley, Corey Pavlich, Colter Ray, Sarah Riforgiate, James Stein, Melissa Tafoya, Lisa van Raalte, Alice Veksler, and Nathan Woo.

I am especially grateful to my co-editor, René Weber, for his hard work, insights, and patience during this process. This was our first collaboration and I'm certain it won't be our last. Finally, I would like to thank my husband Brian for his unwavering love and support.

René's individual acknowledgment: Editing a book is much harder than I thought, so I have to start by thanking my awesome graduate students and research assistants at the University of California Santa Barbara's Media Neuroscience Lab, who motivated and supported me throughout the process. My special thanks go to Jacob Fisher, Frederic Hopp, and Chelsea Lonergan who added extra work to their busy schedules so that I could find the time for editing. I am also grateful to my advisors, mentors, and friends who shaped my professional career in important ways at critical moments and who had, and still have, a lasting effect on my scientific perspectives and thinking; in temporal order, they are Jürgen Bortz, Klaus Mathiak, Peter Vorderer, Jennings Bryant, Ron Tamborini, Charles Atkin, John Sherry, Tim Levine, and Michael Gazzaniga. You are my most important role-models; if there were a scholar who could combine all your talents, it would be the perfect scholar. Your example and your guidance throughout the years have been invaluable for me and are reflected in this book.

I also would like to thank the leadership of the International Communication Association for their early support in founding an interest group that has the same title as our book, Communication Science and Biology. My special thanks go to Michael Haley and Laura Sawyer, who always supported this idea and helped me to master all the organizational issues to get the interest group off the ground. My gratitude also includes all past and current members of the ICA Communication Science and Biology Interest Group. I learned so much from all of you over the years. Your work within the communication discipline is inspiring. Your scientific missions and your support ultimately gave rise to this book.

Finally, I thank my co-editor Kory Floyd for sharing his experiences as a book editor with me, for his always supportive communication style, for his patience in getting all work completed, and for his generosity with his time when helping me to meet deadlines. I hope this will not be our last book project together. Thank you all!

Abbreviations

AAAS	American Association for the Advancement of Science
ACC	anterior cingulate cortex
ACS	acute coronary syndrome
ACTH	adrenocorticotropic hormone
AET	Affection Exchange Theory
AI	anterior insula
AMP	Affect misattribution procedures
ANS	autonomic nervous system
API	Application Programming Interface
AR	augmented reality
ARF	Advertising Research Foundation
ASD	Autism spectrum disorder
BAP	Brain-as-Predictor
BCIs	brain-computer interfaces
BDNF	brain-derived neurotrophic factor
BIS	Barratt Impulsiveness Scale
BLC	brain-like creatures
BOLD	blood-oxygen-level dependent
BP	blood pressure
BPM	beats per minute
BSC	biological sensitivity to context
CAIS	complete androgen insensitivity syndrome
CAP	Communication and Psychophysiology
CARISMA	Center for Avatar Research and Immersive Social Media Applications
CIU	compulsive internet use
CMV	Cytomegalovirus
CNS	central nervous system
CRH	corticotropin releasing hormone
CRP	C-reactive protein
CRQ	Cross-recurrence quantification
CRTT	Competitive Reaction Time Task
CRVP	Copy Research Validity Project
CS	corrugator supercilii
dACC	dorsal portions of the anterior cingulate cortex
DAMs	deceptive affectionate messages
DAT	dopamine transporter
dB	decibels
DHCCST	Dynamic Human-Centered Communication Systems Theory
DHEA-S	dehydroepiandrosterone-sulfate
DLPFC	dorsolateral prefrontal cortex
DMN	default mode network

DMPFC	dorsomedial prefrontal cortex
DSD	disorder of sex development
DSM	diathesis-stress model
DST	Dynamic Systems Theory
DZ	dizygotic
EBV	Epstein-Barr virus
EDA	Electrodermal activity
EEA	environment of evolutionary adaptedness
EEA	equal environments assumption
EEG	electroencephalogram
eMFD	extended, Moral Foundation Dictionary
EMG	electromyography
EPMs	evolved psychological mechanisms
ERPs	event-related potentials
ERSP	event-related spectral perturbations
FEF	frontal eye fields
FFA	fusiform face area
fMRI	functional magnetic resonance imaging
fNIRS	functional near-infrared spectroscopy
FPS	first-person shooter
fTCS	functional transcranial Doppler sonography
GLM	General Linear Model
GWAS	Genome-Wide Association Study
HCCST	Human-centered Communication Systems Theory
HCI	human-computer interaction
HPA	hypothalamic-pituitary-adrenal
HPG	hypothalamic-pituitary-gonadal
HR	heart rate
HRA	heart rate acceleration
HRD	heart rate deceleration
HRF	hemodynamic response function
HRV	heart rate variability
HS	hyperscanning
HSR	human stress response
IAD	internet addiction disorder
IAT	Internet Addiction Test
IBI	inter-beat interval
ICA	International Communication Association
ICD-11	International Classification of Diseases
IED	intermittent explosive disorder
IgA	immunoglobulin A
IgE	immunoglobulin E
IgG	immunoglobulin G
IgM	immunoglobulin M
IIs	imagined interactions
IL-1	interleukin-1
IL-1B	interleukin 1-beta
IL-2	interleukin-2

IL-6	interleukin-6
IPS	Interpersonal synchrony
ISC	Intersubject Correlation Analysis
ISFC	inter-subject functional connectivity
JNR	*Journal of Neuroscience Research*
LC3MP	Limited Capacity Model of Mediated Message Processing
LC4MP	Limited Capacity Model of Motivated Mediated Message Processing
LDA	linear discriminant analysis
LIP	lateral intraparietal area
LIWC	Linguistic Inquiry and Word Count
LPC	late positive component or late positive complex
lPFC	lateral prefrontal cortex
LR	likelihood ratio
MAOA	monoamine oxidase A gene
MEA	Motion Energy Analysis
MF-AMP	moral foundations-affect misattribution procedure
MFC	medial frontal cortex
MFD	Moral Foundations Dictionary
MF-LDT	moral foundations lexical decision-making task
MIDUS	Midlife in the United States
MIME	Model of Intuitive Morality and Exemplars
MMI	Media Multitasking Index
M-MIA	moral measure of intuition accessibility
M-MIMB	moral measure of intuitively motivated behavior
MMPFC	middle medial prefrontal cortex
MoNA	Moral Narrative Analyzer
MPFC	medial prefrontal cortex
MR	mixed reality
MRI	magnetic resonance imaging
MVPA	Multivoxel Pattern Analysis
MVT	Marginal Value Theorem
MZ	monozygotic
NAc	nucleus accumbens
NE	norepinephrine
NGF	nerve growth factor
NIH	National Institutes of Health
NK	natural killer
NLP	natural language processing
NOM	Neuroscience of Messages
nSCRs	nonspecific skin conductance responses
NT-3	neurotrophin-3
NT-4	neurotrophin-4
OFH	Optimal Foraging Hypothesis
OO	orbicularis oculi
OR	orienting response
OXTR	oxytocin receptor
PA	post-auricular
PAG	periaqueductal grey

PC	precuneus
PCC	precuneus
PET	positron emission tomography
PFC	prefrontal cortex
PG-YBOCS	pathological gambling adaptation of the Yale-Brown Obsessive-Compulsive Scale
PIU	problematic internet use
PME	perceived message effectiveness
pMFC	posterior portion of the medial prefrontal cortex
PNS	parasympathetic nervous system
PNS	peripheral nervous system
PPI	psychophysiological interaction
PSA	public service announcement
PSAP	Point Subtraction Aggression Paradigm
QC	quantum cognition
ROI	region of interest
RT	response time
SA	sentiment analysis
sAA	salivary α-amylase
SABV	Sex as a Biological Variable
SAM	sympathetic-adrenal-medullary
SC	skin conductance
SCR	skin conductance response
SDT	self-determination theory
SNP	single nucleotide polymorphism
SNS	sympathetic nervous system
SPL	superior parietal lobule
SPS	sensory processing sensitivity
SR	startle response
SST	sexual strategies theory
STF	Synchronization Theory of Flow
STRT	secondary task reaction times
STS	superior temporal sulcus
SVM	support vector machines
TAP	Taylor Aggression Paradigm
TD	typically developing
TMIM	Theory of Motivated Information Management
TMS	transcranial magnetic stimulation
TP	temporal pole
TPJ	temporoparietal junction
TR	time interval
TRRL	theory of resilience and relational load
TSST	Trier Social Stress Test
U&G	Uses and Gratifications
UHV	universal human values
VBM	voxel-based morphometry
VLPF	ventral lateral prefrontal
VMPFC	ventromedial prefrontal cortex

VPA	vaginal pulse amplitude
VPC	verbal person-centeredness
VR	Virtual Reality
VTA	ventral tegmental area
WEIRD	Western, educated, industrialized, rich, and democratic
WHO	World Health Organization
ZM	zygomaticus major

Part I

COMMUNICATION SCIENCE AND BIOLOGY

1

Introduction

Kory Floyd and René Weber

INTRODUCTION

To call human behavior multifactorial is a profound understatement. Communication and social interaction are shaped by virtually innumerable influences, many of which are regularly adjudicated by social scientists. These influences include, among others, culture (Matsumoto & Hwang, 2016), ethnicity (Bernhold & Giles, 2018), socioeconomic status (Betancourt, Brodsky, & Hurt, 2015), parenting style (Moreno-Ruiz, Martínez-Ferrer, & García-Bacete, 2019), sex (Despins, Turkstra, Struchen, & Clark, 2016), gender (Wilhelm, 2018), sexual orientation (Mark, Garcia, & Fisher, 2015), age (Fitzpatrick, Fox, Hoffman, & Dehlendorf, 2016), religion (Croucher, Sommier, Kuchma, & Melnychenko, 2015), mass media (Andersen, Bjarnøe, Albæk, & De Vreese, 2016), social media (Neubaum & Krämer, 2017), and technology (Cantoni & Danowski, 2015; Rogers, 1986).

As potent as these characteristics of social behavior are, however, none can truly be called universal. Many behaviors vary as a function of culture, socioeconomic status, or media exposure, for instance, but not all do (see, e.g., Russell, 2017). Similarly, women and men differ in some ways but not in others (Zell, Krizan, & Teeter, 2015). As Floyd and Afifi (2012) contend, however, perhaps the only universal characteristic—applying without exception to communication behavior—is biology. No verbal or nonverbal message can be either encoded or decoded, that is, without the direct intervention of multiple anatomical and physiological systems. Natural variation in anatomical and physiological systems also affects the qualities of some communication behaviors. For instance, natural pitch variance during phonation varies as a function of laryngeal size and the length and thickness of the vocal folds (Hollien, 2014), whereas decoding accuracy for facial emotion displays varies as a function of developmental stage (Herba & Phillips, 2004), and all of these influences are shaped by natural selection. Conversely, deficits in anatomical and physiological systems—for example, in cases of traumatic brain injury (Rousseaux, Vérigneaux, & Kozlowski, 2010) or sensory impairments such as deafness and blindness (Damen, Janssen, Ruijssenaars, & Schuengel, 2015)—correspond to limitations in social interaction ability that routinely require remediation through education and therapy.

Absolutely none of these observations renders irrelevant the often-substantial effects of proximal environmental factors, such as culture, economic status, religion, or social media use. As the theorizing in Chapter 3 by Tooby and Cosmides so profoundly demonstrates, species' genome and their environment co-evolve! What these assertions do imply is that a science of human social behavior that is willfully or even unintentionally ignorant of the biological and physiological substrates of such behavior will likely overlook strong and enduring influences altogether, or will misattribute their effects, such as by misinterpreting parent-child similarities in genetically heritable behavior patterns as outcomes of parenting style. It also implies that without biological perspectives, many of our communication theories will likely remain proximate theories, or "theoretical frames" that are inductively built over time to explain some specific aspect of communication behavior. Sometimes, these kinds of theories duplicate past theorizing in communication and other disciplines, and as such only marginally increase knowledge and contribute to what is known as additive science. To be clear, there is nothing wrong with attempts to explain how observed phenomena (concepts) relate to each other and predict other phenomena; most of very successful engineering research is devoted to solving "how" questions. But the core scientific mission is to generate explanations of why phenomena relate to each other and why they may exist in the first place, and then demonstrate these relationships with reproducible empirical evidence (Weber, Sherry, & Mathiak, 2008). Without adding a biological perspective to communication science that is firmly grounded in sound principles of natural selection, we believe that it will be difficult, if not impossible, to enhance our theorizing with ultimate *why* explanations.

The Handbook of Communication Science and Biology articulates and synthesizes the theoretical and empirical work identifying the neurological, genetic, anatomical, hormonal, and peripheral physiological antecedents, correlates, and consequences of human social behavior. As described below, this area of focus has burgeoned within the communication discipline in recent years, supporting the development of new theories, the compilation of special issues of communication journals (including three in 2015 alone), and the organization of a new International Communication Association (ICA) interest group with the same name as this book: The ICA Communication Science and Biology Interest Group (www.commscience.org). Since its foundation in 2016 with René Weber as inaugural chair, the membership of this ICA interest group is steadily increasing and it is predicted that the interest group will become an ICA Division within the next two years. The *Handbook* charts the state of the art in the field by summarizing and critiquing the present and potential applications of the biological perspective to a wide range of communicative and social behaviors.

We begin this introductory chapter with a brief history of the biological perspective on social behavior as it has evolved in the communication discipline, noting both its triumphs and its challenges. More details regarding its history and the philosophical and epistemological foundations of biological reasoning in communication science can be found in Chapter 2 by Capella, in this volume, in Weber, Sherry, and Mathiak (2008), and Weber, Eden, Huskey, Mangus, and Falk (2015). We then offer an overview of this *Handbook*, which was organized intentionally to respond to existing challenges and position the biological perspective to advance communication science in substantive ways in the coming years.

COMMUNICATION AND BIOLOGY: A BRIEF HISTORY

Attention to the biological dimensions of social behavior is not particularly novel in allied social sciences. Theory development, coursework, and empirical research on the biological and physiological correlates and effects of social interaction have a longstanding history in the fields of psychophysiology, biological psychiatry, and behavioral medicine, for instance (see Weber, 2015a, 2015b). A focus on the biological causes, correlates, and outcomes of *communication* behavior has been slower to develop, although it was certainly foreshadowed by seminal writings such as Cappella's (1991) article, "The biological origins of automated patterns of human interaction," published in *Communication Theory*.

Despite drawing attention to the benefits of understanding communication behavior from a biological perspective, Cappella's article did not necessarily have an immediate effect on how communication behavior is conceptualized or measured, at least in the communication discipline. Even by the end of the same decade, Craig's (1999) enumeration of the seven principal theoretical traditions in the communication field had not yet reflected any ontologies acknowledging the role of biology, anatomy, or physiology as a cause, correlate, or consequence of communicative behavior.

In the intervening years, however, researchers, including (in alphabetical order) Falk, Lang, Reeves, Weber, and Zillmann in media studies and Afifi, Beatty, Floyd, and McCroskey in interpersonal communication, have constructed compelling theoretical and empirical bases for considering the neurological, genetic, endocrine, immunological, muscular, hematological, and cardiovascular dimensions of social behavior. Although this research may have been characterized as being in its relative infancy at the turn of the 21st century, it has made substantial strides since that time and has established itself as a viable paradigm for understanding communication behavior.

Several pieces of data support this contention. In 2006, for instance, Floyd edited a special issue of the *Journal of Social and Personal Relationships* focused on the biological dimensions of social behavior. Nine years later, in 2015, three journals—*Communication Monographs*, *Communication Methods and Measures*, and the *Journal of Media Psychology*—published special issues on the topic in the same year. Two focused on original research studies and theory (*Communication Monographs*, *82*, edited by Tamara Afifi; and the *Journal of Media Psychology*, *27*(3), edited by René Weber). One double special issue was dedicated to methodological issues around biological communication research (*Communication Methods and Measures*, *9*(1 & 2), edited by René Weber). In addition, prior to the foundation of ICA's Communication Science and Biology Interest Group, there were three ICA pre-conferences (2013 in London, 2014 in Seattle, and 2015 in San Juan) on the topic of "Communication Science—Evolution, Biology, and Brains," chaired and organized by René Weber and a team of junior communication scientists. With more than 100 participants each, attendance was unusually high for an ICA pre-conference. The conference presence together with the representation of research and theory in special issues of the field's flagship journals is a testament to the increasing relevance and visibility of biological dimensions in communication.

As the biological approach to understanding communication has grown, however, one challenge has been that research focused on media use/media effects and research focused on interpersonal/organizational communication have evolved separately and largely independently, often with little intellectual conversation between the two areas. This gap has shrunk in recent years, thanks in part to the aforementioned special issues of journals and pre-conference programming at ICA that have helped to make each area's work familiar to those in the other area. Before now, however, no substantial repository has existed of the discipline's theoretical and empirical contributions to the study of communication and biology that would bridge the gap between media and interpersonal communication behavior and spur new work, including integrative work, in both areas of focus. *The Handbook of Communication Science and Biology* represents a significant step toward filling that emergent need.

This volume comprehensively charts the field and sets the agenda for future scientific research on the biological dimensions of communication behavior. By highlighting the work of the foremost experts in the area, we anticipate that this *Handbook* will serve as a seminal resource for the training of the current and next generation of communication scientists. It will boost the intellectual foundations for the study of communication and biology and provide readers with an extensive overview of the current state of affairs in this evolving field across its diverse epistemological, theoretical, and methodological traditions.

STRUCTURE OF THE BOOK

We elected to organize the book into five principal Parts: (1) Communication Science and Biology; (2) Evolutionary Perspectives; (3) Communication and Media Neuroscience; (4) Interpersonal Communication; and (5) Integrated Perspectives. Chapters in Part I provide a broad introduction for a diverse readership that will approach this book from different theoretical, methodological, and epistemological perspectives. Besides this introductory chapter, Chapter 2 by Cappella maps biological communication research as a growing field of inquiry in its own right, outlines some of its historical trajectories, and explores its unifying themes and potential for the discipline.

Part II focuses on the use of evolutionary reasoning to explore and explain communication behavior. This section begins with Chapter 3 by Tooby and Cosmides, pioneers in the development of evolutionary psychology. Their chapter on natural selection and the nature of communication outlines the foundational principles that provide *why* explanations for the origin and evolution of communication. Following Tooby and Cosmides, in Chapter 4, Reid et al. apply costly signaling theory—a sub-theory of natural selection—to human communication phenomena, and Bryant explores the evolution, structure, and functions of human laughter in Chapter 5. York discusses the use of behavior genetics and twin studies in communication research in Chapter 6, and Brill and Schwab anchor Part II, in Chapter 7, by exploring the proper use of evolutionary reasoning in communication scholarship.

Parts III and IV are dedicated to research on the biological antecedents, correlates, and outcomes of media and interpersonal communication, respectively. In Part III, Schmälzle and Grall start off in Chapter 8 by explaining why and how mediated messages have the capacity to synchronize brain responses, and they introduce a corresponding, innovative analytical paradigm in neuroimaging—inter-subject correlation analysis—which does not require tightly controlled experimental conditions and allows for multi-modal, complex stimulus material, such as a movie or a public service announcement. These new ideas are followed by the neuroscience of persuasion (Chapter 9 by Baek et al.), social media (Chapter 10 by Meshi & Özdem-Mertens), and political knowledge and misinformation (Chapter 11 by Coronel & Bucy). The next two chapters focus on one of the oldest and perhaps the most central concepts in the cognitive sciences: attention. In Chapter 12, Weber and Fisher advance the synchronization theory of flow experiences which conceptualizes flow as a synchronization of specific attentional and reward networks. Fisher and Keene further explore attentional and working memory networks and their involvement in functional and dysfunctional media multitasking behaviors in Chapter 13. After these ideas, the next two chapters introduce two prominent research areas that are traditionally associated with potential negative consequences of media exposure: compulsive media use and the exposure to media violence. In Chapter 14, Gilbertson and Gentile examine the involvement of reward systems during video gaming and possible impairments of cognitive control networks in players' brains. Klasen and Mathiak, in Chapter 15, address research on the exposure to media violence and its underlying neurocognitive mechanisms. The remaining chapters in Part III address primarily methodological innovations or challenges in the area of communication and media neuroscience. In Chapter 16, Alicea describes the many opportunities of using virtual reality in communication neuroscience. In Chapter 17, Turner argues that studying individual differences in processing media messages is not only of advantage at the behavioral level, but also at the neurocognitive level. This chapter demonstrates that there is an amazingly large amount of individual variation at the level of brain responses and that this variation can be predicted by various environmental influences and individually different "cognitive styles." Bellmann and Varan take communication and media neuroscience into the commercial arena, in Chapter 18, and critically discuss issues of measurement in neuromarketing. The final two chapters anchor the section by addressing state-of-the art methodology in peripheral nervous system measurement (Chapter 19 by Potter & Bolls) and brain imaging (Chapter 20 by Hopp & Weber).

Chapters in Part IV address multiple contemporary interpersonal behaviors and constructs that have been studied from a physiological perspective. These include discussions of seminal interpersonal topics, such as emotion and emotional communication (Chapter 21 by Hesse & Mikkelson), affectionate communication (Chapter 22 by Floyd et al.), and social support (Chapter 23 by Veksler et al.). In Chapter 24, Sawyer examines the stress response during communication; Honeycutt and Rasner, in Chapter 25, explore arousal during rumination about conflict; and Afifi and colleagues explore stress and thriving in close relationships in Chapter 26. Denes et al. apply a biological perspective to the study of sexual communication in Chapter 27, and Part IV ends with a methodologically focused chapter on "wet psychophysiology" by Floyd and Hesse (Chapter 28).

Our aim in Part V, the final section, was to transcend the traditional media/interpersonal divide by including chapters that offered a more integrated approach to understanding human communication. The section begins with Lang's description of how a widely used, biologically grounded theoretical model, the Limited Capacity Model of Motivated Mediated Message Processing (LC4MP), evolved into the more contemporary Dynamic Human-Centered Communication Systems Theory (DHCCST). In Chapter 29, Lang explicitly cautions readers against relying on the recently suggested updates of the LC4MP published by Fisher et al. (2018a, 2018b), and argues that the authors of these updates have misstated aspects of the LC4MP and have drawn incorrect conclusions about what it predicts. Thus, we granted Fisher et al. the opportunity to write a short, constructive commentary in which the authors clarify their conclusions and highlight the continued utility of the LC4MP for investigating human communication behavior (Chapter 30). We hope the commentary will help our readers to better understand the academic disagreement. Next, in Chapter 31, Bente and Novotny explore synchronization mechanisms in biological systems and show their application in both mediated-virtual and interpersonal communication environments. In Chapter 32, Lonergan and Palomares examine different conceptualizations and measurements of sex and gender, and Borghetti and colleagues apply principles of quantum cognition to the study of communication behavior in Chapter 33. Part V ends with Chapter 34 by Tamborini and Weber, advancing the model of intuitive morality and exemplars (MIME). As our readers will see, the theorizing in these final chapters spans across the artificial categorization of our discipline into mediated and interpersonal communication.

We believe that this coverage of biologically focused communication science—although not necessarily exhaustive—offers a broad representation of the research currently being conducted in communication and related disciplines. The observant reader may have noticed that an important core area of human communication is woefully missing in our book: a scientific-biological perspective on language. We firmly believe that a book on the science and biology of human communication must contain a chapter on language. We tried as much as possible, but we ultimately were unable to receive this important chapter. We assure our readers that in possible future editions of this book, there will be a comprehensive chapter on language. Nevertheless, we hope that our selection of chapters in this *Handbook* will prompt new questions, spur new conversations, and energize new investigations into the myriad ways in which anatomy, physiology, genetics, and environments intersect with human communication behavior.

NOTE

The editors' names appear alphabetically to denote equal contributions to this volume.

REFERENCES

Afifi, T. D. (Ed.). (2015). Biological and physiological approaches to communication [Special issue]. *Communication Monographs, 82*(1).

Andersen, K., Bjarnøe, C., Albæk, E., & De Vreese, C. H. (2016). How news type matters: Indirect effects of media use on political participation through knowledge and efficacy. *Journal of Media Psychology: Theories, Methods, and Applications, 28*, 111–122.

Bernhold, Q. S., & Giles, H. (2018). Ethnic differences in grandparent–grandchild affectionate communication. *Communication Reports, 31*, 188–202.

Betancourt, L. M., Brodsky, N. L., & Hurt, H. (2015). Socioeconomic (SES) differences in language are evident in female infants at 7 months of age. *Early Human Development, 91*, 719–724.

Cantoni, L., & Danowski, J. A. (2015). *Communication and technology.* Berlin, Germany: De Gruyter.

Cappella, J. N. (1991). The biological origins of automated patterns of human interaction. *Communication Theory, 1*, 4–35.

Craig, R. T. (1999). Communication theory as a field. *Communication Theory, 9*, 119–161.

Croucher, S. M., Sommier, M., Kuchma, A., & Melnychenko, V. (2015). A content analysis of the discourses of "religion" and "spirituality" in communication journals: 2002–2012. *Journal of Communication & Religion, 38*, 42–79.

Damen, S., Janssen, M. J., Ruijssenaars, W. A. J. J. M., & Schuengel, C. (2015). Communication between children with deafness and deafblindness and their social partners: An intersubjective developmental perspective. *International Journal of Disability, Development and Education, 62*, 215–243.

Despins, E. H., Turkstra, L. S., Struchen, M. A., & Clark, A. N. (2016). Sex-based differences in perceived pragmatic communication ability of adults with traumatic brain injury. *Archives of Physical Medicine and Rehabilitation, 97*, S26–S32.

Fisher, J. T., Huskey, R., Keene, J. R., & Weber, R. (2018a). The limited capacity model of motivated mediated message processing: Looking to the future. *Annals of the International Communication Association, 42*(4), 291–315.

Fisher, J. T., Keene, J. R., Huskey, R., & Weber, R. (2018b). The limited capacity model of motivated mediated message processing: Taking stock of the past. *Annals of the International Communication Association, 42*(4), 270–290.

Fitzpatrick, J., Fox, E., Hoffman, A., & Dehlendorf, C. (2016). Differences in social communication about contraception by age and race/ethnicity: Baseline results from a randomized controlled trial. *Contraception, 94*, 411–412.

Floyd, K. (Ed.). (2006). Physiology and human relationships [Special issue]. *Journal of Social and Personal Relationships, 23*(2).

Floyd, K., & Afifi, T. D. (2012). Biological and physiological perspectives on interpersonal communication. In M. L. Knapp & J. A. Daly (Eds.), *The handbook of interpersonal communication* (4th ed., pp. 87–127). Thousand Oaks, CA: Sage.

Herba, C., & Phillips, M. (2004). Development of facial expression recognition from childhood to adolescence: Behavioural and neurological perspectives. *Journal of Child Psychology and Psychiatry, 45*, 1185–1198.

Hollien, H. (2014). Vocal fold dynamics for frequency change. *Journal of Voice, 28*, 395–405.

Mark, K. P., Garcia, J. R., & Fisher, H. E. (2015). Perceived emotional and sexual satisfaction across sexual relationship contexts: Gender and sexual orientation differences and similarities. *The Canadian Journal of Human Sexuality, 24*, 120–130.

Matsumoto, D., & Hwang, H. C. (2016). The cultural bases of nonverbal communication. In D. Matsumoto, H. C. Hwang, & M. G. Frank (Eds.), *APA handbook of nonverbal communication* (pp. 77–101). Washington, DC: American Psychological Association.

Moreno-Ruiz, D., Martínez-Ferrer, B., & García-Bacete, F. (2019). Parenting styles, cyberaggression, and cybervictimization among adolescents. *Computers in Human Behavior, 93*, 252–259.

Neubaum, G., & Krämer, N. C. (2017). Opinion climates in social media: Blending mass and interpersonal communication. *Human Communication Research, 43*, 464–476.

Rogers, E. M. (1986). *Communication technology: The new media in society.* New York, NY: Free Press.

Rousseaux, M., Vérigneaux, C., & Kozlowski, O. (2010). An analysis of communication in conversation after severe traumatic brain injury. *European Journal of Neurology, 17*, 922–929.

Russell, J. A. (2017). Cross-cultural similarities and differences in affective processing and expression. In M. Jeon (Ed.), *Emotions and affect in human factors and human-computer interaction* (pp. 123–141). San Diego, CA: Academic Press.

Weber, R. (2015a). Brain, mind, and media: Neuroscience meets media psychology. *Journal of Media Psychology, 27*, 89–92.

Weber, R. (2015b). Biology and brains—Methodological innovations in communication science: Introduction to the special issue. *Communication Methods and Measures, 9*(1), 1–4.

Weber, R., Eden, A., Huskey, R., Mangus, J. M., & Falk, E. (2015). Bridging media psychology and cognitive neuroscience: Challenges and opportunities. *Journal of Media Psychology, 27*(3), 146–156.

Weber, R., Sherry, J., & Mathiak, K. (2008). The neurophysiological perspective in mass communication research. Theoretical rationale, methods, and applications. In M. J. Beatty, J. C. McCroskey, & K. Floyd (Eds.), *Biological dimensions of communication: Perspectives, methods, and research* (pp. 41–71). Cresskill, NJ: Hampton Press.

Wilhelm, C. (2018). Gender role orientation and gaming behavior revisited: Examining mediated and moderated effects. *Information, Communication & Society, 21*, 224–240.

Zell, E., Krizan, Z., & Teeter, S. R. (2015). Evaluating gender similarities and differences using metasynthesis. *American Psychologist, 70*, 10–20.

2

Building Communication Science through the Blueprints of D. C. Dennett and Robert Pirsig

Joseph N. Cappella

This chapter will be a glimpse of a personal, intellectual journey—my own—in attempting to pursue a version of communication science and scholarship. I will try to lay out the influences that have led to some of my research choices over my career, which some have seen as laying a basis for the study of more biologically based approaches to communication research (although only a portion of my work can be characterized in that way). In sketching some of the choices and decisions I have made, both consciously and subconsciously, I make no brief for the field of communication research as a whole but rather for a set of approaches—biological and behavioral in orientation—that had received short shrift in a field dominated by cultural and social approaches, at least until recently. The core argument that I make here and have made elsewhere (Cappella, 1996) is that comprehensive scholarship and well-grounded explanation about human communication cannot reduce the biological to an afterthought but rather that human communication occurs centrally in and through and to bodies, with all the wet wear that accompanies.

In what follows, I will sketch some of the influences on my own views of communication science, explain how and why biological approaches play into the study of human communication, and illustrate how some of these perspectives influenced the research that my labs have conducted from their earliest days at the University of Wisconsin to the more collaborative work being conducted now at the University of Pennsylvania. Hopefully, some of these comments will offer readers insights into studying communication processes complementary to the more established and conventional works emphasizing social, psychological, and cultural factors. The chapter concludes with the hope and promise of the future of biological approaches to communication research ushering in an era of fresh insights into human communicative behavior whose shelf life is robust.

CRAIG'S META-MODEL

In characterizing the role of biological approaches in human communication theory and research, it makes sense to begin with the big picture. In his influential meta-model of communication theory and the field of communication more broadly, Robert Craig argues that there is no theorization of communicative practice that would be biological in orientation (Craig, 1999). As he puts it, "I am unaware of any unique biological conceptualization of communicative practice itself" (p. 152). He doesn't rule out the possibility that someone could discover or invent a biological theorization of communication but does not hold out that possibility in part because of the way he thinks about communicative practice as core to the seven approaches[1] that make up communication theory as he sees them.

Craig's description of the field has been influential (Craig, 1999, had 800 Google Scholar citations and 130 Web of Science citations by 2015) and long lived (Craig, 2006, 2007, 2015), so his explicit denial of a role for a biologically based approach to communication theory must be understood and put in context. That context makes two core assumptions (Craig, 1999, 2001) about what would constitute an element in Craig's seven approaches: the theory must (1) imply a specific and distinct approach to the *practice* of communication, and (2) offer a language—a theoretical discourse—for constituting and regulating communication in society. There is certainly communication within biological systems, but these communicative exchanges are not communication practices within what is ordinarily meant by society. So, de facto, Craig's meta-model and the description it offers of the field of communication omit biological approaches not because he denies they could be antecedents or consequents of communicative behavior but rather as inappropriate descriptions of communication practice itself. Without putting too fine a point on it, the explicit omission of biologically oriented theories of communication is not equivalent to disallowing predictions and explanations for communication behavior using biologically based concepts and operational techniques. I do not think that Craig would disagree with this characterization, but neither should researchers doing biologically significant work in communication feel slighted by the explicit omission of the biological from Craig's meta-model.

THE EPISTEMOLOGIES OF PIRSIG AND DENNETT

Despite the omission of the biological from Craig's meta-model of the field, there is certainly a place for biological approaches to communication research and theory about communication. This place is not located in communicative practice itself but in the kinds of causal antecedents to communicative expression, and to explanations and predictions about communicative influence. These might be assimilated in some senses into the social psychological traditions of the field as Craig identifies them, but they are certainly beyond the typical psychological and social antecedents and consequences that reside centrally in the social psychological traditions of communication research and theory.

But how can biologically-based variables (e.g., deactivation of occipital cortex) and approaches (e.g., inclusive fitness) be integrated into the domains of communicative practice that Craig identifies as the defining characteristics of communication theory? My answer to this question goes back to the author Robert Pirsig's (1974) novel *Zen and the Art of Motorcycle Maintenance* and the philosopher of mind Daniel Dennett (Dennett, 1971) and eventually his book *The Intentional Stance* (Dennett, 1987).

Motorcycle Maintenance

When I was an assistant professor at the University of Wisconsin, my first doctoral student, Joseph Folger, encouraged me to read *Zen and the Art of Motorcycle Maintenance*. My first reaction was incredulity. "No, my God, motorcycle maintenance? What does he think? I ride a bicycle but I'm not a motorcycle guy." But he was absolutely right. I found the novel brilliant, and along with some early pieces by Dennett, it formed my epistemology about communication research and theory.

Dennett's work came my way through Donald Cushman. Cushman was a debate coach at Michigan State when I was a graduate student but also the person who taught me a lot about the philosophy of science. On one occasion (of many), he stood over me in my office insisting I finish reading an article by D. C. Dennett published in the *Journal of Philosophy* in 1971, telling me over and over how important it was. He was right.

I took some important lessons from Pirsig and from Dennett helping me to define who I was as a scholar and the way in which I wanted to approach research problems. Pirsig's book, rejected by 121 publishers before it was finally published, eventually becoming a bestseller, is a book about a motorcycle trip, but is really a book about the philosophy of science. The narrator takes his son across the country on a motorcycle trip with another couple, the Sutherlands. The Sutherlands are the romantics, and the narrator (some say Pirsig, suggesting the autobiographical nature of the book) is the classicist.

The core question that Pirsig addresses as the group motors across country is motorcycle repair and maintenance and general treatment. Pirsig—who had an old, beloved machine—wanted to understand the motorcycle fully so when it was not working properly, he would be able to fix it. He wanted to be able to take it apart and tweak what needed to be tweaked, so that the motorcycle would work better.

The Sutherlands, on the other hand, could not care less about how their motorcycle worked. They cared only about the beauty of their much more modern machine and wanted only to derive pleasure from the ride it provided. And so, in some senses, what you have is this contrast between the narrator and the Sutherlands. The narrator insists on understanding how the machine works with sufficient depth and breadth of knowledge to allow him to repair it when it fails, as it surely will. Knowledge also frees the narrator from unscrupulous mechanics, who could take advantage of his ignorance. The lesson I took away from the narrator, at least initially, was that to fully understand the workings of his motorcycle he had to be able to fix it by taking it apart and putting it back together successfully. In short, he needed to be able to build and rebuild his motorcycle from scratch in order to ensure that it did what it was designed to do—its potential in an Aristotelian sense.

The Sutherlands, on the other hand, have an appreciation for the beauty, the pleasure, the delight that come from riding the motorcycle they have; the narrator, less so. And it is that juxtaposition that influenced me, because I identified with Pirsig, the narrator. I adopted the view that "I want to know how things work. I want to be able to fix them when they break. I want to make them a little better if they're not doing what they need to do."

The ability to intervene is part of what has driven the way I think about and understand problems, and the kinds of solutions that I try to bring about in handling those problems. But that does create an interesting separation between subject and the object being studied. As you distance yourself, you separate yourself from the object of inquiry as it becomes apart from you. In an absolutely brilliant speech that he gave as president of the International Communication Association (ICA), Klaus Krippendorff described how it is that scientists do science (Krippendorff, 1989). His answer, and that of other philosophers of science (e.g., Polanyi,

1958), is that scientists create the realities they are trying to understand. And you cannot do that by being apart, totally apart. You cannot objectify in that way, you need to be near, you need to be close, you need to appreciate, as the Sutherlands appreciate the character of the world that they were trying to participate in and from which they took pleasure.

So, in the end what I took from Pirsig's novel about scientific problem-solving was the need for both the classic and the romantic view of motorcycles. To understand them fully requires the ability to repair and improve and even to build them from the ground up, but one must also appreciate in a holistic way the beauty and simplicity of their structure and the experience they provide. Both elements are necessary to intervene when the machine breaks or needs to be redesigned for higher performance. In practice, these metaphors for building a machine are really calls for knowledge of causal process acquired through a science of analysis and synthesis accompanied by involvement with the phenomenon via the art of empathy and gestalt appreciation.

How to Play Chess

Because both the classicist and romantic approaches are kinds of construction, but at very different levels of abstraction, which is to be preferred at which stages of inquiry? Here our philosopher of mind—D. C. Dennett—enters the arena. Dennett too addresses problem-solving in science.

Pirsig identified two very broad "stances" toward problem-solving: the classic and the romantic. Dennett was also concerned with scientists' stances toward the problems they were seeking to solve. In his 1971 paper, which later became a full-fledged book manuscript (Dennett, 1987), Dennett argues that one can approach problems in many different ways. The question is not necessarily "which of these approaches is the correct approach?," but rather "which of these approaches works best with the questions that must be answered?" His most famous example focuses on playing chess. His hypothetical supposes that one is playing chess with an entity whose nature is unknown. Is it a computer, a person, an on-line group?

Which stance should you take with respect to an unknown entity? Should you treat it as if it's a person? Well, if you want to play the game and win, then you had best treat whatever you are interacting with as if it has those purposes and goals of a human chess player, such as defending the center of the board and making sure you can castle when the king is in danger. Basically, treat the system as intentional agent. But if you want to improve the quality of play in the entity with which you are playing, treating it as if it were an intentional being is not going to get you anywhere. You are going to need to get into the programming instructions; you are going to need to be able to alter its playing rules and strategies. For example, one will need to write instructions for how to protect the center of the board in a fuller range of contexts. If you want the machine to run faster, if you want the machine to use less energy, or if you want the machine to be able to play against 100 players simultaneously, you need to take a different stance at the level of the microscopic, the almost subatomic, quantum level in some ways. You need to take a physical stance.

Now, which of these stances is right? They're all right. But the issue is, what questions need to be answered? Pirsig's narrator wants to ensure that his machine is operating at its best, to be able to improve the quality of his machine. Then he at least needs to operate at a design stance, in order to be able to get in there and fix what needs fixing. He may not have to operate at the molecular level to make the machine run properly but needs to be able to intervene to make things work.

Motorcycles, Chess, and the Meta-Model

If the expression or effects of communication are at issue, then influencing these outcomes requires causal knowledge at the design and, in some cases, the physical level to even entertain the possibility of intervention. Given that the practices of communication are social as well as psychological, they are still carried out in embodied ways and by beings with bodies. Intervention in communicative practice will necessarily involve more than just communication qua communication but will require data from and about the bodies enacting communication. In Craig's view of communicative paradigms, the biological and bio-behavioral cannot be theories, but in his conception, interventions to modify communicative practices are not primary ends of theoretical work. When intervention and modification are primary ends—the very definition of practical theory—then the unearthing and verification of causal processes are the *sine qua non* of theory and, I would argue, the bio-behavioral antecedents and consequences are significant elements of a full understanding of communicative expressions and effects.

RESEARCH IN BIO-BEHAVIORAL ASPECTS OF COMMUNICATION

The entire volume in which this chapter appears is a testimony to the vibrancy of biological approaches to communication research. Research with which I have been associated has employed bio-behavioral and biological measures and protocols in some cases. In the early years of my research, the focus was on social interaction primarily on basic human behaviors, such as talk and silence sequences in two-person interactions (Cappella, 1980; Cappella & Planalp, 1981), broadening later to other aspects of nonverbal behaviors and how these more automatic elements of face-to-face communication affected behaviors manifested by partners in adult-adult and infant-adult interactions (Cappella, 1981). After these fact patterns were established, I sought to offer a theoretical account of the proximate causal processes involved in the patterns of mutual influence observed (Cappella & Greene, 1982).

From this base my work ventured into a biological and evolutionary account of how the interactions I had been studying came to be in the first place (Cappella, 1991). This was not about proximate causes, but rather evolutionary pressures and processes molding humans' social interaction in certain directions. This foray into evolutionary reasoning (and biological mechanisms more generally) was unprecedented in the field of communication and so required some justification.

I offered the following lines of argument that I believed justified our biological approaches to the study of communication behavior and communication processes (Cappella, 1991, 1996). These included: (1) the incompleteness of descriptions of behavior when they ignore the biological; (2) a reorientation from what it is that differentiates us one from the other to that which unites us in common biological and cross-cultural function, re-emphasizing what we hold in common and what unifies us as a species; (3) communicative processes rooted in core biological systems are likely to be less susceptible to the ephemera of changing history and geography, decreasing the possibility of findings that are temporally, geographically, and historically bound; (4) orienting researchers to reasoning and evidence that are developmental, cross-species, cross-cultural, psychopharmacological, neural, and psychophysiological, broadening—without contradiction—the more common social and cultural, cognitive and symbolic evidence commonly identified as the *sine qua non* of communicative practices; and (5) offering data that are more covert and less under the purposeful (and too often misleading) control of respondents motivated to manage impressions and to adhere to social desirability.

These early forays from interactional behavior patterns to evolutionary accounts of the necessity of certain interactional patterns were generalized to more verbal behaviors indicative of cooperation and its lack (Cappella, 1995) and, more broadly, to the metaphor of genetic propagation as the basis for the cascading of behavioral memes in the political and social world (Cappella, 2002). The mechanisms of gene propagation were simply radically different from other forms of causal effect fitting the notion of what survives and moves through social systems more appropriately than simple causal sequences (Smith, Niederdeppe, Blake, & Cappella, 2013).

Less grandiose works have focused on more specific bio-behavioral sources of data in the service of understanding the effects of various communications engineered for persuasive effect. Eye-tracking research has dealt with visual attention to motivating stimuli such as smoking cues to smokers (Lochbuehler et al., 2017; Sanders-Jackson et al., 2011) and distracting effects of non-congruent texts and images in tobacco control messages (Lochbuehler et al., 2016; Lochbuehler, Wileyto, Mercincavage et al., 2018). Psychophysiological measures have provided covert evidence of attention to smoking cues in anti-smoking PSAs (Kang, Cappella, Strasser, & Lerman, 2009; Strasser et al., 2009). Neural response data have helped us to understand that messages high in sensation value interfere with deep processing of content (Langleben et al., 2009), the impact of strong arguments on neural responses (Wang et al., 2013), and processes of retransmission of news in the real world (Scholz et al., 2017). One study even identified some of the genetic differences in those more responsive to tobacco control messages (Falcone et al., 2011). Some of our work in the world of self-reports of message effectiveness has been carefully validated against bio-behavioral measures to avoid the methodological circularity of validating self-report measures against other self-report measures (Cappella, 2018; Zhao & Cappella, 2016).

Although not all of our research has involved bio-behavioral data or biological or evolutionary explanations, I expect that the "shelf life" of the work in that category will be longest-lived.

OBJECTIONS TO BIOLOGICAL APPROACHES TO COMMUNICATION THEORY

In addition to Craig's principled rejection of the possibility of a biological-based theory of communication, a number of other criticisms and concerns were raised and addressed in Weber, Sherry, and Mathiak (2008), and in several special issues, including in the *Journal of Communication* (Cappella, 1996), the *Journal of Media Psychology* (Weber, 2015a; Weber, Eden, Huskey, Mangus, & Falk, 2015), and *Communication Methods and Measures* (Weber, 2015b). Degler (1991) and Gould (1981) remind us that there have been grotesque political and social persecutions in the name of the biological sciences of human behavior. These histories are often coupled with assumptions that biological accounts are deterministic, ideologically conservative, and irrelevant to the texts that are central to human communication. The ideas that biological causes are innate and that genetic-phenotypic relationships are determinist are both naïve and counter to modern theories of biological and genetic causation (Arnhart, 1995; Masters 1993; Weber et al., 2008). Just because there is a biological disposition to acquire language does not mean that everyone acquires the same language (Chomsky, 1968). Just because there is a biological basis for a behavior does not make that behavior natural and therefore acceptable morally in modern society. This naturalistic fallacy is too simplistic, just as is the imputed determinism of a genetic-evolutionary explanation.

Are these criticisms still true? I think not. The editors of this volume make clear that biological approaches to core questions in communication research have increased, matured, and taken their rightful place alongside the social, cultural, psychological, legal, regulatory, and political approaches to communication scholarship. It should not be otherwise, as social-cultural considerations are still executed by bodies.

A BRIGHT FUTURE

In putting together the anniversary issue of 25 years of *Communication Theory* in 2016, the ten original articles and commentaries gave virtually no space to more biologically inclined theory development. Despite this myopia and the failure of Craig's wide net to allow biologically based theories of communication, there is little doubt that biologically oriented communication research is alive and well in the early decades of the 21st century. Serious, paradigmatic, consequential, well-funded research in communication can be counted in the column of biological work, including neural response (Falk, Cascio, & Coronel, 2015; Falk, O'Donnell et al., 2015; Weber, Westcott-Baker, & Anderson, 2013; Weber et al., 2015b), physiological reactions (Lang, 2014; Lee & Lang, 2015), affection and health (Floyd & Afifi, 2012; Floyd, Hesse, & Generous, 2015), and eye-tracking behaviors (Lochbuehler, Wileyto, Tang et al., 2018). Perhaps the next generations of theoretic work will incorporate rather than ignore the biological in explaining communication processes.

NOTE

1. Craig suggests a set of categories for communication scholarship that include rhetorical, semiotic, phenomenological, cybernetic, sociopsychological, sociocultural, and critical. These "schemes," as he calls them, are distinguished by the problems and definitions they offer, their vocabularies, and their common assumptions (Craig, 1999).

REFERENCES

Arnhart, L. (1995). The new Darwinian naturalism in political theory. *American Political Science Review, 89*, 389–400.
Cappella, J. N. (1980). Talk and silence sequences in informal social conversations II. *Human Communication Research, 6*, 130–145.
Cappella, J. N. (1981). Mutual influence in expressive behavior: Adult and infant-adult dyadic interaction. *Psychological Bulletin, 89*, 101–132.
Cappella, J. N. (1991). The biological origins of automated patterns of human interaction. *Communication Theory, 1*, 4–35.
Cappella, J. N. (1995). An evolutionary psychology of Gricean cooperation. *Journal of Language and Social Psychology, 14*(1–2), 167–181.
Cappella, J. N. (1996). Why biological explanation? *Journal of Communication, 46*(3), 1–4.
Cappella, J. N. (2002). Cynicism and social trust in the new media environment. *Journal of Communication, 52*(1), 229–241.
Cappella, J. N. (2018). Perceived message effectiveness meets the requirements of a reliable, valid, and efficient measure of persuasiveness. *Journal of Communication, 68*(5), 994–997.
Cappella, J. N., & Greene, J. O. (1982). A discrepancy-arousal explanation of mutual influence in expressive behaviors for adult-adult and infant-adult interactions. *Communication Monographs, 49*, 89–114.

Cappella, J. N., & Planalp, S. (1981). Talk and silence sequences in informal conversations III. Interspeaker influence. *Human Communication Research, 7*, 117–132.

Chomsky, N. (1968). *Language and mind.* New York, NY: Harcourt, Brace.

Craig, R. T. (1999). Communication theory as a field. *Communication Theory, 9*, 119–161.

Craig, R. T. (2001). Minding my metamodel, mending Myers. *Communication Theory, 11*, 231–240.

Craig, R. T. (2006). Communication as a practice. In G. J. Shepherd, J. St. John, & T. Striphas (Eds.), *Communication as ...: Perspectives on theory* (pp. 38–47). Thousand Oaks, CA: Sage.

Craig, R. T. (2007). Pragmatism in the field of communication theory. *Communication Theory, 17*(2), 125–145.

Craig, R. T. (2015). The constitutive metamodel: A 16-year review *Communication Theory, 25*(4), 356–374.

Degler, C. N. (1991). *In search of human nature.* New York, NY: Oxford University Press.

Dennett, D. C. (1971). Intentional systems. *The Journal of Philosophy, 68*(4), 87–106.

Dennett, D. C. (1987). *The intentional stance*. Cambridge, MA: MIT Press.

Falcone, M., Jepson, C., Sanborn, P., Cappella, J. N., Lerman, C., & Strasser, A. A. (2011). Association of BDNF and COMT genotypes with cognitive processing of anti-smoking PSAs. *Genes, Brain & Behavior, 10*(8), 862–867.

Falk, E. B., Cascio, C. N., & Coronel, J. C. (2015). Neural prediction of communication-relevant outcomes. *Communication Methods and Measures, 9*(1–2), 30–54.

Falk, E. B., O'Donnell, M. B., Tompson, S., Gonzalez, R., Dal Cin, S., Strecher, V., ... An, L. (2015). Functional brain imaging predicts public health campaign success. *Social Cognitive and Affective Neuroscience, 11*(2), 204–214.

Floyd, K., & Afifi, T. D. (2012). Biological and physiological perspectives on interpersonal communication. In M. L. Knapp & J. A. Daly (Eds.), *The handbook of interpersonal communication* (4th ed., pp. 87–127). Thousand Oaks, CA: Sage.

Floyd, K., Hesse, C., & Generous, M. A. (2015). Affection exchange theory: A bio-evolutionary look at affectionate communication. In D. O. Braithwaite & P. Schrodt (Eds.), *Engaging theories in interpersonal communication: Multiple perspectives* (2nd ed., pp. 303–314). Thousand Oaks, CA: Sage.

Gould, S. J. (1981). *The mismeasure of man.* New York, NY: W. W. Norton.

Kang, Y., Cappella, J. N., Strasser, A., & Lerman, C. (2009). The effect of smoking cues in antismoking advertisements on smoking urge and psychophysiological reactions. *Nicotine and Tobacco Research, 11*(3), 254–261.

Krippendorff, K. (1989). On the ethics of constructing communication. In B. Dervin, L. Grossberg, B. J. O'Keefe, & E. Wartella (Eds.), *Rethinking communication: Paradigm issues* (Vol. 1, pp. 66–96). Newbury Park, CA: Sage.

Lang, A. (Ed.). (2014). *Measuring psychological responses to media messages*. New York, NY: Routledge.

Langleben, D., Loughead, J. W., Ruparel, K., Hakun, J. G., Holloway, M. B., Busch, S. I., ... Lerman, C. (2009). Reduced prefrontal and temporal processing and recall of high "sensation value" ads. *Neuroimage, 46*, 219–225.

Lee, S., & Lang, A. (2015). Redefining media content and structure in terms of available resources: Toward a dynamic human-centric theory of communication. *Communication Research, 42*(5), 599–625.

Lochbuehler, K., Mercincavage, M., Tang, K. Z., Tomlin, C. D., Cappella, J. N., & Strasser, A. A. (2018). Effect of message congruency on attention and recall in pictorial health warning labels. *Tobacco Control, 27*(3), 266–271.

Lochbuehler, K., Tang, K. Z., Souprountchouk, V., Campetti, D., Cappella, J. N., Kozlowski, L. T., & Strasser, A. A. (2016). Using eye-tracking to examine how embedding risk corrective statements improves cigarette risk beliefs: Implications for tobacco regulatory policy. *Drug and Alcohol Dependence, 164*, 97–105.

Lochbuehler, K., Wileyto, E. P., Mercincavage, M., Souprountchouk, V., Burdge, J. Z., Tang, K. Z., ... Strasser, A. A. (2018). Temporal effects of message congruency on attention to and recall of pictorial health warning labels on cigarette packages. *Nicotine & Tobacco Research, 21*(7), 879–886.

Lochbuehler, K., Wileyto, E. P., Tang, K. Z., Mercincavage, M., Cappella, J. N., & Strasser, A. A. (2018). Do current and former cigarette smokers have an attentional bias for e-cigarette cues? *Journal of Psychopharmacology, 32*(3), 316–323.

Masters, R. D. (1993). *Beyond relativism: Science and human values.* Hanover, NH: University Press of New England.

Pirsig, R. M. (1974). *Zen and the art of motorcycle maintenance.* New York, NY: William Morrow.

Polanyi, M. (1958). *Personal knowledge: Towards a post critical epistemology.* Chicago, IL: University of Chicago Press.

Sanders-Jackson, A. N., Cappella, J. N., Linebarger, D. L., Piotrowski, J., O'Keeffe, M., & Strasser, A. A. (2011). Visual attention to anti-smoking PSAs: Smoking cues versus other attention-grabbing features. *Human Communication Research, 37*, 275–292.

Scholz, C., Baek, E. C., O'Donnell, M. B., Kim, H. S., Cappella, J. N., & Falk, E. B. (2017). A neural model of valuation and information virality. *Proceedings of the National Academy of Sciences, 114*(11), 2881–2886.

Smith, K. C., Niederdeppe, J., Blake, K. D., & Cappella, J. N. (2013). Advancing cancer control research in an emerging news media environment. *Journal of the National Cancer Institute Monographs, 47*, 175–181.

Strasser, A. A., Cappella, J. N., Jepson, C., Fishbein, M., Tang, K. Z., Han, E., & Lerman, C. (2009). Experimental evaluation of anti-tobacco PSAs: Effects of message content and format on physiological and behavioral outcomes. *Nicotine and Tobacco Research, 11*(3), 293–230.

Wang, A. L., Loughead, J. W., Strasser, A. A., Ruparel, K., Romer, D. R., Blady, S. J., … Langleben, D. D. (2013). Content matters: Neuroimaging investigation of brain and behavioral impact of televised anti-tobacco public service announcements. *Journal of Neuroscience, 33*, 7420–7427.

Weber, R. (2015a). Brain, mind, and media: Neuroscience meets media psychology. *Journal of Media Psychology, 27*(3), 89–92.

Weber, R. (2015b). Biology and brains—Methodological innovations in communication science: Introduction to the special issue. *Communication Methods and Measures, 9*, 1–4.

Weber, R., Eden, A., Huskey, R., Mangus, J. M., & Falk, E. (2015a). Bridging media psychology and cognitive neuroscience: Challenges and opportunities. *Journal of Media Psychology, 27*(3), 146–156.

Weber, R., Huskey, R., Mangus, J. M., Westcott-Baker, A., & Turner, B. O. (2015b). Neural predictors of message effectiveness during counterarguing in antidrug campaigns. *Communication Monographs, 82*, 4–30.

Weber, R., Sherry, J., & Mathiak, K. (2008). The neurophysiological perspective in mass communication research. Theoretical rationale, methods, and applications. In M. J. Beatty, J. C. McCroskey, & K. Floyd (Eds.), *Biological dimensions of communication: Perspectives, methods, and research* (pp. 41–71). Cresskill, NJ: Hampton Press.

Weber, R., Westcott-Baker, A., & Anderson, G. (2013). A multilevel analysis of antimarijuana public service announcement effectiveness. *Communication Monographs, 80*(3), 302–330.

Zhao, X., & Cappella, J. N. (2016). Perceived argument strength. In D. K. Kim & J. Dearing (Eds.), *Health communication research measures* (pp. 119–126). New York, NY: Peter Lang.

Part II

EVOLUTIONARY PERSPECTIVES

3

Natural Selection and the Nature of Communication

John Tooby and Leda Cosmides

One of the deeply satisfying things about an evolutionary approach is that one can start at the causal foundations of a topic and work from the ground up to build one's way into a systematic map of some particular conceptual territory. This is especially true of communication, many of whose principles emerge so naturally from the theory of natural selection that the process feels almost effortless. The key insights emerge from an examination of the logical and causal interrelationships among the following concepts: entropy, order, replication, natural selection, replicative order, frames of reference, replicative entropy, replicative work, function, evolved design, regulation, cybernetics, computation, information, learning, inference, and finally, communication and meaning.

SOME PHYSICS ESCAPES THE REALM OF ENTROPY

The vast expanses of the universe are overwhelmingly populated by phenomena embodying eddies of ever-increasing disorder, salted with only the rarest of exceptions. The domain of phenomena that biological evolution applies to—life—is exceptional. Life consists of the set of physical systems that cause the assembly of (near-duplicate) physical systems (Dawkins, 1976; Williams, 1966). To qualify as self-replicating, these systems must, in their turn, maintain the same capacity for self-replication in the offspring-replicant as in the parent. This means that some of the components of the regulatory system that constructs descendant physical systems must in some fashion reliably transmit something crucial: the capacity to further replicate the inherited design *into* each new descendant. Such components are, by definition, called the genetic system of the organism. By evolved design, the genetic system conserves and transmits the information necessary to construct descendants, and this with great and necessary accuracy. Nevertheless, through inevitable physics, entropy injects random changes into this otherwise conserved, design-replicating information as it passes from generation to generation. (*Mutation* is nothing technical: It is simply the Latinate synonym for *change*.)

Hence physics poses an unsparing problem: How is the existence of life forms at all consistent with a physical world that is pervaded by a tendency to move randomly over time from the far rarer, less probable, more organized states (should they come to exist) to the far more probable, far less organized arrangements? Even more puzzling, how can populations of organisms often move uphill against entropy toward more ordered configurations, as they clearly have? Plants and animals flare out as remarkable departures from the rest of the physically normal entities, such as Kuiper Belt objects, solar convection cells, lunar impact craters, particle clouds driven by Martian wind storms, gamma ray bursts scouring the surfaces of planets, and rippling planetary auroras. What sets all organisms—from ash trees and buckthorn to kestrels and radiolarians—apart from all other expressions of the physical universe is that their designs are spectacularly unlikely arrangements of highly calibrated relationships. Life forms reliably achieve inconceivably highly ordered states—and specifically, order that is highly functional with respect to replication. We will scrutinize what *functional* means in this framework, after a longer look at the question posed by the omnipresence of entropy.

It is true that neither our planet nor its organisms are closed systems, because these systems allow interactions between their internal elements and the environment. Thus, thermodynamic entropy can still increase globally (consistent with the second law of thermodynamics) while (sometimes) decreasing locally in populations of organisms lucky enough not to starve or go extinct. Being open systems permits, but does not explain, the high levels of organization and energy concentration found and deployed in life. Nevertheless, as highly ordered physical systems, organisms should tend to slide rapidly back toward a state of high disorder, because disordered states are so much more probable than ones consistent with survival (much less reproduction).

As the physicist Erwin Schrödinger wrote in his book *What Is Life?* (1944/1992), "It is by avoiding the rapid decay into the inert state that an organism appears so enigmatic" (pp. 69–70). Indeed, each organism does decay into an inert state, but instead of doing so in the milliseconds that chemistry makes reasonable, this decay is fought off in an evolutionarily near-perfected rearguard action for days, weeks, months, or years. From one perspective we should all be extinct, but enough lineages escape from terminal entropy through the operation of the defining property of life: replication. Replication paired with entropy itself in the form of mutation (design copying errors) jointly interact to provide the platform out of which the entire four-billion-year tower of elaborating life fountains upwards toward the farthest reaches of high order.

So how is this very high order achieved? Given that random physical processes introduce differences that modify the otherwise conserved design of these physical systems, it is inevitable that at least some will change the operation of the replicating systems in a way that is material to their exact methods of self-reproduction, and hence to their rate of reproduction in a given environment. By far the most probable outcome is that random changes to conserved design would cause the system's ability to replicate to degrade (a case of harmful mutation). This is because successfully carrying out replication requires immensely well-organized, and hence improbable, causal machinery carrying out a dizzying array of exacting processes. If design degradation were all that happened, then replicators (however they came into existence) would decay into increasingly noisy and drifting chains of parents and descendants, until each chain terminated through an inability to complete all the necessary cause-and-effect steps of an offspring assembly.

Given opportunities provided by a large enough set of successful replications, a small residual subset of these changes in inherited design will be accidentally generated that improves the physical system's replication-promoting organization—that is, that constitutes increased replicative order. What does this mean? Improved design features are defined by having the consequence of multiplying themselves over generations. This happens because the favorable design change interacts with relevant, cross-generationally enduring properties of the world in a systematic, repeating, nonrandom method to cause net increases (positive feedback) in the subsequent frequency of their genetic basis. As these modified, better-designed organisms cause themselves to become more numerous, they increase the order to be found in the world—one mystery we are trying to explain. Conversely, design features that cause net decreases in their gene frequencies become less numerous and eventually disappear (negative feedback on bad design), which can also (potentially) increase the order in the world, as material shifts from constituting the less well-organized to being utilized by the better-organized. The better-ordered design self-amplifies and displaces the less-well-ordered design. It is exactly this attribute of physically self-caused differential replication of the better-designed that densely populates the world with (in Darwin's awed phrase) "endless forms most beautiful and most wonderful"—forms manifesting high *replicative order*, whose lineages often tend to become even better organized over long stretches of cross-generational time.

NATURAL SELECTION AND REPLICATIVE FUNCTIONAL ORDER

Natural selection drives the escape from entropy into the world of functional replicative order. It is essential to be clear on precisely what specific type of natural order evolution uniquely produces, namely, *functional replicative order*. One could imagine the world manifesting many types of things that might seem to embody complex functionality—lakes of wine, hedges of bread, watches synchronized with the orbital period of Titan, mountains in the form of houses complete with electricity, air conditioning, and running water. But, these are so inconceivably improbable that natural physical processes (as opposed to intelligently directed processes) obviously never produce them. More importantly, if we use an intuitive human framing derived from our evolved theory of mind system, *functional* means having properties that assist an agent in reaching its valued goal. Using this definition, nothing can be objectively functional per se, absent a frame of reference that specifies an agent and goal. Lakes of wine would be useless to a crater or a whale. However, in the domain of self-replicating physical systems, there are refracted conceptual counterparts corresponding to *agent* and *goal*, namely, *organism* and *replication*. These license a special, biologically restrictive conceptual analog to the intuitive concept of function. Making these substitutions, we can see that the only kind of organized functionality evolution produces (at rates higher than chance) is *replicative functional order*, that is, functionality to be found in sets of characteristics in an *organism* whose interactions contribute (through however many intermediate steps) to systematically causing enhanced *reproduction* (in a specific lineage) over multiple generations.

Because reproduction-promoting characteristics are the only (naturally occurring) characteristics that cause their own increase in frequency, they constitute the only (naturally occurring) kind of functionality that proliferates through time and space. Obviously, intelligent organisms such as humans can artificially produce functional systems, because they constitute agents with valued goals—and predicting the behavior of this animate world is why we evolved our intuitive concepts of *agent* and *goal*. This process is physically self-executing and requires no actual goals, agents, or values, although the evolutionary process might eventually produce them as computational elements in some lineages. This emphatically does not mean that organisms are, in reality, agents (although some may be), much less agents whose represented or teleological goal is replication. It means only that organisms are assemblages of devices whose functional organization was tailored by selection based on whether, under typical ancestral conditions, they caused replication. Moths fly into flames because their organization implements its outputs, not because they are seeking reproduction. It is a seductive and common error, even among biologists, to think of organisms as agents with the goal of reproduction, instead of considering the actual cause-and-effect design of their behavior control machinery. We are designed by evolution to (mis)interpret organisms in terms of our evolved concepts of *agent* and *goal* because (1) humans (and some other species) sometimes operate partly as agents with consciously represented goals; (2) the behavior of many more organisms can be partly predicted by projecting this interpretation to them; (3) the fact that humans make material arrangements (such as tools), and behavioral arrangements (such as a hunt) to pursue valued goals allows the evolved concept *goal* (*function, purpose*) to help provide a useful explanatory framework for them, using what Dennett calls the intentional stance (Dennett, 1987).

Understanding the centrality of replicative functional order leads to a very stark, logically spare framework, fundamentally free of all of the rich, life-like, animate, and mental features we implicitly are inclined to imbue the world with. From a human perspective, organization that causes an increase in the probability of successful gene propagation (the multiplication of some molecular sequences over others) is a very bizarre, humanly meaningless, and restricted kind of functionality. Yet, it is the only complex functionality ("adaptations," "mechanisms," "useful" traits, "beneficial" traits, "favored" traits, etc.) that will non-arbitrarily characterize systems found spontaneously occurring in the natural world. Adaptations or devices that manifest replicative functional order execute *replicative work* for the organism—that is, they tend to move the organism along mechanistic paths that bring it closer to a life history of realized reproduction. The various sub-theories of natural selection, such as kin selection, sexual selection, foraging theory, the theory of animal conflict, and theories of cooperation and communication, can be used to understand why the architectures of organisms have the designs that they do. Replicative functional order is the only kind of functional order that is objectively present in undomesticated organisms (i.e., caused to be present in organisms).

Replicative order can be conceptually benchmarked through considering the matrix of physically and developmentally possible changes in the design of the organism and analyzing the change each would make in the rate of replication for the design given the prevailing conditions. As used here, *design* simply means the replicative order in the organism built and conserved over generations by natural selection, together with its systematic by-products. Greater effective design for reproduction (given a specified environment) constitutes higher replicative order, and reduced design for reproduction represents lower replicative order.

REPLICATIVE ORDER AND FRAME OF REFERENCE

Functional replicative order is a special kind of objective physical order, and is different from other kinds of order. Its recognition allows us to understand the causal architectures of organisms in a powerful new way that link together observed design features to evolved function, enduring environmental conditions, and ancestral event populations. It has been by bringing the causal framework centered on replication into focus that the door to understanding the engineering principles underlying the naturally selected architectures of organisms has been opened, including their systems of communication. On all different scales, organisms necessarily consist of suites of interlocking mechanistic devices, each organized according to some cause-and-effect arrangement designed (tailored) by the feedback processes inherent in natural selection to bring about outcomes which, taken in combination with the outputs of the other devices, increase the probability of a lifespan of successful replication. We can understand the reason that each of these devices (adaptations) has the characteristics it does by understanding how these characteristics interact functionally to cause the solution of their associated adaptive problem.

These design features make replicative functional sense, and understanding their replicative function—how they solve their adaptive problem—makes sense of why their architecture has the organization that it does. That is why inventorying and breaking down the set of evolutionarily recurrent tasks each species must solve, over a life-history of successful reproduction, is central to understanding a species. Like a particularly shaped key fitting into a particular lock, it is the highly ordered mesh between adaptive problem (including task environment) and adaptive problem-solver (adaptation) that allows us to recognize and understand the functional organization of the architecture.

George Williams (1966) called observations of these improbably well-ordered relationships between the design features and the adaptive problem *evidence of special design* that the system was built by natural selection for solving the adaptive problem. (The evidence is particularly compelling when the design features of the adaptation are predicted in advance, and are then discovered based on those predictions. This happens, for example, when the predictions lead to the construction of experiments to detect the existence of previously unknown mechanisms and design features; e.g., Kurzban, Tooby, & Cosmides, 2001; Lieberman, Tooby, & Cosmides, 2007; Sell, Tooby, & Cosmides, 2009.) Like all scientific claims, the claim that something is an adaptation involves an analysis of probability: whether the reliably developing features of an organism constitute a mechanistic adaptive solution to an adaptive problem that is too good to have arisen by chance (too well-coordinated with the problem-space, too effective, too precise, too economical, too reliable, etc.; Williams, 1966).

For example, the water-soluble protein crystallin is found in the cornea and the lens of the eye and has the extraordinarily rare property—for a protein—of increasing the refractive index while not obstructing light (Jester, 2008). Nearly all possible proteins are opaque. Not obstructing light is the lock and crystallin is the key. An extremely long search through ancestor-descendant protein sequences along a decreasing gradient of light obstruction would have been necessary to evolutionarily discover such a transparent protein out of the dense jungle of far worse alternatives.

This coordination represents a specific kind of objective physical order—replicative order—traceable to the hill-climbing process of encounters of replicated subcomponents and their variants with ancestral cause-and-effect event populations. It is an objective physical property, but a property of a particular species or lineage. That is, an increase in the replicative order of one species can represent a decrease in the replicative order of another—such as host and parasite, or predator and prey. Or they can both increase, as in symbiotic relationships. The point is, replicative order is specific to a frame of reference defined by the design of an organism in its environment.

Organisms exploit the distinct and clashing kinds of orders inherent in different physical frames of reference, including replicative order and replicative entropy. Order, replicative or not, must always be defined with respect to a frame of reference. The frame of reference (e.g., books spatially arranged alphabetically by author) categorizes the state of the described system (the physical order of the books on the bookcase) to the degree that it is in correspondence with the kind of order defined by the reference frame. Books on a bookcase can be organized by author, title, topic, or—in the case of an eccentric friend—color. That is, what is orderly (versus what increases disorder) depends on your metric or frame of reference (books arranged by author would be disordered according to the color frame of reference but ordered according to the alphabetical-by-author frame). So, for example, the initial formation of the Earth involved accretion, impacts, and the sorting of its chemical constituents by density, so that the heavier metals sank to the core, the less heavy materials migrated to the mantle, the lighter solids floated up to form the crust, with water on top, and gases emerging to form the atmosphere as the lightest, topmost layer. Now, from the reference frame of chemical homogeneity and density, the Earth spontaneously became more orderly, not less. Yet, at the same time it moved toward greater thermodynamic entropy or energy dispersal (and by that frame of reference moved toward being less well-ordered). As this example shows, whether something is categorized as migrating from greater to lesser order, or the reverse, depends on the frame of reference one chooses to view the dynamics from. Of course, causation—what happens—is objectively determined by physics, and the analyst's choice of reference frame will not change that. But different causal systems—and what one wants to understand about them—may make different reference frames relevant, and opens diverse engineering possibilities both to human engineers, and—of more relevance here—to natural selection functioning as an engineer.

It is indispensable to recognize that the upward climb toward replicative order against what we will call *replicative entropy* references kinds of physical order and entropy that are fundamentally different from either thermodynamic order and thermodynamic entropy, or informational order and Shannon entropy. Replicative order and replicative entropy are proprietary concepts that emerge specifically from the logic of natural selection, and not from thermodynamics or information theory. The concepts of a lineage's replicative frame of reference and its replicative order will turn out to be fundamental to understanding both the evolution of communication and the evolution of meaning.

Organisms can exist only because evolution exploits the potentiality inherent in deploying different entropic frames of reference and their different types of order and marshaling how they interact to accomplish different and complementary kinds of replicative work. That is, what is naturally increasing disorder (moving toward increasingly probable states) for one frame of reference inside one physical regime can be harnessed to decrease disorder and perform replicative work with respect to another frame of reference coupled to another physical regime. For example, ordinary gaseous diffusion, a typical kind of increasing disorder, can be used throughout the body as something that performs replicative work, increasing replicative order. Oxygen in the lungs diffuses across membranes into capillaries, falling toward more probable high entropy states, and hence increasing disorder with respect to statistical mechanics. However, from the frame of reference of replication-promotion, this same process is decreasing order—just a different kind of physical order. Hemoglobin, trapped on one side of the capillary membranes, binds to the oxygen, which carries it to tissues, where dropping pH lowers the binding affinity, causing the oxygen to dissociate and diffuse into the cell; there it is burned, to provide the energy driving the system, including breathing in oxygen, pumping the blood, manufacturing hemoglobin and capillary membranes, and so on. The increasing disorder of gas diffusion is harnessed by natural selection to increase replicative order (resurrecting ATP so it is available for energy). More generally, natural selection picks out and links different entropic domains (e.g., substrates, enzymes, membranes, cells, assemblages) that each impose their own proprietary entropic frames of reference locally. When the right ones are associated with each other in the right way, they interact to perform replicative work. They do this through harnessing various types of increasing entropy (like gas diffusion) to decrease other kinds of entropy in ways (regenerating ATP) that are useful for the organism. The correct flow of energy and substrates allows the organism not only to successfully self-assemble against other kinds of disruptive buffeting but also enact steps to achieve extremely high replicative order (an adult healthy phenotype, successfully engaging over time in an ensemble of actions that increase the probability of reproduction).

When not looking at all the complementary frames of reference, life's Herculean triumphs of self-ordering seem like the action of some miraculous life force—something like Escher's watermill driven by his perpetual closed loop waterfall. The body, as well as intimately associated parts of the environment, consists of an interpenetrating patchwork quilt at all scales—one that weaves these heterogeneous domains into interactions that perform replicative work together, while preventing their distinct disordering gradients from disrupting each other (too much).

GENETIC AND ENVIRONMENTAL INHERITANCE

Evolution coordinates the two parallel inheritances each organism receives—the genetic inheritance and the inheritance of environmental regularities—that interact to cause development. Adaptive problems—that is, the statistical set of cause-and-effect relationships in the world that, over evolutionary time, pushed a species' genes from their initial appearance as rare mutants to near universality (or stable high frequency polymorphism)—can be called the species' *environment of evolutionary adaptedness*. The mutation-selection-new mutation-further selection process requires time to push designs very far toward increasingly effective solutions to adaptive problems. The need for many generations means that species' designs necessarily embody organization—sets of organic devices—that constitute functional responses to the temporally long-enduring structure of the world, rather than unpredictable transient events. That is, something does not qualify as an adaptive problem—even if it kills an organism—if the particular events that affect replication do not occur in persistent enough sets over the species range to durably drive genetic change. For selection to build a complex adaptation—an adaptive problem solver—the adaptation must see the world (define the boundaries of the distribution of events it functionally responds to) in terms of large categories of long enduring conditions or continually recurring instances.

Our evolved architectures make bets based on the long-term evolutionary average properties of the members of the categories they evolved to respond to. For example, the majority of snakes are harmless. However, the cost of severe injury or death from the deadly minority selected the design of the snake phobia system so that (at least initially) fear is evoked to all members of the category *snake*, rather than to the subcategories that are actually dangerous, yet harder to discriminate. Hence, the granularity with which our adaptations evolved to categorize the world is large, approximate, and ancient in scale. Accordingly, our cognitive architectures are more likely to be equipped with conceptual primitives such as *snake*, *mother*, *meat*, *enemy*, *fruit*, *night*, *blade*, *fire*, *mate*, *stranger*, *storm*, *scream of terror*, and *sweet*, rather than (obviously) *Czech*, *dove*, *candle*, *waiter*, *arquebus*, *intertextuality*, and so on. (This principle will be significant in dissecting the evolved foundations of meaning.) Even at this early stage of knowledge, converging lines of evidence indicate that there are computational entities, "conceptual primitives," that reliably develop in all normal humans as part of evolved interpretive systems that inhabit our psychological adaptations, even though we do not know all of their formal properties, nor almost anything about how they are neurally implemented or their genetic basis (Barrett, 2015; Boyer, 2018; Carey, 2009; Pinker, 2007; Tooby & Cosmides, 1992; Tooby, Cosmides, & Barrett, 2005).

There is, of course, no such thing as genetic determinism: All developmental outcomes are the disruptable, dynamic, joint product of gene-environment interactions (Tooby & Cosmides, 1990c, 1992, 2015). Yet, species reliably reproduce their designs across generations just as if there were such a thing as genetic determinism—even seemingly sometimes across tens of millions of years. How can this paradox be resolved? It is commonly overlooked that the developing organism not only receives a genetic inheritance, but also receives a predictable second inheritance logically parallel to the genes—the set of enduring, ancestrally recurring environmental regularities the genes evolved to depend on for development and successful functioning. These are "inherited" simply by persisting as an encompassing envelope of conditions from one generation to the next. This includes everything stable about the species' local world that would negatively affect development if it were changed: the laws of chemistry; the typical temperature range; the reliable initial presence (for mammals) of the mother; the patterned shifts in the spectral composition of terrestrial illumination over the day that our color vision assumes; the stereotypical vocal signals for one's own species (other members of the species constitute a critical part of the environmental inheritance); milk (for mammals); typical species that function as food (e.g., eucalyptus leaves for koala bears); for humans, an adult language community—indeed, one that conforms to Universal Grammar (assuming the correctness of the Chomskyan account); the array of species-typical facial expressions of emotion; the fact that the size and strength of adversaries predict their ability to inflict costs; and perhaps tens of thousands of other aspects of the world (Tooby & Cosmides, 2015).

It is the evolutionarily tuned interaction between these two paired inheritances—a genetic inheritance and an environmental inheritance—that causes the (largely) species-typical design to develop along its functionally coordinated trajectory. Each genetic or environmental element is shaped by ancestral event populations—the summation of incredibly numerous actual individual physical events in the past that filtered some genes over others. For convenience, we term the inheritance of environmental regularities the *environment of evolutionary ontogeny*. This turns out to be a subset of the environment of evolutionary adaptedness, as conceptualized through the lens of the task of successful development.

THE COEVOLUTION OF GENOME AND ENVIRONMENT

Over evolution, natural selection selects not only the genome but also the species' environment of ontogeny. Successful self-assembly is the first adaptive task faced by every organism, and it represents an enormous hill-climbing achievement by moving from a single cell toward amazingly improbable fine-grained arrangements of (often) massive quantities of matter (e.g., trillions of cells). It must be accomplished in the face of the world's continuous change, chaos, and disruption, which constitute ongoing entropic onslaughts to the realization of the organism's replicative order. Obviously, each individual (surviving or not) reflects the compromised intersection of ordering and disordering processes, so perfection or optimality is not to be expected—simply large departures toward functionality from random disorder.

The key to understanding how these challenges are so often surmounted lies in recognizing how natural selection picks the elements of the species' environment as well as the genes that make up its genome. Every time selection picks out a gene, it is also picking out aspects of the environment that the gene's downstream phenotypic consequences interact with. One genetic alternative might make the offspring imprint on the mother's appearance as a guide to mate selection; the alternative might not, and so (for this purpose) the mother's appearance in the second case would not be part of that offspring's environment of ontogeny. That is, the species' developmentally relevant environment becomes different depending on which gene was selected. Selection, by picking one gene over others, constructs the species' environment of ontogeny out of the total superset of total environmental conditions. It also picks out how the developing organism interacts with that aspect of the environment. So, the environment does not just appear unanticipated at the conception of each individual, as a new sculptor might show up, shaping the clay in some individually idiosyncratic way according to her unique vision. Instead, whether and in what way the specific aspects of the environment affect development have been highly engineered by its long and intricate history of ancestral interactions with the evolving genome in past environments. It matters in what ways the particular environment the individual encounters resembles the species' evolved environment of ontogeny—that is, resembles the environment the species evolved in, and therefore the environment its replicative functionality evolved to "assume."

This, surprisingly, makes the lineage's environment just as much a product of natural selection as the genes are. Indeed, natural selection acts *through* genes by retaining or subtracting mutations (or established alleles). But selection goes further, by acting *on* the relationship between the genes and the organism's developmentally relevant environment, so that their interactions are coordinated to produce the species' functional architecture. Over evolutionary time, selection will sift for genes that pick out certain aspects of the environment based on whether they are useful, stable, and organizing for the successful realization of the organism's design. In effect, that is, selection asks, do these aspects of the environment assist in transforming the *latent replicative order* present in the genes into the *expressed functional order* of the potentially replicating phenotype? At the same time, evolution will select against genes that open development to disruption from other aspects of the environment that do not usefully contribute to development. Hence, development is (imperfectly) hardened against the kinds of regularly experienced environmental (and genetic) variations that diminish replicative order. Indeed, developmental adaptations are not only hardened against variation, but are also often designed to harvest environmental "information" present in that variation so that the resulting adaptations are positioned to richly articulate with details of particular environments in ways that perform replicative work. Of course, novel environmental features might disrupt development in completely novel ways, because the genome has not been selected to be prepared for them (Tooby & Cosmides, 1990c, 1992, 2015).

One answer, therefore, as to why organisms can successfully rebuild their designs across generations despite the ever-changing, disordering nature of the world is that things only change (or not) with respect to a replicative frame of reference; natural selection, acting over evolutionary time designs the replicative order in the lineage so that its replicative frame of reference minimizes the degree to which the flow of events is experienced as change rather than continuity. Heraclitus claimed we never step in the same river twice, but the argument here is that developmental adaptations are anti-Heraclitan in design: that is, they were sampled until ones were found that experience the water as being the same, time after time. This helps explain why, say, species-typical anatomical features, or, the facial expressions of emotion (e.g., anger) appear as if they were genetically determined when there can be no such thing. In fact, what masquerades as genetic determination is joint co-determination by the evolutionarily tuned interaction of the inheritance of genetic regularities and the enduring parallel inheritance of environmental regularities. If you change the developmental environment outside of the envelope of conditions defined by the evolutionary environment of ontogeny, then a different phenotype emerges diverging from the species-typical phenotype. If you change it enough, the organism, lineage, or species ceases to exist.

The coevolution of the genome and the environment allows the evolution of developmental adaptations that circumvent the bandwidth limitations of the genome. The consideration of the role that the environment of evolutionary ontogeny plays in development allows researchers to understand how adaptations can encompass enormous bodies of information about the organism's environment, despite the limited bandwidth of the genome itself. Natural selection only cares, so to speak, that the phenotype be caused to develop an organization that solves its adaptive problems, without caring whether the information necessary for this comes from the genes, from the environment or (as always) from their interaction. So, evolved order is not just "in" the genes or "from" the genes, and environmental order is not just "in" or "from" or "determined by" the environment. Developmental questions are never a matter of genes "versus" the environment: Genes enable the method by which particular aspects of the environment are caused to participate in the integrative interactions that produce a phenotypic and computational product of (potentially) remarkable complexity, poised to do incredibly subtle and sophisticated replicative work. The information "from" the environment is assigned evolutionary meaning in the process of co-constructing and organizing the behavior-regulatory adaptations. For example, to take a simple case, a perceptual construct may be assigned the coupled meaning (conceptual primitive) "food" in evolutionarily pre-organized data structures accessed by the appetitive system. In opening the architecture to environmental inputs, the system is not escaping evolved organization, but is further realizing it on ever-increasing and ever more precisely articulated scales.

Hence, it is important to fully appreciate that natural selection, in effect, stores information impartially in the environment just as well as in the genome, so that their integration in the developing organism—say, in the brain—always embodies evolved organization and not just "genetic" or "environmental" effects free of each other's "influence." This evolutionary functional developmental framework is not nativist (although it incorporates aspects of nativism); or environmentalist (although it incorporates aspects of environmentalism); instead, it recognizes that every aspect of every organism is the joint product of an evolutionarily-coordinated gene-environment co-regulated developmental system.

Given that everything develops (from the original zygote), adaptations should be conceptualized as having two functional modes. Researchers generally focus on what can be called the executive mode, in which adaptations are executing their evolved functions (e.g., prudent snake avoidance; successful food choice; successful predator evasion; incest avoidance). What often escapes attention is their organizational mode, in which they are operating in a fashion that assembles them and organizes them to be better positioned to execute their function (turn-taking in motor practice of evasion and chase; sampling potential foods). A great deal of what appears to be nonfunctional behavior when viewed from an executive mode perspective (e.g., play fighting, babbling) seems plausibly to be adaptations operating in a self-organizational mode (Cosmides & Tooby, 2000; Tooby & Cosmides, 2001). One face of aesthetic motivation appears to be the product of adaptations operating their organizational mode, motivating the organism to interact with aspects of the world in ways that increase the power and acuity of the mind's functionality.

Over evolutionary time, a lineage can discover new stores of organization that are reliably present in the environment, and hence available to be exploited once adaptations (in their organizational mode) become tailored by selection to transmute them into replicatively useful somatic or neurocognitive structure. Such stores offer the opportunity for species to evolve adaptations whose designs guide the developing organism to exploit them to become better organized to perform replicative work. For example, the hedonic reward signal of chase play or play fighting, together with the cognitive/decision-making architecture it is a part of, guide the organism in activities that harvest environmentally dispersed fragments of information into developed perceptual motor skills, skills of concealment, exploration of the environment for refuge and evasion paths, etc. Consequently, as the organism plays, it develops a far greater competence in predator detection and evasion, there to be executed when menaced (Barrett, 2015; Symons, 1978).

Others' minds are a particularly pivotal source of information embedded in the environment. For example, the brains of the local population of adult macaques contain useful information about the magnitudes of the local ecological risks of venomous snakes. Macaques recalibrate their initial fear of snakes by watching conspecifics display fear expressions toward snakes (Mineka, Davidson, Cook, & Keir, 1984; Öhman, 2009; Öhman & Mineka, 2001). The fact that the snake phobia system is designed to reset parameters based on others' fear levels frees the fear system from the potentially unrepresentative prison of individual experience—an incipient form of culture.

From the point of view of each child entering the world, the stores of information in minds of the pre-existing members of the social group are a reliably recurring aspect of the environment. Humans, of course, have taken the mining of information to be found in the minds of others to remarkable zoological extremes. Indeed, humans have evolved an entire suite of adaptations that propelled us into what we and others have called the cognitive niche, a way of life involving the intensive capture and use of detailed local information, based on a dramatic drop in the individual cost of acquiring information (Pinker, 2010; Tooby & DeVore, 1987). In the cognitive niche, humans use unprecedented amounts of socially accumulated local cause-and-effect information and expertise, enabling (among other things) tool use and the contingent improvisation of tactics for highly productive ecological exploitation. This has allowed humans to multiply across nearly all terrestrial habitats, successfully diversifying into a remarkable number of subsistence practices.

The computational architecture underlying the cognitive niche includes adaptations supporting information pooling in the social group (culture); the evolution of specializations for the low-cost transfer of (largely) propositional information from mind-to-mind (involving the evolution of language and innovation in other aspects of communication); mind-reading adaptations enabling striking advances in inferring the contents of others' minds (involving theory of mind, pragmatic implicature, epistemic vigilance, scope syntax, etc.); adaptations for causal reasoning supporting tool use, environmental manipulation and instrumental action ("intelligence"); and a rich set of adaptations for cooperation, exchange, and interpersonal negotiation (moral and socioemotional adaptations; Cosmides & Tooby, 2000; Mercier & Sperber, 2011; Pinker, 1994; Sperber & Wilson, 1996; Sznycer, Al-Shawaf, et al., 2017; Sznycer, Tooby, et al., 2016; Sznycer, Xygalatas, Agey, et al., 2018; Sznycer, Xygalatas, Alami, et al., 2018; Tooby & Cosmides, 1992, 2008, 2010; Tooby & DeVore, 1987).

COMMUNICATION AND SYSTEMS OF EVOLVED MEANING

Innovations in communication lie at the heart of the human information-intensive cognitive and behavioral revolution. These innovations create and support the immeasurably enriched human inner worlds that constitute our species' strange existence. This is because low-cost, high-bandwidth communication spreads the costs of acquiring individual pieces of information across all the individuals in the social group (or communicating population), radically lowering the *per capita* cost of information (Tooby & DeVore, 1987). This reduction in price fueled striking increases in the quantity that can be cost-efficiently used. Moreover, the payoff of discovering new information is not just reaped by the discoverer (as it would be without communication), but also by kin, cooperators, and descendants, explaining the evolutionary increase in the human lineage in the motivational intensity of curiosity, exploratory imperatives, and life-long propensities to play.

Traditional approaches to the social and human sciences (blank slate views, or the Standard Social Science Model) have reasonably foregrounded culture, language, learning, intelligence, rationality, and sociality as central to human uniqueness (Tooby & Cosmides, 1992). However, we want to briefly caution that the traditional conceptions—the implicit and often explicit models of what these are like computationally—contain some assumptions that need to be revised, especially that (1) the dominant learning architectures are prepared to face all contents equally; and (2) they contain no reliably developing content native to their systems. This contrasts, for example, with Chomsky's (1957, 1959, 1975) arguments that for the human cognitive architecture to be successful in acquiring the syntax of the local language, it needed a specialized language acquisition device. Specifically, its principles of language induction contentfully reflected in some form correct limiting assumptions about universal grammar—that is, grammatical patterns and principles characteristic of all human languages. In the terms used here, the linguistic environment of evolutionary ontogeny included the cognitive architectures of language-competent speakers. Species-typical universals characteristic of these computational architectures (what Chomsky called *universal grammar*) selected for language learning adaptations that presume that the local language conformed to universal grammar, a step which made this cryptographic-like decoding problem (the child inducing the local grammar) computationally possible (Chomsky, 1975; Pinker, 1989, 1994; Tooby & Cosmides, 1990b). This was the strongest early case study indicating "learning" was not what cognitive and psychological scientists had thought it was for almost a century—a powerful but content-free system that could solve any learning problem without needing any content-specialized computational principles built in. On the blank slate view, the human mind was conceptualized as something like a tape recorder or video camera (or blank slate)—the mechanism of recording added no signal of its own; the only content was supplied by the external environment. This was Aristotle's theory that, in Aquinas' distillation, "There is nothing in the mind that was not first in the senses." Plato famously argued that the mind was full of innate ideas derived from past lives.

It is important to identify two related, mutually reinforcing arguments which, rightly or wrongly, summarize key points of departure of an evolutionary functional view of the mind from the Standard Social Science Model paradigm for the behavioral sciences (including the psychological, social, neural, and allied biological sciences). The first is that evolution would not have produced a blank slate mind, because it would have been hopelessly computationally inefficient, having to consider endless sterile possibilities while starving or being eaten by predators, etc. (Tooby & Cosmides, 1992). The second is that evolution would not have produced a blank slate mind, because (as Chomsky argued in the case of language) the organism faces a diverse number of adaptive problems that must be solved but that are computationally intractable to a blank slate mind, because the necessary information is not out in the environment (Chomsky, 1957, 1959, 1975; Tooby, Cosmides, & Barrett, 2005; Tooby & Cosmides, 1992). For example, Hume famously pointed out that one cannot derive an ought from an is, which means that for every domain where the definition of biological success is different (e.g., foraging, mating) the organism must have values or definitions of biological success supplied to it from its architecture (Tooby et al., 2005).

The field of evolutionary psychology is largely based on the recognition (or some argue, misconception) that the human neurocomputational architecture is full of replicative order in the form of neural programs that are functionally specialized just like the language acquisition device is. These programs cover an enormous range of adaptive problems and functional activities: child rearing, foraging, navigation, altruism toward kin, negotiating welfare tradeoffs to social others, forging group identity, participating in collective actions, resolving conflicts of interest cost-effectively (managed by the anger system), effort-allocation, alliance detection, preventing infidelity, incest avoidance, mate choice, sexual motivation, status management, collective aggression, animacy detection, food evaluation, cooperation, exchange, friendship, coalitional action, and so on (for an overview of the logic, see Tooby & Cosmides, 1992, 2015; for reviews of the field, see Buss, 2015).

On this view, each of these functionally specialized programs contains their own evolved proprietary computational architecture. While the computational specifics of particular adaptations are becoming increasingly resolved as research progresses, our neuroscientific ignorance about how exactly information-processing is neurally implemented makes modesty appropriate. Nevertheless, these programs appear to have conceptual primitives (*mother*, *food*, *conditional exchange*, *goal*, *agent*, *harm*, *kinship*, *contamination*, *sexually attractive*, *own-offspring*, *enemy*, *ally*, *us*, *them*, *health*, *sickness*, etc.); interpretive frameworks (which link primitives into systems with specialized inference engines, including often motivational triggers); specialized inferential elements (pollution by contact, cheater detection, grammar-morphology induction, alliance cue signals alliance group, strength predicts dominance, playface initiates pretend play, seeing is knowing, etc.); motivational programs (punitive sentiment toward free riders; erotic attraction; familial love, etc.); value-concept gradient systems connected to specific motivation and emotion programs (setting, e.g., the magnitude of fear related to closeness to spider; the magnitude of the sacrifice one is willing to make for the welfare of a specific known other; magnitude of rivalrous dislike given the magnitude of the threat of loss or displacement; the magnitude of hunger); and so on (Buss, 2015).

If one accepts that there is a rich, evolved set of content-specialized programs in the human mind, then this raises the question of how the contingent elements of each individual's life become correctly mapped into these pre-existing cognitive structures so that they can perform their evolved functions. Each adaptation faces the problem of binding or attaching represented local contingent facts to its appropriate evolutionarily meaningful conceptual or regulatory proxies: The grammar system must detect that specific words in the local language are *verbs*. The kin detection program must attach a perceptually defined person-representation to a kinship index—i.e., who is your *sister*?, what is the *name* of the *stranger*?, what substances around you are *food*?, what faces are *enemies*? Systems for binding ontogenetic specifics have distinct methods specific to the nature of the interpretive problem, but in general these subcomponents can be called psychophysical front ends associated with programs for assigning meaning and motivational tags. This *face* is your *mother*; this person who supported you in the argument is on *your side* and is your *ally; you* are both *members* of your *coalition*, and in *conflict* with *them*, the *opposing coalition*. This is the world our brains construct for us, in interaction with local inputs.

Each ancestral human was born into a particular place and situation full of contingent specificities: These trees, my mother, my sister, this fig in the hand, this snake on the branch, these utterances, these facial expressions, these tools, this set of band-mates, these neighboring groups, and so on. If *verb* is a part of universal grammar, then instances of verbs must be recognized in the lexicon as verbs. Evolutionary theory suggests that genetic kin constitutes a special category of persons that the actor benefits from treating altruistically, and benefits by avoiding them as sex partners. To perform these functions, the kin detection and motivation system must discover and map which familiar others are genetic kin. Research indicates humans have such an evolved program for detecting genetic kin; it uses the cues of seeing one's mother care for a newborn, and length of co-residence during childhood to bind regulatory magnitudes of *genetic kinship* (the higher the index, the more sexual desire toward them decreases and altruism increases) to particular people (Lieberman, Oum, & Kurzban, 2008). Similarly, an alliance detection program uses behavioral cues of alliance observed in one's social world to sort people into one or more alliances (Kurzban, Tooby, & Cosmides, 2001; Pietraszewski, Curry, Petersen, Cosmides, & Tooby, 2015). For an adaptation to be activated and guide behavior, this process must happen for every active psychological adaptation. Therefore, the mind experiences and represents the person's situation and flow of experiences at two levels simultaneously: The first level is the world of unique particularities; in the second level, these particularities are clothed in or mapped to meaningful evolved categories (predator; beloved child; weapon), and weighted with motivational magnitudes (close family member; desirable game animal; annoying free-rider). In aggregate, the current activation of these programs provides the organism a *situation representation*—something the organism always needs, so that she always knows how to act in the present moment.

Understanding the reality of these interpenetrating systems of evolved meaning throughout our species' neurocomputational architecture is crucial to understanding the nature and operation of human-to-human communication. To understand fully why, it is necessary to take a step back and explore the relationship between the perspective that emerges from the theory of replicative functionality, and Shannon's (1948) separately developed theory of information and communication, embedded in Weiner's (1948) conceptualization of cybernetics.

EVOLUTION, REPLICATIVE ORDER, AND SHANNON'S THEORY OF COMMUNICATION

Evolution provides a framework that grounds Claude Shannon's theory of communication and Norbert Weiner's science of cybernetics in replicative order. A key question is how Shannon's theory of communication, and especially information theory, relates to the evolution of life. To answer this, we need to identify exactly when information entered the world, where it exists, and how it exists (leaving aside the different meaning of information in quantum mechanics). For Shannon, information is a concept analytically embedded in his theory of communication. "The fundamental problem of communication is that of reproducing at one point, either exactly or approximately, a message selected at another point" (Shannon, 1948, p. 379). This model posits two entities, the source and the receiver, which are already coordinated on a pre-existing set of possible messages. In any particular instance of communication "[t]he significant aspect is that the actual message is one selected from a set of possible messages" (Shannon, 1948, p. 379; see also Shannon & Weaver, 1949). At the risk of stating the obvious, it follows that information in Shannon's sense only comes into being with respect to the existence of a source and receiver sharing such a set of pre-established messages. In the nonliving physical world, there is feed-forward causation and causal structure, but no shared sets of messages, and so no information in Shannon's sense.

With the emergence of life with its replicative order, things no longer merely happen; instead, a new kind of physical order, replicative order, differentially accumulates. Replicative order in the design of organisms is designed to bring about some outcomes (those doing replicative work) over other outcomes. Living order benefits from, and is selected to accumulate, organization for the targeted regulation of functional processes, including behavior. With the passage of evolutionary time, it is easy to see that the design of an organism might evolve regulatory elements designed to exercise control over other functional subsystems—elements which benefit by such regulation. General cells evolve into nerve and other cell types. Within the organism, it is easy to see how the system benefits from components which send, and other components that receive, evolutionarily coordinated messages from a pre-existing set, such as a pain receptor in the foot triggering muscle contraction. Although we have been culturally and technologically shepherded by our digital tools towards thinking increasingly in terms of Alan Turing's conception of computation, Norbert Weiner's proposal of a more inclusive field of cybernetics is a more illuminating framework. He defined cybernetics as "the scientific study of control and communication in the animal and the machine" (Weiner, 1948). Fitting regulation (control), information, communication, and then computation together inside an evolutionarily informed cybernetics puts them together logically and historically in the wider scientific landscape of the evolution of the physical universe.

Although we now associate information with computation, and often consider computation in its most general and abstract form with mathematical analysis taken independently of real-world consequences, much is lost (as well as gained) with this Turing-style framing. Adversaries in war naturally tried to defeat others' understanding of one's codes by injecting maximum unpredictability into the signal. For this reason, computational approaches to encryption and hence decryption are designed to minimize fallible assumptions about what the incoming signal might be. Decryption—depending on the detection of the deviation from noise—pushes for computational generality in a way other engineering problems do not. The resulting implicit and often explicit model of a computational system, generalized to be ready for anything is, we argue, fundamentally misleading for understanding the emergence and nature of natural computational systems.

In contrast, Norbert Weiner was recruited to work on the problem of fire control systems in anti-aircraft guns. Such particular engineering problems are intrinsically computationally specialized, wherein the actual embodiment of the guns, the properties of ammunition, the dynamics of the targets, and the atmosphere manifest regularities that can be treated as a stable background; the computational elements developed to solve the fire control problem can implicitly presume these regularities in their computational implementations. This vastly simplifies the computational problem of inputting or estimating the remaining open parameters (target speed, altitude, wind direction, temperature, air pressure, distance) that must be integrated for the target to be hit. (Interestingly, Claude Shannon started his career at Bell Labs also working on fire control systems.) In approaching control problems, Weiner did not need a general-purpose computational system that implemented a totally flexible conceptualization of the situation. Instead, the methods required for regulatory guidance of a system to the goal are the residual set of those not already solved by the regularities of the task environment and the system embedded in it. "Representation" of the total situation can be largely dispensed with. These often allow elegant minimal hacks to solve a specific engineering problem. For example, to catch a baseball or intersect with any other projectile, you only need to move to keep the object at the same point in your visual field—you do not need a computation of the ballistics of the projectile and yourself with respect to a three-dimensional environment. (We suggest that, despite the apparent flexibility of human intelligence, the reality of natural computational systems is far more analogous to fire control systems than to Turing machine implementations on von Neuman architectures; Tooby & Cosmides, 1992.)

As a consequence of the evolutionary accumulation of replicative order in living systems, adaptations for systems control that perform replicative work evolve computational and signaling systems. In these systems, information comes into being with the evolution of coordinated sets of messages between senders and receivers internal to the organism. Replicative order provides frames of reference with respect to which information exists, rather than just physical causation. Hence, information in Shannon's sense first enters the world embedded in evolved regulatory systems. Natural computation enters the world as specialized regulatory elements designed to solve particular adaptive problems for the organism (at all scales—even single-celled organisms have large numbers of regulatory information-processing elements; Bray, 2011). All exist within frames of reference provided by the replicative order of the respective species or lineages involved.

It is important to recognize that information per se in Shannon's sense does not exist in the external physical world. Physical causation exists in the external world, but information does not. There is no pre-existing objective parsing of the world into an exhaustive set or superset of messages. Despite how our senses present the world to us, the world simply exists as an endless uncarved flow of physics, and each cause-and-effect relationship is just an infinitesimal part of the limitless ocean of unparsed structure. (It is important to note that DNA involves communication in Shannon's sense, and is both information-bearing and also computationally system-regulating—merged in one superlatively functionally ordered replicative system. DNA replicates and enters or forms a new cell, which constitutes the transmission of a message written in a code both within and between individuals. It also is the source of regulation for the cell, or a cascade of larger entities.)

In contrast, whereas information does not exist in the non-living world, the precursors to information—the raw material out of which information is extracted—obviously do exist. This allows us to be more precise about what learning is (in the broadest possible sense) and how information in our brains about the world is produced through interaction with a world—a world that merely contains precursors to information but not information itself: An organism benefits by regulating its behavior in functional accordance with selected aspects of the actual state of affairs in the external world (it needs to detect and flee predators; identify and eat foods, etc.). Hence, the regulatory architecture evolves to detect those states of the world that the regulatory system needs to discriminate and differentially respond to implement functional regulatory responses (i.e., perform replicative work). Brains are infinitesimal compared to the magnitude of the world, so brains evolve to cost-effectively assay those limited aspects of the world that might be useful to coordinate their behavior in conjunction with. It throws away or does not detect the rest.

Learning can be defined in the least restrictive way as the assessment of states of the world that one or more regulatory systems in the organism developed to discriminate in order to perform replicative work. Under this broad definition, even perception is a form of learning. The replicative order of the organism (usually in interaction with prior developmental processes) provides a computationally implemented frame of reference designed to interpret these input assessments. These computational systems interpret them in terms of the functional regulatory outputs they enable (i.e., flight driven by the detection of a predator; mastication driven by the recognition of a food). In fact, a set of information carried by signals to a target in a regulatory system coevolves with an interpretive system that frames physical patterns as information. That is, information is physical (as Shannon and other information theorists emphasize), but exists only with respect to an interpretive system in an organism—a system that provides a frame of reference (derived from replicative order) that makes the information meaningful to the regulation of the system (and other subsequent computational steps the organism might be designed to make). It is always important to remember that information only exists with respect to an interpretive system, and natural interpretive systems only came into existence as aspects of evolved organisms. Of course, multiple organisms (especially of the same species) can and commonly do share the same interpretive systems in terms of their abstract properties (such as object recognition, predator detection, social hierarchy, or phonological processing). These are shared in the sense that each member of the species (or a developmentally coordinated group) has its own instantiation of the interpretive system (e.g., a local dialect's phonemic boundaries) that is paralleled in others. Those who share interpretive systems will therefore interpret external situations in the same way. As we will see, this sharing of interpretive systems—and hence potential coordination—between individuals are important, because they are what allows Shannon's theory of communication to be applied to communication between organisms.

In Shannon's conceptualization of communication, there is a shared set of messages, and the problem is one of reconstruction in the receiver of the message that was sent. In an evolved system of sending and receiving within an organism, this needs to be analyzed somewhat differently: There must be a principled mapping between the message sent and an interpretation in the receiver, but the receiver does not need to reconstruct the "same" message—just the useful next step in the usefully managed regulatory process. In murine rodents, olfactory cues to cats trigger predator evasion, but the receipt of the signal by the behavioral system is triggered evasion, not the reconstruction of a smell template (Kinderman, Siemers, & Fendt, 2009). The point here is that the organism's regulatory system provides (1) a frame of reference that (2) assesses some physically detected signal in order to (3) characterize some external state of the world that (4) discriminates it from other states of the world, in terms of (5) an interpretation that (6) the regulatory system uses to produce or improve behavioral (or physiological) regulatory outputs. So, more precisely, natural information only exists in the world paired with or relative to an interpretive system—a system that provides a frame of reference that gives an interpretation to the informational substrate that turns the substrate into information. These (naturally occurring, as opposed to artificially built) information-interpretative system pairs exist only inside organisms (or between organisms), as a result of the organisms' replicative order.

In the case of evolved instead of artificial communications systems, the properties of the receiver's communicative and interpretive interface are the lock (the recurrent adaptive problem) that the sender's signal production interface evolves to unlock. Likewise, the properties of the sender's signal production interface are the lock that the receiver's communications and interpretive system evolves to unlock. At its simplest, when conflict is absent, senders evolve architectures capable of being understood by receivers, and receivers evolve architectures capable of understanding—each constituting an adaptive problem for the other and each coevolving to constitute an adaptive solution to the adaptive problem posed by the other.

Shannon separates off the question of meaning from the analysis of physically instantiated information, a useful and prudent step, given that it takes an evolutionary framework to tackle the question about what natural systems of meaning are in evolved organisms like humans. Here, however, we suggest that scientific progress on questions concerning the nature of meaning—at least specific types of meaning—can be facilitated by considering how each specific type of information generated in one part of the architecture is used by other components as inputs, intermediate computational products, or regulatory outputs in other functional parts of the system: all to drive replicative work. Meanings in our architecture are not Shannon-information decoded from messages sent; instead, meaning systems emerge in evolved human psychological adaptations as they assemble themselves in their organizational modes. For example, *mother* may have propositional linkages in the conceptual and language systems, but it is also linked in implicit regulatory ways to motivated proximity management, welfare trade-off valuation, as triggers of kin detection for her other children, and many other systems which are quite functionally specifiable but fall outside a lexical database (Bowlby, 1969; Lieberman, Tooby, & Cosmides, 2007; Tooby et al., 2008; Tooby & Cosmides, 2008). Meaning is generated by the programs in our brains that evolved to serve our regulatory systems. However, precisely because we are not unitary agents pursuing the goal of reproduction but assemblages of quasi-autonomous programs, we embody large, diverse sets of evolved, haphazardly developed interpretive frameworks that assign affect-laden meanings.

COMMUNICATION AND LEARNING

Communication between different organisms is based on learning—that is, adaptations designed to detect states of the world. Natural communication first evolves within organisms. Obviously, however, eventually communication evolves between different organisms (more precisely, as we will see, between subcomponents of different organisms). It is not, though, always appreciated that the driver of the evolution of inter-individual communication is located in the learning systems and associated interpretive systems of the receiver. That is, the organism evolves to capture those types of information from the world that assist it in improving its different systems of behavior regulation (foraging, mating, alliance, aggression, predator evasion, etc.). That is, it evolves learning systems that correspond to what it needs to know. The world happens to supply the great majority of what the organism needs to know, but the world per se was not generally selected to supply it.

The designs of the learning systems (such as its specification of the states of the world it benefits from registering, discriminating, and interpreting) evolve to match the organism's learning problems. However, for a subset of what the organism is designed to assess, there are payoffs to the supplier of assisting the receiver in capturing this information. In most cases of learning, there are information seekers ("receivers") but not information senders. Here, we are not speaking about the intentions of the sender (whatever that might mean), but whether selection has shaped the monitored organisms to assist in delivery of informative signals to the information seeker. So, we have information seekers, which attempt to capture information that is useful to them; there are incidental information suppliers, which the seekers evolve to capture information about and from. For example, trees are opaque, and therefore animals avoid colliding with them: They supply the information about their locations, but they do not send it, because they have not been selected to make supplying this information easier.

Of course, there are organisms that benefit when the information seekers receive information about them. They become subject to selection pressures that modify their designs to more readily supply this information to information seekers. In this case, we have finally arrived at what is describable as inter-individual communication, with actual information, actual senders (those designed to supply information), and actual receivers (those designed to capture this information). The characteristics of the adaptations for sending are driven by the characteristics of the receiver—that is, what states of the world the receiver is designed to capture. With the emergence of this system there is genuine inter-individual (as opposed to within-individual) communication of information. Having come so far, it is still important to recognize that this is only unidirectional communication, from the sender to the receiver, and not bidirectional communication. Such unidirectional situations are common. For example, toxic organisms, such as certain species of butterflies, are brightly colored; potential predators benefit by being warned before they sample toxic foods; toxic prey benefits by not being sampled. Prey evolves to signal toxicity to predators, but predators are not (in this case) selected to send any corresponding signal back.

Finally, among many animal species there is the evolution of full reciprocal or bidirectional communication. Under some circumstances, both parties benefit from capturing information from the other, and both are selected to supply it. These systems of communication may or may not share a full set of messages between sender and receiver, but that do not mirror the sender's message. What can serve as a substitute for Shannon's shared set of messages is a set of messages potentially sent by the sender, matched with a set of functionally useful interpretations derived by the receiver—interpretations that do replicative work for the receiver. The butterfly has warning coloration, but the avian predator does not also have warning coloration as a possible return signal; what the avian predator has is a weighting on prey choice by appearance that causes it to avoid the butterfly—that is, that gives a regulated functional response to the signal.

In contrast, in the case of human language, there is the full Shannon communication system, wherein the sender and receiver share a set of messages and the receiver reconstructs the message sent by the sender. Moreover, language is typically bidirectional (but obviously can be unidirectional, in which language production is in one of two individuals only, so that an aphasic or a very young child may understand the speaker but not be able to speak in turn). It is arresting that the form of communication that seems so easy and natural to us should in fact have such an extensive set of layered properties, and should be at the upper end of the evolutionary process.

With no hope of its being acted on, we gently suggest that it might be useful when discussing evolved communications systems, to sometimes use the term supplier rather than sender, since the supplier has not always been shaped by selection to send signals; instead, in many cases the source happens to supply to observation the precursor or information substrate to what becomes transformed into information when interpreted by the information-seeking interpreter and regulatory system. So, one has suppliers and seekers rather than senders and receivers.

CONFLICT AND DECEPTION AS FEATURES OF INTER-INDIVIDUAL COMMUNICATION

Conflict is a ubiquitous feature of inter-individual communication. Signaling theory and the evolutionary dynamics of communication are still debated, and papers outlining key issues are readily available, so we will not extensively review these issues (Dawkins & Krebs, 1978; Grafen, 1990; Krebs & Dawkins, 1984; Maynard Smith & Harper, 2003; Searcy & Nowicki, 2005; see also Chapter 4 by Reid, Zhang, Anderson, & Keblusek, in this volume). For those unfamiliar with these issues (but familiar with the nature of social life), it will come as no surprise that the primary dynamics arise from the fact that different replicators have different fitness interests, and so are often selected to pursue conflicting agendas, generating conflict, and over generations antagonistic coevolution. Organisms (typically animals) for reasons already described may be selected to produce signals that supply information to information seekers. Because fitness interests are rarely perfectly aligned, inter-individual communication is challenged by the possibility that signals may sometimes be deceptive. If signals of a given type are never accurate, then information seekers will evolve to disregard them, and there will cease to be communication in that channel. Hence, the only signals that animals stably monitor are ones that continue to have some information in them. The question is what keeps signaling honest under conditions where senders are benefitted by supplying signals that are sometimes honest, sometimes not. Of course, there is ongoing selection in the information seekers to distinguish honest from deceptive signals.

One theory is that, under the right envelope of conditions, the honest signals are costly to send. If the signals are costlier for deceivers than for honest signalers (for example, because the honest signalers are in better condition (Grafen, 1990; Zahavi, 1975), then this can place a limit on the invasion of deceit into a system of honest signaling. According to this view, for example, only the healthiest peacocks can afford the largest and most colorful tail (Zahavi, 1975). Index theories propose, in contrast, that in many cases of communication, the nature of what is being assessed cannot be faked by virtue of the causal pathways involved: For example, if what is being advertised is a low level of mutations, and the mutations express themselves if they are present, it is not matter of cost to distinguish low- from high-quality genotypes (Maynard Smith & Harper, 2003). The apparent *joie de vivre* of a male whale breaching the water and falling back is explained on this theory by the fact that the sound of his impact is strongly physically correlated with his size, and therefore with his aggressive formidability—something hard to assess otherwise in the visually challenging underwater murk. It is a satisfying complement to index theory that the causal structure of the world is such that organisms often cannot help but supply information simply by pursuing their replicative activities. Information seekers evolve to track whatever useful information may be supplied in such observations, which sets the stage for a gradualism of the seeker focusing on, and developing more sensitive detectors for those physical indices which reliably convey information in a difficult to fake way.

Often, evolutionary dynamics of signaling for a set of organisms can be sorted out by asking the following series of questions: What sets of information does each set of organisms benefit by observing? For each set of organisms and each body of information, who is benefitted by the truth being revealed? Who is harmed by the truth being known? Who knows which players would be helped or hurt? Who is in a position to release or broadcast the truth (about a given set of information)? Who is in a position to clarify or obscure the truth being known? What are the costs and benefits for each player of changing the knowledge states of others, and their own? For example, in a system of sexual selection, a high quality male might benefit by his mate quality becoming known to local females, while other males might be harmed. Other females might benefit from discovering his mate quality, whereas his present mate might be harmed. In contrast, if senders and receivers (such as mothers and offspring) are both benefitted by the truth being known (i.e., is there a predator?), then signals might evolve to become very reliable, and highly trusted. If a young male who has been rapidly increasing in strength knows he has become relatively strong, then he can advantageously recalibrate others by arranging a display (like chimpanzee tree-branch shaking) that advertises his strength. Using these questions, communications dynamics for any given situation can be broken down and analyzed. They apply to illuminating the dynamics of rivals, cooperators, those like potential mates engaged in consequential choices, predators, prey, parasites, allies, outgroup members, and mixtures, and the distribution of camouflage, the economics of display, and numerous other questions. The basic principles are nearly self-evident: The agents who benefit by knowing the truth are in favor of transparency; those harmed will discourage displays, produce anti-information, and attempt to increase noise.

It is important to note that neither costly signaling theory nor index theory of honest signaling seem well positioned to explain the two major human cases of human communication: language, and the (often involuntary) expression of emotions. It does not seem to require more calories to tell a lie or to move a deceptive set of muscles in your face than to tell the truth. Nor does it seem the truth is compelled by physical necessity (as with the sound of the whale's impact). We have suggested that in these cases another principle is at play (Tooby & Cosmides, 2008). The argument is that what keeps these signaling systems honest is the relative downstream consequences of broadcasting true versus untrue information among close cooperators over time (Tooby & Cosmides, 2008). This analysis operates more strongly the longer a piece of information is used in the group. Some pieces of information once released will be used repeatedly. It seems plausible that an individual can anticipate how that item of information will be used the first time, in circumstances she can foresee. But second and subsequent uses may be harder to anticipate. The average return on releasing versus withholding information will be, other things being equal, given by the average degree that others take the releasers' values into account in their decision-making. Antagonists will use accurate information to the detriment of an individual, while close cooperators will often use information in the interest of the information source. The argument is that rather than attempt costly strategic analysis of the payoff of every piece of information that might or might not be broadcast, some categories of information on average will have a positive expected return if released, because you are broadcasting to those who place a high weight on your welfare. If so, evaluative information, for example, signaled by certain emotional expressions such as happiness, disgust, fear, and so on, will allow others to better take into account your welfare when they act, by knowing your values. This model not only predicts emotional signaling's most surprising characteristic—its automaticity—but it also predicts that people will be more emotionally expressive, the more they are together with people they know and like. In contrast, the more they are with people who are antagonistic, or strangers, the more shy, reserved, impassive, and withdrawn they should be.

The explanation for why language is reliable against invasion by deceit may rest on two factors, one similar to the argument for the automaticity of emotional broadcasting. The most basic point is that there is no problem explaining the reliability in communication where there is a reliable harmony or convergence of interests among the communicators. Indeed, humans evolved in small, closely related, highly interdependent groups, which is consistent with the high level of cooperation, and intensive communication mandated by the hominin entry into the cognitive niche—a way of life that depends on high levels of information sharing. Also, it is not an overstatement to say that humans evolved to be nearly obligately group-living in our mutual interdependence. Cross-cultural evidence shows that humans on every inhabited continent, and in small-scale societies as well as developed societies, track in a detailed way the specific values of others in our community. We feel strong shame when we are seen to violate their values (Sznycer, Al-Shawaf, et al., 2017; Sznycer, Tooby, et al., 2016). We feel strongly attracted to winning approval by publicly upholding our group's behavioral evaluations (Sznycer, Xygalatas, Agey, et al., 2018; Sznycer, Xygalatas, Alami, et al., 2018). Thus, our pre-existing plateau of cooperation, based on our obligate small group-based ancestry, is a first-tier explanation for our levels of deceit being lower than they could be.

Second, the kinds of complex propositional information transmitted linguistically often carries its own credibility tests in the way with which it meshes, or fails to mesh, with the immense stores of information we already have. As Mercier and Sperber (2011) and others have written about in enormous detail, humans have a large set of adaptations for epistemic vigilance that are deployed when processing others' utterances (e.g., Mercier, 2020; Sperber & Wilson, 1996). Indeed, we have argued that humans come equipped with what might be called a scope syntax, which is necessary for the human entry into the cognitive niche, because in our information-drenched way of life, individuals are bombarded with vast amounts of information of varying and hard-to-evaluate truth value; that is, our minds need systems for evaluating, and quarantining sets of information so that undetected false representations will not, by inferential propagation, spread incoherence or falsehood throughout one's knowledge bases (Cosmides & Tooby, 2000).

LANGUAGE ACQUISITION INVOLVES EVOLVED SYSTEMS OF MEANING

Language acquisition involves evolved systems of meaning and situation representation. Chomsky's argument about the poverty of the stimulus, and the computational or learnability problems posed by syntax had an immediate and decisive impact on the development of the cognitive sciences (Chomsky, 1957, 1959, 1975; Pinker, 1994; Skinner, 1957). The pre-theoretical commitment to the computational generality of learning systems, especially human learning, meant that almost all attention became stalled at the debate between what one might call (as a convenient name rather than a sophisticated or fair theoretical characterization) partisans of blank slate learning, versus Chomskyans, who focused primarily on syntax. Conveniently, the syntaxes of various specific languages had a formal structure that could (perhaps optimistically) be characterized as at least well enough to test whether, say, finite state grammars could succeed at acquiring them (as they could not).

What few people realized at the time was that Shannon's theory of communication indicated that arguments analogous to Chomsky's about learning syntax applied with even more of a vengeance to learning the semantics of languages. Cryptography is possible only when cryptographers have *a priori* statistical knowledge about a pre-existing set of incoming messages drawn from a shared set. The child's task is parallel to the cryptographer's. The child's task of discovering word and sentence meanings involves isolating, out of an infinite (or indefinitely large) set of possible meanings, the actual meanings intended by other speakers. The less information the child (or cryptographer) has about likely messages, the more messages the child must receive to converge on answers. At the limit, if any message is possible—the system is completely general—then even an infinite amount of information would not allow the child to converge on the local meanings of words. If the utterance "Do you want some soup?" could just as well mean "Martin Luther told me he stole Gaius Marius's dignity when he found a hole in his atomistic argument" or "I hope the clouds sing tomorrow" as it could mean "Do you want some soup?," then the developing child is in trouble.

Fortunately, natural selection, unlike philosophers and psychologists, would never allow the evolution of such a fatally crippled system. Culturally, we are all attracted to the idea of a mind free of content because it (falsely) invites the idea that such a mind would leave us free to think anything without limitation. (However, such an architecture would actually prevent us from thinking.) The developing child, in order to be able to solve the adaptive problem of learning the meanings of words, must reliably have in her evolved cognitive repertoire enough interpretations she shares in common with competent speakers that she can guess at the likely meanings of the utterances she hears. Even more fundamentally, from an information-theoretic point of view, the more general the architecture, the more possibilities it has to compute over—the higher its starting entropy—and the more inefficient it is. Our traditional expectation that brains should be general purpose is upended: Natural selection as an engineer would move brains in the opposite direction, collapsing unnecessary or inefficient dimensions.

As discussed, our evolved psychological architecture must necessarily be permeated with evolved systems of meaning as components or aspects of adaptive specializations that regulate our behavior adaptively (theory of mind; theory of objects; skeletal tool core knowledge; social exchange logic; the logic of aggression; the logic of alliance; child-caretaking; food concepts and motivations; pretend play; syntactic core knowledge; spatial representation; causation, core emotion expression interpretation with the ability to understand the associated evaluative signals; and so on). These systems of meaning supply the interpretations that must be shared between the learning child and speakers necessary for the child to acquire meanings: communication depends on a shared list of messages Moreover, these systems of meaning do something even more basic: All organisms, in order to generate behavior more or less appropriate to their circumstances, have to construct an ongoing situation interpretation (even if it is the emotion mode of confusion—which motivates the suspension of action and attempts to rebuild coherent representations). So, the child herself must always be generating situation representations, which is the central element of constructing what Sperber and Wilson call the mutually manifest between two communicators, as well as the ground of the acquisition word meaning. Situation representation and its sharing are, it turns out, also crucial to understanding the social assignment of meaning in group processes.

Before turning to that final issue, it is important to identify, on the topic of learnability and computational tractability, the relationship between Shannon's theory of information and communication and the efficient design of regulatory and computational architectures. If a human engineer or evolution is building a regulatory system, it obviously needs to be as computationally powerful as necessary to solve its adaptive or control problem. This means its data structures, frames of reference, and interpretive system need to be as large as necessary to accommodate the number of states to be discriminated and differentially responded to. So, for every dimension in the adaptive problem or task environment that needs bandwidth, one might expect (given cost-performance tradeoffs) for the architecture to have sufficient capacity. This corresponds to what people intuitively call "flexibility" in the cognitive architecture.

Reciprocally, however, for dimensions that have never been needed, one would expect (if somehow they came to exist) for selection (or the canny engineer) to remove them for efficiency and cost considerations. Unencountered dimensions of "flexibility" should have been stripped out or never built, and their occurrence should not be expected to be more than accidental or as a by-product aspect of our cognitive architectures. Evolved systems have only encountered past conditions, so there is no selection to prepare them for all possible conditions, even ones never encountered, or encountered so rarely it would not pay to dedicate resources to the possibility. So, in this sense one expects dimensions of capacity or flexibility with respect to evolutionarily recurrent problems (or their structural isomorphs), but not to Turing generality. In this respect, our minds should bear what Darwin called the stamp of their lowly origins. This is not as depressing as it might sound, because as Chomskyans are fond of pointing out, combinatorics can generate very large outputs from finite elements. Also, it bears pointing out that these capacity issues are somewhat independent of the other deeply contested claim: That our neurocomputational architectures are imbued with large amounts of contentful computational structure, which reconstruct in our minds large systems of evolutionary saturated or evolutionarily inflected meaning. This is no longer as controversial as it once was, when many anthropologists are documenting universal moral sentiments, universal emotion programs, core knowledge systems, and so on.

COMMUNICATION AND HUMAN COALITIONS

For individuals to act together in a group toward shared goals, the members must be coordinated with each other—something that depends on communication. Coalitions are sets of individuals interpreted by their members and/or by others as sharing a common abstract identity (including propensities to act as a unit, to defend joint interests, and to have shared mental states and other properties of a single human agent, such as status, prerogatives, and aggressive power). Underneath is a set of programs that evolved out of pre-existing individual social adaptations, but that also now induce us to form, maintain, join, support, recognize, defend, defect from, factionalize, exploit, resist, subordinate, distrust, dislike, oppose, and attack (other) coalitions. Communication is central to this entire social system, because while an individual is in total control of her own behavior, a set of individuals can only act jointly to the extent they are dynamically communicating and coordinating. This depends on a shared situation interpretation of what they are doing, and a joint motivational commitment to carrying out joint actions.

Most species do not and cannot see and feel about the social world in this way. Among elephant seals, for example, an alpha can reproductively exclude other males, even though beta and gamma are physically capable of beating alpha—if only they could coordinate. The fitness payoff is enormous for solving the constellation of regulatory and motivational problems inherent in acting in groups. Beta and gamma get no matings in a world without coalitions, whereas if they were capable of teaming up, they could drive off alpha and each get half of the matings. Two can beat one, three can beat two, etc., and so once the computational machinery evolves to act in groups, solitary animals are wiped out, and the species becomes a set of contending coalitions rather than struggling individuals. We are descended from some of the vanishingly few species to have solved these problems by evolving coalitional instincts. In this transformed world, power shifted from the solitary alphas to larger numbers of the cognitively coordinated non-alphas. This gave rise to the human world of politics, coalitions, ingroups and outgroups, factions, and war; our ancestors lived in a world where other groups expanded at their expense, or shrunk as a result of their group's collective dominance. Humans, with our high levels of ingroup cooperation and closeness have our pluses as well, compared to most other species—especially in times of peace. But our minds, even in peacetime, are saturated by the shadows of factions, allies, menaces, and the joy of moving collectively against our adversaries.

The first set of interpretations that our minds construct is intersubjective agreement on the existence of a set of groups. To be successful in the landscape of groups, you need to represent their existence. They exist to the extent to which individuals ally themselves with each other, so public acts of individual alliance are interpreted in our minds in ways that construct a map of our world as being populated with groups or alliances. Significant research shows that humans have an automatic alliance detector that takes individual acts of alliance, and produces group representations. For example, the propensity to categorize by race is one output of these system, and so seeing race in the social world is a function of the extent to which acts of alliance cross or maintain racial lines (Kurzban, Tooby, & Cosmides, 2001; Pietraszewski, Cosmides, & Tooby, 2014).

During our evolution, reproductive resources were often limited, in a zero-sum fashion. Organisms, such as humans, evolved adaptations that take advantage of opportunities to capture enhanced shares of these resources, leaving more copies of these design features than those without lacking adaptations as well-designed. Individual competition exists, but has been largely displaced by coalitions, because coalitions with their greater power in the last analysis tend to determine final outcomes. Coalitions form and compete in a collective, zero-sum fashion for status (relative entitlement to determine outcomes). In more developed areas, government institutions have largely pre-empted violence as the final determiner, but politics becomes the way to govern peaceful societies. In any case, the evolved adaptations still populate our minds with coalitional motivations, constructs, and emotions.

Humans have an evolved, group-directed motivational system that is designed to link individuals together to act as a unit to enhance their status, or initiate aggression in the interest of seizure, exploitive supremacism, or self-defensive deterrence (Tooby & Cosmides, 1988, 2010; Tooby, Cosmides & Price, 2006; Wrangham, 2019). The system is volatilely sensitive to contagious coordination in other group members, as the increases or decreases in the number of individuals who act together volatilely change the power of the group. Indeed, ad hoc mobs may materialize to strip victims of their property, homes, or lives.

Individuals, factions, and groups have two primary avenues of social negotiation: (1) threatening or inflicting harms to the target (aggressive formidability in the Asymmetric War of Attrition); and (2) conferring or threatening to withhold benefits (conferral power). Groups' divisions of social or material resources are determined by representations of status (formidability/conferral power) in the brains of two sets of agents (set A and set B). Where these representations are mutually consistent, there is no (overt) conflict, and these relationships are exhibited in flows of acts that are expressed in welfare tradeoffs between members of the two groups. The expected welfare tradeoffs between the two groups can be mutually consistent and mutually manifest. For example, in the American South under Jim Crow Laws in the 1930s, the relative status of whites was high, and blacks was low, and whites expected black behavior to reflect this. They enforced this by violence and other sanctions. Welfare tradeoffs are proportionate to relative status of the two sets (e.g., social dominance). In such a world, the acts of one or more individuals towards one or more members of another group are interpreted as communicating the relative status of the two groups.

When the weight placed on the welfare of one or more members of the ingroup is perceived as being too low (i.e., less than what the agents implicitly compute they can enforce given their joint formidability or power to withhold or confer benefits), then this becomes the internal trigger for the anger program. The emotion of anger evolved to orchestrate the agent's (individual or group) bargaining behavior during conflicts of interest (Sell, Tooby, & Cosmides, 2009; Sell et al., 2017). The function of the anger system is to leverage through prospective or actual bargaining actions (harming, or withholding help) to recalibrate upward the weight the other party places on their welfare. When groups enter the picture, anger potentially becomes an entrained part of group psychology. Coalitional psychology includes evolved circuits designed to link together the emotion programs of the individuals co-participating in the coalition. The status of the group is a public good shared among them that they all benefit from or suffer from. It is important to recognize this is not a model of rational agents or irrational agents, or of justice-seeking. The communication going on need not be conscious, and need not be directed at each other as agents under executive control. The motivational system involved is designed to assess things such as the formidability of the group, the formidability of the individuals from the other group, and then trigger motivations based on automatic assessments. As people who have been involved in mobs might recall, perhaps it makes sense to say that signals are being sent from computational subsystems inside one or more individuals to subsystems inside other individuals, within and across groups.

Groups negotiate relative rank and entitlement through (1) registration of cues (number, formidability, cohesion, etc.); (2) broadcast signals (expressed anger or outrage, menace); and (3) actions that communicate the intention to incentivize the other party (through directing violence, rioting, withholding or destroying valued things, etc.) until one or both sides recalibrates sufficiently. This is the point where a new equilibrium of mutually consistent welfare tradeoffs is reached (co-registered by all concerned). Permanent, durable groups (like individuals) should be imputed to have a relatively stable bargaining power, implicitly based on their numbers, cohesion, aggressive skills, and ability to grant or withhold benefits. Because of demography, health, maturation and other factors, groups need to assert their deterrent power through managing challenges, or invite attack.

Events in which one or more members of one group injure the welfare of one or more members of another group (defined here as an *outrage*) are implicitly viewed by all aware of them as proposing a change in the intergroup welfare tradeoff relationship—a new precedent that reflects the new relative power of the two groups. This helps to explain something that could be abstractly seen as strange—why should conflict between two individuals come to involve all the individuals of the two groups? This is because group status is a public good that applies to all members of a group. If the group cannot defend member A, then it equally cannot defend member J. Should this new welfare tradeoff precedent be accepted, this would reset expectations for future interactions between members of the two groups to the detriment of one and the benefit of the other. If the group whose member(s) have been injured feels stronger and entitled to more deference than the proposed welfare tradeoff implies (with its potentially undeterred mistreatment precedent), mutual awareness by ingroup members of the outrage mobilizes others to come join a coordinated aggressive action to attempt to reset the other group's welfare tradeoffs toward the ingroup. Here the messages sent back and forth between the two groups are easily and mutually interpreted because everyone shares the same evolved interpretive system. Often these messages are not even articulated—or even lexically accessible.

Typically, outrages and the joint attention they summon trigger collective responses, and so representations of outrages and grievances function as group-mobilizing resources, and are nurtured, embroidered, and exaggerated for their utility in advancing the group's interests, including in subordinating outgroup members. They trigger cohesion, turning uncoordinated individuals into joint actors.

For this reason, representations of outrages are group resources. Outrages by the outgroup provide a threat to the public good of the ingroup coalition's status entitlement representations; this advertises the possibility of a massively increased payoff for a coordinated group response. Because it is normally difficult to get individuals to set aside competing agendas within groups, yet group power increases with coordination, outrages (real or fabricated) become a resource ingroup individuals strategically deploy to mobilize joint action they (as individuals) benefit from. Nearly all wars are precipitated (rather than caused) by outrages, as well as many social movements (Hitler staged an invasion of Germany by German officers wearing Polish army uniforms immediately before invading Poland; Nazi irregular government power was greatly increased after the Reichstag fire; the Civil War in the United States began after southern troops fired on Fort Sumpter; the modern civil rights movement was significantly triggered by the torture and murder of the boy, Emmett Till, etc.).

CONCLUSION

Having built all this way from first principles, it is important to recognize the centrality that communication plays in cognitively populating the world with motivated groups, often in conflict. Moreover, it is not just communication in general that is involved, but specialized systems of communication that are serving their evolved functions. Consequently, it is important to recognize that groups can only exist because specialized adaptations exist that interpret the world so that we see events in terms of the actions of groups, even when it would be just as accurate to see mere individuals. These groups can only exist because specialized communication links—linked not between individuals, but from subsystems within individuals to other subsystems in other people. These communication links provide us with interpretive systems that impose an evolved set of functions on our choices, shepherd us into acting in groups in highly conflictual ways, activate largely pre-existing content that evolved to be native to our psychologies, and that provide us with a shared hallucinatory reality not often to our good.

REFERENCES

Barrett, H. C. (2015). *The shape of thought: How mental adaptations evolve*. New York, NY: Oxford University Press.

Bowlby, J. (1969). *Attachment: Attachment and loss*, Vol. 1. New York, NY: Basic Books.

Boyer, P. (2018). *Minds make societies: How cognition builds the world humans create*. New Haven, CT: Yale University Press.

Bray, D. (2011). *Wetware: A computer in every living cell*. New Haven, CT: Yale University Press.

Buss, D. M. (Ed.). (2015). *The handbook of evolutionary psychology* (2nd ed.). Hoboken, NJ: John Wiley & Sons.

Carey, S. (2009). *The origin of concepts*. New York, NY: Oxford University Press.

Chomsky, N. (1957). *Syntactic structures*. The Hague, The Netherlands: Mouton & Co.

Chomsky, N. (1959). Review of Skinner's "verbal behavior." *Language, 35*, 26–58.

Chomsky, N. (1975). *Reflections on language*. New York, NY: Random House.

Cosmides, L., & Tooby, J. (1994). Origins of domain-specificity: The evolution of functional organization. In L. Hirschfeld & S. Gelman (Eds.), *Mapping the mind: Domain-specificity in cognition and culture* (pp. 85–116). New York, NY: Cambridge University Press.

Cosmides, L., & Tooby, J. (2000). Consider the source: The evolution of adaptations for decoupling and metarepresentation. In D. Sperber (Ed.), *Metarepresentations: A multidisciplinary perspective* (pp. 53–115). New York, NY: Oxford University Press.

Dawkins, R. (1976). *The selfish gene*. Oxford, England: Oxford University Press.

Dawkins, R., & Krebs, J. (1978). Animal signals: Information or manipulation. In J. Krebs & N. Davies (Eds.), *Behavioural ecology: An evolutionary approach* (pp. 282–309). Oxford, England: Blackwell.

Dennett, D. (1987). *The intentional stance*. Cambridge, MA: MIT Press.

Grafen, A. (1990). Biological signals as handicaps. *Journal of Theoretical Biology, 144*, 517–546.

Jester, J. (2008). Corneal crystallins and the development of cellular transparency. *Seminars in Cell & Developmental Biology, 19*(2), 82–93.

Kinderman, T., Siemers, B., & Fendt, M. (2009). Innate or learned acoustic recognition of avian predators in rodents? *Journal of Experimental Biology, 212*, 506–513.

Krebs, J., & Dawkins, R. (1984). Animal signals: Mind-reading and manipulation. In J. Krebs & N. Davies (Eds.), *Behavioural ecology: An evolutionary approach* (pp. 380–402). Oxford, England: Blackwell.

Kurzban, R., Tooby, J., & Cosmides, L. (2001). Can race be erased?: Coalitional computation and social categorization. *Proceedings of the National Academy of Sciences, 98*(26), 15387–15392.

Lieberman, D., Oum, R., & Kurzban, R. (2008). The family of fundamental social categories includes kinship: Evidence from the memory confusion paradigm. *European Journal of Social Psychology, 38*, 998–1012.

Lieberman, D., Tooby, J., & Cosmides, L. (2007). The architecture of human kin detection. *Nature, 445*, 727–731.

Maynard Smith, J., & Harper, D. (2003). *Animal signals*. Oxford, England: Oxford University Press.

Mercier, H. (2020). *Not born yesterday: The science of who we trust and what we believe*. Princeton, NJ: Princeton University Press.

Mercier, H., & Sperber, D. (2011). Why do humans reason? Arguments for an argumentative theory. *Behavioral and Brain Sciences, 34*(2), 94–111.

Mineka, S., Davidson, M., Cook, M., & Keir, R. (1984). Observational conditioning of snake fear in rhesus monkeys. *Journal of Abnormal Psychology, 93*(4), 355–372.

Öhman, A. (2009). Of snakes and faces: An evolutionary perspective on the psychology of fear. *Scandinavian Journal of Psychology, 50*(6), 543–552.

Öhman, A., & Mineka, S. (2001). Fear, phobias and preparedness: Toward an evolved module of fear and fear learning. *Psychological Review, 108*, 483–522.

Pietraszewski, D., Cosmides, L., & Tooby, J. (2014). The content of our cooperation, not the color of our skin: Alliance detection regulates categorization by coalition and race, but not sex. *PLoS One, 9*(2), e88534.

Pietraszewski, D., Curry, O., Petersen, M., Cosmides, L., & Tooby, J. (2015). Constituents of political cognition: Race, party politics, and the alliance detection system. *Cognition, 140*, 24–39.

Pinker, S. (1989). *Learnability and cognition: The acquisition of argument structure*. Cambridge, MA: MIT Press.

Pinker, S. (1994). *The language instinct*. New York, NY: HarperCollins.

Pinker, S. (2007). *The stuff of thought: Language as a window into human nature*. New York, NY: Viking.

Pinker, S. (2010). The cognitive niche: Coevolution of intelligence, sociality, and language. *Proceedings of the National Academy of Sciences, 107*(2), 8993–8999.

Searcy, W., & Nowicki, S. (2005). *The evolution of animal communication: Reliability and deception in signaling systems*. Princeton, NJ: Princeton University Press.

Sell, A., Sznycer, D., Al-Shawaf, L., Lim, J., Krauss, A., Feldman, A., … Tooby, J. (2017). The grammar of anger: Mapping the computational architecture of a recalibrational emotion. *Cognition, 168*, 110–128.

Sell, A., Tooby, J., & Cosmides, L. (2009). Formidability and the logic of human anger. *Proceedings of the National Academy of Sciences, 106*(35), 15073–15078.

Shannon, C. (1948). A mathematical theory of communication. *Bell System Technical Journal, 27*, 379–423 (July); *27*, 623–656 (October).
Shannon, C., & Weaver, W. (1949). *The mathematical theory of communication*. Urbana, IL: University of Illinois Press.
Skinner, B. F. (1957). *Verbal behavior*. Acton, MA: Copley Publishing Group.
Sperber, D., & Wilson, D. (1996). *Relevance: Communication and cognition* (2nd ed.). Hoboken, NJ: Wiley-Blackwell.
Symons, D. (1978). *Play and aggression: A study of Rhesus monkeys*. New York, NY: Columbia University Press.
Sznycer, D., Al-Shawaf, L., Bereby-Meyer, Y., Curry, O. S., De Smet, D., Ermer, E., ... Tooby, J. (2017). Cross-cultural regularities in the cognitive architecture of pride. *Proceedings of the National Academy of Sciences, 114*(8), 1874–1879.
Sznycer, D., Tooby, J., Cosmides, L., Porat, R., Shalvi, S., & Halperin, E. (2016). Shame closely tracks the threat of devaluation by others, even across cultures. *Proceedings of the National Academy of Sciences, 113*(10), 2625–2630.
Sznycer, D., Xygalatas, D., Agey, E., Alami, S., An, X.-F., Ananyeva, K., et al. (2018). Cross-cultural invariances in the architecture of shame. *Proceedings of the National Academy of Sciences, 115*(39), 9702–9707.
Sznycer, D., Xygalatas, D., Alami, S., An, X.-F., Ananyeva, K., Fukushima, S., & Tooby, J. (2018). Invariances in the architecture of pride across small-scale societies. *Proceedings of the National Academy of Sciences, 115*(33), 8322–8327.
Tooby, J., & Cosmides, L. (1988). The evolution of war and its cognitive foundations. *Institute for Evolutionary Studies Technical Report*, 88–1.
Tooby, J., & Cosmides, L. (1990a). The past explains the present: Emotional adaptations and the structure of ancestral environments. *Ethology and Sociobiology, 11*, 375–424.
Tooby, J., & Cosmides, L. (1990b). Toward an adaptationist psycholinguistics. *Behavioral and Brain Sciences, 13*, 760–762.
Tooby, J., & Cosmides, L. (1990c). On the universality of human nature and the uniqueness of the individual: The role of genetics and adaptation. *Journal of Personality, 58*, 17–67.
Tooby, J., & Cosmides, L. (1992). The psychological foundations of culture. In J. Barkow, L. Cosmides, & J. Tooby (Eds.), *The adapted mind: Evolutionary psychology and the generation of culture* (pp. 19–136). New York, NY: Oxford University Press.
Tooby, J., & Cosmides, L. (2001). Does beauty build adapted minds? Toward an evolutionary theory of aesthetics, fiction and the arts. *SubStance, 30*(1), 6–27.
Tooby, J., & Cosmides, L. (2008). The evolutionary psychology of the emotions and their relationship to internal regulatory variables. In M. Lewis, J. M. Haviland-Jones, & L. Feldman Barrett (Eds.), *Handbook of emotions* (3rd ed., pp. 114–137). New York, NY: Guilford.
Tooby, J., & Cosmides, L. (2010). Groups in mind: Coalitional psychology and the roots of war and morality. In H. Høgh-Olesen (Ed.), *Human morality and sociality: Evolutionary and comparative perspectives* (pp. 191–234). London, England: Palgrave Macmillan.
Tooby, J., & Cosmides, L. (2015). The theoretical foundations of evolutionary psychology. In D. M. Buss (Ed.), *The handbook of evolutionary psychology* (2nd ed., Vol. 1, pp. 3–87). Hoboken, NJ: John Wiley & Sons.
Tooby, J., Cosmides, L., & Barrett, H. C. (2005). Resolving the debate on innate ideas: Learnability constraints and the evolved interpenetration of motivational and conceptual functions. In P. Carruthers, S. Laurence, & S. Stich (Eds.), *The innate mind: Structure and content* (pp. 305–337). New York, NY: Oxford University Press.
Tooby, J., Cosmides, L., & Price, M. (2006). Cognitive adaptations for n-person exchange: The evolutionary roots of organizational behavior. *Managerial and Decision Economics, 27*, 103–129.
Tooby, J., Cosmides, L., Sell, A., Lieberman, D., & Sznycer, D. (2008). Internal regulatory variables and the design of human motivation: A computational and evolutionary approach. In A. J. Elliot (Ed.) *Handbook of approach and avoidance motivation* (pp. 251–271). Mahwah, NJ: Lawrence Erlbaum Associates.

Tooby, J., & DeVore, I. (1987). The reconstruction of hominid behavioral evolution through strategic modeling. In W. Kinzey (Ed.), *Primate models of hominid behavior* (pp. 183–237). New York, NY: SUNY Press.

Weiner, N. (1948). *Cybernetics: Or control and communication in the animal and the machine.* Paris, France: Hermann & Cie.

Williams, G. (1966). *Adaptation and natural selection.* Princeton, NJ: Princeton University Press.

Wrangham, R. (2019). *The goodness paradox: The strange relationship between virtue and violence in human evolution.* New York, NY: Pantheon Books.

Zahavi, A. (1975). Mate selection: A selection for a handicap. *Journal of Theoretical Biology, 53*, 205–214.

4

Costly Signaling in Human Communication

Scott A. Reid, Jinguang Zhang, Grace L. Anderson, and Lauren Keblusek

In nature, an extraordinary array of communication systems is found. Many systems appear unrelated at first glance, but closer inspection reveals surprising cross-species commonalities. For example, many species can settle conflicts of interest over fitness-enhancing resources (e.g., mates, food, territories) without fighting because honest signals of aggressive intent and/or formidability exist. Such signals are mutually beneficial because the more motivated or stronger gain access to resources without engaging in costly fights, and the less motivated or weaker avoid the physical costs of fighting superior opponents. These signaling mechanisms are found in cuttlefish (Adamo & Hanlon, 1996), octopuses (Scheel, Godfrey-Smith, & Lawrence, 2016), and chameleons (Ligon & McGraw, 2016), who all communicate aggressive intent by darkening their skin. Other species have fixed traits (e.g., status badges) that advertise their formidability. Examples include paper wasps that have yellow/black facial patterns that vary with formidability (Tibbetts & Dale, 2004), and Harris's sparrows that vary rank with bib and crown darkening (Rohwer & Ewald, 1981). Vocal signals are also common; humans signal aggressive intent by modulating their voices to a lower pitch (Zhang & Reid, 2017), ortolan buntings use songs (Osiejuk, Łosak, & Dale, 2007), and red deer stags' roars signal formidability via formant frequency (Reby & McComb, 2003).

But how have all of these species evolved a means to communicate aggressive intentions and/or formidability when the incentive to cheat—i.e., bluff intentions and/or formidability to gain access to fitness-enhancing resources—is palpably strong? A generalized form of this problem is present whenever we encounter communications that mediate conflict over fitness-enhancing resources.

Part of the answer is that signal honesty exists because it is difficult to fake—but why is it difficult to fake? Biologists who study animal communication have produced the theoretical advances that explain communication systems that are susceptible to cheating, and their work has influenced the social sciences mainly through costly signaling theory. Costly signaling theory concerns the transmission of reliable (i.e., honest) information between signalers and receivers and the cost mechanisms that maintain information reliability. The theory began with the application of the handicap principle to sexually selected traits (Zahavi, 1975), was refined by Zahavi (1977), generalized to naturally selected adaptations (see Zahavi & Zahavi, 1997), and mathematically formalized by Grafen (1990a, 1990b).

There is abundant evidence that nature employs costly signals. Costly signaling theory has been used by biologists to explain aggression, begging, sexual advertisements, alarm calls, and food calls (see Searcy & Nowicki, 2005); by anthropologists to explain costly ritual behaviors, such as scarification, tattooing, and genital mutilation (Sosis, Kress, & Boster, 2007), hazing rituals (Cimino, 2011), and non-violent risk-taking (Fessler, Tiokhin, Holbrook, Gervais, & Snyder, 2014); and by social psychologists to explain how romantic motives produce sexual dimorphism in conspicuous consumption (Griskevicius et al., 2007; Wang & Griskevicius, 2014) and competitive altruism (van Vugt & Hardy, 2010). Despite the fact that costly signaling mechanisms are at play in human social relations, the theory has had scant influence on our field.

This is surprising because human communication is likely to contain many signals. For example, from a brief verbal exchange human listeners can discern at a better-than-chance rate a speaker's sex, age, health, ethnicity, and sexual orientation; psychological characteristics including intelligence, personality traits, emotional state and psychiatric status; and social category memberships, including social class, education, and nationality (see Kreiman & Sidtis, 2011). Moreover, some variants yield greater payoffs than others. For example, men who have relatively low-pitched voices have higher threat potential (Puts, Apicella, & Cárdenas, 2012) and reproductive success (Apicella, Feinberg, & Marlowe, 2007), whereas people who produce voiced, prosodic ("Duchenne") laughter are more favorably evaluated and attract more friendly intentions from listeners than people who produce non-Duchenne snorts, grunts, and cackles (Provine & Fischer, 1989). When communicating in one fashion yields greater social or reproductive advantages, what prevents cheaters from flourishing?

We describe how costly signaling theory solves this type of problem, and we provide detailed examples (voice pitch as aggressive intention signals and the role of laughter in managing social bonds) of costly signaling mechanisms. We hope to enhance the accessibility of costly signaling theory for our field and demonstrate its generative potential. We conclude by discussing avenues for research on signaling in human communication.

THE COSTLY SIGNALING MECHANISM

A signal is "an act or structure that alters the behavior of another organism, which evolved because of that effect, and which is effective because the receiver's response has also evolved" (Maynard Smith & Harper, 2003, p. 15). Signals evolve when they provide net fitness benefits to signalers *and* receivers, otherwise receivers would ignore signals and they would not evolve. Signals require a communication medium (e.g., vocalizations, facial expressions, bodily ornaments and armaments, or chemicals), psychological adaptations for their production and reception, and well-mapped behavioral responses. Signals are distinguished from indexing mechanisms, in which information cannot be faked because it is a by-product of a physiological process (e.g., one might discern that a bear fully prepared for hibernation is less willing and able to chase prey because it is encumbered in proportion to the size of its fat stores).

Zahavi (1975) conceived the handicap principle to explain the exaggerated forms of sexually selected ornaments and armaments. For example, Zahavi argued that peacocks with more massive tails are more attractive to females because they had passed a survival test (i.e., larger tails are a handicap to survival), which reliably communicated superior genetic quality. Honest signals of quality are of mutual benefit because better quality males get more matings and females can reliably select the best quality males. However, Maynard Smith (1976) showed that the male trait and the female preference could not co-evolve because females could not bear the cost of losing their phenotypically weak sons to predation. Zahavi (1977) responded that the handicap principle could work if the costs paid by signalers were proportional to their phenotypic condition. Such "revealing" handicaps work mathematically because only the best quality males can afford the most intense displays, and lesser males display less intensely. In this elaboration of the handicap principle, a "signal is reliable when the investment required for its use is greater than the potential gain a cheater would make from using it improperly. The investment should be acceptable to an honest signaler and prohibitive to a cheater" (Zahavi, 1979, p. 227).

Grafen's (1990a, 1990b) mathematical formalization of Zahavi's (1977) verbal model demonstrated that the handicap principle was evolutionarily stable and a special case of a more general costly signaling theory. Namely, a signal will remain honest—that is, it will communicate reliable information about a signaler—when (1) more intense signals are more effective (e.g., a more conspicuous trait is more attractive); (2) more intense signals are also more costly (e.g., a more conspicuous trait consumes more energy); and (3) broadcasting more intense signals incurs greater marginal costs for lower than higher-quality individuals. Because more intense signals are more effective, everyone is motivated to signal more intensely. However, the differential cost for different quality individuals constrains their signal to the intensity they can afford given their condition (e.g., genetic quality, fighting ability, motivation, wealth). Grafen (1990a) showed that the maximum level of intensity for a signal given a signaler's condition is not only optimal but also evolutionarily stable—no other strategies can yield a larger payoff, making the signal immune to invasion by other strategies.

Johnstone (1997) represented this model graphically (see Figure 4.1). Signal intensity is on the x-axis, costs and benefits are represented on the y-axis, and cost lines are included for

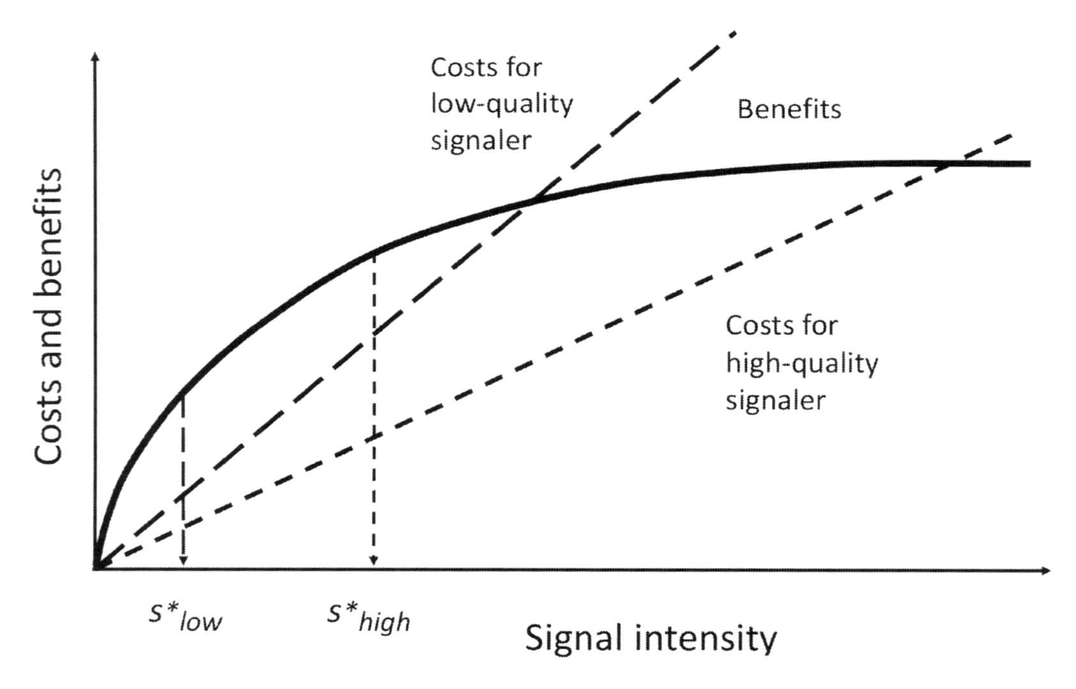

Figure 4.1 Zahavi's (1977) Model as Graphically Represented in Johnstone (1997)

low- and high-quality individuals. Benefits are assumed to be identical for both individuals and tail off with increases in signal intensity. To illustrate, imagine that wealth is signaled via the purchase of a car and the payoffs are reproductive opportunities. All individuals are motivated to signal at an intensity that maximizes the payoff they would receive, given their wealth. For a wealthy individual, a $100k car may be optimal because it maximizes the ratio between costs and benefits given wealth, whereas for a poor individual a $10k car may be optimal. For a wealthy individual, a $10k car is less costly than for a poor individual because it would represent a smaller proportion of total wealth—in this sense, more intense signals are more costly for lower-quality individuals. Although low-quality individuals could cheat by signaling with an intensity beyond their wealth, this is unlikely because all individuals are incentivized to signal at the level they can afford given their quality (i.e., cheating is possible but rare). Just like peacocks' tails, human costly consumption functions as a signal in mate competition (see Griskevicius et al., 2007).

COST TYPES

Costs are expressed in the currency of biological fitness and are assumed to be consistent with the selfish gene approach (Hamilton, 1963; Williams, 1966; Dawkins, 1976). Multiple cost types may apply to any given signal and may vary across the life course of an individual. Nonetheless, it is informative to subdivide costs into types because doing so aids in hypothesis generation.

Receiver-Independent Costs

- *Production costs* typically occur when signaling consumes energy and impedes engagement in other fitness-enhancing activities. For example, red deer stags' roaring rate is an honest signal of formidability, with stags who roar faster being more successful in repelling mating rivals (Clutton-Brock & Albon, 1979). However, older stags' roars show faster decline in frequencies than younger stags' roars because roaring is energetically demanding (Clutton-Brock & Albon, 1979). Additional examples include the duration and/or intensity of bird song and the reckless acts of human male teenagers. In each case, there are energetic costs, and costs associated with engaging in activities that might be directed at other fitness-enhancing activities, such as foraging or doing homework.
- *Developmental costs* are paid when the formation of a signal involves a developmental trade-off. For example, Nowicki, Peters, and Podos (1998) argued that song complexity is a costly signal because the formation of the necessary neural structures occurs during a developmental stage in which birds are vulnerable to nutritional stress. Only birds who have superior genetic quality and/or access to better quality nutrition could afford to develop those elaborate neural structures necessary to produce complex songs in adulthood. It is possible that similar models will account for variations in human vocal and creative (e.g., musical, comedic, scientific, and literary) abilities (Miller, 2001).
- *Maintenance costs* are paid for the retention of a display structure independent of whether a signal is issued. Searcy and Nowicki (2005) offer the example of long tails in birds that compromise flight and thus increase the risk of predation. In humans, an example is male musculature, heavy jaw, and high cheekbones, which come with metabolic costs and the cost of depressed immunity (Lassek & Gaulin, 2009).

Receiver-Dependent Costs

- *Retaliation costs* are paid when signaling induces attacks from signal receivers (hence "receiver dependent"; Enquist, 1985). In this case, receivers aggressively test signalers' true intent and/or formidability upon receiving an intense signal, which ensures that less motivated or formidable individuals do not cheat. Retaliation costs are crucial in maintaining the honesty of signals that are not physically constrained or incur little cost to produce, such as most threat displays (Enquist, 1985; Hurd & Enquist, 2005). If threat displays are made despite high risks of retaliation, those displays are likely honest signals of aggressive intent. Retaliation costs also apply to cases of human nonverbal aggression, including vocal pitch shifts, yells, and angry facial expressions.
- Signalers pay a *vulnerability cost* when the act of signaling renders signalers more prone to injuries (Zahavi, 1982) or possibly a loss of status. As in the case of retaliation costs, if an animal performs intense threat displays despite the fact that those displays make the animal more prone to injury, those displays must signal aggressive intent. Laidre (2007) tested the vulnerability cost hypothesis by varying the degree to which hermit crabs' soft abdomens were exposed. Crabs whose abdomens were more exposed (and thus more vulnerable) were less likely to display threats. This finding suggests that the degree of vulnerability constrains signal intensity, just as energy availability places constraints on stags' roaring rates. Similar processes are likely involved in acquiescence signals, such as when a dog or wolf rolls over, making its neck and abdomen vulnerable to a bite. Vulnerability costs may explain Duchenne laughter (discussed in detail below), pupil dilation in the presence of sexually attractive people, crying, smiling (a homologue of acquiescence signals in other primates), and patterns of gossip and self-disclosure. In each case, there are benefits to signaling, cheating is possible, and physical or social vulnerabilities are likely concurrent with the signal.

APPLICATIONS OF COSTLY SIGNALING TO HUMAN COMMUNICATION

The application of costly signaling theory is a three-part enterprise. First, signal content is reverse-engineered using the strategies of evolutionary psychology (see Chapter 3 by Tooby & Cosmides, in this volume). Second, a complementary hypothesis regarding the cost(s) that maintain signal honesty is developed. Third, predictions are deduced from these general hypotheses and are empirically tested.

Aggressive Intent Signals Maintained by Retaliation Costs

Since Darwin (1872), researchers have identified a set of facial and vocal expressions of anger (e.g., lowering brow-ridges, pressing lips against each other, use of a deep, monotonic voice or yell) that develop during infancy and are observed among, and recognized by, people across cultures (Ekman, 2006; Scherer, Johnstone, & Klasmeyer, 2003). Anger expressions produce fear, defense physiology, and withdrawal tendencies in observers (Dimberg & Öhman, 1996; Marsh, Ambady, & Kleck, 2005), suggesting that signalers can use anger expressions to enhance access to fitness-enhancing resources. Thus, anger expressions present a signaling problem. If people who use expressions of anger can gain access to fitness-enhancing resources, what accounts for the honesty of the signal?

Signal Content

Sell, Tooby, and Cosmides (2014) postulated that anger is an emotion program designed to resolve conflicts of interest in favor of the angry person, and that it does so by enhancing the angry person's ability to inflict physical costs. Consistent with this "physicality-enhancing" hypothesis, a man who displays an angry face (e.g., with lowered brow-ridges, raised cheekbones, and pressed lips) (Sell et al., 2014) or an intense yell (i.e., with an elevated pitch level and amplitude) (Sell, Tooby, & Cosmides, 2007) is perceived to be physically stronger than the same man who displays an emotionally neutral face or a non-violent yell.

However, anger expressions may also communicate signalers' aggressive intent (a state) in addition to formidability (a trait). That is, by appearing angry, a person is telling his or her opponent that he or she is willing to deploy the physical force being advertised. Studies suggest that there are two kinds of vocal correlates of human anger: (1) a monotonic voice with low mean fundamental frequency (f_0), or (2) yelling (Banse & Scherer, 1996). Zhang and Reid (2012) found that male and female participants perceived a male speaker who either lowered his voice pitch to address an opponent or yelled to be significantly angrier than when he raised his pitch or presented an emotionless statement. In turn, perceived anger predicted receivers' perceptions of the speaker's likelihood to be aggressive, and did so independent of the speaker's perceived physical strength. It appears that physical strength and aggressive intentions are independent pieces of information that receivers extract from vocal expressions of anger.

Signal Costs

The honesty of anger expressions needs an explanation because such expressions enhance the angry person's chance of obtaining fitness-enhancing resources, and the signal itself is not physically constrained. That is, people of different physicality and/or with different levels of aggressive intent can all appear or sound angry, and that means that anger expressions are readily faked. Zhang (2018) postulated that the first mechanism that maintains the honesty of anger expressions is the production costs of activating an anger response, which involves muscle contraction, blood vessel dilation, and heart rate increases. As such, compared to men in poorer physical condition, men in better physical condition would be more capable of effectively mobilizing physical strength, rendering anger expressions genuine and threatening. It is also likely that other mechanisms maintain the honesty of trait formidability. Namely, low-pitched male voices can function as a dominance signal (Puts et al., 2012) through an indexing mechanism (i.e., taller and stronger men tend to have a lower F_0; Hodges-Simeon, Gurven, Puts, & Gaulin, 2014; Pisanski et al., 2014), or a developmental cost because of higher testosterone, which is immunosuppressive (Hodges-Simeon, Gurven, & Gaulin, 2015). The existence of these receiver-independent mechanisms does not preclude the operation of receiver-dependent costs, especially considering that men can fake dominance by lowering their pitch (Fraccaro et al., 2013), which presumably does not consume much energy.

Specifically, the honesty of human anger expressions may also be maintained by a retaliation cost. Enquist (1985) analyzed a situation wherein signaling in aggressive interactions is free of indexing, production and developmental cost mechanisms, such that all individuals, regardless of their physical strength or valuation of a contested resource (i.e., motivation), are capable of using more intense signals to repel rivals. Through mathematical modeling, Enquist showed that signal honesty (i.e., individuals use a more intense signal only when physically stronger or more motivated compared to a rival) is maintained when more intense signals induce aggression from physically strong or highly motivated signal receivers. When the cost of being attacked by those signal receivers outweighs the benefit of bluffing, a signaling system can be evolutionarily stable.

Much research on nonhuman animals has supported Enquist's (1985) retaliation-cost model (e.g., Anderson, Searcy, Hughes, & Nowicki, 2012; Molles & Vehrencamp, 2001; Popp, 1987). Recently, Zhang and Reid (2017) conducted the first test of the model with humans using low-pitched male voices as stimuli. For the retaliation-cost mechanism to work, more intense signaling (e.g., high-threat displays) must incur aggressive responses from physically strong or highly motivated signal receivers. Given that a low-pitched male voice (relative to a high-pitched voice) is perceived as more physically dominant (thus constituting a more intense signal), the retaliation-cost model predicts that it should induce aggression from men high in threat potential. As hypothesized, Zhang and Reid (2017) found that heterosexual male participants showed enhanced aggressive cognitions (in a word completion task) or reported stronger aggressive behavioral tendencies upon hearing low-pitched male voices when primed with intrasexual competition. Importantly, these effects were moderated by receiver formidability, with more trait-aggressive males being more aggressive, and physically stronger males reporting more aggressive tendencies. These findings provide what we believe is the first direct evidence for the retaliation-cost model in human aggressive signaling.

Laughter as a Trust Barometer Maintained by Vulnerability Costs

Laughter has the hallmarks of having evolved. Laughter in all of its characteristic forms is found in all human groups; laughter is observed in people who are born deaf and blind; laughter is observed in very young babies (Darwin, 1872); laughter has homologous forms in primates (Davila-Ross, Owren, & Zimmerman, 2009), which suggests that laughter is an ancient adaptation that evolved in a common ancestor who lived 10–16 million years ago; and laughter appears to be involved in the creation and management of social bonds (Dunbar et al., 2011; Provine & Fischer, 1989). Human laughter can take a "voiced" Duchenne form, which is unforced and prosodic, or an unvoiced or non-Duchenne form, such as nasal snorts, cackles, and grunts. Laughter is also used to regulate the flow of conversation (Provine, 1993; Vettin & Todt, 2004) by punctuating (usually unfunny) statements and questions: "Most pre-laugh dialogue is like that of an interminable television situation comedy scripted by an extremely ungifted writer" (Provine, 1993, p. 296). Laughter appears to be a more complex signal than voice pitch shifts, but we are nonetheless left with the same type of puzzle: Why do humans laugh, what is signaled, and, if it is a signal, how is honesty maintained?

Signal Content

There is evidence that Duchenne laughter signals trustworthiness (we use the term "friendly intentions" interchangeably) (cf. Dunbar et al., 2011; Provine & Fischer, 1989). Of course, trustworthiness or friendly intentions are attributions—a conscious proximal mechanism that links beliefs about others to the likelihood that they will behave in ways that provide us with fitness costs or benefits. Because friendly intentions cannot be directly observed, and because there are strong potential fitness consequences associated with knowing who is or is not a reliable ally, we can expect that friendly intention signals are important in evolution. In fact, signals of friendly intentions may make alliances possible among non-genetic relatives, and once an alliance is established, people can regulate a host of mutually beneficial behaviors, such as social exchange, cooperation, and coordinated actions in the face of shared threats.

Consistent with the hypothesis that laughter signals friendly intentions, research demonstrates that laughter is elicited almost exclusively in social situations, such as when people are in a group versus when alone (Provine & Fischer, 1989), among people who are in close proximity (Freedman & Perlick, 1979), or a shared social situation (Chapman, 1975). The social bond between people is also important for eliciting laughter. Foot, Chapman, and Smith (1977) found that laughter among children who were friends was of greater duration and more frequent than that among children who were unacquainted, and this was particularly true for girls who were interacting with boys. There is also comparative evidence for the social bond hypothesis. Infant humans, chimps, siamangs, gorillas, bonobos, and orangutans all laugh when tickled by humans and conspecifics that they are familiar with and bonded to (Davila-Ross et al., 2009). It appears that laughter is elicited in social situations among individuals who share a social bond.

But if primates have established a social bond, why would they continue to laugh together sometimes intensely and at other times not at all? We suspect that laughter does not simply reflect established social bonds; rather, laughter functions as a barometer of dynamically changing social bonds because social alliances and situations are by nature fluid. For example, Davila-Ross, Allcock, Thomas, and Bard (2011) found that laughter was common among orphaned chimps in newly established groups, which suggests that laughter can probe others' friendly intentions. Anecdotal evidence also suggests that humans are particularly likely to laugh in threatening situations. For example, Darwin (1872) described a case of German soldiers readily laughing at "small" jokes during the siege of Paris. Our barometer hypothesis suggests that this laughter is a test of the strength of social bonds when dissolution of the alliance could have severe fitness costs. There is also evidence for a similar hypothesis for the production of and reactions to humor in romantic relationships (Li, Griskevicius, Durante, Jonason, Pasisz, & Aumer, 2009). Li et al. (2009) argued that humor can monitor interest in romantic relationships, such as relationship initiation, as well as monitor existing relationships. Consistent with their hypothesis, Li et al. found that both men and women are more likely to initiate humor with people they are attracted to and rate the humor of attractive people as being more funny.

If laughter contains a signal, then there must be benefits for signalers. Indeed, compared with non-Duchenne laughter, listeners are more interested in meeting people who produce Duchenne laughter, experience more favorable emotions upon hearing it, think it more appropriate as an accompaniment for a laugh track, and perceive the laugher as more friendly and more sexually attractive (Li et al., 2009), particularly if the laugher is female (Bachorowski, Smoski, & Owren, 2001). Laughter appears to be a signal to ascertain who would make a good alliance or romantic partner.

Signal Costs

If Duchenne laughter reliably indicates a signaler's trust in receivers, then signal reliability is likely maintained by costs. We propose that honesty is maintained by another receiver-dependent cost: increased vulnerability to the aggressive acts of others (Zahavi, 1982). Common experience suggests that laughing intensely can be painful, hence the terms "side-splitting laughter" and "I was in stitches." At the same time, more intense laughter comes with tears ("I laughed until I cried") and Duchenne smiles that involve much involuntary use of the *orbicularis oculi* muscles. Tears (rendering vision impaired) and heavily taxed abdominal muscles (increasing vulnerability to abdominal blows) would presumably make the laughing person less able to respond effectively to physical threats. Thus, people would risk intense Duchenne laughter only when in the presence of people they trust. Receivers can therefore be sure that

Duchenne laughter is an honest signal of a social bond, and thereby trust the signaler as a committed coalitional partner.

But does that mean vulnerability costs maintain the honesty of only intense Duchenne laughter? Even moderate-intensity Duchenne laughter may be physically taxing, especially when repeated. Duchenne laughter involves constant exhalation and rhythmic movements of abdominal muscles. Further, the acoustic features of laughter show that it is typically high in amplitude and has an abrupt onset with a high and oscillating f_0 compared with regular speech (Bachorowski et al., 2001; Provine & Yong, 1991; Vettin & Todt, 2004). Direct evidence that laughter represents a high energetic cost is found in research on laughter and pain tolerance. Cogan, Cogan, Waltz, and McCue (1987) exposed participants to 20-minute audio tapes designed to induce laughter, relaxation, or boredom, and then measured discomfort thresholds using a blood pressure cuff. Both men and women in the laughter group tolerated greater pain than those in the other conditions. Further still, in six experiments conducted in the laboratory and a natural context, Dunbar et al. (2011) found that pain tolerance was increased by social laughter. These findings all suggest that laughter is a taxing physical activity, because it would not otherwise release endogenous opiates, and that moderate intensity laughter may be sufficiently costly for vulnerability costs to apply.

If the vulnerability cost hypothesis is correct, a number of predictions follow. Due to increased vulnerability to attack (because of distraction and/or physical depletion), people who have recently engaged in a substantial bout of Duchenne laughter should be less able to respond quickly to a startling stimulus, and the more the laughter, the lower the ability to respond. The speed and effectiveness with which people can mount a fight-or-flight response should decrease with increases in the amount of laughter they have engaged in. We should also find that physical strength is compromised by a recent bout of laughter, and the extent of the strength decrease should be greater for those who laughed more.

CONCLUSION

This chapter outlined the contours of costly signaling theory. Many substantive issues regarding the nature of signals and mechanisms remain a source of new research and have not been hinted at here (see Maynard Smith & Harper, 2003; Searcy & Nowicki, 2005).

There are countless avenues for new research on costly signaling. For example, in both of our detailed examples we made use of receiver-dependent costs—specifically, retaliation and vulnerability cost mechanisms. These kinds of mechanisms are most readily applicable when the phenomenon being explored involves situational and motivational signals that are modulated by signalers. This potentially includes the six basic emotional expressions identified by Darwin (1872), as well as pride and guilt, and yawn contagion, among others. When it comes to traits rather than states, however, developmental, production, and maintenance costs are more likely involved. For example, languages vary in complexity (Lupyan & Dale, 2010), and some individuals are more adept at producing creative, complex language than others (see Miller, 2001). It is possible that languages and individual differences in linguistic ability are amenable to a signaling analysis. Indeed, recent research suggests that more complex language can be sexually attractive, which invites cheating (see Lange, Zaretsky, Schwarz, & Euler, 2014). Other traits such as health, sexual orientation, and personality are also likely affected by developmental, and maintenance costs. Costly signaling theory will ultimately serve as an organizing framework for much of our understanding of human nonverbal communication.

Other work may consider the multi-modal nature of communication signals. For example, sexually selected ornaments communicate information to potential mates, and more elaborate ornaments are more sexually attractive. But what explains why humans consider large breasts, full lips, and a waist-to-hip ratio of 0.7 as most physically attractive in women? Do these different traits communicate unique or redundant information about the quality of the signaler? Disentangling multiple-message and redundant message hypotheses is likely to be a source of much ongoing theorizing and research.

Deception research in communication science is also likely to be viewed in a new light when a signaling analysis is applied. The majority of current work is focused on the proximal processes involved in deception detection without considering how deception could evolve, who might be best at it, when, and why. Costly signaling theory requires deception to co-exist with honesty, and states that the level of deception found is proportional to the costs that receivers are willing to tolerate. In some areas, deception is likely to be highly tolerable and difficult to detect, simply because the costs of deception are low. When the costs are relatively high, however, we expect stringent penalties to be applied to deception, and that this would tend to keep the rate of deception in check.

Costly signaling theory will not only cast our field's current questions in a new light. It is also likely to lead to new questions and new discoveries.

REFERENCES

Adamo, S. A., & Hanlon, R. T. (1996). Do cuttlefish (Cephalopoda) signal their intentions to conspecifics during agonistic encounters? *Animal Behaviour, 52*, 73–81.

Anderson, R. C., Searcy, W. A., Hughes, M., & Nowicki, S. (2012). The receiver-dependent cost of soft song: A signal of aggressive intent in songbirds. *Animal Behaviour, 83*, 1443–1448.

Apicella, C. L., Feinberg, D. R., & Marlowe, F. W. (2007). Voice pitch predicts reproductive success in male hunter-gatherers. *Biology Letters, 3*, 682–684.

Bachorowski, J. A., Smoski, M. J., & Owren, M. J. (2001). The acoustic features of laughter. *Journal of the Acoustical Society of America, 110*, 1581–1597.

Banse, R., & Scherer, K. R. (1996). Acoustic profiles in vocal emotion expression. *Journal of Personality and Social Psychology, 70*, 614–636.

Chapman, A. J. (1975). Humorous laughter in children. *Journal of Personality and Social Psychology, 31*, 42–49.

Cimino, A. (2011). The evolution of hazing: Motivational mechanisms and the abuse of newcomers. *Journal of Cognition and Culture, 11*, 241–267.

Clutton-Brock, T. H., & Albon, S. D. (1979). The roaring of red deer and the evolution of honest advertisement. *Behaviour, 69*, 145–170.

Cogan, R., Cogan, D., Waltz, W., & McCue, M. (1987). Effects of laughter and relaxation on discomfort thresholds. *Journal of Behavioral Medicine, 10*, 139–144.

Darwin, C. (1872). *The expression of the emotions in man and animals*. London, England: John Murray.

Davila-Ross, M., Allcock, B., Thomas, C., & Bard, K. A. (2011). Aping expressions? Chimpanzees produce distinct laugh types when responding to laughter of others. *Emotion, 11*, 1013–1020.

Davila-Ross, M., Owren, M. J., & Zimmermann, E. (2009). Reconstructing the evolution of laughter in great apes and humans. *Current Biology, 19*, 1106–1111.

Dawkins, R. (1976). *The selfish gene*. Oxford, England: Oxford University Press.

Dimberg, U., & Öhman, A. (1996). Behold the wrath: Psychophysiological responses to facial stimuli. *Motivation and Emotion, 20*, 149–182.

Dunbar, R. I. M., Baron, R., Frangou, A., Pearce, E., van Leeuwin, E. J. C., Stow, J., ... van Vugt, M. (2011). Social laughter is correlated with an elevated pain threshold. *Proceedings of the Royal Society B: Biological Sciences, 279*, 1161–1167.

Ekman, P. (2006). Cross-cultural studies of facial expression. In P. Ekman (Ed.), *Darwin and facial expression: A century of research in review* (pp. 169–220). Los Altos, CA: Malor Books.

Enquist, M. (1985). Communication during aggressive interactions with particular reference to variation in choice of behaviour. *Animal Behaviour, 33*, 1152–1161.
Fessler, D. M., Tiokhin, L. B., Holbrook, C., Gervais, M. M., & Snyder, J. K. (2014). Foundations of the Crazy Bastard Hypothesis: Nonviolent physical risk-taking enhances conceptualized formidability. *Evolution and Human Behavior, 35*, 26–33.
Foot, H. C., Chapman, A. J., & Smith, J. R. (1977). Friendship and social responsiveness in boys and girls. *Journal of Personality and Social Psychology, 35*, 401–411.
Fraccaro, P. J., O'Connor, J. J., Re, D. E., Jones, B. C., DeBruine, L. M., & Feinberg, D. R. (2013). Faking it: Deliberately altered voice pitch and vocal attractiveness. *Animal Behaviour, 85*, 127–136.
Freedman, J. L., & Perlick, D. (1979). Crowding, contagion and laughter. *Journal of Experimental Social Psychology, 15*, 295–303.
Grafen, A. (1990a). Sexual selection unhandicapped by the Fisher process. *Journal of Theoretical Biology, 144*, 473–516.
Grafen, A. (1990b). Biological signals as handicaps. *Journal of Theoretical Biology, 144*, 517–546.
Griskevicius, V., Tybur, J. M., Sundie, J. M., Cialdini, R. B., Miller, G. F., & Kenrick, D. T. (2007). Blatant benevolence and conspicuous consumption: When romantic motives elicit strategic costly signals. *Journal of Personality and Social Psychology, 93*, 85–102.
Hamilton, W. D. (1963). The evolution of altruistic behavior. *The American Naturalist, 97*, 354–356.
Hodges-Simeon, C. R., Gurven, M., & Gaulin, S. J. (2015). The low male voice is a costly signal of phenotypic quality among Bolivian adolescents. *Evolution and Human Behavior, 36*, 294–302.
Hodges-Simeon, C. R., Gurven, M., Puts, D. A., & Gaulin, S. J. (2014). Vocal fundamental and formant frequencies are honest signals of threat potential in peripubertal males. *Behavioral Ecology, 25*, 984–988.
Hurd, P. L., & Enquist, M. (2005). A strategic taxonomy of biological communication. *Animal Behaviour, 70*, 1155–1170.
Johnstone, R. A. (1997). The evolution of animal signals. In J. R. Krebs & N. B. Davies (Eds.), *Behavioural ecology: An evolutionary approach* (4th ed., pp. 155–178). Malden, MA: Blackwell.
Kreiman, J., & Sidtis, D. (2011). *Foundations of voice studies: An interdisciplinary approach to voice production and perception*. Chichester, England: Wiley.
Laidre, M. E. (2007). Vulnerability and reliable signaling in conflicts between hermit crabs. *Behavioral Ecology, 18*, 736–741.
Lange, B. P., Zaretsky, E., Schwarz, S., & Euler, H. A. (2014). Words won't fail: Experimental evidence on the role of verbal proficiency in mate choice. *Journal of Language and Social Psychology, 33*, 482–499.
Lassek, W. D., & Gaulin, S. J. (2009). Costs and benefits of fat-free muscle mass in men: Relationship to mating success, dietary requirements, and native immunity. *Evolution and Human Behavior, 30*, 322–328.
Li, N. P., Griskevicius, V., Durante, K. M., Jonason, P. K., Pasisz, D. J., & Aumer, K. (2009). An evolutionary perspective on humor: Sexual selection or interest indication? *Personality and Social Psychology Bulletin, 35*, 923–936.
Ligon, R. A., & McGraw, K. J. (2016). Social costs enforce honesty of a dynamic signal of motivation. *Proceedings of the Royal Society B: Biological Sciences, 283*, 20161873.
Lupyan, G., & Dale, R. (2010). Language structure is partly determined by social structure. *PloS One, 5*, e8559.
Marsh, A. A., Ambady, N., & Kleck, R. E. (2005). The effects of fear and anger facial expressions on approach and avoidance-related behavior. *Emotion, 5*, 119–124.
Maynard Smith, J. (1976). Sexual selection and the handicap principle. *Journal of Theoretical Biology, 57*, 239–242.
Maynard Smith, J., & Harper, D. (2003). *Animal signals*. Oxford, England: Oxford University Press.
Miller, G. F. (2001). *The mating mind: How sexual choice shaped the evolution of human nature*. London, England: Vintage/Ebury.
Molles, L. E., & Vehrencamp, S. L. (2001). Songbird cheaters pay a retaliation cost: Evidence for auditory conventional signals. *Proceedings of the Royal Society of London B: Biological Sciences, 268*, 2013–2019.

Nowicki, S., Peters, S., & Podos, J. (1998). Song learning, early nutrition and sexual selection in songbirds. *American Zoologist, 38*, 179–190.

Osiejuk, T. S., Łosak, K., & Dale, S. (2007). Cautious response of inexperienced birds to conventional signal of stronger threat. *Journal of Avian Biology, 38*, 644–649.

Pisanski, K., Fraccaro, P. J., Tigue, C. C., O'Connor, J. J., Röder, S., Andrews, P. W., … Feinberg, D. R. (2014). Vocal indicators of body size in men and women: A meta-analysis. *Animal Behaviour, 95*, 89–99.

Popp, J. W. (1987). Risk and effectiveness in the use of agonistic displays by American goldfinches. *Behaviour, 103*, 141–156.

Provine, R. R. (1993). Laughter punctuates speech: Linguistic, social and gender contexts of laughter. *Ethology, 95*, 291–298.

Provine, R. R., & Fischer, K. R. (1989). Laughing, smiling and talking: Relation to sleeping and social context in humans. *Ethology, 83*, 295–305.

Provine, R. R., & Yong, Y. L. (1991). Laughter: A stereotyped human vocalization. *Ethology, 89*, 115–124.

Puts, D. A., Apicella, C. L., & Cárdenas, R. A. (2012). Masculine voices signal men's threat potential in forager and industrial societies. *Proceedings of the Royal Society of London B: Biological Sciences, 279*, 601–609.

Reby, D., & McComb, K. (2003). Anatomical constraints generate honesty: Acoustic cues to age and weight in the roars of red deer stags. *Animal Behaviour, 65*, 519–530.

Rohwer, S., & Ewald, P. W. (1981). The cost of dominance and advantage of subordination in a badge signaling system. *Evolution, 35*, 441–454.

Scheel, D., Godfrey-Smith, P., & Lawrence, M. (2016). Signal use by octopuses in agonistic interactions. *Current Biology, 26*, 377–382.

Searcy, W. A., & Nowicki, S. (2005). *The evolution of animal communication: Reliability and deception in signaling systems*. Princeton, NJ: Princeton University Press.

Sell, A., Tooby, J., & Cosmides, L. (2007, May). Violent yells dissected: Physical strength is revealed in the voice and enhanced during anger. Paper presented at the annual convention of the Human Behavior and Evolution Society, Williamsburg, VA.

Sell, A., Tooby, J., & Cosmides, L. (2009). Formidability and the logic of human anger. *Proceedings of the National Academy of Sciences, 106*, 15073–15078.

Sell, A., Tooby, J., & Cosmides, L. (2014). The human anger face evolved to enhance cues of strength. *Evolution and Human Behavior, 35*, 425–429.

Scherer, K. R., Johnstone, T., & Klasmeyer, G. (2003). Vocal expression of emotion. In R. J. Davidson, K. R. Scherer, & H. H. Goldsmith (Eds.), *Handbook of affective sciences* (pp. 433–456). New York, NY: Oxford University Press.

Sosis, R., Kress, H. C., & Boster, J. S. (2007). Scars for war: Evaluating alternative signaling mechanisms for cross-cultural variance in ritual costs. *Evolution and Human Behavior, 28*, 234–247.

Tibbetts, E. A., & Dale, J. (2004). A socially enforced signal of quality in a paper wasp. *Nature, 432*, 218–222.

Van Vugt, M., & Hardy, C. L. (2010). Cooperation for reputation: Wasteful contributions as costly signals in public goods. *Group Processes & Intergroup Relations, 13*, 101–111.

Vettin, J., & Todt, D. (2004). Laughter in conversation: Features of occurrence and acoustic structure. *Journal of Nonverbal Behavior, 28*, 93–115.

Wang, Y., & Griskevicius, V. (2014). Conspicuous consumption, relationships, and rivals: Women's luxury products as signals to other women. *Journal of Consumer Research, 40*, 834–854.

Williams, G. C. (1966). *Adaptation and natural selection: A critique of some current evolutionary thought.* Princeton, NJ: Princeton University Press.

Zahavi, A. (1975). Mate selection: A selection for a handicap. *Journal of Theoretical Biology, 53*, 205–214.

Zahavi, A. (1977). The cost of honesty (further remarks on the handicap principle). *Journal of Theoretical Biology, 67*, 603–605.

Zahavi, A. (1979). The fallacy of conventional signaling. *Philosophical Transactions of the Royal Society of London B: Biological Sciences, 340*, 227–230.
Zahavi, A. (1982). The pattern of vocal signals and the information they convey. *Behaviour, 80*, 1–8.
Zahavi, A., & Zahavi, A. (1997). *The handicap principle*. Oxford, England: Oxford University Press.
Zhang, J. (2018). The human anger face likely carries a dual-signaling function. *Frontiers in Behavioral Neuroscience, 12*, article 26.
Zhang, J., & Reid, S. A. (2012). Lowering voice-pitch and yelling make men sound more likely to aggress: Evidence for an aggressive-intention signaling hypothesis of vocal anger expressions. Unpublished manuscript, University of California, Santa Barbara.
Zhang, J., & Reid, S. A. (2017). Aggression in young men high in threat potential increases after hearing low-pitched male voices: Two tests of the retaliation-cost model. *Evolution and Human Behavior, 38*, 513–521.

5

Evolution, Structure, and Functions of Human Laughter

Gregory A. Bryant

Vocal communication is fundamental to human and nonhuman social interaction and is remarkably ancient and widespread across taxa (Bass, Gilland, & Baker, 2008). It is not particularly mysterious why this form of communication is so pervasive in nature—vocalizations allow animals to rapidly communicate, sometimes at great distances, as well as in coordination with one another and to potentially large audiences simultaneously. Vocal emotion production is highly conserved in mammals, meaning that selection has retained underlying brain circuitry responsible for affective vocal control, making it quite similar across many otherwise dissimilar species (Ackermann, Hage, & Ziegler, 2014; Jürgens, 2002). Humans maintain a large repertoire of vocalizations that can be studied effectively through a comparative lens, but few afford such an analysis as well as laughter. This chapter will describe human laughter from an evolutionary perspective, including discussion of its basic acoustic structure, phylogenetic origins, and social communicative functions. The study of laughter is a burgeoning research topic that provides a window into not only human vocal communication and evolution, but also the complex social cognitive niche that humans have created.

Like all biological adaptations, vocal signals follow basic principles of form and function (Owren & Rendall, 2001). An analysis of the physical properties of vocalizations can help elucidate the adaptive problems the signals are designed to solve. Some structural features are more obvious than others. For example, the loud and acute dissonance of infant crying is easy to understand as a sound that motivates adaptive caretaker action. By causing a parent to induce crying cessation, the signal enhances the fitness of both parent and offspring, driving the signaling system's evolution and design (Lummaa, Vuorisalo, Barr, & Lehtonen, 1998). If crying sounded like gentle birdsong, it would not be nearly as effective. Similarly, we can understand laughter as being a signal shaped by selection to induce positive affect for mutual benefit in senders and receivers on average, but the form-function relationship is not quite as obvious as in crying. Before we explore this issue, we will consider the specific acoustic properties of laughter, which, along with its phylogenetic history, provide clues regarding its multiple possible communicative functions.

THE ACOUSTIC STRUCTURE OF LAUGHTER

What is a laugh? Most basically, laughter is a nonverbal vocalization characterized by a specific coordination of rhythmic respiratory and laryngeal activity (Bachorowski, Smoski, & Owren, 2001; Citardi, Yanagisawa, & Estill, 1996; Luschei, Ramig, Finnegan, Bakker, & Smith, 2006; Provine, 2000; Titze, Finnegan, Laukkanen, Fuja, & Hoffman, 2008). Laughter typically occurs as a sequence of acoustic bursts, collectively known as a bout—but single laugh bursts occur with some regularity. The energy underlying laugh bouts arises from rhythmic pulsing of abdominal and intercostal muscles, forcing air through the glottis, situated in the larynx, housing the vocal folds. Initial bursts often contain the greatest energy, which then decay over time in a near-linear fashion, often affecting both loudness and perceived pitch (Titze et al., 2008). A unique feature of laughter is the rapid opening and closing of the glottis (and hence the vocal folds), resulting in oscillating bursts of energy containing tonal and nontonal (i.e., voiced and unvoiced) components. The tonal elements arise due to the brief closures of the vocal folds and subsequent vibration regimes that typically generate a fundamental frequency (f_o) associated with perceived pitch. Along with harmonic structure associated with the voice source, resonances of the vocal tract result in formant structure in laughs that determines perceived vowel sounds. Laughter generally consists of stable vowel configurations within bouts (*ha-ha-ha* rather than *ha-he-ha*) (Bachorowski, Smoski, & Owren, 2001; Provine, 2000). One cycle of closing and opening the glottis constitutes a single laugh burst, with the process often continuing for several seconds, manifesting as a structured series of complex vocal sounds. Complexity here refers to the multiple dimensions of acoustic variability we see in laughter, including not only the tonal (i.e., periodic) components such as f_o and formants, but also nontonal spectral features associated with perceived voice quality. Because of occasional high intensity moments, laugh bouts can contain chaotic nonlinear phenomena such as subharmonics (i.e., bands of energy between harmonics due to period doubling) and deterministic chaos (broadband noise) (Bachorowski, Smoski, & Owren, 2001; Fitch, Neubauer, & Herzel, 2002). These features can contribute to a variety of subjective sounds such as hoarseness, roughness, breathiness, and of course loudness.

Figure 5.1 shows the acoustic similarities between a chimpanzee play vocalization (recording courtesy of M. Davila-Ross) and an ingressive (i.e., inward airflow) laugh produced by an actor (Bill Paxton as Chet in the 1985 movie *Weird Science*). When slowed down two and a half times, this ingressive laugh sounds like a nonhuman animal, much like the spontaneous laughs tested by Bryant and Aktipis (2014). High-intensity, ingressive vocal effort often results in noisy features, such as deterministic chaos, which is evident in the example in Figure 5.1 (right column, first row). A completely ingressive laugh in humans such as the one displayed here is highly unusual in ordinary discourse. But ingressive components tied with breathing are quite common at laugh onsets and offsets. An example in Figure 5.1 shows a typical manifestation of an ingressive laugh offset (left column, third row). Laughs composed of predominantly voiced components are often perceived as being volitional (or "fake"), likely due to their apparent production by the speech system (Bryant & Aktipis, 2014). The infamous laughter of Hillary Clinton illustrates this well (right column, third row). The laugh bursts are regularly spaced, highly voiced, and contain relatively low acoustic variation across bursts, including in f_o, spectral, and duration properties. Volitional laughs functioning as backchannels (i.e., brief utterances while listening in conversation) are often low in variation both within and between bouts.

The portions of laughter linked to the closing stage of the glottal cycle are important for the sound of a laugh (e.g., f_o dynamics), but the intervals between those bursts are also crucial, and

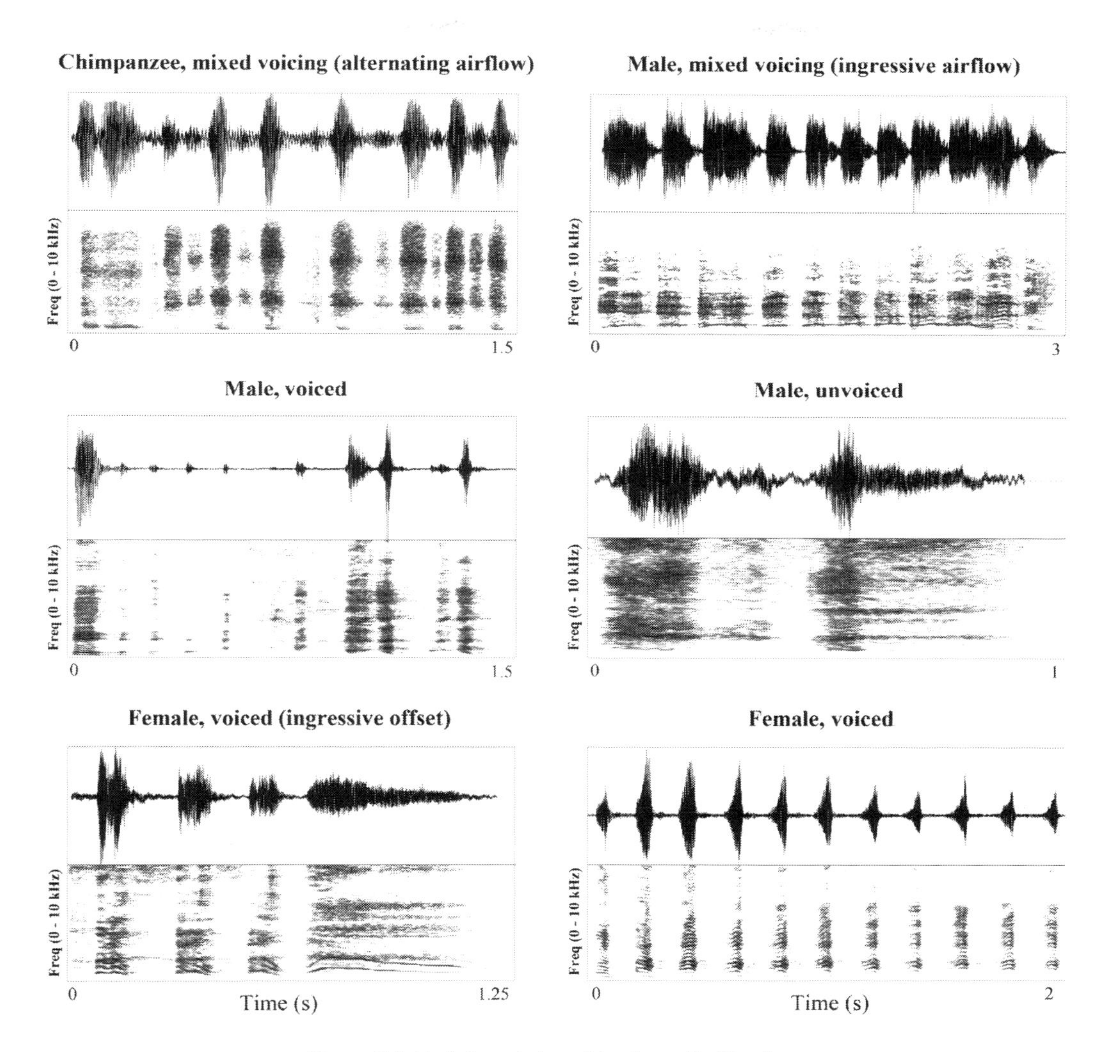

Figure 5.1 Variations in Laughter Acoustic Structure

can provide unique information regarding the nature of a laugh. One measure we developed—rate of intervoicing interval—quantifies the averaged proportion of time that a laugh spends in the open portion of the glottal stage across a laugh bout, and captures an important dimension linked to breath control. Perceptually, the measure is positively associated with judgments of laughter being spontaneous, as well as listeners confusing slowed versions of spontaneous laughter with nonhuman vocalizations (Bryant & Aktipis, 2014). Breath control is a crucial mechanistic difference between the evolutionarily conserved vocal emotion system and the human species-specific speech system. At some point during relatively recent hominin evolution, increased thoracic innervation afforded fine control over breathing dynamics that facilitated the production of long linguistic phrases with well-timed inspiratory breaks (MacLarnon & Hewitt, 1999). Consequently, vocalizations produced by the speech system contain the signature of speech-timed glottal dynamics. We will return to this issue below.

Beyond the immediate physiology of laugh bursts, researchers have described the ways that laughter production coordinates with other kinds of vocalizations, namely, speech. It appears that laughter often follows simple production rules. For example, Provine (1993) described the ways people laugh during ordinary talk, including the ways laughs are placed relative to words and sentences. Speakers punctuate sentences with laughs instead of placing them randomly within them. This rule-governed organization is due primarily to the different cortical systems underlying vocal emotion production and speech—systems that must share connections to the vocal apparatus. This fact explains phenomena such as people's inability to speak while their breathing is labored or while they are involuntarily laughing, crying, or screaming.

LAUGHTER PHYLOGENY

We can learn much about laughter by looking at evolutionarily related vocal behaviors in other primates, especially the great apes. The comparative method affords important insights into the physical forms of signals and their connections to communicative functions. Human laughter is well established as homologous to nonhuman play vocalizations (Gervais & Wilson, 2005; Panksepp & Burgdorf, 2003; Provine, 2000; Ruch & Ekman, 2001; van Hooff, 1972). Many social mammals produce vocalizations during physical play, including rats (Knutson, Burgdorf, & Panksepp, 1998), dogs (Feddersen-Petersen, 2000), chimpanzees (Davila-Ross, Allcock, Thomas, & Bard, 2011; Matsusaka, 2004), squirrel monkeys (Biben & Symmes, 1986), and other primate species (Masataka & Kohda, 1988). Play vocalizations provide an excellent example of the concept of ritualization—a process of signal evolution proposed by behavioral biologists (Krebs & Dawkins, 1984; Tinbergen, 1952). Many ritualized signals begin as cues, which can be any behavior that reliably provides some kind of predictive information to perceivers. Mutual recurrent benefits between producers and perceivers from a given pattern of predictable interaction can result in an escalation and ratcheting of signal properties and corresponding response biases (Maynard Smith & Harper, 2003). The ritualization process follows the dynamics of arms races. For example, costly signals of genetic quality in competitive mating contexts can become increasingly exaggerated in response to female resistance and choosiness.

In the case of play vocalizations, the heavy breathing associated with excessive play provides honest information about the physical state of the participants and reveals vulnerability. The continued play indicates a willingness to be vulnerable, trust in the social partner, and an investment in the action itself. Play can function in a variety of ways to prepare animals for situations they are likely to encounter as adults. For example, play fighting helps calibrate muscle systems for physical conflict (Byers & Walker, 1995), and play chasing can tune systems designed for predatory skills of hunting and killing prey (Smith, 1982). But as animals are engaged in the playful encounters—preparing for adult contexts—some of their behaviors could be potentially construed as aggressive and not playful. The exaggerated production of a by-product behavior, such as heavy breathing and panting that become ritualized into affective voicing, can reassure a co-participant during play that one's intentions are benign. The ritualized transformation of by-product vocal cues into laughter signals can contribute to increases in the frequency and intensity of play, benefitting participants over evolutionary time.

The connection between human laughter and nonhuman animal play vocalizations is most evident in acoustic analyses. Davila-Ross, Owren, and Zimmermann (2009) examined audio recordings of six ape species, including humans, producing tickle-induced vocalizations, and developed a phylogeny (i.e., an evolutionary history) of the vocal behavior based on acoustic comparisons. Their reconstructed phylogeny of laughter suggests that the last common ancestor of all extant ape species, who lived approximately 25 million years ago, likely produced a play vocalization characterized by long, noisy bursts, mostly through eggressive (i.e., outward) breath flow. Most generally, play vocalizations have evolved to be more tonal; that is, they increasingly include vocal vibration regimes (perceived as pitch), and the calls have become shorter and sometimes alternating in airflow, such as in chimpanzees and bonobos. But this class of vocalizations in humans has become incorporated into much broader communicative contexts that include verbal interaction, deeply interwoven with signaling affiliation and cooperative intentions. Of course, this includes the so-called darker side of laughter, such as taunting and ironic laughing that can be quite hostile and aggressive to particular targets. The origins of human laughter in the ritualized labored breathing of our ape ancestors help explain its unique vocal features and its affective source.

An additional factor in humans that dramatically complicates the theoretical picture for laughter is the evolution of the speech capacity that works in conjunction with vocal emotions. The dual pathway model has consequences for understanding all vocal behavior (Owren, Amoss, & Rendall, 2011). In social mammals that produce affective vocalizations, the same basic circuit underlies vocal control: a projection from the anterior cingulate, routed through the periaqueductal grey (PAG) area into laryngeal muscles (Jürgens, 2002). But one of the defining characteristics of modern humans is the ability to produce articulated speech sounds, attributable to species-specific neural projections connecting motor cortical areas to laryngeal musculature (Ackermann, Hage, & Ziegler, 2014). Having volitional control over the vocal apparatus affords the capability to mimic the repertoire of vocalizations originally produced by the emotional vocal system. As humans, we can produce "fake" cries, pain shrieks, fear screams, orgasm calls, and laughs. Volitional forms of vocalizations can be produced for deceptive and cooperative reasons, and the game-theoretic dynamics of those strategies must work in the context of evolved linguistic and emotional signaling systems. Difficulties in neuroimaging techniques have contributed to uncertainties regarding the independence of these vocal production routes, and research suggests these systems might interact in complex ways (Pisanski, Cartei, McGettigan, Raine, & Reby, 2016), further complicating the theoretical issues.

Although all vocalizations generated by the vocal emotion system have volitional counterparts, laughter is the only one that has been examined empirically for physical characteristics and perceptual correlates of different production types. Many proposed taxonomies of laughter distinguish between spontaneous (i.e., genuine/involuntary) and volitional (i.e., deliberate/voluntary) forms (e.g., Gervais & Wilson, 2005; Keltner & Bonanno, 1997). A growing body of research reveals that listeners are quite adept at distinguishing these laugh types (Brown, Sacco, & Young, 2018; Bryant & Aktipis, 2014; Lavan, Rankin, Lorking, Scott, & McGettigan, 2017; Lavan, Scott, & McGettigan, 2016; McGettigan et al., 2013; Wood, Martin, & Niedenthal, 2017), including across widely disparate cultures (Bryant et al., 2018). Moreover, neuroimaging research reveals differential activation of brain regions during both the production and perception of spontaneous and volitional laughter (Lavan et al., 2017; McGettigan et al., 2013; Szameitat et al., 2010). Because the laugh types are generated using different vocal production systems, there are predictable and documented acoustic differences, many attributable to differences in arousal during production, but also due to differential breath control in the vocal emotion versus speech systems. Spontaneous laughs typically have higher f_o, shorter burst duration, as well as fewer voiced elements, including a higher rate of intervoicing intervals (Bryant & Aktipis, 2014; Lavan, Scott, & McGettigan, 2016).

In sum, laughter is clearly evolved from ancestral ape play vocalizations, and now manifests itself as a complex signal retaining the ancestral function of signaling positive affect and cooperative intent. The evolution of volitional control over vocal behavior in conjunction with language has resulted in a diversification of laughter's functions as a strategic tool in humans' cognitive niche. We now turn to the issue of how laughter can function in social life.

SOCIAL FUNCTIONS OF LAUGHTER

Human laughter occurs most typically in the context of conversational turn-taking (Jefferson, Sacks, & Schegloff, 1987; Vettin & Todt, 2004). Despite extensive description at this level of analysis, its specific functions remain poorly understood. In social contexts, laughter is often generated by groups of people, and sometimes directed at specific individuals or other groups. The well-known sentiment, "I'm laughing *with* you, not *at* you," speaks to the important social dynamics that unfold related to laughter episodes, even within dyads. A substantive body of research in social psychology has explored the role of in-groups and out-groups in how people are affected by laughter, generally suggesting that laughter is functioning as a means to coordinate sentiment and appraisals of social situations (for a review, see Platow et al., 2005). Most treatments of laughter center on the idea of positive emotions and prosocial intent. The positive subjective feelings behind laughter are likely the product of endogenous opioid activity (Manninen et al., 2017; Wild, Rodden, Grodd, & Ruch, 2003), a process shown even to elevate pain thresholds (Dunbar et al., 2011). The proximate rewards of laughter point to an adaptive communication system, one likely functioning in social groups for the purposes of social assortment and alliance formation (Flamson & Bryant, 2013).

Signaling ongoing affiliation or cooperative intent is widely considered to be an important pragmatic function of laughter and is directly related to the affiliative function of nonhuman play vocalizations described earlier (Brown et al., 2018; Bryant et al., 2016; Bryant & Aktipis, 2014; Curran et al., 2018; Dezecache & Dunbar, 2012; Mehu & Dunbar, 2008; Morisseau et al., 2017; Owren & Bachorowski, 2003; Panksepp, 2005; Platow et al., 2005; Provine, 2016, 2000; Ruch & Ekman, 2001; Scott, Lavan, Chen, & McGettigan, 2014; Wood & Niedenthal, 2018). There are many forms of play in human behavior and most involve laughter at some level, with tickling being the classic, and likely primordial, form. In conversation, verbal play elicits laughter in a variety of contexts, including volitional laughter that can mark recognition or understanding of speakers' intentions, often with play and humor (e.g., Bryant, 2011; Jefferson, 1979). Humor and laughter are traditionally linked, but their relationship is far from simple. Provine (1993) argued that much of what people laugh at is not obviously funny, and more often than not, speakers are the first to laugh in verbal exchanges. This goes against the folk notion that people laugh primarily in response to something funny, and points to more complex signaling dynamics. The encryption theory of humor explains this nicely. According to this approach, a central reason why people generate intentional humor, whether through words, behavior, or other representations such as literature and art, is to allow social agents to assess knowledge, preferences, and beliefs indirectly (Flamson & Barrett, 2008). When humor is used as a social assortment device, laughter can function as a covert signal of shared information, affording adaptive strategies of social alignment through the mutual recognition of subtle, spontaneous signals within groups (Flamson & Bryant, 2013; Lynch, 2011; Smaldino, Flamson, & McElreath, 2018).

Research examining the darker side of laughter resonates well with a social assortment approach. Titze (1996) and others (e.g., Ruch & Proyer, 2008) have described a clinical condition (gelotophobia) in which individuals have a pathological fear of being the target of ridicule indicated specifically by laughter. Although a pathology, this hypersensitivity is an exaggeration of a phenomenon nobody wants to experience. Klages and Wirth (2014) asked people to recall personal incidents when they were the target of exclusive laughter. Relative to recalling being a part of inclusive laughter, or the activity of a typical weekday, exclusive laughter victims were more motivated to aggress, felt more social pain, had more negative moods, and experienced reduced self-evaluation. Papousek et al. (2014) examined responses to laughter as a possible social rejection cue in individuals with gelotophobia. Participants were systematically insulted through an intercom during an arithmetic task, and sometimes the insults were followed by incidental laughter (i.e., made to look accidental and not intentionally tied to the insults). When experiencing the laughter, gelotophobic individuals had marked heart rate deceleration relative to controls. Slowed heart rate is consistent with the notion that social fear can create a "freezing" response that could contribute to stopping an ongoing behavior, as opposed to the fight-or-flight response that generates heart rate acceleration and other responses preparatory for action. Overall, gelotophobes often seem to process laughter as socially relevant in ambiguous situations, revealing a hypersensitivity to the social signal. But even normal individuals have stronger and more elongated neurocognitive emotional processing of insults when they are paired with laughter (Otten, Mann, van Berkum, & Jonas, 2017).

People's disposition and social standing can affect the way they hear laughter, as well as how they produce it. For example, Oveis et al. (2016) found that individuals in a dominant social position (fraternity members versus pledges) produced more dominant sounding laughs, whereas pledges produced more submissive sounding laughs. The dominant laughter was coded as more disinhibited, including higher in pitch and faster in burst rate, both indicative of greater arousal. The difference here could be reflecting a difference in spontaneous versus volitional production, a possibility that the authors do not discuss. Submissive social actors might often refrain from producing genuine social signals, and instead feign emotional engagement deemed as more socially appropriate. Moreover, producing laughter that potentially sounds like an attempt to be dominant could have negative social repercussions. Polite laughs are one strategic option for individuals desiring social approval, as well as a means for signaling one's acceptance of their own current social position. Consistent with this idea, compared to typically developing individuals, boys at risk for antisocial behavior showed lower neural responses to laughter in the supplementary motor area, a region associated with readiness for social interaction (O'Nions et al., 2017). Moreover, boys at risk for psychopathy showed lowered activation in the anterior insula, a region connected to auditory-motor processing and integrating action tendencies with subjective affective feelings. These individuals also had a lowered motivation to colaugh. Clinical data such as these point toward the normal functioning of social-cognitive processes that crucially involve the integration of multiple social signals including laughter.

COLAUGHTER

An important but understudied aspect of laughter is the fact that most laughing occurs in the context of a group, and that people often laugh together. Research typically focuses on the role of laughter between individuals, looking at how people signal to one another, or at least that approach is implied. Theorists often either consider laughter to be some kind of reflection of an internal affective state that is then conveyed to others, or they focus on the proximate factors triggering the laughter, usually having to do with humor and what people consider to be funny. An evolutionary approach requires a functional explanation—when we are talking about expressive behavior in general, we need to think in terms of signals and cues. From this perspective, what is the evidence for colaughter being a group signal? There are many examples of group-produced signals in humans and nonhuman animals alike. Many nonhuman species collectively produce territorial advertisements, alarm calls, and signals of mateships, and humans generate group emotional signals, organized chanting and other choreographed vocal displays, as well as musical performances (Bryant, 2013; Hagen & Bryant, 2003; Hagen & Hammerstein, 2009). Are colaughing individuals inadvertently providing information regarding their social relationship to overhearers (i.e., a cue), or is the colaughter shaped by selection to signal to those outside of the group?

Although not in abundance, research has explored colaughter in a few different ways. People laugh together, and not only is this behavior associated with particular states of relationships, but listeners are sensitive to it. Smoski and Bachorowski (2003a, 2003b) examined antiphonal laughter, which they defined as laughter in response to a social partner's laughter. They looked at how this pattern of laughing occurred in developing friendships as well as between friends and strangers. Women tended to start producing antiphonal laughter in their relationships earlier than men did, and women laughed together more than men overall. Not surprisingly, friends produced antiphonal laughs more than strangers, and females produced them more in mixed-sex pairs than did males. The authors proposed that the laughter reinforced mutual positive emotions between laughers, but never suggested any kind of signaling function outside of the dyads.

More recently, Kurtz and Algoe (2017, 2015) examined the effect of "shared laughter" on personal relationships. Using both video coding for romantic couples (Kurtz & Algoe, 2015), and surveys of people online (Kurtz & Algoe, 2017), the authors reported that shared laughter (minimally defined as simply co-occurring) was associated with increased affiliative feelings and higher perceived personal similarity. Romantic couples who produced more shared laughter in recordings also reported feelings of greater support and closeness with their partners. Dezecache and Dunbar (2012) explored group laughter as a means of social grooming. They observed naturally forming conversational groups and assessed the typical conversational and laughter group size, finding that members of a social group most often laughed in groups of three or four, similar to documented conversational group sizes. By laughing in groups, individuals can essentially groom others in a way that is analogous to physical grooming exhibited by many primate species. These findings, as well as those of Smoski and Bachorowski, point to a basic function of laughter in its proximate role in relationship development. There are endogenous rewards for laughing with others, and we can signal information about our cooperative intentions and feelings of positive affect. But the acoustic design features of laughter suggest there is more to it.

Laughter appears to be well designed for intergroup communication, as it contains acoustic features traditionally associated with wide broadcast and the penetration of noisy environments. This is especially true for colaughter. Wiley (1983) described four features of vocalizations that ensure detectability and reliability: (1) alerting components; (2) conspicuousness; (3) small repertoires; and (4) redundancy. Consider the typical sound features of laughter described earlier. Laughs often begin with a loud, abrupt onset, including high frequency and loudness that can function to alert listeners. The overall sound of laughter stands out and is unique among human vocalizations, making it highly recognizable and conspicuous. Whereas a variety of sounds can constitute laughter, vocalizers generally produce limited inventories of sounds that follow some basic production rules. Common repertoires generally incorporate voiceless glottal fricatives (e.g., /h/) along with central vowels (e.g., /ə/) that together maximize airflow through the vocal tract (Stevens, 1998). Finally, laughter is repetitious with the same components being repeated on short and long timescales. Two other important features of laughter that speak to its communicative function are its contagiousness and overall loudness. One of the most effective triggers of laughter is another person's laughter (Provine, 1992), and when multiple people are laughing together, it is often significantly louder than the sound of ordinary conversation. In Figure 5.2, we see a little over 10 seconds of conversation between two female friends. Twice in this time they laugh together, and the increases in loudness during the laugh bouts relative to their talk are quite visible (and audible). Taken together, a form-function account strongly points toward a broadcast function that is different from the quiet, personal signaling that occurs in play contexts among nonhuman primates. When we laugh together, we appear to be signaling broadly to those outside the immediate communicative group.

Two recent large cross-cultural studies examined the perception of both colaughter and individually produced laughs. How well can listeners extract meaningful information about social relationships from hearing only brief isolated clips of two people laughing together? Indeed, people from all over the world, ranging from hunter-gatherers in Africa to college students across numerous industrialized countries, perform rather well on this task, with some variation—listeners everywhere could reliably identify friends versus strangers from just one-second segments of colaughter (Bryant et al., 2016). The colaughter in this study was extracted from real conversations between either established friends who knew each other on

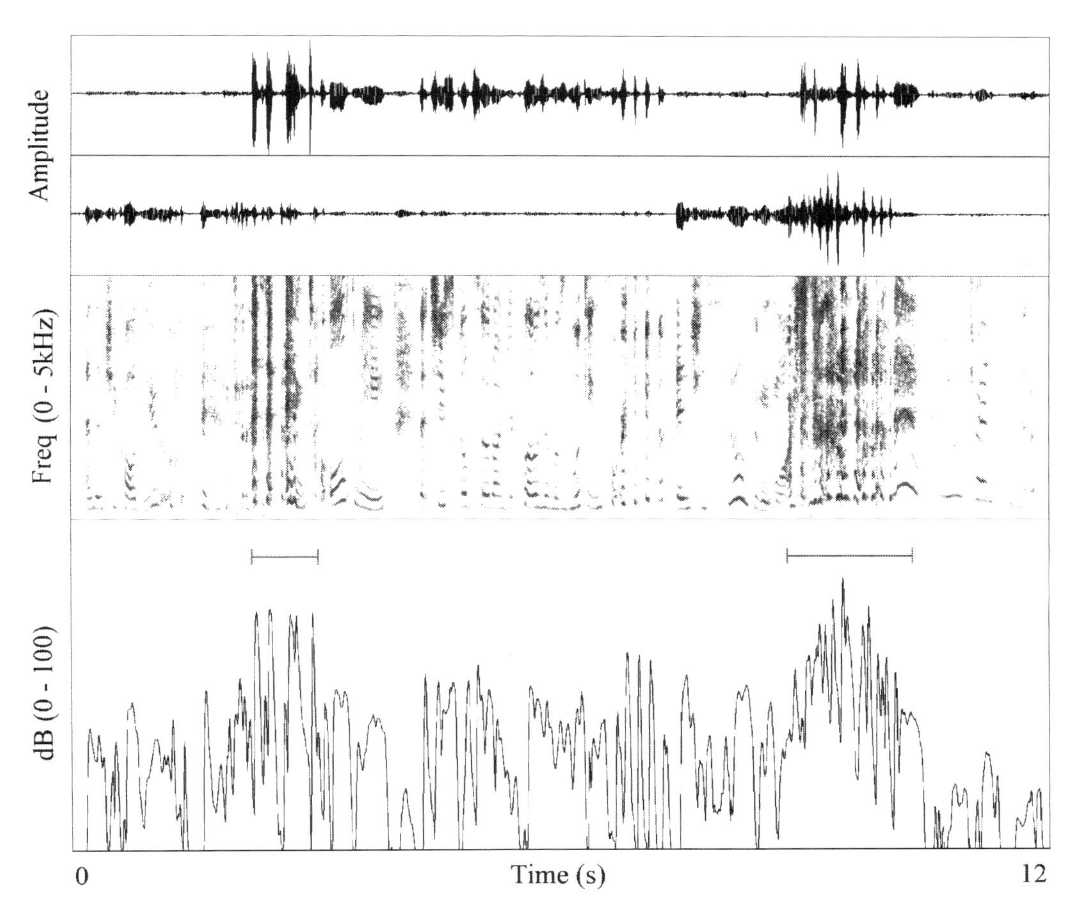

Figure 5.2 Colaughter in Conversational Context

average for about two years, to newly acquainted strangers who had just met moments prior to the recording of the conversations. One somewhat surprising finding was that people everywhere had a slight response bias to judge two women laughing together as more likely to be friends, resulting in accuracy for identifying female friends to be highest in every society. People's intuitions about female colaughter seem to hold across quite disparate societies, and as described earlier, research on antiphonal laughter found that women not only exhibit this behavior more than men, but they begin doing it earlier in their relationships.

Interestingly, listeners everywhere tended to use the same acoustic features in making their judgments. Colaughter that was faster and had irregular dynamics in pitch and loudness was often associated with friends. In a follow-up study, participants from 21 societies (many of the same study sites as the colaughter study) were able to reliably distinguish between spontaneous and volitional laughter (Bryant et al., 2018). In this study, the acoustic correlates of people's judgments were also examined, and similar acoustic features predicted judgments of laughter being spontaneous as in the previous study of colaughter between friends. What is the connection here? The most parsimonious view is that, in both cases, listeners are tracking acoustic correlates of physical arousal in the speakers. When people are in the presence of their friends, arousal is heightened. We speak with more enthusiasm and our emotions are more expressive. If, during interactions with friends, our vocal emotions are triggered more reliably, we will generate more spontaneous expressions, including spontaneous laughter. Moreover, colaughter seems to be particularly efficient in transmitting social information; better than, for example, overlapping talk (i.e., cospeech). A recent study found that listeners were able to more accurately assess whether pairs of interactants were friends or strangers using colaughter than cospeech from the same speakers that was over double in length (Bryant, Wang, & Fusaroli, 2019).

Another source of evidence for the social functions of colaughter comes from infants. Although fairly little developmental research has been done on laughter in children, a few studies have examined the social nature of laughter development. For example, work by Chapman (1973) found that 7-year-old children were much more likely to laugh in the presence of other children than by themselves. Recent work has also shown this effect in younger children, finding that 3- to 4-year-olds laugh significantly more at a cartoon when at least one other child is present, and not really much more if the number of children increases (Addyman, Fogelquist, Levakova, & Rees, 2018). Moreover, when asked about how funny they thought the videos were after watching, children's subjective ratings of funniness were associated with neither how much they laughed or smiled nor whether other children were present or not. These results show that group laughter is calibrated early and it is independent, to some degree, of external proximate triggers beyond other people laughing.

Another recent study explored infants' perception of colaughter. Using the same colaughter recordings as the large cross-cultural study described above, Vouloumanos and Bryant (2019) found that 5-month-old infants preferred to listen to streams of colaughter between friends over the same from strangers. In a second experiment, a different group of 5-month-olds observed pairings of short videos and colaughter. The videos showed two adult women acting either in an affiliative manner (i.e., silent greeting with a wave and then standing facing one another) or a non-affiliative manner (i.e., no greeting, then turning away from one another). These videos were paired with either colaughter between friends or colaughter between strangers. Infants were relatively surprised by incongruent pairs (i.e., they looked longer)—that is, affiliative interactions paired with colaughter between strangers, or non-affiliative interactions with friends' colaughter. Even in the first months of life, infants are prepared to link colaughter to social action. Such an early sensitivity suggests the action of an adaptive perceptual system.

Colaughter appears to be a reliable signal of affiliation that is readily perceived across disparate cultures and detectable by very young infants. It appears more efficient at transmitting social information than other kinds of dynamic information produced between interlocutors, such as overlapping talk. But it also appears that laughter is not well designed to provide specific group-identifying information (Lavan, Short, Wilding, & McGettigan, 2018; Ritter & Sauter, 2017), suggesting that the signaling functions are about affective intentions and potentially group alliance structure as opposed to person identification, or cultural and linguistic background. Future work here should explore the social dimensions that reliably affect colaughter behavior and examine more closely any possible coordinated dynamical features in the colaughter itself. Bryant et al. (2016) failed to find evidence that coordinated acoustic features between the individual laughs making up colaughter pairs contributed to listeners' judgments of affiliation, possibly due to the extreme brevity of the stimulus clips (~1 sec). But behavior matching during interpersonal interaction (e.g., Louwerse, Dale, Bard, & Jeuniaux, 2012) suggests coordinated multimodal features, potentially including laughter, are informative.

CONCLUSION

Human laughter constitutes a family of nonverbal social vocalizations homologous to play vocalizations in nonhuman animals. The evolution of language and articulated speech introduced a second vocal production pathway that afforded multiple interactive functions manifesting in conversational turn-taking and group contexts. Most communicative functions of laughing stem from complex social strategies of coalition formation and maintenance, including humans' universal deep motivation to develop extended social ties for successful long-term cooperative relationships. Fueled by a proximate endogenous reward system, laughter provides one of the many tools people use to signal social information, such as implicit knowledge and preferences. Recent research documents the salient social information available in temporally coincident colaughter, and the widespread ability of listeners to glean rich social information from very thin slices of colaughter stimuli. Laughter appears to function both within dyads and larger interactive groups as well as between groups. A group signaling approach explains certain ubiquitous features of laughter, most notably its loud overall sound and other features that ensure its detectability, including alerting components and conspicuous acoustic structure.

Future research should further explore the fine-grained social-pragmatic functions of laughter and how these signals occur in the context of conversation and social group dynamics. Acoustically, more work is needed to explore cultural universals and variations in laughter structure. For example, we might expect more variation in volitional laughter across languages due to the linguistic constraints on speech development, while spontaneous laughter might be relatively unaffected by that developmental process, and instead manifest itself quite similarly across disparate language and cultural groups. Laughter provides a unique window into human vocal signaling and cooperative behavior, as well as an example of how ancestral communicative behaviors become integrated with later evolving systems. The story of the evolution of laughter is central to the development of our cooperative nature and complex social lives.

REFERENCES

Ackermann, H., Hage, S. R., & Ziegler, W. (2014). Brain mechanisms of acoustic communication in humans and nonhuman primates: An evolutionary perspective. *Behavioral and Brain Sciences, 37*, 529–546.

Addyman, C., Fogelquist, C., Levakova, L., & Rees, S. (2018). Social facilitation of laughter and smiles in preschool children. *Frontiers in Psychology, 9*, 1048.

Bachorowski, J. A., Smoski, M. J., & Owren, M. J. (2001). The acoustic features of human laughter. *Journal of the Acoustical Society of America, 110*, 1581–1597.

Bass, A. H., Gilland, E. H., & Baker, R. (2008). Evolutionary origins for social vocalization in a vertebrate hindbrain–spinal compartment. *Science, 321*(5887), 417–421.

Biben, M., & Symmes, D. (1986). Play vocalizations of squirrel monkeys (Saimiri sciureus). *Folia Primatologica, 46*(3), 173–182.

Brown, M., Sacco, D. F., & Young, S. G. (2018). Spontaneous laughter as an auditory analog to affiliative intent. *Evolutionary Psychological Science, 4*, 1–7.

Bryant, G. A. (2011). Verbal irony in the wild. *Pragmatics & Cognition, 19*, 291–309.

Bryant, G. A. (2013). Animal signals and emotion in music: Coordinating affect across groups. *Frontiers in Psychology, 4*, 990.

Bryant, G. A., & Aktipis, C. A. (2014). The animal nature of spontaneous human laughter. *Evolution and Human Behavior, 35*(4), 327–335.

Bryant, G. A., Fessler, D. M. T., Fusaroli, R., Clint, E., Aarøe, L., Apicella, C. L., … Zhou, Y. (2016). Detecting affiliation in colaughter across 24 societies. *Proceedings of the National Academy of Sciences, 113*(17), 4682–4687.

Bryant, G. A., Fessler, D. M. T., Fusaroli, R., Clint, E., Amir, D., Chávez, B., … Zhou, Y. (2018). The perception of spontaneous and volitional laughter across 21 societies. *Psychological Science, 29*, 1515–1525.

Bryant, G. A., Wang, C. S., & Fusaroli, R. (2019). *Recognizing affiliation in colaughter and cospeech.* Manuscript submitted for publication.

Byers, J. A., & Walker, C. (1995). Refining the motor training hypothesis for the evolution of play. *The American Naturalist, 146*(1), 25–40.

Chapman, A. J. (1973). Social facilitation of laughter in children. *Journal of Experimental Social Psychology, 9*, 528–541.

Citardi, M. J., Yanagisawa, E., & Estill, J. (1996). Videoendoscopic analysis of laryngeal function during laughter. *Annals of Otology, Rhinology, and Laryngology, 105*, 545–549.

Curran, W., McKeown, G. J., Rychlowska, M., André, E., Wagner, J., & Lingenfelser, F. (2018). Social context disambiguates the interpretation of laughter. *Frontiers in Psychology, 8*, 2342.

Davila-Ross, M., Allcock, B., Thomas, C., & Bard, K. A. (2011). Aping expressions? Chimpanzees produce distinct laugh types when responding to laughter of others. *Emotion, 11*, 1013–1020.

Dezecache, G., & Dunbar, R. I. (2012). Sharing the joke: The size of natural laughter groups. *Evolution and Human Behavior, 33*(6), 775–779.

Dunbar, R. I., Baron, R., Frangou, A., Pearce, E., van Leeuwin, E. J., Stow, J., … Van Vugt, M. (2011). Social laughter is correlated with an elevated pain threshold. *Proceedings of the Royal Society of London B: Biological Sciences*, rspb20111373.

Feddersen-Petersen, D. U. (2000). Vocalization of European wolves (*Canis lupus lupus* L.) and various dog breeds (*Canis lupus* f. fam.). *Archives Animal Breeding, 43*(4), 387–398.

Fitch, W. T., Neubauer, J., & Herzel, H. (2002). Calls out of chaos: The adaptive significance of nonlinear phenomena in mammalian vocal production. *Animal Behaviour, 63*(3), 407–418.

Flamson, T., & Barrett, H. C. (2008). The encryption theory of humor: A knowledge-based mechanism of honest signaling. *Journal of Evolutionary Psychology, 6*(4), 261–281.

Flamson, T. J., & Bryant, G. A. (2013). Signals of humor: Encryption and laughter in social interaction. In M. Dynel (Ed.), *Developments in linguistic humour theory* (pp. 49–74). Amsterdam, the Netherlands: John Benjamins.

Gervais, M., & Wilson, D. S. (2005). The evolution and functions of laughter and humor: A synthetic approach. *Quarterly Review of Biology, 80*, 395–430.

Hagen, E. H., & Bryant, G. A. (2003). Music and dance as a coalition signaling system. *Human Nature, 14*(1), 21–51.

Hagen, E. H., & Hammerstein, P. (2009). Did Neanderthals and other early humans sing? Seeking the biological roots of music in the territorial advertisements of primates, lions, hyenas, and wolves. *Musicae Scientiae, 13*(2, suppl.), 291–320.

Jefferson, G. (1979). A technique for inviting laughter and its subsequent acceptance/declination. In G. Psathas (Ed.), *Everyday language: Studies in ethnomethodology* (pp. 79–96). New York, NY: Irvington Publishers.

Jefferson, G., Sacks, H., & Schegloff, E. (1987). Notes on laughter in the pursuit of intimacy. In G. Button & J. R. E. Lee (Eds.), *Talk and social organisation* (pp. 152–205). Clevedon, England: Multilingual Matters.

Jürgens, U. (2002). Neural pathways underlying vocal control. *Neuroscience & Biobehavioral Reviews, 26*, 235–238.

Keltner, D., & Bonanno, G. A. (1997). A study of laughter and dissociation: Distinct correlates of laughter and smiling during bereavement. *Journal of Personality and Social Psychology, 73*(4), 687–702.

Klages, S. V., & Wirth, J. H. (2014). Excluded by laughter: Laughing until it hurts someone else. *The Journal of Social Psychology, 154*(1), 8–13.

Knutson, B., Burgdorf, J., & Panksepp, J. (1998). Anticipation of play elicits high-frequency ultrasonic vocalizations in young rats. *Journal of Comparative Psychology, 112*(1), 65.

Krebs, J. R., & Dawkins, R. (1984). Animal signals: Mind-reading and manipulation. In J. R. Krebs & N. B. Davies (Eds.), *Behavioral ecology: An evolutionary approach* (pp. 380–402). Oxford, England: Blackwell Scientific.

Kurtz, L. E., & Algoe, S. B. (2015). Putting laughter in context: Shared laughter as behavioral indicator of relationship well-being. *Personal Relationships, 22*(4), 573–590.

Kurtz, L. E., & Algoe, S. B. (2017). When sharing a laugh means sharing more: Testing the role of shared laughter on short-term interpersonal consequences. *Journal of Nonverbal Behavior, 41*(1), 45–65.

Lavan, N., Rankin, G., Lorking, N., Scott, S., & McGettigan, C. (2017). Neural correlates of the affective properties of spontaneous and volitional laughter types. *Neuropsychologia, 95*, 30–39.

Lavan, N., Scott, S. K., & McGettigan, C. (2016). Laugh like you mean it: Authenticity modulates acoustic, physiological and perceptual properties of laughter. *Journal of Nonverbal Behavior, 40*, 133–149.

Lavan, N., Short, B., Wilding, A., & McGettigan, C. (2018). Impoverished encoding of speaker identity in spontaneous laughter. *Evolution and Human Behavior, 39*(1), 139–145.

Louwerse, M. M., Dale, R., Bard, E. G., & Jeuniaux, P. (2012). Behavior matching in multimodal communication is synchronized. *Cognitive Science, 36*(8), 1404–1426.

Luschei, E. S., Ramig, L. O., Finnegan, E. M., Bakker, K. K., & Smith, M. E. (2006). Patterns of laryngeal EMG and the activity of the respiratory system during spontaneous laughter. *Journal of Neurophysiology, 96*, 442–450.

Lummaa, V., Vuorisalo, T., Barr, R. G., & Lehtonen, L. (1998). Why cry? Adaptive significance of intensive crying in human infants. *Evolution and Human Behavior, 19*(3), 193–202.

Lynch, R. (2011). It's funny because we think it's true: Laughter is augmented by implicit preferences. *Evolution and Human Behavior, 31*, 141–148.

MacLarnon, A. M., & Hewitt, G. P. (1999). The evolution of human speech: The role of enhanced breathing control. *American Journal of Physical Anthropology, 109*(3), 341–363.

Manninen, S., Tuominen, L., Dunbar, R., Karjalainen, T., Hirvonen, J., Arponen, E., … Nummenmaa, L. (2017). Social laughter triggers endogenous opioid release in humans. *Journal of Neuroscience, 37*, 6125–6131.

Masataka, N., & Kohda, M. (1988). Primate play vocalizations and their functional significance. *Folia Primatologica, 50*(1–2), 152–156.

Matsusaka, T. (2004). When does play panting occur during social play in wild chimpanzees? *Primates, 45*, 221–229.

Maynard Smith, J., & Harper, D. (2003). *Animal signals*. Oxford, England: Oxford University Press.

McGettigan, C., Walsh, E., Jessop, R., Agnew, Z. K., Sauter, D. A., Warren, J. E., & Scott, S. K. (2013). Individual differences in laughter perception reveal roles for mentalizing and sensorimotor systems in the evaluation of emotional authenticity. *Cerebral Cortex, 25*, 246–257.
Mehu, M., & Dunbar, R. I. M. (2008). Naturalistic observations of smiling and laughter in human group interactions. *Behaviour, 145*(12), 1747–1780.
Morisseau, T., Mermillod, M., Eymond, C., Van Der Henst, J. B., & Noveck, I. A. (2017). You can laugh at everything, but not with everyone. *Interaction Studies, 18*(1), 116–141.
O'Nions, E., Lima, C. F., Scott, S. K., Roberts, R., McCrory, E. J., & Viding, E. (2017). Reduced laughter contagion in boys at risk for psychopathy. *Current Biology, 27*(19), 3049–3055.
Otten, M., Mann, L., van Berkum, J. J., & Jonas, K. J. (2017). No laughing matter: How the presence of laughing witnesses changes the perception of insults. *Social Neuroscience, 12*, 182–193.
Oveis, C., Spectre, A., Smith, P. K., Liu, M. Y., & Keltner, D. (2016). Laughter conveys status. *Journal of Experimental Social Psychology, 65*, 109–115.
Owren, M. J., Amoss, R. T., & Rendall, D. (2011). Two organizing principles of vocal production: Implications for nonhuman and human primates. *American Journal of Primatology, 73*, 530–544.
Owren, M. J., & Bachorowski, J. A. (2003). Reconsidering the evolution of nonlinguistic communication: The case of laughter. *Journal of Nonverbal Behavior, 27*, 183–200.
Owren, M. J., & Rendall, D. (2001). Sound on the rebound: Bringing form and function back to the forefront in understanding nonhuman primate vocal signaling. *Evolutionary Anthropology: Issues, News, and Reviews, 10*(2), 58–71.
Panksepp, J. (2005). Beyond a joke: From animal laughter to human joy? *Science, 308*(5718), 62–63.
Panksepp, J., & Burgdorf, J. (2003). "Laughing" rats and the evolutionary antecedents of human joy? *Physiology & Behavior, 79*(3), 533–547.
Papousek, I., Aydin, N., Lackner, H. K., Weiss, E. M., Bühner, M., Schulter, G., ... Freudenthaler, H. H. (2014). Laughter as a social rejection cue: Gelotophobia and transient cardiac responses to other persons' laughter and insult. *Psychophysiology, 51*(11), 1112–1121.
Pisanski, K., Cartei, V., McGettigan, C., Raine, J., & Reby, D. (2016). Voice modulation: A window into the origins of human vocal control? *Trends in Cognitive Sciences, 20*(4), 304–318.
Platow, M. J., Haslam, S. A., Both, A., Chew, I., Cuddon, M., Goharpey, N., ... Grace, D. M. (2005). "It's not funny if they're laughing": Self-categorization, social influence, and responses to canned laughter. *Journal of Experimental Social Psychology, 41*(5), 542–550.
Provine, R. R. (1992). Contagious laughter: Laughter is a sufficient stimulus for laughs and smiles. *Bulletin of the Psychonomic Society, 30*(1), 1–4.
Provine, R. R. (1993). Laughter punctuates speech: Linguistic, social and gender contexts of laughter. *Ethology, 95*, 291–298.
Provine, R. R. (1996). Laughter. *American Scientist, 84*(1), 38–45.
Provine, R. R. (2000). *Laughter: A scientific investigation.* New York, NY: Penguin Press.
Provine, R. R. (2016). Laughter as a scientific problem: An adventure in sidewalk neuroscience. *Journal of Comparative Neurology, 524*, 1532–1539.
Provine, R. R. (2017). Laughter as an approach to vocal evolution: The bipedal theory. *Psychonomic Bulletin & Review, 24*(1), 238–244.
Ritter, M., & Sauter, D. A. (2017). Telling friend from foe: Listeners are unable to identify in-group and out-group members from heard laughter. *Frontiers in Psychology, 8*, 2006.
Ruch, W., & Ekman, P. (2001). The expressive pattern of laughter. In A. W. Kaszniak (Ed.), *Emotions, qualia, and consciousness* (pp. 426–443). Tokyo, Japan: World Scientific Publisher.
Ruch, W., & Proyer, R. T. (2008). The fear of being laughed at: Individual and group differences in Gelotophobia. *Humor, 21*(1), 47–67.
Scott, S. K., Lavan, N., Chen, S., & McGettigan, C. (2014). The social life of laughter. *Trends in Cognitive Sciences, 18*, 618–620.
Smaldino, P. E., Flamson, T. J., & McElreath, R. (2018). The evolution of covert signaling. *Scientific Reports, 8*(1), 4905.

Smith, P. K. (1982). Does play matter? Functional and evolutionary aspects of animal and human play. *Behavioral and Brain Sciences, 5*(1), 139–155.

Smoski, M. J., & Bachorowski, J. A. (2003a). Antiphonal laughter between friends and strangers. *Cognition and Emotion, 17*(2), 327–340.

Smoski, M. J., & Bachorowski, J. A. (2003b). Antiphonal laughter in developing friendships. *Annals of the New York Academy of Sciences, 1000*(1), 300–303.

Stevens, K. N. (1998). *Acoustic phonetics*. Cambridge, MA: MIT Press.

Szameitat, D. P., Kreifelts, B., Alter, K., Szameitat, A. J., Sterr, A., Grodd, W., & Wildgruber, D. (2010). It is not always tickling: Distinct cerebral responses during perception of different laughter types. *Neuroimage, 53*(4), 1264–1271.

Tinbergen, N. (1952). Derived activities: Their causation, biological significance, origin and emancipation during evolution. *Quarterly Review of Biology, 27*, 1–32.

Titze, I. R., Finnegan, E. M., Laukkanen, A. M., Fuja, M., & Hoffman, H. (2008). Laryngeal muscle activity in giggle: A damped oscillation model. *Journal of Voice, 22*(6), 644–648.

Titze, M. (1996). The Pinocchio Complex: Overcoming the fear of laughter. *Humor & Health Journal, 5*(1), 1–11.

van Hooff, J. A. (1972). A comparative approach to the phylogeny of laughter and smiling. In R. A. Hinde (Ed.), *Nonverbal communication* (pp. 209–241). Cambridge, England: Cambridge University Press.

Vettin, J., & Todt, D. (2004). Laughter in conversation: Features of occurrence and acoustic structure. *Journal of Nonverbal Behavior, 28*, 93–115.

Vouloumanos, A., & Bryant, G. A. (2019). Five-month-old infants detect affiliation in colaughter. *Scientific Reports, 9*, 4158.

Wild, B., Rodden, F. A., Grodd, W., & Ruch, W. (2003). Neural correlates of laughter and humour. *Brain, 126*(10), 2121–2138.

Wiley, R. W. (1983). The evolution of communication: Information and manipulation. In T. R. Halliday & P. J. B. Slater (Eds.), *Communication* (pp. s156–215). Oxford, England: Blackwell Scientific Publications.

Wood, A., Martin, J., & Niedenthal, P. (2017). Towards a social functional account of laughter: Acoustic features convey reward, affiliation, and dominance. *PLoS ONE, 12*(8), e0183811.

Wood, A., & Niedenthal, P. (2018). Developing a social functional account of laughter. *Social and Personality Psychology Compass, 12*(4), e12383.

6

Behavior Genetics and Twin Studies

Principles, Analytical Techniques, and Data Resources for Innovative Communication Research

Chance York

Social scientists are often interested in identifying factors that lead to individual differences in traits and behaviors such as intelligence, obesity, criminal delinquency, political attitudes, and media use. Traditionally, individual differences in such traits and behaviors have been attributed to the influence of parents, culture, and other social-environmental forces of "nurture." Yet, almost a century of research in the field of behavior genetics has shown that nurture plays only a small role in what makes people distinct psychologically and behaviorally. Instead, variation in hundreds of complex traits and behaviors, from personality characteristics to tobacco use, is primarily rooted in genes, or "nature" (Plomin, 2018; Polderman et al., 2015).

The logical implication is that genes should also account for variability in the expression of *communication* traits and behaviors. However, few researchers have used behavior genetic approaches to investigate the biological origins of traits and behaviors of interest to the communication discipline. This chapter attempts to address this gap by providing readers with a brief introduction to behavior genetics and twin studies.

PRINCIPLES OF BEHAVIOR GENETICS AND TWIN STUDIES

Behavior genetics is a field of scientific research in which investigators pursue fundamental explanations for individual differences. The central goal of behavior genetics is to trace variation in the way organisms appear, think, feel, and act to root genetic and environmental causes.[1] To do so, researchers rely on genetically informative research designs that allow them to isolate the effects of genes and the environment on traits and behaviors of interest. Using these designs, researchers have repeatedly shown that variation at the genetic level partly explains why many complex traits and behaviors—or, phenotypes—differ so extensively within populations of organisms.

In behavior genetics, the term phenotype is used as a catch-all that refers to any observable physical, physiological, cognitive, psychological, or behavioral characteristic of an organism. Hair color is a phenotype, as is a complex psychological trait such as extraversion. Critically, from a behavior genetic perspective, all phenotypic variation within populations of organisms can be traced back to a relative combination of genetic and environmental factors or an interaction between the two (see Figure 6.1). This perspective makes intuitive sense: all phenotypes can only originate in and be shaped by genetic and environmental forces, which in behavior genetics are treated as the only sources of phenotypic variation.

For example, consider Tay-Sachs disease. Phenotypic variation in the symptoms of a rare inherited illness such as Tay-Sachs is almost entirely explained by underlying genetic mechanisms. The reason why one person develops Tay-Sachs symptoms and another does not is entirely due to genetic variation between individuals, not variation in the social or physical environment. Conversely, a facial scar that results from an idiosyncratic childhood trauma is a phenotype that is almost entirely accounted for by random environmental variation, not genes. For personality traits, IQ, educational attainment, political ideology, voter turnout, and hundreds of other complex phenotypes, both genetic and environmental factors play a role (see Lockyer & Hatemi, 2018; Plomin, DeFries, Knopik, & Neiderhiser, 2016). The question is: *To what degree* do genetic and environmental factors explain variation in the expression of complex psychological and behavioral phenotypes? One behavior genetic research design, the twin study, is uniquely positioned to answer this question.

A twin study is a genetically informative research design used to determine the extent to which genetic and environmental factors contribute to individual differences in traits and behaviors. Twin studies represent only one behavior genetic design that relies on biological relationships among kin to disentangle genetic and environmental influences on phenotypes.[2] However, twin studies are considered an important research design because they can capitalize on two unique properties associated only with twin research participants.

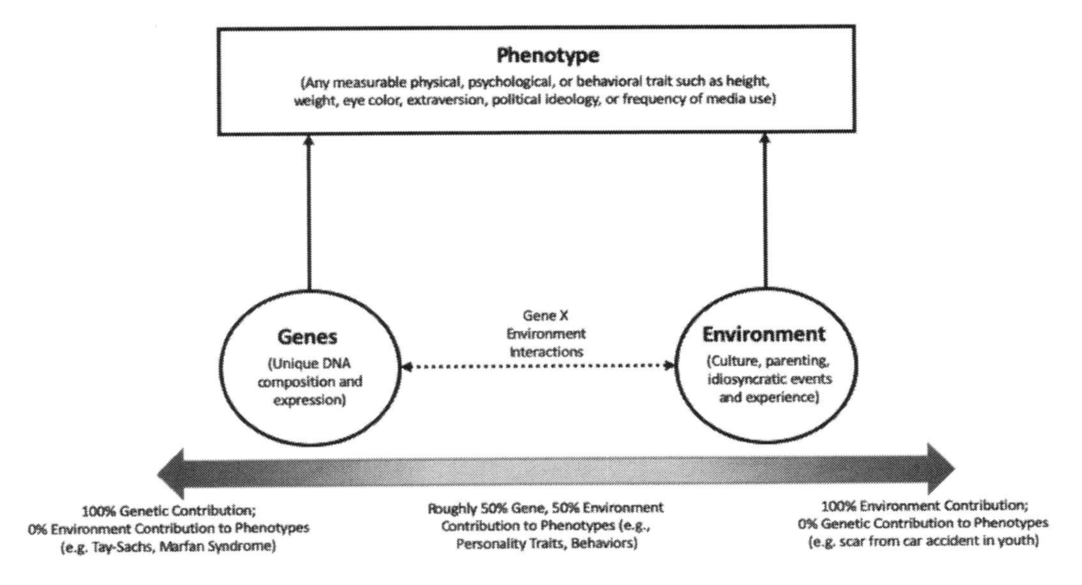

Figure 6.1 The Behavior Genetic Perspective

Source: Adapted from Blokland, Mosing, Verweij, & Medland (2013).

The first property relates to how twins develop *in utero* (Figure 6.2). Specifically, it is well established that identical or monozygotic (MZ) twins are conceived from a single zygote, or fertilized ovum, that divides into two separate yet indistinguishable embryos, ultimately leading to the production of two genetically identical same-sex individuals. Fraternal or dizygotic (DZ) twins, on the other hand, develop from separate zygotes that ultimately become same- or opposite-sex twins who share roughly half of their DNA in common just as full siblings do.

Unlike full siblings, however, both fraternal and identical twins share highly similar social and physical environments. This is the second property. All twins share the same uterine conditions during their initial development. All twins are born at roughly the same time, and all twins presumably experience a comparable upbringing by the same parents in the same culture.

Thus, twins act as a kind of "natural experiment" in which shared aspects of their environment is the factor that is "held constant" or "controlled" within pairs while shared DNA (roughly 50% for fraternal twins and 100% for identical twins) is the factor that varies between identical and non-identical pairs.[3] Other first-degree relatives, as well as grandparents, cousins, and so on may share certain degrees of genetic similarity, but none experience a similar upbringing in the way twins do. In fact, it is these two properties unique to twins—shared genetics and environments—that make twin studies a powerful and effective research design, and a design that is increasingly popular in the social sciences.

COLLECTING AND ANALYZING TWIN DATA

Methodologically, twin studies are executed by collecting survey or biometric data from twins and their co-twins. Each twin's observation on a single measure of a physical, psychological, or behavioral phenotype (e.g., frequency of eating high-fat foods) is then compared. In a very basic sense, researchers can then use what is known about the unique genetic and environmental

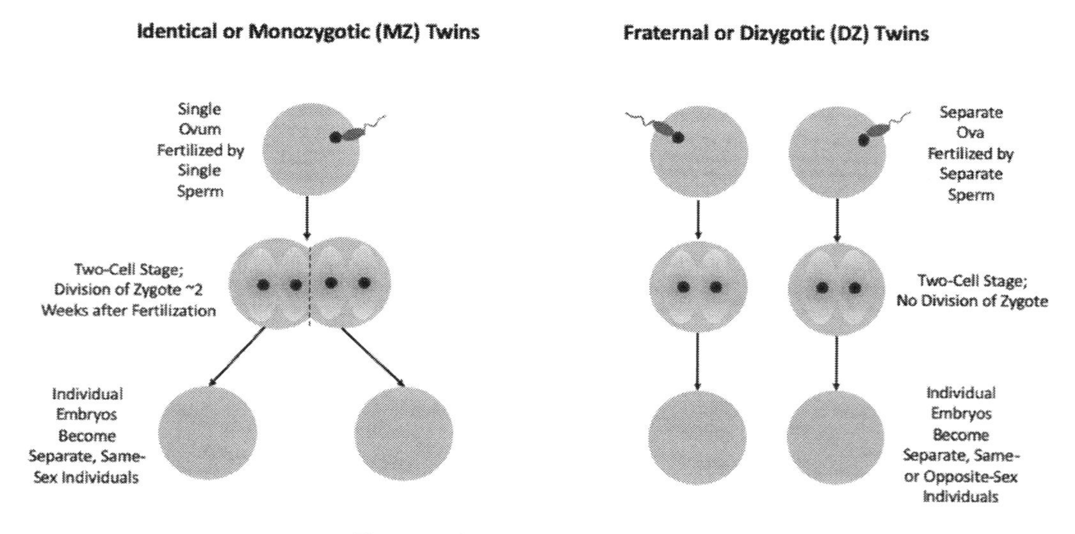

Figure 6.2 Twin Prenatal Development

Source: Adapted from Blokland et al. (2013).

properties shared by twins to determine why observations on a measure of the same phenotype are so similar, and if they differ, why they differ.

Analytical Approaches

The first step in twin data analysis is to simply use correlations to assess similarity in phenotypes among twin dyads. In theory, stronger correlations among identical twin pairs compared to fraternal twins is indicative of genetic influence owing to identical twins' greater level of shared DNA (Knopik, Neiderheiser, DeFries, & Plomin, 2016). For example, suppose a researcher fields a survey that asks twin respondents to report their "years of formal education." If after fielding the survey the researcher discovers that within-pair correlations on the educational attainment measure are stronger among identical twins than among fraternal twins, this result would be indicative of latent genetic influence on educational attainment. The same would be true of dyadic correlations on communication behavior or media use phenotypes—stronger within-pair correlations among more closely related kin can be interpreted as a preliminary sign of latent genetic influence.

Simple correlations, however, cannot establish the relative contribution of genes and environments to phenotypic variation. To do that, structural equation models are required.

Structural equation models can be used to estimate how much total variance in a phenotype measured independently for both twins in a pair is accounted for by the genetic and environmental attributes of twins, which are modeled as latent factors. These latent factors are intended to represent the level of additive genetic traits shared by twins (notated *A*), the common environmental conditions that should presumably be shared by both twins in a pair, such as being raised by the same parents in the same household (*C*), and any unique environmental conditions specific to an individual twin in the pair, such as when one twin belongs to a distinct peer group in childhood (*E*).[4] The logic of this latent modeling approach is that the total variance (σ^2) in an observed phenotype (*P*) can ultimately be accounted for by just these three latent factors (*A*), (*C*), and (*E*), which also gives the approach its nickname: *ACE* modeling.

How these three latent factors relate conceptually to a phenotype in a standard univariate *ACE* model is illustrated in Figure 6.3. Note that the model is considered "univariate" because a single phenotype is measured for each twin in a dyad.[5]

Importantly, when researchers employ *ACE* models to analyze twin data *the objective is to explain how much of the total variance in a phenotype is accounted for by each latent factor*, not to compare means, as is the case with analysis of variance, or predict a phenotype's average value or probability of occurrence, as is the case in regression analysis. It is imperative to emphasize this point as it is prone to misconceptions. *ACE* modeling is only concerned with decomposing the total variance in a phenotype by latent genetic and environmental factors.

In theory, any statistics package with a structural equation module can be used to estimate an *ACE* model. In practice, however, only two statistics packages—*MPlus* and *R's OpenMX* program—are uniquely adapted to estimate *ACE* models. The developers of these packages also provide many useful resources on their homepages, including example syntax for modeling twin data, user manuals, tutorials, and discussion boards.[6]

Regardless of the statistics package researchers use for twin analysis, it is instructive to understand how univariate *ACE* models are specified. Below, an outline of specifications required for univariate *ACE* modeling is provided. It should be noted that far more technical explanations of *ACE* model specifications exist elsewhere (e.g., Neale & Maes, 2004).

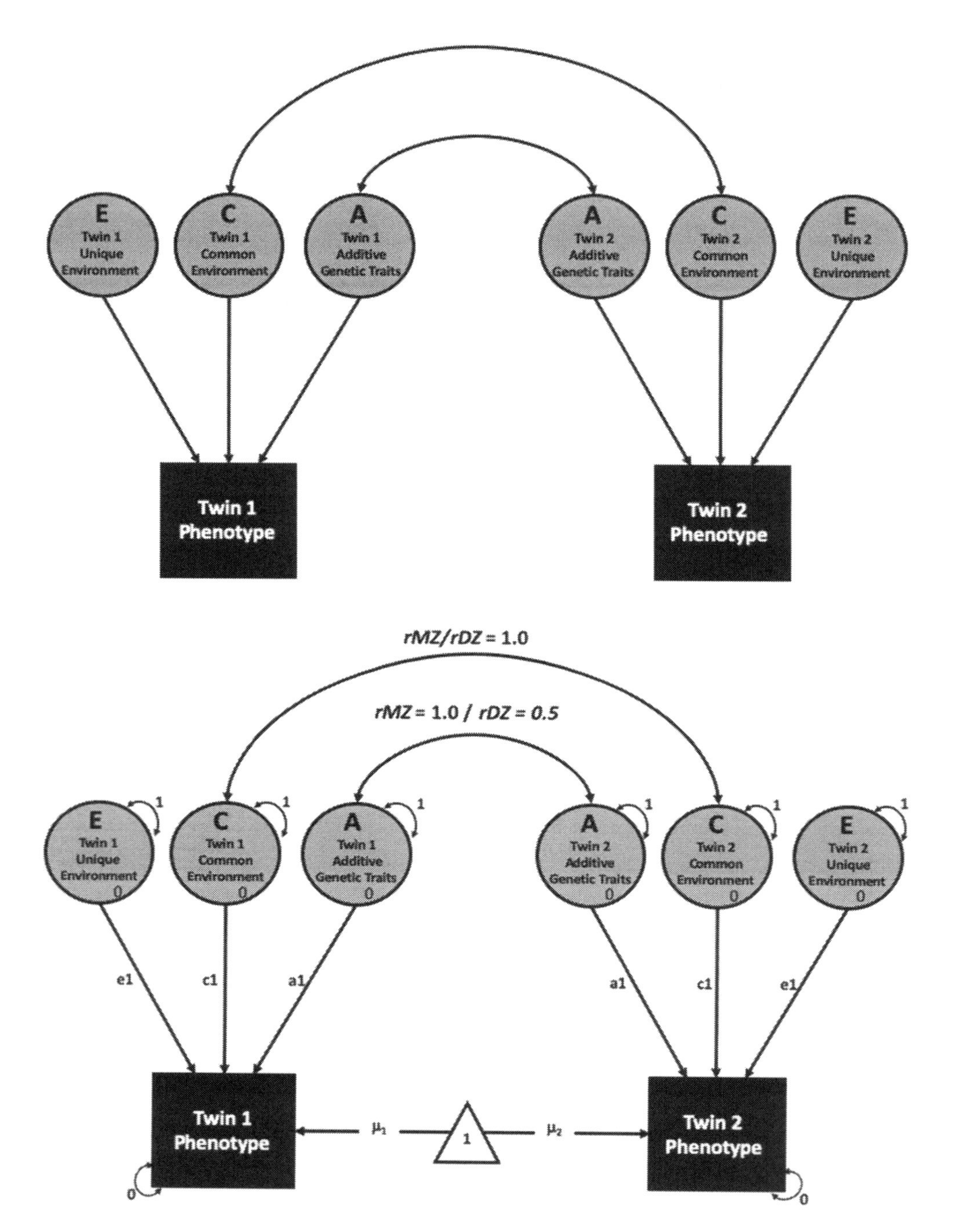

Figure 6.3 Conceptual and Functional Univariate *ACE* Model

ACE Model Specifications and Fit

Researchers must specify *ACE* models in a way that explicitly accounts for the unique properties of twins. To accomplish this, the genetic resemblance between twins should be accounted for by fixing the covariance between latent factors representing additive genetic traits (*A*) at 1.0 for identical twins who share all their DNA in common and at 0.5 for fraternal twins who share roughly half (see Figure 6.3(b)). Similarly, covariance between latent factors that represent the common environment (*C*) should be fixed at 1.0 for all twin respondents, as this factor is assumed to represent the "constant" aspects of the environment experienced by both twins in a pair. The latent factor (*E*) theoretically represents any aspect of the environment that makes each twin in a pair unique and should therefore be freely estimated.

In addition, model paths linking latent factors to a phenotype measured for both members of the twin pair (*a1*, *c1*, and *e1*) should be specified as equal across co-twins because "in most models we do not expect the magnitude of genetics effects, or the environmental effects, to differ between first and second twins" (Neale & Maes, 2004, p. 110). To enable model identification, the latent variable means are required to be fixed at 0 and variances at 1.0. Likewise, the residual variance for the phenotype is set to 0 as it is confounded with variance in (*E*). Estimates of (*E*) will therefore sequester error in the observed phenotype along with any unique environmental effects on the phenotype (see Neale & Maes, 2004, p. 111; Neale, 2009).

Estimating the *ACE* model using these specifications will produce path coefficients for *a1*, *c1*, and *e1*. Again, because paths are set to be equal across twins, coefficients should not differ between co-twins (e.g., *a1* should be the same for both twins). It is then possible—though not required—to use the resulting coefficients to manually determine the proportion of total variance each latent factor explains in the phenotype. This can be accomplished by simply tracing up a path (*a*), for example, and then multiplying by 1.0 (the variance of latent variable *A*), and then tracing back down the path (multiplying by *a*). This results in ($a*1*a$) or more simply (a^2). As a thought exercise, the same procedure could be performed for paths *c* and *e*, which would result in ($c*1*c$) and ($e*1*e$). The total variance in the phenotype can thus be stated mathematically as the sum of squared path coefficients, or

$$\sigma^2_P = a^2 + c^2 + e^2 = 1.0 \text{ or } 100\%$$

Covariance values can also be found by manually tracing back across double-headed arrows. For reference, a full variance-covariance matrix for a univariate *ACE* model is shown in Figure 6.4.[7]

The final *ACE* modeling consideration regards fit. Post-estimation likelihood ratio (LR) tests should be used to determine if the full *ACE* model or a nested model with factor (*A*) or (*C*) dropped provides the most parsimonious fit to the data. If dropping a model parameter (*A*) or

$$\mathbf{Matrix} = \begin{bmatrix} var\ \text{twin1} & cov\ \text{twin1 \& twin2} \\ cov\ \text{twin1 \& twin2} & var\ \text{twin2} \end{bmatrix}$$

$$\mathbf{MZ} = \begin{bmatrix} a^2 + c^2 + e^2 & a^2 + c^2 \\ a^2 + c^2 & a^2 + c^2 + e^2 \end{bmatrix}$$

$$\mathbf{DZ} = \begin{bmatrix} a^2 + c^2 + e^2 & .5 * a^2 + c^2 \\ .5 * a^2 + c^2 & a^2 + c^2 + e^2 \end{bmatrix}$$

Figure 6.4 Variance-Covariance Matrix for a Univariate *ACE* Model

(*C*) significantly worsens fit at some prespecified critical value (typically $p<0.05$), the full *ACE* model should be preferred. In cases in which dropping a model parameter (*A*) or (*C*) does not significantly worsen fit, then the nested *AE* or *CE* model should be preferred. In such cases, fit statistics for a nested model *AE* or *CE* should be comparable to that of the *ACE* model. The associated chi-square statistic for the LR test should be close to zero and the *p*-value close to 1.00, all of which signifies no significant change in model fit as a result of dropping a parameter.[8]

Application of *ACE* Modeling

What does *ACE* modeling look like in practice as it is applied to communication- and media-related phenotypes? In this section, an example of *ACE* modeling using secondary panel survey data from the University of Wisconsin's Midlife in the United States (MIDUS) study is provided.[9] The MIDUS study is composed of a national probability sample of unrelated non-twin adults or "singletons," full siblings, identical twins, and fraternal twins. Due to its unique sample composition, the MIDUS study has been frequently used by social scientists to answer questions about genetic and environmental effects on a wide array of phenotypes.

As an initial cut of these data, we simply plot within-pair correlations on a variety of communication- and media-related phenotypes by level of genetic relatedness (see Figure 6.5). On the left-hand side of each panel in Figure 6.5 are dyadic correlations for genetically identical twins. On the far right-hand side are dyadic correlations for genetically unrelated pairs matched at random. In between are correlations for fraternal twins and full siblings.

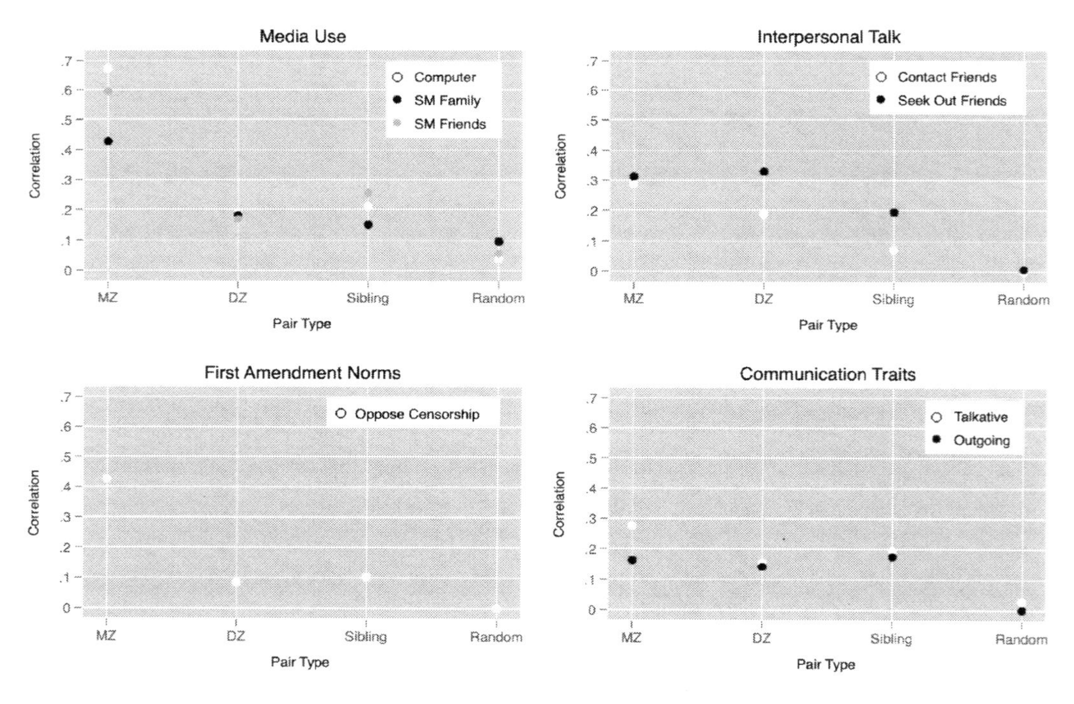

Figure 6.5 Within-Pair Correlations

What is immediately striking about Figure 6.5 is the strength of correlations among dyads of differing degrees of genetic relatedness. Moving across each panel from left to right, from genetically identical to genetically unrelated pairs, within-pair correlation coefficients decline from approximately ($r = 0.20$ to 0.70) for identical twins on the left-hand side of each graph to approximately ($r = 0.00$) for random pairs on the right-hand side. Overall, identical twins were far more similar on communication- and media-related phenotypes than were their non-identical comparison groups. This is a preliminary indicator of possible genetic influence on phenotypes.

Next, we used *OpenMx* syntax adapted from Maes (2016) and the model specifications outlined above to estimate univariate *ACE* models for each MIDUS variable depicted in Figure 6.5. *ACE* model results are shown in Table 6.1. Note that the results shown in Table 6.1 provide the reader with estimates of the *total proportion (or percentage) of variance explained* in each

Table 6.1 *ACE* Model Results

Variable	*Model*	a^2	c^2	e^2	*–2LL*	X^2
Media Use						
Computer Use	ACE	0.82 (0.36, 0.90)	0.00 (0.00, 0.39)	0.18 (0.00, 0.32)	–1,791.88	Base
	AE	**0.82 (0.68, 0.90)**	–	**0.18 (0.09, 0.32)**	**1,791.88**	**0.00**
	CE	–	0.59 (0.43, 0.71)	0.41 (0.29, 0.57)	1,802.93	11.05
SM Family	ACE	0.41 (0.00, 0.56)	0.00 (0.00, 0.32)	0.59 (0.44, 0.78)	–2,825.96	Base
	AE	**0.41 (0.22, 0.56)**	–	**0.59 (0.44, 0.79)**	**–2,825.96**	**0.00**
	CE	–	0.28 (0.13, 0.41)	0.72 (0.59, 0.87)	–2,829.47	3.51
SM Friends	ACE	0.57 (0.30, 0.70)	0.00 (0.00, 0.19)	0.43 (0.30, 0.60)	–2,819.98	Base
	AE	**0.57 (0.40, 0.70)**	–	**0.43 (0.30, 0.60)**	**–2,819.98**	**0.00**
	CE	–	0.37 (0.23, 0.49)	0.63 (0.51, 0.76)	–2,854.22	10.79
Interpersonal Talk						
Contact Friends	**ACE**	**0.10 (0.00, 0.45)**	**0.20 (0.00, 0.28)**	**0.70 (0.55, 0.86)**	**–2,908.20**	**Base**
	AE	0.32 (0.17, 0.46)	–	0.68 (0.54, 0.83)	–2,909.15	0.95
	CE	–	0.27 (0.14, 0.38)	0.73 (0.62, 0.86)	–2,908.34	0.15
Seek Out Friends	**ACE**	**0.14 (0.00, 0.51)**	**0.20 (0.00, 0.42)**	**0.66 (0.49, 0.84)**	**–2,013.79**	**Base**
	AE	0.37 (0.19, 0.52)	–	0.63 (0.48, 0.81)	–2,014.56	0.77
	CE	–	0.30 (0.15, 0.43)	0.70 (0.55, 0.85)	–2,014.04	0.25
First Amendment Norms						
Oppose Censorship	ACE	0.53 (0.23, 0.68)	0.00 (0.00, 0.21)	0.47 (0.32, 0.66)	–2,126.33	Base
	AE	**0.53 (0.34, 0.68)**	—	**0.47 (0.32, 0.66)**	**–2,126.33**	**0.00**
	CE	–	0.32 (0.17, 0.45)	0.68 (0.55, 0.83)	–2,134.41	8.07
Traits						
Talkative	**ACE**	**0.31 (0.00, 0.53)**	**0.04 (0.00, 0.37)**	**0.65 (0.47, 0.88)**	**–2,030.54**	**Base**
	AE	0.36 (0.16, 0.54)	—	0.64 (0.46, 0.84)	–2,030.58	0.04
	CE	—	0.25 (0.10, 0.39)	0.75 (0.61, 0.90)	–2,031.57	1.03
Outgoing	**ACE**	**0.21 (0.00, 0.44)**	**0.02 (0.00, 0.29)**	**0.77 (0.57, 0.97)**	**–1,919.47**	**Base**
	AE	0.24 (0.03, 0.44)	—	0.76 (0.56, 0.97)	–1,919.48	0.01
	CE	—	0.16 (0.02, 0.30)	0.84 (0.70, 0.99)	–1,924.59	0.45

Notes

$n = 456$, or 228 pairs (132 DZ; 96 MZ). DZ = fraternal twin pairs. MZ = identical twin pairs. a^2 = additive genetic traits, c^2 = common environment, e^2 = unique environment. Estimates for a^2, c^2, and e^2 represent the proportion of variance in each observed variable explained by additive genetic, common environment, and unique environment factors. The 95% confidence intervals are in parentheses. The –2 log-likelihood (–2LL) statistic is an indicator of model fit. The change in the –2LL statistic as a result of dropping (*A*) or (*C*) from the base model can be tested with a X^2 test. Models shown in bold are the best-fitting models based on these tests. Because variables were ordinal, all estimates were derived from liability threshold models for ordered categorical outcomes (Neale, 2009).

phenotype by latent genetic traits (a^2), the common environment (c^2), and the unique environment (e^2). These estimates should always sum to 1.0 or 100% of the variance in the phenotype. Importantly, the estimate of (a^2) is often referred to as the estimate of heritability because it directly indicates the proportion of phenotypic variation that is attributable to the effects of latent genetic variation. LR tests of model fit are shown on the right. Confidence intervals around estimates indicate the possible range of phenotypic variation explained in the population.

Consistent with previous twin study research, the results shown in Table 6.1 reveal an overall pattern of genetic influence across phenotypes and little or no common environmental influence (Plomin, 2018; Polderman et al., 2015). Averaged across all models, the estimate of additive genetics (a^2) accounted for 38.63% of the variance in communication- and media-related phenotypes. Estimates of the common environment (c^2) were null in four of the eight analyses. In these analyses, the *AE* model with the (C) parameter dropped provided a more parsimonious fit to these data and was therefore preferred over the *ACE* model. Critically, when estimates of the common environment (c^2) were non-zero, they never explained more than 20% of the total variance in a phenotype.

Environmental factors, however, still played a major role in explaining variance in phenotypes. In fact, the unique environment factor (E), which theoretically represents anything related to the environment that is unshared between twins (a spouse, a friend, experiencing an idiosyncratic event, etc.), consistently explained between 17% and 77% of variance in phenotypes. This is also consistent with previous twin studies. It suggests that a considerable amount of variance in phenotypes is explained neither by shared genetic or environmental factors, but by random environmental factors (e.g., having a special teacher in grade school).

In sum, the *ACE* model results shown here reveal latent genetic traits as an important factor that explains considerable variance in phenotypes relevant to the communication discipline. This is remarkable, as there are no existing theories of communication that anticipate such effects.[10] The implications for existing communication theories are perhaps just as striking.

THEORETICAL IMPLICATIONS

Consider the implications for Uses and Gratifications (U&G), which is widely considered the preeminent audience-centered theory of communication behavior and effects. U&G suggests "communication behavior, including the selection and use of the media, is goal-directed, purposive, and motivated" by felt needs (Rubin, 2009, p. 167). These needs and any related communication behaviors that result from them are thought to be preceded and ultimately constrained by background characteristics specific to individuals, namely, psychological "predispositions [and] the environment" (Rubin, 2009, p. 167). However, the results presented in this chapter alone imply psychological predispositions and the environment are not the only individual background characteristics that regulate U&G processes. The results presented here clearly imply that "genes" should be added to the list of background factors that regulate U&G processes. Alternatively, it may be more accurate to simply replace "psychological predispositions" with "biological predispositions," given individual differences in personality traits and many other psychological phenotypes are themselves informed by genes (see Plomin et al., 2016).

Two further implications for communication research are apparent. First, it has been theorized that communication behavior and media use patterns develop in part from social learning, which occurs when children observe and later re-enact the behavior of parents (e.g., Edgerly, Thorson, Thorson, Vraga, & Bode, 2018). Consequently, any similarities between frequency of parent media use and child media use, for example, are typically attributed to the social transmission of behavior from parent to child through social learning or direct discussion ("nurture") mechanisms. In fact, such relationships may actually reflect the genetic transmission of behavior (see York & Scholl, 2015), as genetic traits shared between parents and children ("nature") may partially explain similarities in communication behavior and media use.

Second, because the analysis presented in this chapter demonstrates that latent genetic traits explain considerable variance in media use and communication behavior, it is plausible that genetic traits consequently shape patterns of media and communication effects. Audience-centered communication theories that link behavior to effects—U&G, Selective Exposure, Mood Management—identify only psychological and social antecedents to behavior-effects processes within their frameworks, not genetic antecedents. As such, it is understandable that genes have not been theorized as an antecedent mechanism regulating effects processes. However, mounting empirical evidence, including evidence shown in this chapter, suggests future research should explore this possibility.

INNOVATIVE APPLICATIONS IN COMMUNICATION RESEARCH

Behavior genetic approaches have been underutilized in communication research, but a handful of existing twin studies have shown promising initial results. These studies have shown that latent genetic traits explain substantial proportions of variance in television viewing and interpersonal communication (Kirzinger et al., 2012; York, 2019), social media use (Ayorech, Von Stumm, Haworth, Davis, & Plomin, 2017; York, 2017), mobile phone use (Miller, Zhu, Wright, Hansell, & Martin, 2012), and multiple dimensions of problematic internet use (Deryakulu & Ursavaş, 2014), among other communication- and media-related phenotypes (see Table 6.2). Few as they are, these studies almost uniformly point to biological in addition to environmental explanations for individual differences in phenotypes of interest to the discipline.

Moving forward, it will be critical for researchers to continue this line of inquiry, exploring the relative contribution of genetic and environmental factors to individual differences in communication traits, behaviors, media use, and effects responses. This is the first step in understanding the genetic foundations of phenotypes of broad interest to the field. Additionally, future research could use longitudinal twin designs and multivariate modeling techniques (see Hatemi et al., 2010; Medland & Hatemi, 2009) to explore more complex questions, such as:

1. How do genetic and environmental factors explain individual differences in communication behavior and media use phenotypes over time? For example, do genetic factors explain more variation in news-reading as individuals age and become further removed from the socializing pressures of the childhood home?
2. Are there interactions between latent genetic factors and specific environmental factors (*GxE interactions*) that explain individual differences in communication- and media-related phenotypes? For instance, do genetic traits explain more variation in affectionate communication among individuals raised in families that encourage affection (triggering a genetic propensity for affectionate communication) vs. families that suppress affection?
3. To what extent do genetic and environmental factors explain the *relationship* between two or more phenotypes? For example, do genetic traits explain more covariance between playing video games and aggressive behavior than does the shared environment?

Table 6.2 Twin Studies of Communication- and Media-Related Phenotypes

Citation	*Design*	*Traits examined*	n *(pairs)*	*Data Sources*	*Analysis*	*Genetic Effect*
Ayorech et al. (2017)	Twin	Social media use	1,978 MZ; 1,674 DZ	Twins early development study	SEM	Yes
Beatty et al. (2001)	Twin	Communication adaptability	62 MZ; 43 DZ	Original data	Correlation	Yes
Deryakulu et al. (2016)	Twin	Computer self-Efficacy	55 MZ; 110 DZ	Original data	Correlation	No
Deryakulu & Ursavaş (2014)	Twin	Problematic internet use	80 MZ; 157 DZ	Original data	Correlation	Yes
Hazel et al. (2006)	Twin	Communication traits	15 MZ; 22 DZ	Original data	Correlation	No
Horvath (1995)	Twin	Communication traits	62 MZ; 42 DZ	Original data	Correlation	Yes
Kirzinger et al. (2012)	Twin	Media use and interpersonal discussion	(1) 221 MZ, 206 DZ; (2) 220 MZ, 216 DZ; (3) 342 MZ, 349 DZ	Add Health III, Add Health IV; MIDUS I	SEM	Yes
Miller et al. (2012)	Twin	Mobile phone use	(1) 88 MZ, 185 DZ; (2) 112 MZ, 201 DZ	Original data	SEM	Yes
York (2017)	Twin	Social media use	94 MZ; 134 DZ	MIDUS III	DF Regression	Yes
York (2019)	Twin	Political discussion	138 MZ; 29 DZ 347 MZ; 231 DZ	Original data; Minnesota Twins Political Survey	SEM	Yes

Notes
MZ = monozygotic or identical twin pairs. DZ = dizygotic or fraternal twin pairs. The table presents a comprehensive list of articles published between January 1950 and May 2018 that used twin studies to investigate communication- and media-related phenotypes. To obtain this list, combinations of the keywords "behavior genetics," "behavioral genetics," "twin study," "twin studies," "genetics," "genes," "biology" and "biological" were sought in the databases Communication Mass Media Complete, Academic Search Complete, Communication Abstracts, PsychInfo, and Google Scholar. This process yielded just 10 twin studies of relevant phenotypes.

Answers to these questions and many more can be readily addressed by pairing secondary twin data with advanced behavior genetic modeling techniques (see Table 6.3). Alternatively, national and international gatherings of twins held annually at sites such as the Twinsburg, Ohio "Twins Days Festival" allow researchers relatively inexpensive opportunities to collect original twin survey or biometric data.[11] Twin databases, such as the Washington Twin Registry, likewise offer cost-effective solutions to original twin data collection.[12]

In the future, communication researchers should also triangulate twin study results with candidate gene and genome-wide association study (GWAS) approaches that are designed to link the expression of specific genes to the expression of phenotypes. Candidate gene and GWAS approaches are increasingly used in psychology, sociology, and political science (e.g., Deppe, Stoltenberg, Smith, & Hibbing, 2013) and have generated important conclusions about how individual genotypes, or sets of genes in human DNA, contribute to phenotypic expression. One of the most significant conclusions from this research is that "many genes of small effect" (Plomin et al., 2016, p. 6), not any one gene, shape individual differences in pheno-

Table 6.3 Examples of Twin Study Datasets Containing Relevant Variables

Repository	*Design*	*Respondents*	*n (pairs)*	*Variables*
Midlife in the United States (MIDUS) Study[1]	3 wave panel, 1995–present	Twins, siblings	357 MZ/584 DZ; 475 Siblings	Social media use, computer use, frequency of reading and writing, opposition to censorship, interpersonal discussion, communication traits
Add Health[2]	5 wave panel, 1994–present	Twins, siblings, adoptees, parents	307 MZ/452 DZ; 1,251 Full Sibling; 442 Half Sibling; 560 Adopted	Use non-English media, student newspaper participation, TV use, parental mediation of TV, video game use, internet use
Minnesota Twins Political Survey[3]	Cross-sectional, 2008–2009	Twins	356 MZ/240 DZ	Censor newspapers, political argumentation, political talk (current/childhood), political agreement and disagreement
Swedish Adoption/Twin Study on Aging[4]	7 Wave panel, 1983–2010	Twins, adoptees	279 MZ/546 DZ	Interpersonal communication, telephone use
TwinLife[5]	Multi-wave panel, 2014–present	Twins	1,441 MZ/1,633 DZ	Computer use, internet use, computer games, game consoles, television, videos and DVDs

Notes
MZ = monozygotic or identical twin pairs. DZ = dizygotic or fraternal twin pairs. To obtain data from the sources listed in the table, see: [1]http://midus.wisc.edu/ [2]www.cpc.unc.edu/projects/addhealth [3]https://doi.org/10.7910/DVN/0F2FJ0 [4]www.icpsr.umich.edu/icpsrweb/NACDA/studies/3843. [5]www.twin-life.de/en/data. The author also maintains a repository of original twin data at: https://dataverse.harvard.edu/dataverse/chanceyork.

types.[13] However, little is known about the genes responsible for promoting variation in communication- and media-related phenotypes.

DNA research designs might also help flesh out assumptions regarding the gene-brain pathways that ultimately influence phenotypic variation. That is, it is widely assumed that genes influence the expression of psychological traits and behaviors by directly regulating neurophysiological responses to stimuli (Hatemi, Byrne, & McDermott, 2012). Alternatively, certain genes may indirectly influence phenotypes by triggering a "propensity toward a temperament" that promotes patterns of trait and behavior expression (Mukherjee, 2017, p. 386). In either case, genes are assumed to influence phenotypes through neural pathways, and in a probabilistic rather than a deterministic manner. However, the exact gene-brain-phenotype causal process "remains unknown even to the most experienced" geneticists (Hatemi et al., 2012, p. 308) and is thus a critical area for future research in genetics, behavior genetics, and the social sciences.

CONCLUSION

Behavior genetics research is thriving. Over 11,000 behavior genetic studies involving human participants were published between 1960 and 2014, and more than 3,000 of those studies (about 27%) were published between 2010 and 2014 alone (Ayorech et al., 2016). Part of the recent popularity in behavior genetics can be attributed to psychologists, sociologists, and political scientists increasingly implementing behavior genetic research designs to answer questions of interest to their respective disciplines. Still, despite this increasing popularity, very few communication researchers have utilized behavior genetic approaches. Why?

As this chapter demonstrates, the implementation of twin studies and other behavior genetic approaches in communication has not been limited by analytical hurdles or available data—resources for both have been publicly available for years. Instead, the lack of behavior genetic studies in communication likely is due to an absence of postgraduate training in behavior genetic approaches as well as theoretical frameworks that can be tested using genetically informative research designs. It is imperative that communication scholars work diligently to address these deficits. Provided the appropriate training and theory frameworks, behavior genetics and twin studies could usher in a new and exciting chapter of innovative communication research dedicated to exploring the foundations of traits, behaviors, and effects processes.

NOTES

1. "Organisms" is used here to indicate that, in behavior genetics, variation in traits and behaviors is also studied among non-human animal populations. For examples, see Gewirtz and Kim (2016).
2. Other designs include adoption studies, sibling studies, and extended family studies with parents and grandparents.
3. Twin studies are built on several important conceptual assumptions (see Medland & Hatemi, 2009). One of the most important is the equal environments assumption (EEA), which refers to the idea that phenotypic similarity among identical twins is not due to identical twins being treated more similarly on average than are fraternal twins. Although there is evidence that identical twins are treated more similarly than fraternal twins in certain contexts, some empirical tests show the influence of violating EEA has only a minimal effect on estimates of genetic influence (e.g., Littvay, 2012). Twin studies also rely on the assumption of no assortative mating.
4. Another way to think about the latent factor (E) is that it represents anything unique to an individual twin in a pair not explained by the first two latent factors that represent shared genes and environments. That is, (E) represents what is unshared between twins that makes each twin in a pair unique.

5. Note that there are also bivariate and multivariate *ACE* models that examine genetic and environmental contributions to relationships between multiple phenotypes (see Medland & Hatemi, 2009).
6. I also provide *ACE* model syntax for *R's OpenMx* program and tutorials on my website: www.chanceyork.com.
7. Thankfully, programs such as *OpenMX* perform these calculations automatically, providing estimates of variance explained in the phenotype based on the underlying variance-covariance matrix. Path tracing procedures are mentioned here only to illustrate the logic behind *ACE* modeling.
8. It is important to note that *ACE* models rely on several analytical assumptions. For example, measures of phenotypes are assumed to be multivariate normal. For non-normally distributed measures of phenotypes, alternative models with different distributional assumptions must be specified (e.g., threshold models). In addition, *ACE* models rely on the assumptions of no interactions between genetic and environmental factors and additive genetic effects. In cases in which nonadditive genetic effects such as genetic dominance (*D*) are detected, alternative *ADE* models should be specified. Finally, *ACE* modeling requires researchers to pick starting values for model paths or thresholds. Although opinions vary on how to select starting values, one approach is simply to allow the model to select starting values at random (for an extended discussion of *ACE* assumptions, see Medland & Hatemi, 2009).
9. For more information about the MIDUS study, including sampling methods, interview procedures, and question wording, see http://midus.wisc.edu/.
10. However, it is important to acknowledge that the "communibiological" paradigm popularized by McCroskey and Beatty in the 1990s offered a theoretical framework in which genes were posited to influence communication phenotypes indirectly, via neurophysiological pathways and temperament (see Beatty, McCroskey, & Pence, 2009; Sherry, 2001; Weber, Sherry, & Mathiak, 2009).
11. More information is available at www.twinsdays.org/.
12. See the Washington Twin Registry homepage at https://wstwinregistry.org/.
13. This occurs when traits or behaviors are explained by polygenic inheritance.

REFERENCES

Ayorech, Z., Selzam, S., Smith-Woolley, E., Knopik, V. S., Neiderhiser, J. M., DeFries, J. C., & Plomin, R. (2016). Publication trends over 55 years of behavioral genetic research. *Behavior Genetics, 46*(5), 603–607.

Ayorech, Z., Von Stumm, S., Haworth, C. M., Davis, O. S., & Plomin, R. (2017). Personalized media: A genetically informative investigation of individual differences in online media use. *PLoS One, 12*(1), e0168895.

Beatty, M. J., Marshall, L. A., & Rudd, J. E. (2001). A twins study of communicative adaptability: Heritability of individual differences. *Quarterly Journal of Speech, 87*(4), 366–377.

Beatty, M. J., McCroskey, J. C., & Pence, M. E. (2009). Communibiological paradigm. In M. J. Beatty, J. C. McCroskey, & K. Floyd (Eds.), *Biological dimensions of communication: Perspectives, methods, and research* (pp. 3–16). Cresskill, NJ: Hampton Press.

Blokland, B. A. M., Mosing, M. A., Verweij, K. J. M., & Medland, S. E. (2013). Twin studies and behavior genetics. In T. D. Little (Ed.), *The Oxford handbook of quantitative methods* (Vol. 2, pp. 198–218). Oxford, England: Oxford University Press.

Deppe, K. D., Stoltenberg, S. F., Smith, K. B., & Hibbing, J. R. (2013). Candidate genes and voter turnout: Further evidence on the role of 5-HTTLPR. *American Political Science Review, 107*(2), 375–381.

Deryakulu, D., McIlroy, D., Ursavaş, Ö. F., & Çalışkan, E. (2016). Intrapair similarity of computer self-efficacy in Turkish adolescent twins. *Journal of Educational Computing Research, 54*(6), 840–862.

Deryakulu, D., & Ursavaş, Ö. F. (2014). Genetic and environmental influences on problematic Internet use: A twin study. *Computers in Human Behavior, 39*, 331–338.

Edgerly, S., Thorson, K., Thorson, E., Vraga, E. K., & Bode, L. (2018). Do parents still model news consumption? Socializing news use among adolescents in a multi-device world. *New Media & Society, 20*(4), 1263–1281.

Gewirtz, J. C., & Kim, Y. K. (Eds.). (2016). *Animal models of behavior genetics*. New York, NY: Springer.

Hatemi, P. K., Byrne, E., & McDermott, R. (2012). Introduction: What is a 'gene' and why does it matter for political science? *Journal of Theoretical Politics, 24*(3), 305–327.

Hatemi, P. K., Hibbing, J. R., Medland, S. E., Keller, M. C., Alford, J. R., Smith, K. B., … Eaves, L. J. (2010). Not by twins alone: Using the extended family design to investigate genetic influence on political beliefs. *American Journal of Political Science, 54*(3), 798–814.

Hazel, M., Wongprasert, T. K., & Ayres, J. (2006). Twins: How similar are fraternal and identical twins across four communication variables? *Journal of the Northwest Communication Association, 35*, 46–59.

Horvath, C. W. (1995). Biological origins of communicator style. *Communication Quarterly, 43*(4), 394–407.

Kirzinger, A. E., Weber, C., & Johnson, M. (2012). Genetic and environmental influences on media use and communication behaviors. *Human Communication Research, 38*(2), 144–171.

Knopik, V. S., Neiderheiser, J., DeFries, J. C., & Plomin, R. (2016). *Behavioral genetics*. New York, NY: Palgrave Macmillan.

Littvay, L. (2012). Do heritability estimates of political phenotypes suffer from an equal environment assumption violation? Evidence from an empirical study. *Twin Research and Human Genetics, 15*(1), 6–14.

Lockyer, A., & Hatemi, P. K. (2018). Genetics and politics: A review for the social scientist. In R. L. Hopcroft (Ed.), *Oxford handbook of evolution, biology, and society* (pp. 281–304). New York, NY: Oxford University Press.

Maes, H. (2016). Program: oneACEo.R. Retrieved from: http://ibg.colorado.edu/cdrom2016/maes/UnivariateAnalysis/one/oneACEc.R.

Medland, S., & Hatemi, P. (2009). Political science, behavior genetics, and twin studies: A methodological primer. *Political Analysis, 17*, 191–214.

Miller, G., Zhu, G., Wright, M. J., Hansell, N. K., & Martin, N. G. (2012). The heritability and genetic correlates of mobile phone use: A twin study of consumer behavior. *Twin Research and Human Genetics, 15*(1), 97.

Mukherjee, S. (2017). *The gene: An intimate history*. New York, NY: Simon & Schuster.

Neale, M. (2009). Biometrical models in behavioral genetics. In Y.-K. Kim (Ed.), *Handbook of behavior genetics* (pp. 15–33). New York, NY: Springer.

Neale, M. C., & Maes, H. M. (2004). *Methodology for genetic studies of twins and families*. Dordrecht, The Netherlands: Kluwer Academic.

Plomin, R. (2018). *Blueprint: How DNA makes us who we are*. Cambridge, MA: MIT Press.

Plomin, R., DeFries, J. C., Knopik, V. S., & Neiderhiser, J. M. (2016). Top 10 replicated findings from behavioral genetics. *Perspectives on Psychological Science, 11*(1), 3–23.

Polderman, T. J., Benyamin, B., De Leeuw, C. A., Sullivan, P. F., Van Bochoven, A., Visscher, P. M., & Posthuma, D. (2015). Meta-analysis of the heritability of human traits based on fifty years of twin studies. *Nature Genetics, 47*(7), 702.

Rubin. A. M. (2009). Uses-and-gratifications perspective on media effects. In J. Bryant & M. B. Oliver (Eds.), *Media effects: Advances in theory and research* (pp. 165–184). New York, NY: Routledge.

Sherry, J. L. (2001). Toward an etiology of media use motivations: The role of temperament in media use. *Communication Monographs, 68*(3), 274–288.

Weber, R., Sherry, J., & Mathiak, K. (2009). The neurophysiological perspective in mass communication research. In M. J. Beatty, J. C. McCroskey, & K. Floyd (Eds.), *Biological dimensions of communication: Perspectives, methods, and research* (pp. 41–71). Cresskill, NJ: Hampton Press.

York, C. (2017). A regression approach to testing genetic influence on communication behavior: Social media use as an example. *Computers in Human Behavior, 73*, 100–109.

York, C. (2019). Genetic influence on political discussion: Results from two twin studies. *Communication Monographs*. Advance online publication.

York, C., & Scholl, R. M. (2015). Youth antecedents to news media consumption: Parent and youth newspaper use, news discussion, and long-term news behavior. *Journalism & Mass Communication Quarterly, 92*(3), 681–699.

7

Evolutionary Reasoning in Communication Scholarship

Generating and Testing Sound Hypotheses

Michael Brill and Frank Schwab

There is a wide consensus in science that Darwin's theory of evolution by means of natural selection (Darwin, 1859) offers a plausible explanation for the phylogenetic development of modern day humans: 98% of members of the American Association for the Advancement of Science (AAAS) agree with the statement that "humans and other living beings have evolved over time" (Pew Research, 2015). Several assumptions with relevance for media research could be derived from the theory of evolution: that, for example, the modern human brain is a product of selection processes within the environments of our phylogenetic ancestors; that this brain produces human behavior, but also artifacts, such as media; and, finally, that human communication and media use are also vastly controlled by this evolved brain (Weber, 2015). Given the widespread, fundamental acceptance of the evolution theory, one might expect that researchers in communication and psychology would frequently engage in empirical testing of such derived assumptions. Yet, relatively few studies have considered the biological foundations of human emotion, cognition, and behavior.

This chapter argues in favor of a biologically grounded perspective with an evolutionary theory background. To illustrate the expected benefits, we refer to Tinbergen's Four Whys and his categorization of causes of behavior into proximate and ultimate causes, an approach that has become standard in biology (Tinbergen, 1963). We will then detail how considering the aspect of ultimate causes can increase our understanding of why humans think, feel, and act as they do, particularly in such complex domains as human communication and media use. At first glance, these domains may apparently focus on products of culture and socialization and are thus often exclusively discussed as an output of modern society and technologies, and not so much from an evolutionary perspective (Hennighausen & Schwab, 2015).

However, this potentially beneficial perspective is under-used: moreover, research employing this perspective is frequently subject to fundamental criticism regarding its scientific validity. Such criticism may in part result from a lack of understanding of why and how evolutionarily grounded hypotheses are tested and can thus be countered by explaining how scholars in evolutionary psychology adhere to well-established scientific paradigms. Naturally, it is all the more important to adhere to such standards, so that a sound, rigorous testing of hypotheses on the biological foundations of human behavior can be ensured. This chapter summarizes scholarly advice on how to do so. First, we will provide an introduction on how communication scholars can benefit from evolutionarily grounded explanations. As a guideline, we have relied on Tinbergen's Four Whys, which serve as guiding questions in the tradition of Aristotle for biological research. On this foundation, we will discuss directives by Holcomb (1998) and Buss (1995) on how researchers can perform testing of evolutionarily grounded concepts and discuss why a Lakatosian approach (Ketelaar & Ellis, 2000) especially benefits evolutionary research. We will conclude with brief examples of how an evolutionary perspective could advance research on some common phenomena of human communication and media use.

NATURE BLINDNESS IN MEDIA RESEARCH

For a long time, research on the micro level—that is, on the level of individuals—has focused on proximate and rather biology-free approaches and has been driven by neo-behaviorism, social learning theory, cognitive approaches, modern psychoanalysis, and postmodern radical constructivist approaches (Hennighausen & Schwab, 2015; Schwab, 2010). Concerning the influence of biology, a certain "nature blindness" (Sherry, 2004) could be found in media research, where biological influence is rarely considered in theory development and hypothesizing. Instead, the human mind is often treated solely as a product of culture, socialization, or environment (Hennighausen & Schwab, 2015). In this respect, research frequently neglects the importance of the interaction between an evolved phenotype and the current environment for the production of a given behavior (Buss, 1995; Confer et al., 2010; Hennighausen & Schwab, 2015). Speaking metaphorically, researchers make extensive use of their "nurture eye" when focusing on cognitive learning-only approaches (or the "software" component of behavior production), while at the same time not using their "nature eye" for evolved biological influences (or the component for "hardware" and some additional "operating system software" for behavior production). When biology has been considered, then mostly aspects such as genetics, neurophysiology (Weber, 2015), and peripheral physiology (Zillmann, 2000) have been taken into account, thereby focusing on proximate explanations for human perception or behavior. From an evolutionary perspective, however, the important ultimate question, namely, why exactly these mechanisms have evolved and which purpose they initially served, remains disregarded far too often.

A DISTORTED IMAGE OF BIOLOGY

Although evolutionary psychology is considered to be a promising research paradigm (e.g., Barkow, Cosmides, & Tooby, 1992; Buss, 1995, 1999; Confer et al., 2010), it has likewise been criticized (e.g., Buller, 2005; Fodor, 2005; Lewontin, 1990, 1998; Schlinger, 2002; Silvers, 2010; Trafimow & Gambacorta, 2012). As we have stated above, criticism of evolutionary approaches may result in part from a lack of understanding regarding its underlying concepts or even an inaccurate conception of biology or evolutionary explanations. To address some common misconceptions (see Schwab, 2007), we can first clarify that the evolutionary process itself is far from being merely a theater for brutal competition and pure egoism, as the concept "survival of the fittest" may imply. Although reproductive success is certainly the ultimate purpose to accomplish during evolution, many strategies have developed in nature to achieve this goal. Especially for organisms that live in social groups, this includes mechanisms that could be regarded as extremely desirable from a cultural perspective, such as helping behavior and altruism, cooperation, and affection toward kin and significant others (Trivers, 1971).

Second, biology is far from proposing an absolute genetic determinism and preformism. Heritable traits can mostly be described as a learning bias or disposition, and most genetically influenced behaviors are not fixed, innate, or impossible to modify. Most behaviors instead develop and mature as a result of the interaction between an evolved phenotype and the current environment (Bischof, 1989; Brill, Lange, & Schwab, 2018).

Third, biology and evolutionary approaches do not follow a "pan-adaptationist" program. They do not claim that every observable aspect of an organism is an adaptation, nor that all observable aspects of human behavior are free of evolved influences or adaptations, as many scholars who favor the Standard Social Science Model do (Tooby & Cosmides, 1992). Approaches informed by biology or evolutionary considerations instead ask to what extent organismic behavior or experience results from, or is influenced by, evolved aspects.

Fourth, evolutionary reasoning is no telling of "just-so-stories." Questions regarding the adaptive value and the phylogenetic development of an observed phenomenon are necessary to provide biologically complete explanations. Careful evolutionary reasoning relies on an established body of research, makes a priori predictions instead of *ex post facto* explanations, and evaluates research concepts based on accumulated empirical evidence. Because evolutionary reasoning as a guideline requires a plausible connection between hypothesized evolutionary aspects and known proximate explanations, it can serve to guide thinking about proximate explanations.

Fifth, evolutionary thinking is descriptive and does not aim to promote or support naturalistic fallacies. An exact description of human behavior, including an understanding of its ultimate functions in the sense of Tinbergen (1963), can rather serve societal discourse about how our modern environment can or should be shaped by society. On the other hand, it is advisable to exert care concerning restrictions arising through moralistic fallacies, when aspects of nature with socially difficult consequences are negated, or when research on such aspects is labeled as undesirable. After all, such restrictions would impede societal emancipation from our evolved heritage in a modern environment. Finally, a constant source for conflict seems to be the relation between nature and culture. A host of theories in evolutionary thinking addresses this issue (Schwab & Lange, 2017) and agrees upon a culture-by-nature perspective: humans, as well as some other species, have an evolved capacity to develop something like culture.

After this brief excursion about what an evolutionary approach does not claim, we can now move on to several core assumptions made in biology and evolutionary psychology.

TINBERGEN'S FOUR WHYS

The ethologist and Nobel Prize-winner Nikolaas Tinbergen (1963) proposed four guiding questions for biological behavioral research, which are based on Aristotle's four causes. The four causes aim at providing different answers to the question "why?," and for Aristotle, the four necessary answers should cover the aspects of matter, form, source, and end. Tinbergen derived four complementary categories of explanations for animal behavior from these philosophical considerations (see Table 7.1). The four questions address different levels of analysis and can be grouped into proximate causes and ultimate causes. Proximate causes address (1) the specific mechanism that is necessary to produce a given behavior, including neurophysiological, physiological, and cognitive mechanisms, or stimuli in the environment that serve as a trigger for

Table 7.1 Tinbergen's Concept of the Four Whys as Guiding Questions for Biological Research, Organized in Proximate and Ultimate, and Dynamic and Static Aspects

	Dynamic View	*Static View*
Proximate view	Ontogeny (development) Developmental explanations for changes in *individuals*, from DNA to their current form	Mechanism (causation) Mechanistic explanations for how an organism's structures work
Ultimate view	Phylogeny (evolution) The history of the evolution of sequential changes in a *species* over many generations	Function (adaptation) A species trait that solves a reproductive or survival problem in the current environment

behavior production. As an example, researchers could address the effects of testosterone on aggressive behavior. Further, proximate causes address (2) ontogenetic aspects, focusing on the entire lifespan of an individual organism, from the embryonic state onwards. For example, researchers could ask what the necessary developmental steps are for a given behavior to emerge. It is clear that communication science and psychology usually focus on one (or both) of these proximate causes. The two additional categories proposed by Tinbergen address ultimate causes of behavior and are considered less frequently outside of evolutionary psychology or ethology. The questions for ultimate causes ask for (3) the ultimate function of a given behavior, that is, the behavior's adaptive value for the organism in the process of natural or sexual selection, and (4) phylogenetic aspects, thus focusing on the evolutionary history of an organism. For example, researchers might ask what the adaptive value of different forms of aggression within a social group is, when or under which circumstances in evolutionary history this behavior might have occurred for the first time, and when it was presumably established in the population.

Addressing all four causes for a given research question can be seen as the gold standard for developing a biological answer to a phenomenon and can also be used in media research. Although this approach may not always readily present all answers to all aspects of a research question, it will still guide researchers by leading them to ask the right questions. For example, researchers could be interested in a biologically grounded explanation for parasocial behavior toward a media persona (e.g., Schramm, 2008). To answer all four questions for this phenomenon, researchers would address proximate and ultimate causes. The proximate cause of mechanism leads to an understanding of what sort of physical process (e.g., neurological changes), mental-cognitive processes (e.g., reaction time changes), and social behaviors (e.g., observational data) result in or accompany the phenomenon. This proximate domain is probably the usual focus in media research, with scholars focusing on mental-cognitive functioning, social behavior, neuropsychological processes (Weber, 2015) or physiological causes, such as hormonal processes. Beyond this scope of research, the additional questions by Tinbergen can help produce more complete explanations for the behavior at hand. The proximate ontogenetic causes address how this behavior develops and changes over the course of life (see also life history theory; Alexander, 1987), a domain usually addressed by developmental psychologists (Bjorklund & Pellegrini, 2002). Parasocial relations may, for example, develop along a theory of mind (Premack & Woodruff, 1978; Tomasello, 2002) in children at the age of 3 or 4, and the behavior could occur more prominently during different developmental stages. For example, a hypothesis from an evolutionary point of view could be that cross-gender parasocial interaction effects (Schramm, 2008) should correlate with periods of increased mating investment as a life-history aspect. The question for the ultimate cause (that is, the question for the behavior's adaptive value) would use an evolutionary perspective and would address the question of where the adaptive benefit is, that is, how exactly it should increase reproductive success across generations. Finally, the question for the ultimate cause of phylogenetic development would be what evolutionary mechanisms have caused the behavior or experience to evolve throughout phylogeny. Reconstructing the phylogeny of a species often makes it possible to understand the "uniqueness" of recent characteristics: Why are parasocial interactions and relationships working the way they do? Identifying earlier phylogenetic stages of a behavior, or identifying necessary preconditions for the behavior to occur (e.g., processing of social and emotional cues), can help determine the form of more modern characteristics, such as human behavior when using media with fictional narratives.

DERIVING AND TESTING SOUND HYPOTHESES IN EVOLUTIONARY MEDIA RESEARCH

Integrating ultimate aspects from Tinbergen's guidelines into research can lead to a better understanding of human behavior. However, for media researchers with experimental paradigms in particular, the question is how they can integrate Tinbergen's ultimate perspective into their established research processes, and at the same time avoid telling "just-so-stories" or using evolutionary approaches for *ex post facto* explanations of observed data (Schwab 2010; Schwab & Schwender, 2010). We have organized scholarly advice on the appropriate incorporation of ultimate questions into research procedures into three steps: (1) considering the assumed evolved functioning of the human mind; (2) using a Lakatosian philosophy of science, in contrast to the usual Popperian philosophy (Ketelaar & Ellis, 2000); and (3) following guidelines for developing and testing evolutionary theories and hypotheses (Buss, 1995).

Biology of the Human Mind

When looking for biologically grounded explanations for human behavior, researchers actually observe organisms that live in an environment that is very different from the environment evolution has equipped them to live in. Thus, when making assumptions about the functioning of the evolved human mind, researchers in ethology and evolutionary psychology put particular focus on the Pleistocene, a period of time when our ancestors inhabited a relatively stable environment for many generations (Barkow et al., 1992; Cosmides & Tooby, 1992; Schwab, 2006; Schwab & Schwender, 2010). During the Pleistocene, the *Homo* species recurrently faced adaptive problems, and it is assumed that evolution shaped a human mind that is equipped to deal effectively with the adaptive problems encountered in this ancient environment. Researchers assume that this has been achieved by a series of specialized, domain-specific mechanisms that facilitate the solution of recurrent adaptive problems: the evolved psychological mechanisms (EPMs; Barkow et al., 1992; Buss, 1999; Schwab, 2010). These EPMs represent an adaptive toolbox (Gigerenzer & Selten, 2001; Schwab & Schwender, 2010) attuned to the Pleistocene environment. They process a limited amount of specific information and produce "physiological activity, information to other psychological mechanisms, or manifest behavior" (Buss, 1999, p. 64), all of which is aimed at the solution of a specific adaptive problem of survival or reproduction.

When a successful EPM emerged in individuals within the population, it increased their chance of survival and reproduction, so an increasing number of members of a species was equipped with this new tool in their psychological toolbox (Buss, 1995; Confer et al., 2010;

Schwab & Schwender, 2010). For example, if some human individuals developed (sex-specific) mechanisms for jealousy, their reproductive success would exceed the success of individuals who care less about mate-guarding or their mate's sexual activities (Buss, Larsen, Westen, & Semmelroth, 1992). Consequently, this specific EPM would be established in the population over time.

These evolutionary processes that occurred during the Pleistocene formed the human mind to perform well in the Pleistocene environment. However, the timespan between the end of the Pleistocene and modern times is—in evolutionary terms—very short, and since *Homo sapiens* formed the first settlements about 40,000 years ago (Cosmides & Tooby, 1994), the environment humans inhabit has changed dramatically. As, in addition, probably no selective pressures have surfaced against the Pleistocene EPMs, it can be assumed that the evolved psychological mechanisms for the solution of Pleistocene adaptive problems are still present in today's psychological toolbox, but they are now active in a very different environment. There is a mismatch of evolutionary design (Workman & Reader, 2008) between the environment the mechanisms evolved for and the environment in which they are currently performing.

One prominent example of a domain of mismatch is the media, which is a very recent achievement of human activity in environment-modification, especially when considering technologically sophisticated media. As a consequence, the human mind did not have enough time to develop media-specific EPMs, and media can serve as a dummy stimulus that triggers the archaic EPMs. Just as a scarecrow is designed to deter a bird by appealing to its EPMs, events and figures in media are designed to appeal to human EPMs (Schwender, 2006). Media can present multi-sensory cues and bind the recipient's attention, evoke emotions, or even lead to parasocial relationships with fictional characters (Hennighausen & Schwab, 2015; Horton & Wohl, 1956; Schramm, 2008; Schwab & Schwender, 2010).

Evolved Psychological Mechanisms as an Access Route for Empirical Testing

Whether researchers explicitly aim to examine evolutionary theories or are merely looking for a research framework to understand human behavior, EPMs are the component of the evolutionary process that offer an approach for empirical testing (Holcomb, 1998). But how can researchers in communication and psychology test the respective concepts with their established empirical research methods? Holcomb (1998) organized the consideration of the evolutionary background into three steps: (1) an adaptive problem of the Pleistocene environment is identified; (2) a specific EPM is hypothesized that would have been able to facilitate solving this adaptive problem within the Pleistocene environment; and (3) the derived hypotheses are tested empirically. This helps to determine if the hypothesized EPM can be found in modern-day populations, and whether it can be regarded as part of the human mind's evolved adaptive toolbox. This process should take into consideration that EPMs may manifest with different effects when interacting with the modern-day environment (Hennighausen & Schwab, 2015; Holcomb, 1998). For example, a preference for food with high amounts of sugar and fat may have been very adaptive in the Pleistocene environment, because it motivated adequate caloric intake when nutrient-rich food was scarce. However, in the modern environment with its abundance of high-calorie, processed food, such an EPM can have maladaptive effects. One can speculate that some preferences of media characteristics (e.g., for content such as violence, crime, love, and sex) are—like food preferences (see the "cheesecake argument" below)—influenced by evolved mechanisms.

In any case, it is important to understand that the theory of evolution and the assumed phylogenetic processes are not tested directly (Ketelaar & Ellis, 2000). Researchers instead use

Darwin's theory of evolution and related modern evolutionary theories as a meta-theoretical framework to construct hypotheses (Buss, 1995). In this respect, evolutionary media research can follow a Lakatosian philosophy of science (Ketelaar & Ellis, 2000), which is slightly different from the commonly employed Popperian philosophy of science.

Philosophy of Science in Evolutionary Psychology: Popper or Lakatos

In Popperian epistemology, science should use the method of falsification (Popper, 1959). In this process, theories are developed, hypotheses are deduced from these theories as specific predictions, and conformity of these predictions to reality is tested in empirical studies. If the hypotheses' predictions are in line with collected data, this is considered a preliminary verification of the hypotheses, and they are accepted. If, on the other hand, the collected data do not support the hypotheses' predictions, then the hypotheses are rejected. If empirical testing shows that hypotheses that have been deduced from a given theory are rejected, this theory is seen as falsified because it does not adequately describe reality.

Ketelaar and Ellis (2000) argue that this rather strict approach would, especially in the field of psychology, result in "too many empirical findings being cast in terms of their support or refutation of binary oppositions, such as nature versus nurture, central versus peripheral, serial versus parallel" (Newell, 1973, as cited in Ketelaar and Ellis, 2000, p. 3). Thus, more and more negative knowledge about the human mind would be collected, whereas positive knowledge from not-yet falsified theories would be comparably small. Ketelaar and Ellis (2000) thus consider a Lakatosian philosophy of science a more efficient research strategy (Hennighausen & Schwab, 2015).

Lakatos (1970, 1978) proposed that researchers first agree on a set of basic assumptions as a "hard core," such as Newton's four principles in physics. The hard core serves as a metatheoretical framework and guides research in development and testing of theories and hypotheses (Ketelaar & Ellis, 2000). Smaller, more specific theories can be added to the metatheoretical framework to explain certain aspects in more detail. By developing and testing such an extended theoretical body, new aspects of the metatheory are challenged in empirical studies. The metatheory's guiding function directs researchers to test assumptions that are necessarily related to the overall framework, so over time, more empirical evidence on the underlying metatheory is gathered. As the derivative theories and hypotheses of the metatheories are either confirmed or falsified over time, the core metatheory either makes progressive or degenerative changes. Together with constant refinement and reformulation, researchers can choose between rival metatheories. Those competitors can be in a process of approximating reality; that is, they may themselves not offer perfect explanatory value, but researchers can choose the theory that best explains observed phenomena (Hennighausen & Schwab, 2015; Ketelaar & Ellis, 2000).

In the Lakatosian approach, the derivative theories form a "protective belt" of middle-level theories around the core metatheory, and these mid-level concepts are then tested empirically. This offers the chance to test specific aspects of the metatheory empirically. If observed data support the predictions of the middle-level concepts, then the core metatheoretical concept is supported as well. The "protective" function of the surrounding belt comes into play once a middle-level theory is not supported by empirical data. This does not lead researchers to immediately reject the core metatheory, because only the middle-level theory has been tested. Although repeated rejection of many derived middle-level theories would ultimately lead to rejection of the hard core, as well, this indirect testing approach offers researchers more breathing room for testing complex phenomena. For example, it is possible to derive rival middle-level explanations for observed phenomena, where each explanation is perfectly in line with the guiding metatheory, while the explanations contradict each other, however. This is frequently the case in evolutionary research when researchers ask if an observed behavior is an actual adaptation or merely a by-product of a different adaptation. Within the protective belt, both theories can be tested; the concept with better scientific value can replace the other concept, so the overall understanding of the core concept—in this case, the theory of evolution—has been improved (Hennighausen & Schwab, 2015).

When transferring the Lakatosian approach to a research heuristic for evolutionary media research, four levels of analysis can be identified (Buss, 1995; Ketelaar & Ellis, 2000): (1) the core metatheory; (2) the middle-level theories that have been derived from the metatheory; (3) the hypotheses that have been derived from the middle-level theory; and (4) specific predictions used to test the hypotheses (see Figure 7.1). For evolutionary media research, the metatheoretical framework consists of Darwin's theory of evolution through natural and sexual selection, and related assumptions made by modern evolutionary theories (e.g., Hamilton, 1964; Tooby & Cosmides, 1992; see Hennighausen & Schwab, 2015). From these core concepts, middle-level theories on specific aspects of the metatheoretical framework are derived. As we have already

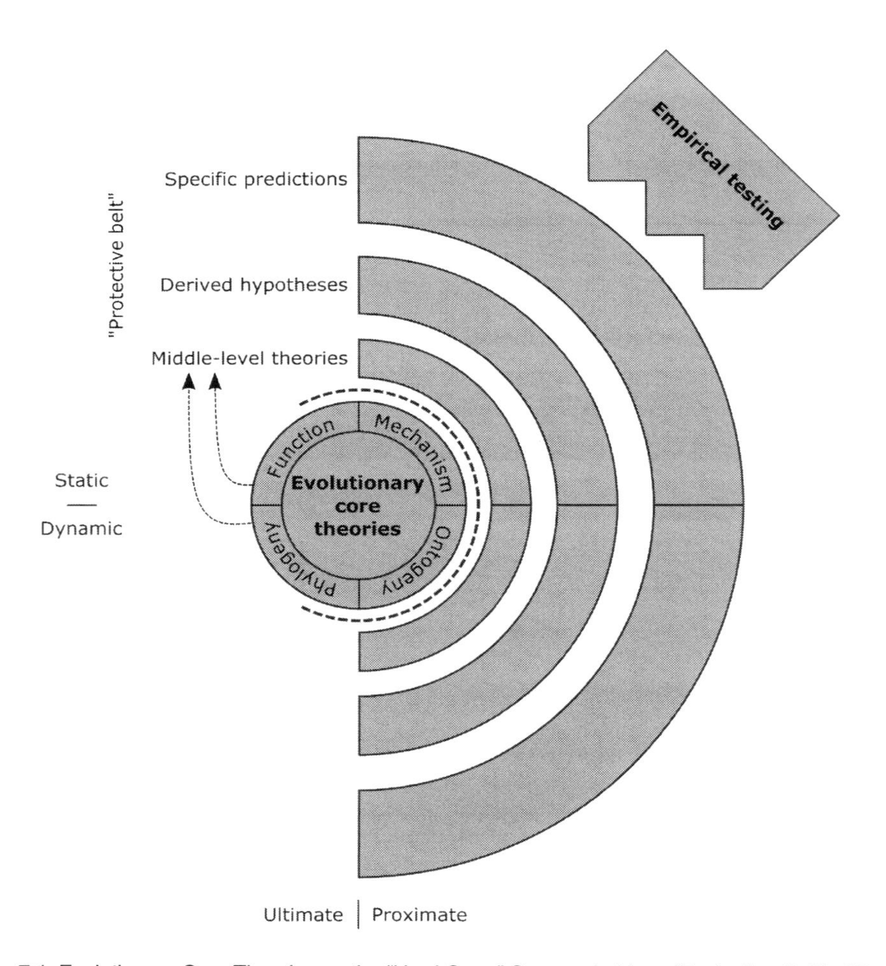

Figure 7.1 Evolutionary Core Theories as the "Hard Core," Surrounded by a "Protective Belt" of Derived Middle-Level Theories and Hypotheses

seen, these middle-level theories are all consistent with the metatheory (otherwise, they could, e.g., be part of a different, rival metatheoretical framework), but it is certainly possible—and even beneficial—that they can make conflicting predictions. Regarding the choice of middle-level theories, Buss (1995) states that not only theories from the evolutionary framework can be introduced into the protective belt, but that theories and explanations from non-evolutionary approaches can be used as well. This is an important aspect: as we have stated, evolutionary approaches explicitly acknowledge that behavior results from an interaction between phenotype and environment. That means that cultural aspects—being an important product of the evolved mind—are not necessarily neglected, and neither is theorizing on cultural aspects, as long as it can be linked to the metatheory. For example, mate selection preferences observed in the human species can be explained from an evolutionary perspective, as in sexual strategies theory (SST; Buss & Barnes, 1986), or from a non-evolutionary perspective, as in social roles theory (Eagly et al., 2000).

The next step is to formulate the specific hypotheses and predictions for empirical testing, which can be carried out using one or more middle-level theories and can reflect contradicting predictions from the theories from which the hypotheses were deduced.

They can either be derived from middle-level theories in a top-down process, or they can be constructed from observed phenomena in a bottom-up process (Buss, 1995). According to Buss (1995), the latter approach can be used when encountering phenomena that are suspected to be triggered by evolved mechanisms: then the metatheoretical framework's accumulated knowledge needs to be taken into account, for example, on the evolved functioning of the human mind, including its adaptive toolbox of EPMs. Following the three steps formulated by Holcomb (1998), the identification of a Pleistocene adaptive problem helps to illuminate observed behavior. As a guideline, Buss (1995) offers a selection of adaptive problems for the domains of survival, sex and mating, parenting and kinship, and adaptive problems of living in social groups. In Holcomb's (1998) second and third steps, assumed EPMs and empirical testing on modern-day populations lead to empirical insights. For empirical media researchers in the social sciences, the research paradigms used to empirically test these assumptions will in many cases be experimental group comparisons on data gathered through self-report or observational methods. However, Buss (1995) states that additional paradigms and sources of data can corroborate evolutionary research, with paradigms including comparison of men and women, within-sex comparisons, and comparison of individuals in different contexts (Buss, 1995, p. 65), which can, for this type of media research, include experimentally modified settings. For the data that should be used to test the derived assumptions, Buss (1995, p. 65) suggests the use of additional sources of data to be beneficial for evolutionary research, such as archeological records, data collected in contemporary hunter-gatherer societies, life-history data available in public records, and analysis of cultural artifacts (like media). With continued testing of the specific predictions derived from hypotheses, the scientific value of the middle-level theories can be evaluated, which ultimately leads to progression or degradation of the core metatheoretical framework (Hennighausen & Schwab, 2015).

CONCLUSION

The consideration of evolutionary aspects, and in particular of ultimate causes for observed phenomena, offers new pathways for theorizing and hypothesis testing. When focusing exclusively on, for example, learning-only approaches as explanations, researchers would effectively neglect a considerable part of reality, such as the biological influences on the production of behavior. Arguing with Tinbergen's Four Whys, reality could be better explained when all known aspects are considered for theorizing.

A few examples from media research and adjacent areas are given to illustrate the potential benefit of evolutionarily grounded explanations. For example, many theories in communication include assumptions of almost unbounded rationality, for instance, in media selection theories such as media richness. From a biological point of view, researchers could instead be guided by the assumption that rationality very often functions in the form of heuristics (Gigerenzer & Selten, 2001). When explaining sex differences in media research, "sex-only explanations" rely exclusively on the effects of culture and socialization (Eagly et al., 2000). Beyond such cultural aspects, a biological perspective would also include sex-typical (or even sex-specific) aspects, such as hormonal or neuropsychological differences (Weber, 2015), evolved biases, or differences in parental investment (Trivers, 1972). In research on uses and gratifications, diverse motives for media use have been identified (Katz & Foulkes, 1962). An evolutionary perspective could provide an organizing framework for the different motives, for example, along evolved solutions for species-specific adaptive problems (see Bischof, 1989; Buss, 1995). For the explanation of emotional media aspects, assumptions have been made that consider biological aspects (Meadowcroft & Zillmann, 1987; Zillmann, 2000), given that most emotion theories understand emotions as a bodily phenomenon. These proximate biological considerations could be corroborated by an evolutionary perspective on ultimate causes, and assumptions on evolved emotional mechanisms (Tooby & Cosmides, 2008). As a final example, media narrations are understood by most communication scholars as a cultural phenomenon, whereas evolutionarily informed research may regard narrations either as evolved adaptations or as the by-product of adaptations (Ohler & Nieding, 2006; Pinker, 2002; Schwab, 2010; Schwender, 2006), a perspective that offers new approaches for empirical testing.

Evolution as a guiding metatheoretical framework may also benefit interdisciplinary research. If scholars from different disciplines, such as communication, psychology, and life sciences, adhere to the same conceptional standards, their theories and concepts may become more interconnectable, leading to more diverse approaches for empirical testing. An overview of how different branches of science are addressing different facets and mechanisms of organisms has been proposed by Medicus (2005), and a slightly modified version can provide guidance for scholars in evolutionary media research, as well (see Table 7.2). The overview table can then serve as a map to identify gaps in knowledge and can inform researchers who are looking for interdisciplinary cooperation, both within social sciences, or between life sciences and social sciences.

For explanations on experiences and behavior of the human species in particular, evolution and biology seem to be natural options for an interdisciplinary guiding framework.

Table 7.2 Modified Version of the Map of Gaps of Knowledge by Medicus (2005), Illustrating which Aspects of Human Evolved Functioning Are Addressed in Different Branches of Life Sciences and Social Sciences

		Ontogeny	*Mechanism*	*Phylogeny*	*Function*
		Generating Hypothesis		*Guiding Theorizing*	
Organ (brain)	Evolutionary neuroscience				
Individual	Evolutionary psychology				
Family/group	Evolutionary social psychology				
Society	Evolutionary sociology				

REFERENCES

Alexander, R. D. (1987). *The biology of moral systems*. New York, NY: Aldine de Gruyter.

Barkow, J. H., Cosmides, L., & Tooby, J. (1992). *The adapted mind: Evolutionary psychology and the generation of culture*. New York, NY: Oxford University Press.

Bischof, N. (1989). Emotionale Verwirrungen—Oder: Von den Schwierigkeiten im Umgang mit der Biologie [Emotional confusions—or: On the difficulties of dealing with biology]. *Psychologische Rundschau, 40*, 188–205.

Bjorklund, D. F., & Pellegrini, A. D. (2002). *The origins of human nature: Evolutionary developmental psychology*. Washington, DC: American Psychological Association.

Brill, M., Lange, B. P., & Schwab, F. (2018). Digital sandboxes for Stone Age minds. Virtual worlds as Bischofian fitness potential landscapes. In J. Breuer, D. Pietschmann, B. Liebold, & B. P. Lange (Eds.), *Evolutionary psychology and digital games: Digital hunter-gatherers* (pp. 32–48). New York, NY: Routledge.

Buller, D. J. (2005). Evolutionary psychology: The emperor's new paradigm. *Trends in Cognitive Sciences, 9*, 277–283.

Buss, D. M. (1995). Evolutionary psychology: A new paradigm for psychological science. *Psychological Inquiry, 6*, 1–30.

Buss, D. M. (1999). *Evolutionary psychology: The new science of the mind*. Needham Heights, MA: Allyn & Bacon.

Buss, D. M., & Barnes, M. (1986). Preferences in human mate selection. *Journal of Personality and Social Psychology, 50*, 559–570.

Buss, D. M., Larsen, R. J., Westen, D., & Semmelroth, J. (1992). Sex differences in jealousy: Evolution, physiology, and psychology. *Psychological Science, 3*, 251–256.

Confer, J. C., Easton, J. A., Fleischman, D. S., Goetz, C. D., Lewis, D. G., Perilloux, C., & Buss, D. M. (2010). Evolutionary psychology: Controversies, questions, prospects, and limitations. *American Psychologist, 65*, 110–126.

Cosmides, L., & Tooby, J. (1992). Cognitive adaptations for social exchange. In J. H. Barkow, L. Cosmides, & J. Tooby (Eds.), *The adapted mind: Evolutionary psychology and the generation of culture* (pp. 163–228). New York, NY: Oxford University Press.

Cosmides, L., & Tooby, J. (1994). Beyond intuition and instinct blindness: Toward an evolutionarily rigorous cognitive science. *Cognition and Emotion, 50*, 41–77.

Darwin, C. (1859). *On the origin of species by means of natural selection, or the preservation of the favoured races in the struggle for life*. London, England: John Murray.

Eagly, A. H., Wood, W., & Diekman, A. B. (2000). Social role theory of sex differences and similarities: A current appraisal. In T. Eckes & H. M. Trautner (Eds.), *The developmental social psychology of gender* (pp. 123–174). New York, NY: Psychology Press.

Fodor, J. (2005). Reply to Steven Pinker "So how does the mind work?" *Mind & Language, 20*, 25–32.

Gigerenzer, G., & Selten, R. (2001). *Bounded rationality: The adaptive toolbox*. Cambridge, MA: MIT Press.

Hamilton, W. D. (1964). The genetical evolution of social behavior. II. *Journal of Theoretical Biology, 7*, 17–52.

Hennighausen, C., & Schwab, F. (2015). Evolutionary media psychology and its epistemological foundation. In T. Breyer (Ed.), *Epistemological dimensions of evolutionary psychology* (pp. 131–158). New York, NY: Springer.

Holcomb, H. (1998). Testing evolutionary hypotheses. In C. B. Crawford & D. L. Krebs (Eds.), *Handbook of evolutionary psychology: Ideas, issues, and applications* (pp. 303–334). Mahwah, NJ: Lawrence Erlbaum Associates.

Horton, D. W. R., & Wohl, R. (1956). Mass communication and parasocial interaction: Observation on intimacy at a distance. *Psychiatry, 19*, 215–229.

Katz, E., & Foulkes, D. (1962). On the use of mass media as "escape": Clarification of a concept. *Public Opinion Quarterly, 26*, 377–388.

Ketelaar, T., & Ellis, B. J. (2000). Are evolutionary explanations unfalsifiable? Evolutionary psychology and the Lakatosian philosophy of science. *Psychological Inquiry, 11*, 1–21.

Lakatos, I. (1970). Falsifications and the methodology of scientific research programmes. In I. Lakatos & A. Musgrave (Eds.), *Criticism and the growth of knowledge* (pp. 91–196). Cambridge, England: Cambridge University Press.

Lakatos, I. (1978). *The methodology of scientific research programs: Philosophical papers* (Vol. 1). Cambridge, England: Cambridge University Press.

Lewontin, R. C. (1990). The evolution of cognition. In D. N. Osherson & E. E. Smith (Eds.), *Thinking: An invitation to cognitive science* (Vol. 3, pp. 229–246). Cambridge, MA: MIT Press.

Lewontin, R. C. (1998). The evolution of cognition: Questions we will never answer. In D. Scarborough & S. Sternberg (Eds.), *Methods, models, and conceptual issues: An invitation to cognitive science* (Vol. 4, pp. 106–132). Cambridge, MA: MIT Press.

Meadowcroft, J. M., & Zillmann, D. (1987). Women's comedy preferences during the menstrual cycle. *Communication Research, 14*, 204–218.

Medicus, G. (2005). Mapping transdisciplinarity in human sciences. In J. W. Lee (Ed.), *Focus on gender identity* (pp. 95–114). New York, NY: Nova Science Publishers.

Newell, A. (1973). You can't play 20 questions with nature and win: Projective comments on the papers in this symposium. In W. G. Chase (Ed.), *Visual information processing* (pp. 283–308). New York, NY: Academic.

Ohler, P., & Nieding, G. (2006). An evolutionary perspective on entertainment. In J. Bryant & P. Vorderer (Eds.), *Psychology of entertainment* (pp. 423–433). Mahwah, NJ: Lawrence Erlbaum Associates.

Pew Research. (2015). An elaboration of AAAS scientists' views. Retrieved from http://assets.pewresearch.org/wp-content/uploads/sites/14/2015/07/.Report-AAAS-Members-Elaboration_FINAL.pdf.

Pinker, S. (2002). *The blank slate: The modern denial of human nature*. New York, NY: Viking.

Popper, K. R. (1959). *The logic of scientific discovery*. Oxford, England: Basic Books.

Premack, D., & Woodruff, G. (1978). Does the chimpanzee have a theory of mind? *Behavioral and Brain Sciences, 1*, 515–526.

Schlinger, H. D. (2002). Not so fast, Mr. Pinker: A behaviorist looks at the blank slate. A review of Steven Pinker's *The blank slate: The modern denial of human nature*. *Behavior and Social Issues, 12*, 75–79.

Schramm, H. (2008). Parasocial interaction, Parasoziale Interaktion. In N. C. Krämer, S. Schwan, D. Unz, & M. Suckfüll (Eds.), *Media psychology: Key terms and concepts, Medienpsychologie: Schlüsselbegriffe und Konzepte* (pp. 253–258). Stuttgart, Germany: Kohlhammer.

Schwab, F. (2006). Are we amusing ourselves to death? Answers from evolutionary psychology. *SPIEL. Siegener Periodikum zur internationalen empirischen Literaturwissenschaft, 22*, 329–338.

Schwab, F. (2007). Evolutionäres Denken: Missverständnisse, Trugschlüsse und Richtigstellungen. *Zeitschrift für Medienpsychologie. Themenheft, 19*, 140–144.

Schwab, F. (2010). *Moving pictures: An evolutionary media psychology of entertainment, Lichtspiele: Eine evolutionäre Medienpsychologie der Unterhaltung*. Stuttgart, Germany: Kohlhammer.

Schwab, F., & Lange, B. P. (2017). Evolutionäre Kulturtheorien. In G. Jüttemann (Ed.), *Psychogenese. Das zentrale Erkenntnisobjekt einer integrativen Humanwissenschaft* (pp. 83–94). Lengerich, Germany: Pabst Publishers.

Schwab, F., & Schwender, C. (2010). The descent of emotions in media: Darwinian perspectives. In K. Doveling, C. V. Scheve, & E. A. Konijn (Eds.), *The Routledge handbook of emotion and mass media* (pp. 15–36). New York, NY: Routledge.

Schwender, C. (2006). *Media and emotions: Evolutionary psychological components of a media theory, Medien und Emotionen. Evolutionspsychologische Bausteine einer Medientheorie* (2nd ed.). Wiesbaden, Germany: DUV.

Sherry, J. L. (2004). Media effects theory and the nature/nurture debate: A historical overview and directions for future research. *Media Psychology, 6*, 83–109.

Silvers, S. (2010). Methodological and moral muddles in evolutionary psychology. *Journal of Mind and Behavior, 31*, 65–84.

Tinbergen, N. (1963). On the aims and methods of ethology. *Zeitschrift für Tierpsychologie, 20*, 410–433.

Tomasello, M. (2002). *Die kulturelle Entwicklung des menschlichen Denkens*. Frankfurt, Germany: Suhrkamp.

Tooby, J., & Cosmides, L. (1992). The psychological foundations of culture. In J. H. Barkow, L. Cosmides, & J. Tooby (Eds.), *The adapted mind: Evolutionary psychology and the generation of culture* (pp. 19–136). New York, NY: Oxford University Press.

Tooby, J., & Cosmides, L. (2001). Does beauty build adapted minds? Toward an evolutionary theory of aesthetics, fiction and the arts. *SubStance, 30*, 6–27.

Tooby, J., & Cosmides, L. (2008). The evolutionary psychology of the emotions and their relationship to internal regulatory variables. In M. Lewis, J. M. Haviland-Jones, & L. F. Barrett (Eds.), *Handbook of emotions* (3rd ed., pp. 114–137). New York, NY: Guilford Press.

Trafimow, D., & Gambacorta, D. (2012). How obvious are hypotheses in evolutionary psychology? *The Journal of Social, Evolutionary, and Cultural Psychology, 6*, 1–12.

Trivers, R. (1971). The evolution of reciprocal altruism. *Quarterly Review of Biology, 46*, 35–57.

Trivers, R. L. (1972). Parental investment and sexual selection. In B. Campbell (Ed.), *Sexual selection and the descent of man: 1871–1971* (pp. 136–179). Chicago, IL: Aldine.

Weber, R. (2015). Brain, mind, and media. Neuroscience meets media psychology. *Journal of Media Psychology, 27*, 89–92.

Workman, L., & Reader, W. (2008). *Evolutionary psychology: An introduction* (2nd ed.). New York, NY: Cambridge University Press.

Zillmann, D. (2000). The coming of media entertainment. In D. Zillmann & P. Vorderer (Eds.), *Media entertainment: The psychology of its appeal* (pp. 1–20). Mahwah, NJ: Lawrence Erlbaum Associates.

Part III

COMMUNICATION AND MEDIA NEUROSCIENCE

8

Mediated Messages and Synchronized Brains

Ralf Schmälzle and Clare Grall

Every morning thousands of commuters listen to the same radio program, and every evening millions watch the same news on TV. Large crowds gather to see the same movies, and readers flock around the same top-selling books. In each of these mass communication situations, the messages are the same for every recipient. Thus, the same stream of images, sounds, or letters will enter each recipients' eyes and ears, where it gets converted into neural impulses. Through their various pathways in the brain, these impulses produce activities at multiple levels of the neural hierarchy ranging from basic sensory-perceptual responses to higher-order computations involved in comprehension, attention, memory, and emotion (Gazzaniga, Ivry, & Mangun, 2013). This chapter discusses how inter-subject correlation analysis (ISC) of neuroimaging data can be used to uncover cross-receiver similarities as well as individual differences in how audience members respond to mediated messages (Hasson et al., 2008; Hasson, Malach, & Heeger, 2010). Put simply, we ask whether, where, and how strongly a message synchronizes[1] brain activity across receivers. This speaks to the issue highlighted in the famous quote by Claude Shannon, the creator of information theory, who stated that "the fundamental problem of communication is that of reproducing at one point, either exactly or approximately, the message selected at another point" (Shannon, 1948, p. 379).

In a seminal study, Hasson et al. exposed a group of viewers to the movie *The Good, the Bad, and the Ugly* while recording their brain activity using functional magnetic resonance imaging (fMRI) (Hasson, Nir, Levy, Fuhrmann, & Malach, 2004). In their analysis, they extracted the response time-course in a given brain region from one viewer and compared it to the response time-course obtained in the corresponding brain region from another viewer (Figure 8.1).

The time-course of brain activity in a given region of one recipient's brain is compared with the time-course from the corresponding region in another brain exposed to the same movie. The results reveal commonalities in how brains respond to mediated messages. Conceptually, this is like listening to the voice signal of individual brain regions and measuring how "in unison" the same regions from different brains are responding to a particular message. Empirically, similar brain responses emerge for regions that subserve specific aspects of message processing: visual cortex for processing elemental visual features, face-sensitive cortex (e.g., fusiform face area, FFA) for processing structural aspects of faces, or voice-sensitive cortex for processing vocal

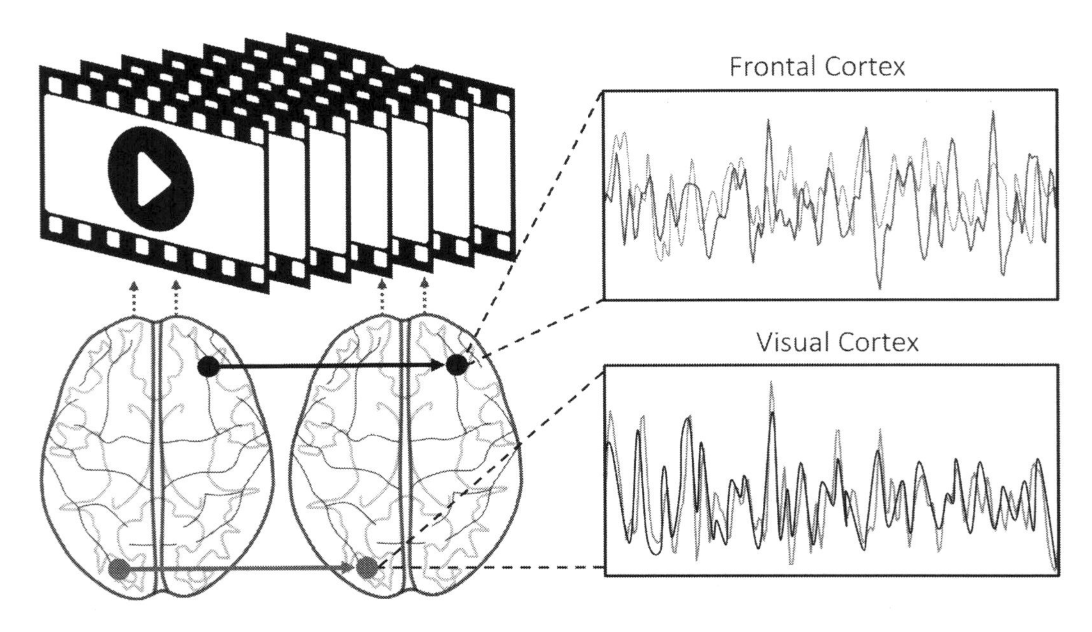

Figure 8.1 Principle of Inter-Subject Correlation Analysis

Source: Figure adapted from Hasson et al. (2004) and brain images from Abraham et al. (2014).

features in the case of spoken messages. More abstract psychological processes (e.g., related to comprehension) are rarely associated with single brain regions, but extensions of ISC analysis can also help to understand commonalities between spatially distributed processes.

By computing this analysis for many brain regions and across all pairs of viewers, they identified brain regions that responded similarly to the movie.[2] Their results revealed highly similar responses in visual and auditory cortices, but also in frontal and parietal regions. This suggests that beyond collectively stimulating each receiver's sensory and perceptual brain apparatus, the movie also engaged systems involved in comprehension, following the narrative, and responding emotionally to its events in a similar way across individual recipients.

Following this seminal study, the basic ISC logic has been applied to a broad variety of mediated messages, including popular movies (e.g., *Indiana Jones*, *Forrest Gump*, and Charlie Chaplin films), TV series (e.g., *Alfred Hitchcock Presents*, *Curb Your Enthusiasm*, *The Office*), written messages and narratives, speeches, documentaries and educational programs, health prevention messages and advertisements, podcasts, and music (Barnett & Cerf, 2017; Byrge, Dubois, Tyszka, Adolphs, & Kennedy, 2015; Cantlon & Li, 2013; Dmochowski et al., 2014; Dmochowski, Sajda, Dias, & Parra, 2012; Finn, Corlett, Chen, Bandettini, & Constable, 2017; Hanke et al., 2014; Hasson et al., 2008; Haxby et al., 2011; Imhof, Schmälzle, Renner, & Schupp, 2017; Jääskeläinen et al., 2008; Kauppi, Jääskeläinen, Sams, & Tohka, 2010; Nummenmaa et al., 2012; Richardson, Lisandrelli, Riobueno-Naylor, & Saxe, 2018; Schmälzle, Häcker, Honey, & Hasson, 2015; Schmälzle, Häcker, Renner, Honey, & Schupp, 2013; Wilson, Molnar-Szakacs, & Iacoboni, 2008, see overview in Vanderwal, Eilbott, & Castellanos, 2019).

An ISC approach is highly beneficial to the study of communication processes for several reasons. First, it yields an objective biological metric that is inherently social because it summarizes brain activity measured in an entire audience. Theoretically, this is highly relevant because it exposes brain-to-brain similarities that can be understood as commonalities between observers at the level of regional brain processes. In addition to the notion of "commonalities"—obviously central to communication[3]—the word "processes" is chosen deliberately here because ISC analysis is computed on temporal data. Considering time is indispensable whenever the input signal is temporal in nature, which is clearly the case for speech, movies, and narratives. Further, ISC analysis can be applied to messages of any complexity, and this paves the way for experiments with mediated messages that contain a mixture of images, sounds, and language. In fact, although complex, these messages are in a way perfectly controlled because they are the same for all receivers. This makes it possible to assess functional similarities across brains during complex cognitive processing, even when the functions of individual regions and how exactly different stimulus properties drive regional activity are not yet fully understood. Finally, the method is equally applicable to multiple measurement modalities, such as fMRI, electroencephalography (EEG), functional near-infrared spectroscopy (fNIRS), psychophysiological data, eye-tracking data, and continuous response measures. In the following sections, we review how this approach can be used to gain new insights into how mediated messages (1) collectively engage audiences; (2) transmit common meaning; and (3) underlie or produce socially shared processes.

CAPTURING COLLECTIVE ENGAGEMENT

ISC holds great promise to assess the common effects of messages on multiple recipients, which are central to mass communication, media effects, and persuasion research. For example, one of communication science's oldest effects is the phenomenon that the rhetorical quality of a speech strongly impacts the degree to which the audience is engaged (Aristotle, 1939). Powerful speeches "draw the audience in" and make everybody listen intently, whereas weaker speeches fail to captivate their listeners' minds, letting them drift away. Schmälzle and colleagues (2015) examined this phenomenon by exposing a group of listeners to rhetorically strong and weak political speeches while recording fMRI data. As predicted, listening to any speech induced similar neural response time-courses in brain regions involved in spoken language processing, but this effect was stronger for rhetorically powerful speeches. In particular, auditory regions as well as the medial prefrontal cortex, a region associated with processing self-relevant stimuli (Schmitz & Johnson, 2007) and following the social content of a story (Mar, 2011), exhibited increased ISC.

These patterns of correlated brain responses across listeners, which are amplified during high-quality speeches, bear resemblance to metaphors of resonance that are often brought up in the context of audience research. The emerging picture suggests that ISC can be used to measure the "strength of the grip" (Hasson et al., 2008) a given message has on an audience, revealing the extent to which people become involved with the message or how it "resonates" with them (Greenwald & Leavitt, 1984; Schmälzle et al., 2015; Sherif & Sargent, 1947). Similar findings have been obtained using a variety of messages, including emotional stories, movies, and effective health messages (Hermans et al., 2011; Imhof et al., 2017; Nummenmaa et al., 2012). From a neuroscientific perspective, such resonance-like effects can be linked to the literature on motivated or selective attention (Lang, Bradley, & Cuthbert, 1997; Schupp, Flaisch, Stockburger, & Junghöfer, 2006). Of note, while ISC can be used to gauge the level of audience engagement, the quality of engagement is a variable that needs further study. Given an appropriate experimental design and by triangulating brain and self-report data, it may become possible to assess whether all recipients are enthusiastic about a speech or collectively hate it.

In sum, this new line of work clearly invites further research on the neural mechanisms underlying entertaining or persuasive messages (Barnett & Cerf, 2017; Dmochowski et al., 2012, 2014; Dudai, 2012; Hasson et al., 2008), topics that have occupied center-stage in communication science since the field's earliest days (Lazarsfeld & Stanton, 1949).

EXAMINING HOW MESSAGES CONVEY MEANING

Beyond examining engagement with a stimulus, which can be encouraged through editing various substantive and stylistic features, a large body of evidence shows that brain activity correlations (ISC) are sensitive to the transmission of meaning. Upon first view, one may be tempted to attribute brain similarities in response to the same messages to purely sensory-driven processes that are not psychologically relevant. Indeed, it is hardly surprising that a message prompts similar responses across recipients if the identical physical properties command similar auditory and visual processes. However, an elegant study of narratives demonstrated that this view is too simplistic (Honey, Thompson, Lerner, & Hasson, 2012). In brief, the authors compared the fMRI response time-courses of Anglophone and Russophone listeners who heard a Russian narrative and its English translation. Thus, the same conceptual content was delivered to the listeners' brains via different sensory-perceptual input, namely, sequences of English or Russian words. When analyzing the brain activity between groups of Anglophone listeners exposed to the English story or Russophone listeners exposed to the Russian story, the early auditory regions, linguistic regions, as well as higher-order regions involved in story processing exhibited reliable ISC, suggesting that responses in these regions were similar across viewers from each language group. However, when Anglophone listeners were exposed to the Russian story, it was unintelligible to them and ISC across their brains was confined to early auditory regions. This demonstrates that ISC in post-sensory regions, rather than being sensory-driven, depends on learned knowledge structures (i.e., knowledge of the Russian language), and similar brain responses across listeners can only emerge if they share common knowledge as a group (Figure 8.2).

In Figure 8.2 (A), ISC can measure the degree to which a given message commands similar processes in different brains. The schematic graphic illustrates overall ISC (computed across the entire brain), but more refined metrics in individual brain regions or systems are possible. In Figure 8.2 (B), Anglophone listeners exposed to a story told in English exhibit ISC because they can successfully decode the message. Similarly, Russophone listeners understand the story when it is conveyed in Russian, but Anglophone listeners exposed to Russian stories showed no ISCs in higher-order regions. In Figure 8.2 (C), measuring inter-brain similarities during exposure to message content opens up new possibilities for multilevel integrative studies of social phenomena and may explain social network phenomena.

Notably, the authors also compared the brain responses of Anglophone listeners exposed to English to the responses from Russophone listeners exposed to Russian. Results showed that despite dissimilarities in auditory regions, similarities emerged in higher-order regions. Thus, these regions responded similarly to the same story being conveyed in different languages. This suggests that conceptual aspects of the story were represented in a similar fashion across all listeners, regardless of the language in which the story was initially delivered. Understanding how this similarity at the conceptual rather than the perceptual level relates to situation models, working memory, and comprehension (Kintsch, 1988; Zwaan, 1999) is currently a vibrant area of research (Hasson, Chen, & Honey, 2015; Lerner, Honey, Silbert, & Hasson, 2011).

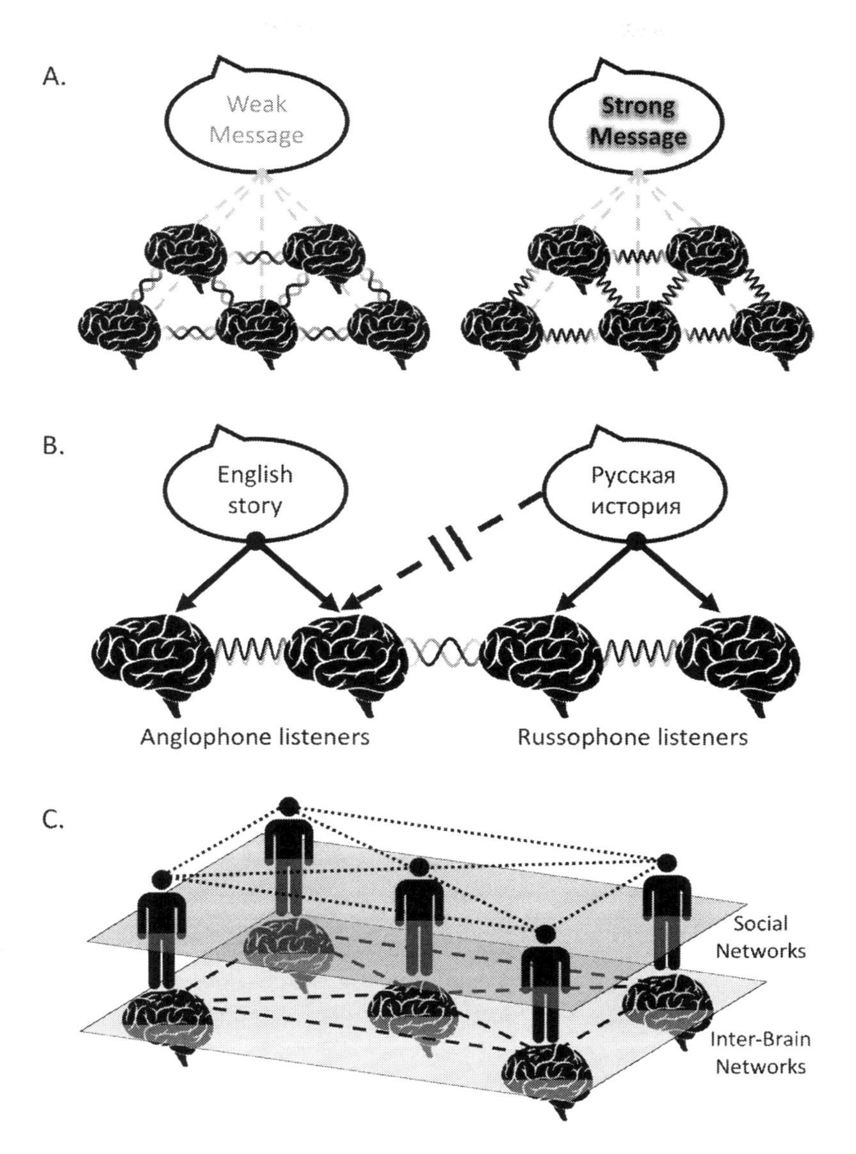

Figure 8.2 Applications of ISC Analysis

Several other studies have shown the great potential of using complex stories to understand higher linguistic and communicative functions related to narrative processing, event segmentation, and natural discourse (Baldassano et al., 2017; Chen et al., 2016; Dikker, Silbert, Hasson, & Zevin, 2014; Farbood, Heeger, Marcus, Hasson, & Lerner, 2015; Stephens, Silbert, & Hasson, 2010; Wilson et al., 2008; Zadbood, Chen, Leong, Norman, & Hasson, 2017). For example, manipulating receiver expectations or interpretations for the same message leads to similar brain response patterns between those with similar interpretations, and different responses between those with different interpretations (Lahnakoski et al., 2014; Yeshurun, Nguyen, & Hasson, 2017; Yeshurun, Swanson et al., 2017). These studies complement the findings from the Russian-English study by demonstrating that not only can different messages produce similar brain responses, but that the same message can also produce divergent responses depending on short-term (e.g., using framing to manipulate expectations) or long-term (e.g., pre-existing language ability) contextual factors.

Altogether, cross-receiver similarities during message processing extend from sensory-driven auditory and visual responses to similarities associated with comprehension and mutual understanding. In this regard, ISC can be used to measure how messages "stimulate meaning" in the minds of others, which is one way that communication has been defined (McCroskey, 2015). This important feat is perhaps best illustrated by a study of brain-injured patients (Naci, Cusack, Anello, & Owen, 2014) in which one patient, although being behaviorally completely unresponsive, exhibited a brain response that tracked with a movie's narrative in a similar way as the brains of healthy viewers. This provides strong evidence for conscious experience of the brain-injured patient, who was unable to communicate, but showed signs of functional executive processing.

SHARED PROCESSING AMONG SOCIAL GROUPS

A classic adage from psychology holds that at some level every person is *like all others, like some others, like no other person* (Kluckhohn. Murray, & Schneider, 1948). This continuum from universality, to group-specificity, to individuality provides a useful framework for thinking about why shared and nonshared brain responses emerge, how they can be manipulated, and what we can learn by studying them. Although important individual differences and culture-specific influences do exist, there are well-researched universalities in genetics, brain architecture, and function, as well as many shared ontogenetic factors that provide everyone with a vast amount of similarly structured sensory, perceptual, and conceptual "training data." Examples of this include seeing faces, using language to communicate, as well as emotional and socio-moral experiences (Brothers, 1990; Haidt, 2001; Panksepp, 2004). The sum of these commonalities, when addressed through message content, gives rise to similar brain responses. For example, every person's visual system will respond roughly similarly to changes in brightness and contrast (Purves et al., 2008; Tootell et al., 1998), and all humans have a biological disposition to like sweet and dislike bitter tastes (Berridge & Kringelbach, 2013). In fact, a recent study demonstrated substantial inter-species correlations between in particular the visual cortices of monkey and human brains while watching the same movie (Mantini et al., 2012). This points to commonalities at least at the level of visual sensory and perceptual processes, if not beyond (Sliwa & Freiwald, 2017). From a communication perspective, similar brain responses to the same messages represent message main effects, or the effects that messages evoke uniformly in all receivers (Littlejohn & Foss, 2010). Such shared functions arguably provide the common protocol that enables communication between human brains (Hasson, Ghazanfar, Galantucci, Garrod, & Keysers, 2012; Weber, Sherry, & Mathiak, 2008; Zadbood et al., 2017). Similar ideas come from different disciplines, such as denotational meaning in linguistics, the field of semiotics (Chandler, 2007; Galantucci & Garrod, 2012), and philosophy of mind more broadly (Wittgenstein, 2013).

Beyond the main effects of a message, receiver-side factors can interact with content variables. Such message–receiver interactions are nicely illustrated by the examples from above, where the same story was interpreted differently and thus produced different brain responses (Lahnakoski et al., 2014; Schmälzle et al., 2013; Yeshurun, Nguyen et al., 2017; Yeshurun, Swanson et al., 2017). These phenomena are well studied at the behavioral level and associated with a plethora of theories, such as motivated reasoning, attribution, and others (Donsbach, 2008; Heider & Simmel, 1944; Kunda, 1990). For instance, in their classic study of selective perception entitled "They saw a game," Hastorf and Cantril (1954) examined how spectators of a contested football match interpreted identical events in strikingly different ways based on team affiliation. A clear next step would be to use ISC-based approaches to pinpoint neural correlates of such group differences in interpretation of the same physical input. In this sense, ISC can serve as a tool to study message processing in dyads, small teams, groups, or cultures. For instance, one can look at commonalities and differences during political messages (where recipients can be categorized into pro- and anti-groups, e.g., Turner et al., 2018) or any situation where recipients diverge on internal attitudinal or other knowledge structures. For example, during the outbreak of the H1N1 pandemic, Schmälzle et al. (2013) exposed recipients who were selected to be extreme in terms of risk perceptions (i.e., either considered H1N1 to be a major concern or no concern at all) to an authentic half-hour TV documentary about H1N1. ISC analysis of fMRI data showed that both groups, regardless of risk perception level, responded similarly in visual and auditory cortices. The high-risk perception group, however, showed more strongly aligned responses in the anterior cingulate cortex, a region associated with processing salient information and anticipatory anxiety. This indicates a message-receiver interaction in how the affective significance of a message is modulated based on internal factors, in this case, pre-existing risk perception. Of note, the previously mentioned study of Russian and English listeners is another example of this type of design by using a quasi-experimental manipulation of the cognitive variable language knowledge. The same logic has been applied to examine processing differences associated with clinical diagnoses (e.g., autism and depression; Byrge et al., 2015; Hasson et al., 2009; Salmi et al., 2013; Wang et al., 2017) and can be used to examine whether any type of social group membership—stable or experimentally induced—affects how people respond to the same content.

The previous examples conceived of the social group as the independent variable and compared the degree of similarity of neural processing (dependent variable) between groups to test if group status makes a difference. In a recent study, Parkinson and colleagues (Parkinson, Kleinbaum, & Wheatley, 2018) turned this logic on its head, asking if the degree of brain similarity between individuals could predict a social outcome. The researchers had a large cohort view videos and recorded their brain activity. Additionally, they obtained information about the participant's social networks, which allowed them to compute the degree of separation between each pair of participants. They then tested if the similarity of neural processing would be associated with social network proximity. The results showed that as neural similarity decreased, real-world social network distance increased. In other words, these findings provide initial evidence of homophily at the neural level. Clearly, there is much more to explore in how *social brain networks* are linked to classical social networks at the interpersonal level (Falk & Bassett, 2017; Schmälzle et al., 2017). One intriguing question, for example, is whether messages preferably propagate along diffusion paths that are "carved" by similarities of brain responding.

In sum, approaches based on ISC group difference or individual difference provide a versatile tool to identify responses to messages that are shared among some but not all recipients, and thus can either define group membership (e.g., in the case of diagnostic applications or audience segmentation) or provide a manifestation of its effects (e.g., in the case of motivated political reasoning). Given that only a handful of studies exist to date, there are still many important questions that have yet to be asked. This research direction holds great potential for message-tailoring and targeting, conflict and risk communication, audience segmentation, or for identifying the influence of factors such as age, ability, or interest on message processing.

FUTURE RESEARCH ON MEDIATED MESSAGES AND SYNCHRONIZED BRAINS

The field of media neuroscience (Hasson et al., 2008; Weber et al., 2008) has rapidly expanded and is currently one of the major growth areas of neuroimaging, including initiatives to scan the brains of large populations (Campbell et al., 2015; Chen et al., 2016; Cohen et al., 2017; Dubois & Adolphs, 2016). For communication scholars, this increase in the availability of large neuroimaging datasets that use media stimuli allows for investigations into the message content that drives increased alignment in the brain function of audiences. This includes looking at content characteristics, such as affect, incongruency, or similar variables (Grall & Schmälzle, 2018), or classic variables from the persuasion literature, such as message sensation value, argument strength, or general message strength (Schmälzle et al., 2015).

In parallel with topical expansion, several methodological developments have taken place. New methods have been introduced to improve the alignment of different brains (Conroy, Singer, Guntupalli, Ramadge, & Haxby, 2013; Haxby et al., 2011) and improve the statistical inference on ISC matrices, which are challenging due to the relational and nested structure of the data (Chen et al., 2016, 2019). As discussed above, the basic ISC framework assumes a one-to-one mapping between responses, but although human brains are quite similar at a coarse spatial scale, they differ in their meso- and micro-structure just like the shapes of our ears and faces. Newer methods hold promise to be more sensitive to fine-grained differences that wash out during conventional spatial alignment procedures. Another area is the integration of ISC analysis with network neuroscience approaches (Bassett & Sporns, 2017). In brief, these approaches typically focus on response similarities within brains, or *intra-brain* connectivity, whereas inter-subject correlation methods assess *between-brain* connectivity. A recently proposed variant of ISC analysis, called inter-subject functional connectivity (ISFC) (Simony et al., 2016), blends these approaches. Last, but not least, a very important push comes from studies that apply ISC-based analyses to other methods than fMRI, particularly EEG with its excellent temporal resolution, fNIRS, and psychophysiological measures (Bridwell, Roth, Gupta, & Calhoun, 2015; Cohen et al., 2018; Dikker et al., 2017; Dmochowski et al., 2014; Haufe et al., 2017).

These methodological developments are certainly important, but what really matters is whether ISC as a uniquely audience-focused approach to neural data analysis can stimulate new insights in substantive domains of communication science. In that regard, beyond the three main topics outlined above (engagement, message meaning, social influences), we see the following avenues as most promising: First, the majority of studies have treated message-evoked brain synchrony (ISC) as the dependent variable. Relatively few studies have used a brain-as-predictor approach (Berkman & Falk, 2013), in which the brain serves as the independent variable. However, given that ISC offers a group-based metric for audience engagement, it seems ideal to link this measure to outcomes at a larger scale (Barnett & Cerf, 2017; Dmochowski et al., 2014; Parkinson et al., 2018). Second, hardly any work examines how collective brain engagement waxes and wanes over the course of a message, such as during engaging moments of a speech, interesting parts of a lecture, or suspenseful scenes of a movie. These questions can be addressed with dynamic, time-resolved ISC analyses (Dmochowski et al., 2012; Grall & Schmälzle, 2018; Naci et al., 2014), essentially providing a brain-based version of Lazarsfeld and Stanton's famous Program Analyzer (Levy, 1982) for measuring how audiences respond to messages.

CONCLUSION

Mass-mediated messages carry content from a sender via a medium to the receiver. When a given message is processed by many receivers, a natural question to ask is how similarly their brains will respond to the same content. ISC analysis provides the answer to this question. In this sense, the ISC approach provides a good example that "there is nothing so theoretical as a good method" (Greenwald, 2012; Weber, Fisher, Hopp, & Lonergan, 2018) because the correlated brain responses indicate that communication has been successful, namely, that a message has been transmitted and evokes shared informational states across multiple recipients.

NOTES

1. The term synchronization is often associated with coupled oscillators in physics (Nolte, 2014). However, the original word meaning derives from Greek "*syn*"—together, with, and "*chronos*"—time, temporal, and may be translated as *concurrent* or *simultaneous*. Here we conceptualize message-evoked brain synchronization as the common effect of messaging on brain dynamics, which would not exhibit similarities if they were not exposed to the same message. However, the mechanism that elicits these similarities, or synchronicities, is not the same as in the physics of coupled oscillators, but rather works via commonalities in brain function and shared knowledge structures, which enable communication.
2. Of note, the viewers did not watch the movie simultaneously, but sequentially. The critical aspect is that the movie presentation was time-locked across viewers, which is the key prerequisite for running ISC analysis. Time-locked presentation guarantees that the message-evoked brain responses can be compared across the audience. Single-scanner environments are typically not equipped to study "true" audience effects, which require simultaneous processing of messages. However, it is technically possible to measure the brains of two people in a scanner, or to perform so-called hyper-scanning (see Chapter 20 by Hopp & Weber, in this volume). Furthermore, techniques such as EEG, fNIRS, and psychophysiological measurement can also be used to study simultaneous audience responses.
3. Latin "*communis*"—shared, common.

REFERENCES

Abraham, A., Pedregosa, F., Eickenberg, M., Gervais, P., Mueller, A., Kossaifi, J., & Varoquaux, G. (2014). Machine learning for neuroimaging with scikit-learn. *Frontiers in Neuroinformatics, 8*, 14.

Aristotle. (1939). *The "art" of rhetoric* (Trans. J. H. Freese). Cambridge, MA: Harvard University Press.

Baldassano, C., Chen, J., Zadbood, A., Pillow, J. W., Hasson, U., & Norman, K. A. (2017). Discovering event structure in continuous narrative perception and memory. *Neuron, 95*, 709–721.

Barnett, S. B., & Cerf, M. (2017). A ticket for your thoughts: Method for predicting content recall and sales using neural similarity of moviegoers. *The Journal of Consumer Research, 44*, 160–181.

Bassett, D. S., & Sporns, O. (2017). Network neuroscience. *Nature Neuroscience, 20*, 353–364.

Berkman, E. T., & Falk, E. B. (2013). Beyond brain mapping: Using neural measures to predict real-world outcomes. *Current Directions in Psychological Science, 22*, 45–50.

Berridge, K. C., & Kringelbach, M. L. (2013). Neuroscience of affect: Brain mechanisms of pleasure and displeasure. *Current Opinion in Neurobiology, 23*, 294–303.

Bridwell, D. A., Roth, C., Gupta, C. N., & Calhoun, V. D. (2015). Cortical response similarities predict which audiovisual clips individuals viewed, but are unrelated to clip preference. *PloS One, 10*, e0128833.

Brothers, L. (1990). The neural basis of primate social communication. *Motivation and Emotion, 14*, 81–91.

Byrge, L., Dubois, J., Tyszka, J. M., Adolphs, R., & Kennedy, D. P. (2015). Idiosyncratic brain activation patterns are associated with poor social comprehension in autism. *The Journal of Neuroscience, 35*, 5837–5850.

Campbell, K. L., Shafto, M. A., Wright, P., Tsvetanov, K. A., Geerligs, L., & Cusack, R. (2015). Idiosyncratic responding during movie-watching predicted by age differences in attentional control. *Neurobiology of Aging, 36*, 3045–3055.

Cantlon, J. F., & Li, R. (2013). Neural activity during natural viewing of *Sesame Street* statistically predicts test scores in early childhood. *PLoS Biology, 11*, e1001462.

Chandler, D. (2007). *Semiotics: The basics*. New York, NY: Routledge.

Chen, G., Shin, Y.-W., Taylor, P. A., Glen, D., Reynolds, R. C., Israel, R. B., & Cox, R. W (2016). Untangling the relatedness among correlations, part I: Nonparametric approaches to inter-subject correlation analysis at the group level. *NeuroImage, 142*, 248–259.

Chen, G., Taylor, P. A., Qu, X., Molfese, P. J., Bandettini, P. A., Cox, R. W., & Finn, E. S. (2019). Untangling the relatedness among correlations, part III: Inter-subject correlation analysis through Bayesian multilevel modeling for naturalistic scanning. *bioRxiv*, 655738.

Chen, J., Leong, Y. C., Honey, C. J., Yong, C. H., Norman, K. A., & Hasson, U. (2016). Shared memories reveal shared structure in neural activity across individuals. *Nature Neuroscience, 20*, 115–125.

Cohen, J. D., Daw, N., Engelhardt, B., Hasson, U., Li, K., Niv, Y., … Willke, T. L. (2017). Computational approaches to fMRI analysis. *Nature Neuroscience, 20*, 304–313.

Cohen, S. S., Madsen, J., Touchan, G., Robles, D., Lima, S. F. A., Henin, S., & Parra, L. C. (2018). Neural engagement with online educational videos predicts learning performance for individual students. *bioRxiv*, 253260.

Conroy, B. R., Singer, B. D., Guntupalli, J. S., Ramadge, P. J., & Haxby, J. V. (2013). Inter-subject alignment of human cortical anatomy using functional connectivity. *NeuroImage, 81*, 400–411.

Dikker, S., Silbert, L. J., Hasson, U., & Zevin, J. D. (2014). On the same wavelength: Predictable language enhances speaker-listener brain-to-brain synchrony in posterior superior temporal gyrus. *The Journal of Neuroscience, 34*, 6267–6272.

Dikker, S., Wan, L., Davidesco, I., Kaggen, L., Oostrik, M., McClintock, J., … Poeppel, D. (2017). Brain-to-brain synchrony tracks real-world dynamic group interactions in the classroom. *Current Biology, 27*, 1375–1380.

Dmochowski, J. P., Bezdek, M. A., Abelson, B. P., Johnson, J. S., Schumacher, E. H., & Parra, L. C. (2014). Audience preferences are predicted by temporal reliability of neural processing. *Nature Communications, 5*, 4567.

Dmochowski, J. P., Sajda, P., Dias, J., & Parra, L. C. (2012). Correlated components of ongoing EEG point to emotionally laden attention—a possible marker of engagement? *Frontiers in Human Neuroscience, 6*, 112.

Donsbach, W. (2008). Selective perception and selective retention. In W. Donsbach (Ed.), *The international encyclopedia of communication* (Vol. 64, p. 123). Chichester, England: John Wiley & Sons.

Dubois, J., & Adolphs, R. (2016). Building a science of individual differences from fMRI. *Trends in Cognitive Sciences, 20*, 425–443.

Dudai, Y. (2012). The cinema-cognition dialogue: A match made in brain. *Frontiers in Human Neuroscience, 6*, 248.

Falk, E. B., & Bassett, D. S. (2017). Brain and social networks: Fundamental building blocks of human experience. *Trends in Cognitive Sciences, 21*, 674–690.

Farbood, M. M., Heeger, D. J., Marcus, G., Hasson, U., & Lerner, Y. (2015). The neural processing of hierarchical structure in music and speech at different timescales. *Frontiers in Neuroscience, 9*, 157.

Finn, E. S., Corlett, P. R., Chen, G., Bandettini, P. A., & Constable, R. T. (2017). Trait-level paranoia shapes inter-subject synchrony in brain activity during an ambiguous social narrative. *bioRxiv*, 231738.

Galantucci, B., & Garrod, S. (2012). *Experimental semiotics: Studies on the emergence and evolution of human communication*. Philadelphia, PA: John Benjamins Publishing.

Gazzaniga, M. S., Ivry, R. B., & Mangun, G. R. (2013). *Cognitive neuroscience: The biology of the mind* (4th ed.). New York, NY: W. W. Norton & Company.

Grall, C., & Schmälzle, R. (2018, May). The coupled brains of captivated audiences: How suspense in a movie modulates collective brain dynamics. Abstract presented at the 11th Annual Meeting of the Social and Affective Neuroscience Society, Brooklyn, NY.

Greenwald, A. G. (2012). There is nothing so theoretical as a good method. *Perspectives on Psychological Science, 7*, 99–108.

Greenwald, A. G., & Leavitt, C. (1984). Audience involvement in advertising: Four levels. *The Journal of Consumer Research, 11*, 581–592.

Haidt, J. (2001). The emotional dog and its rational tail: A social intuitionist approach to moral judgment. *Psychological Review, 108*, 814–834.

Hanke, M., Baumgartner, F. J., Ibe, P., Kaule, F. R., Pollmann, S., Speck, O., ... Stadler, J. (2014). A high-resolution 7-Tesla fMRI dataset from complex natural stimulation with an audio movie. *Scientific Data, 1*, 140003.

Hasson, U., Avidan, G., Gelbard, H., Vallines, I., Harel, M., Minshew, N., & Behrmann, M. (2009). Shared and idiosyncratic cortical activation patterns in autism revealed under continuous real-life viewing conditions. *Autism Research, 2*, 220–231.

Hasson, U., Chen, J., & Honey, C. J. (2015). Hierarchical process memory: Memory as an integral component of information processing. *Trends in Cognitive Sciences, 19*, 304–313.

Hasson, U., Ghazanfar, A. A., Galantucci, B., Garrod, S., & Keysers, C. (2012). Brain-to-brain coupling: A mechanism for creating and sharing a social world. *Trends in Cognitive Sciences, 16*, 114–121.

Hasson, U., Landesman, O., Knappmeyer, B., Vallines, I., Rubin, N., & Heeger, D. J. (2008). Neurocinematics: The neuroscience of film. *Projections, 2*, 1–26.

Hasson, U., Malach, R., & Heeger, D. J. (2010). Reliability of cortical activity during natural stimulation. *Trends in Cognitive Sciences, 14*, 40–48.

Hasson, U., Nir, Y., Levy, I., Fuhrmann, G., & Malach, R. (2004). Intersubject synchronization of cortical activity during natural vision. *Science, 303*, 1634–1640.

Hastorf, A. H., & Cantril, H. (1954). They saw a game: A case study. *Journal of Abnormal Psychology, 49*, 129–134.

Haufe, S., DeGuzman, P., Henin, S., Arcaro, M., Honey, C. J., Hasson, U., & Parra, L. C. (2017). Elucidating relations between fMRI, ECoG and EEG through a common natural stimulus. *bioRxiv*, 207456.

Haxby, J. V., Guntupalli, J. S., Connolly, A. C., Halchenko, Y. O., Conroy, B. R., Gobbini, M. I., ... Ramadge, P. J. (2011). A common, high-dimensional model of the representational space in human ventral temporal cortex. *Neuron, 72*, 404–416.

Heider, F., & Simmel, M. (1944). An experimental study of apparent behavior. *The American Journal of Psychology, 57*, 243–259.

Hermans, E. J., van Marle, H. J. F., Ossewaarde, L., Henckens, M. J. A. G., Qin, S., van Kesteren, M. T. R., ... Fernández, G. (2011). Stress-related noradrenergic activity prompts large-scale neural network reconfiguration. *Science, 334*, 1151–1153.

Honey, C. J., Thompson, C. R., Lerner, Y., & Hasson, U. (2012). Not lost in translation: Neural responses shared across languages. *The Journal of Neuroscience, 32*, 15277–15283.

Imhof, M. A., Schmälzle, R., Renner, B., & Schupp, H. T. (2017). How real-life health messages engage our brains: Shared processing of effective anti-alcohol videos. *Social Cognitive and Affective Neuroscience, 12*, 1188–1196.

Jääskeläinen, I. P., Koskentalo, K., Balk, M. H., Autti, T., Kauramäki, J., Pomren, C., & Sams, M. (2008). Inter-subject synchronization of prefrontal cortex hemodynamic activity during natural viewing. *The Open Neuroimaging Journal, 2*, 14–19.

Kauppi, J.-P., Jääskeläinen, I. P., Sams, M., & Tohka, J. (2010). Inter-subject correlation of brain hemodynamic responses during watching a movie: Localization in space and frequency. *Frontiers in Neuroinformatics, 4*, 5.

Kintsch, W. (1988). The role of knowledge in discourse comprehension: A construction-integration model. *Psychological Review, 95*, 163–182.

Kluckhohn, C., Murray, H. A., & Schneider, D. M. (Eds.). (1948). *Personality in nature, society, and culture* (2nd ed. rev.). Oxford, England: Knopf.

Kunda, Z. (1990). The case for motivated reasoning. *Psychological Bulletin, 108*, 480–498.

Lahnakoski, J. M., Glerean, E., Jääskeläinen, I. P., Hyönä, J., Hari, R., Sams, M., & Nummenmaa, L. (2014). Synchronous brain activity across individuals underlies shared psychological perspectives. *NeuroImage, 100*, 316–324.

Lang, P. J., Bradley, M. M., & Cuthbert, B. N. (1997). Motivated attention: Affect, activation, and action. In P. J. Lang, R. F. Simons, & M. Balaban (Eds.), *Attention and orienting; Sensory and motivational processes* (pp. 97–135). New York, NY: Routledge.

Lazarsfeld, P. F., & Stanton, F. N. (Eds.). (1949). *Communications research*. Oxford, England: Harper.

Lerner, Y., Honey, C. J., Silbert, L. J., & Hasson, U. (2011). Topographic mapping of a hierarchy of temporal receptive windows using a narrated story. *The Journal of Neuroscience, 31*, 2906–2915.

Levy, M. R. (1982). The Lazarsfeld-Stanton program analyzer: An historical note. *The Journal of Communication, 32*, 30–38.

Littlejohn, S. W., & Foss, K. A. (2010). *Theories of human communication* (10th ed.). Long Grove, IL: Waveland Press.

Mantini, D., Hasson, U., Betti, V., Perrucci, M. G., Romani, G. L., Corbetta, M., … Vanduffel, W. (2012). Interspecies activity correlations reveal functional correspondence between monkey and human brain areas. *Nature Methods, 9*, 277–282.

Mar, R. A. (2011). The neural bases of social cognition and story comprehension. *Annual Review of Psychology, 62*, 103–134.

McCroskey, J. C. (2015). *An introduction to rhetorical communication: A western rhetorical perspective* (9th ed.). New York, NY: Routledge.

Naci, L., Cusack, R., Anello, M., & Owen, A. M. (2014). A common neural code for similar conscious experiences in different individuals. *Proceedings of the National Academy of Sciences, 111*, 14277–14282.

Nolte, D. D. (2014). *Introduction to modern dynamics: Chaos, networks, space and time*. Oxford, England: Oxford University Press.

Nummenmaa, L., Glerean, E., Viinikainen, M., Jääskeläinen, I. P., Hari, R., & Sams, M. (2012). Emotions promote social interaction by synchronizing brain activity across individuals. *Proceedings of the National Academy of Sciences, 109*, 9599–9604.

Panksepp, J. (2004). *Affective neuroscience: The foundations of human and animal emotions*. Oxford, England: Oxford University Press.

Parkinson, C., Kleinbaum, A. M., & Wheatley, T. (2018). Similar neural responses predict friendship. *Nature Communications, 9*, 332.

Purves, D., Cabeza, R., Huettel, S. A., LaBar, K. S., Platt, M. L., Woldorff, M. G., & Brannon, E. M. (2008). *Cognitive neuroscience*. Sunderland: Sinauer Associates, Inc.

Richardson, H., Lisandrelli, G., Riobueno-Naylor, A., & Saxe, R. (2018). Development of the social brain from age three to twelve years. *Nature Communications, 9*, 1027.

Salmi, J., Roine, U., Glerean, E., Lahnakoski, J., Nieminen-von Wendt, T., Tani, P., … Sams, M. (2013). The brains of high functioning autistic individuals do not synchronize with those of others. *NeuroImage, 3*, 489–497.

Schmälzle, R., Brook O'Donnell, M., Garcia, J. O., Cascio, C. N. C., Bayer, J., Vettel, P., … Falk, E. (2017). Brain connectivity dynamics during social interaction reflect social network structure. *Proceedings of the National Academy of Sciences, 114*, 5153–5158.

Schmälzle, R., Häcker, F., Honey, C. J., & Hasson, U. (2015). Engaged listeners: Shared neural processing of powerful political speeches. *Social, Cognitive, and Affective Neurosciences, 1*, 168–169.

Schmälzle, R., Häcker, F., Renner, B., Honey, C. J., & Schupp, H. T. (2013). Neural correlates of risk perception during real-life risk communication. *The Journal of Neuroscience, 33*, 10340–10347.

Schmitz, T. W., & Johnson, S. C. (2007). Relevance to self: A brief review and framework of neural systems underlying appraisal. *Neuroscience & Biobehavioral Reviews, 31*, 585–596.

Schupp, H. T., Flaisch, T., Stockburger, J., & Junghöfer, M. (2006). Emotion and attention: Event-related brain potential studies. *Progress in Brain Research, 156*, 31–51.

Shannon, C. E. (1948). A mathematical theory of communication. *Bell System Technical Journal, 27*, 379–423.

Sherif, M., & Sargent, S. S. (1947). Ego-involvement and the mass media. *The Journal of Social Issues, 3*, 8–16.

Simony, E., Honey, C. J., Chen, J., Lositsky, O., Yeshurun, Y., Wiesel, A., & Hasson, U. (2016). Dynamic reconfiguration of the default mode network during narrative comprehension. *Nature Communications, 7*, 12141.

Sliwa, J., & Freiwald, W. A. (2017). A dedicated network for social interaction processing in the primate brain. *Science, 356*, 745–749.

Stephens, G. J., Silbert, L. J., & Hasson, U. (2010). Speaker-listener neural coupling underlies successful communication. *Proceedings of the National Academy of Sciences, 107*, 14425–14430.

Tootell, R. B., Hadjikhani, N. K., Vanduffel, W., Liu, A. K., Mendola, J. D., Sereno, M. I., & Dale, A. M. (1998). Functional analysis of primary visual cortex (V1) in humans. *Proceedings of the National Academy of Sciences, 95*, 811–817.

Turner, B. O., Huskey, R., Amir, O., & Weber, R. (2018, May). The immoral opponent in political attack ads: Intersubject correlations across the moral brain differ by party affiliation. Abstract presented at the 11th Annual Meeting of the Social and Affective Neuroscience Society, Brooklyn, NY.

Vanderwal, T., Eilbott, J., & Castellanos, F. X. (2019). Movies in the magnet: Naturalistic paradigms in developmental functional neuroimaging. *Developmental Cognitive Neuroscience, 36*, 10600.

Wang, J., Ren, Y., Hu, X., Nguyen, V. T., Guo, L., Han, J., & Guo, C. C. (2017). Test-retest reliability of functional connectivity networks during naturalistic fMRI paradigms. *Human Brain Mapping, 38*, 2226–2241.

Weber, R., Fisher, J. T., Hopp, F. R., & Lonergan, C. (2018). Taking messages into the magnet: Method–theory synergy in communication neuroscience. *Communication Monographs, 85*, 81–102.

Weber, R., Sherry, J., & Mathiak, K. (2008). The neurophysiological perspective in mass communication research. In M. J. Beatty, J. C. McCroskey, & K. Floyd (Eds.), *Biological dimensions of communication: Perspectives, methods, and research* (pp. 41–71). Cresskill, NJ: Hampton Press.

Wilson, S. M., Molnar-Szakacs, I., & Iacoboni, M. (2008). Beyond superior temporal cortex: Intersubject correlations in narrative speech comprehension. *Cerebral Cortex, 18*, 230–242.

Wittgenstein, L. (2013). *Tractatus logico-philosophicus*. New York, NY: Routledge.

Yeshurun, Y., Nguyen, M., & Hasson, U. (2017). The butterfly effect: Amplification of local changes along the temporal processing hierarchy. *bioRxiv*, 102590.

Yeshurun, Y., Swanson, S., Simony, E., Chen, J., Lazaridi, C., Honey, C. J., & Hasson, U. (2017). Same story, different story: The neural representation of interpretive frameworks. *Psychological Science, 28*, 307–319.

Zadbood, A., Chen, J., Leong, Y. C., Norman, K. A., & Hasson, U. (2017). How we transmit memories to other brains: Constructing shared neural representations via communication. *Cerebral Cortex, 27*, 4988–5000.

Zwaan, R. A. (1999). Situation models: The mental leap into imagined worlds. *Current Directions in Psychological Science, 8*, 15–18.

9

The Neuroscience of Persuasion and Information Propagation

The Key Role of the Mentalizing System

Elisa C. Baek, Christin Scholz, and Emily B. Falk

What are the psychological drivers that lead communicators to share information with and influence others? How do receivers process shared content that leads to successful information propagation? How do contextual factors moderate the mechanisms at play? In this chapter, we focus on mentalizing (i.e., considering the mental states of other people). Among many processes involved in social influence, mentalizing is one key input to decision-making in communicators and receivers of influence (Baek, Scholz, O'Donnell, & Falk, 2017; Cascio, O'Donnell, Bayer, Tinney, & Falk, 2015; van Hoorn, van Dijk, Meuwese, Rieffe & Crone, 2016; Welborn et al., 2015), and it is theorized to play a central role in whether efforts to persuade and influence are successful (Cascio et al., 2015; Falk, Morelli, Welborn, Dambacher, & Lieberman, 2013; Welborn et al., 2015). Messages that lead to successful propagation of ideas are distinguished by higher levels of mentalizing in both the communicator and receiver. We also highlight key contextual factors, including development, culture, and social networks, that moderate the association between mentalizing and influence, and conclude by highlighting future research opportunities in these domains.

WHAT IS MENTALIZING?

Mentalizing, defined as considering the mental states of others, plays a critical role in the human social experience (Adolphs, 2009; Frith & Frith, 2006). Without the ability to interpret and understand the attitudes, beliefs, and feelings of others that precede their actions, we would be lost in navigating the social world. The ability to mentalize is not only necessary to explicitly understand that others might hold different beliefs than we do (e.g., "My brother never saw me eat the last banana, so he might look for it"), but also to navigate complex social situations through implicit cues (e.g., "Susie feels embarrassed after falling, and she might feel better if I reassure her"). Indeed, deficits in the ability to mentalize characterize psychosocial disorders including autism (Chung, Barch, & Strube, 2014), schizophrenia (Chung et al., 2014), and borderline personality disorder (Bateman & Fonagy, 2012).

Behavioral tasks are often used to test and measure mentalizing abilities of individuals. In one classic task, participants are tested in their ability to understand that others might hold an incorrect belief about the state of the world (Wimmer & Perner, 1983). In this false-belief task, participants read through a narrative such as the following: *Mark puts his toy inside his drawer and leaves the room. Meanwhile, Katy moves the toy to under the bed. Upon returning, where will Mark look for the toy?* In order to identify that Mark will look for the toy in the drawer, participants must be able to understand that Mark holds a belief about the location of the toy that is both incorrect and different from their own. This simple false-belief task has mostly been utilized to test mentalizing abilities in children, and more complex versions of the task have been adapted for adults that test the ability to understand more complex mental states and intentions (Happé, 1994; Wang & Su, 2013). Findings from such tasks shed light on developmental stages in mentalizing ability (Frith & Frith, 2003) and mentalizing deficits that characterize psychosocial disorders (Chung et al., 2014).

Neuroscience evidence has also identified neural systems that support mentalizing. In particular, portions of the medial prefrontal cortex (MPFC), particularly subregions in the middle and dorsomedial prefrontal cortex (MMPFC, DMPFC), as well as bilateral temporoparietal junction (TPJ), precuneus (PC/PCC), superior temporal sulcus (STS), and temporal poles are robustly engaged when people are actively considering the mental state of others, as identified in a meta-analytic map retrieved from Neurosynth (see Figure 9.1; see also Dufour et al., 2013; Frith & Frith, 2006). These regions involved in mentalizing overlap substantially with the brain's "default-mode network" (i.e., regions of the brain active when participants of fMRI studies are not given a specific task), which is thought to generally support continuous processing of the environment in order to anticipate future events and enact appropriate behavioral trajectories, including efficient social judgments (Meyer, Davachi, Ochsner, & Lieberman, 2018). Although we review evidence from prior work that interprets activity in these regions as involved in mentalizing, the subregions that form the mentalizing network have also been implicated in other cognitions as part of the default-mode network, including reinforcement learning and goal-directed action (for a review, see Raichle, 2015). Thus, mentalizing may be one specific type of the broader class of processes

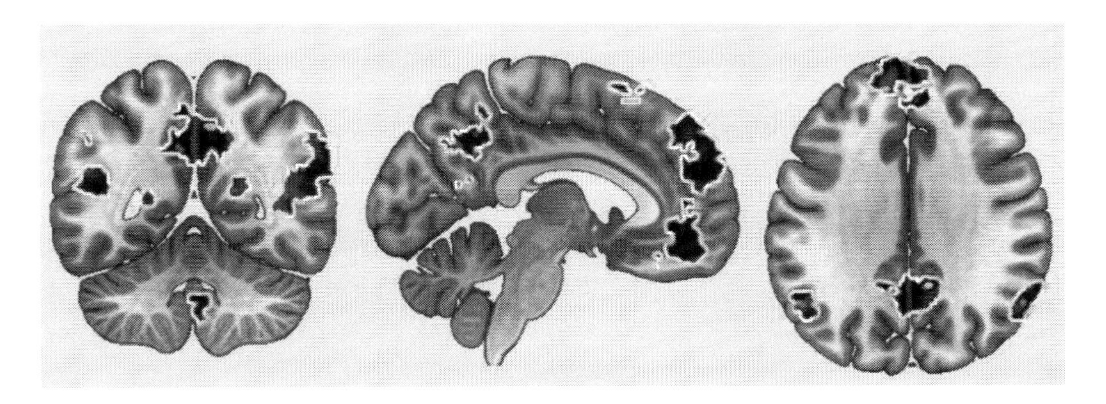

Figure 9.1 Brain Regions Associated with Mentalizing

Note: A reverse inference meta-analytic map of the functional neuroimaging literature on "mentalizing" retrieved from Neurosynth shows that subregions in the middle and dorsal medial prefrontal cortex (MMPFC, DMPFC), bilateral temporoparietal junction (TPJ), precuneus (PC/PCC), superior temporal sulcus (STS), and temporal poles have been implicated in mentalizing.

that are implicated in interpreting the environment and predicting future events to optimize behavioral response.

With these caveats in mind (see Poldrack, 2011, for a review), neuroimaging allows for activity in the mentalizing system to be tracked in real time as individuals are actively processing stimuli, and thereby does not rely on individuals' ability to introspect on their thought processes. This is an advantage because mentalizing is often an implicit, instinctive process (Frith & Frith, 2003); in many social situations, people understand others' perspectives without intentional effort. The ability to assess these processes implicitly has enabled researchers to study the role of mentalizing in non-verbal communication (Kampe, Frith, & Frith, 2003), predicting behavioral outcomes ranging from susceptibility to peer influence (Cascio et al., 2015; Welborn et al., 2015), and encoding of successful ideas (Falk et al., 2013; Falk, O'Donnell, & Lieberman, 2012).

This chapter focuses on how mentalizing contributes to decision-making in one aspect of human communication: social influence. We highlight recent findings from neuroscience that suggest that mentalizing contributes to decision-making processes that determine whether specific forms of social influence, such as persuasion and information propagation, are successful. Findings highlight that neural measures of mentalizing activity can be used to predict behavioral outcomes of interest, such as the success of social influence at the message, individual person, and population levels.

THE INVOLVEMENT OF MENTALIZING IN PERSUASION, SOCIAL INFLUENCE, AND INFORMATION PROPAGATION

Depending on the context, people can be either communicators or receivers of social influence, influencing or being influenced by others to change behaviors or preferences. Social influence can take numerous forms such as active persuasion attempts (e.g., ordering a salad because your friend convinced you to eat healthy) or more subtle influences on behavior (e.g., ordering a salad because you saw your friend eat healthy). In order for social influence to successfully take place, a message of some form (e.g., verbal, written, non-verbal) from a communicator must be successfully transmitted to the receiver. We highlight neuroscientific evidence that implicates mentalizing in two types of social influence: (1) persuasion (i.e., explicit attempts to influence others); and (2) information propagation (i.e., transfer of information between communicators and receivers).

Mentalizing in Communicators Increases Success of Social Influence

Mentalizing is associated with communicators' decisions to share information with others and predicts whether such shared information leads to success in persuading and influencing others (for reviews, see Baek & Falk, 2018; Falk & Scholz, 2018). This may be in part because to effectively share information and influence others, communicators should consider the audience of their message, taking into account the receiver's attitude, knowledge, and preferences. Mentalizing is one example of the broader human ability to learn about the world around us, predict future events, and reinforce previous knowledge. Specifically, mentalizing allows a communicator to estimate how a message might be received by the receiver and, thus, what the probable social consequences of a given interaction will be (Berger, 2014). For instance, sharing information may lead to social rewards through opportunities to bond and obtain others' approval; on the other hand, it can lead to social punishment, for instance, when a message is ridiculed by the receiver. Indeed, one account suggests that communicators adjust their strategies based on the characteristics of the receivers of their messages (Barasch & Berger, 2014; Scholz, Baek, O'Donnell, & Falk, 2019).

Further, communicators recruit portions of the mentalizing network (i.e., sub-regions of MPFC, PC, bilateral TPJ, and right STS defined based on a meta-analysis of mentalizing) when making decisions about sharing news articles compared to decisions to read the articles themselves or considering the content (Baek et al., 2017). Activity in mentalizing regions also positively scales with communicators' self-reported intentions to share the news articles; this suggests that the mentalizing system is recruited not only when communicators think about whether to share, but also that increased mentalizing is associated with increased sharing intention (Baek et al., 2017). This predictive power of activity in the mentalizing system extends beyond individual study participants to population-level behavior. Brain activity in a small number of participants as they considered news articles was associated with how often each article was actually shared by larger groups of people in real life (Scholz et al., 2017). Specifically, across two studies, brain activity in mentalizing regions showed an indirect effect on population-level sharing of the articles that was mediated through neural value-related activity; this suggests that articles that encouraged mentalizing activity also increased value-related brain activity in communicators (Scholz et al., 2017). In this way, mentalizing activity is associated with increased likelihood that information is shared.

Individual Differences in Mentalizing Are Associated with Successful Influence

Given that a communicator decides to share information, what sets successful communicators apart from their less successful counterparts? Communicators who show heightened consideration of the thoughts of their receivers can be more likely to be successful (for a review, see Baek & Falk, 2018). Successful communicators are more likely to be socially flexible, with higher levels of self-monitoring, or the tendency to adjust their behavior based on interpersonal cues (Baek & Falk, 2018; Snyder, 1974). Further, successful communicators are characterized by environmental contexts that encourage a greater ability to take the perspective of others, such as occupying social network positions that span many structural holes, meaning that more of their contacts rely on them to communicate with one another (Burt, 2004). Within the brain, salespeople who self-reported higher levels of mentalizing also showed increased recruitment of brain regions associated with mentalizing, including the TPJ and MPFC, and were more successful at adapting their strategies to meet customers' needs (Dietvorst et al., 2009). Similarly, non-professional participants who were more successful at convincing others to accept or reject TV show ideas based on their own preferences showed greater recruitment of the TPJ during initial exposure to the ideas (Falk et al., 2013). Thus, the level of mentalizing in communicators is one indication of the likelihood that information is shared, as well as whether the shared information leads to successful social influence.

Mentalizing in Receivers Leads to Increased Likelihood of Social Influence

Receivers of social influence also engage in mentalizing, and consideration of others' perspectives can increase a receiver's susceptibility to influence and persuasion. Social belonging and the maintenance of social relationships are fundamental human needs (Baumeister & Leary, 1995). As such, the value of complying with a message is influenced by social norms and by the potential social rewards of compliance (Cialdini & Goldstein, 2004). For instance, agreeing with a message shared by a friend can lead to positive relational outcomes, increasing the value of compliance (Berger, 2014). Likewise, acting to be in compliance with a perceived social norm can lead to a feeling of belongingness (Cialdini & Goldstein, 2004). In order for receivers to find conforming to social influence valuable, it is helpful to understand others' mindsets, perspectives, and values. For instance, receivers may be more likely to find adopting an idea valuable if they are able to understand the perspectives of the communicators. The use of social norms in persuasive communication increases the likelihood of an individual shifting their behavior or attitude toward the norm (Cialdini et al., 2006; Goldstein, Cialdini, & Griskevicius, 2008). For instance, hotel room signs promoting reuse of towels that invoke social norms and group belongingness (e.g., "the majority of guests reuse their towels") were more successful at gaining compliance from guests compared to signs solely focusing on environmental appeals (e.g., "you can show your respect for nature and help save the environment by reusing your towels during your stay"; Goldstein et al., 2008). Although not explicitly measured, invoking a social norm likely activated mentalizing in the receivers of the influence as they considered the behavior of others in calculating the value of complying.

Neural evidence provides additional insights into this theory. Text-based persuasive messages that invoked higher activity in regions of the brain implicated in mentalizing (DMPFC, posterior STS, temporal pole) were rated to be higher in persuasiveness across two cultural contexts (i.e., American and Korean participants) and message modalities (i.e., text and video messages; Falk et al., 2010). Thus, messages that elicit thoughts of social implications, even if not explicitly from a social source, may be more likely to lead to successful influence.

The mentalizing system is also associated with successful influence during dyadic interactions. Receivers engage the mentalizing system when considering the opinions of others, and this activity is predictive of whether the message will lead to successful social influence (Cascio et al., 2015; Welborn et al., 2015). Increased activity in the mentalizing system while receivers were reading and considering peer recommendations of mobile game apps was associated with increased likelihood that receivers would change their own recommendations toward that of their peers, leading to successful propagation of recommendations (Baek et al., 2019). Likewise, in adolescents, greater neural activity in regions of the mentalizing system (DMPFC, rTPJ, left temporal pole) was associated with increased likelihood that adolescent participants shifted their preference of art based on social feedback from both their peers and parents (Welborn et al., 2015). In this way, mentalizing activity may be one signal that contributes to whether receivers are socially influenced.

Given that mentalizing is associated with increased likelihood that a receiver will be successfully influenced, can it also distinguish individual differences in susceptibility to social influence? Indeed, individuals more susceptible to social influence (i.e., more likely to change their own opinion to match that of their peers) also showed greater recruitment of the brain's mentalizing system while they were exposed to the opinions of others (Cascio et al., 2015; Welborn et al., 2015). Thus, the extent to which a receiver considers the social implications of incorporating a communicator's opinion may be a key input to an overall value calculation that determines whether the receiver will be socially influenced (for a review, see Falk & Scholz, 2018).

Mentalizing Supports Information Transfer Between Communicators and Receivers

Evidence reviewed thus far has highlighted the role of mentalizing in both communicators and receivers of messages, demonstrating that social relevance is a key consideration when a communicator determines the value of sharing information with others, and when a receiver determines the value of incorporating such shared information. Further, the extent to which communicators and receivers consider social relevance contributes to whether a message from a communicator will lead to successful propagation in the receiver. In this section, we highlight evidence suggesting that these processes in communicators and receivers do not exist in isolation but interact with one another to lead to successful communication.

Communicating dyads show behavioral and biological coupling during interactions, coordinating with their partner in the use of language (e.g., linguistic style matching, word matching) (Gonzales, Hancock, & Pennebaker, 2010), non-verbal signals (e.g., physical gestures, vocalizations, eye movements; Cappella, 1997; Richardson & Dale, 2005), and brain activity (e.g., similar time course of neural activations; Stephens, Silbert, & Hasson, 2010). Higher levels of coupling and coordination between communicating dyads across these measures predict the success of various interpersonal outcomes, including conversational satisfaction (Cappella, 1997), discourse comprehension (Richardson & Dale, 2005), and group performance (Gonzales et al., 2010). Indeed, the large literature on behavioral mimicry (see also Chapter 31 by Bente & Novotny, in this volume) suggests that this human tendency to match or mimic one's interaction partner occurs automatically without conscious effort, and has evolutionary advantages in supporting the need of humans to socially affiliate with one another for survival (Lakin & Chartrand, 2003; Lakin, Jefferis, Cheng, & Chartrand, 2003).

Neuroscience evidence corroborates and extends such findings, showing that the human brain has not only evolved systems dedicated to support social coordination for individuals in isolation, but also that the brains of communicating dyads display coupling that supports coordination during social interactions (Schippers, Roebroeck, Renken, Nanetti, & Keysers, 2010; Stephens et al., 2010). In particular, neural coupling has not only been observed in low-level sensory regions of the brain but also in high-level cognitive systems, including those that support mentalizing (Schippers et al., 2010; Scholz et al., 2019; Stephens et al., 2010), with increased coupling in these regions predicting communication success (Scholz et al., 2019; Stephens et al., 2010).

Brain activity while a set of communicators told real-life stories was associated with brain activity in receivers who listened to these stories; specifically, significant temporally-coupled activity was found in brain regions within the mentalizing system (the dorsolateral prefrontal cortex, the precuneus, MPFC, TPJ), as well as in low-level auditory regions and the mirror neuron system (the ventral premotor area and the intraparietal area; Stephens et al., 2010). Further, greater neural coupling in these regions between communicators and receivers was associated with higher receiver scores on story comprehension, suggesting that biological coordination between communicating dyads may be a critical component of successful communication (Stephens et al., 2010). Similarly, significant neural coupling was found in brain regions previously implicated in mentalizing (the bilateral TPJ, the right superior temporal lobe, PC, and portions of the MPFC), as well as in self-related processing (the posterior cingulate cortex, MPFC), and subjective valuation (the ventral striatum, VMPFC) while communicators were considering news articles to share, and receivers who saw the written shared messages (Scholz et al., 2019). In addition, neural coupling in all three brain systems was associated with coupling in self-reported perceived benefits of sharing in the communicator-receiver pairs (Scholz et al., 2019), further supporting the theory that neural coupling in the mentalizing system may contribute to the propagation of ideas in a communicating dyad. Coupling also occurs during non-verbal social interactions. Communicators and receivers in a non-verbal game of charades showed significant neural coupling in VMPFC and the putative mirror neuron system (Schippers et al., 2010).

Combined, the findings highlight that the brain's mentalizing system supports the transfer of ideas between communicators and receivers, with neural coupling in systems relevant to mentalizing robustly implicated across different forms of communication, including verbal, non-verbal, and short text-based interactions on social media. This coupling in the brain's mentalizing system may be necessary to promote the successful propagation of ideas through facilitating mutual understanding of the perspectives, thoughts, and emotions of others, and increasing the value of social coordination. This coordination in the brain's mentalizing system may reflect communicators transferring the social value of information to their receivers through their messages, providing cues to which information may be socially valued (Scholz et al., 2017). This transfer may rely on the communicator and receiver sharing a social context in which a collective sense of identity, norms, and values exists.

MODERATORS OF MENTALIZING AND FUTURE DIRECTIONS

With few exceptions, extant studies linking mentalizing activity with successful social influence, persuasion, and message propagation have focused on highly educated young adult populations in the West, primarily observing individuals (and their brains) removed from their environmental and social contexts. However, the effect of mentalizing on social influence may be moderated by various contextual variables. For instance, social norms vary across populations (e.g., cultures that emphasize interdependence), and the same content may be processed differently depending on how it is framed (e.g., when described in more vs. less social terms). Likewise, the vast majority of studies utilizing neuroscience have studied brain systems in isolation, extracting mean activation using a region-of-interest approach.

Growing bodies of research in recent years have begun to address this gap, suggesting that incorporating contextual variables in conjunction with neural networks can further our understanding of how the brain interacts dynamically with its environment. These new methods afford opportunities for increased precision in understanding brain activity beyond mean activation in a large brain system, by measuring patterns of brain activity and interactions between and within brain systems. In the present section, we highlight key developments in the study of social influence, and communication processes more broadly, that have incorporated these contextual considerations. Given these promising findings, we suggest that new research in each of these areas may be fruitful in moving the field forward.

Message-Level and Contextual Considerations

Recent findings suggest that certain message-level features may increase the likelihood of a message being shared, such as novelty, perceived utility, and emotionality (Berger & Milkman, 2012; Kim, 2015). Less is known, however, about how these message-level features may moderate the relationship between brain activity and successful propagation of ideas as explored in previous sections, or whether mentalizing is the key mediating factor between these message variables and their effects. In one recent study (Baek et al., 2019), participants first saw generic descriptions of numerous mobile game apps and provided initial ratings on their likelihood to recommend each mobile game app. Next, in the fMRI scanner, participants read peer recommendations about the same mobile game apps and had the opportunity to update their initial ratings after incorporating this social feedback. Corroborating previous findings (Cascio, et al., 2015), greater mentalizing activity while participants considered social recommendations was associated with successful recommendation propagation, in that participants shifted their initial ratings toward that of the peer recommender. In this case, the relationship between mentalizing and opinion change was particularly strong when people responded to negative recommendations. One possibility is that understanding the mental states of others is particularly critical in situations involving negative feedback. More generally, certain message-level characteristics may lead to a greater consideration of social implications that drive social influence, and future studies are needed to explore these interactions further.

Development

The mentalizing system undergoes changes throughout the course of human development, and processes underlying social influence are also differently implicated based on a person's age. Research in this area may be particularly important given that the mentalizing system develops significantly over childhood and adolescence, with implicit abilities to take others' mental states developing at around 18 months of age and explicit abilities emerging between 4 and 6 years of age (Frith & Frith, 2003). The mentalizing system continues to change throughout adolescence (Blakemore, 2008; Harenski, Harenski, Shane, & Kiehl, 2012) and into older adulthood (Harenski et al., 2012; Moran, Jolly, & Mitchell, 2012; Wang & Su, 2013). Regions of the mentalizing system undergo structural and functional changes during adolescence (Blakemore, 2008; Steinberg, 2008), and one account suggests that this leads to heightened sensitivity to social cues and rewards during this time (Blakemore, 2008; Blakemore & Mills, 2014). Corroborating this theory, adolescents showed stronger functional connectivity within regions of the mentalizing system than young adults during a task that required processing of social emotions (Burnett & Blakemore, 2009). Further, older adulthood is marked with decreased functionality of the brain's mentalizing system (Moran et al., 2012). Older adults showed behavioral impairments in social judgment tasks, and these deficits were reflected by decreased activity in the brain's mentalizing system during social cognition tasks (Moran et al., 2012). Combined, these findings suggest that age should be carefully considered in studying the processes involved in social influence, especially given that extant studies have primarily recruited young adult participants.

Recent research has begun to address some of these considerations, with researchers investigating the neural signatures of social influence in adolescents (Cascio et al., 2015; Falk et al., 2014; Peake et al., 2013; van Hoorn et al., 2016; Welborn et al., 2015). Adolescent male participants who showed greater engagement in regions of the mentalizing system (DMPFC, rTPJ, PCC) while being socially excluded were also more likely to conform to social influence, showing an increased likelihood of more risky driving when paired with a peer passenger who encouraged risk-taking in a simulated driving task (Falk et al., 2014). This suggests that although similar mechanisms may be in play in adolescents and adults during social influence, additional research investigating how such effects may be moderated by age (as well as contexts such as during exclusion) may be fruitful.

Cultural and Environmental Considerations

The majority of research investigating social influence, persuasion, and propagation has focused on Western samples. Recent evidence, however, suggests the importance of considering cultural variables that may moderate these effects (Tompson, Lieberman, & Falk, 2015). For one, persuasive messages promoting healthy behavior that align with individuals' cultural and social norms are more effective (Kalichman & Coley, 1995; Uskul & Oyserman, 2010). Europeans and Americans were more likely to find messages on health risks of caffeine more persuasive after reading messages emphasizing individualism, whereas Asian Americans were more likely to find the messages persuasive after reading messages emphasizing relational obligations (Uskul & Oyserman, 2010). Further, a meta-analysis comparing neural activity in East Asians and Westerners in social and non-social processes reported that East Asians show increased recruitment of brain systems associated with mentalizing (DMPFC, TPJ) compared to Westerners during social compared to non-social processing (Han & Ma, 2014).

Neural patterns associated with social influence and persuasion might also be moderated by environmental variables, such as socioeconomic status (Cascio, O'Donnell, Simons-Morton, Bingham, & Falk, 2017). Critically, socioeconomic status moderated the effects of mentalizing activity during social exclusion on subsequent susceptibility to social influence, such that increased mentalizing activity during social exclusion was associated with less likelihood of conformity to social influence in adolescents higher in socioeconomic status (Cascio et al., 2017). Combined, these studies emphasize the importance of considering cultural and environmental contexts in the study of social influence in the brain.

Dynamic Brain and Social Networks

A growing body of research has begun to leverage recent developments in the analysis of social networks and brain network dynamics to understand how an individual's real-life social network position might influence brain patterns associated with social influence, persuasion, and information propagation. For instance, adolescents who hold positions of brokerage in their online social networks (i.e., higher ego-betweenness centrality, meaning that individuals have more opportunities to "broker" information between others who are otherwise not directly connected to one another) also showed greater mentalizing activity (MPFC, PC, TPJ) when making recommendations about products to their peers (O'Donnell, Bayer, Cascio, & Falk, 2017). Indeed, the human brain is dynamically associated with an individual's social network, and the brain has mechanisms to track and update the social network positions of others (Parkinson, Kleinbaum, & Wheatley, 2017; Zerubavel, Bearman, Weber, & Ochsner, 2015; for a review, see Falk & Bassett, 2017).

Highlighting the value of incorporating real-life social networks, brain dynamics, and social context, adolescents who had real-life social networks that were less dense (i.e., had less friends who were connected with one another) also showed greater changes in functional connectivity within the mentalizing system when being socially excluded (Schmälzle et al., 2017). Further, adolescents who showed greater global functional connectivity within the brain's mentalizing system during social exclusion were also more susceptible to peer influence in a driving simulation task (Wasylyshyn et al., 2018). Thus, susceptibility to social influence may not only involve mean engagement of the brain's mentalizing system (Falk et al., 2014), but also functional coordination within the system. Accordingly, incorporating variables from real-life social networks and dynamic neural patterns is a promising avenue to uncover variables that lead to social influence.

CONCLUSION

Extant research highlights mentalizing as one signal that contributes to decision-making that leads to successful social influence, persuasion, and information propagation. Both communicators and receivers of social influence engage in mentalizing as they consider the social implications of their actions, and the degree to which they engage in these processes is predictive of whether attempts at social influence, persuasion, and message propagation will be successful. Recent research has also begun to uncover contextual variables that may moderate the relationship between mentalizing and social influence. Future research that explores these avenues may be productive in understanding the variables that lead to social influence.

REFERENCES

Adolphs, R. (2009). The social brain: Neural basis of social knowledge. *Annual Review of Psychology, 60*, 693–716.

Baek, E. C., & Falk, E. B. (2018). Persuasion and influence: What makes a successful persuader? *Current Opinion in Psychology, 24*(1), 53–57.

Baek, E. C., Scholz, C., O'Donnell, M. B., & Falk, E. B. (2017). The value of sharing information: A neural account of information transmission. *Psychological Science, 28*(7), 851–861.

Baek, E. C., Scholz, C., O'Donnell, M. B., & Falk, E. B. (2019). The involvement of mentalizing and valuation brain networks in social influence on message receivers. Manuscript submitted for publication.

Barasch, A., & Berger, J. (2014). Broadcasting and narrowcasting: How audience size affects what people share. *Journal of Marketing Research, 51*, 286–299.

Bateman, A. W., & Fonagy, P. (Eds.). (2012). *Handbook of mentalizing in mental health practice*. Arlington, VA: American Psychiatric Publishing.

Baumeister, R. F., & Leary, M. R. (1995). The need to belong: Desire for interpersonal attachments as a fundamental human motivation. *Psychological Bulletin, 117*(3), 497–529.

Berger, J. (2014). Word of mouth and interpersonal communication: A review and directions for future research. *Journal of Consumer Psychology, 24*(4), 586–607.

Berger, J., & Milkman, K. L. (2012). What makes online content viral? *Journal of Marketing Research, 49*(2), 192–205.

Blakemore, S. J. (2008). The social brain in adolescence. *Nature Reviews Neuroscience, 9*(4), 267–277.

Blakemore, S. J., & Mills, K. L. (2014). Is adolescence a sensitive period for sociocultural processing? *Annual Review of Psychology, 65*(1), 187–207.

Burnett, S., & Blakemore, S. J. (2009). Functional connectivity during a social emotion task in adolescents and in adults. *European Journal of Neuroscience, 29*(6), 1294–1301.

Burt, R. S. (2004). Structural holes and good ideas. *American Journal of Sociology, 110*, 349–399.

Cappella, J. N. (1997). Behavioral and judged coordination in adult informal social interactions: Vocal and kinesic indicators. *Journal of Personality and Social Psychology, 72*(1), 119–131.

Cascio, C. N., O'Donnell, M. B., Bayer, J., Tinney, F. J., & Falk, E. B. (2015). Neural correlates of susceptibility to group opinions in online word-of-mouth recommendations. *Journal of Marketing Research, 52*(4), 559–575.

Cascio, C. N., O'Donnell, M. B., Simons-Morton, B. G., Bingham, C. R., & Falk, E. B. (2017). Cultural context moderates neural pathways to social influence. *Culture and Brain, 5*(1), 50–70.

Chung, Y. S., Barch, D., & Strube, M. (2014). A meta-analysis of mentalizing impairments in adults with schizophrenia and autism spectrum disorder. *Schizophrenia Bulletin, 40*(3), 602–616.

Cialdini, R. B., Demaine, L. J., Sagarin, B. J., Barrett, D. W., Rhoads, K., & Winter, P. L. (2006). Managing social norms for persuasive impact. *Social Influence, 1*(1), 3–15.

Cialdini, R. B., & Goldstein, N. J. (2004). Social influence: Compliance and conformity. *Annual Review of Psychology, 55*, 591–621.

Dietvorst, R. C., Verbeke, W. J. M., Bagozzi, R. P., Yoon, C., Smits, M., & van der Lugt, A. (2009). A sales force-specific theory-of-mind scale: Tests of its validity by classical methods and functional magnetic resonance imaging. *Journal of Marketing Research, 46*, 653–668.

Dufour, N., Redcay, E., Young, L., Mavros, P. L., Moran, J. M., Triantafyllou, C., … Saxe, R. (2013). Similar brain activation during false belief tasks in a large sample of adults with and without autism. *PLoS One, 8*(9), e75468.

Falk, E. B., & Bassett, D. S. (2017). Brain and social networks: Fundamental building blocks of human experience. *Trends in Cognitive Sciences, 21*(9), 674–690.

Falk, E. B., Cascio, C. N., Brook O'Donnell, M., Carp, J., Tinney, F. J., Bingham, C. R., … Simons-Morton, B. G. (2014). Neural responses to exclusion predict susceptibility to social influence. *The Journal of Adolescent Health: Official Publication of the Society for Adolescent Medicine, 54*(5 Suppl.), S22–S31.

Falk, E. B., Morelli, S. A., Welborn, B. L., Dambacher, K., & Lieberman, M. D. (2013). Creating buzz: The neural correlates of effective message propagation. *Psychological Science, 24*(7), 1234–1242.

Falk, E. B., O'Donnell, M., & Lieberman, M. (2012). Getting the word out: Neural correlates of enthusiastic message propagation. *Frontiers in Human Neuroscience, 6*, 313.

Falk, E. B., Rameson, L., Berkman, E. T., Liao, B., Kang, Y., Inagaki, T. K., & Lieberman, M. D. (2010). The neural correlates of persuasion: A common network across cultures and media. *Journal of Cognitive Neuroscience, 22*(11), 2447–2459.

Falk, E. B., & Scholz, C. (2018). Persuasion, influence and value: Perspectives from communication and social neuroscience. *Annual Review of Psychology, 69*, 329–356.

Frith, C. D., & Frith, U. (2006). The neural basis of mentalizing. *Neuron, 50*(4), 531–534.

Frith, U., & Frith, C. D. (2003). Development and neurophysiology of mentalizing. *Philosophical Transactions of the Royal Society B: Biological Sciences, 358*(1431), 459–473.

Goldstein, N. J., Cialdini, R. B., & Griskevicius, V. (2008). A room with a viewpoint: Using social norms to motivate environmental conservation in hotels. *Journal of Consumer Research, 35*(3), 472–482.

Gonzales, A. L., Hancock, J. T., & Pennebaker, J. W. (2010). Language style matching as a predictor of social dynamics in small groups. *Communication Research, 37*(1), 3–19.

Han, S., & Ma, Y. (2014). Cultural differences in human brain activity: A quantitative meta-analysis. *NeuroImage, 99*, 293–300.

Happé, F. G. E. (1994). An advanced test of theory of mind: Understanding of story characters' thoughts and feelings by able autistic, mentally handicapped, and normal children and adults. *Journal of Autism and Developmental Disorders, 24*(2), 129–154.

Harenski, C. L., Harenski, K. A., Shane, M. S., & Kiehl, K. A. (2012). Neural development of mentalizing in moral judgment from adolescence to adulthood. *Developmental Cognitive Neuroscience, 2*(1), 162–173.

Kalichman, S. C., & Coley, B. (1995). Context framing to enhance HIV-antibody-testing messages targeted to African American women. *Health Psychology, 14*(3), 247–254.

Kampe, K. K. W., Frith, C. D., & Frith, U. (2003). "Hey John": Signals conveying communicative intention toward the self activate brain regions associated with "mentalizing," regardless of modality. *The Journal of Neuroscience, 23*(12), 5258–5263.

Kim, H. S. (2015). Attracting views and going viral: How message features and news-sharing channels affect health news diffusion. *Journal of Communication, 65*, 512–534.

Lakin, J. L., & Chartrand, T. L. (2003). Using nonconscious behavioral mimicry to create affiliation and rapport. *Psychological Science, 14*(4), 334–339.

Lakin, J. L., Jefferis, V. E., Cheng, C. M., & Chartrand, T. L. (2003). The chameleon effect as social glue: Evidence for the evolutionary significance of nonconscious mimicry. *Journal of Nonverbal Behavior, 27*(3), 145–161.

Meyer, M. L., Davachi, L., Ochsner, K. N., & Lieberman, M. D. (2018). Evidence that default network connectivity during rest consolidates social information. *Cerebral Cortex, 29*(5), 1910–1920.

Moran, J. M., Jolly, E., & Mitchell, J. P. (2012). Social-cognitive deficits in normal aging. *Journal of Neuroscience, 32*(16), 5553–5561.

O'Donnell, M. B., Bayer, J. B., Cascio, C. N., & Falk, E. B. (2017). Neural bases of recommendations differ according to social network structure. *Social Cognitive and Affective Neuroscience, 12*(1), 61–69.

Parkinson, C., Kleinbaum, A. M., & Wheatley, T. (2017). Spontaneous neural encoding of social network position. *Nature Human Behaviour, 1*(5), 0072.

Peake, S. J., Dishion, T. J., Stormshak, E. A., Moore, W. E., & Pfeifer, J. H. (2013). Risk-taking and social exclusion in adolescence: Neural mechanisms underlying peer influences on decision-making. *NeuroImage, 82*(7), 23–34.

Poldrack, R. A. (2011). Inferring mental states from neuroimaging data: From reverse inference to large-scale decoding. *Neuron, 72*(5), 692–697.

Raichle, M. E. (2015). The brain's default mode network. *Annual Review of Neuroscience, 38*, 433–447.

Richardson, D. C., & Dale, R. (2005). Looking to understand: The coupling between speakers' and listeners' eye movements and its relationship to discourse comprehension. *Cognitive Science, 29*(6), 1045–1060.

Schippers, M. B., Roebroeck, A., Renken, R., Nanetti, L., & Keysers, C. (2010). Mapping the information flow from one brain to another during gestural communication. *Proceedings of the National Academy of Sciences, 107*(20), 9388–9393.

Schmälzle, R., Brook O'Donnell, M., Garcia, J. O., Cascio, C. N., Bayer, J., Bassett, D. S., ... Falk, E. B. (2017). Brain connectivity dynamics during social interaction reflect social network structure. *Proceedings of the National Academy of Sciences, 114*(20), 5153–5158.

Schmälzle, R., Häcker, F. E. K., Honey, C. J., & Hasson, U. (2014). Engaged listeners: Shared neural processing of powerful political speeches. *Social Cognitive and Affective Neuroscience, 10*(8), 1137–1143.

Scholz, C., Baek, E. C., O'Donnell, M. B., & Falk, E. B. (2019). Decision-making about broad- and narrowcasting: A neuroscientific perspective. *Media Psychology*. Advance online publication.

Scholz, C., Baek, E. C., O'Donnell, M. B., Kim, H. S., Cappella, J. N., & Falk, E. B. (2017). A neural model of valuation and information virality. *Proceedings of the National Academy of Sciences, 114*(11), 2881–2886.

Scholz, C., Doré, B. P., Baek, E. C., O'Donnell, M. B., & Falk, E. B. (2019). A neural propagation system: Neurocognitive and preference synchrony in information sharers and their receivers. Manuscript submitted for publication.

Snyder, M. (1974). Self-monitoring of expressive behavior. *Journal of Personality and Social Psychology, 30*(4), 526–537.

Steinberg, L. (2008). A social neuroscience perspective on adolescent risk-taking. *Developmental Review, 28*(1), 78–106.

Stephens, G. J., Silbert, L. J., & Hasson, U. (2010). Speaker-listener neural coupling underlies successful communication. *Proceedings of the National Academy of Sciences, 107*(32), 14425–14430.

Tompson, S., Lieberman, M. D., & Falk, E. B. (2015). Grounding the neuroscience of behavior change in the sociocultural context. *Current Opinion in Behavioral Sciences, 5*, 58–63.

Uskul, A. K., & Oyserman, D. (2010). When message-frame fits salient cultural-frame, messages feel more persuasive. *Psychology and Health, 25*(3), 321–337.

van Hoorn, J., van Dijk, E., Meuwese, R., Rieffe, C., & Crone, E. A. (2016). Peer influence on prosocial behavior in adolescence. *Journal of Research on Adolescence, 26*(1), 90–100.

Wang, Z., & Su, Y. (2013). Age-related differences in the performance of theory of mind in older adults: A dissociation of cognitive and affective components. *Psychology and Aging, 28*(1), 284–291.

Wasylyshyn, N., Falk, B. H., Garcia, J. O., Cascio, C. N., O'Donnell, M. B., Bingham, C. R., ... Falk, E. B. (2018). Global brain dynamics during social exclusion predict subsequent behavioral conformity. *Social Cognitive and Affective Neuroscience, 13*(2), 182191.

Welborn, B. L., Lieberman, M. D., Goldenberg, D., Fuligni, A. J., Galván, A., & Telzer, E. H. (2015). Neural mechanisms of social influence in adolescence. *Social Cognitive and Affective Neuroscience, 11*(1), 100–109.

Wimmer, H., & Perner, J. (1983). Beliefs about beliefs: Representation and constraining function of wrong beliefs in young children's understanding of deception. *Cognition, 13*(1), 103–128.

Zerubavel, N., Bearman, P. S., Weber, J., & Ochsner, K. N. (2015). Neural mechanisms tracking popularity in real-world social networks. *Proceedings of the National Academy of Sciences, 112*(49), 15072–15077.

10

Social Media in Neuroscience Research

Dar Meshi and Ceylan Özdem-Mertens

Over 2.5 billion people around the world regularly use social media platforms on their desktop computers, tablets, and mobile phones (Statista, 2018). This number increases every day as more people gain access to the internet. Currently, 88% of 18-to-29-year-olds in the United States use at least one form of social media, and typical users have profiles on multiple platforms, such as Facebook, Instagram, Snapchat, and Twitter (Pew Research Center, 2018). Social media users aren't just creating profiles, they're spending substantial, recurring periods of time on these platforms observing and interacting with others. For example, in 2016, Facebook reported that the average user spends 50 minutes a day on its sites (Stewart, 2016). With this widespread and time-consuming use, people are establishing behaviors and social norms concerning the use of these platforms, as well as generating a tremendous wealth of behavioral data.

Social scientists have conducted myriad research projects that attempt to better understand behaviors on social media platforms, as well as capitalize on this boon of generated behavioral data. These scientists come from a variety of fields, such as communication, psychology, economics, and sociology. Their work has resulted in copious journal publications on the topic—as of April 2018, there have been almost 20,000 research articles published that contain the term "social media." This number is up from 10,000 just three years prior (Meshi, Tamir, & Heekeren, 2015). The field of neuroscience stands in stark contrast to the abovementioned fields and has greatly lagged behind the social sciences—as of April 2018, there have been only 18 published primary research articles that related a behavioral or self-reported measure obtained from social media use to a measure of brain structure or function obtained with a magnetic resonance imaging (MRI) scanner. This chapter reviews these 18 publications, comprehensively surveying the history of neuroscience research with measures of social media use.

SOCIAL NETWORKS

In the real world, people develop and cultivate a social network of friends, colleagues, and family. People also develop and cultivate social networks in the online world, on certain social media sites, such as Facebook and Instagram. These online networks can consist of both people who are known in the real world and also people who are only known online (Ellison, Steinfield, & Lampe, 2007). Importantly, information about one's real-world social network is not easy to obtain, as process usually consists of a lengthy survey that requires the respondent to recall each individual in their network. Conversely, obtaining one's online social network consists of simply consulting their social media profile, as one's social network size is usually quantified and displayed by social media platforms. Research has demonstrated that the structure of one's online social network (e.g., layers and frequency of contacts within layers) closely mirrors the structure of one's offline network (Dunbar, Arnaboldi, Conti, & Passarella, 2015). Therefore, using online social network data as a proxy for real-world social network data is ideal for investigation into the neural correlates of social network size and structure (Meshi et al., 2015).

Social Network Size

In 2012, Kanai and colleagues conducted the first ever study to relate a measure of social media use to a measure of brain structure (Kanai, Bahrami, Roylance, & Rees, 2012). They investigated whether individual differences in social network size correlated with regional gray matter density of the brain. To achieve their aim, Kanai and colleagues obtained individuals' online social network size from participants' Facebook accounts and then used voxel-based morphometry (VBM) to examine gray matter density across the whole brain, as well as specifically looking at the amygdala. They examined the amygdala because a prior study using offline social network size had already indicated that the amygdala volume correlates with real-world social network size (Bickart, Wright, Dautoff, Dickerson, & Barrett, 2010). Kanai and colleagues found that when they conducted their whole brain analysis, gray matter density in the left middle temporal gyrus, right superior temporal sulcus, and right entorhinal cortex all positively correlated with number of Facebook friends. Interestingly, only when conducting a region of interest analysis in the amygdala did they reveal a significant correlation with online social network size. Kanai and colleagues also used a survey to assess offline social network size and analyses revealed a correlation with the gray matter density of the amygdala, but not the regions that were significant for online social network size. This disparity between regions was addressed by the next study published on the topic.

Von der Heide and colleagues built upon the above work, further investigating both online and offline social network size (Von der Heide, Vyas, & Olson, 2014). These researchers collected participants' Facebook friend number as a measure of online social network size, as well as two measures of real-world social network size. They used VBM to examine individual differences in regional gray matter density of the brain. Rather than conducting a whole-brain analysis, they focused on specific brain regions previously indicated to correlate with either online or offline social network size (Bickart et al., 2010; Kanai et al., 2012). These regions consisted of the bilateral amygdala, the superior temporal sulcus and middle temporal gyrus, the entorhinal cortex, and the orbitofrontal cortex. Their analyses revealed positive correlations for *both* online and offline social network size in the bilateral amygdala, the superior frontal gyrus, the entorhinal cortex, and the orbitofrontal cortex. Note, they did not replicate Kanai and colleagues' correlations with online social network size in the middle temporal gyrus or the superior temporal sulcus.

Von der Heide and colleagues also conducted a functional magnetic resonance imaging (fMRI) experiment to examine individual differences in brain activation during a social closeness task. In this task, participants viewed pictures of two people, a friend or an unknown person, and indicated which was more socially close to them. The results demonstrated that brain activation in the right amygdala when viewing faces positively correlates with both online and offline social network size, across participants. Interestingly, brain activation in the right entorhinal cortex positively correlated with the number of Facebook friends, but just for participants' friends, not for unfamiliar faces.

Social Network Density

In 2015, Bayer and colleagues investigated the relationship between one's social network density and how one responds to social exclusion (Bayer, O'Donnell, Cascio, & Falk, 2015). The authors obtained adolescent participants' social network data, including the connections between participants' friends, via the Facebook Application Programming Interface (API)—note, information concerning user's friends is no longer available through the Facebook API, as of April 2014. The authors analyzed this information and were able to ascertain the participants' full online social network, as well as their recent interaction network. They related these network measures to neuroimaging data collected while participants experienced social exclusion.

To induce social exclusion, Bayer and colleagues had participants play Cyberball while they were in the MRI scanner. Cyberball is essentially a computerized game of catch, where the participant passes the ball to two other people represented by online avatars. Importantly, even though the participant believes these avatars represent real people, they do not—the game is controlled by the experimenters. In the first session, the avatars play nicely, passing the ball around equally, but in a second session, the avatars do not pass the ball back to the participant, socially excluding the participant from the game. The authors examined regions of interest that were previously demonstrated to process social exclusion, such as the anterior cingulate cortex and anterior insula, and found that the density of participants' full online network did not predict their brain response, but their recent interaction network did. In other words, there was a positive relationship between the density of participants' recent interaction network and their brain's response to social exclusion. These findings demonstrate that individuals who have communicated recently with a highly dense social network display a greater physiological reactivity when being socially excluded.

Following this work, Schmälzle and colleagues conducted a similar experiment with adolescents, again capitalizing on the social network data available via the Facebook API and the Cyberball social exclusion task (Schmälzle et al., 2017). In this experiment, the authors focused on the social exclusion brain system discussed above, as well as the brain's mentalizing system (see Chapter 9 by Baek, Scholz, & Falk, in this volume), which consists of the medial prefrontal cortex, precuneus, and the bilateral temporoparietal junction. The authors found greater connectivity within the mentalizing network during social exclusion compared to social inclusion. This connectivity difference was not present in the social exclusion network. Furthermore, the authors found that individuals who have a highly dense, full online social network display greater bilateral temporoparietal junction connectivity during social exclusion.

INTENSITY OF ONLINE SOCIAL NETWORKING

Users of online social network sites, like Facebook, display different intensities of use—some people frequently check these platforms and have a strong psychological dependence on them, while others demonstrate less frequent checking and a lack of psychological dependence (Ellison et al., 2007; Griffiths, Kuss, & Demetrovics, 2014). To provide an explanation for these individual differences in social media use intensity, Meshi and colleagues conducted the first study to relate a measure of social media use to a measure of functional brain activation (Meshi,

Morawetz, & Heekeren, 2013). The authors adapted a previously published task (Izuma, Saito, & Sadato, 2008) that provided study participants with reputation-related social rewards (positive social feedback) about both their character and another person's character. In this task, participants were video-recorded during a brief interview while believing that after the interview, a group of judges would view their video and select words that described them. Importantly, participants also believed that an "other" participant was also interviewed and evaluated by these judges. The participants then entered the MRI scanner to observe what the judges thought of them and this "other" person. Meshi and colleagues measured activation in a key region of the brain's reward system, the nucleus accumbens, relating it to participants' self-reported intensity of Facebook use—as revealed by the Facebook Intensity Scale (Ellison et al., 2007). The results demonstrated that the nucleus accumbens' response to simple, self-related social reward did not correlate with the intensity of social media use across participants. Importantly, however, the nucleus accumbens' response to socially compared reward (the self-related reward minus the other-related reward) did correlate with the intensity of social media use across participants. In other words, the bigger the difference in the nucleus accumbens' response to self-related vs. other-related social rewards, the more that individual uses Facebook.

Another study followed up on this research, also examining the nucleus accumbens and intensity of social media use, but by specifically assessing brain structure and relating it to an actual measure of mobile Facebook use, not just self-report (Montag et al., 2017). Montag and colleagues tracked Facebook use on 62 participants' mobile phones for five weeks, monitoring both duration of use and frequency of checking. They also collected a survey on participants' intensity of use. They then correlated these measures with the gray matter volume of various brain structures. They found significant negative correlations between both duration and frequency of mobile Facebook use and the gray matter volume of the bilateral nucleus accumbens, while the survey revealed a significant negative relationship only with the right nucleus accumbens. Furthermore, they found no significant associations between any measure of Facebook use and amygdala or hippocampal gray matter volumes. Taken together, the above study by Meshi and colleagues and the study by Montag and colleagues both demonstrate a clear connection between the intensity of online social networking and a key region of the brain's reward system.

Problematic Online Social Network Use

Intensity of online social networking can also be framed as excessive or problematic use, with some researchers using the term "addiction" (Griffiths et al., 2014)—although using this term is hotly debated (Carbonell & Panova, 2017). Some researchers categorize problematic online social networking as an addiction because of behavioral similarities with substance use and other behavioral addictive disorders, namely, salience (preoccupation), mood modification, tolerance, withdrawal, conflict, and relapse (Griffiths et al., 2014; see also Chapter 14 by Gilbertson & Gentile, in this volume). Furthermore, case studies exist for individuals whose online social networking significantly impaired their personal and professional lives (Karaiskos, Tzavellas, Balta, & Paparrigopoulos, 2010). Along these lines, Turel and colleagues had 20 participants complete a questionnaire regarding addictive symptomology for Facebook use and then, one week later, these participants were given Facebook specific go/no-go tasks while in the scanner (Turel, He, Xue, Xiao, & Bechara, 2014). The authors' whole-brain analysis revealed a correlation between ventral striatum activation during Facebook-go trials and Facebook "addiction" scores across participants. Aberrant function of the ventral striatum has been previously demonstrated for individuals with substance use and behavioral addictive disorders (Luijten, Schellekens, Kühn, Machielse, & Sescousse, 2017), so Turel and colleagues' finding creates a parallel between problematic use of social media and these other disorders.

This group of researchers then performed three follow-up studies specifically examining the relationship between brain structure and surveys assessing excessive online social networking. First, the authors again used 20 participants and correlated their results on the same Facebook questionnaire with gray matter volume, using VBM (He, Turel, & Bechara, 2017). This analysis revealed negative correlations between the volume of both the right and left amygdala and excessive online social networking across participants. This analysis also revealed a positive correlation between the volume of the anterior cingulate cortex and excessive online social networking. Interestingly, no correlation was observed with the left or right ventral striatum, in contrast to the study by Montag and colleagues cited above. Next, these researchers conducted a similar analysis, but compared two groups of 25 participants each—the authors gave 50 individuals their survey of excessive Facebook use and performed a median split for comparisons between groups (He, Turel, Brevers, & Bechara, 2017). The group of individuals who scored high on the excessive use survey demonstrated reduced volumes of both the right and left amygdala, supporting the authors' previous results. Interestingly, at the whole-brain level, the authors' analysis again revealed no result with the ventral striatum, but upon conducting a region of interest analysis, the right ventral striatum was revealed to be smaller in excessive Facebook users. Finally, this group of researchers again conducted a VBM analysis with gray matter volumes, but this time they used a different survey method (Turel, He, Brevers, & Bechara, 2017). They asked 33 participants two open-ended questions: (1) about the number of days per week they use Facebook, and (2) about the hours per day that they use Facebook. The authors then multiplied these numbers to obtain participants' estimated hours per week using Facebook. The authors also asked the same participants the same two questions, but participants responded on a 5-point Likert scale. The authors then conducted a principal component analysis on the open-ended and Likert responses to reveal a single measure of estimated Facebook use. This self-reported measure positively correlated with several regions of gray matter in the temporal lobe: the bilateral middle temporal gyrus, right superior temporal gyrus, and left fusiform gyrus. The results in these regions appear to overlap with the previous reports regarding social network size discussed above. Overall, these studies demonstrate the complexity of excessive online social networking, indicating the involvement of not only the brain's reward system, but also brain regions previously indicated in social cognitive functions.

SOCIAL INFLUENCE

Social media platforms provide arenas for individuals to express their opinions and influence others. For example, platforms such as Yelp allow users to review restaurants and businesses, and these reviews are intended to explicitly shape the opinions and behaviors of others. More implicit social influence may also occur on social media sites, for example, through the effects of popularity. YouTube provides an example of this by displaying the number of views of each video on its site. This number may influence how much people like the video and share it with others. With these explicit and implicit mechanisms in mind, several studies have investigated the neuroscience underlying social influence as instantiated on social media.

In 2015, Cascio and colleagues built on previous research into offline social influence (e.g., Klucharev, Rijpkema, Smidts, Hytönen, & Fernández, 2009) to examine the neural correlates of explicit online social influence (Cascio, O'Donnell, Bayer, Tinney, & Falk, 2015). These researchers simulated an online software application (app) recommendation platform, similar to the app reviews one sees in the iTunes App Store. While outside the scanner, 65 adolescents (either 16 or 17 years old) saw descriptions of game apps and rated them on how likely they'd be to recommend the apps. Then, while inside the scanner, participants were reminded of their rating and were then provided a rating that they believed was the aggregate of a group of their peers—this rating was randomly either identical to, or higher or lower than, their initial rating. Participants were then allowed to update their rating if they wanted to. Analyses revealed that when participants changed their opinions, there was more activation in the ventral striatum and ventromedial prefrontal cortex. Moreover, the authors correlated the percent recommendation change for each participant across trials with activity in the right temporoparietal junction. Individuals who changed their responses to a greater extent had more activation in this area. These brain regions have previously been implicated in reward processing and social cognition, such as mentalizing about others' thoughts (Bartra, McGuire, & Kable, 2013; Saxe & Kanwisher, 2003).

This group of researchers followed up on this study, using the same app recommendation task to investigate the relationship between social influence and one's social network characteristics (O'Donnell, Bayer, Cascio, & Falk, 2017). Fifty adolescents (again either 16 or 17 years old) performed the task, and similar to the above social network size experiments, their online social network was obtained with the Facebook API. O'Donnell and colleagues' analyses revealed that social network size was not related to behavioral influence on the task, but there was a marginal relationship between influence and ego betweenness centrality. Ego betweenness centrality measures how important actors are in a network (Everett & Borgatti, 2005), and in the present study the authors operationalized the measure to represent the ability to broker information within one's network. Specifically, individuals low in information brokerage were more likely to change their opinion when presented with a peer opinion different from their own. The authors then focused on the mentalizing network of the brain. They examined if activity within this network, while receiving peer feedback and adjusting one's opinion, correlated with social network size or ego betweenness centrality. Interestingly, brain activation in the mentalizing network did not correlate with social network size, but there was a positive correlation with ego betweenness centrality. In other words, adolescents who have more information brokerage capability had more activation in their mentalizing network while seeing opinions of others and adjusting their own opinion.

The above studies specifically looked at explicit influence on social media when discovering that others' recommendations were either in line with, or contrary to, one's own recommendation. However, another more implicit type of influence may occur online on social media—it could be that the previous popularity of a social media post can influence a user to like the content more. In other words, if something is already well-liked online, will it cause individuals to like it more, and vice versa? To investigate this type of social influence, Sherman and colleagues simulated the online social networking platform Instagram and conducted two studies, one with adolescents and the other with both adolescents and young adults (Sherman, Greenfield, Hernandez, & Dapretto, 2017; Sherman, Payton, Hernandez, Greenfield, & Dapretto, 2016). In their first study, conducted with 32 adolescents in high school between 13 and 18 years of age, Sherman and colleagues had participants view photos in the MRI scanner while they had the opportunity to either "like" them or move to the next photo (Sherman et al., 2016). Importantly, participants believed that 50 other participants had already viewed these photos and had the opportunity to "like" them as well. The researchers manipulated the number of likes each photo had already received and the content of the photos (risky behaviors vs. neutral images). The results demonstrated that adolescents will like photos more often if the photos have already received a large number of likes, regardless of the content of the image.

With regard to neuroimaging, the ventral striatum displayed greater activation when participants saw that their own images had received a high number of likes as compared to a low number of likes. Furthermore, a region of interest analysis revealed that the left nucleus accumbens was also more active when participants simply saw neutral photos that had received a high number of likes as compared to a low number of likes.

To replicate their results and compare age differences, Sherman and colleagues conducted a second study. In this study, they had a group of 27 university students perform the same task (Sherman et al., 2017). Indeed, the researchers replicated their previous behavioral and neuroimaging findings with this older sample. Sherman and colleagues then analyzed the neuroimaging data of both the young, high school sample from the previous study and the new university sample with respect to individuals' age. The results demonstrated a positive correlation, across participants in the high school sample, between age and left nucleus accumbens' activation when seeing that one's own photos received a high number of likes, as compared to a low number of likes. This correlation was not present in the university sample, leading the authors to conclude that nucleus accumbens' sensitivity to social rewards peaks at around 16 to 17 years of age.

SHARING INFORMATION

Social media platforms allow users to broadcast information about themselves and others (Meshi et al., 2015). Recent research has demonstrated that sharing information about oneself and others is associated with both behavioral and neural signatures of value (Tamir & Mitchell, 2012; Tamir, Zaki, & Mitchell, 2015). With this in mind, two studies thus far have investigated the act of sharing information on social media platforms.

In 2016, Meshi and colleagues conducted a resting state neuroimaging study examining self-related information sharing on Facebook (Meshi et al., 2016). Meshi and colleagues used a 6-item survey to assess degree of self-related sharing in 35 participants, and they then related individual survey scores with the intrinsic (resting) functional connectivity of participants' brains. To do this, the authors placed seeds for analysis in three regions of the brain that were generated from a meta-analysis on self-related cognition: the medial prefrontal cortex, the central precuneus, and the caudal anterior cingulate cortex. The authors conducted whole-brain correlations across participants with their individual degree of self-related sharing on Facebook. The analysis revealed a network of brain regions involved in self-disclosure on social media. Connectivity of both the medial prefrontal cortex and the central precuneus to the dorsolateral prefrontal cortex correlated with degree of self-related information sharing on Facebook. Furthermore, connectivity between the central precuneus and the left lateral orbitofrontal cortex also correlated with the degree of self-related information sharing. Interestingly, analysis with seeds in the nucleus accumbens did not reveal any significant correlations.

In 2017, Scholz and colleagues also investigated the sharing of information on social media, but they used an fMRI task and actual news articles from the *New York Times* (Scholz et al., 2017). Importantly, these researchers had real data regarding the number of times these articles had been shared online (via Facebook, Twitter, and email). In their task, which they conducted twice—once with 41 participants and again with 39 participants—individuals were shown news article headlines and abstracts (in one condition, individuals also heard a recorded reading of these articles). The authors used neuroimaging data from three previously defined networks for analysis: the self-related cognition network, the social cognition network, and the reward system. The authors conducted a path analysis linking the neuroimaging data from within these networks, while reading article headlines and abstracts, to the actual amount these articles had been shared online by others. Neural activity in the self-related cognition network and the social cognition network was not directly related to population-level sharing (i.e., how often the news articles were shared). Rather, activity in these networks was indirectly related to population-level sharing specifically through increased activation in the brain's reward system. Furthermore, the brain data predicted population-level sharing better than simple article characteristics or the study participants' self-reported intention to share.

FINANCIAL DECISION-MAKING

Social media websites like Kiva and Kickstarter allow users to provide financial support for people or companies in need. Visitors to these sites can browse profiles that make appeals for money and then decide whether, and to what extent, they would like to fund the person or project. These platforms provide a novel environment to investigate financial decision-making, especially to predict decisions made by the overall market. For example, Genevsky and Knutson showed Kiva profiles to 28 participants while they were in the scanner and asked them if they would fund these financial appeals (Genevsky & Knutson, 2015). Similar to the above study with *New York Times* articles, this group of researchers obtained data from Kiva on the actual rates at which these financial appeals were funded. The authors analyzed the neuroimaging data when participants saw the appeals, revealing that their nucleus accumbens' response predicted the amount of money that the overall market provided to these Kiva profiles. Importantly, the neuroimaging data predicted the market behavior above and beyond the participants' self-reported decisions on whether to loan money or not.

In another study, the same group of researchers focused on the funding of Kickstarter projects (Genevsky, Yoon, & Knutson, 2017). Genevsky and colleagues showed 30 participants Kickstarter funding appeals and asked if they would fund these projects. Again, the nucleus accumbens' response outperformed the participants' behavioral report, predicting the market-level funding outcomes of these Kickstarter appeals. This result was independently replicated with another sample of 32 participants. Taken together, these studies demonstrate the ability of neuroimaging data, obtained from a relatively small number of participants, to predict market-level metrics of financial decisions on social media.

CONCLUSION

In this chapter, we reviewed primary research that related a behavioral or self-reported measure obtained from social media use (or simulated social media use) to a measure of brain structure or function obtained in an MRI scanner. This research utilized measures of online social network size and density, intensity of social media use, social influence, information sharing, and financial decision-making to provide a better understanding of our brain and behavior. These studies, conducted over the last 6 years, comprise the first wave of neuroscientific investigations with social media.

Note, these studies appear to have taken one of two general approaches. In the first approach, researchers answer questions about the brain by capitalizing on measures obtained from social media. In this approach, a measure from social media is used as a proxy for an offline social behavior, for example, one's social network size. The research focus is not on understanding social media use, per se, but on understanding the neural mechanisms of human social behavior. With the second approach, researchers answer questions directly related to the use of social media. In this approach, the research focus is on understanding brain processes when using social media or understanding the effects of social media use on the brain.

Future research will continue with these two approaches, or apply new approaches, on a variety of topics (Meshi & Deters, 2015), for example, this research may address questions regarding adolescents (Crone & Konijn, 2018) and the elderly, interpersonal relationships, multitasking, mental health and well-being, and further investigating the neuroscience underlying excessive use of online social networking platforms (Meshi et al., 2015). This next wave of neuroimaging research will provide a better understanding of our social world and the neuroscience underlying our complex, contemporary social behaviors.

REFERENCES

Bartra, O., McGuire, J. T., & Kable, J. W. (2013). The valuation system: a coordinate-based meta-analysis of BOLD fMRI experiments examining neural correlates of subjective value. *NeuroImage, 76*, 412–427.

Bayer, J., O'Donnell, M. B., Cascio, C. N., & Falk, E. B. (2015, November). Linking Facebook network structure to neural responses during social exclusion. Paper presented at the annual meeting of the National Communication Association, Las Vegas, NV.

Bickart, K. C., Wright, C. I., Dautoff, R. J., Dickerson, B. C., & Barrett, L. F. (2010). Amygdala volume and social network size in humans. *Nature Neuroscience, 14*(2), 163–164.

Carbonell, X., & Panova, T. (2017). A critical consideration of social networking sites' addiction potential. *Addiction Research and Theory, 25*(1), 48–57.

Cascio, C. N., O'Donnell, M. B., Bayer, J., Tinney, F. J., & Falk, E. B. (2015). Neural correlates of susceptibility to group opinions in online word-of-mouth recommendations. *Journal of Marketing Research, 52*(4), 559–575.

Crone, E. A., & Konijn, E. A. (2018). Media use and brain development during adolescence. *Nature Communications, 9*(1), 588.

Dunbar, R. I. M., Arnaboldi, V., Conti, M., & Passarella, A. (2015). The structure of online social networks mirrors those in the offline world. *Social Networks, 43*, 39–47.

Ellison, N. B., Steinfield, C., & Lampe, C. (2007). The benefits of Facebook "friends:" Social capital and college students' use of online social network sites. *Journal of Computer-Mediated Communication, 12*(4), 1143–1168.

Everett, M., & Borgatti, S. P. (2005). Ego network betweenness. *Social Networks, 27*, 31–38.

Genevsky, A., & Knutson, B. (2015). Neural affective mechanisms predict market-level microlending. *Psychological Science, 26*(9), 1411–1422.

Genevsky, A., Yoon, C., & Knutson, B. (2017). When brain beats behavior: Neuroforecasting crowdfunding outcomes. *The Journal of Neuroscience, 37*(36), 8625–8634.

Griffiths, M. D., Kuss, D. J., & Demetrovics, Z. (2014). Social networking addiction: An overview of preliminary findings. In K. P. Rosenberg & L. C. Feder (Eds.), *Behavioral addictions: Criteria, evidence, and treatment* (pp. 119–141). London, England: Academic Press.

He, Q., Turel, O., & Bechara, A. (2017). Brain anatomy alterations associated with social networking site (SNS) addiction. *Scientific Reports, 7*, 1–8.

He, Q., Turel, O., Brevers, D., & Bechara, A. (2017). Excess social media use in normal populations is associated with amygdala-striatal but not with prefrontal morphology. *Psychiatry Research: Neuroimaging, 269*, 31–35.

Izuma, K., Saito, D. N., & Sadato, N. (2008). Processing of social and monetary rewards in the human striatum. *Neuron, 58*(2), 284–294.

Kanai, R., Bahrami, B., Roylance, R., & Rees, G. (2012). Online social network size is reflected in human brain structure. *Proceedings of the Royal Society, Biological Sciences, 279*(1732), 1327–1334.

Karaiskos, D., Tzavellas, E., Balta, G., & Paparrigopoulos, T. (2010). Social network addiction: A new clinical disorder? *European Psychiatry, 25*, 855.

Klucharev, V., Rijpkema, M., Smidts, A., Hytönen, K., & Fernández, G. (2009). Reinforcement learning signal predicts social conformity. *Neuron, 61*, 140–151.

Luijten, M., Schellekens, A. F., Kühn, S., Machielse, M. W. J., & Sescousse, G. (2017). Disruption of reward processing in addiction: An image-based meta-analysis of functional magnetic resonance imaging studies. *JAMA Psychiatry, 74*(4), 387–398.

Meshi, D., & Deters, F. G. (2015). The need to connect, predicting personality, and the brain's reward system: Emerging themes from the Berlin Social Media Research Workshop. *International Journal of Developmental Sciences, 9*(2), 53–56.

Meshi, D., Mamerow, L., Kirilina, E., Morawetz, C., Margulies, D. S., & Heekeren, H. R. (2016). Sharing self-related information is associated with intrinsic functional connectivity of cortical midline brain regions. *Scientific Reports, 6*, 22491.

Meshi, D., Morawetz, C., & Heekeren, H. R. (2013). Nucleus accumbens response to gains in reputation for the self relative to gains for others predicts social media use. *Frontiers in Human Neuroscience, 7*, 439.

Meshi, D., Tamir, D. I., & Heekeren, H. R. (2015). The emerging neuroscience of social media. *Trends in Cognitive Sciences, 19*(12), 771–782.

Montag, C., Markowetz, A., Blaszkiewicz, K., Andone, I., Lachmann, B., Sariyska, R., ... Markett, S. (2017). Facebook usage on smartphones and gray matter volume of the nucleus accumbens. *Behavioural Brain Research, 329*, 221–228.

O'Donnell, M. B., Bayer, J. B., Cascio, C. N., & Falk, E. B. (2017). Neural bases of recommendations differ according to social network structure. *Social Cognitive and Affective Neuroscience, 12*(1), 61–69.

Pew Research Center. (2018). Social media use in 2018. Retrieved from www.pewinternet.org/2018/03/01/social-media-use-in-2018/.

Saxe, R., & Kanwisher, N. (2003). People thinking about thinking people: The role of the temporo-parietal junction in "theory of mind." *NeuroImage, 19*, 1835–1842.

Schmälzle, R., O'Donnell, M. B., Garcia, J. O., Cascio, C. N., Bayer, J., Bassett, D. S., ... Falk, E. B. (2017). Brain connectivity dynamics during social interaction reflect social network structure. *Proceedings of the National Academy of Sciences, 114*(20), 5153–5158.

Scholz, C., Baek, E. C., O'Donnell, M. B., Kim, H. S., Cappella, J. N., & Falk, E. B. (2017). A neural model of valuation and information virality. *Proceedings of the National Academy of Sciences, 114*(11), 2881–2886.

Sherman, L. E., Greenfield, P. M., Hernandez, L. M., & Dapretto, M. (2017). Peer influence via Instagram: Effects on brain and behavior in adolescence and young adulthood. *Child Development, 89*(1), 37–47.

Sherman, L. E., Payton, A. A., Hernandez, L. M., Greenfield, P. M., & Dapretto, M. (2016). The power of the like in adolescence: Effects of peer influence on neural and behavioral responses to social media. *Psychological Science, 27*(7), 1027–1035.

Statista. (2018). Number of global social network users 2010–2021. Retrieved from www.statista.com/statistics/278414/number-of-worldwide-social-network-users/.

Stewart, J. B. (2016). Facebook has 50 minutes of your time each day. It wants more. Retrieved from www.nytimes.com/2016/05/06/business/facebook-bends-the-rules-of-audience-engagement-to-its-advantage.html.

Tamir, D. I., & Mitchell, J. P. (2012). Disclosing information about the self is intrinsically rewarding. *Proceedings of the National Academy of Sciences, 109*(21), 8038–8043.

Tamir, D. I., Zaki, J., & Mitchell, J. P. (2015). Informing others is associated with behavioral and neural signatures of value. *Journal of Experimental Psychology: General, 144*(6), 1114–1123.

Turel, O., He, Q., Brevers, D., & Bechara, A. (2017). Social networking sites use and the morphology of a social-semantic brain network. *Social Neuroscience, 13*(5), 628–636.

Turel, O., He, Q., Xue, G., Xiao, L., & Bechara, A. (2014). Examination of neural systems sub-serving Facebook "addiction." *Psychological Reports, 115*(3), 675–695.

Von der Heide, R., Vyas, G., & Olson, I. R. (2014). The social network-network: Size is predicted by brain structure and function in the amygdala and paralimbic regions. *Social Cognitive and Affective Neuroscience, 9*(12), 1962–1972.

11

A Cognitive Neuroscience Perspective on Political Knowledge, Misinformation, and Memory for "Facts"

Jason C. Coronel and Erik P. Bucy

Memory is a fundamental concept in the fields of public opinion and political communication research. For example, the seminal studies in political science that examined what voters knew (or didn't know) about political candidates were based on assessments of memory for information that was disseminated over the course of an election campaign (Berelson, Lazarsfeld, & McPhee, 1954). Since the establishment of media research as a field of study, survey-based attempts to estimate media use have relied heavily on how well respondents can remember utilizing a specific news source (for criticisms of this approach, see Prior, 2009a, 2009b). Moreover, experimental work in media information processing has demonstrated that attention to certain features of news, such as negative compelling images, may retroactively inhibit memory for the preceding content (Newhagen & Reeves, 1992), distorting the accuracy of recollections about news stories that viewers have just seen. Taken together, insights about the nature of memory are critical to understanding how citizens make sense of the political world.

The central role of memory in explanations of political phenomena, then, begs the question of the extent to which public opinion and political communication scholars possess an accurate understanding of the nature and organization of human memory. This chapter argues that researchers in our field have much to gain by incorporating concepts and methods from studies of the cognitive neuroscience of memory.[1] In the following sections, we demonstrate how adopting a cognitive neuroscience perspective on memory can advance theoretical and empirical work in public opinion and political communication research. We proceed by discussing two important domains in which utilizing a cognitive neuroscience perspective can pay enormous dividends: (1) political learning and knowledge; and (2) processing misinformation.

This chapter is organized as follows. First, we discuss how public opinion and communication scholars have conceptualized memory. Different conceptualizations of memory have distinct implications for scholarly assessments of how effectively voters are able to understand their political world. Then, we describe a prominent view of memory from cognitive neuroscience: the notion of multiple memory systems. Adopting the theoretical view of multiple memory systems has profound implications for how scholars conceptualize the manner in which political learning from the media environment occurs. Second, we turn to the domain of political misinformation. In particular, how neuroscience studies of false memories can contribute to a more nuanced understanding of why political misinformation persists and is difficult to eradicate. The chapter ends by briefly observing the implications of a cognitive neuroscience perspective on memory on assessments of how well equipped we assume citizens are to take on the challenges of democratic life.

MEMORY IN PUBLIC OPINION AND POLITICAL COMMUNICATION RESEARCH

Much of the foundational work in public opinion did not articulate an explicit conceptualization of memory. Early media effects research similarly ignored information processing issues, casting the "human processor as an impenetrable 'black box' with unknowable processes taking place between message reception and the traditional outcomes of learning, attitudes, or behaviors" (Geiger & Newhagen, 1993, p. 42). However, one could make inferences regarding how scholars implicitly thought about memory based on their central research questions and the methods they used in answering them. In particular, public opinion research in its formative years was focused on estimating what citizens knew about politics and the kinds of information they gleaned from media (Berelson et al., 1954; Campbell, Converse, Miller, & Stokes, 1960; Converse, 1964). Early research was heavily reliant on the survey method and respondent self-reports to determine what voters remembered about candidates, parties, policies, and other political constructs. An implicit assumption of this body of work is that voters fail tests of political knowledge because they possess no memory of an event or fact to which they were previously exposed (e.g., information disseminated from a political campaign).[2] From this perspective, the ability to verbally describe a past political event or fact served as an indicator for memory for political information.

This survey-based approach had a large influence on scholars' assessments of the ability of voters to function as informed citizens. Critically, this influential work led to the conclusion that citizens possessed little political knowledge because they performed poorly on survey questions (Berelson et al., 1954; Campbell et al., 1960; Converse, 1964; Lazarsfeld, Berelson, & Gaudet, 1944)—a problem that had as much to do with the way questions were presented as it did with how much voters actually knew about candidates and issues (see Prior, 2014). Many theorists viewed such findings as troubling (Delli-Carpini & Keeter, 1997; Lupia & McCubbins, 1998), given that they challenged participatory models of democracy that cast an informed electorate as a prerequisite for a well-functioning democracy (see Dewey, 1927).

Furthermore, prominent theories of political learning assumed that political information could only exert an influence on attitudes and behavior if voters were able to explicitly remember it in a narrow, test-like format. The seminal Columbia and Michigan studies assumed that people's ability to think in ideological terms and engage in policy-based voting required that they successfully retrieve and apply information about parties, issues, and nuances of political ideology (Berelson et al., 1954; Campbell et al., 1960; Converse, 1964). In addition, the formation of political attitudes about issues was conceptualized as involving a process of retrieving considerations associated with each issue, including recollection about official party positions, arguments made by political figures, and so on (Zaller, 1992; Zaller & Feldman, 1992). In these "fixed memory" studies of political knowledge and sophistication, scholars ultimately concluded that the mass of voters were largely unable to think in ideological terms, at least as operationally defined in survey instruments, and were incapable of making policy-based judgments.

In the late 1980s, however, a group of public opinion scholars challenged the conclusions of the static model of memory by introducing a new conceptualization. Borrowing from the information processing approach in social psychology, Lodge and colleagues distinguished between "memory-based" and "on-line" processing of political information (Lodge et al., 1989, 1995). In this model, memory-based representations were conceptualized as political information that citizens could consciously remember or recite verbally when tested, a view of memory similar to one adopted by early work in public opinion research. In contrast, Lodge and colleagues introduced the concept of an on-line affective tally, an overall summary evaluation based on affect that is derived, but independent from, memory-based political information. Through a series of experimental studies, they showed that rather than storing specific information about candidates and retrieving relevant bits at the time of decision-making (i.e., memory-based processing), people instead acted like on-line processors; that is, they could not report the specifics, but these details still affected their summary judgments via an affective tally.

For example, in a typical political campaign, voters are exposed to various types of information associated with the candidates running for office. Voters may encode (in classic "memory-based" models, the process of storing information in memory is often referred to as "encoding") and then later use this information to decide whether they like or dislike a particular candidate. Scholars have traditionally viewed this candidate evaluation process as driven primarily by deeply ingrained specific facts about the candidate that can be consciously remembered. That is, scholars have assumed that in order for voters to render reasonable like or dislike judgments about candidates, they must first be able to recall specific information associated with candidates (e.g., the candidates' issue positions; Enelow & Hinich, 1984; Kelley & Mirer, 1974). This recalled information is then compared with political preferences and drives affective judgments about the candidate. An important implication of this memory-based model is that meaningful political evaluations and voting decisions depend entirely on the successful recall of issue information.

On-line processing specifies a different means of representing and storing information. According to the on-line model of candidate evaluation, voters extract affective information from exposure to candidate issue positions and public statements (Lodge et al., 1989, 1995), as when negative emotion is elicited from a candidate whose issue positions, values, or even communication style diverge from the voter's political preferences. This negatively stamped affective information is then incorporated into an accumulated affective tally for that candidate. Critically, the updated tally endures *irrespective* of whether information about candidates and the specific issue positions they advocate are also stored in memory. When the voter is later queried about that particular candidate, she only has to reactivate the affective tally to render an evaluation. As a consequence, voters can render reliable judgments even if they cannot remember any specific issue stands or policy positions. Because the average voter will likely not remember much specific, policy-related information previously learned during a campaign, as documented by public opinion survey data (e.g., Delli-Carpini & Keeter, 1997) on-line processing represents an efficient method of voter alignment with preferred candidates.

The on-line processing model has had substantial impact on public opinion and political communication research. First, the model has revised how scholars conceptualize the storage of political knowledge in memory. According to this model, political knowledge does not just specify representations of information that voters must retrieve from long-term memory and express verbally but can also take the form of general, affective representations. Second, these studies reached a substantially different conclusion about voter capacities to engage in informed citizenship than the earlier literature. Voters, they argued, can be responsive to campaign information even though they are unable to remember the specific considerations that went into their political evaluations. In the next section, we discuss the extent to which the on-line model of political learning and evaluation comports with modern views of memory from cognitive neuroscience.

A COGNITIVE NEUROSCIENCE PERSPECTIVE ON MEMORY

Modern research on the cognitive neuroscience of memory began in 1957 when two neuropsychologists described a patient who became known by his initials, H.M. (Scoville & Milner, 1957). H.M. underwent brain surgery and had the medial portion of his temporal lobe removed (these included brain regions such as the hippocampus, amygdala, and the parahippocampal gyrus) in order to treat his epilepsy—a medical procedure considered severe and inhumane today. Though the surgery decreased the occurrence of his epileptic seizures, it also severely impaired his capacity to form new long-term memories, a condition referred to as "anterograde amnesia." Indeed, when exposed to new people, places, or events, H.M. and other amnesic patients cannot explicitly state that they have learned new information.

Importantly, however, later studies showed that although amnesic patients were incapable of identifying new information to which they had just been exposed, their behavior showed evidence of prior exposure (Cohen & Eichenbaum, 1993; Eichenbaum & Cohen, 2001). The evidence for this type of learning has been shown in a variety of domains and contexts. For example, amnesic patients have the capacity to learn new motor skills (Milner, Corkin, & Teuber, 1968), show enhanced performance in perceptually identifying objects or words to which they were previously exposed (Warrington & Weiskrantz, 1968), and are cautious toward individuals with whom they previously have had negative interactions (Claparède, 1911; Feinstein, Duff, & Tranel, 2010; Tranel & Damasio, 1993).

Similar to the on-line processing model, this work has shown that the capacity to have one's performance shaped by learning experiences is distinct from the capacity to consciously remember the learning experiences themselves. Many memory researchers attributed this dissociation to the operation of distinct forms of memory that are mediated by different brain systems. In particular, researchers theorized that the memory system responsible for implementing the encoding, storage, and retrieval of facts and events was dependent on the medial temporal lobe structures (e.g., the hippocampus) that are damaged in amnesic patients. Thus, studies of amnesic patients provide one of the key insights of modern memory research: memory is not a monolithic process but consists of distinct and separate capacities, each mediated by different brain networks. The notion that memory is expressed in many ways by multiple brain networks is often referred to as *multiple memory systems* (Squire, 1992).

In the decades following research on patient H.M., the multiple systems view of memory has been strongly supported by findings from a variety of cognitive neuroscience approaches. Indeed, reviews of the literature reporting multiple, converging results from both humans and animals has provided robust evidence for the existence of multiple memory systems (e.g., Eichenbaum & Cohen, 2001). In particular, memory researchers have made a key distinction between two broad classes of memory: declarative and nondeclarative (Cohen & Squire, 1980; Cohen & Eichenbaum, 1993; Squire, 2004). Declarative memory refers to knowledge of facts and events that can be accessed and expressed consciously. It can support all manner of arbitrary relations and has a high degree of flexibility. That is, such memories can be manipulated and used in a wide range of novel contexts. In contrast, nondeclarative memory refers to a broad collection of unconscious learning capacities that are expressed through performance (Squire, 2004). They neither require nor necessarily permit conscious access for expression.

Nondeclarative memory includes capacities such as procedural learning, emotional memory, conditioning, and priming. Unlike declarative memory, expression of nondeclarative memory most often influences automatic, procedural, and habitual behavior.

Within the framework of multiple memory systems, the concepts of "memory-based" and "on-line" processing can be viewed, respectively, as forms of declarative and nondeclarative memory processes. Indeed, recent research has provided evidence for the claim that political evaluations comport with a multiple memory systems model of political learning (Coronel et al., 2012). This study examined the extent to which amnesic patients could vote for candidates whose issue positions come closest to their own political views, despite not remembering any of the candidates' issue positions. The study exposed amnesic patients and neurologically intact comparison participants (i.e., matched on age, education, and sex) to fictitious political candidates who endorsed issues that either converged or diverged with participants' political preferences. The researchers then assessed whether amnesic patients had the capacity to identify candidates whose issue positions came closest to their own views. The study found that the amnesic patients did vote for candidates whose issues positions matched their own despite not consciously remembering their associated issue stands.

These findings support the claim in the on-line processing literature that voters can be responsive to campaign information even though they are unable to consciously remember the information that influenced their political evaluations. The distinction made by public opinion and political communication scholars between "memory" and "on-line" based processing also seems to map onto the distinction between declarative and nondeclarative memory made by cognitive neuroscientists. However, despite the seemingly potent capacity of nondeclarative memory to reliably support advantageous political decisions, findings from cognitive neuroscience may lead to less than optimistic conclusions about the capacity of nondeclarative memory systems, such as the affective tally, to aid voters in navigating the political world.

Multiple Memory Systems and Political Performance

The memory-based and on-line processing models have been highly influential in the public opinion and political communication literatures. Evidence suggests that Lodge and colleagues' on-line processing framework can be viewed as a nondeclarative, emotional learning account of political evaluations. This work has been presented as a rebuttal to critical depictions of the democratic citizen. However, prior work on on-line processing neglects a critical feature from multiple memory systems research. In particular, evidence from memory research suggests that we should *not expect* nondeclarative memory to aid people's ability to effectively navigate the political world as much as declarative memory. The learning experience via nondeclarative memory is not consciously accessible and, as a consequence, nondeclarative memory is unlikely to be as robust as declarative memory in its use across multiple, political learning and decision-making contexts. Nondeclarative memory may aid political decision-making but is likely to do so only under a narrow set of conditions. Given that information contained in nondeclarative memory is largely inaccessible to conscious awareness, it is less amenable to scrutiny in political decision-making.

For instance, the claim that the on-line tally can compensate for lack of declarative memory for political facts in the context of political decision-making is based almost entirely on results from studies in which participants are exposed only to issue-oriented political information that is accurately tied to a specific candidate. The real-world information landscape, however, is quite "noisy" and increasingly requires voters to distinguish between accurate and inaccurate political information. This problem is compounded by the growing use of digital and social media for election campaigns and governance. Devolving traditional editorial decisions about what qualifies as newsworthy to anyone with a social media account means that the individual must now navigate an expanding universe of content choices and delivery platforms, deciding what is important, believable, and true and what is not (Bucy & Newhagen, 2019).

The rise of "fake news" calls the assertion that the affective tally can support advantageous political decisions into question because such noisy messaging can also convey affective information. Suppose a voter encounters an article claiming that Hillary Clinton poisoned an FBI agent who leaked her emails. Although the voter recognizes the news as "fake," exposure to it could still generate a negative emotional response toward Clinton, particularly if the message is delivered in a menacing or skeptical tone that reinforces the negativity. If the voter remembers the declarative source of these affective associations (i.e., a non-credible source), then she may simply discount these negative feelings since it's clear that they have an unreliable pedigree. However, if the voter does not remember the source of such affective associations, then she may treat the negative affective information derived from "fake news" the same way as affective information derived from a credible source. In other words, the voter may incorporate this emotional information into her affective tally.

Taken together, a cognitive neuroscience view of memory leads us to a more skeptical perspective on the extent to which nondeclarative processes such as the affective tally can, without the aid of declarative memory, effectively aid voters in performing some important functions in their role as citizens (evaluating candidates, aligning candidates with their policy positions based on accurate political information, making voting decisions, and so on). Although our arguments are admittedly speculative, future work employing a multiple memory systems perspective can test such predictions regarding the capacity of nondeclarative memories to aid political decision-making in real-world information environments. Finally, work on the cognitive neuroscience of memory is continually evolving. Emerging theoretical views have begun to move away from the consciousness/unconscious distinction that has been previously used to characterize declarative and nondeclarative memories (Hannula & Greene, 2012; Hasson, Chen, & Honey, 2015; Henke, 2010; Ranganath & Ritchey, 2012). These new ways of characterizing memory systems (i.e., differences in processing modes, support for relational memories) will likely have implications for future work in political communication and public opinion.

A FALSE MEMORY PERSPECTIVE ON POLITICAL MISINFORMATION

Work on false memory over the past two decades suggests that much of the American public is probably misinformed about key issues (Kuklinski, Quirk, Jerit, Schwieder, & Rich, 2000; Pasek, Sood, & Krosnick, 2015). That is, many citizens confidently hold beliefs that are demonstrably false. These false beliefs range from inaccurate views about scientific facts to who benefits from social policies, to health-related misnomers, to propaganda about political candidates (Kuklinski et al., 2000; Pasek et al., 2015). Political communication scholars theorize that the popularity of convenient media sources that disseminate fake news and questionable reports, such as blogs, apps, and social media platforms, has played a major role in misinforming members of the mass public (Bessi et al., 2015; Mocanu, Rossi, Zhang, Karsai, & Quattrociocchi, 2015). Indeed, this concern gained international prominence in 2016 during the U.S. presidential election and the U.K. Brexit referendum campaign with the proliferation of fabricated news stories across legions of misleading websites and social media accounts (Lapowsky, 2018).

Here, we focus on another important source of misinformation: instances in which individuals are exposed to accurate information from a source (e.g., news websites), but biases inherent in memory cause them to misremember information. In particular, we focus on instances in which individuals can form false memories about politics. A "false memory" refers to the vivid recollection of an event that did not occur (for a review, see Gallo, 2014). The mechanisms that lead to the generation of false memories are not commonly considered in the political communication literature as potential sources of misinformation but are likely part of the dynamic that deters citizens from factual understandings of important events.

In a now-classic study, Gilliam and Iyengar (2000) presented participants with a local television news story about a violent crime. In one of the conditions, the story did not show or even mention the word "suspect." After exposure to the story, participants were asked if the story showed a suspect. Strikingly, 44% of participants inaccurately recalled seeing a black "suspect" in the story that did not show or mention a suspect. There are at least two possible explanations for why these errors occurred. First, participants may have relied on stereotypic or schema-based judgments about groups to make a strategic guess about the presumed race of the suspect in the news story. This is largely the explanation provided by the authors of the study. Under this strategic guessing account, participants had no memory of the suspect in the news story. Instead, they used their knowledge of stereotypes—that African Americans are often associated with crime in news stories (Dixon & Linz, 2000)—to infer that the suspect was likely an African American.

These errors, however, can also stem from a different source: false memories. In this scenario, an individual's stereotypic beliefs about African Americans may be sufficiently developed that exposure to information stereotypically associated with African Americans (e.g., crime) can lead to the implicit generation of other, related concepts (e.g., a black person). According to this account, information is stored in the form of schemas consisting of an organized network of semantically related concepts. For certain voters, for example, the concept of "African American" may be linked with "crime." When a concept is encountered, its associated representation in memory becomes active and that activation spreads to surrounding concepts within the network (Collins & Loftus, 1975). Incidental activation of a related, non-encountered concept could form a long-lasting memory representation. A false memory occurs when an individual retrieves this memory representation and misattributes its source, mistakenly thinking that they encountered information that was, instead, internally activated (Johnson, Hashtroudi, & Lindsay, 1993; Johnson & Raye, 1981; Roediger & McDermott, 1995).

Thus, the critical difference between the strategic guessing and false memory account of false beliefs is that there is presumably a memory trace of a black suspect in the latter whereas there is no memory trace in the former. That is, a false memory account would predict that people possess an actual memory of the black suspect. In contrast, the strategic guessing account would predict that people do not possess an actual memory of the black suspect. Strategic guessing and false memory are two qualitatively distinct processes and distinguishing them is critical. Indeed, a large body of work suggests that individuals treat false memories the same as true memories in that they report similar levels of confidence in real and falsely remembered events (see Gallo, 2014). Thus, political misinformation is potentially harder to correct if it is the outcome of a false memory process than if it is the result of strategic guessing. In the next section, we discuss how cognitive neuroscience techniques, specifically, event-related potentials, can be used to investigate false memories in the context of political misinformation.

Using Event-Related Potentials to Investigate False Memories

Event-related potentials (ERPs) are well suited to investigating false memories as they can provide information specifically linked to different types of memory processes. ERPs are measures of electrical brain activity recorded from the scalp that are time-locked to a stimulus event (photo, phrase, individual word, etc.) and, therefore, index the information processing operations engaged as a result of exposure to the stimulus (for a general introduction to ERPs, see Amodio, Bartholow, & Ito, 2014). Within the more global ERP signal are components that can be described on the basis of polarity (positive or negative), amplitude (measured in microvolts), latency (time in milliseconds from stimulus onset to peak amplitude), topography (distribution of amplitude across the scalp), and functional sensitivity (i.e., the types of perceptual, motor, cognitive, and affective factors to which they respond). Based on functional sensitivity, as well as an understanding of the underlying neural generators (when known), these components have come to be associated with cognitive processes of interest to communication research, namely, attention and memory.

There are two ERP components that reflect different aspects of memory processing: N400 and the LPC.[3] The N400 has been linked to the nondeclarative aspects of memory processing (for a review, see Kutas & Federmeier, 2011). For example, N400 memory effects are preserved in amnesic patients with compromised declarative memory systems (Olichney et al., 2000). In contrast, the LPC has been associated with declarative memory-related processes (Friedman & Johnson, 2000). Indeed, the LPC disappears in patients with anterograde amnesia, suggesting that the generation of the LPC requires an intact declarative memory system (Düzel, Vargha-Khadem, Heinze, & Mishkin, 2001).

A study in the domain of candidate evaluation has examined how one could use the N400 and LPC to disentangle strategic guessing from false memory-based processes (Coronel, Federmeier, & Gonsalves, 2014). The study examined people's tendency to misattribute issue positions that are consistent with a candidate's party affiliation, even when the candidate has never explicitly stated or endorsed such issue positions. For example, during the 2008 presidential election, half of Americans surveyed believed that it was Barack Obama and not John McCain who was in favor of embryonic stem cell research, when in fact, both candidates supported this position. These errors can stem from strategic guessing or false memories. Under the strategic guessing account, voters might reason that since McCain is a Republican and most Republicans are against stem cell research, then McCain must also be against stem cell research. Under this explanation, voters might assume that because their knowledge is incomplete, they should guess where candidates stand on some issues. In contrast, a false memory account suggests that voters' stereotypes or schemas of major party candidates can be so strong that voters create false memories about the issues that candidates support or oppose. Because false memories are indistinguishable from real memories in consciousness, voters believe candidates support issues they actually oppose, or vice versa.

In the study, participants learned about fictitious political candidates and their issue positions. In a subsequent test phase, brain activity in the form of ERPs were recorded. During this memory test phase, participants were shown issue positions that they were previously exposed to ("old") or positions they were not shown during the study phase ("new"). Individuals were asked to classify the items as old or new. The memory test generated four types of trials: hits (old issue positions correctly classified as "old"), misses (old issue positions incorrectly classified as "new"), correct rejections (new issue positions correctly classified as "new"), and false alarms (new issue positions incorrectly classified as old). The study generated the classic old/new effect, the finding that hits generate distinct ERPs (N400 and LPC) from correct rejections.

Given the functional interpretations of the N400 and LPC as indexing memory processes, this difference in ERPs between hits and correct rejections is theorized to arise from individuals possessing a memory signal for hits but possessing no memory for correct rejections.

The critical items are the false alarms—issue positions that were not shown in the study phase but that were incorrectly identified as shown. The strategic guessing account predicts that ERP responses to false alarms should resemble ERP responses to correct rejections since no memory signal should exist for these items. With the absence of a memory signal, false alarms are likely the product of strategic guessing. In contrast, the false memory account predicts that ERP responses to false alarms should be similar to ERP responses to hits. This would suggest that there are memory signals for false alarms that are indistinguishable from true memories (hits) (for studies examining the neural networks associated with false recognition, see Dennis, Bowman, & Vandekar, 2012; Garoff-Eaton, Slotnick, & Schacter, 2006). The study found that ERP responses to false alarms were indistinguishable from hits (Coronel et al., 2014). In other words, falsely attributed issue positions exhibited similar brain patterns to *true* memories of issue positions. In addition, participants reported high levels of confidence when making false alarm choices.

The relationship between false memories and high confidence levels in recognition has implications for the behavioral consequences of misinformation. Indeed, previous work in different domains shows that confidence can influence behaviors. For instance, individuals who are uncertain about the validity of their beliefs are more likely to seek out additional information to reduce their feelings of uncertainty (Locander & Hermann, 1979). Under strategic guessing, voters are likely aware that their knowledge is incomplete and recognize that they are making educated guesses. With more information, voters likely would update their views, replacing incorrect guesses with facts.

In contrast, because false memories are indistinguishable from real memories, voters can possess stronger confidence in their beliefs. Unlike educated guesses, false memories are potentially harder to correct even in light of new information given that voters believe this information to be valid or the events in question to have really occurred. Although we are not claiming that all types of misinformation can be explained by false memory-based processes, false memories generated by schemas or stereotypes may explain why some patently false beliefs are held in high confidence and why they are hard to correct.

CONCLUSION

The capacity of voters to retrieve political information about candidates, parties, and policies from memory has been a central criterion for assessing the competence and performance of democratic citizens since the earliest empirical research in public opinion. Indeed, over the last half century, memory and memory-based processes have played a key role in theories of political behavior and decision-making, sometimes as an implicit or assumed process rather than a fully articulated construct. Historically, public opinion and political communication scholars have conceptualized voters' memories for political information in different ways. As discussed, early attempts by public opinion and political communication scholars to conceptualize memory generated multiple and sometimes conflicting conclusions regarding the competence and political performance of citizens. Thus, insights into the organization and function of memory is critical to understanding the nature of citizen decision-making in democratic governance.

Although nondeclarative memory, via on-line processing, has been used as a refutation to critical portrayals of voters, memory research suggests that we should have a skeptical view of such a claim. In particular, nondeclarative memory is characterized by its lack of conscious access and scrutiny. Thus, nondeclarative memory should not be as robust as declarative memory in its use in political decision-making across different information contexts.

The work in cognitive neuroscience on false memory suggests that some types of political information may be the product of false memories rather than strategic guessing. This perspective on misinformation is important, given the evidence that people treat false memories the same as true memories and false memories tend to elicit high levels of confidence (while generating a memory signal that is indistinguishable from a true memory). An important implication of the false memory account of information processing is that some types of misinformation may be harder to correct than previously thought.

In summary, the use of concepts and frameworks from cognitive neuroscience should be useful in advancing theoretical and empirical work in public opinion and political communication research. The role of memory in assessing citizen competence and performance is an example of how knowledge about the brain can revise concepts central to communication. Indeed, different conceptualizations and characterizations about memory can lead to vastly different assessments of citizens' political performance in democratic governance. More broadly, a cognitive neuroscience approach to questions about citizen competence has the potential to change how political communication scholars think about the underlying mechanisms of citizenship—including memory, attention, emotion, and cognition—in a way that increases our understanding of how citizens make sense of the political world.

NOTES

1. The arguments and ideas from this entry were also developed in the first author's doctoral dissertation (Coronel, 2012).
2. It was unclear from these early studies whether the inability to correctly answer survey questions was due to memory failures (people were exposed to relevant political information but just forgot them) or failures in exposure (people were never exposed to relevant political information).
3. The N400 is so named because it describes a negative deflection that peaks around 400 milliseconds after stimulus onset. The LPC (late positive component or late positive complex) is a positive deflection that peaks 500 and 800 milliseconds post-stimulus onset.

REFERENCES

Amodio, D. M., Bartholow, B. D., & Ito, T. A. (2014). Tracking the dynamics of the social brain: ERP approaches for social cognitive and affective neuroscience. *Social Cognitive and Affective Neuroscience, 9*(3), 385–393.

Berelson, B. R., Lazarsfeld, P. F., & McPhee, W. N. (1954). *Voting: A study of opinion formation in a presidential campaign.* Chicago, IL: University of Chicago Press.

Bessi, A., Coletto, M., Davidescu, G. A., Scala, A., Caldarelli, G., & Quattrociocchi, W. (2015). Science vs conspiracy: Collective narratives in the age of misinformation. *PLoS One, 10*(2), e0118093.

Brader, T. (2006). *Campaigning for hearts and minds: How emotional appeals in political ads work.* Chicago, IL: University of Chicago Press.

Bucy, E. P., & Newhagen, J. E. (2019). Fake news finds an audience. In J. E. Katz & K. K. Mays (Eds.), *Journalism and truth in an age of social media* (pp. 201–222). New York, NY: Oxford University Press.

Campbell, A., Converse, P. E., Miller, W. E., & Stokes, D. E. (1960). *The American voter.* Chicago, IL: University of Chicago Press.

Claparède, E. (1911). Recognition et moïté. *Archives de Psychologie, 11*, 79–90.
Cohen, N. J., & Eichenbaum, H. (1993). *Memory, amnesia, and the hippocampal system*. Cambridge, MA: MIT Press.
Cohen, N. J., & Squire, L. (1980). Preserved learning and retention of pattern-analyzing skill in amnesia: Dissociation of knowing how and knowing that. *Science, 210*(4466), 207–210.
Collins, A. M., & Loftus, E. (1975). A spreading activation theory of semantic processing. *Psychological Review, 82*, 407–428.
Converse, P. E. (1964). The nature of belief systems in mass publics. In D. E. Apter (Ed.), *Ideology and discontent* (pp. 164–193). New York, NY: Free Press.
Coronel, J. (2012). Memory and voting: Neuropsychological and electrophysiological investigations of voters remembering political events. Unpublished doctoral dissertation, University of Illinois, Champaign-Urbana, IL.
Coronel, J. C., Duff, M. C., Warren, D. E., Federmeier, K. D., Gonsalves, B. D., Tranel, D., & Cohen, N. J. (2012). Remembering and voting: Theory and evidence from amnesic patients. *American Journal of Political Science, 56*(4), 837–848.
Coronel, J. C., Federmeier, K. D., & Gonsalves, B. D. (2014). Event-related potential evidence suggesting voters remember political events that never happened. *Social Cognitive and Affective Neuroscience, 9*(3), 358–366.
Delli-Carpini, M. X., & Keeter, S. (1997). *What Americans know about politics and why it matters*. New Haven, CT: Yale University Press.
Dennis, N. A., Bowman, C. R., & Vandekar, S. N. (2012). True and phantom recollection: An fMRI investigation of similar and distinct neural correlates and connectivity. *Neuroimage, 59*, 2982–2993.
Dewey, J. (1927). *The public and its problems*. Chicago, IL: Swallow Press.
Dixon, T., & Linz, D. (2000). Overrepresentation and underrepresentation of African Americans and Latinos as lawbreakers on television news. *Journal of Communication, 50*(2), 131–154.
Düzel, E., Vargha-Khadem, F., Heinze, H. J., & Mishkin, M. (2001). Brain activity evidence for recognition without recollection after early hippocampal damage. *Proceedings of the National Academy of Sciences, 98*(14), 8101–8106.
Eichenbaum, H., & Cohen, N. J. (2001). *From conditioning to conscious recollection: Memory systems of the brain*. New York, NY: Oxford University Press.
Enelow, J. M., & Hinich, M. J. (1984). *The spatial theory of voting: An introduction*. Cambridge, England: Cambridge University Press.
Feinstein, J. S., Duff, M. C., & Tranel, D. (2010). Sustained experience of emotion after loss of memory in patients with amnesia. *Proceedings of the National Academy of Sciences, 107*(17), 7674–7679.
Friedman, D., & Johnson, R. (2000). Event-related potential (ERP) studies of memory encoding and retrieval: A selective review. *Microscopy Research and Technique, 51*(1), 6–28.
Gallo, D. A. (2014). *Associative illusions of memory: False memory research in DRM and related tasks* (2nd ed.). New York, NY: Psychology Press.
Garoff-Eaton, R. J., Slotnick, S. D., & Schacter, D. L. (2006). Not all false memories are created equal: The neural basis of false recognition. *Cerebral Cortex, 16*(11), 1645–1652.
Geiger, S., & Newhagen, J. E. (1993). Revealing the black box: Information processing and media effects. *Journal of Communication, 43*(4), 42–50.
Gilliam, F. D., & Iyengar, S. (2000). Prime suspects: The influence of local television news on the viewing public. *American Journal of Political Science, 44*, 560–573.
Hannula, D. E., & Greene, A. J. (2012). The hippocampus reevaluated in unconscious learning and memory: At a tipping point? *Frontiers in Human Neuroscience, 6*(80), 1–20.
Hasson, U., Chen, J., & Honey, C. J. (2015). Hierarchical process memory: memory as an integral component of information processing. *Trends in Cognitive Sciences, 19*(6), 304–313.
Henke, K. (2010). A model for memory systems based on processing modes rather than consciousness. *Nature Reviews Neuroscience, 11*(7), 523.
Johnson, M. K., & Raye, C. L. (1981). Reality monitoring. *Psychological Review, 88*(1), 67–85.

Johnson, M. K., Hashtroudi, S., & Lindsay, D. S. (1993). Source monitoring. *Psychological Bulletin, 114*(1), 3–28.

Kelley, S., & Mirer, T. W. (1974). The simple act of voting. *American Political Science Review, 68*(2), 572–591.

Knowlton, B. J., & Squire, L. R. (1996). Artificial grammar learning depends on implicit acquisition of both abstract and exemplar-specific information. *Journal of Experimental Psychology: Learning, Memory, and Cognition, 22*(1), 169–181.

Kuklinski, J. H., Quirk, P. J., Jerit, J., Schwieder, D., & Rich, R. F. (2000). Misinformation and the currency of democratic citizenship. *Journal of Politics, 62*, 790–816.

Kutas, M., & Federmeier, K. D. (2011). Thirty years and counting: Finding meaning in the N400 component of the event-related brain potential (ERP). *Annual Review of Psychology, 62*(1), 621–647.

Lapowsky, I. (2018, April 4). Facebook exposed 87 million users to Cambridge Analytica. *Wired.* Retrieved from www.wired.com/story/facebook-exposed-87-million-users-tocambridge-analytica.

Lazarsfeld, P. F., Berelson, B., & Gaudet, H. (1944). *The people's choice: How the voter makes up his mind in a presidential campaign.* New York, NY: Columbia University Press.

Lewandowsky, S., Ecker, U. K. H., Seifert, C. M., Schwarz, N., & Cook, J. (2012). Misinformation and its correction: Continued influence and successful debiasing. *Psychological Science in the Public Interest, 13*(3), 106–131.

Locander, W. B., & Hermann, P. W. (1979). The effect of self-confidence and anxiety on information seeking in consumer risk reduction. *Journal of Marketing Research, 16*, 268–274.

Lodge, M., McGraw, K. M., & Stroh, P. (1989). An impression-driven model of candidate evaluation. *American Political Science Review, 83*(2), 399–419.

Lodge, M., Steenbergen, M. R., & Brau, S. (1995). The responsive voter: Campaign information and the dynamics of candidate evaluation. *American Political Science Review, 89*(2), 309–326.

Lupia, A., & McCubbins, M. D. (1998). *The democratic dilemma: Can citizens learn what they need to know?* New York, NY: Cambridge University Press.

Milner, B., Corkin, S., & Teuber, H. L. (1968). Further analysis of the hippocampal amnesic syndrome: 14-year follow-up study of H.M. *Neuropsychologia, 6*, 215–234.

Mocanu, D., Rossi, L., Zhang, Q., Karsai, M., & Quattrociocchi, W. (2015). Collective attention in the age of (mis)information. *Computers in Human Behavior, 51*, 1198–1204.

Newhagen, J. E., & Reeves, B. (1992). This evening's bad news: Effects of compelling negative television news images on memory. *Journal of Communication, 42*(2), 25–41.

Olichney, J. M., Van Petten, C., Paller, K. A., Salmon, D. P., Iragui, V. J., & Kutas, M. (2000). Word repetition in amnesia: Electrophysiological measures of impaired and spared memory. *Brain: A Journal of Neurology, 123*, 1948–1963.

Pasek, J., Sood, G., & Krosnick, J. A. (2015). Misinformed about the Affordable Care Act? Leveraging certainty to assess the prevalence of misperceptions. *Journal of Communication, 65*, 660–673.

Prior, M. (2009a). Improving media effects research through better measurement of news exposure. *Journal of Politics, 71*(3), 893–908.

Prior, M. (2009b). The immensely inflated news audience: Assessing bias in self-reported news exposure. *Public Opinion Quarterly, 73*(1), 130–143.

Prior, M. (2014). Visual political knowledge: A different road to competence? *Journal of Politics, 76*(1), 41–57.

Ranganath, C., & Ritchey, M. (2012). Two cortical systems for memory-guided behaviour. *Nature Reviews Neuroscience, 13(*10), 713.

Roediger, H. L., & McDermott, K. (1995). Creating false memories: Remembering words not presented in lists. *Journal of Experimental Psychology: Learning, Memory & Cognition, 21*(4), 803–814.

Scoville, W. B., & Milner, B. (1957). Loss of recent memory after bilateral hippocampal lesions. *Journal of Neurology, Neurosurgery, and Psychiatry, 20*(1), 11–21.

Squire, L. R. (1992). Declarative and nondeclarative memory: Multiple brain systems supporting learning and memory. *Journal of Cognitive Neuroscience, 4*(3), 232–243.

Squire, L. R. (2004). Memory systems of the brain: A brief history and current perspective. *Neurobiology of Learning and Memory, 82*(3), 171–177.

Squire, L. R., Knowlton, B., & Musen, G. (1993). The structure and organization of memory. *Annual Review of Psychology, 44*(1), 453–495.
Tranel, D., & Damasio, A. R. (1993). The covert learning of affective valence does not require structures in hippocampal system or amygdala. *Journal of Cognitive Neuroscience, 5*(1), 79–88.
Warrington, E. K., & Weiskrantz, L. (1968). New method of testing long-term retention with special reference to amnesic patients. *Nature, 217*(5132), 972–974.
Zaller, J. (1992). *The nature and origins of mass opinion*. New York, NY: Cambridge University Press.
Zaller, J., & Feldman, S. (1992). A simple theory of the survey response: Answering questions versus revealing preferences. *American Journal of Political Science, 36*(3), 579–616.

12

Advancing the Synchronization Theory of Flow Experiences

René Weber and Jacob T. Fisher

Flow is a psychological state characterized by full absorption in the task at hand (Csikszentmihalyi, 1990; Nakamura & Csikszentmihalyi, 2014). Individuals in a state of flow lose track of time, experience reduced self-consciousness, and feel a sense that their skills are matched to the challenges of the task. Flow states have been investigated in a broad variety of domains, including sports (Harris, Vine, & Wilson, 2017; Jackson, Ford, Kimiecik, & Marsh, 1998; Stein, Kimiecik, Daniels, & Jackson, 1995), music and art (MacDonald, Byrne, & Carlton, 2006), work (Hallberg & Schaufeli, 2006; Salanova, Bakker, & Llorens, 2006), and media (Ghani & Deshpande, 1994; Sherry, 2004; Weber, Tamborini, Westcott-Baker, & Kantor, 2009). In each of these domains, flow research has been characterized by conceptual broadness and methodological heterogeneity, leading to a flood of findings regarding the precursors, correlates, and outcomes of flow, but little clarity as to how these diverse findings can be integrated into a general understanding of the state itself (Harris et al., 2017).

The Synchronization Theory of Flow (STF) (Weber et al., 2009) aimed to rectify this issue by proposing specific neurophysiological processes underlying the experience of flow, leading to greater theoretical and methodological clarity (Weber, Sherry, & Mathiak, 2008). Nearly twelve years have now passed since the publication of the original premises and predictions of STF. In the intervening time, STF has garnered theoretical and empirical support from within communication and cognate fields, bolstering its core propositions. At the same time, advances in neuroscience have augmented our understanding of the neurophysiological mechanisms underlying flow experiences, allowing for more refined predictions.

These two developments necessitate a return to the original theory to review its premises and to refine its core hypotheses. In this chapter, we begin with a brief overview of the core tenets of STF, followed by a review of the burgeoning empirical evidence for its predictions. On this foundation we propose a selection of theoretical and methodological advances that clarify and extend the theory, and we provide future research avenues for scholars. Finally, we discuss a selection of domains in which STF is particularly useful for guiding the design of messages, games, and tools that contribute to health and well-being.

THE SYNCHRONIZATION THEORY OF FLOW

Much of communication scholarship adopts the perspective that psychological and communicative phenomena of interest can be understood using theories and methods that solely consider a portion of human experience (Floyd, 2014; Sherry, 2004). Recent work in communication science considers behavioral, sociocultural, and psychological phenomena that are the traditional provenance of communication research, but also emphasizes relevant chemical, neural, physiological, and evolutionary factors. These approaches more accurately capture the dynamic, multilevel nature of media psychology phenomena than do approaches that ignore the human processing system (Lang, 2013, 2014; Lang & Ewoldsen, 2010; Sherry, 2015; Weber et al., 2008). STF follows in this tradition. In prying open "the black box" of the brain, STF leverages a wealth of research from biology, neuroscience, and communication to generate accurate and falsifiable predictions and explanations at multiple levels (Weber et al., 2009).

Communication phenomena of interest are often emergent properties of complex, networked, dynamic, multilevel systems (Lang & Ewoldsen, 2010). Observing and attempting to explain these high-level phenomena often result in a preponderance of difficult-to-falsify theories that may even be logically inconsistent with one another (Watts, 2017; Weber et al., 2008; Yarkoni & Westfall, 2017). Specifying the neural substrates of these phenomena constrains the sorts of predictions available to researchers and affords an increased depth of understanding related to how the brain enables certain behaviors. These factors increase explanatory and predictive power as well as the likelihood that independent researchers will arrive at similar conclusions regarding the same behavioral phenomena (Ashby & Helie, 2011).

Weber and colleagues propose five key premises of STF regarding the neurophysiological substrates of flow (Weber, Huskey, & Craighead, 2017; Weber et al., 2009):

1. *Brains are oscillating systems* that rely on metabolic energy. Information is encoded in the firing of single neurons, but also in the synchronized activity of large networks acting in unison (Breakspear, Heitmann, & Daffertshofer, 2010; Buzsáki, 2006; Buzsáki, Logothetis, & Singer, 2013).
2. *Synchronized systems are energy-efficient.* Oscillating systems that are out of sync with one another expend more metabolic energy (and are therefore less efficient) than oscillating systems that are in sync (Arenas, Díaz-Guilera, Kurths, Moreno, & Zhou, 2008; Strogatz, 2004).
3. *Brain states are discrete.* Transitions into and out of specific neural states are not continuous and as such cannot be characterized as "more" or "less" of that state. Brain systems are either in sync or they are not—there is no "in between." Transitions between phases follow a power law distribution, meaning that there are many more small-phase changes than large ones. The dynamic coupling of these systems comprises a quasi-critical system (Beggs, 2008; Priesemann et al., 2014; Wilting & Priesemann, 2018).
4. *Brains are functionally organized.* Different brain regions and networks are semi-specialized for the performance of certain tasks and are dynamically recruited for processing tasks based on a set of simple rules (Betzel & Bassett, 2017; Kanwisher, 2010; Khambhati, Mattar, Wymbs, Grafton, & Bassett, 2018).
5. *Brains are hierarchically organized.* Higher-order cognitive states emerge from lower-order cognitive processes (Bassett & Gazzaniga, 2011). These processes unfold across the brain on multiple timescales. Emergence of higher-order cognitive states and representations from these lower-level processes are associated with conscious attention (Regev et al., 2018).

These premises lead to three primary predictions regarding the biological, behavioral, and phenomenological aspects of flow (for more detail, see Weber et al., 2009, 2017). First, from a phenomenological perspective, attention and reward networks should be recruited during flow experiences and, given premise 1, they should be synchronized with one another. Second, according to premises 2 and 3, the synchronization of attention and reward networks observed during flow states should be sudden at critical time points and correspond to energetic optimization in brain networks. Finally, premises 4 and 5 suggest that the observed network synchronization spans large-scale attention and reward networks that are hierarchically organized and that this pattern of network synchronization should be predictive of behavioral aspects of flow, including enjoyment and increased performance in the task at hand.

Networks of Attention

Before reviewing extant evidence for these predictions, we provide a brief overview of the neurobiological underpinnings of attention and reward in the human brain. Attention in STF was initially conceptualized as a tripartite process relying on separable but closely interlinked neural substrates (Petersen & Posner, 2012; Posner & Petersen, 1990; Raz & Buhle, 2006). More recent treatments of STF focus on connectivity between large-scale cortical and subcortical networks involved in task performance, reward, and task disengagement (Huskey, Craighead, Miller, & Weber, 2018; Huskey, Wilcox, & Weber, 2018).

The first of these processes, *alerting*, is conceptualized as a generalized state of vigilance that increases the brain's readiness to perform an impending task. Alerting is primarily dependent on the neurotransmitter norepinephrine (NE), and involves a network of brain regions centered on the locus coeruleus (Aston-Jones & Cohen, 2005) and extending to prefrontal and parietal areas. Alerting is related to (but not commensurate with) arousal, an emotional state associated with broad activation in the parasympathetic nervous system leading to physiological responses such as increased heart rate, greater skin conductance levels, and wide-ranging modulations to perceptual and cognitive processes (Dolcos, Wang, & Mather, 2015; Lee, Itti, & Mather, 2012; Mather & Sutherland, 2011). The flow state is associated with increased physiological arousal and greater self-reported alertness (Peifer, Schulz, Schächinger, Baumann, & Antoni, 2014) but also with reduced activation in the amygdala, a pattern often associated with reduced arousal (Ulrich, Keller, & Grön, 2016; Ulrich, Keller, Hoenig, Waller, & Grön, 2013).

The *orienting* subprocess involves selecting information from the sensory field and modulating sensory and cognitive systems in order to maximize processing of the selected information (Corbetta & Shulman, 2002; Sokolov, 1963). A rich research area within media psychology is concerned with elucidating how and when the human processing system orients to stimuli within a mediated environment (Francuz & Zabielska-Mendyk, 2013; Lang, Bradley, Park, Shin, & Chung, 2006; Potter, Lang, & Bolls, 2008; Thorson & Lang, 1992). In the brain, orienting is primarily associated with a cholinergic network involving the frontal eye fields (FEF), the temporoparietal junction (TPJ), the superior parietal lobule (SPL), and ventral portions of the frontal lobes (Corbetta & Shulman, 2002; Davidson & Marrocco, 2000; Petersen & Posner, 2012).

The *executive control* subprocess is implicated in situations wherein stimuli must be selected or inhibited based on the parameters of an individual's goals. Executive control is perhaps the most widely researched of the three subprocesses, as it is heavily involved in theoretically related processes and models such as the dual mechanisms of cognitive control framework (Braver, Gray, & Burgess, 2007; Miller & Cohen, 2001), working memory (Baddeley, 1992; Baddeley & Hitch, 1974), and the Limited Capacity Model of Motivated Mediated

Message Processing (LC4MP; Fisher, Huskey, Keene, & Weber, 2018; Fisher, Keene, Huskey, & Weber, 2018; Lang, 2000, 2009) (see Chapter 29 by Lang, in this volume). The executive control network in the brain consists of frontoparietal regions and the anterior cingulate cortex (ACC; Petersen & Posner, 2012). Frontoparietal structures—especially the lateral and inferior portions of the prefrontal cortex and the lateral intraparietal area (LIP)—seem to work together to control attention in important ways (Aron, Robbins, & Poldrack, 2004, 2014; Buschman & Miller, 2007). Most notably, evidence suggests that these regions work together to create a salience map of the visual field that directs top-down or bottom-up orienting processes (Bisley & Goldberg, 2010; Buschman & Miller, 2007; Sestieri, Shulman, & Corbetta, 2009).

EMERGING EMPIRICAL EVIDENCE

Since the publication of STF, multiple studies have been undertaken to test its core predictions and to compare them to predictions made by competing models of the neuronal underpinnings of flow (such as the hypofrontality model proposed by Dietrich, 2004). These studies have largely bolstered the predictions of STF. Using state-of-the-art methodological tools afforded by network neuroscience (Bassett & Sporns, 2017; Medaglia, Lynall, & Bassett, 2015), this research has found compelling evidence for the presence of synchronization during flow states and for the energetic efficiency of these states. Furthermore, emerging research suggests promising future avenues for advancing STF into new theoretical and methodological domains and illuminates pathways for potential applications in media design and health communication contexts.

Hypothesis 1: Synchronization

STF predicts that attention and reward networks in the brain will be active during flow states and that these networks will operate in sync with one another (Weber et al., 2009). It has long been suggested that synchronization of neural oscillations is the core means through which the brain integrates influences from disparate brain regions. The synchronous activation of cortical regions during perceptual processing is theorized to be a solution to the *binding problem*—how the brain combines perceived object features into a coherent representation of the object itself (Crick & Koch, 1990). Various neurons throughout the cortex and subcortical regions are specialized for the processing of certain object features, such as color, shape, motion, and orientation (for an overview of visual processing, see Nassi & Callaway, 2009). These neurons oscillate at certain frequencies and are connected either directly to one another or indirectly via various nuclei in the thalamus and hippocampus (Buzsáki, 1991, 2006; Crick & Koch, 1990; Singer, 2001). This network synchronization is proposed to be the means by which disparate perceptual representations are "linked" to one another to form a coherent representation of an object (Stryker, 1989).

Network synchronization is critical for the effective performance of a number of cognitive tasks, including sustained attention (Clayton, Yeung, & Cohen Kadosh, 2015). This research has shown that a network composed of the posterior portion of the medial prefrontal cortex (pMFC), the lateral prefrontal cortex (lPFC), and the lower-level sensory processing regions serves to direct attentional processing (MacDonald, Cohen, Stenger, & Carter, 2000; Ridderinkhof, Ullsperger, Crone, & Nieuwenhuis, 2004). This network largely overlaps with the executive control network discussed earlier in this chapter (Buschman & Miller, 2007). Importantly, the nodes within this network have been found to phase-synchronize within the theta oscillatory band (4–7 Hz) following commission of errors in an attentional task (Cavanagh, Cohen, & Allen, 2009; van de Vijver, Ridderinkhof, & Cohen, 2011; van Driel, Ridderinkhof, & Cohen, 2012). This theta-synchronization between nodes in the executive control network is theorized to subserve error-correction processes and has been shown to predict learning and cognitive control outcomes (Cavanagh et al., 2009; Cavanagh, Zambrano-Vazquez, & Allen, 2012).

STF does not specify which oscillatory band(s) will be used for internode communication within attention and reward networks, but it is likely that synchronization within different bands is associated with different communication processes between nodes. For example, in attention research, gamma oscillations have been linked to active but unfocused attention, beta oscillations with focused attention, and alpha oscillations with passive unfocused attention (Başar, 2012; Ward, 2003). Interestingly, a recent study found a significant power increase only in the alpha band during flow (Núñez Castellar, Antons, Marinazzo, & Van Looy, 2019). However, these separate oscillations also seem to function as separate attentional "refresh rates" for stimuli in central and peripheral attention, rhythmically sampling various portions of the visual field for the presence of salient stimuli (Fiebelkorn, Saalmann, & Kastner, 2013; Lakatos, Karmos, Mehta, Ulbert, & Schroeder, 2008; Landau & Fries, 2012). These varying oscillatory refresh rates may serve as a mechanism for flexibly gating the reallocation of attentional resources (Buschman & Kastner, 2015).

Research in this area has produced broad support for Hypothesis 1 of STF. Several studies have shown that the neural correlates of flow include robust activation in frontoparietal attention and reward networks (Huskey, Craighead et al., 2018; Huskey, Wilcox et al., 2018; Klasen, Weber, Kircher, Mathiak, & Mathiak, 2012; Ulrich et al., 2016, 2013; Yoshida et al., 2014). Furthermore, activation patterns in these networks are correlated with one another during flow states, a pattern not observed when skills and reward are mismatched (Huskey, Craighead et al., 2018). These findings have been augmented by recent work leveraging methods from network neuroscience (Bassett & Sporns, 2017) to investigate many-to-many connectivity within the frontoparietal attention network during flow states. This work has revealed that during flow, the average degree of nodes within the frontoparietal attention network is highest during flow (Huskey, Wilcox et al., 2018).

Hypothesis 2: Optimization

The second core hypothesis of STF is that the flow state should be reflected in an energetic optimization of cortical networks in the brain. This hypothesis is based on extensive evidence that synchronized systems in nature are more energy-efficient than are non-synchronized systems (Laufs et al., 2003; Sporns, 2011; Strogatz, 2004). Neural systems have structurally evolved to maximize information processing ability while minimizing wiring costs (Achard & Bullmore, 2007; Bassett et al., 2009). Short-range connections are energetically cheaper than long-range connections, but long-range connections are required in order to effectively integrate information from different parts of the network (Sporns, 2011). This has resulted in widespread "small-world" topology wherein most connections in the brain are short-range (within a small brain region) but a small number of connections extend to greater distances (Bassett & Sporns, 2017).

The cost efficiency of functional connections can be measured as the average of the inverse of the shortest path length between all nodes in a brain network (Bassett et al., 2009). The cost-efficiency of functional connectivity is associated with cognitive performance in a variety of domains (Bassett et al., 2009; Gießing, Thiel, Alexander-Bloch, Patel, & Bullmore, 2013) and with general intelligence (Langer et al., 2012). Less cost-efficient network topologies are associated with high performance in effortful tasks that demand high levels of control whereas more cost-efficient topologies are associated with high performance in less control-demanding, more automatic tasks (Bullmore & Sporns, 2012). In STF, cost-efficient synchronization of brain networks is proposed to be an underlying factor in the perceived effortlessness of flow experiences, despite the objective difficulty of the tasks being performed (Huskey, Wilcox et al., 2018; Weber, Huskey, & Craighead, 2016).

Emerging support for the energetic efficiency hypothesis is provided in one recent study (Huskey, Wilcox et al., 2018). In this study, participants engaged in a media task (playing a video game) while undergoing brain imaging. The video game stimulus was programmed to be either very easy (skill>>difficulty), very difficult (skill<<difficulty), or approximately matched to participant ability (skill≈difficulty). As matching of difficulty and ability is proposed to be a precursor of flow (Nakamura & Csikszentmihalyi, 2014), it was hypothesized that brain networks would be energetically optimized in the skill ≈ ability condition as compared to the other two conditions. This was found to be the case across various edge thresholding cutoffs, supporting the second hypothesis of STF.

Hypothesis 3: Outcomes

The final prediction of STF is that synchronization of brain networks observed during flow experiences should be associated with behavioral and self-reported outcomes such as higher performance and greater enjoyment (Weber et al., 2009). This hypothesis has been broadly supported in a variety of different research paradigms, including video games (Huskey, Craighead et al., 2018; Huskey, Wilcox et al., 2018; Keller & Bless, 2008; Klasen et al., 2012), and cognitive tasks (Ulrich et al., 2013, 2016). In each of these paradigms, the matching of skill and difficulty leads to activation in flow-related brain regions as well as higher self-reported flow and enjoyment.

Flow experiences in video games have been measured using both fMRI and functional near-infrared spectroscopy (fNIRS). In one study (Yoshida et al., 2014), brain activity was measured using fNIRS during video game play in a boredom condition (skill»difficulty) and in a flow condition (skill≈difficulty). The flow condition was associated with activation in the frontal pole, ventrolateral PFC, and ventromedial PFC (nodes in the frontoparietal attentional network) as well as with higher self-reported flow (Yoshida et al., 2014). Huskey and colleagues (Huskey, Craighead et al., 2018; Huskey, Wilcox et al., 2018) used *Asteroid Impact*,[1] an open-source video game stimulus, to manipulate the balance of difficulty and reward during three behavioral experiments and during fMRI scanning. In these studies, self-reported rewardingness of the video game was highest when skill≈difficulty. Flow experiences were also associated with activation in cognitive control and reward-related brain regions. Ulrich and colleagues (2013, 2016) observed brain activation while participants solved math problems that were either easy (skill»difficulty), difficult (skill«difficulty), or matched to their skill level (skill≈difficulty). In both of these studies, the flow condition was associated with activation in cognitive control and reward-related brain regions as well as higher self-reported enjoyment and willingness to do the task again.

ADVANCING THE SYNCHRONIZATION THEORY OF FLOW

Evidence is mounting for all three core hypotheses of STF. Flow experiences have been shown to be associated with activity in cognitive control and reward networks, and to result in efficient synchronization in the brain. In addition, these brain activation patterns are associated with behavioral outcomes such as enjoyment and performance. Now, almost 12 years after the publication of the original theory, theoretical and methodological advances in neuroscience and communication research allow for refinement of the assumptions and predictions of STF. In this section, we introduce four key ways in which STF can be pushed forward. Two of these advances are theoretical, reflecting key insights from neuroscience regarding the dynamic networks undergirding attention, reward, and cognitive control. The last two proposed advances are methodological: (1) introducing novel methods from network neuroscience, and (2) incorporating recent work concerning the social nature of flow-inducing tasks.

Theoretical Advance: Beyond the Tripartite Network Model

In the original conceptualization of STF, attention was proposed to rely on three primary subprocesses: alerting, orienting, and executive control (Petersen & Posner, 2012; Posner & Petersen, 1990). More recent treatments of the model (Weber et al., 2017) expand STF beyond the tripartite network model to consider the role of *cognitive control* in flow experiences. Cognitive control is an overarching term for a set of neural processes that enable the brain to focus on certain stimuli in the environment that are relevant for individual goals while inhibiting responses to less-relevant stimuli (Botvinick, Braver, Barch, Carter, & Cohen, 2001; Miller & Cohen, 2001). The process of cognitive control is enabled by three large and well-researched networks in the brain: the task-positive network, the task-negative network, and the salience network (Menon, 2011; see also Chapter 13 by Fisher & Keene, in this volume). In this section, we review these three networks, discussing their relevance for understanding the neural underpinnings of flow.

Cognitive control requires recruitment of excitatory and inhibitory neurons throughout wide swaths of the cortex and subcortical regions (Botvinick & Braver, 2015). These connections comprise the *task-positive* (central executive) network. The task-positive network is anchored in the dorsolateral prefrontal cortex (DLPFC) and lateral portions of the posterior parietal cortex (Seeley et al., 2007). Nodes in this network are activated during flow and also highly synchronized with other nodes in the task-positive network (Huskey, Craighead et al., 2018; Huskey Wilcox, et al., 2018). Consistent activation of the task-positive network during flow points to its importance for the maintenance of flow states in the presence of potentially salient distractors. Thus, it can be expected that the task-positive network will exhibit three primary characteristics during flow: (1) *activation*—nodes in the network should exhibit higher activation during flow than during conditions of skill/difficulty/reward mismatch; (2) *synchronization*—edge weights between nodes in the network should be high, reflecting large correlation coefficients between nodes; and (3) *allegiance*—connections between nodes in the task-positive network should be more connected to one another than to nodes outside of the network.

In contrast to the task-positive network, the *task-negative network* (also called the *default mode network; DMN*) consists of a collection of brain regions that have been observed to become more active whenever someone is *not* engaged in the task at hand (Greicius, Krasnow, Reiss, & Menon, 2003; Raichle et al., 2001; Seeley et al., 2007). This network is involved in mind-wandering, self-referential thought, planning, mentalizing, reward evaluation, and a wide selection of other behaviors (Raichle, 2015; Spreng, Mar, & Kim, 2009). Flow is associated with reduced activation of nodes in the DMN (Huskey, Craighead et al., 2018; Ulrich et al., 2013). An important exception to this pattern is the precuneus. The precuneus is a core member of the task-negative network but is also highly active during flow (Huskey, Craighead et al., 2018). The precuneus seems to serve as a hub node connecting the DMN and the task-positive network (Cavanna & Trimble, 2006; Utevsky, Smith, & Huettel, 2014). Thus, during flow, the DMN seems to serve to update representations of cognitive effort and commensurate rewards, cueing the individual that the current behavior is rewarding and should be continued.

A third network—the *salience network*—serves as a bridge between the task-positive and the default mode network (Menon, 2011, 2015; Seeley et al., 2007). The salience network serves to detect important stimuli in the internal or external environment and direct attention toward those stimuli as a function of their sensory salience and motivational priority. The primary hubs of this network are the anterior insula (AI; Menon & Uddin, 2010) and dorsal portions of the anterior cingulate cortex (dACC). Activation in the salience network is often coupled with activity in the dopaminergic midbrain, including the nucleus accumbens (NAc), the caudate nucleus, and the putamen (Pessoa, 2009). These brain regions have long been known to be involved in the processing of rewards and threats, and are of critical importance for network models of emotion and motivation (Pessoa, 2017). The salience network is proposed to undergird the initiation of the flow experience, serving as the inflection point between non-flow and flow experiences. Drawing on the metaphor of car driving in previous work (Weber, Sherry, & Mathiak, 2008), we envision the salience network serving as a clutch that transitions an idling car (default mode) into a driving car (executive processing) at system-critical time-points. System criticality is determined by the value of external stimuli and once reached leads to a sudden change of the system (from idle to driving). Huskey et al. (2018) provide preliminary evidence for the clutch model of flow experiences.

Theoretical Advance: Individual Differences in Capacity and Control

In STF, flow experiences are proposed to be reliant on synchronized activity between cognitive control networks and reward networks in the brain. The reward network consists of a web of connections between dopaminergic regions of the midbrain and the cortex. These connections transmit dopamine both tonically (over a long period of time) and phasically (over shorter periods of time; Bromberg-Martin, Matsumoto, & Hikosaka, 2010; Grace, Floresco, Goto, & Lodge, 2007). Tonic dopamine release serves to maintain homeostasis of dopamine levels in cortical regions, whereas phasic dopamine release results in sharp increases/decreases in dopamine levels that influence specific behaviors (Schultz, 1998, 2007; Schultz, Dayan, & Montague, 1997). These dopamine afferents have been shown to exhibit substantial individual differences in both tonic levels of activation and responsivity to motivational stimuli, and are implicated in cognitive disorders such as ADHD (LaHoste et al., 1996; Swanson et al., 2007). Since the original publication of STF, research in this area has progressed rapidly, allowing for more informed predictions as to how individual differences in flow proneness and in the robustness of flow states are contingent on tonic and phasic activation in dopaminergic circuits.

Tonic levels of dopamine in the brain are regulated by spontaneously firing neurons in the midbrain (Grace et al., 2007). These neurons act like a pacemaker, firing at regular intervals regardless of external or internal stimuli in order to maintain homeostasis of dopamine levels and to enact changes in dopamine concentrations at relatively long time periods (seconds to minutes). In a recent model it has been proposed that tonic changes in dopamine levels in the midbrain serve to "prepare" a set of responses to the environment that would likely result in reward based on sensory inputs and on memory. This creates a limited workspace wherein more fine-grained actions can be chosen and enacted (Keeler, Pretsell, & Robbins, 2014). Thus, individual differences in tonic dopamine levels are likely associated with proneness to certain behavioral patterns such as flow. Support for this idea has been found in one study showing that flow proneness is correlated with the availability of dopamine D_2 receptors in the ventral portion of the striatum (de Manzano et al., 2013).

In contrast, phasic dopamine responses are mediated by rapid bursts of neuronal firing regulated by inputs from the hippocampus, the frontal regions, and other areas (Grace et al., 2007). These bursts occur in response to motivationally relevant stimuli and release a large amount of dopamine in a short time (Berridge, 2000; Dayan & Balleine, 2002; Schultz et al., 1997). Phasic dopamine responses bias the action selection process by interacting with the task-positive, default mode, and salience networks to guide the processing system toward actions that are predicted to be the most rewarding (Braver et al., 2014; Westbrook & Frank, 2018). In this framework, phasic changes in dopamine levels help the brain "select" appropriate responses from the set prepared by tonic activity.

Incorporating a more thorough account of the reward processes that undergird action preparation and selection in the brain allows for several additional falsifiable predictions regarding individual differences in flow proneness and the robustness of flow in the presence of salient distractors. First, flow proneness should be associated with firing rates of dopaminergic neurons in the ventral tegmental area (VTA) of the midbrain as well as receptor availability. Emerging support for this prediction has already been identified (de Manzano et al., 2013). Second, those with known deficits in dopaminergic pathways (such as in ADHD) should evidence higher variance in the task parameters that induce flow and should have less efficient network activation during flow. Those with ADHD have been shown to have abnormally low dopamine receptor availability in the VTA, as well as genetic mutations that produce underactivity in cortico-midbrain dopamine pathways (Ilgin, Senol, Gucuyener, Gokcora, & Sener, 2001; LaHoste et al., 1996). Those with ADHD are also known to exhibit high variance in attention behaviors, ranging from extreme inability to focus attention to a hyper-focused state (wherein ADHD individuals perform a repetitive task without interruption for hours on end; Bush, 2010).

Methodological Advance: Dynamic Brain Networks

Another methodological update that is changing the face of STF is the development and rapid expansion of network neuroscience, especially methods that use dynamic and multilayer networks (Bassett & Sporns, 2017; Sporns, 2014). The hypotheses of STF are undeniably network hypotheses. Testing the primary hypotheses of STF involves measuring the synchronization between brain regions and networks, measuring metabolic efficiency of active brain networks, and predicting behavioral outcomes using network models. These hypotheses could be tested in a rudimentary fashion using methods available at the time, including psychophysiological interaction analysis (PPI; Friston et al., 1997), and GLM-based brain imaging methods, but rigorous testing of these hypotheses required development of more refined computational and statistical techniques.

The field of network neuroscience has developed rapidly since the publication of STF, especially in the creation and application of tools and models that can be used in naturalistic environments, such as video games and media viewing tasks (Vanderwal et al., 2017; Weber, Alicea, Huskey, & Mathiak, 2018). The most widespread of these approaches utilizes a parcellation atlas to divide the brain into a certain number of regions (e.g., Power et al., 2011). Edges between nodes are typically defined as the average correlation between activity patterns of each node over time (Fornito, Zalesky, & Bullmore, 2016). Dynamic approaches consider edge weights as the correlation or coherence between nodes within a sliding window, allowing for comparisons of network characteristics (such as average path length, node degree, or edge weight) over time during a task (Zalesky, Fornito, Cocchi, Gollo, & Breakspear, 2014).

In the tradition of method-theory synergy in communication neuroscience (Weber, Fisher, Hopp, & Lonergan, 2018), these methodological developments have allowed for testing existing hypotheses in new ways, but also for the development of more refined predictions. To this end, an especially salient metric for flow researchers is that of *network allegiance*—the amount of times a node within a network changes module membership over time (Bassett, Wymbs, & Porter, 2011; Cooper et al., 2018; Khambhati et al., 2018). As highly modular brain networks are metabolically efficient (Bullmore & Sporns, 2012), one would expect that during flow, nodes in the task-positive network would exhibit high connection with one another, but also high allegiance within the network. The allegiance of nodes within the task-positive network during flow states should be compromised in those with cognitive disorders such as ADHD and should dynamically predict variables of interest such as enjoyment and performance.

Methodological Advance: Social Demand Tasks vs. Visuo-Motor Demand Tasks

In the original conceptualization of flow theory, Csikszentmihalyi (Csikszentmihalyi, 1990) described flow as an almost completely domain-general concept. Flow could happen in essentially any task so long as it contained the right balance of difficulty and reward. In most flow research, investigation of flow states has been limited to a narrow range of cognitive and visuomotor tasks such as solving mathematical problems or playing puzzle games. The neural substrates and cognitive outcomes of flow in these sorts of tasks have been well explicated (Harris et al., 2017; Huskey, Craighead et al., 2018; Huskey, Wilcox et al., 2018; Ulrich et al., 2013, 2016). Less is known, though, about how social demands and rewards may contribute to flow experiences. The social aspect of many media experiences is frequently posited to be one of the primary motivations for using media (Sherry, 2004) and social rewards have been shown to be a primary motivator guiding choices to play particular video games (Weber & Shaw, 2009).

Recent work has developed a framework to understand how the social demands and rewards present in video games and other tasks can be incorporated into STF (Kryston, Novotny, Schmälzle, & Tamborini, 2018). In this framework, tasks that incorporate a social element—such as collaborating with a team or competing against another individual—have different sets of demands/rewards than do simple visuomotor or cognitive tasks. This could lead to differences in the precursors or outcomes of flow in social vs. nonsocial tasks. Behavioral work has shown that teams that had higher interaction and higher levels of flow outperformed teams that interacted less or had lower levels of flow (Aubé, Brunelle, & Rousseau, 2014). It has been proposed that social interaction may catalyze the neural synchrony that undergirds flow experiences (Harris et al., 2017), but this assertion has not yet been tested.

Cooperative success in social tasks requires dynamically adapting one's behavior based on the behaviors and perceived intentions of others (Dale, Fusaroli, Duran, & Richardson, 2013). As in neural systems, evidence suggests that increased synchrony of actors within social systems serves to increase the efficiency and effectiveness of the system as a whole (Fusaroli & Tylén, 2016; Miles, Nind, & Macrae, 2009) and to increase cooperative performance on specific tasks. Interpersonal synchrony also has social rewards. Interaction partners who show increased gestural, lexical, or neural synchrony show increased positive affect and bonding (Koehne, Hatri, Cacioppo, & Dziobek, 2016; Wheatley, Kang, Parkinson, & Looser, 2012), and are more likely to become friends (Parkinson, Kleinbaum, & Wheatley, 2018).

Methodological developments in brain imaging and in network science have allowed for a proliferation of recent studies investigating synchrony both within and between brains (Hasson, Ghazanfar, Galantucci, Garrod, & Keysers, 2012). This emerging field at the intersection of brain networks and social networks provides key tools for scaffolding predictions regarding flow in social environments (Falk & Bassett, 2017). Rather than considering one brain network in isolation, the brains of those co-participating in a task (e.g., playing a video game) are considered as a four-dimensional hypergraph. Edge weights are calculated at each time point for each node within one brain, but also for the corresponding nodes in other brains (see Schmälzle et al., 2017). This allows for calculations of synchrony both within and between brains. Emerging evidence has shown that more effective media messages elicit higher synchrony between individuals (Schmälzle, Häcker, Honey, & Hasson, 2015) and that synchrony between individual brains during media viewing can predict shared memories during a recall task (Chen et al., 2017).

With these things in mind, the three core hypotheses of STF can be outlined in view of social tasks. First, flow within cooperative partners or teams should be reflected as synchronization within and between task-positive and reward networks in each individual brain, but also as increased synchronization *across* brains. This should be observable at the neural level (e.g., as increased average coherence between the brain networks of interactants) but also at the behavioral level (e.g., as increased reciprocity, gestural entrainment, or lexical similarity; see Fusaroli, Konvalinka, & Wallot, 2014). Second, this flow state should be energetically optimized within each brain, and should also be observable at the behavioral level using measures of communication efficiency (as a ratio of words used or time spent communicating per unit of task performance; see Fay, Arbib, & Garrod, 2013). Finally, this synchronized state within and between brains should be associated with cognitive, affective, and social outcomes of interest, such as higher performance, more positive affect, and a greater desire to perform the same sorts of tasks with the same sorts of individuals in the future.

APPLYING SYNC THEORY

The concept of flow has been widely utilized to illuminate the antecedents and consequences of flow experiences in a variety of contexts, including education (Shernoff, Csikszentmihalyi, Schneider, & Shernoff, 2014), business (Ghani & Deshpande, 1994; Salanova et al., 2006), art (MacDonald et al., 2006), athletics (Harris et al., 2017; Jackson et al., 1998; Stein et al., 1995), and many others. The clear physiological operationalization of flow employed in STF provides further specification as to how, when, and why the human brain enters these discrete periods of efficient, synchronized, goal-focused activity. This makes STF uniquely amenable to understanding flow states, but also for contributing to practical and humanitarian outcomes outside of the "ivory tower." In this section we will review three potentially fruitful applications of STF: (1) designing media that are challenging, engaging, and enjoyable; (2) creating innovative treatments for cognitive processing disorders, such as ADHD; and (3) understanding how and why teams are successful in fast-paced cooperative tasks.

First, STF can be used to design media that are engaging and enjoyable while still being challenging. Media that induce efficient synchronization of cognitive control and reward networks can be said to be more flow-inducing than media that do not elicit this physiological response pattern. Thus, physiological responses can be used to index flow states in situations when self-report measures may be undesirable due to their intrusiveness (as in a video game or other interactive media) or their lack of reliability. Media designers can use STF to construct pilot tests of films, video games, or other stimuli to understand which structural or content features elicit flow states reliably and which do not. This is especially salient for those designing educational multimedia (such as learning games or educational television shows) or public service announcements. These efforts are informed by the *brain-as-predictor* approach (Berkman & Falk, 2013), wherein brain activation patterns in a small group of individuals are used to predict real-world outcomes.

Second, STF can be used to explore and develop novel treatments for cognitive and behavioral disorders, such as ADHD. Brain imaging approaches have provided notable insights into individual differences in the structure and function of human brain networks in health and disease (Braun, Muldoon, & Bassett, 2015). These insights allow for the development of brain-imaging-based approaches to cognitive and behavioral change that are personalized to the brains of those undergoing treatment (Gabrieli, Ghosh, & Whitfield-Gabrieli, 2015), and provide a promising way forward for remedying the notable ineffectiveness of "brain training" games (Lindenberger, Wenger, & Lövdén, 2017). For example, flow states are characterized by high levels of synchronization, efficiency, and network allegiance in cognitive control networks. This synchronization is robust to distraction in a curvilinear fashion (Weber, Alicea et al., 2018). Current work in our lab investigates how structural features of video games (e.g., perceptual load) can be used to elicit flow (and thus increased performance outcomes) in those with ADHD (Fisher, Hopp, & Weber, 2018). Individual differences in the robustness of this attentional network in the face of distraction can be used to inform and adapt treatments aimed at ameliorating symptoms of distractibility and impulsiveness observed in ADHD.

Finally, STF—with the social components outlined in this manuscript—can be used to predict and explain performance differences in pairs and small teams. If flow in team environments is characterized by a state of increased synchronization both within and between brains, this increased synchrony can be used to predict performance on complex tasks. Brain imaging and electrophysiological approaches are more cost-effective and widely available than in the past. Technologies such as functional near-infrared spectroscopy (fNIRS; Ayaz et al., 2013) and electroencephalography (EEG; Cohen, 2017) allow for measurement of brain activity outside of an MRI scanner, allowing for measurement of interactive dynamics in a much more flexible fashion. These measures have also been shown to be useful for measurement of interpersonal synchrony (Cacioppo et al., 2014; Liu et al., 2017). Thus, STF would predict that optimal team performance would be associated with high synchrony both between and within brains.

CONCLUSION

By defining the neurophysiological substrates of flow experiences, STF moves beyond descriptive research and provides a way forward to investigate the "why" of flow experiences. In this manuscript, we have reviewed an upwelling of recent support for its core hypotheses driven by advances in network science and computational methodologies. We have also outlined five ways in which STF can be pushed forward into the next ten years of research: (1) redefinition of the core networks of cognitive control and reward; (2) incorporating an increased understanding of cognitive and neural individual differences; (3) updating core methodologies in light of advances in network modeling and dynamic approaches; (4) including a proper conceptualization and operationalization of the social components of flow experiences; and (5) outlining a path toward conducting sync theory research on a broad scale.

We also included a selection of applications of STF in the non-academic sphere: informing the design of challenging and engaging media messages, contributing to the development of more effective cognitive and behavioral treatments and training applications, and increasing practical understanding of how and why effective teams synchronize with one another to complete complex tasks. These methodological, theoretical, and practical advances improve the predictive and explanatory power of STF and pave the way for future decades of research into the neural underpinnings of flow experiences.

NOTE

1. https://github.com/medianeuroscience/asteroid_impact.

REFERENCES

Achard, S., & Bullmore, E. (2007). Efficiency and cost of economical brain functional networks. *PLoS Computational Biology, 3*(2), e17.

Arenas, A., Díaz-Guilera, A., Kurths, J., Moreno, Y., & Zhou, C. (2008). Synchronization in complex networks. *Physics Reports, 469*(3), 93–153.

Aron, A. R., Robbins, T. W., & Poldrack, R. A. (2004). Inhibition and the right inferior frontal cortex. *Trends in Cognitive Sciences, 8*(4), 170–177.

Aron, A. R., Robbins, T. W., & Poldrack, R. A. (2014). Inhibition and the right inferior frontal cortex: One decade on. *Trends in Cognitive Sciences, 18*(4), 177–185.

Ashby, G. F., & Helie, S. (2011). A tutorial on computational cognitive neuroscience: Modeling the neurodynamics of cognition. *Journal of Mathematical Psychology, 55*(4), 273–289.

Aston-Jones, G., & Cohen, J. D. (2005). An integrative theory of locus coeruleus-norepinephrine function: Adaptive gain and optimal performance. *Annual Review of Neuroscience, 28*, 403–450.

Aubé, C., Brunelle, E., & Rousseau, V. (2014). Flow experience and team performance: The role of team goal commitment and information exchange. *Motivation and Emotion, 38*(1), 120–130.

Ayaz, H., Onaral, B., Izzetoglu, K., Shewokis, P. A., McKendrick, R., & Parasuraman, R. (2013). Continuous monitoring of brain dynamics with functional near infrared spectroscopy as a tool for neuroergonomic research: Empirical examples and a technological development. *Frontiers in Human Neuroscience, 7*, 871.

Baddeley, A. D. (1992). Working memory. *Science, 255*(5044), 556–559.

Baddeley, A. D., & Hitch, G. J. (1974). Working memory. In G. H. Bower (Ed.), *Psychology of learning motivation* (pp. 48–79). New York, NY: Academic Press.

Başar, E. (2012). *Brain function and oscillations*: Vol II: *Integrative brain function: Neurophysiology and Cognitive Processes*. Berlin, Germany: Springer Science & Business Media.

Bassett, D. S., Bullmore, E. T., Meyer-Lindenberg, A., Apud, J. A., Weinberger, D. R., & Coppola, R. (2009). Cognitive fitness of cost-efficient brain functional networks. *Proceedings of the National Academy of Sciences of the U.S.A., 106*(28), 11747–11752.

Bassett, D. S., & Gazzaniga, M. S. (2011). Understanding complexity in the human brain. *Trends in Cognitive Sciences, 15*(5), 200–209.

Bassett, D. S., & Sporns, O. (2017). Network neuroscience. *Nature Neuroscience, 20*(3), 353–364.

Bassett, D. S., Wymbs, N. F., & Porter, M. A. (2011). Dynamic reconfiguration of human brain networks during learning. *Proceedings of the National Academy of Sciences, 108*, 7641–7646.

Beggs, J. M. (2008). The criticality hypothesis: How local cortical networks might optimize information processing. *Philosophical Transactions. Series A, Mathematical, Physical, and Engineering Sciences, 366*(1864), 329–343.

Berkman, E. T., & Falk, E. B. (2013). Beyond brain mapping: Using neural measures to predict real-world outcomes. *Current Directions in Psychological Science, 22*(1), 45–50.

Berridge, K. C. (2000). Reward learning: Reinforcement, incentives, and expectations. *Psychology of Learning and Motivation, 40*, 223–278.

Betzel, R. F., & Bassett, D. S. (2017). Multi-scale brain networks. *NeuroImage, 160*, 73–83.

Bisley, J. W., & Goldberg, M. E. (2010). Attention, intention, and priority in the parietal lobe. *Annual Review of Neuroscience, 33*, 1–21.

Botvinick, M. M., & Braver, T. S. (2015). Motivation and cognitive control: From behavior to neural mechanism. *Annual Review of Psychology, 66*, 82–113.

Botvinick, M. M., Braver, T. S., Barch, D. M., Carter, C. S., & Cohen, J. D. (2001). Conflict monitoring and cognitive control. *Psychological Review, 108*(3), 624–652.

Braun, U., Muldoon, S., & Bassett, D. (2015). On human brain networks in health and disease. In *eLS* (Vol. 26, pp. 1–9). Chichester, England: John Wiley & Sons, Ltd.

Braver, T. S., Gray, J. R., & Burgess, G. C. (2007). Explaining the many varieties of working memory variation: Dual mechanisms of cognitive control. In A. R. Conway, C. Jarrold, M. J. Kane, A. Miyake, & J. N. Towse (Eds.), *Variation in working memory* (pp. 76–106). New York, NY: Oxford University Press.

Braver, T. S., Krug, M. K., Chiew, K. S., Kool, W., Westbrook, J. A., Clement, N. J., ... Somerville, L. H. (2014). Mechanisms of motivation-cognition interaction: Challenges and opportunities. *Cognitive, Affective & Behavioral Neuroscience, 14*(2), 443–472.

Breakspear, M., Heitmann, S., & Daffertshofer, A. (2010). Generative models of cortical oscillations: Neurobiological implications of the Kuramoto model. *Frontiers in Human Neuroscience, 4*, 190.

Bromberg-Martin, E. S., Matsumoto, M., & Hikosaka, O. (2010). Dopamine in motivational control: Rewarding, aversive, and alerting. *Neuron, 68*(5), 815–834.

Bullmore, E., & Sporns, O. (2012). The economy of brain network organization. *Nature Reviews Neuroscience, 13*(5), 336–349.

Buschman, T. J., & Kastner, S. (2015). From behavior to neural dynamics: An integrated theory of attention. *Neuron, 88*(1), 127–144.

Buschman, T. J., & Miller, E. K. (2007). Top-down versus bottom-up control of attention in the prefrontal and posterior parietal cortices. *Science, 315*(5820), 1860–1862.

Bush, G. (2010). Attention-deficit/hyperactivity disorder and attention networks. *Neuropsychopharmacology, 35*(1), 278–300.

Buzsáki, G. (1991). The thalamic clock: Emergent network properties. *Neuroscience, 41*(2–3), 351–364.

Buzsáki, G. (2006). *Rhythms of the brain*. Oxford, England: Oxford University Press.

Buzsáki, G., Logothetis, N., & Singer, W. (2013). Scaling brain size, keeping timing: Evolutionary preservation of brain rhythms. *Neuron, 80*(3), 751–764.

Cacioppo, S., Zhou, H., Monteleone, G., Majka, E. A., Quinn, K. A., Ball, A. B., ... Cacioppo, J. T. (2014). You are in sync with me: Neural correlates of interpersonal synchrony with a partner. *Neuroscience, 277*, 842–858.

Cavanagh, J. F., Cohen, M. X., & Allen, J. J. B. (2009). Prelude to and resolution of an error: EEG phase synchrony reveals cognitive control dynamics during action monitoring. *The Journal of Neuroscience, 29*(1), 98–105.

Cavanagh, J. F., Zambrano-Vazquez, L., & Allen, J. J. B. (2012). Theta lingua franca: A common midfrontal substrate for action monitoring processes. *Psychophysiology, 49*(2), 220–238.

Cavanna, A. E., & Trimble, M. R. (2006). The precuneus: A review of its functional anatomy and behavioural correlates. *Brain, 129*(Pt 3), 564–583.

Chen, J., Leong, Y. C., Honey, C. J., Yong, C. H., Norman, K. A., & Hasson, U. (2017). Shared memories reveal shared structure in neural activity across individuals. *Nature Neuroscience, 20*(1), 115–125.

Clayton, M. S., Yeung, N., & Cohen Kadosh, R. (2015). The roles of cortical oscillations in sustained attention. *Trends in Cognitive Sciences, 19*(4), 188–195.

Cohen, M. X. (2017). Where does EEG come from and what does it mean? *Trends in Neurosciences, 40*(4), 208–218.

Cooper, N., Garcia, J. O., Tompson, S., O'Donnell, M. B., Falk, E. B., & Vettel, J. M. (2018). Time-evolving dynamics in brain networks forecast responses to health messaging. *Network Neuroscience, 3*, 1–40.

Corbetta, M., & Shulman, G. L. (2002). Control of goal-directed and stimulus-driven attention in the brain. *Nature Reviews Neuroscience, 3*(3), 201–215.

Crick, F., & Koch, C. (1990). Towards a neurobiological theory of consciousness. *Seminars in the Neurosciences, 2*, 263–275.

Csikszentmihalyi, M. (1990). *Flow: The psychology of optimal performance*. New York, NY: Harper & Row.

Dale, R., Fusaroli, R., Duran, N. D., & Richardson, D. C. (2013). The self-organization of human interaction. In B. H. Ross (Ed.), *Psychology of learning and motivation* (Vol. 59, pp. 43–95). Cambridge, MA: Academic Press.

Davidson, M. C., & Marrocco, R. T. (2000). Local infusion of scopolamine into intraparietal cortex slows covert orienting in rhesus monkeys. *Journal of Neurophysiology, 83*(3), 1536–1549.
Dayan, P., & Balleine, B. W. (2002). Reward, motivation, and reinforcement learning. *Neuron, 36*(2), 285–298.
de Manzano, Ö., Cervenka, S., Jucaite, A., Hellenäs, O., Farde, L., & Ullén, F. (2013). Individual differences in the proneness to have flow experiences are linked to dopamine D_2-receptor availability in the dorsal striatum. *NeuroImage, 67*, 1–6.
Dietrich, A. (2004). Neurocognitive mechanisms underlying the experience of flow. *Consciousness and Cognition, 13*(4), 746–761.
Dolcos, F., Wang, L., & Mather, M. (2015). Current research and emerging directions in emotion-cognition interactions. *Frontiers in Integrative Neuroscience, 1*(83), 1–11.
Falk, E. B., & Bassett, D. S. (2017). Brain and social networks: Fundamental building blocks of human experience. *Trends in Cognitive Sciences, 21*(9), 674–690.
Fay, N., Arbib, M., & Garrod, S. (2013). How to bootstrap a human communication system. *Cognitive Science, 37*(7), 1356–1367.
Fiebelkorn, I. C., Saalmann, Y. B., & Kastner, S. (2013). Rhythmic sampling within and between objects despite sustained attention at a cued location. *Current Biology, 23*(24), 2553–2558.
Fisher, J. T., Hopp, F., & Weber, R. (2018, May). "Look harder!" ADHD, media multitasking, and the effects of cognitive and perceptual load on resource availability and processing performance. Paper presented at the 68th annual conference of the International Communication Association, Prague, the Czech Republic.
Fisher, J. T., Huskey, R., Keene, J. R., & Weber, R. (2018). The limited capacity model of motivated mediated message processing: Looking to the future. *Annals of the International Communication Association, 42*(4), 291–315.
Fisher, J. T., Keene, J. R., Huskey, R., & Weber, R. (2018). The limited capacity model of motivated mediated message processing: Taking stock of the past. *Annals of the International Communication Association, 42*(4), 270–290.
Floyd, K. (2014). Humans are people, too: Nurturing an appreciation for nature in communication research. *Review of Communication Research, 2*, 1–29.
Fornito, A., Zalesky, A., & Bullmore, E. (2016). *Fundamentals of brain network analysis*. New York, NY: Academic Press.
Francuz, P., & Zabielska-Mendyk, E. (2013). Does the brain differentiate between related and unrelated cuts when processing audiovisual messages? An ERP study. *Media Psychology, 16*(4), 461–475.
Friston, K. J., Buechel, C., Fink, G. R., Morris, J., Rolls, E., & Dolan, R. J. (1997). Psychophysiological and modulatory interactions in neuroimaging. *NeuroImage, 6*(3), 218–229.
Fusaroli, R., Konvalinka, I., & Wallot, S. (2014). Analyzing social interactions: The promises and challenges of using cross recurrence quantification analysis. In N. Marwan, M. Riley, & A. Giuliani (Eds.), *Translational recurrences* (pp. 137–155). Cham, Switzerland: Springer International.
Fusaroli, R., & Tylén, K. (2016). Investigating conversational dynamics: Interactive alignment, Interpersonal synergy, and collective task performance. *Cognitive Science, 40*(1), 145–171.
Gabrieli, J. D. E., Ghosh, S. S., & Whitfield-Gabrieli, S. (2015). Prediction as a humanitarian and pragmatic contribution from human cognitive neuroscience. *Neuron, 85*(1), 11–26.
Ghani, J. A., & Deshpande, S. P. (1994). Task characteristics and the experience of optimal flow in human-computer interaction. *The Journal of Psychology, 128*(4), 381–391.
Gießing, C., Thiel, C. M., Alexander-Bloch, A. F., Patel, A. X., & Bullmore, E. T. (2013). Human brain functional network changes associated with enhanced and impaired attentional task performance. *The Journal of Neuroscience, 33*(14), 5903–5914.
Grace, A. A., Floresco, S. B., Goto, Y., & Lodge, D. J. (2007). Regulation of firing of dopaminergic neurons and control of goal-directed behaviors. *Trends in Neurosciences, 30*(5), 220–227.
Greicius, M. D., Krasnow, B., Reiss, A. L., & Menon, V. (2003). Functional connectivity in the resting brain: A network analysis of the default mode hypothesis. *Proceedings of the National Academy of Sciences, 100*(1), 253–258.

Hallberg, U. E., & Schaufeli, W. B. (2006). "Same same" but different? Can work engagement be discriminated from job involvement and organizational commitment? *European Psychologist, 11*(2), 119–127.
Harris, D. J., Vine, S. J., & Wilson, M. R. (2017). Neurocognitive mechanisms of the flow state. In M. R. Wilson, V. Walsh, & B. Parkin (Eds.), *Progress in brain research* (Vol. 234, pp. 221–243). Cambridge, MA: Academic Press.
Hasson, U., Ghazanfar, A. A., Galantucci, B., Garrod, S., & Keysers, C. (2012). Brain-to-brain coupling: A mechanism for creating and sharing a social world. *Trends in Cognitive Sciences, 16*(2), 114–121.
Huskey, R., Craighead, B., Miller, M. B., & Weber, R. (2018). Does intrinsic reward motivate cognitive control? A naturalistic-fMRI study based on the synchronization theory of flow. *Cognitive, Affective & Behavioral Neuroscience, 18*(5), 902–924.
Huskey, R., Wilcox, S., & Weber, R. (2018). Network neuroscience reveals distinct neuromarkers of flow during media use. *The Journal of Communication, 68*(5), 872–895.
Ilgin, N., Senol, S., Gucuyener, K., Gokcora, N., & Sener, S. (2001). Is increased D_2 receptor availability associated with response to stimulant medication in ADHD? *Developmental Medicine and Child Neurology, 43*(11), 755–760.
Jackson, S. A., Ford, S. K., Kimiecik, J. C., & Marsh, H. W. (1998). Psychological correlates of flow in sport. *Journal of Sport and Exercise Psychology, 20*(4), 358–378.
Kanwisher, N. (2010). Functional specificity in the human brain: A window into the functional architecture of the mind. *Proceedings of the National Academy of Sciences, 107*(25), 11163–11170.
Keeler, J. F., Pretsell, D. O., & Robbins, T. W. (2014). Functional implications of dopamine D_1 vs. D_2 receptors: A "prepare and select" model of the striatal direct vs. indirect pathways. *Neuroscience, 282*, 156–175.
Keller, J., & Bless, H. (2008). Flow and regulatory compatibility: An experimental approach to the flow model of intrinsic motivation. *Personality & Social Psychology Bulletin, 34*(2), 196–209.
Khambhati, A. N., Mattar, M. G., Wymbs, N. F., Grafton, S. T., & Bassett, D. S. (2018). Beyond modularity: Fine-scale mechanisms and rules for brain network reconfiguration. *NeuroImage, 166*, 385–399.
Klasen, M., Weber, R., Kircher, T. T. J., Mathiak, K. A., & Mathiak, K. (2012). Neural contributions to flow experience during video game playing. *Social Cognitive and Affective Neuroscience, 7*(4), 485–495.
Koehne, S., Hatri, A., Cacioppo, J. T., & Dziobek, I. (2016). Perceived interpersonal synchrony increases empathy: Insights from autism spectrum disorder. *Cognition, 146*, 8–15.
Kryston, K., Novotny, E., Schmälzle, R., & Tamborini, R. (2018). Social demand in video games and the synchronization theory of flow. In N. D. Bowman (Ed.), *Video games: A medium that demands our attention* (pp. 161–177). New York, NY: Routledge.
LaHoste, G. J., Swanson, J. M., Wigal, S. B., Glabe, C., Wigal, T., King, N., & Kennedy, J. L. (1996). Dopamine D4 receptor gene polymorphism is associated with attention deficit hyperactivity disorder. *Molecular Psychiatry, 1*(2), 121–124.
Lakatos, P., Karmos, G., Mehta, A. D., Ulbert, I., & Schroeder, C. E. (2008). Entrainment of neuronal oscillations as a mechanism of attentional selection. *Science, 320*(5872), 110–113.
Landau, A. N., & Fries, P. (2012). Attention samples stimuli rhythmically. *Current Biology, 22*(11), 1000–1004.
Lang, A. (2000). The limited capacity model of mediated message processing. *Journal of Communication, 50*(1), 46–70.
Lang, A. (2009). The limited capacity model of motivated mediated message processing. In R. Nabi & M. B. Oliver (Eds.), *The Sage handbook of media processes and effects* (pp. 193–204). Thousand Oaks, CA: Sage.
Lang, A. (2013). Discipline in crisis? The shifting paradigm of mass communication research. *Communication Theory, 23*(1), 10–24.
Lang, A. (2014). Dynamic human-centered communication systems theory. *The Information Society, 30*(1), 60–70.
Lang, A., Bradley, S. D., Park, B., Shin, M., & Chung, Y. (2006). Parsing the resource pie: Using STRTs to measure attention to mediated messages. *Media Psychology, 8*(4), 369–394.

Lang, A., & Ewoldsen, D. (2010). Beyond effects: Conceptualizing communication as dynamic, complex, nonlinear, and fundamental. In S. Allen (Ed.), *Rethinking communication: Keywords in communication research* (pp. 109–120). New York, NY: Hampton Press.

Langer, N., Pedroni, A., Gianotti, L. R. R., Hänggi, J., Knoch, D., & Jäncke, L. (2012). Functional brain network efficiency predicts intelligence. *Human Brain Mapping, 33*(6), 1393–1406.

Laufs, H., Kleinschmidt, A., Beyerle, A., Eger, E., Salek-Haddadi, A., Preibisch, C., & Krakow, K. (2003). EEG-correlated fMRI of human alpha activity. *NeuroImage, 19*(4), 1463–1476.

Lee, T., Itti, L., & Mather, M. (2012). Evidence for arousal-biased competition in perceptual learning. *Frontiers in Psychology, 3*(241), 1–9.

Lindenberger, U., Wenger, E., & Lövdén, M. (2017). Towards a stronger science of human plasticity. *Nature Reviews Neuroscience, 18*(5), 261–262.

Liu, Y., Piazza, E. A., Simony, E., Shewokis, P. A., Onaral, B., Hasson, U., & Ayaz, H. (2017). Measuring speaker-listener neural coupling with functional near infrared spectroscopy. *Scientific Reports, 7*, 43293.

MacDonald, A. W., Cohen, J. D., Stenger, V. A., & Carter, C. S. (2000). Dissociating the role of the dorsolateral prefrontal and anterior cingulate cortex in cognitive control. *Science, 288*(5472), 1835–1838.

MacDonald, R., Byrne, C., & Carlton, L. (2006). Creativity and flow in musical composition: An empirical investigation. *Psychology of Music, 34*(3), 292–306.

Mather, M., & Sutherland, M. R. (2011). Arousal-biased competition in perception and memory. *Perspectives on Psychological Science, 6*(2), 114–133.

Medaglia, J. D., Lynall, M.-E., & Bassett, D. S. (2015). Cognitive network neuroscience. *Journal of Cognitive Neuroscience, 27*(8), 1471–1491.

Menon, V. (2011). Large-scale brain networks and psychopathology: A unifying triple network model. *Trends in Cognitive Sciences, 15*(10), 483–506.

Menon, V. (2015). Salience network. In A. Toga (Ed.), *Brain mapping: An encyclopedic reference* (pp. 597–611). Waltham, MA: Academic Press.

Menon, V., & Uddin, L. Q. (2010). Saliency, switching, attention and control: A network model of insula function. *Brain Structure & Function, 214*(5–6), 655–667.

Miles, L. K., Nind, L. K., & Macrae, C. N. (2009). The rhythm of rapport: Interpersonal synchrony and social perception. *Journal of Experimental Social Psychology, 45*(3), 585–589.

Miller, E. K., & Cohen, J. D. (2001). An integrative theory of prefrontal cortex function. *Annual Review of Neuroscience, 24*, 167–202.

Nakamura, J., & Csikszentmihalyi, M. (2014). The concept of flow. In M. Csikszentmihalyi (Ed.), *Flow and the foundations of positive psychology: The collected works of Mihaly Csikszentmihalyi* (pp. 239–263). Dordrecht, The Netherlands: Springer.

Nassi, J. J., & Callaway, E. M. (2009). Parallel processing strategies of the primate visual system. *Nature Reviews Neuroscience, 10*(5), 360–372.

Núñez Castellar, E. P., Antons, J.-N., Marinazzo, D., & Van Looy, J. (2019). Mapping attention during gameplay: Assessment of behavioral and ERP markers in an auditory oddball task. *Psychophysiology*, e13347.

Parkinson, C., Kleinbaum, A. M., & Wheatley, T. (2018). Similar neural responses predict friendship. *Nature Communications, 9*(1), 332.

Peifer, C., Schulz, A., Schächinger, H., Baumann, N., & Antoni, C. H. (2014). The relation of flow-experience and physiological arousal under stress — Can u shape it? *Journal of Experimental Social Psychology, 53*, 62–69.

Pessoa, L. (2009). How do emotion and motivation direct executive control? *Trends in Cognitive Sciences, 13*(4), 160–166.

Pessoa, L. (2017). A network model of the emotional brain. *Trends in Cognitive Sciences, 21*(5), 357–371.

Petersen, S. E., & Posner, M. I. (2012). The attention system of the human brain: 20 years after. *Annual Review of Neuroscience, 35*(1), 73–89.

Posner, M. I., & Petersen, S. E. (1990). The attention system of the human brain. *Annual Review of Neuroscience, 13*, 25–32.

Potter, R. F., Lang, A., & Bolls, P. D. (2008). Identifying structural features of audio: Orienting responses during radio messages and their impact on recognition. *Journal of Media Psychology, 20*(4), 168–177.

Power, J. D., Cohen, A. L., Nelson, S. M., Wig, G. S., Barnes, K. A., Church, J. A., ... Petersen, S. E. (2011). Functional network organization of the human brain. *Neuron, 72*(4), 665–678.

Priesemann, V., Wibral, M., Valderrama, M., Pröpper, R., Le Van Quyen, M., Geisel, T., ... Munk, M. H. J. (2014). Spike avalanches in vivo suggest a driven, slightly subcritical brain state. *Frontiers in Systems Neuroscience, 8*, 108.

Raichle, M. E. (2015). The brain's default mode network. *Annual Review of Neuroscience, 38*, 433–447.

Raichle, M. E., MacLeod, A. M., Snyder, A. Z., Powers, W. J., Gusnard, D. A., & Shulman, G. L. (2001). A default mode of brain function. *Proceedings of the National Academy of Sciences, 98*(2), 676–682.

Raz, A., & Buhle, J. (2006). Typologies of attentional networks. *Nature Reviews Neuroscience, 7*(5), 367–379.

Regev, M., Simony, E., Lee, K., Tan, K. M., Chen, J., & Hasson, U. (2018). Propagation of information along the cortical hierarchy as a function of attention while reading and listening to stories. *Cerebral Cortex*. Advance online publication.

Ridderinkhof, K. R., Ullsperger, M., Crone, E. A., & Nieuwenhuis, S. (2004). The role of the medial frontal cortex in cognitive control. *Science, 306*(5695), 443–447.

Salanova, M., Bakker, A. B., & Llorens, S. (2006). Flow at work: Evidence for an upward spiral of personal and organizational resources. *Journal of Happiness Studies, 7*(1), 1–22.

Schmälzle, R., Brook O'Donnell, M., Garcia, J. O., Cascio, C. N., Bayer, J., Bassett, D. S., ... Falk, E. B. (2017). Brain connectivity dynamics during social interaction reflect social network structure. *Proceedings of the National Academy of Sciences, 114*(20), 5153–5158.

Schmälzle, R., Häcker, F. E. K., Honey, C. J., & Hasson, U. (2015). Engaged listeners: Shared neural processing of powerful political speeches. *Social Cognitive and Affective Neuroscience, 10*(8), 1137–1143.

Schultz, W. (1998). Predictive reward signal of dopamine neurons. *Journal of Neurophysiology, 80*(1), 1–27.

Schultz, W. (2007). Multiple dopamine functions at different time courses. *Annual Review of Neuroscience, 30*, 259–288.

Schultz, W., Dayan, P., & Montague, P. R. (1997). A neural substrate of prediction and reward. *Science, 275*(5306), 1593–1599.

Seeley, W. W., Menon, V., Schatzberg, A. F., Keller, J., Glover, G. H., Kenna, H., ... Greicius, M. D. (2007). Dissociable intrinsic connectivity networks for salience processing and executive control. *The Journal of Neuroscience, 27*(9), 2349–2356.

Sestieri, C., Shulman, G. L., & Corbetta, M. (2009). Top-down attention for memory and perception: Segregated networks in the parietal lobe. *NeuroImage, 47*, S43.

Shernoff, D. J., Csikszentmihalyi, M., Schneider, B., & Shernoff, E. S. (2014). Student engagement in high school classrooms from the perspective of flow theory. In M. Csikszentmihalyi (Ed.), *Applications of flow in human development and education* (pp. 475–494). Dordrecht, The Netherlands: Springer.

Sherry, J. L. (2004). Flow and media enjoyment. *Communication Theory, 14*(4), 328–347.

Sherry, J. L. (2015). The complexity paradigm for studying human communication: A summary and integration of two fields. *Review of Communication Research, 3*(1), 22–54.

Singer, W. (2001). Consciousness and the binding problem. *Annals of the New York Academy of Sciences, 929*, 123–146.

Sokolov, E. N. (1963). Higher nervous functions: The orienting reflex. *Annual Review of Physiology, 25*, 545–580.

Sporns, O. (2011). The non-random brain: Efficiency, economy, and complex dynamics. *Frontiers in Computational Neuroscience, 5*, 5.

Sporns, O. (2014). Contributions and challenges for network models in cognitive neuroscience. *Nature Neuroscience, 17*(5), 652–660.

Spreng, R. N., Mar, R. A., & Kim, A. S. N. (2009). The common neural basis of autobiographical memory, prospection, navigation, theory of mind, and the default mode: A quantitative meta-analysis. *Journal of Cognitive Neuroscience, 21*(3), 489–510.

Stein, G. L., Kimiecik, J. C., Daniels, J., & Jackson, S. A. (1995). Psychological antecedents of flow in recreational sport. *Personality & Social Psychology Bulletin, 21*(2), 125–135.

Strogatz, S. (2004). *Sync: The emerging science of spontaneous order*. London, England: Penguin.

Stryker, M. P. (1989). Is grandmother an oscillation? *Nature, 338*(6213), 297–298.

Swanson, J. M., Kinsbourne, M., Nigg, J., Lanphear, B., Stefanatos, G. A., Volkow, N., ... Wadhwa, P. D. (2007). Etiologic subtypes of attention-deficit/hyperactivity disorder: Brain imaging, molecular genetic and environmental factors and the dopamine hypothesis. *Neuropsychology Review, 17*(1), 39–59.

Thorson, E., & Lang, A. (1992). The effects of television videographics and lecture familiarity on adult cardiac orienting responses and memory. *Communication Research, 19*(3), 346–369.

Ulrich, M., Keller, J., & Grön, G. (2016). Neural signatures of experimentally induced flow experiences identified in a typical fMRI block design with BOLD imaging. *Social Cognitive and Affective Neuroscience, 11*(3), 496–507.

Ulrich, M., Keller, J., Hoenig, K., Waller, C., & Grön, G. (2013). Neural correlates of experimentally induced flow experiences. *NeuroImage, 86*(1), 194–202.

Utevsky, A. V., Smith, D. V., & Huettel, S. A. (2014). Precuneus is a functional core of the default-mode network. *The Journal of Neuroscience, 34*(3), 932–940.

Vanderwal, T., Eilbott, J., Finn, E. S., Craddock, R. C., Turnbull, A., & Castellanos, F. X. (2017). Individual differences in functional connectivity during naturalistic viewing conditions. *NeuroImage, 157*, 521–530.

van de Vijver, I., Ridderinkhof, K. R., & Cohen, M. X. (2011). Frontal oscillatory dynamics predict feedback learning and action adjustment. *Journal of Cognitive Neuroscience, 23*(12), 4106–4121.

van Driel, J., Ridderinkhof, K. R., & Cohen, M. X. (2012). Not all errors are alike: Theta and alpha EEG dynamics relate to differences in error-processing dynamics. *The Journal of Neuroscience, 32*(47), 16795–16806.

Ward, L. M. (2003). Synchronous neural oscillations and cognitive processes. *Trends in Cognitive Sciences, 7*(12), 553–559.

Watts, D. J. (2017). Should social science be more solution-oriented? *Nature Human Behaviour, 1*, 0015.

Weber, R., Alicea, B., Huskey, R., & Mathiak, K. (2018). Network dynamics of attention during a naturalistic behavioral paradigm. *Frontiers in Human Neuroscience, 12*, 182.

Weber, R., Fisher, J. T., Hopp, F. R., & Lonergan, C. (2018). Taking messages into the magnet: Method-theory synergy in communication neuroscience. *Communication Monographs, 85*(1), 81–102.

Weber, R., Huskey, R., & Craighead, B. (2016). Flow experiences and well-being: A media neuroscience perspective. In M. B. Oliver & L. Reinecke (Eds.), *Handbook of media use and well-being: International perspectives on theory and research on positive media effects* (pp. 183–196). New York, NY: Routledge.

Weber, R., Huskey, R., & Craighead, B. (2017). Flow experiences and well-being: A media neuroscience perspective. In L. Reinecke & M. B. Oliver (Eds.), *The Routledge handbook of media use and well-being* (pp. 183–196). New York, NY: Routledge.

Weber, R., & Shaw, P. (2009). Player types and quality perceptions: A social cognitive theory based model to predict video game playing. *International Journal of Gaming and Computer-Mediated Simulations, 1*(1), 66–89.

Weber, R., Sherry, J., & Mathiak, K. (2008). The neurophysiological perspective in mass communication research: Theoretical rationale, methods, and applications. In M. J. Beatty, J. C. McCroskey, & K. Floyd (Eds.), *Biological dimensions of communication: Perspectives, methods, and research* (pp. 41–71). Cresskill, NJ: Hampton Press.

Weber, R., Tamborini, R., Westcott-Baker, A., & Kantor, B. (2009). Theorizing flow and media enjoyment as cognitive synchronization of attentional and reward networks. *Communication Theory, 19*(4), 397–422.

Westbrook, A., & Frank, M. (2018). Dopamine and proximity in motivation and cognitive control. *Current Opinion in Behavioral Sciences, 22*, 28–34.

Wheatley, T., Kang, O., Parkinson, C., & Looser, C. E. (2012). From mind perception to mental connection: Synchrony as a mechanism for social understanding: Mind perception and mental connection. *Social and Personality Psychology Compass, 6*(8), 589–606.

Wilting, J., & Priesemann, V. (2018). Inferring collective dynamical states from widely unobserved systems. *Nature Communications, 9*(1), 2325.

Yarkoni, T., & Westfall, J. (2017). Choosing prediction over explanation in psychology: Lessons from machine learning. *Perspectives on Psychological Science, 12*(6), 1100–1122.

Yoshida, K., Sawamura, D., Inagaki, Y., Ogawa, K., Ikoma, K., & Sakai, S. (2014). Brain activity during the flow experience: A functional near-infrared spectroscopy study. *Neuroscience Letters, 573*, 30–34.

Zalesky, A., Fornito, A., Cocchi, L., Gollo, L. L., & Breakspear, M. (2014). Time-resolved resting-state brain networks. *Proceedings of the National Academy of Sciences, 111*(28), 10341–10346.

13

Attention, Working Memory, and Media Multitasking

Jacob T. Fisher and Justin Robert Keene

Multitasking with media has become a nearly ubiquitous phenomenon. Individuals multitask up to 92 percent of the time when consuming certain forms of media (Deloitte, 2015), and switch between screens up to 2.5 times per minute (Brasel & Gips, 2017). In addition, multitasking on a single screen is becoming more common as devices such as smartphones and laptop computers incorporate "picture-in-picture" capabilities and more efficient task-switching mechanisms. Studies of single-screen multitasking have shown that the median time spent on any one task is less than 20 seconds (Yeykelis, Cummings, & Reeves, 2014). Media multitasking is an often-problematized phenomenon, implicated in bringing about shortening attention spans, shallower information processing, and greater mindlessness (see, e.g., Carr, 2011; Gazzaley & Rosen, 2016; Small & Vorgan, 2008). A core thesis in this research is that media multitasking induces neuroplastic changes in the brain that encourage more "breadth-biased" forms of information processing and lead to greater distractability (Lin, 2009; Loh & Kanai, 2016; Ophir, Nass, & Wagner, 2009).

Several potentially concerning findings have emerged in recent years regarding the association between media multitasking and suboptimal cognitive, social, educational, and health outcomes. These include higher rates of internet addiction (Brand, Young, & Laier, 2014), lower grades (Cain, Leonard, Gabrieli, & Finn, 2016; Junco & Cotten, 2012), greater self-reported distractibility (Baumgartner, Weeda, van der Heijden, & Huizinga, 2014; Ralph, Thomson, Cheyne, & Smilek, 2013), more impulsivity (Sanbonmatsu, Strayer, Medeiros-Ward, & Watson, 2013), higher rates of depression (Becker, Alzahabi, & Hopwood, 2013), and reduced social success (Xu, Wang, & David, 2016). One of the best-supported correlations in this domain is between habitual media multitasking and attentional control problems. Numerous studies have found associations between media multitasking and self-reported deficits in working memory, attentional control, and personal achievement (Baumgartner et al., 2014; Bowman, Waite, & Levine, 2015; Jeong & Hwang, 2016; Nikkelen, Valkenburg, Huizinga, & Bushman, 2014; Uncapher et al., 2017; Uncapher, Thieu, & Wagner, 2015). A recently reported longitudinal cross-lagged panel study suggests that media multitasking may even have a causal influence on the development of attentional problems (Baumgartner, van der Schuur, Lemmens, & te Poel, 2018).

Interestingly, despite these relatively consistent associations between media multitasking behaviors and self-reported cognitive deficits, experimental evidence for this effect has been far less clear (Wiradhany & Niewwenstein, 2017). A groundswell of research in recent years has investigated the relationship between media multitasking behaviors as measured by the Media Multitasking Index (MMI; Ophir et al., 2009) and experimental measures of attentional control (Baumgartner et al., 2014; Cain & Mitroff, 2011; Lin, 2009; Moisala et al., 2016), working memory (Cain et al., 2016; Cardoso-Leite et al., 2016; Minear, Brasher, McCurdy, Lewis, & Younggren, 2013; Uncapher et al., 2015), dual-tasking (Alzahabi & Becker, 2013; Minear et al., 2013; Sanbonmatsu et al., 2013), task-switching (Alzahabi & Becker, 2013), and multisensory integration abilities (Lui & Wong, 2012).

These studies have produced mixed findings regarding the purported negative influence of media multitasking. Some have found that media multitaskers perform worse in cognitive tasks (Moisala et al., 2016; Ralph & Smilek, 2017; Sanbonmatsu et al., 2013), whereas others have found no difference in performance between heavy and light media multitaskers (Minear et al., 2013). Some studies report that heavy media multitaskers actually perform *better* than light media multitaskers (e.g., Alzahabi & Becker, 2013; Fisher, Hopp, & Weber, 2018; Lui & Wong, 2012) or even that intermediate media multitaskers outperform both light and heavy multitaskers (Cardoso-Leite et al., 2016). A recent meta-analysis in this research area found that the effect of individual differences in reported media multitasking on performance in lab-based cognitive tasks was nearly zero (Wiradhany & Nieuwenstein, 2017).

The clear disconnection between self-reported and experimentally observed cognitive processing deficits suggests that further theoretical and operational clarity is needed in this research area. A large proportion of media multitasking studies treat the brain as a "black box" between proposed inputs (trait media multitasking) and outputs (cognitive processing differences/deficits), bifurcating survey respondents into "heavy" and "light" media multitasking groups with little attempt made to explicate a biological mechanism that could predict or explain why each group would be expected to score differently on these measures. The primary measure of media multitasking used in the literature (the Media Multitasking Index; Ophir et al., 2009) simply asks participants to report which media they multitask with and how often they multitask. This measure ignores pertinent cognitive differences between different forms of media multitasking, limiting the explanatory power of these "high" and "low" multitasking groups (Wang, Irwin, Cooper, & Srivastava, 2015). For example, a participant who spends eight hours typing an essay in a word processing program while listening to music (tasks that use separate modalities and that can be efficiently parallelized, Cassidy & MacDonald, 2009) would have the same score on the MMI as another student who spends eight hours watching online videos while concurrently playing video games (tasks that use the same modality and cannot be efficiently done in parallel).

We argue that to understand the cognitive effects of media multitasking, researchers must take care to elucidate the cognitive and neural dimensions of multitasking behaviors. Initial progress in this area has come through the work of Wang and colleagues, who conceptualized media multitasking as varying along four main axes: task inputs, task outputs, task relations, and user differences (Wang et al., 2015). In this chapter, we outline a framework for predicting and explaining cognitive effects of media multitasking centered on understanding the neural substrates involved in multitasking with media. In doing so, we undertake three primary objectives: (1) outlining the neurophysiological substrates of attention, how they are relevant for understanding multitasking behavior, and how they may change with continued multitasking; (2) discussing how digital media is uniquely amenable to multitasking behaviors very different from those possible in a non-digital environment; and (3) providing predictions as to how these potential neuroplastic changes may be expected to lead to variance in self-reported or lab-based measures of cognitive processing.

NEUROPHYSIOLOGICAL MECHANISMS OF ATTENTION AND MULTITASKING

Understanding the potential effects of multitasking behaviors on cognitive processing in the short and long term depends on clearly defining the neurophysiological substrates implicated in these behaviors. This allows media and psychology researchers to leverage findings from neuropsychology, model systems, computational neuroscience, and other fields to increase understanding of the mechanism through which habitual media multitasking seems to lead to self-reported and behaviorally observed differences in attentional control and working memory. In this section, we define the neurophysiological substrates implicated in the maintenance and shifting of attention, and discuss the relevance of these substrates for understanding how various forms of multitasking may impact the brain.

Attention is notoriously difficult to define and perhaps even more difficult to measure (Chun, Golomb, & Turk-Browne, 2011; Treisman, 1964). Early models of attention focused on its role in selectively filtering objects from the external world, directing attention to certain items while ignoring others (James, 1896). As psychological research has developed, "attention" has become a catch-all term applying to any of the numerous mechanisms through which the brain controls its own information processing. These mechanisms include (conservatively) vigilance, orienting, sensing and recognizing stimuli, memory, emotion, executive control and many others (Chun et al., 2011; Rosenberg, Finn, Scheinost, Constable, & Chun, 2017). Furthermore, attention seems to be intricately intertwined with—and perhaps inseparable from—consciousness (Cohen, Cavanagh, Chun, & Nakayama, 2012), making it unwieldy and perhaps impossible to study as a monolithic construct. For this reason, many of the most productive models in the field divide attention into various subcomponents, illuminating separable neurophysiological mechanisms that support different facets of attentional processing.

As the neurophysiological substrates of attention have become better characterized, researchers have begun to push past strictly descriptive or correlational approaches in favor of constructing multivariate predictive models of brain activation during task performance (Kragel, Koban, Barrett, & Wager, 2018). These network models have been successfully used to predict performance in attention and working memory tasks (Rosenberg et al., 2017) and to better characterize individual cognitive differences (Finn et al., 2017, 2015). Investigating the dynamics of these large-scale networks during media multitasking behavior can afford scientists an understanding of how various forms of multitasking may recruit different neural substrates and facilitate different cognitive outcomes. It is these networks we focus on in this section, highlighting the utility of predictive multivariate network models in seeking to better characterize the neurophysiological dimensions of media multitasking and to more accurately predict and explain cognitive individual differences between those with different media multitasking habits.

TOP-DOWN AND BOTTOM-UP

A wide selection of neuroimaging, behavioral, and model systems research suggests that attention relies on different neural substrates, depending on whether the initial act of attending is "goal-driven" or "stimulus-driven" (Corbetta & Shulman, 2002; Fox, Corbetta, Snyder, Vincent, & Raichle, 2006). Goal-driven attention is a "top-down" process wherein an individual selectively attends to stimuli in the environment that are relevant for the completion of a given goal while ignoring less relevant stimuli. Stimulus-driven orienting is a "bottom-up" process wherein an individual's attention is pulled to a stimulus as a result of its sensory or affective salience (Connor, Egeth, & Yantis, 2004; Itti, 2005).

Multitasking can be driven by top-down processes in which an individual chooses to attend to a secondary task, but it can also be driven by bottom-up processes. Consider for example the case of someone picking up a phone while watching a television show. This process of attending to the secondary task could be driven by a goal in working memory (such as remembering to call to make a reservation for dinner that evening) but it could also be driven by a more bottom-up process (such as the phone ringing or providing other push notifications).

This top-down/bottom-up dichotomy within the orienting network has become influential within media psychology, due in large part to its inclusion in the Limited Capacity Model for Motivated Mediated Message Processing (LC4MP; Lang, 2006, 2009; Lang et al., 2000) (see Chapter 29 by Lang, in this volume). In recent years, though, it has become clear that the dichotomy between top-down and bottom-up influence isn't as clear-cut as was originally thought (Anderson, 2013; Awh, Belopolsky, & Theeuwes, 2012; Chelazzi et al., 2014; Chelazzi, Perlato, Santandrea, & Della Libera, 2013).

A wealth of research has illuminated the neural substrates of goal-driven and stimulus-driven orienting networks (Fox et al., 2006; Sarter, Givens, & Bruno, 2001). This research has served to localize goal-driven orienting to a "dorsal stream" of brain regions running along the cortex near the crown of the skull. Activity in this network increases when stimuli are presented that are relevant to an individual's goals or when individuals are instructed as to where or when they will be required to direct their attention to a given stimulus (Shulman et al., 2003). Stimulus-driven orienting seems to depend on brain regions organized in a "ventral stream" located in cortical regions beneath the sides of the skull. This network is activated when an individual is presented with visually salient (high contrast, moving, or high luminance) stimuli, especially when those targets are presented in novel locations or with unexpected features (Astafiev et al., 2003; Kincade, Abrams, Astafiev, Shulman, & Corbetta, 2005).

Importantly, these systems interact and bias each other's behavior in a substantial way. For example, the ventral system seems to be much more responsive to novel or unexpected stimuli when those stimuli match with learned goals (Anderson, Laurent, & Yantis, 2011; Serences et al., 2005). Individual goals and learned incentive values have also been shown to bias bottom-up processing in relation to monetary (Anderson et al., 2011) and social (Anderson, 2016b) rewards, and these biasing effects persist even when task parameters have changed (e.g., when incentives are removed from previously rewarded stimuli, Anderson, 2016a). Furthermore, it seems that the dorsal network can down-regulate the ventral network—especially the temporoparietal junction—during focused attention, but that this biasing effect is overcome whenever stimuli are salient enough to result in a change of current goals (Corbetta, Patel, & Shulman, 2008).

These findings have led to calls for discarding the top-down/bottom-up dichotomy in favor of an approach that considers goal-driven and stimulus-driven influences on orienting behavior in tandem with one another (Awh et al., 2012). The idea that goal-driven patterns of orienting can develop bottom-up qualities is especially interesting for understanding media multitasking. The frequently incentive-laden nature of certain forms of media use (discussed in detail in the next section) could modulate these attention mechanisms such that the salience of media stimuli is increased, making attentional capture by these media highly automatic, even preconscious.

LARGE-SCALE DYNAMIC NETWORKS

Emerging evidence suggests that bottom-up and top-down influences on attention converge in a large-scale, dynamic "salience network" anchored in the anterior insula (Menon & Uddin, 2010; Seeley et al., 2007). The elucidation of this network is an example of a wider body of more recent work in the neuroscience of attention that has adopted tools from network science (Bassett & Sporns, 2017; Newman, 2018) to more precisely map the large-scale, dynamic networks that undergird cognitive processing. In this approach, brain voxels or regions of interest are treated as "nodes" in a network. The weight of the connections (edges) between the nodes in the network can be conceptualized as either correlations between activation patterns, directed causal influence, wavelet coherence, or other metrics (Fornito, Zalesky, & Bullmore, 2016). Frequently, these networks are analyzed using time-varying approaches, allowing for the creation of models that can predict how the network varies over time in relation to task parameters (Braun et al., 2015; Zalesky, Fornito, Cocchi, Gollo, & Breakspear, 2014), how network dynamics can predict individual characteristics (Rosenberg et al., 2017; Schmälzle et al., 2017), and several other outcomes of interest (Cooper et al., 2018; Kucyi, Hove, Esterman, Hutchison, & Valera, 2016; Liu et al., 2017; Wasylyshyn et al., 2018).

Researchers in this area have demonstrated that the brain is organized into numerous complex functional networks that oscillate in sync with one another in a task-dependent manner (Bassett & Sporns, 2017; Bullmore & Sporns, 2009; Petersen & Sporns, 2015). Dynamic changes in these networks can predict factors, such as learning (Bassett, Wymbs, & Porter, 2011) and psychopathology (Xia et al., 2018). This research has highlighted three networks especially important for attention and task-switching behaviors: (1) the "task-positive" network; (2) the "task-negative" (default mode) network; and (3) the "task-switching" (salience) network (Fox et al., 2006, 2005; Greicius, Krasnow, Reiss, & Menon, 2003; Menon & Uddin, 2010; Raichle et al., 2001).

The Task-Positive Network

The task-positive (also described as the *central executive*) network is activated whenever an individual is engaged in a task that requires *cognitive control*—paying attention to certain inputs while ignoring others (Clayton, Yeung, & Kadosh, 2015; Rosenberg et al., 2017; Seeley et al., 2007). This network is also implicated in maintaining and manipulating information in working memory, applying flexible rules to solve problems, and making decisions in view of goals (Koechlin & Summerfield, 2007; Miller & Cohen, 2001). Connectivity within the task-positive network is tied to attentional control abilities (Clayton et al., 2015) and regions in the network are consistently implicated in ADHD (Cortese et al., 2012; Dickstein, Bannon, Xavier Castellanos, & Milham, 2006; Tegelbeckers et al., 2018).

Recent work has shown that the task-positive network is activated whenever reward is approximately equal to skill in a video game task (Huskey, Craighead, Miller, & Weber, 2018), and that activation in this network is associated with greater proactive cognitive control and slower task switching (Botvinick & Braver, 2015). Other work has found that connectivity within the task-positive network increases in a nonlinear fashion as distraction decreases in a naturalistic task (Weber, Alicea, Huskey, & Mathiak, 2018). This work suggests that activation in the task-positive network could predict an individual's ability to maintain focus in the presence of potentially distracting secondary media streams, and that disruptions in robustness of the connectivity within this task-positive network may be a candidate mechanism for the distractability frequently reported by media multitaskers.

The Task-Negative Network

The task-negative (default mode) network is a strongly interconnected collection of regions that seems to come on-line whenever attention is disengaged–reflecting a state of non-specific monitoring of the internal or external environment (Bressler & Menon, 2010; Fox et al., 2005; Raichle et al., 2001). Activity within this network has been associated with mind-wandering (Fox, Spreng, Ellamil, Andrews-Hanna, & Christoff, 2015), distraction (Christoff, Gordon, Smallwood, Smith, & Schooler, 2009; Mason et al., 2007), and task-disengagement as a result of effort-reward mismatches (Huskey et al., 2018).

The default mode network (DMN) is one of the most widely researched brain networks in recent years. This research has reliably demonstrated that activity within the DMN is the most energy-consuming process in the brain (Raichle, 2015). In addition, many have found that aberrant connectivity in this network is associated with various psychopathologies, most notably ADHD (Broyd et al., 2009; Whitfield-Gabrieli & Ford, 2012). Furthermore, insufficient differentiation between the task-positive network and the DMN is also highly associated with distractibility and deficits in cognitive control (Kelly, Uddin, Biswal, Castellanos, & Milham, 2008; Uddin et al., 2008). Recent work has shown that certain tasks can be used to highlight individual differences in connectivity in the task-positive and task-negative networks during task performance, and that these differences can be used to predict performance on new tasks (Finn et al., 2017; Rosenberg et al., 2017).

In media multitasking research, activity in the task-negative network is especially salient as a measure of distraction and task-nonspecific thought. It is likely that certain forms of multitasking behavior will induce more activation in the task-negative network than others (e.g., tasks that are characterized by an imbalance between skill and difficulty; Huskey et al., 2018). Activation within the task-negative network is associated with lower performance on tasks that require cognitive control, and thus, we posit that this network may be used to predict deficits in performance due to task disengagement or focus on a secondary task.

The Task-Switching Network

The "task-switching" (salience) network subserves the detection of internal (interoceptive) or external (sensory) events that may require attention and reallocation of cognitive resources toward processing these novel or motivationally relevant stimuli (Menon, 2011; Seeley et al., 2007). In the beginning stages of research into this network it was thought that there were several specialized brain systems that performed slightly different processing tasks (Botvinick, Cohen, & Carter, 2004; Singer, Critchley, & Preuschoff, 2009). However, more recent research suggests that these networks can best be thought of as working as one to judge the comparative priority of stimuli in the internal or external milieu and guide behavior to effectively respond to the stimuli which are most prioritized (Menon & Uddin, 2010).

The task-switching network is especially interesting for media multitasking research, as it seems to serve as a "bridge" between the task-positive and task-negative networks and between successive activations of the task-positive network (Goulden et al., 2014). This network is also implicated as the neural mechanism of "reactive cognitive control" (Botvinick & Braver, 2015). An over-reliance on reactive cognitive control mechanisms has been proposed to be one of the core differences between high media multitaskers and low media multitaskers (Fisher, Hopp, & Weber, 2018; Lin, 2009). Thus, the "breadth bias" that is frequently referred to within media multitasking literature may be measurable in the brain as increased activation within the task-switching network during various tasks.

Plasticity in Large-Scale Networks

Although research in this area is still in its incipient stages, it seems that aberrant connectivity within and between these large-scale cortical networks could be a candidate mechanism for explaining observed cognitive differences in relation to multitasking behaviors. It is clear that these cortical networks reconfigure themselves during the performance of behaviors (Bassett et al., 2011; Braun et al., 2015; Khambhati, Mattar, Wymbs, Grafton, & Bassett, 2018) and that this reconfiguration persists over time (Bassett & Mattar, 2017; Bassett, Yang, Wymbs, & Grafton, 2015; Baum et al., 2017). Thus, it is our prediction that the continued persistence of certain behaviors (e.g., continually interrupting a control-demanding task with an unrelated media task) could facilitate reconfiguration of these dynamic networks to adapt to these behaviors, especially if the interruption is rewarding.

THE ROLE OF MEDIA IN MEDIA MULTITASKING

Thus far, we have highlighted the neurophysiological substrates that undergird focused attention, distraction, and task-switching. Specifically, we placed emphasis on their relevance for understanding the unique behavioral patterns that take place during everyday media multitasking. It is worth noting that the task-positive, task-negative, and task-switching networks are by no means unique to media processing behavior. For much of evolutionary history, humans and their ancestors have been required to attend to and multitask between a myriad of sensory and interoceptive inputs to hunt for food, avoid threats, and forge social relationships. The successful performance of these adaptive behaviors depends on the ability of these networks to attend to salient inputs, tune out perceptually salient but goal-irrelevant stimuli, and monitor the internal and external environment.

Tasks that do not involve media tend to have inputs and outputs that are largely inflexible—bound to the physical environment in which they take place. Said a different way, the "state space" of cognitive processes required by these tasks is constrained by the parameters of the task at hand and by the perceptual limitations of the human processing system. However, when we introduce media into such a situation, the state space of cognitive demands can increase markedly. Unmoored from the physical constraints of the media-absent world, media can provide easily accessible (even intrusive) *extensions* to the physical environment that introduce numerous and idiosyncratic cognitive demands.

These demands may be quite motivationally relevant, such as updates on the lives of close friends, vivid accounts of violence and harm, moral outrage, appetitive imagery, and so on. They may be task-relevant, such as looking up a recipe or cooking video, or quite task-irrelevant, such as watching cat videos. Furthermore, there is often no "end" to the information present in a mediated environment, making it what evolutionary theorists would describe as "a non-depleting 'information' resource" (Sandstrom, 1994). Finally, media are also quite flexible in the modality (visual, auditory, etc.) in which they are presented. For example, a harried parent cooking dinner and helping a child with homework may choose to forgo multitasking with a visual medium (such as television) and turn on music, allowing him or her to maintain focus on the task at hand while still multitasking across two modalities. In this section we introduce media as *motivational*, *accessible*, *non-exhaustive*, and *cross-modal and inter-modal*, which we argue cause media to be particularly prone to being used in multitasking situations.

Motivational

First, media are *motivational*. Indeed, in the conceptualization of media multitasking presented here it is possible to consider motivation as the umbrella under which the other three characteristics reside. A wealth of studies point to the motivational nature of a myriad of media-use behaviors including watching television (Keene & Lang, 2016; Lang, Zhou, Schwartz, Bolls, & Potter, 2010; Wang, Lang, & Busemeyer, 2011), sharing and seeking information on social media (Alhabash, Almutairi, Lou, & Kim, 2018; Sherman, Payton, Hernandez, Greenfield, & Dapretto, 2016; Tamir & Mitchell, 2012), playing video games (Huskey et al., 2018; Sherry, 2004), using a smartphone (Bayer, Campbell, & Ling, 2015; Jung & Sundar, 2018), and many others. Importantly, media seem to be processed at the preconscious level largely isomorphic to that of real-world information (Reeves & Nass, 1996). This means that motivational information presented through digital means is likely to have the same attention-capturing effects as rewarding—or threatening—information presented in an analog fashion. At its core, this characteristic is built upon the idea that interactions—whether mediated or not—elicit specific activation patterns in the appetitive and the aversive systems (Bradley, Codispoti, Cuthbert, & Lang, 2001; Cacioppo, Berntson, Norris, & Gollan, 2011).

The appetitive and the aversive systems derive from functionally separable dynamic networks of neuronal activation (Norris, Gollan, Berntson, & Cacioppo, 2010). The appetitive system, which responds to pleasant information or situations that present potential reward (e.g., food or sex), is more active when a situation is neutral and activates in a relatively positively linear fashion. As the appetitive system activates, allocation of cognitive resources also increases in a linear fashion (Lang, 2009). The aversive system responds to unpleasant information or situations of potential risk or threat (e.g., predators, weapons, etc.). Activation in the aversive system causes increasing allocation of cognitive resources up to a tipping point of high aversive activation where the allocation decreases significantly as the individual proceeds down the defensive cascade (Bradley et al., 2001). Whereas the aversive system is less active than the appetitive system in neutral situations, the aversive system does activate much more quickly (i.e., a steeper slope of activation) than the appetitive system as situations become more intense.

These motivations can best be thought of as judgments of the association between a particular stimulus and the likelihood, intensity, and proximity of a given risk/threat or reward (Braver et al., 2014; Fisher, Huskey, Keene, & Weber, 2018). These estimates bias the planning and execution of cognitive and emotional processes related to motivation. Some of these judgments have likely been passed down over evolutionary time (Pessoa, 2010), resulting in associations between the stimulus and reward/threat likelihood, intensity, and proximity that are *rigid*, context-agnostic, and inflexible. This means that the preconscious reaction to these stimuli is always in the same direction with the same relative intensity. Reactions to situations involving sex (Sparks & Lang, 2015), food (Yegiyan & Bailey, 2016), predators (Löw, Lang, Smith, & Bradley, 2008), disease (Leshner, Vultee, Bolls, & Moore, 2010), and a host of other potential threats or rewards would fall into this category.

Other associations between a stimulus and an outcome are more *flexible*, context-specific, and task- or goal-oriented (Fisher, Huskey et al., 2018). These flexible associations are likely highly individualized and are activated within particular contexts (e.g., a particular cue in a video game that would hold no motivational value outside of that specific mediated environment). These flexible associations must be held in working memory to bias attentional processing, and thus tasks which load working memory should interfere with their modulatory power (Anderson & Yantis, 2012). Importantly, flexible associations can, with enough reinforcement, become more rigidly encoded in the cognitive system (Anderson, 2013), which reduces their reliance on working memory and increases their power to bias preconscious attentional processing.

It is our argument here that many forms of media multitasking are driven by associations between media behaviors and reward/threat and that individuals seek to balance effort expenditure and reward across multiple tasks (Braver, 2015; Westgate & Wilson, 2018). For example, if people are in a state of cognitive depletion, they may choose to turn on the television and watch an enjoyable and unchallenging movie while they complete their housework, but if they feel more cognitively energized, they may choose a more challenging but also more rewarding movie (Eden, Johnson, & Hartmann, 2017).

Accessible

Another characteristic of media that makes it particularly amenable to multitasking behaviors is that it is *accessible*. Often, the performance of multiple tasks that do not involve media is subject to severe physical limitations. For example, an individual cannot choose to strike up a conversation with a friend if the friend is not co-present, and cannot listen to a favorite tune without traveling to a location where it is being played. Media afford the opportunity to engage in high reward interactions with very little physical exertion.

Furthermore, media are readily available and adaptable into personal formats of varying size from home theater systems to smartwatches. In the age of smartphones, an increasing variety of formerly separate varieties are now available in our pockets at all times. One unique outcome of the accessibility of media is that it affords faster task switching to more diverse tasks than might otherwise be possible in an environment without media tools. These tasks likely vary in both value and in their relevance to the primary task. For example, while concurrently attending to a lecture, a university student might search the internet for a phrase used by the lecturer in order to better understand the material. Conversely, the same student might go to a shopping website and begin browsing for ideas for a gift for an upcoming holiday or may converse with a friend through a chat app. Each of these secondary tasks can be executed with minimal physical effort by the student, but one is closely related to the primary task while the others are orthogonal to it.

This highlights a key issue within media multitasking research: the idea that an individual may choose a low-value but also low-effort (high proximity) task over a task that is higher in value but also higher in effort (low proximity). Because media tasks are frequently motivational and are highly proximal to an individual (e.g., they take little effort to access), they are likely to be chosen in situations of cognitive *offloading* wherein a primary task becomes either too unrewarding or too effortful to continue to draw full allocation of cognitive resources (Fisher, Huskey, et al., 2018). A promising area of recent research in this area also concerns cognitive *underload* wherein it has been shown that understimulated individuals choose to undergo cognitively loading tasks even without associated reward (Bench & Lench, 2013), and choose more stimulating media over less stimulating ones (Bryant & Zillmann, 1984).

Cross- and Inter-Modal

In addition to being motivationally relevant and accessible, media are frequently adaptable within and between modalities (e.g., visual, auditory). The modality of these information streams can be manipulated by the user in order to balance effort expenditure and reward. This flexibility allows for more breadth in the construction of an environment that optimally balances effort and reward in relation to current goals and states. For example, an individual working on a resource-intensive task in a primarily visual modality (such as writing a book chapter) could choose to add a media task in the auditory modality (such as listening to classical music).

This modal flexibility is especially salient when considering research findings that higher-order cognitive processes seem to operate most effectively when perceptual processing channels are "plugged" by sensory information (Forster & Lavie, 2009; Lavie, Hirst, de Fockert, & Viding, 2004). The idea that selective attention can be *augmented* by the introduction of load in perceptual channels points to promising future research directions in media multitasking research aimed at elucidating the quite common multitasking pattern wherein individuals use media (such as familiar television shows or music as a backdrop while completing other tasks). Research from cognitive neuroscience suggests that individuals exposed to extraneous background noise (either auditory or visual) actually *improved* their cognitive performance relative to controls (see e.g., Gleiss & Kayser, 2014; Rausch, Bauch, & Bunzeck, 2014; Söderlund, Sikström, & Smart, 2007). This provides support for the proposition that some forms of multitasking may actually be beneficial for the completion of cognitively-taxing tasks and emphasizes the necessity of considering the neurophysiological substrates of different forms of multitasking and how tonic engagement in different forms of multitasking may result in diverging outcomes.

Non-Exhaustive

A final aspect of media environments that needs to be considered is its *non-exhaustive* nature. Evidence from the biological sciences strongly suggests that humans and other organisms treat information resources in much the same way as they treat nutritive resources. Information is a crucial currency for survival, especially in social and tool-making animals (Dall, Giraldeau, Olsson, McNamara, & Stephens, 2005) and informational rewards have been shown to rely on the same valuation network in the brain as nutritive rewards (Adams, Watson, Pearson, & Platt, 2012; Pearson, Watson, & Platt, 2014). Thus, it has been proposed that goal-driven information foraging behaviors (such as those common in media use) should follow an optimization function similar to that of food foraging.

The Optimal Foraging Hypothesis (OFH; Charnov, 1976) and the associated Marginal Value Theorem (MVT) have served as a scaffold for particularly productive approach to understanding how and when animals switch from "patch" to "patch" in a food foraging environment based on judgments of the value and accessibility of various patches of food. In these models, food (or information) foraging follows an optimization function wherein organisms take into account the value and proximity of a food patch, and also the diminishment factor of the current food patch (i.e., how much can be eaten before the food here runs out).

It is our suggestion here that information foraging within a mediated environment works in a similar way. Prior communication research has demonstrated that some media behavior—such as channel changing—is driven by consistent re-evaluation of the potential value that a change might present over staying with the current information stream (Wang et al., 2011). Indeed, the amount of and the value of information in a particular "patch" may vary widely between successive visits, which leads to consistent monitoring of the patches to assess their potential motivational value. Furthermore, many forms of media are functionally inexhaustible (e.g., there is no "end" to a social media feed or an online video queue), which results in a diminishment factor near zero. Emerging evidence suggests that at diminishment factors approaching zero, foraging behaviors become less efficient, resulting in suboptimal switches between patches (Bloomfield, Harbison, Campbell, Bradley, & Saner, 2016).

NEUROPHYSIOLOGICAL DIMENSIONS OF MEDIA MULTITASKING

The motivational, accessible, multimodal, and inexhaustible nature of media environments interacts with the task-positive, task-negative, and task-switching networks of the brain to facilitate numerous and idiosyncratic forms of multitasking behavior. As such, attempting to develop a typology of media multitasking from a media-centric perspective (such as trying to document every potential combination of device, app, context, etc.) would be all but impossible. For this reason, researchers have begun to categorize multitasking behaviors in relation to their cognitive dimensions (Wang et al., 2015). This research has made great strides in bridging the divide between lab-based studies of multitasking and studies that consider multitasking as an individual difference variable.

Wang and colleagues suggest considering media multitasking as varying along 11 primary dimensions: task hierarchy, task switch, task relevance, information modality, shared modality, task contiguity, control over information flow, emotional content, response requirements, time pressure, and user differences (Wang et al., 2015). These dimensions have been shown to guide individual behavior and to correspond to cognitive outcomes. In a meta-analysis of multitasking research using these dimensions, three of these proposed dimensions (user control, task-relevance, and task-contiguity) were found to be significant moderators of the effects of multitasking on cognitive/attitudinal outcomes (Jeong & Hwang, 2016).

We propose that the adoption of a neurophysiological perspective on the question of multitasking allows researchers to further simplify these dimensions in view of what the brain is doing while an individual is multitasking with media. We propose that the 11 cognitive dimensions of media multitasking may be condensed into three neurophysiological dimensions: (1) *multiplexing*—the extent to which two tasks share resources in the brain; (2) *motivation*—the incentive salience of each task combined with the effort that it takes to switch between tasks; and (3) *control*—the amount of control an individual has over their multitasking behavior (e.g., whether multitasking is voluntary or induced by an experimenter).

Multiplexing

Various combinations of tasks load the human processing system in different ways. From the early days of attention research, it has been proposed that the cognitive system is of limited capacity (Kahneman, 1973). The exact nature of this capacity is still hotly debated, but evidence suggests that it is related to the dynamic availability of metabolic resources (such as oxygen and glucose) and structural constraints within specialized subnetworks (e.g., such as executive processing or auditory/visual sensory processing; Buschman, Siegel, Roy, & Miller, 2011; Todd & Marois, 2004). These constraints place limits on the sorts of tasks that can be concurrently pursued. For example, it is rather simple to listen to the radio while driving a vehicle. It is rather hard to watch a movie while driving a vehicle.

Tasks that do not rely on the same resources are amenable to true multitasking—concurrent performance of both tasks with little impairment of either task. On the other hand, tasks that share resources must be *multiplexed* (Salvucci & Taatgen, 2008). Multiplexing, also known as task-switching, is the process of switching back and forth between two or more tasks within a short window of time. Importantly, the processes of multitasking and multiplexing are likely associated with differentiable patterns of brain activation over time (Crone, Wendelken, Donohue, & Bunge, 2005; Feng, Schwemmer, Gershman, & Cohen, 2014). Multitasking—the true performance of two concurrent tasks—is to be associated with high network integrity in the task-positive network. Multiplexing, on the other hand, would be expected to be associated with reduced integrity in the task-positive network, and greater activation in the task-switching network during switches between tasks. Importantly, given the premise that tonic differences in media multitasking behaviors should lead to differentiable cognitive outcomes, individual differences in cognitive abilities should be expected to vary based on (1) the *quantity* of multitasking (as measured using the MMI), but also (2) the relative quantity of multitasking vs. multiplexing behaviors. Those who multiplex more would (all other things being equal) be expected to have lower robustness in the task-positive network and be more prone to switching. Conversely, those who multitask more would not be expected to show these differences. These predictions are similar for studies in which multitasking is induced in the lab. Tasks that rely on separate resources in the cognitive system would be expected to lead to a smaller decrement in performance when concurrently performed as compared to tasks that must be multiplexed.

Motivation

A second proposed neurophysiological dimension of media multitasking is the relative *motivation* associated with the primary and the secondary task. Recall that the human cognitive system seeks to balance effort and reward, choosing combinations of tasks that maximize the tradeoff between these two variables. Thus, the cognitive system can be said to make judgments as to where to allocate cognitive resources as a function of rigid/flexible associations with reward/threat likelihood, magnitude, and proximity. If a secondary task becomes superior to the primary task in this calculation (e.g., the rewards are higher, the costs are lower, and switching requires little effort) and the two tasks are unable to be concurrently performed, a switch would be expected to occur. If the two tasks are able to be concurrently performed, then it is likely that an individual will choose to combine tasks in a way that optimizes this subjective balance (e.g., listening to an enjoyable song while performing a cognitively underwhelming task). This also leads to the prediction that the more highly rewarding and easily accessible a task is, the more likely it will be to be a secondary task in self-reported or observed multitasking.

Control

The final neurophysiological dimension proposed here is that of *control*. In many studies of multitasking, participants are told to concurrently perform two tasks with little control over the modality, rewardingness, or difficulty of either the primary or secondary task. A substantial amount of evidence suggests that decrements in performance observed in media multitasking conditions are reduced whenever individuals are given control over aspects of the secondary task that they perform, such as the background music that they listen to (Cassidy & MacDonald, 2009) or whether to toggle a background video on or off (Ralph, 2018). It has also been shown that performance decrements in multitasking are removed if participants are given freedom to choose when they switch back and forth between the primary and secondary task (Kononova, Joo, & Yuan, 2016). These studies suggest that in real-world multitasking, individuals may dynamically adapt the sorts of secondary tasks that are performed in order to maintain optimal performance in the primary task. Thus, conclusions drawn from evidence regarding degraded performance while engaging in forced multitasking in the lab may not cleanly generalize to real-world situations unless control is taken into account.

MULTITASKING FROM A NEW PERSPECTIVE

Multitasking, particularly involving media, is becoming a larger part of our lives in an increasingly connected society. As these behaviors increase, so too do the concerns about the potentially deleterious effects of multitasking. In this chapter, we presented a neurophysiological perspective on media multitasking research. To accomplish this goal, we first presented an overview of the large-scale dynamic networks that enable focused task performance, non-specific monitoring of the internal and external environment, and task-switching. We then highlighted aspects of media that uniquely facilitate multitasking behaviors, homing in on the motivational, accessible, non-exhaustive, and modality-flexible nature of digital environments. Finally, we presented three candidate neurophysiological dimensions along which media multitasking tasks may vary, highlighting key predictions regarding in-lab performance and potential individual differences in regard to these dimensions.

In a recent paper, a large consortium of clinical and academic researchers issued a clarion call to conduct rigorous research into the cognitive, psychosocial, neural, and academic ramifications of the proliferation of media multitasking—especially in children (Uncapher et al., 2017). It is critical that research that seeks to advance understanding in this area be conducted from a common understanding of neural function, the nature of media, and the dimensions along which multitasking can vary. We argue that approaching media multitasking from a neurophysiological perspective provides a structure upon which to construct this understanding. By highlighting how different forms of media-enabled multitasking may rely on differentiable neural substrates and be expected to lead to different outcomes, media multitasking researchers can make more precise predictions regarding the immediate and long-term effects of media multitasking and can more effectively explain the mechanism that may undergird these effects. Findings from research undertaken using this framework are well situated to inform research, policy, and digital design decisions in the private, public, and academic spheres, that pave the way for the creation of digital technologies that enhance well-being and performance in numerous domains.

REFERENCES

Adams, G. K., Watson, K. K., Pearson, J., & Platt, M. L. (2012). Neuroethology of decision-making. *Current Opinion in Neurobiology, 22*(6), 982–989.

Alhabash, S., Almutairi, N., Lou, C., & Kim, W. (2018). Pathways to virality: Psychophysiological responses preceding likes, shares, comments, and status updates on Facebook. *Media Psychology*. Advance online publication.

Alzahabi, R., & Becker, M. W. (2013). The association between media multitasking, task-switching, and dual-task performance. *Journal of Experimental Psychology: Human Perception and Performance, 39*(5), 1485–1495.

Anderson, B. A. (2013). A value-driven mechanism of attentional selection. *Journal of Vision, 13*(3), 7.

Anderson, B. A. (2016a). The attention habit: How reward learning shapes attentional selection. *Annals of the New York Academy of Sciences, 1369*(1), 24–39.

Anderson, B. A. (2016b). Social reward shapes attentional biases. *Cognitive Neuroscience, 7*(1–4), 30–36.

Anderson, B. A., Laurent, P. A., & Yantis, S. (2011). Value-driven attentional capture. *Proceedings of the National Academy of Sciences, 108*(25), 10367–10371.

Anderson, B. A., & Yantis, S. (2012). Value-driven attentional and oculomotor capture during goal-directed, unconstrained viewing. *Attention, Perception, & Psychophysics, 74*(8), 1644–1653.

Astafiev, S. V., Shulman, G. L., Stanley, C. M., Snyder, A. Z., Van Essen, D. C., & Corbetta, M. (2003). Functional organization of human intraparietal and frontal cortex for attending, looking, and pointing. *Journal of Neuroscience, 23*(11), 4689–4699.

Awh, E., Belopolsky, A. V., & Theeuwes, J. (2012). Top-down versus bottom-up attentional control: A failed theoretical dichotomy. *Trends in Cognitive Sciences, 16*(8), 437–443.
Bassett, D. S., & Mattar, M. G. (2017). A network neuroscience of human learning: Potential to inform quantitative theories of brain and behavior. *Trends in Cognitive Sciences, 21*(4), 250i264.
Bassett, D. S., & Sporns, O. (2017). Network neuroscience. *Nature Neuroscience, 20*(3), 353.
Bassett, D. S., Wymbs, N. F., & Porter, M. A. (2011). Dynamic reconfiguration of human brain networks during learning. *Proceedings of the National Academy of Sciences, 108*(18), 7641–7646.
Bassett, D. S., Yang, M., Wymbs, N. F., & Grafton, S. T. (2015). Learning-induced autonomy of sensorimotor systems. *Nature Neuroscience, 18*(5), 744–751.
Baum, G. L., Ciric, R., Roalf, D. R., Betzel, R. F., Moore, T. M., Shinohara, R. T., … Satterthwaite, T. D. (2017). Modular segregation of structural brain networks supports the development of executive function in youth. *Current Biology, 27*(11), 1561–1572.
Baumgartner, S. E., van der Schuur, W. A., Lemmens, J. S., & te Poel, F. (2018). The relationship between media multitasking and attention problems in adolescents: Results of two longitudinal studies. *Human Communication Research, 12*(1), 116–127.
Baumgartner, S. E., Weeda, W. D., van der Heijden, L. L., & Huizinga, M. (2014). The relationship between media multitasking and executive function in early adolescents. *Journal of Early Adolescence, 34*(8), 1120–1144.
Bayer, J. B., Campbell, S. W., & Ling, R. (2015). Connection cues: Activating the norms and habits of social connectedness. *Communication Theory, 26*(2), 128–149.
Becker, M. W., Alzahabi, R., & Hopwood, C. J. (2013). Media multitasking is associated with symptoms of depression and social anxiety. *Cyberpsychology, Behavior, and Social Networking, 16*(2), 132–135.
Bench, S. W., & Lench, H. C. (2013). On the function of boredom. *Behavioral Sciences, 3*(3), 459–472.
Bloomfield, A., Harbison, J. I., Campbell, S. G., Bradley, P., & Saner, L. (2016, July). Information search with depleting and non-depleting resources. Paper presented at the annual meeting of the Cognitive Science Society, Philadelphia, PA.
Botvinick, M., & Braver, T. (2015). Motivation and cognitive control: From behavior to neural mechanism. *Annual Review of Psychology, 66*, 83–113.
Botvinick, M. M., Cohen, J. D., & Carter, C. S. (2004). Conflict monitoring and anterior cingulate cortex: An update. *Trends in Cognitive Sciences, 8*(12), 539–546.
Bowman, L. L., Waite, B. M., & Levine, L. E. (2015). Multitasking, and attention: Implications for college students. In L. D. Rosen & N. Cheever (Eds.), *The Wiley handbook of psychology, technology, and society* (pp. 388–403). Chichester, England: John Wiley & Sons.
Bradley, M. M., Codispoti, M., Cuthbert, B. N., & Lang, P. J. (2001). Emotion and motivation I: Defensive and appetitive reactions in picture processing. *Emotion, 1*(3), 276–298.
Brand, M., Young, K. S., & Laier, C. (2014). Prefrontal control and internet addiction: A theoretical model and review of neuropsychological and neuroimaging findings. *Frontiers in Human Neuroscience, 8*, 375.
Brasel, S. A., & Gips, J. (2017). Media multitasking: How visual cues affect switching behavior. *Computers in Human Behavior, 77*, 258–265.
Braun, U., Schäfer, A., Walter, H., Erk, S., Romanczuk-Seiferth, N., Haddad, L., … Bassett, D. S. (2015). Dynamic reconfiguration of frontal brain networks during executive cognition in humans. *Proceedings of the National Academy of Sciences, 112*(37), 11678–11683.
Braver, T. S. (2015). Cognitive effort: A neuroeconomic approach. *Cognitive, Affective, & Behavioral Neuroscience, 12*(182), 395–415.
Braver, T. S., Krug, M. K., Chiew, K. S., Kool, W., Westbrook, J. A., Clement, N. J., … Somerville, L. H. (2014). Mechanisms of motivation-cognition interaction: Challenges and opportunities. *Cognitive, Affective & Behavioral Neuroscience, 14* (2), 443–472.
Bressler, S. L., & Menon, V. (2010). Large-scale brain networks in cognition: Emerging methods and principles. *Trends in Cognitive Sciences, 14*(6), 277–290.
Broyd, S. J., Demanuele, C., Debener, S., Helps, S. K., James, C. J., & Sonuga-Barke, E. J. (2009). Default-mode brain dysfunction in mental disorders: A systematic review. *Neuroscience & Biobehavioral Reviews, 33*(3), 279–296.

Bryant, J., & Zillmann, D. (1984). Using television to alleviate boredom and stress: Selective exposure as a function of induced excitational states. *Journal of Broadcasting & Electronic Media, 28*(1), 1–20.

Bullmore, E., & Sporns, O. (2009). Complex brain networks: Graph theoretical analysis of structural and functional systems. *Nature Reviews Neuroscience, 10*(3), 186–198.

Buschman, T. J., Siegel, M., Roy, J. E., & Miller, E. K. (2011). Neural substrates of cognitive capacity limitations. *Proceedings of the National Academy of Sciences, 108*(27), 11252–11255.

Cacioppo, J. T., Berntson, G. G., Norris, C., & Gollan, J. (2011). The evaluative space model. In P. A. M. Van Lange, A. W. Kruglanski, & E. T. Higgins (Eds.), *Handbook of theories of social psychology* (Vol. 1., pp. 50–72). London, England: Sage.

Cain, M. S., Leonard, J. A., Gabrieli, J. D. E., & Finn, A. S. (2016). Media multitasking in adolescence. *Psychonomic Bulletin & Review, 23*(6), 1–10.

Cain, M. S., & Mitroff, S. R. (2011). Distractor filtering in media multitaskers. *Perception, 40*(10), 1183–1192.

Cardoso-Leite, P., Kludt, R., Vignola, G., Ma, W. J., Green, C. S., & Bavelier, D. (2016). Technology consumption and cognitive control: Contrasting action video game experience with media multitasking. *Attention, Perception, & Psychophysics, 78*(1), 218–241.

Carr, N. (2011). *The shallows: What the internet is doing to our brains*. New York, NY: W.W. Norton & Company.

Cassidy, G., & MacDonald, R. (2009). The effects of music choice on task performance: A study of the impact of self-selected and experimenter-selected music on driving game performance and experience. *Musicae Scientiae, 13*(2), 357–386.

Charnov, E. L. (1976). Optimal foraging, the marginal value theorem. *Theoretical Population Biology, 9*(2), 129136.

Chelazzi, L., Eštočinová, J., Calletti, R., Lo Gerfo, E., Sani, I., Della Libera, C., & Santandrea, E. (2014). Altering spatial priority maps via reward-based learning. *Journal of Neuroscience, 34*(25), 8594–8604.

Chelazzi, L., Perlato, A., Santandrea, E., & Della Libera, C. (2013). Rewards teach visual selective attention. *Vision Research, 85*, 58–72.

Christoff, K., Gordon, A. M., Smallwood, J., Smith, R., & Schooler, J. W. (2009). Experience sampling during fMRI reveals default network and executive system contributions to mind wandering. *Proceedings of the National Academy of Sciences, 106*(21), 8719–8724.

Chun, M. M., Golomb, J. D., & Turk-Browne, N. B. (2011). A taxonomy of external and internal attention. *Annual Review of Psychology, 62*(1), 73–101.

Clayton, M. S., Yeung, N., & Kadosh, R. C. (2015). The roles of cortical oscillations in sustained attention. *Trends in Cognitive Sciences, 19*(4), 188–195.

Cohen, M. A., Cavanagh, P., Chun, M. M., & Nakayama, K. (2012). The attentional requirements of consciousness. *Trends in Cognitive Sciences, 16*(8), 411–417.

Connor, C. E., Egeth, H. E., & Yantis, S. (2004). Visual attention: Bottom-up versus top-down. *Current Biology, 14*(19), R850-R852.

Cooper, N., Tompson, S., O'Donnell, M. B., Vettel, J. M., Bassett, D. S., & Falk, E. B. (2018). Associations between coherent neural activity in the brain's value system during antismoking messages and reductions in smoking. *Health Psychology, 37*(4), 375–384.

Corbetta, M., Patel, G., & Shulman, G. L. (2008). The reorienting system of the human brain: From environment to theory of mind. *Neuron, 58*(3), 306–324.

Corbetta, M., & Shulman, G. L. (2002). Control of goal-directed and stimulus-driven attention in the brain. *Nature Reviews Neuroscience, 3*(3), 201–215.

Cortese, S., Kelly, C., Chabernaud, C., Proal, E., Di Martino, A., Milham, M. P., & Castellanos, F. X. (2012). Toward systems neuroscience of ADHD: A meta-analysis of 55 fMRI studies. *American Journal of Psychiatry, 169*(10), 1038–1055.

Crone, E. A., Wendelken, C., Donohue, S. E., & Bunge, S. A. (2005). Neural evidence for dissociable components of task-switching. *Cerebral Cortex, 16*(4), 475–486.

Dall, S. R., Giraldeau, L.-A., Olsson, O., McNamara, J. M., & Stephens, D. W. (2005). Information and its use by animals in evolutionary ecology. *Trends in Ecology & Evolution, 20*(4), 187–193.

Deloitte. (2015). *Digital democracy survey: A multi-generational view of consumer technology, media and telecom trends* (9th ed.). London, England: Author.

Dickstein, S. G., Bannon, K., Xavier Castellanos, F., & Milham, M. P. (2006). The neural correlates of attention deficit hyperactivity disorder: An ALE meta-analysis. *Journal of Child Psychology and Psychiatry, 47*(10), 1051–1062.

Eden, A., Johnson, B. K., & Hartmann, T. (2017). Entertainment as a creature comfort: Self-control and selection of challenging media. *Media Psychology, 21*(3), 352–376.

Feng, S. F., Schwemmer, M., Gershman, S. J., & Cohen, J. D. (2014). Multitasking versus multiplexing: Toward a normative account of limitations in the simultaneous execution of control-demanding behaviors. *Cognitive, Affective, & Behavioral Neuroscience, 14*(1), 129–146.

Finn, E. S., Scheinost, D., Finn, D. M., Shen, X., Papademetris, X., & Constable, R. T. (2017). Can brain state be manipulated to emphasize individual differences in functional connectivity? *NeuroImage, 160*, 140–151.

Finn, E. S., Shen, X., Scheinost, D., Rosenberg, M. D., Huang, J., Chun, M. M., Constable, R. T. (2015). Functional connectome fingerprinting: identifying individuals using patterns of brain connectivity. *Nature Neuroscience, 18*(11), 1664–1671.

Fisher, J. T., Hopp, F. R., & Weber, R. (2018, May). "Look harder!" ADHD, media multitasking, and the effects of cognitive and perceptual load on resource availability and processing performance. Paper presented at the annual conference of the International Communication Association, Prague, the Czech Republic.

Fisher, J. T., Huskey, R., Keene, J. R., & Weber, R. (2018). The limited capacity model of motivated mediated message processing: Looking to the future. *Annals of the International Communication Association, 42*, 291–315.

Fornito, A., Zalesky, A., & Bullmore, E. (2016). *Fundamentals of brain network analysis*. London, England: Academic Press.

Forster, S., & Lavie, N. (2009). Harnessing the wandering mind: The role of perceptual load. *Cognition, 111*(3), 345–355.

Fox, K. C., Spreng, R. N., Ellamil, M., Andrews-Hanna, J. R., & Christoff, K. (2015). The wandering brain: Meta-analysis of functional neuroimaging studies of mind-wandering and related spontaneous thought processes. *NeuroImage, 111*, 611–621.

Fox, M. D., Corbetta, M., Snyder, A. Z., Vincent, J. L., & Raichle, M. E. (2006). Spontaneous neuronal activity distinguishes human dorsal and ventral attention systems. *Proceedings of the National Academy of Sciences, 103*(26), 10046–10051.

Fox, M. D., Snyder, A. Z., Vincent, J. L., Corbetta, M., Van Essen, D. C., & Raichle, M. E. (2005). The human brain is intrinsically organized into dynamic, anticorrelated functional networks. *Proceedings of the National Academy of Sciences, 102*(27), 9673–9678.

Gazzaley, A., & Rosen, L. D. (2016). *The distracted mind: Ancient brains in a high-tech world*. Cambridge, MA: MIT Press.

Gleiss, S., & Kayser, C. (2014). Acoustic noise improves visual perception and modulates occipital oscillatory states. *Journal of Cognitive Neuroscience, 26*(4), 699–711.

Goulden, N., Khusnulina, A., Davis, N. J., Bracewell, R. M., Bokde, A. L., McNulty, J. P., & Mullins, P. G. (2014). The salience network is responsible for switching between the default mode network and the central executive network: Replication from DCM. *NeuroImage, 99*, 180–190.

Greicius, M. D., Krasnow, B., Reiss, A. L., & Menon, V. (2003). Functional connectivity in the resting brain: A network analysis of the default mode hypothesis. *Proceedings of the National Academy of Sciences, 100*(1), 253–258.

Huskey, R., Craighead, B., Miller, M. B., & Weber, R. (2018). Does intrinsic reward motivate cognitive control? A naturalistic-fMRI study based on the synchronization theory of flow. *Cognitive, Affective, & Behavioral Neuroscience*. Advance online publication.

Itti, L. (2005). Models of bottom-up attention and saliency. In L. Itti, G. Rees, & J. K. Tsotsos (Eds.), *Neurobiology of attention* (pp. 576–582). Burlington, MA: Elsevier.

James, W. (1896). *The principles of psychology* (vols. 1 and 2). New York, NY: Henry Holt and Company.

Jeong, S.-H., & Hwang, Y. (2016). Media multitasking effects on cognitive vs. attitudinal outcomes: A meta-analysis. *Human Communication Research, 42*(4), 599–618.
Junco, R., & Cotten, S. R. (2012). No A 4 U: The relationship between multitasking and academic performance. *Computers & Education, 59*(2), 505–514.
Jung, E. H., & Sundar, S. S. (2018). Status update: Gratifications derived from Facebook affordances by older adults. *New Media & Society*. Advance online publication.
Kahneman, D. (1973). *Attention and effort*. Englewood Cliffs, NJ: Prentice-Hall.
Keene, J. R., & Lang, A. (2016). Dynamic motivated processing of emotional trajectories in public service announcements. *Communication Monographs, 83*(4), 468–485.
Kelly, A. M. C., Uddin, L. Q., Biswal, B. B., Castellanos, F. X., & Milham, M. P. (2008). Competition between functional brain networks mediates behavioral variability. *NeuroImage, 39*(1), 527–537.
Khambhati, A. N., Mattar, M. G., Wymbs, N. F., Grafton, S. T., & Bassett, D. S. (2018). Beyond modularity: Fine-scale mechanisms and rules for brain network reconfiguration. *NeuroImage, 166*, 385–399.
Kincade, J. M., Abrams, R. A., Astafiev, S. V., Shulman, G. L., & Corbetta, M. (2005). An event-related functional magnetic resonance imaging study of voluntary and stimulus-driven orienting of attention. *Journal of Neuroscience, 25*(18), 4593–4604.
Koechlin, E., & Summerfield, C. (2007). An information theoretical approach to prefrontal executive function. *Trends in Cognitive Sciences, 11*(6), 229–235.
Kononova, A., Joo, E., & Yuan, S. (2016). If I choose when to switch: Heavy multitaskers remember online content better than light multitaskers when they have the freedom to multitask. *Computers in Human Behavior, 65*, 567–575.
Kragel, P. A., Koban, L., Barrett, L. F., & Wager, T. D. (2018). Representation, pattern information, and brain signatures: From neurons to neuroimaging. *Neuron, 99*(2), 257–273.
Kucyi, A., Hove, M. J., Esterman, M., Hutchison, R. M., & Valera, E. M. (2016). Dynamic brain network correlates of spontaneous fluctuations in attention. *Cerebral Cortex, 27*, 1831–1840.
Lang, A. (2006). Using the limited capacity model of motivated mediated message processing to design effective cancer communication messages. *Journal of Communication, 56*(s1), S57–S80.
Lang, A. (2009). The limited capacity model of motivated mediated message processing. In R. L. Nabi & M. B. Oliver (Eds.), *The Sage handbook of media processes and effects* (pp. 193–204). Thousand Oaks, CA: Sage.
Lang, A., Basil, M. D., Bolls, P. D., Bradley, S. D., Lee, S., Potter, R. F., … Sparks, J. V. (2000). The limited capacity model of mediated message processing. *Journal of Communication, 50*(1), 46–70.
Lang, A., Zhou, S., Schwartz, N., Bolls, P. D., & Potter, R. F. (2010). The effects of edits on arousal, attention, and memory for television messages: When an edit is an edit, can an edit be too much? *Journal of Broadcasting and Electronic Media, 44*(1), 94–109.
Lavie, N., Hirst, A., de Fockert, J. W., & Viding, E. (2004). Load theory of selective attention and cognitive control. *Journal of Experimental Psychology: General, 133*(3), 339–354.
Leshner, G., Vultee, F., Bolls, P. D., & Moore, J. (2010). When a fear appeal isn't just a fear appeal: The effects of graphic anti-tobacco messages. *Journal of Broadcasting & Electronic Media, 54*(3), 485–507.
Lin, L. (2009). Breadth-biased versus focused cognitive control in media multitasking behaviors. *Proceedings of the National Academy of Sciences, 106*(37), 15521–15522.
Liu, Y., Hong, X., Bengson, J. J., Kelley, T. A., Ding, M., & Mangun, G. R. (2017). Deciding where to attend: Large-scale network mechanisms underlying attention and intention revealed by graph-theoretic analysis. *NeuroImage, 157*, 45–60.
Loh, K. K., & Kanai, R. (2016). How has the Internet reshaped human cognition? *The Neuroscientist, 22*(5), 506–520.
Löw, A., Lang, P. J., Smith, J. C., & Bradley, M. M. (2008). Both predator and prey: Emotional arousal in threat and reward. *Psychological Science, 19*(9), 865–873.
Lui, K. F. H., & Wong, A. C. N. (2012). Does media multitasking always hurt? A positive correlation between multitasking and multisensory integration. *Psychonomic Bulletin & Review, 19*(4), 647–653.
Mason, M. F., Norton, M. I., Van Horn, J. D., Wegner, D. M., Grafton, S. T., & Macrae, C. N. (2007). Wandering minds: The default network and stimulus-independent thought. *Science, 315*(5810), 393–395.

Menon, V. (2011). Large-scale brain networks and psychopathology: A unifying triple network model. *Trends in Cognitive Sciences, 15*(10), 483–506.

Menon, V., & Uddin, L. Q. (2010). Saliency, switching, attention and control: A network model of insula function. *Brain Structure and Function, 214*(5–6), 655–667.

Miller, E. K., & Cohen, J. D. (2001). An integrative theory of prefrontal cortex function. *Annual Review of Neuroscience, 24*(1), 167–202.

Minear, M., Brasher, F., McCurdy, M., Lewis, J., & Younggren, A. (2013). Working memory, fluid intelligence, and impulsiveness in heavy media multitaskers. *Psychonomic Bulletin & Review, 20*(6), 1274–1281.

Moisala, M., Salmela, V., Hietajärvi, L., Salo, E., Carlson, S., Salonen, O., … Alho, K. (2016). Media multitasking is associated with distractibility and increased prefrontal activity in adolescents and young adults. *NeuroImage, 134*, 113–121.

Newman, M. (2018). *Networks* (2nd ed.). Oxford, England: Oxford University Press.

Nikkelen, S. W. C., Valkenburg, P. M., Huizinga, M., & Bushman, B. J. (2014). Media use and ADHD-related behaviors in children and adolescents: A meta-analysis. *Developmental Psychology, 50*(9), 2228–2241.

Norris, C. J., Gollan, J., Berntson, G. G., & Cacioppo, J. T. (2010). The current status of research on the structure of evaluative space. *Biological Psychology, 84*(3), 422–436.

Ophir, E., Nass, C., & Wagner, A. D. (2009). Cognitive control in media multitaskers. *Proceedings of the National Academy of Sciences, 106*(37), 15583–15587.

Pearson, J. M., Watson, K. K., & Platt, M. L. (2014). Decision making: The neuroethological turn. *Neuron, 82*(5), 950–965.

Pessoa, L. (2010). Embedding reward signals into perception and cognition. *Frontiers in Neuroscience, 4*, article 17.

Petersen, S. E., & Sporns, O. (2015). Brain networks and cognitive architectures. *Neuron, 88*(1), 207–219.

Raichle, M. E. (2015). The brain's default mode network. *Annual Review of Neuroscience, 38*(1), 433–447.

Raichle, M. E., MacLeod, A. M., Snyder, A. Z., Powers, W. J., Gusnard, D. A., & Shulman, G. L. (2001). A default mode of brain function. *Proceedings of the National Academy of Sciences, 98*(2), 676–682.

Ralph, B. C. W. (2018). Volitional media multitasking: Awareness of performance costs and modulation of media multitasking as a function of task demand. *Psychological Research.* Advance online publication.

Ralph, B. C. W., & Smilek, D. (2017). Individual differences in media multitasking and performance on the *n*-back. *Attention, Perception, & Psychophysics, 79*, 582–592.

Ralph, B. C. W., Thomson, D. R., Cheyne, J. A., & Smilek, D. (2013). Media multitasking and failures of attention in everyday life. *Psychological Research, 78*(5), 661–669.

Rausch, V. H., Bauch, E. M., & Bunzeck, N. (2014). White noise improves learning by modulating activity in dopaminergic midbrain regions and right superior temporal sulcus. *Journal of Cognitive Neuroscience, 26*(7), 1469–1480.

Reeves, B., & Nass, C. (1996). *The media equation: How people treat computers, television, and new media like real people and places*. Cambridge, England: Cambridge University Press.

Rosenberg, M. D., Finn, E. S., Scheinost, D., Constable, R. T., & Chun, M. M. (2017). Characterizing attention with predictive network models. *Trends in Cognitive Sciences, 21*(4), 290–302.

Salvucci, D. D., & Taatgen, N. A. (2008). Threaded cognition: An integrated theory of concurrent multitasking. *Psychological Review, 115*(1), 101–130.

Sanbonmatsu, D. M., Strayer, D. L., Medeiros-Ward, N., & Watson, J. M. (2013). Who multi-tasks and why? Multi-tasking ability, perceived multi-tasking ability, impulsivity, and sensation seeking. *PLoS ONE, 8*(1), e54402–e54408.

Sandstrom, P. E. (1994). An optimal foraging approach to information seeking and use. *The Library Quarterly, 64*(4), 414–449.

Sarter, M., Givens, B., & Bruno, J. P. (2001). The cognitive neuroscience of sustained attention: Where top-down meets bottom-up. *Brain Research Reviews, 35*(2), 146–160.

Schmälzle, R., O'Donnell, M. B., Garcia, J. O., Cascio, C. N., Bayer, J., Bassett, D. S., ... Falk, E. B. (2017). Brain connectivity dynamics during social interaction reflect social network structure. *Proceedings of the National Academy of Sciences, 114*(20), 5153–5158.

Seeley, W. W., Menon, V., Schatzberg, A. F., Keller, J., Glover, G. H., Kenna, H., ... Greicius, M. D. (2007). Dissociable intrinsic connectivity networks for salience processing and executive control. *Journal of Neuroscience, 27*(9), 2349–2356.

Serences, J. T., Shomstein, S., Leber, A. B., Golay, X., Egeth, H. E., & Yantis, S. (2005). Coordination of voluntary and stimulus-driven attentional control in human cortex. *Psychological Science, 16*(2), 114–122.

Sherman, L. E., Payton, A. A., Hernandez, L. M., Greenfield, P. M., & Dapretto, M. (2016). The power of the like in adolescence: Effects of peer influence on neural and behavioral responses to social media. *Psychological Science, 27*(7), 1027–1035.

Sherry, J. L. (2004). Flow and media enjoyment. *Communication Theory, 14*(4), 328–347.

Shulman, G. L., McAvoy, M. P., Cowan, M. C., Astafiev, S. V., Tansy, A. P., d'Avossa, G., & Corbetta, M. (2003). Quantitative analysis of attention and detection signals during visual search. *Journal of Neurophysiology, 90*(5), 3384–3397.

Singer, T., Critchley, H. D., & Preuschoff, K. (2009). A common role of insula in feelings, empathy and uncertainty. *Trends in Cognitive Sciences, 13*(8), 334–340.

Small, G. W., & Vorgan, G. (2008). *iBrain: Surviving the technological alteration of the modern mind.* New York, NY: HarperCollins.

Söderlund, G., Sikström, S., & Smart, A. (2007). Listen to the noise: Noise is beneficial for cognitive performance in ADHD. *Journal of Child Psychology and Psychiatry, 48*(8), 840–847.

Sparks, J. V., & Lang, A. (2015). Mechanisms underlying the effects of sexy and humorous content in advertisements. *Communication Monographs, 82*(1), 134–162.

Tamir, D. I., & Mitchell, J. P. (2012). Disclosing information about the self is intrinsically rewarding. *Proceedings of the National Academy of Sciences, 109*(21), 8038–8043.

Tegelbeckers, J., Kanowski, M., Krauel, K., Haynes, J.-D., Breitling, C., Flechtner, H.-H., & Kahnt, T. (2018). Orbitofrontal signaling of future reward is associated with hyperactivity in attention-deficit/hyperactivity disorder. *Journal of Neuroscience, 38*, 6779–6786.

Todd, J. J., & Marois, R. (2004). Capacity limit of visual short-term memory in human posterior parietal cortex. *Nature, 428*(6984), 751–754.

Treisman, A. M. (1964). Selective attention in man. *British Medical Bulletin, 20*, 12–16.

Uddin, L. Q., Kelly, A. M. C., Biswal, B. B., Margulies, D. S., Shehzad, Z., Shaw, D., ... Milham, M. P. (2008). Network homogeneity reveals decreased integrity of default-mode network in ADHD. *Journal of Neuroscience Methods, 169*(1), 249–254.

Uncapher, M. R., Lin, L., Rosen, L. D., Kirkorian, H. L., Baron, N. S., Bailey, K., ... Wagner, A. D. (2017). Media multitasking and cognitive, psychological, neural, and learning differences. *Pediatrics, 140*(Suppl. 2), S62–S66.

Uncapher, M. R., Thieu, M. K., & Wagner, A. D. (2015). Media multitasking and memory: Differences in working memory and long-term memory. *Psychonomic Bulletin & Review, 23*(2), 483–490.

Wang, Z., Irwin, M., Cooper, C., & Srivastava, J. (2015). Multidimensions of media multitasking and adaptive media selection. *Human Communication Research, 41*(1), 102–127.

Wang, Z., Lang, A., & Busemeyer, J. R. (2011). Motivational processing and choice behavior during television viewing: An integrative dynamic approach. *Journal of Communication, 61*, 71–93.

Wasylyshyn, N., Hemenway Falk, B., Garcia, J. O., Cascio, C. N., O'Donnell, M. B., Bingham, C. R., ... Falk, E. B. (2018). Global brain dynamics during social exclusion predict subsequent behavioral conformity. *Social Cognitive and Affective Neuroscience, 13*(2), 182–191.

Weber, R., Alicea, B., Huskey, R., & Mathiak, K. (2018). Network dynamics of attention during a naturalistic behavioral paradigm. *Frontiers in Human Neuroscience, 12*, 182.

Westgate, E., & Wilson, T. (2018). Boring thoughts and bored minds: The MAC model of boredom and cognitive engagement. *Psychological Review.* Advance online publication.

Whitfield-Gabrieli, S., & Ford, J. M. (2012). Default mode network activity and connectivity in psychopathology. *Annual Review of Clinical Psychology, 8*, 49–76.

Wiradhany, W., & Niewwenstein, M. R. (2017). Cognitive control in media multitaskers: Two replication studies and a meta-analysis. *Attention, Perception, & Psychophysics, 79*(8), 2620–2641.

Xia, C. H., Ma, Z., Ciric, R., Gu, S., Betzel, R. F., Kaczkurkin, A. N., … Satterthwaite, T. D. (2018). Linked dimensions of psychopathology and connectivity in functional brain networks. *Nature Communications, 9*(1), 3003.

Xu, S., Wang, Z. J., & David, P. (2016). Media multitasking and well-being of university students. *Computers in Human Behavior, 55*, 242–250.

Yegiyan, N. S., & Bailey, R. L. (2016). Food as risk: How eating habits and food knowledge affect reactivity to pictures of junk and healthy foods. *Health Communication, 31*(5), 635–642.

Yeykelis, L., Cummings, J. J., & Reeves, B. (2014). Multitasking on a single device: Arousal and the frequency, anticipation, and prediction of switching between media content on a computer. *Journal of Communication, 64*(1), 167–192.

Zalesky, A., Fornito, A., Cocchi, L., Gollo, L. L., & Breakspear, M. (2014). Time-resolved resting-state brain networks. *Proceedings of the National Academy of Sciences, 111*(28), 10341–10346.

14

Video Gaming

A Challenge for the Brain's Reward System?

Rebecca J. Gilbertson and Douglas A. Gentile

Experience alters brain function. This observation is supported by the scientific literature showing that areas of the cortex devoted to motor function expand with repetitive practice such as in professional musicians (Elbert, Pantev, Wienbruch, Rockstroh, & Taub, 1995). Indeed, the brain is particularly sensitive to environmental influence during periods of intense brain development, such as early childhood, and adolescence. Given previous work showing evidence of brain plasticity (i.e., brain change due to experience), perhaps the emerging literature showing functional differences in diffuse areas of the brain, between problematic video gamers and non-gamers should not be a surprise. All learning is evidence of plasticity. However, emerging literature suggests that brain changes in video gamers are noted in circuitry that could perpetuate excessive, problematic video gaming behavior. With Gaming Disorder increasingly classified as a serious health issue (World Health Organization, 2018), an emerging literature is beginning to describe similarities in neural circuitry between substance-related and non-substance-related (behavioral) addictive disorders (with some notable differences). Although a comprehensive review is beyond the scope of this chapter, evidence of possible involvement of reward system, emotional memory, and motor systems is discussed. Also discussed are the cycle of addiction and possible implications for scientists studying excessive, problematic video gamers.

CLASSIFYING PROBLEMATIC VIDEO GAMING BEHAVIOR

Video gaming is immensely popular both in the United States and worldwide. As evidence of this popularity, in 2017, the US video gaming industry reported $36 billion of revenue, an 18% increase from 2016 (Entertainment Software Association, 2018). Increasingly, video gaming is normative, particularly for adolescents, with most individuals reporting that they play video games for recreation or enjoyment (Carras, Van Rooij, Mheen, Musci, & Xue, 2017; Yee, 2006). However, a rapidly expanding literature suggests that for some individuals, video gaming causes significant impairment and/or distress. Thus, the gaming behavior persists in spite of daily consequences such as poor sleep quality and quantity, decreased physical activity, eyesight or hearing problems, physical pain (e.g., back, neck, hands), decreased psychosocial function (i.e., loneliness, depression), and/or poor academic performance (Choo, Gentile, Sim,

Li, Khoo, & Liau, 2010; Dworak, Schierl, Burns, & Struder, 2007; Lemmens, Valkenburg, & Peter, 2011). Although most gamers do not experience severe daily consequences, scientific discourse surrounds how best to classify individuals who do.

Because excessive, problematic video game play affects multiple areas of an individual's life including psychosocial function and physical health, many methods of classification exist in the literature. Most recently, the World Health Organization (WHO) included diagnostic criteria for gaming disorder in the International Classification of Diseases (ICD-11). Examining the scientific literature over the past 30 years, we find a wide range of related terms such as problematic internet use (PIU), internet addiction disorder (IAD), video game addiction, pathological gaming, and compulsive internet use (CIU; Cash, Rae, Steel, & Winkler, 2012). Although they may sound different, each of these tends to use similar criteria focused on significant dysfunction caused by technology use. Recent reviews modernize the classification either through separation of online gaming from problematic internet use (see Grant, Potenza, Weinstein, & Gorelick, 2010), or through consideration of specific internet use disorders (i.e., gaming, gambling, pornography viewing, shopping, or communication; see Brand, Young, Laier, Wölfling, & Potenza, 2016). Of relevance to this chapter and the neuroimaging data presented in it, is the classification of video gaming as an internet use problem (see Weinstein & Lejoyeux, 2015; Weinstein, Livny, & Weizman, 2017). That is, most of the neuroimaging studies include participants who meet the criteria for "internet addiction." Some studies also include an additional psychiatric interview or self-reported measure specific to online gaming disorder (for examples, see Cai et al., 2016; Kaess et al., 2017). Consistent study eligibility criteria based on ICD-11 criteria[1] are needed to allow for direct comparison of neurobiological findings across studies. Continued scholarship of problematic internet users who access specific content similar to Brand, Young, Laier, Wölfling, and Potenza (2016) is also needed.

Consideration of excessive, problematic internet use as an addiction was first suggested by Young in 1998. Scientists typically use the term addiction to refer to a chemical substance that, when taken in excess, causes tolerance, withdrawal symptoms, loss of control, craving and/or psychological, physical or relational problems.[2] Behavioral addictions were first suggested in editorials between 1990 and 2001 and have historically included gambling disorder. Also included were food, shopping, and sex addiction (Marks, 1990). "Internet abuse" was later added in a well-cited editorial as the fastest-growing "behavioral addiction" (Holden, 2001, p. 982).

In 2013, the American Psychiatric Association included Internet Gaming Disorder in the appendix of the DSM-5. The appendix is where emerging problems are included, where the early scientific research looks strong but not yet sufficient to determine whether it should be included as a full diagnosis. Thus, internet gaming disorder was included as a condition for further study (see American Psychiatric Association, 2013; Petry & O'Brien, 2013). Nine criteria were proposed:

1. preoccupation with gaming;
2. tolerance;
3. psychological withdrawal symptoms;
4. unsuccessful attempts to control or limit gaming;
5. loss of interest in previous hobbies;
6. continued use despite knowing that it is a problem;
7. deceiving family members and/or therapists;
8. use of internet games to escape a negative mood;
9. have jeopardized or lost a relationship, job, or educational opportunity (American Psychiatric Association, 2013).

A 27-item scale is now available to measure internet gaming disorder symptoms (Lemmens, Valkenburg, Gentile, & Reynolds, 2015). A 9-item scale is also available (Lemmens et al., 2015). Most recently, WHO adopted criteria for gaming disorder (2018) including offline and online games (as did the DSM-5, despite the name). The suggested diagnostic criteria are not without criticism, particularly concerning over-reliance on substance use disorder criteria (Aarseth et al., 2017; Griffiths, Kuss, Lopez-Fernandez, & Pontes, 2017; Rumpf et al., 2018).

Thus, there is emerging consensus that problematic video gaming behavior may best be classified as addiction by both the American Psychological Association and the World Health Organization. Previous scholarship delineating problem gamers from heavy gamers suggests that the criteria, "playing to escape," may be of use in identifying individuals with problematic internet behavior (Caplan, Williams, & Yee, 2009; Kuss, Louws, & Wiers, 2012).[3] A recent study found preliminary evidence that young adults who are stressed and use games as a coping mechanism may be at greater risk for developing gaming disorder (Plante, Gentile, Groves, Modlin, & Blanco-Herrera, 2018). In the applied settings, tolerance (more time or a new game needed to achieve a desired effect), loss of control (greater amount of time than intended), withdrawal symptoms (negative affect, irritability) and loss of educational opportunities including educational achievement also are supported by the literature (Beutel, Hoch, Wölfing, & Müller, 2011; Cash et al., 2012). Application of the DSM-like criteria for internet gaming disorder in a large sample of 19,000 individuals showed a prevalence of less than 1%. Generally, most studies report the prevalence in Europe and the United States as between 1% and 10%.[4]

NEUROANATOMY OVERVIEW

Many brain regions and their connections are implicated in the contingency learning characteristics of the brain. Prior to discussion of this neural circuitry, an understanding of the organizational structure of the brain is helpful. The brain is often divided into three main sections: the hindbrain (consisting of the brain stem, the area closest to the spine), the midbrain (a very small area in humans), and the forebrain (very large in humans). Entering the brain from the base of the skull, brain stem structures first encountered include those most important for sustaining life. These include the medulla (respiration, heart rate, vomiting, coughing), the pons (arousal), and the cerebellum.

The midbrain contains the tectum and tegmentum. The tectum contains the superior and inferior colliculus that process visual and auditory information, respectively, as it pertains to reflexive behavior and survival of the organism. The midbrain also contains areas of relevance in the study of addiction including the ventral tegmental area and the substantia nigra. These areas contain dopamine neurons and they project to the nucleus accumbens of the ventral striatum in the forebrain (i.e., the mesocorticolimbic pathway) and basal ganglia of the dorsal striatum (i.e., the nigrostriatal pathway). Functions associated with the mesocorticolimbic pathway include motivation, reward, emotion, negative affect, and memory (i.e., the "accelerator system"). The nigrostriatal pathway is associated with motor function.

As the forebrain is large in humans, terms used in the description of brain development, the diencephalon and the telecephalon, are best used to further subdivide this area. The diencephalon includes the hypothalamus, which is involved in basic motivated behaviors, such as seeking food, shelter and reproductive functions, and the thalamus for relaying sensory information to the cortex. Forebrain structures included in the telencephalon include those structures regulating emotion, memory, motor function, attention, and executive control.

Covering the brain is a 1–3 mm-thick area called the cortex. The structure of the human cortex, wrinkled in appearance, allows for more brain tissue to fit into the confined space of the skull. Functional areas of the cortex include those important for judgement, planning, assessment of behavior in social setting/impulse control (i.e., the prefrontal cortex, the "braking system"), visual and auditory processing, motor function, and touch. Additionally, subcortical structures of the brain are connected to the cortex via white matter projections creating neural circuitries. Brain areas involved in addiction, particularly concerning the midbrain and basal forebrain areas, are discussed next.

The Reward System

Studies of the reward system in the brain date to the 1950s (Olds & Milner, 1954). Importantly, much of the early work on the reward system in the brain was completed in animal models. Animal models are advantageous as they allow precise measurement of neurotransmitters within structures involved in the mesolimbic dopamine system (i.e., the reward system). Coordinated with the measurement of neurotransmitter levels was drug-taking behavior (drug self-administration) in the animals. From these studies, two predominant theories of addiction were proposed. The incentive sensitization theory of addiction (Robinson & Berridge, 1993) discussed below, was one of these theories. A second theory, the "hedonic homeostatic dysregulation" model, includes a cycle of addiction (Koob & Le Moal, 1997), and is discussed in the emotional memory section of this chapter. Prior to description of the early rodent work contributing to these theories, a few observations are offered for consideration. First, pre-clinical models approximate human behavior and thus, important species differences exist. One species difference is that substances commonly abused in humans such as nicotine and alcohol are not readily self-administered in rodents. While the neural substrates discussed below were all documented with multiple drugs of abuse (e.g., cocaine, opiates, alcohol), rodents sometimes administer these drugs (i.e., alcohol) differently as compared to humans, particularly during the acquisition phase. Perhaps more relevant to this particular section is that motivational reward circuitry ("the accelerator system") is tempered by the prefrontal cortex in humans ("the braking system"), allowing cognitive control and influencing future decision-making (Jentsch & Taylor, 1999). Thus, while pre-clinical models are crucial to the study of addiction, human studies may have different or added focus, given the higher-order brain structures with a unique role in addiction in humans (e.g., the anterior cingulate, the insula, the prefrontal cortex). While these systems seem to work in parallel in healthy individuals (see Botvinick & Braver, 2015), one theory of addiction describes "hedonic homeostatic dysregulation" as the accelerator system overwhelms the braking system, contributing to loss of control over the drug-seeking and drug-using behavior (Koob & Le Moal, 1997). Notably, in their review of resting-state functional connectivity, Sutherland and colleagues (2012) focus on the insula, the anterior cingulate cortex (a possible "gatekeeper" between the "accelerator" and "braking systems") and the prefrontal cortex, using nicotine dependence as an exemplar. Some of this literature, as it pertains to internet gaming, is discussed below.

Reward in the brain was serendipitously discovered by Olds and Milner in 1954 (Norgren, 2015). Interested in the neural substrates of learning, they discovered that rodents were more likely to return to locations on a table where they had previously received small amounts of electrical current from an electrode implanted directly into the brain. The location of the electrode was assumed to be the midbrain, given the role of the nearby hypothalamus in motivated behavior. However, later imaging showed that Olds had accidentally overshot the midbrain and the electrode was serendipitously placed in the forebrain in the septum pallicidum, an area adjacent to the nucleus accumbens. Thus, the electrode was stimulating the area of the brain surrounding the nucleus accumbens, making a location on the table more salient to the rodents as compared to other locations. Subsequent studies in an operant conditioning chamber showed that rodents would press for electrical stimulation of the same area, a procedure known as intracranial self-stimulation.[5] Approximately four decades after this important discovery, Wise (1996; 2004) showed in a pre-clinical model that dopamine release in the nucleus accumbens was critical for contingency learning (i.e., associating greater value to certain outcomes), for example, associated with drugs of abuse (e.g., cocaine, heroin).

Described another way, studies conducted by Wise and colleagues supported that activation of the mesolimbic dopamine ("reward") pathway assisted in memory formation of the drug-using experience (Wise, 2004). That is, dopamine release in the nucleus accumbens functions to alert the organism (humans, rodents, monkeys, etc.) following behaviors necessary for survival (i.e., food, sex, seeking shelter to stay warm). Dopamine release is pleasurable, influencing motivation and ensuring that the behavior is repeated. Notably, food and sex also cause release of dopamine in the nucleus accumbens. Dopamine, therefore, acts as a signal, alerting and orienting the organism. However, drugs of abuse, including alcohol, cause greater dopamine release than these so-called "natural rewards" (Hernandez & Hoebel, 1988). If seeking, using, and recovering from acute intoxication are repeated, this high level of dopamine release can supplant all other motivated behaviors. Thus, addicted individuals become preoccupied with the substance (i.e., "want"/"crave") almost to the exclusion of every other activity because of its increased salience to their everyday functioning (Robinson & Berridge, 1993). Drug "wanting" versus drug "liking" was theorized to maintain the cycle of addictive behavior (Robinson & Berridge, 1993). Activation of motor circuitry allows for the behavior to be completed and emotional memory allows the experience to be remembered (alternatively, it allows the outcome when the experience doesn't occur, such as the withdrawal symptoms, to be remembered; Koob, 2009).

Although scientists were aware that other neurotransmitters influenced addictive behavior (including opioid and endocannabinoid systems), studies were largely constrained to the study of the mesolimbic dopamine pathway. A series of studies aimed to extend findings to humans concerning dopamine release in the nucleus accumbens being important for maintaining the cycle of addiction. Specifically, a type of brain imaging technology, positron emission tomography (PET) was used in the mid-1990s by Nora Volkow and colleagues to characterize dopamine release in humans with stimulant addiction (cocaine; Volkow et al., 1997). In particular, studies were conducted using the classic dopamine radioligand [^{11} C] raclopride. The method is an indirect measurement of dopamine release as it allows for quantification of dopamine receptor occupancy. Curiously, shortly after these seminal findings by Volkow and colleagues, Koepp and colleagues (1998) demonstrated dopamine release during video game play. Specifically, the study by Koepp and colleagues also used the classic dopamine radioligand [^{11}C] raclopride to demonstrate that playing a video game caused dopamine release (i.e., decreased radioligand binding following video gaming as compared to baseline) in the ventral striatum, an area including the nucleus accumbens (Koepp et al., 1998). To our knowledge, this study has not yet been replicated.[6]

Some evidence suggests variability in dopamine neurotransmitter release or receptor occupancy in internet or gaming addicted individuals. That is, studies may suggest dopamine release in the striatum (ventral or dorsal), reduced dopamine receptor occupancy (dorsal striatum, see Tian et al., 2014; Weinstein & Lejoyeux, 2010), or dopamine transporter (DAT) reduction (Weinstein, Livny, & Weizman, 2017; see also Kim, 2011, for reduced D_2 receptor binding in the striatum). These findings have yet to be replicated and are constrained by heterogeneous selection criteria (also including participants with internet addiction, however; see Tian et al., 2014). Findings are also constrained given somewhat stronger evidence for involvement of dopamine release in the dorsal striatum (motor circuitry; see Kim, 2011; Weinstein et al., 2010) versus ventral striatum (reward), a dissimilarity to substance use disorder literature. Finally, the observations of Clark, director of the Centre of Gambling Research at the University of British Columbia (Clark, 2014) are of relevance when interpreting these findings. In his review of the neurobiology of addiction of gambling disorder, Clark (2014) wrote that studies of gambling disordered individuals sometimes suggest opposite effects to those found in substance use disordered individuals, given the absence of the neurotoxic effects of alcohol and drugs. For instance, concerning dopamine release in the ventral striatum, an area of the brain that includes the nucleus accumbens, individuals with gambling disorder show greater dopamine release following exposure to relevant stimuli. These findings are disparate from the substance use disorder literature where there is a *blunted* dopamine release in the ventral striatum and down-regulated dopamine (D_2) receptor occupancy, a counterintuitive finding indicative of possible withdrawal symptoms (Koob & Le Moal, 2005; Volkow et al., 1997; Volkow, Wang, Fowler, Tomasi, & Telang, 2011). Thus, it is curious that the results of the video gaming literature to date seem to follow that of the substance abuse literature versus gambling disorder, arguably a better comparison given the tendency to classify both as behavioral addictions.

Brain imaging studies (fMRI) have also considered the reward system in video gamers. Findings from these studies suggest differences in both structure and function of reward processing areas in frequent versus infrequent gamers. In the study by Kühn and colleagues (2011), 145 adolescents completed a task designed to measure brain activity during reward anticipation (delayed gratification task). They found slight structural differences between frequent versus infrequent gamers in the striatum, suggesting a possible pre-morbid or pre-existing condition. Specifically, the left ventral striatum was larger in frequent gamers versus non-gamers. Functional differences were also noted as frequent gamers displayed a greater blood-oxygen-level dependent (BOLD) response following feedback regarding small and large losses versus no loss as compared to infrequent gamers. They interpreted their results as showing involvement of the ventral striatum in frequent gamers, similar to substance use disorders. Notably, the direction of findings was opposite that of the substance use disorder literature (some studies report volume reduction), possibly because of the neurotoxic effects of the drugs.

The results of Kühn and colleagues (2011) were also associated with the behavioral performance (reduced time to place a bet) on a gambling task completed outside the scanner. Interestingly, Cai and colleagues (2016) showed that increased ventral striatum volume was associated with scores on Young's Internet Addiction Test in individuals with internet gaming disorder.[7] In the following section, the dorsal striatum, often mentioned in studies of gamers versus non-gamers, is briefly discussed as it often appears in papers with the ventral striatum.

The Motor System

Several studies have noted differences in the dorsal striatum of gamers versus non-gamers (Erickson et al., 2010; Vo et al., 2011). Indeed, often if the study mentions "striatal differences" in structure or function, the dorsal striatum (motor circuitry) versus ventral striatum (reward) is implicated in studies of video gamers. The dorsal striatum consists of motor circuitry, namely, the caudate and putamen. The globus pallidus, also involved in motor circuitry (but not a part of the dorsal striatum), inhibits the thalamus. Movement is allowed to occur when the caudate and putamen inhibit the globus pallidus, allowing an excitatory signal to occur. Considered with reward circuitry, the motor circuitry allows for motivation to become action (Mogenson, Jones,

& Yim, 1980). In the context of addiction, the dorsal striatum is implicated in "habit learning" of repetitive motion (Volkow et al., 2011, p. 15039). Concerning the gaming literature, volumetric differences in the dorsal striatum are positively related to learning to play a video game created for research, so that video game performance could be objectively quantified. This suggests that the dorsal striatum may be a predisposing factor rather than a consequence of excessive game play. For instance, Erickson and colleagues (2010) scanned individuals prior to video game training. *A priori* regions of interest were the hippocampus and the striatum (ventral and dorsal). Following 20 hours of video game training, participants were scanned again. Findings showed that the ventral striatum (reward) was not related to video game performance following training. However, the dorsal striatum (motor circuitry) was positively correlated with video game performance following training. A second study by Vo and colleagues (2011) showed that dorsal striatum white matter at baseline predicted later video game performance. Findings were not significant for gray matter in the dorsal striatum. Thus, the dorsal striatum is an area that may be particularly relevant in prospective studies focused on the etiology of video gaming disorder. Findings concerning the dorsal striatum should, however, be discussed in the context of motor pathway involvement and habit learning versus involvement with reward. Although the substance use literature (cocaine-dependent individuals) shows involvement of both ventral and dorsal striatum (Volkow et al., 2011), the preponderance of the evidence in the gaming literature shows null findings for ventral striatum (reward) and involvement of the dorsal striatum (motor circuitry; Erickson et al., 2010; Vo et al., 2011) as well as for behavioral data (Cai et al., 2016).

Emotional Memory

Also relevant is another complementary theory of addiction that focuses on the maintenance of behaviors to alleviate negative withdrawal symptoms (Koob & Le Moal, 1997). This model is often referred to as the allostasis model and describes an addictive cycle, including negative affect associated with withdrawal. The neural circuitry involved in emotional memory includes the amygdala and surrounding structures, and subsequent projections to the medial prefrontal cortex, the cingulate gyrus, the hippocampus, and the insula (Koob, 2009; Sutherland et al., 2012). This circuitry is activated not only during positive emotional events, but relevant to the current discussion, during negative emotional events. That is, the negative affect and irritability associated with *not* using a drug (or playing video games; Lemmens et al., 2015) is learned over time and remembered (i.e., emotional memory for aversive events). Thus, the *cycle of addiction*, consisting of *preoccupation/anticipation*, *binge/intoxication*, *withdrawal/negative affect* is perpetuated through negative reinforcement, that is, engaging in dysfunctional behavior to avoid the negative aversive event (i.e., withdrawal syndrome; Koob & Le Moal, 1997). As individuals progress in their disease, they binge less for pleasure (positive reinforcement) and more to alleviate negative affect/withdrawal symptoms (negative reinforcement). An analogous model for behavioral addiction suggests that with escalating loss of control, the behavior of gaming for pleasure (i.e., "gratification") diminishes and is replaced by "compensation" (Brand et al., 2016), such as use of the internet in place of social interaction or following a consequence to alleviate a negative mood.

Compensatory neurobiological changes designed to maintain homeostasis during addiction exist, and without the presence of the drug, typically manifest as the withdrawal syndrome. Koob (2009) theorizes that consistent with most biological systems, a regulatory process exists to offset reward. Thus, anti-reward, or the negative affect, irritability, and loss of motivation for natural reward during withdrawal are not only mediated by the amygdala and surrounding structures. The stress system also becomes activated during the withdrawal syndrome.

Stress system responsivity, an example of neurobiological change, is dysregulated in substance use disordered individuals, and predicts relapse (Adinoff, Junghanns, Kiefer, & Krishnan-Sarin, 2005; Lovallo, 2006; Sinha et al., 2011). Given the evidence of dysregulated stress system function following challenge in substance use disorder, studies have recently considered the stress system response in non-substance-related addictive disordered individuals (Bibbey, Phillips, Ginty, & Carrol, 2015; Geisel, Panneck, Hellweg, Wiedemann, & Muller, 2015; Kaess et al., 2017; Lemieux & al'Absi, 2016). Geisel and colleagues (2015) conducted a pilot study of male pathological gamblers and internet gaming disordered individuals and measured serum levels of hypothalamic pituitary adrenal axis function (copeptin, ACTH, cortisol) from a single morning blood sample. They found that although stress system function did not differ between non-substance-related addictive disordered individuals and matched controls, cortisol levels were negatively correlated with the pathological gambling adaptation of the Yale-Brown Obsessive-Compulsive Scale (PG-YBOCS).

Evidence of dysregulated stress system response following an acute stressor in individuals meeting suggested DSM-5 criteria for internet gaming disorder (i.e., addictive disorder criteria) were recently reported by Kaess and colleagues (2017). This study showed that males with internet gaming disorder displayed a lesser cortisol response, but greater negative affect, following a psychosocial stressor (i.e., Trier Social Stress Test) as compared to control participants. Thus, findings are mixed regarding stress system response in problematic online video gamers following a psychosocial stressor (see Bibbey et al., 2015), and additional studies are needed.

COGNITIVE FUNCTION

Loss of control over substance use or video gaming contributes to the maintenance of addictive behavior and may be a distinguishing characteristic between casual and pathological users (APA, 2013; Lemmens et al., 2015; Starcevic, 2013). It also is included in diagnostic criteria for addictive disorders (APA, 2013; WHO, 2018). Thus, the neural circuitry involved in cognitive control, including the anterior cingulate cortex and the lateral prefrontal cortex, are the focus of numerous studies of chemical and behavioral addiction (for reviews, see Goldstein & Volkow, 2011; Meng, Deng, Wang, Guo, & Li, 2015; Sutherland et al., 2012). One function of the anterior cingulate cortex is to monitor for salient events. This structure is particularly sensitive to behavior that may be error-prone, or that may have already caused an error (Sutherland et al., 2012). When the anterior cingulate cortex detects a salient event, it alerts the lateral prefrontal cortex for a "reorientation of attention" (p. 2284). The lateral prefrontal cortex functions to change processing such that relevant events including objects associated with drug use (or video gaming) are processed over irrelevant events (neutral objects; see Sutherland et al., 2012, for a review). Cues (drug or gaming paraphernalia), or previously neutral objects come to be associated with and signify the addictive behavior through associative learning processes and attentional bias. Activation of the lateral prefrontal cortex (specifically, the dorsolateral prefrontal cortex) and the anterior cingulate gyrus is studied in addiction using cue exposure paradigms. A third area of the brain implicated in cognitive control, or rather, the lack of it, is the dorsal anterior cingulate cortex. Deactivation of the dorsal anterior cingulate cortex is implicated in inhibition failure. Tasks measuring inhibition failure include go/no-go and stop signal tasks. Finally, the ventromedial prefrontal cortex (including the orbitofrontal cortex and portions of the anterior cingulate cortex) changes activation with the lateral prefrontal cortex (specifically, the dorsolateral prefrontal cortex) during advantageous decision-making (see

Goldstein & Volkow, 2011, for a review). Evidence of activation changes of neural circuitry involved in cognitive control during cue exposure, and decreases in activation during inhibitory failure and decision making, is briefly discussed below.

Cue Exposure

Cognitive control is often studied in the presence of drug and alcohol cues. An early example of a cue exposure study was conducted by Childress and colleagues in the late 1990s. In the study, Childress and colleagues showed that images of drug paraphernalia, remarkably even those shown at levels below conscious awareness (<100 ms), also cause limbic system activation using fMRI (Childress et al., 1999; Young et al., 2014). Cue exposure has also been studied somewhat extensively in individuals with video gaming disorder. Notably, a recent meta-analysis considered 61 studies for inclusion (Meng et al., 2015). Of the 61 studies, ten were included with focus on two regions typically associated with cognitive control, the anterior cingulate cortex and the dorsolateral prefrontal cortex. Tasks included in the scans included attention, working memory, and cue exposure. In comparison to healthy controls, the meta-analysis found evidence of greater bilateral medial frontal gyrus (Broadmann Area 9) and left cingulate gyrus activation (Broadmann Areas 23, 24). These also significantly correlated with on-line time. It should be noted that findings in substance-dependent individuals typically include larger areas of the cortex, including such areas as the dorsolateral prefrontal cortex including Broadmann Areas 6, 8, 9, and 46 (Goldstein & Volkow, 2011). While the meta-analysis found limited findings in individuals with internet gaming disorder, in a recent study, cue reactivity was assessed in 15 participants with internet gaming addiction and 15 remitted patients (Ko et al., 2013). The results in that study showed greater activation in the dorsolateral prefrontal cortex following cue exposure, an area of the brain associated with attention and craving, specifically in inhibition of the behavioral response to a cue. Earlier studies by Ko and colleagues showed activation in the right orbitofrontal cortex, right nucleus accumbens, bilateral anterior cingulate cortex, medial frontal cortex, right dorsolateral prefrontal cortex, and right caudate nucleus in response to cues (Ko et al., 2013). Thus, although a meta-analysis found narrow areas of activation differences in individuals with IGD versus healthy controls, other studies find limited support for activation differences in cue exposure paradigms.

Attentional Bias

Attentional bias to cues may be studied using visual probe methodology. Briefly, visual probe methodology sometimes involves the presentation of two or more images simultaneously. Typically, one image is neutral and the other image contains substance-related stimuli. Following, the images rapidly disappear and participants respond quickly to a visual probe (e.g., a dot) that is either on the side of the neutral image or the substance-related stimuli. Substance use disordered individuals in comparison to non-users respond more quickly to the dot on the same side of the screen as the substance-related stimuli (Ehrman, Robbins, Bromwell, Lankford, Monterosso, & O'Brien, 2002; Field & Cox, 2008). As applied to video gaming, Lorenz and colleagues (2013) recently demonstrated that "pathological computer game players" display attentional bias to gaming images and control images with positive valence. This attentional bias was associated with greater activation in the medial prefrontal cortex, the anterior cingulate cortex, the orbitofrontal cortex, and the midbrain areas associated with reward. Importantly, this study assessed a variety of demographic variables. Pathological computer game players had higher depression, higher amount of play during the week and weekends, and lower alcohol consumption as compared to healthy controls. Impulsivity, as measured by the Barratt Impulsiveness Scale (BIS), and decision-making were not different between pathological computer game players and controls (although several other studies have found differences on impulsiveness, e.g., Gentile et al., 2011).

Lorenz and colleagues (2013) also showed that although short presentation images (~200 ms) caused activation in motivation areas associated with addiction, participants may have had sufficient time to inhibit these responses in the long presentation of the images. Exploratory functional connectivity analyses tentatively supported that an area of the brain associated with inhibition, the right inferior frontal gyrus, was associated with signal strength modulation in the left orbitofrontal cortex, the left ventral and the bilateral dorsal striatum, as measured in long versus short trials, respectively. These findings are interesting on several levels. First, the study includes a control group and is methodologically sound. Images in the study are specific to gaming and are controlled using positive and negatively valenced images. Also, stimulus presentation time varied and included both long and short presentations. Finally, results were interpreted within the current scholarship of substance use disorders and the authors' conclusions were supported by a subsequent exploratory data analysis. Limitations of the study include a small sample size and minor image presentation methodological issues mentioned by the authors.

Inhibition

Inhibition of behavioral responses to gaming-relevant cues is typically studied using go/no-go tasks. Cued go/no-go tasks assess if participants can inhibit a response following a cue (i.e., a vertical rectangle) that typically is associated with a go-target (i.e., vertical box turning green). Findings in individuals with internet gaming disorder showed activation of more brain regions associated with response inhibition, particularly frontal and caudate, and lower insula activation when processing an impulsive mistake, as compared to controls (Ko et al., 2013). Relatedly, executive functioning was assessed with a risky decision-making task (Dong & Potenza, 2016). When making risky decisions, individuals with internet gaming disorder do so more quickly without needed activation of brain areas associated in decision-making. Interestingly, the insula is also implicated in gambling disorder (Clark, 2014). Individuals with gambling disorder show similar activation to wins and near-misses in a slot machine task as compared to control participants. As Ko and colleagues (2013) showed lower insula activation when processing an impulsive mistake, and Clark (2014) described no change in insula activity when processing near-misses, additional studies are needed to assess insula function during risky decision-making.

CONCLUSION

Findings from imaging studies tentatively suggest that some areas of activation are shared between individuals with video gaming disorder and those with substance use disorders. However, future studies in internet gamers could be improved by continuing to assess the direction of the activation changes (greater versus less), the recruitment of additional areas to complete tasks when one area should be deactivated for efficient brain function (such as the ventromedial prefrontal cortex), and bidirectional task performance indicators (improvement, decrement). Furthermore, there is a need for more prospective studies of individuals with video gaming disorder, such as enrolling participants in a longitudinal study and systematically assessing brain function using multimodal techniques. Participants meeting criteria for chemical or behavioral addiction, as assessed throughout the study, could be compared to those not meeting criteria. This type of study is particularly needed, as much remains to be learned about the etiology of and risk factors for video gaming disorder.

Regarding imaging studies specifically, conclusions are difficult to reach as inclusion criteria, experimental tasks, and the neural regions of interest differ between studies. This heterogeneity makes it difficult to draw conclusions from the neuroimaging literature alone, although numerous behavioral studies generally support the themes found in the neuroimaging literature (e.g., Gentile et al., 2017; Petry et al., 2014). Results, therefore, are still emerging. In studies where there are effects, the effects are generally of less magnitude (in terms of regions affected in the prefrontal cortex) than in the substance abuse literature.

Now that Gaming Disorder has been defined by the World Health Organization, studies should begin to include clinically-identified participants with the disorder. Most studies to date have identified people by the number of symptoms on a screening tool, which is an appropriate first approximation. Recruiting participants using a consistent set of criteria across studies, however, may provide a better sample for finding neurological differences (and allow us to examine the neurobiology underlying different patterns of co-morbidity with Gaming Disorder).

NOTES

1. Symptom counts similar to the mild, moderate, and severe specifiers found in the DSM-5 Substance-Related and Addictive Disorders or polytomous scales assessing the extent to which participants experience a relevant symptom, are additional considerations that could provide useful demographic information regarding a neuroimaging/neurobiological participant sample (see Lemmens, Valkenburg, Gentile, & Reynolds, 2015 for discussion of normal, risky, and disordered groups).
2. Addictive disorder (previously, substance use disorder) criteria include: Tolerance (increased amount to feel the same effect; same amount has less effect than before), withdrawal (a characteristic syndrome possibly involving negative mood, agitation/irritability, physical symptoms; relieved by use), loss of control over activity (cannot quit or cut down on use; more hours or days spent on use than anticipated), or severe physical, psychological, or social/relational consequences (APA, 2013; Starcevic, 2013).
3. However, immersion/escapism is not considered a criterion for substance use disorders/addictive disorders, according to the DSM, thus suggesting that its discriminant validity may be limited.
4. Prevalence estimates for internet gaming disorder are less than 5% in Germany (Festl, Scharkow, & Quandt, 2013), 6.7% in the Netherlands (Lemmens et al., 2015), and 5.9–10.8% in a regional US sample (Eichenbaum, Kattner, Bradford, Gentile, & Green, 2015), and a large study of almost 19,000 international participants (the United States, Canada, the United Kingdom, Germany) employing DSM-5 criteria reported less than 1% (Przybylski, Weinstein, & Murayama, 2017).
5. Electrical stimulation in the nucleus accumbens in humans during implantation of a deep brain stimulation device for obsessive compulsive disorder also causes euphoria (Haq et al., 2011).
6. Of interest in current studies is the specificity of the radioligand used. Some studies report findings with a radioligand specific to monoamines, including both catecholamines (dopamine, norephinphrine) and indolamines (serotonin, melatonin), making data difficult to interpret (e.g., Tian et al., 2014).
7. The selection criteria for the Cai et al. (2016) study included an Internet Addiction Test (IAT) score of over 50 and endorsing five of nine of the suggested DSM criteria for internet gaming disorder.

REFERENCES

Aarseth, E., Bean, A. M., Boonen, H., Colder Carras, M., Coulson, M., Das, D., ... Van Rooij, A. J. (2017). Scholars' open debate paper on the World Health Organization ICD-11 Gaming Disorder proposal. *Journal of Behavioral Addictions, 6*(3), 267–270.

Adinoff, B., Junghanns, K., Kiefer, F., & Krishnan-Sarin, S. (2005). Suppression of the HPA axis stress-response: Implications for relapse. *Alcoholism: Clinical and Experimental Research, 29*, 1351–1355.

APA (American Psychiatric Association). (2013). *Diagnostic and statistical manual of mental disorders* (5th ed.). Arlington, VA: Author.

Beutel, M. E., Hoch, C., Wölfing, K., & Müller, K. W. (2011). Clinical characteristics of computer game and internet addiction in persons seeking treatment in an outpatient clinic for computer game addiction. *Zeitschrift für Psychosomatische Medizin und Psychotherapie, 57*(1), 77–90.

Bibbey, A., Phillips, A. C., Ginty, A. T., & Carrol, D. (2015). Problematic internet use, excessive alcohol consumption, their comorbidity and cardiovascular and cortisol reactions to acute psychological stress in a student population. *Journal of Behavioral Addictions, 4*, 44–52.

Botvinick, M., & Braver, T. (2015). Motivation and cognitive control: From behavior to neural mechanism. *Annual Reviews of Psychology, 66*, 83–113.

Brand, M., Young, K. S., Laier, C., Wölfling, K., & Potenza, M. N. (2016). Integrating psychological and neurobiological considerations regarding the development and maintenance of specific Internet-use disorders: An Interaction of Person-Affect-Cognition-Execution (I-PACE) model. *Neuroscience & Biobehavioral Reviews, 71*, 252–266.

Cai, C., Yuan, K., Yin, J., Feng, D., Bi, Y., Li, Y., ... Tian, J. (2016). Striatum morphometry is associated with cognitive control deficits and symptom severity in internet gaming disorder. *Brain Imaging and Behavior, 10*(1), 12–20.

Caplan, S., Williams, D., & Yee, N. (2009). Problematic internet use and psychosocial well-being among MMO players. *Computers in Human Behavior, 25*(6), 1312–1319.

Carras, M. C., Van Rooij, A. J., Mheen, D. V., Musci, R., & Xue, Q.-L. (2017). Video gaming in a hyperconnected world: A cross-sectional study of heavy gaming, problematic gaming symptoms, and online socializing in adolescents. *Computers in Human Behavior, 68*, 472–479.

Cash, H., Rae, C. D., Steel, A. H., & Winkler, A. (2012). Internet addiction: A brief summary of research and practice. *Current Psychiatry Reviews, 8*(4), 292–298.

Childress, A. R., Mozley, P. D., McElgin, W., Fitzgerald, J., Reivich, M., & O'Brien, C. P. (1999). Limbic activation during cue-induced cocaine craving. *American Journal of Psychiatry, 156*(1), 11–18.

Choo, H., Gentile, D. A., Sim, T., Li, D., Khoo, A., & Liau, A. K. (2010). Pathological video-gaming among Singaporean youth. *Annals Academy of Medicine, 39*, 822–829.

Clark, L. (2014). Disordered gambling: The evolving concept of behavioral addiction. *Annals of the New York Academy of Sciences, 1327*(1), 46–61.

Dong, G., & Potenza, M. N. (2016). Risk-taking and risky decision-making in Internet gaming disorder: Implications regarding online gaming in the setting of negative consequences. *Journal of Psychiatry Research, 73*, 1–8.

Dworak, M., Schierl, T., Bruns, T., & Struder, H. K. (2007). Impact of singular excessive computer game and television exposure on sleep patterns and memory performance of school-aged children. *Pediatrics, 120*, 978–985.

Ehrman, R. N., Robbins, S. J., Bromwell, M. A., Lankford, M. E., Monterosso, J. R., & O'Brien, C. P. (2002). Comparing attentional bias to smoking cues in current smokers, former smokers, and non-smokers using a dot-probe task. *Drug and Alcohol Dependence, 67*, 185–191.

Eichenbaum, A., Kattner, F., Bradford, D., Gentile, D. A., & Green, C. S. (2015). Role-playing and real-time strategy games associated with greater probability of Internet gaming disorder. *Cyberpsychology, Behavior, and Social Networking, 18*(8), 480–485.

Elbert, T., Pantev, C., Wienbruch, C., Rockstroh, B., & Taub, E. (1995). Increased cortical representation of the fingers of the left hand in string players. *Science, 270*, 305–307.

Entertainment Software Association. (2018, January 18). U.S. video game industry revenue reaches $36 billion in 2017. Retrieved from www.theesa.com/article/us-video-game-industry-revenue-reaches-36-billion-2017.

Erickson, K. I., Boot, W. R., Basak, C., Neider, M. B., Prakash, R. S., Voss, M. W., … Kramer, A. F. (2010). Striatal volume predicts level of video game skill acquisition. *Cerebral Cortex, 20*(11), 2522–2530.

Festl, R., Scharkow, M., & Quandt, T. (2013). Problematic computer game use among adolescents, younger and older adults. *Addiction, 108*(3), 592–599.

Field, M., & Cox, W. M. (2008). Attentional bias in addictive behaviors: A review of its development, causes, and consequences. *Drug and Alcohol Dependence, 97*, 1–20.

Geisel, O., Panneck, P., Hellweg, R., Wiedemann, K., & Muller, C. (2015). Hypothalamic-pituitary-adrenal axis activity in patients with pathological gambling and internet use disorder. *Psychiatry Research, 226*, 97–102.

Gentile, D. A., Bailey, K., Bavelier, D., Brockmyer, J. F., Cash, H., Coyne, S. M., … Young, K. (2017). Internet gaming disorder in children and adolescents. *Pediatrics, 140*, S81–S85.

Gentile, D. A., Choo, H., Liau, A., Sim, T., Li, D., Fung, D., & Khoo, A. (2011). Pathological video game use among youth: A two-year longitudinal study. *Pediatrics, 127*, 319–329.

Goldstein, R. Z., & Volkow, N. D. (2011). Dysfunction of the prefrontal cortex in addiction: Neuroimaging findings and clinical implications. *Nature Reviews Neuroscience, 12*(11), 652–669.

Grant, J. E., Potenza, M. N., Weinstein, A., & Gorelick, D. A. (2010). Introduction to behavioral addictions. *The American Journal of Drug and Alcohol Abuse, 36*(5), 233–241.

Griffiths, M. D., Kuss, D. J., Lopez-Fernandez, O., & Pontes, H. M. (2017). Problematic gaming exists and is an example of disordered gaming. *Journal of Behavioral Addictions, 6*(3), 296–301.

Haq, I. U., Foote, K. D., Goodman, W. G., Wu, S. S., Sudhyadhom, A., Ricciuti, N., ... Okun, M. S. (2011). Smile and laughter induction and intraoperative predictors of response to deep brain stimulation for obsessive-compulsive disorder. *NeuroImage, 54*(Suppl. 1), S247–S255.

Hernandez, L., & Hoebel, B. G. (1988). Food reward and cocaine increase extracellular dopamine in the nucleus accumbens as measured by microdialysis. *Life Science, 42*(18), 1705–1712.

Holden, C. (2001). 'Behavioral' addictions: Do they exist? *Science, 294*(5544), 980–982.

Jentsch, J. D., & Taylor, J. R. (1999). Impulsivity resulting from frontostriatal dysfunction in drug abuse: Implications for the control of behavior by reward-related stimuli. *Psychopharmacology, 146*, 373–390.

Kaess, M., Parzer, P., Mehl, L., Weil, L., Strittmatter, E., Resch, F., & Koenig, J. (2017). Stress vulnerability in male youth with Internet Gaming Disorder. *Psychoneuroendocrinology, 77*, 244–251.

Kim, S. H. (2011). Reduced striatal dopamine D_2 receptors in people with Internet addiction. *Neuroreport, 22*, 407–411.

Ko, C. H., Liu, G. C., Yen, J. Y., Chen, C. Y., Yen, C. F., & Chen, C. S. (2013). Brain correlates of craving for online gaming under cue exposure in subjects with Internet gaming addiction and in remitted subjects. *Addiction Biology, 18*, 559–569.

Koepp, M. J., Gunn, R. N., Lawrence, A. D., Cunningham, V. J., Dagher, A., Jones, T., … Grasby, P. M. (1998). Evidence for striatal dopamine release during a video game. *Nature, 393*(6682), 266–268.

Koob, G. F. (2009). Dynamics of neuronal circuits in addiction: Reward, anti-reward, and emotional memory. *Pharmacopsychiatry, 42*, S32–S41.

Koob, G. F., & Le Moal, M. (1997). Drug abuse: Hedonic homeostatic dysregulation. *Science, 278*(5335), 52–58.

Koob, G. F., & Le Moal, M. (2005). Plasticity of reward neurocircuitry and the 'dark side' of drug addiction. *Nature Neuroscience, 8*(11), 1442–1444.

Kühn, S., Romanowski, A., Schilling, C., Lorenz, R., Mörsen, N., Seiferth, T., … Gallinat, J. (2011). The neural basis of video gaming. *Translational Psychiatry, 1*, e53.

Kuss, D. J., Louws, J., & Wiers, R. W. (2012). Online gaming addiction? Motives predict addictive play behavior in massively multiplayer online role-playing games. *Cyberpsychology Behavior and Social Networking, 15*(9), 480–485.

Lemieux, A., & al'Absi, M. (2016). Stress psychobiology in the context of addiction medicine: From drugs of abuse to behavioral addictions. *Progress in Brain Research, 223*, 43–62.
Lemmens, J. S., Valkenburg, P. M., Gentile, D. A., & Reynolds, C. R. (2015). The internet gaming disorder scale. *Psychological Assessment, 27*(2), 567–582.
Lemmens, J. S., Valkenburg, P. M., & Peter, J. (2011). Psychosocial causes and consequences of pathological gaming. *Computers and Human Behavior, 27*, 144–152.
Lorenz, R. C., Kruger, J. K., Neumann, B., Schott, B. H., Kaufmann, C., Heinz, A., & Wüstenberg, T. (2013). Cue reactivity and its inhibition in pathological computer game players. *Addictive Biology, 18*(1), 134–146.
Lovallo, W. R. (2006). Cortisol secretion patterns in addiction and addiction risk. *International Journal of Psychophysiology, 59*(3), 195–202.
Marks, I. (1990). Behavioral (non-chemical) addictions. *Addiction, 85*(11), 1389–1394.
Meng, Y., Deng, W., Wang, H., Guo, W., & Li, T. (2015). The prefrontal dysfunction in individuals with Internet gaming disorder: A meta-analysis of functional magnetic resonance imaging studies. *Addiction Biology, 20*(4), 799–808.
Mogenson, G. J., Jones, D. L., & Yim, C. Y. (1980). From motivation to action: Functional interface between the limbic system and the motor system. *Progress in Neurobiology, 14*(2–3), 69–97.
Norgren, R. (2015, June). Ingestive classics: James Olds and pleasure in the brain. Retrieved from www.ssib.org/web/classic11.php.
Olds, J., & Milner, P. (1954). Positive reinforcement produced by electrical stimulation of septal area and other regions of rat brain. *Journal of Comparative and Physiological Psychology, 47*(6), 419–427.
Petry, N. M., & O'Brien, C. P. (2013). Internet gaming disorder and the DSM-5. *Addiction, 108*(7), 1183–1354.
Petry, N. M., Rehbein, F., Gentile, D. A., Lemmens, J. S., Rumpf, H. J., Mößle, T., … O'Brien, C. P. (2014). An international consensus for assessing internet gaming disorder using the new DSM-5 approach. *Addiction, 109*, 1399–1406.
Plante, C. N., Gentile, D. A., Groves, C. L., Modlin, A., & Blanco-Herrera, J. (2018). Video games as coping mechanisms in the etiology of video game addiction. *Psychology of Popular Media Culture*. Advance online publication.
Przybylski, A. K., Weinstein, N., & Murayama, K. (2017). Internet gaming disorder: Investigating the clinical relevance of a new phenomenon. *American Journal of Psychiatry, 3*(1), 230–236.
Robinson, T. E., & Berridge, K. C. (1993). The neural basis of drug craving: An incentive-sensitization theory of addiction. *Brain Research Reviews, 18*(3), 247–291.
Rumpf, H. J., Achab, S., Billieux, J., Bowden-Jones, H., Carragher, N., Demetrovics, Z., … Poznyak, V. (2018). Including Gaming Disorder in the ICD-11: The need to do so from a clinical and public health perspective. *Journal of Behavioral Addictions, 7*(3), 556–561.
Sinha, R., Fox, H. C., Hong, K. A., Hansen, J., Tuit, K., & Kreek, M. J. (2011). Effects of adrenal sensitivity, stress- and cue-induced craving, and anxiety on subsequent alcohol relapse and treatment outcomes. *Archives of General Psychiatry, 68*, 942–952.
Starcevic, V. (2013). Is internet addiction a useful concept? *Australian & New Zealand Journal of Psychiatry, 47*(1), 16–19.
Sutherland, M. T., McHugh, M. J., Pariyadath, V., & Stein, E. A. (2012). Resting state functional connectivity in addiction: Lessons learned and a road ahead. *NeuroImage, 62*(4), 2281–2295.
Tian, M., Chen, Q., Zhang, Y., Du, F., Hou, H., Chao, F., & Zhang, H. (2014). PET imaging reveals brain functional changes in internet gaming disorder. *European Journal of Nuclear Medicine and Molecular Imaging, 41*(7), 1388–1397.
Vo, L. T. K., Walther, D. B., Kramer, A. F., Erickson, K. I., Boot, W. R., Voss, M., … Wang, W. Y. (2011). Predicting individuals' learning success from patterns of pre-learning MRI activity. *PLoS One, 6*, e16093.
Volkow, N. D., Wang, G., Fowler, J. S., Logan, J., Gatley, S. J., Hitzemann, R., … Pappas, N. (1997). Decreased striatal dopaminergic responsiveness in detoxified cocaine-dependent subjects. *Nature, 386*(6627), 830–833.

Volkow, N. D., Wang, G., Fowler, J. S., Tomasi, D., & Telang, F. (2011). Addiction: Beyond dopamine reward circuitry. *Proceedings of the National Academy of Sciences, 108*(37), 15037–15042.

Weinstein, A., & Lejoyeux, M. (2010). Internet addiction or excessive internet use. *American Journal of Drug and Alcohol Abuse, 36*(5), 277–283.

Weinstein, A., & Lejoyeux, M. (2015). New developments on the neurobiological and pharmaco-genetic mechanisms underlying internet and videogame addiction. *American Journal of Addiction, 24*(2), 117–125.

Weinstein, A., Livny, A., & Weizman, A. (2017). New developments in brain research of internet and gaming disorder. *Neuroscience and Biobehavioral Reviews, 75*, 314–330.

WHO (World Health Organization). (2018). International Classification of Diseases – 11. Retrieved from https://icd.who.int/browse11/l-m/en#/http://id.who.int/icd/entity/338347362.

Wise, R. A. (1996). Addictive drugs and brain stimulation reward. *Annual Review of Neuroscience, 19*, 319–340.

Wise, R. A. (2004). Dopamine, learning and motivation. *Nature Reviews Neuroscience, 5*(6), 483–494.

Yee, N. (2006). Motivations for play in online games. *CyberPsychology & Behavior, 9*(6), 772–775.

Young, K. S. (1998). Internet addiction: The emergence of a new clinical disorder. *CyberPsychology & Behavior, 1*(3), 237–244.

Young, K. A., Franklin, T. R., Roberts, D. C., Jagannathan, K., Suh, J. J., Wetherill, R. R., ... Childress, A. R. (2014). Nipping cue reactivity in the bud: Baclofen prevents limbic activation elicited by subliminal drug cues. *Journal of Neuroscience, 34*(14), 5038–5043.

15

Biological Perspectives of Media Violence and Aggression

Martin Klasen and Klaus Mathiak

Aggression and violence are prominent elements of modern entertainment media forms such as movies and video games. Although there has been a long and controversial debate about the psychological and behavioral effects of media violence, research has only recently begun to address the neurobiological aspects. In contrast, research on the neurobiology of human aggression has a long-standing tradition of more than a century. Experience from this field can be a valuable contribution also for research on media violence. On one hand, aggression research provides empirically supported neurobiological models that can be applied to media research; on the other hand, the established standardized and validated paradigms provide a methodical framework for research on media violence. The aim of the present chapter is to provide a link between these fields. First, we provide an overview of the neurobiology of aggressive personality traits by reviewing structural and functional brain abnormalities in overly aggressive individuals. Second, we describe experimental designs for state aggression research and review corresponding findings from neuroimaging. Third, we outline current knowledge on the neurobiology of media violence and aggression. We conclude this chapter with a discussion of neurobiological aggression models and highlight their relevance for further media research.

NEUROBIOLOGY OF TRAIT AGGRESSION

Structural Brain Anomalies of Aggressive Individuals

From a historical perspective, the first evidence for a relationship between brain regions and aggression comes from case studies. It was observed that after a traumatic brain injury or a tumor at specific locations, patients developed aggressive personality traits. One of the most famous reports in literature is the case of Phineas Gage (1823–1860). Working as a foreman in railway construction, Gage prepared blasting holes with explosives and sealed them with sand, using a tamping iron. On September 13, 1848, the blasting powder exploded prematurely, shooting the tamping iron through the left side of his head and removing a substantial portion of the frontal lobe of his brain (see Figure 15.1 for an overview).

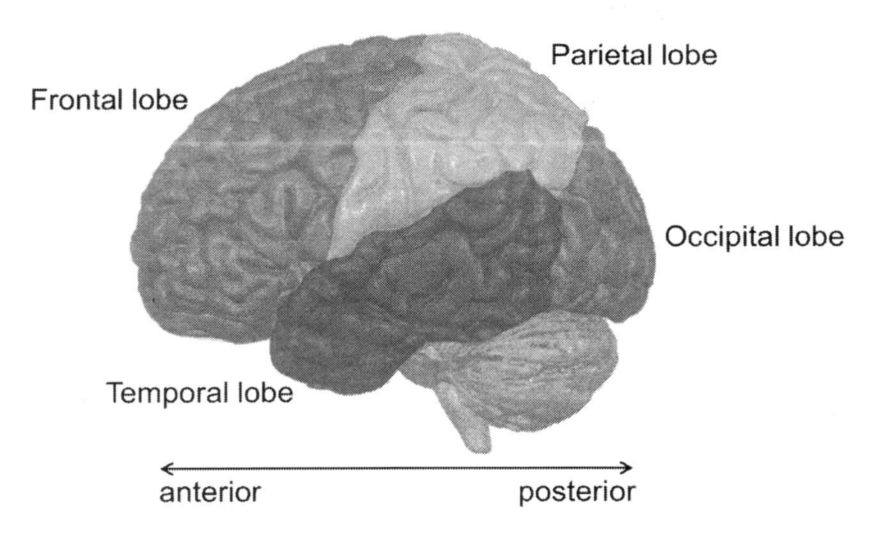

Figure 15.1 The Lobes of the Brain

Apart from the loss of his left eye, he seemed to fully recover physically, but his personality underwent a substantial change in the years after the accident. After his death, his physician described him as "fitful, irreverent, ... impatient of restraint or advice when it conflicts with his desires, at times pertinaciously obstinate, yet capricious and vacillating ..." (Harlow, 1868). Since these days, the case has been employed to illustrate the relevance of the prefrontal cortex (PFC; the anterior part of the frontal lobe) for impulsive-aggressive personality traits. However, secure knowledge about the extent of the brain injury and the nature of Gage's personality changes is sparse. Nonetheless, modern case studies have confirmed the assumption that damage to the PFC leads to aggressive behavior (Malloy, Bihrle, Duffy, & Cimino, 1993). Specifically, PFC lesions seem to impair the ability to foresee negative consequences of aggressive actions (Bechara, Damasio, Tranel, & Damasio, 1997), leading to a behavioral disinhibition. Aggression-enhancing effects have also been observed for damages to structures of the temporal lobes, such as the amygdala. For instance, it is frequently observed that temporal lobe epilepsy affecting the amygdala is accompanied by the development of antisocial behavior (Kiehl, 2006).

In line with these clinical findings, neuroimaging studies have reported structural brain anomalies in aggressive individuals. Magnetic resonance imaging (MRI) allows the non-invasive investigation of brain tissue in these individuals and can identify brain areas with lower gray matter volumes, corresponding to a thinner cerebral cortex. A major focus of aggression research lies on psychopathy, a personality disorder which is characterized by various antisocial and aggressive traits, such as manipulative behavior, pathological lying, a parasitic lifestyle, poor behavioral controls, and criminal versatility (Hare & Neumann, 2008). Psychopaths have reduced PFC and inferior temporal lobe volumes (Gregory et al., 2012), which is related to the extent of psychopathy (Ermer, Cope, Nyalakanti, Calhoun, & Kiehl, 2012). Although most data on brain anomalies and aggression are correlational, recent findings suggest that reduced gray matter in frontotemporal networks is causal for aggressive traits (Lam et al., 2017). Taken together, there is evidence that structural impairments of PFC and temporal structures can serve as a neurobiological model of psychopathic aggression. However, this seems not to be the case for aggression and violence in general. A recent meta-analysis by Lamsma et al. (2017) found evidence for structural aberrations in all four lobes of the brain. The diversity of findings thus reflects the complexity of the aggression concept, both phenomenologically and biologically.

Neuroimaging in Aggressive Individuals

Structural imaging delivers only very limited insight into the function of an area for cognition, emotion, and behavior. This limitation can be overcome by functional magnetic resonance imaging (fMRI). This method measures brain activity via oxygenation of blood hemoglobin in a non-invasive fashion. Analogous to structural findings, fMRI studies have identified aberrant functional patterns in individuals with psychopathic traits. Anatomically, these impairments are well in line with structural findings and can be primarily observed in medial prefrontal cortex regions such as the anterior cingulate cortex (ACC) and temporal structures such as the amygdala, which are core regions for emotion processing. Psychopaths show reduced functioning in these areas during emotional tasks (Kiehl et al., 2001), and it has been concluded that emotional deficits are a central element of psychopathic aggression (Anderson & Kiehl, 2014). Specifically, psychopaths show reduced emotional involvement when others are in distress, which is reflected in a hypoactivation of brain empathy networks (Decety, Chen, Harenski, & Kiehl, 2013), whereas the cognitive ability of moral understanding—knowing right from wrong—is intact (Cima et al., 2010). Moreover, it seems that psychopaths are unable to form associations between environmental cues and aversive consequences. This is particularly evident in fear conditioning paradigms, where an association is formed between an originally neutral cue (tone, picture, etc.) and a subsequent painful stimulation (heat, pressure, etc.), making the cue a fear-inducing stimulus. During this form of "fear learning," psychopaths show reduced activation in the prefrontal cortex, the amygdala, and the anterior insula (Birbaumer et al., 2005), regions that are relevant for the emotional evaluation of a stimulus (Seeley et al., 2007). Thus, the anticipation of aversive consequences does not evoke an emotional response in the brains of psychopaths. This is also supported by clinical reports showing that aggression in psychopaths is mostly unemotional and instrumental (Glenn & Raine, 2009). According to this model, psychopaths use aggression to achieve their goals because they do not empathize with their victims and are not afraid of potential negative consequences.

A somewhat different picture emerges in individuals with intermittent explosive disorder (IED). IED is characterized by anger and emotional outbursts, in contrast to the unemotional and instrumental aggression in psychopathy. In emotion processing tasks, IED patients show brain activation that is more or less the opposite of the pattern in psychopathy, with enhanced activation in amygdala and medial prefrontal areas (Coccaro, McCloskey, Fitzgerald, & Phan, 2007). Thus, one can distinguish two fundamentally different types of pathological aggression: a "cool" aggression, which is unemotional and characterized by a low activity of the salience network, and a "hot" aggression, which is highly emotional and goes along with salience network hyperactivity.

Neurogenetics of Aggression

Aggression is highly heritable (Nelson, 2005). Specifically, a gene-environment interaction has been established as a risk factor for aggression, for low expressing variants of the monoamine oxidase A gene (*MAOA*-L; Pavlov, Chistiakov, & Chekhonin, 2012). In *MAOA*-L carriers, experiencing childhood maltreatment increases the risk for developing an aggressive personality to a much stronger degree than in carriers of other gene variants (Caspi et al., 2002). Genotype effects on emotion processing networks seem to mediate this interaction. In *MAOA*-L carriers, a hyperactivity pattern can be observed in the amygdala, the insula, and the ACC during aggression-related tasks (Alia-Klein et al., 2009), resembling the "hot" aggression pattern in IED. This seems to make the brains of *MAOA*-L carriers more vulnerable to the detrimental effects of childhood maltreatment.

A neurobiological model has been proposed for this gene-environment interaction (Buckholtz & Meyer-Lindenberg, 2008). In this model, higher brain serotonin levels in *MAOA*-L carriers destabilize the ACC-amygdala system. This is partially compensated by coupling of the ACC with the ventromedial prefrontal cortex (VMPFC), a region supporting emotion regulation. Thus, without the experience of childhood maltreatment, the behavioral consequences remain subtle and with no relevance for aggression. In traumatized *MAOA*-L children, in turn, the maturation of the vulnerable ACC-amygdala circuit is disrupted. As a consequence, the development of social skills and emotion control is severely impaired, which leads to impulsive aggression and violence.

Empirical evidence for this neurogenetic aggression model is mixed. *MAOA*-L-specific associations of the VMPFC-ACC-amygdala system with the aggression-related traits, reward dependence, and harm avoidance have been reported (Buckholtz & Meyer-Lindenberg, 2008). However, the model could not be confirmed for genotype effects on trait aggression directly (Klasen et al., 2018). Further research is needed to clarify the neurogenetic mechanisms for trait aggression.

NEUROBIOLOGY OF STATE AGGRESSION

Paradigms for Aggression Research

A major challenge for investigating state aggression is finding an appropriate experimental design for neuroimaging research. The setup should make the participant angry and increase the probability of aggressive responses. Moreover, it should be applicable in an MRI environment, where the participant lies in the scanner and is only indirectly in contact with other persons via microphone, headphones, and a video screen. Most approaches induce aggression via provocation (i.e., aggression of another person toward the participant). The strength of provocation approaches lies in their ecological validity, particularly for "hot" aggression forms, which are impulsive and reactive in nature. Provoking a person is by far less effective when the person knows that the task is designed to provoke. Thus, the participant is not informed that the paradigm is supposed to induce aggression; instead, the ostensible motivation for the experiment is something else, such as measuring reaction times. This "cover story" is a potential weakness of provocation paradigms since it has to be believed by the participant. Asking the participants *a posteriori* whether they believed the cover story may be an option to circumvent this problem, but bears the risk of social desirability effects that may bias the responses. However, there are also other approaches, such as the use of imagined aggression, that do not rely on a cover story. Finally, aggression research has used frustration as a frequent precursor of aggression (frustration-aggression hypothesis; Miller et al., 1941) for frustration-based designs.

Taylor Aggression Paradigm

A well-established neuroimaging aggression paradigm using the provocation approach is the Taylor Aggression Paradigm (TAP; synonym: Competitive Reaction Time Task (CRTT); Taylor, 1967). The cover story in this experiment is an alleged reaction time task. Participants are told that they are playing against an opponent in a reaction game. The winner of a trial is whoever presses a button first when a cue appears. In each trial, the winner can punish the loser with an electric shock. The shock intensity can be adjusted before each trial and serves as a measure for aggression. In reality, there is no opponent, and the frequency of wins and losses as well as the shock intensity are predetermined. This way, the aspect of provocation can be implemented in the TAP via the level of punishment from the alleged opponent. Moreover, the design can distinguish between a decision phase (selecting a punishment for the opponent) and an outcome phase (seeing if one has won or lost the trial). The TAP is considered a valid measure of direct physical aggression (Giancola & Parrott, 2008) and can easily be applied in a neuroimaging setting. As such, it can also be used to measure transfer effects from other aggression-inducing tasks or stimuli, such as violent media (Bushman, 1995). For ethical reasons, many newer versions of the paradigm have replaced the electric shock by a noise blast.

Neuroimaging

Krämer et al. (2007) investigated brain activation while participants performed the TAP in the scanner. The decision phase of the TAP was associated with activity in the ACC and the anterior insula, which was stronger when the provocation was higher in the previous trial. High and low punishment selections had differential influence on regions of the brain reward system. For the outcome phase, winning a trial activated the reward system. Moreover, higher activity in the insula and the ACC was observed for high provocation compared to low provocation trials. Thus, brain systems of aggression depend on cognitive and emotional processes, such as perceived unfairness or satisfaction from revenge. Aggression-related personality traits modulate brain activation in the TAP (Lotze, Veit, Anders, & Birbaumer, 2007). Furthermore, there is evidence for the influence of neurotransmitter levels (Krämer, Riba, Richter, & Munte, 2011) and social exclusion (Beyer, Munte, & Kramer, 2014).

Point Subtraction Aggression Paradigm

A conceptually related aggression paradigm is the Point Subtraction Aggression Paradigm (PSAP; Cherek, 1981). Similar to the TAP, participants play against a fictitious opponent. They are told that their task is to collect as many points as possible via consecutive button presses (e.g., one point for 100 presses). Participants are provoked via a loss of already earned points, which are stolen by the alleged opponent. They can either protect their points for a certain time by pressing a second key, or they can also steal points from their opponent by pressing a third key. Moreover, participants are told that stealing does not increase their own savings but will just reduce the opponent's points. Pressing the stealing button is thus considered a measure of aggression. Like the TAP, the PSAP is a valid and easily applicable measure of physical aggression with good psychometric properties (Geniole, MacDonell, & McCormick, 2017) and can distinguish between different phases of reactive aggression which can be analyzed separately in neuroimaging experiments. Enhanced anger responses of violent compared to non-violent individuals in the PSAP have frequently been reported (Skibsted et al., 2017), but could not be found in all studies (Gan et al., 2016).

Neuroimaging

Aggression toward the opponent in the PSAP leads to activation of the brain reward system (the ventral striatum). Being provoked, in turn, activates the anterior insula and the ACC, along with other areas such as the frontoparietal attention network (Skibsted et al., 2017). Violent offenders show enhanced activity in amygdala and striatum after provocation, which goes along with enhanced aggression in the PSAP and a reduced functional coupling of these areas with the medial prefrontal cortex (da Cunha-Bang et al., 2017).

Verbal Provocation Tasks

An alternative approach to induce aggression is a direct verbal provocation of the participant (i.e., a verbal insult) by the experimenter. As an example, Denson et al. (2009) confronted participants with an anagram task and instructed them to say the answer out loud. The experimenter then interrupted several times, asking the participant to speak louder. Interruptions were getting more and more unfriendly over time, resulting in shouting that the participant was seemingly unable to follow a simple instruction. Similar to TAP and PSAP, provocation increased brain activation in the ACC and the insula, with ACC activity being correlated to both state anger and trait aggression. This approach benefits from a high ecological validity; however, credibility of the setup depends on the acting skills of the experimenter, and standardization may thus be difficult.

Frustration Tasks

A more indirect contribution to the neurobiology of aggression comes from frustration research. According to the well-established frustration-aggression hypothesis (Miller et al., 1941), frustration is frequently a precursor of aggression. In an experimental setting, frustration can be induced by blocking goal achievement. As an example, Yu et al. (2014) blocked participants from obtaining a reward, which resulted in activation in the dorsal ACC, the amygdala, the anterior insula, and the midbrain regions. These findings are in line with an earlier study by Abler et al. (2005) with a similar design. The neural signatures of frustration and state aggression thus seem to be remarkably similar.

Imagined Aggression

An entirely different approach to the concept is the use of imagined aggression and violence. Pietrini et al. (2000) asked their participants to imagine various violent scenarios (e.g., attacking and injuring a person) and compared their brain activity to non-violent scenarios. Imagined violence affected medial PFC areas; specifically, it increased activity in the dorsal parts and decreased activity in the rostral parts. Remarkably, imagined violence also evoked other emotional and physiological signatures of real-life aggression such as anger, frustration, anxiety, and an increase in blood pressure. The imagination approach can also be combined with virtual reality stimulation (e.g., Molenberghs et al., 2015).

NEUROBIOLOGY OF MEDIA AGGRESSION AND VIOLENCE

Media Violence and Aggression Networks

Violence is a prominent element of modern video games. In first-person shooter games (FPS), violent player actions account for about 7% of the total playing time (Weber, Behr, Tamborini, Ritterfeld, & Mathiak, 2009). Given the often very explicit depiction of violence in games, there has been a controversial debate about a possible connection between virtual and real aggression (Anderson & Bushman, 2002; Ferguson, 2007). Functional imaging studies have thus addressed the impact of video game violence on neural networks related to aggression. Mathiak and Weber (2006) measured brain activity in experienced players while they played an FPS in an unrestricted manner. In a quasi-experimental approach, they employed a high-resolution content analysis to identify different types of game events on a frame-by-frame basis. Neuroimaging data was analyzed based on the content categories. In-game violence led to an increased activity in the dorsal ACC regions and to decreased activity in the rostral ACC and the amygdala, with remarkably large effect sizes (Figure 15.2). Moreover, virtual aggression enhanced the functional coupling in the ACC-amygdala network (Weber et al., 2006). The involved brain networks are thus remarkably similar to those reported in trait and state aggression studies. There are two conclusions from this notion. First, violence against a virtual character is not only conceptually related to real-life aggression, it also seems to share the same neural networks. This is also supported by Cheetham et al. (2009), where applying pain to an innocent virtual character involved brain networks of personal distress. Second, video game aggression can serve as a model for real-life aggression in neurobiological research. As such, it

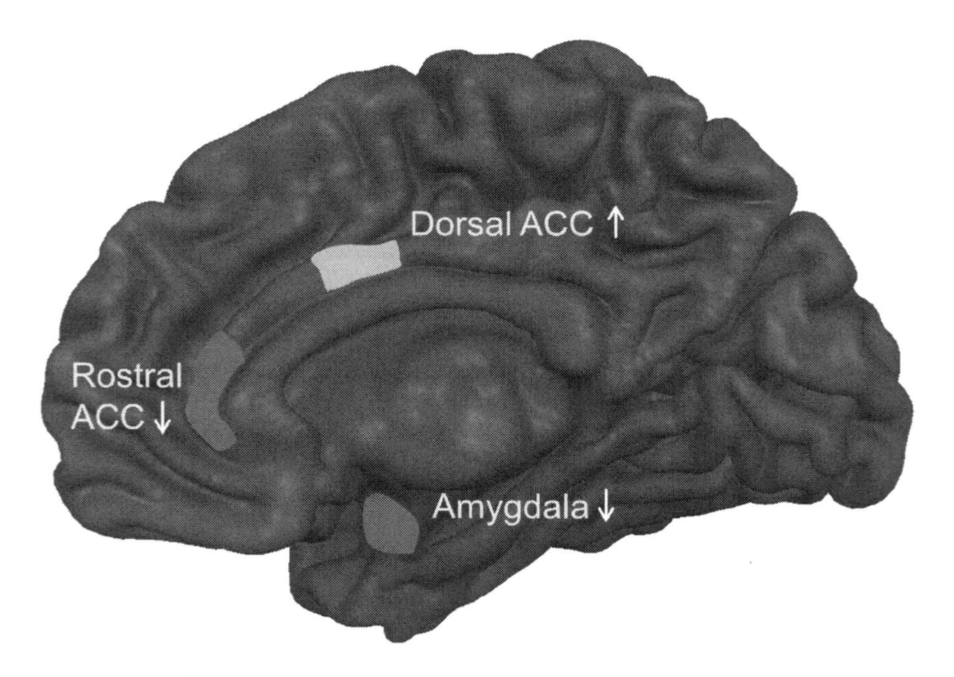

Figure 15.2 Network Activity During Violence in a Video Game

Source: Mathiak & Weber (2006).

has been employed in a pharmacological fMRI study by Klasen et al. (2013). This study tested the neural impact of quetiapine, an atypical antipsychotic with anti-aggressive properties (Comai et al., 2012), in a violent car-racing game as compared to a violence-free control game. During virtual aggression, quetiapine enhanced the connectivity in a circuit of the amygdala, the dorsal ACC, and the dorsolateral PFC, which was directly related to an attenuation of game-induced state aggression, another indication for the similarity between virtual and real-life aggression.

Effects of Media Violence Exposure

For half a century, researchers have investigated potential adverse behavioral effects of violent media exposure (Anderson & Bushman, 2002). In contrast, research on violent media effects on brain systems is a relatively new topic. Thus, only a few studies have addressed this issue so far. Most of them deal with the consequences of video game play; effects of TV violence exposure thus remain hardly investigated and a subject for further studies.

Two studies have investigated neurobehavioral effects of short-term media violence exposure. Kelly et al. (2007) measured brain activity while subjects watched short film clips with violent, fearful, or neutral content. A repeated exposure to violence led to a diminished response of an emotion regulation circuit in the orbitofrontal cortex and the amygdala, which was not the case for any of the other conditions. Neural signatures of reduced emotional responses were also observed by Guo et al. (2013). In a pain empathy task, participants showed diminished activity in the mid-cingulate cortex and the anterior insula after viewing a violent compared to a non-violent movie excerpt. Thus, there is evidence that short-term exposure to media violence diminishes the responsiveness of brain circuits underlying emotion regulation and empathy. However, these studies do not permit any conclusion about how long these effects may last.

Long-term effects of repeated violent media exposure have been a subject of public debate. In particular, it has been suggested that playing violent games may lead to a decrease in empathy and prosocial behavior and an increase in aggression (Anderson & Bushman, 2001). Evidence for this assumption from brain imaging studies is mixed. A recent study by Kühn et al. (2018) could not find any effect of a 16-week violent video game exposure on neural networks for pain empathy. Szycik et al. (2017) also found no such effects in a similar task when comparing regular players and non-players of violent games, in line with findings from an earlier study by Regenbogen et al. (2010). The same seems to be true for resting state brain activity (Pan et al., 2017). In contrast, Montag et al. (2012) reported diminished lateral PFC activity in players of violent games during the processing of negative emotions. Diminished PFC and parietal activity was also associated with TV violence consumption when participants viewed videos with aggressive content (Strenziok et al., 2011). In aggressive adolescents with disruptive behavior disorder, media violence experience increased amygdala activity in an emotional task, indicating that violent media may foster impulsive responses in individuals with high aggressive traits (Kalnin et al., 2011). Besides empathy and aggression, consumption of violent TV content has also been associated with reduced executive functioning and frontoparietal white matter in young adults (Hummer et al., 2014).

Media Violence and the Brain Reward System

Violence and aggression in video games have been associated with activity in the brain reward system. The reward system, encompassing the midbrain areas, the basal ganglia, and the medial PFC, codes the rewarding value of an action or stimulus. In an early study, Koepp et al. (1998)

first described enhanced reward system activity while participants played a violent video game. Whereas this pioneering study was not yet able to distinguish violence effects from game play effects in general, recent findings have confirmed enhanced reward system involvement during a violent as compared to non-violent game (Zvyagintsev et al., 2016). Remarkably, the rewarding value of violent games cannot be reduced to effects of game success only. Using a high-resolution content analysis, Mathiak et al. (2011) investigated the impact of game events on reward system activity. In this study, all events ending the game flow (virtual killing and dying) led to a decrease in reward system activity as compared to the rest of the game play, albeit to a lesser degree for virtual killing events. These findings were replicated in a later study by Kätsyri et al. (2013b). Accordingly, the main reward seems to be the game itself, which is reflected by a tonic increase in reward system activity during the whole game play. Significant activation differences between killing and dying indicate a reward prediction error for violent failure events (dying), i.e., the player expects a reward from the violent situation but does not receive it. Comparing active play to passive watching, Kätsyri et al. (2013b) showed that this violence-related reward prediction error depends on the active role of the player. Reward system activity during virtual violence also depends on the social context, with a stronger response for a human as compared to a computer opponent (Kätsyri et al., 2013a).

Moreover, there is evidence that the affective evaluation of violent game events depends not only on the reward system, but on distributed social processing networks (Mathiak et al., 2011, 2013). The same is true for the emergence of flow experience, a positive mental state of being completely absorbed by an activity (Csikszentmihalyi, 1988; see Chapter 12 by Weber and Fisher, in this volume). In a neuroimaging study by Klasen et al. (2012), flow experience was associated with synchronous activation in reward system and a sensorimotor network, similar to the pattern underlying "virtual presence" (i.e., the feeling of being a part of the virtual environment; Baumgartner et al., 2008). Thus, violence per se seems not to be the rewarding game aspect; instead, it seems to contribute to an overall narrative supporting flow and reward.

CONCLUSION

Structural and functional investigations show a complex neurobiological picture of aggression. Nonetheless, some brain regions have been associated with aggression in a remarkably consistent manner. There is strong evidence that a network of medial PFC (in particular, the ACC), the amygdala, and the anterior insula plays a major role in both trait and state aggression. Blair (2016) highlights their contribution to aggression in a neurobiological model. In this model, ventromedial PFC regions code the value of actions and objects for optimal response selection. Dorsomedial PFC regions, including the dorsal ACC, integrate this information with action outcomes, whereas the anterior insula adjusts behavioral responses based on input from both PFC regions. The amygdala responds to acute threat, together with the midbrain regions and the hypothalamus. The regions of this model show a strong overlap with the "salience network" of the brain (Seeley et al., 2007; see Figure 15.3). Broadly speaking, the salience network processes information that is personally relevant or emotionally engaging and integrates aspects of emotion and cognition for subsequent actions (Menon, 2015). Thus, it can be concluded that there is no specific "aggression network" in the brain. Instead, aggression seems to arise from more fundamental brain networks underlying emotion generation, cognitive assessment, and action selection. This notion is well in line with the phenomenological distinction of "hot" and "cool" aggression, both on the behavioral and on the neurobiological level. Psychopaths with their unemotional aggression show a reduced activity in the salience network, whereas

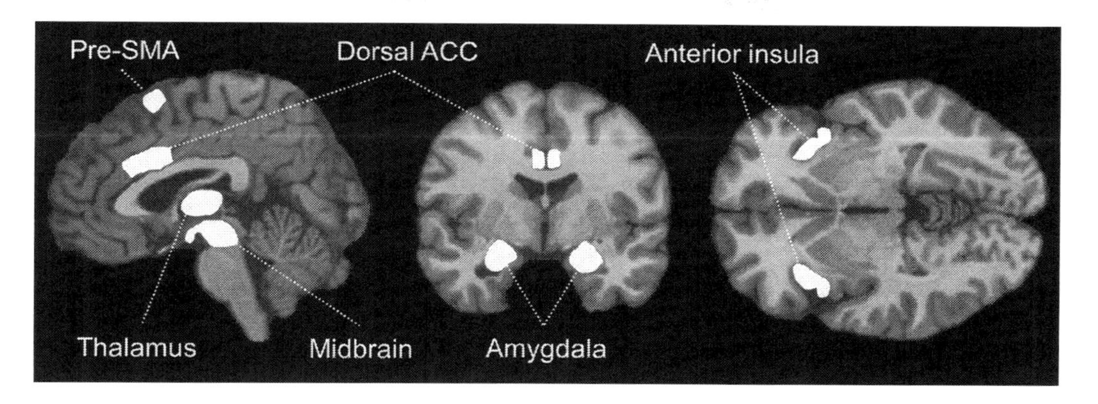

Figure 15.3 The Salience Network

Source: Seeley et al. (2007).

aggression in IED with high anger levels goes along with enhanced activation. A similar picture emerges in individuals from the normal population without aggression problems. Reactive aggression with anger resembles the "hot" aggression pattern. Video game violence, in turn, shows a somewhat different picture. Here, amygdala and rACC deactivation seems to be a characteristic feature. In line with the model of Blair (2016), amygdala deactivation indicates a suppression of feelings of threat. Moreover, similar to the dysfunctions in psychopaths, reduced rACC activity points at reduced emotional involvement. Implications of the ongoing action (i.e., killing the virtual opponent) are suppressed in favor of the cognitive task, a pattern that resembles more of the "cool" instrumental aggression in psychopathy. However, we do not know to which degree this neural signature is related to aggressive personality traits or aggressive behavior in real life, which may be a task for future studies. Nonetheless, it seems justified to consider video game violence as a valid neurobiological aggression model.

Most neurobiological aggression studies have investigated neural systems of media violence in experienced users. However, systematic comparisons on the neural processing of virtual violence in gamers and non-gamers are missing, although they would provide valuable insights into violence effects on the brain. A methodological challenge in comparing gamers and non-gamers may be the motor performance; active execution of in-game violence will always depend on experience and motor skills that may differ considerably between the groups. Possible solutions may lie in simplified game designs or in previous training with a game that is comparable in motor demands, but violence-free.

In a similar vein, long-term effects of media violence on the brain are an understudied topic. Behavioral research has employed longitudinal studies to investigate long-term effects of media violence (Anderson & Bushman, 2002). However, longitudinal designs are de facto absent in the field of neuroimaging research on media violence and aggression. Thus, we do not know how the interplay of media violence and brain systems develops over time, especially concerning aggressive traits and behavioral outcomes. In a similar vein, developmental effects of media violence on the maturation of the adolescent brain are largely unknown, which is a major challenge for future research. This is true not only for video games, but also for other forms of media violence, e.g., in movies.

Recent studies highlight a rewarding effect of media violence. Playing a violent game increases activity in the brain reward system. This seems to be the case for violent game events, such as killing, but there is also evidence that the game, as a whole, leads to a tonic increase in reward system activity. Although it seems plausible to conclude that violence adds to the rewarding value of the game, this issue needs further research. In particular, reward system effects of violent compared to non-violent games and their relationship to subjective experience dimensions may be a target for future research. Here, a special focus may be research on excessive or even pathological video game use. Video game addiction goes along with reward system hyperactivation for game-related cues (Ko et al., 2009), a pattern very similar to the experience of video game violence. Thus, it seems possible that violence may promote the development of game addiction. Further media research may clarify the role of violence in the development of addictive game consumption, or may even help to develop novel therapeutic approaches.

REFERENCES

Abler, B., Walter, H., & Erk, S. (2005). Neural correlates of frustration. *Neuroreport, 16*, 669–672.

Alia-Klein, N., Goldstein, R. Z., Tomasi, D., Woicik, P. A., Moeller, S. J., Williams, B., ... Volkow, N. D. (2009). Neural mechanisms of anger regulation as a function of genetic risk for violence. *Emotion, 9*, 385–396.

Anderson, N. E., & Kiehl, K. A. (2014). Psychopathy and aggression: When paralimbic dysfunction leads to violence. *Current Topics in Behavioral Neuroscience, 17*, 369–393.

Anderson, C. A., & Bushman, B. J. (2001). Effects of violent video games on aggressive behavior, aggressive cognition, aggressive affect, physiological arousal, and prosocial behavior: A meta-analytic review of the scientific literature. *Psychological Science, 12*, 353–359.

Anderson, C. A., & Bushman, B. J. (2002). The effects of media violence on society. *Science, 295*, 2377–2379.

Baumgartner, T., Speck, D., Wettstein, D., Masnari, O., Beeli, G., & Jancke, L. (2008). Feeling present in arousing virtual reality worlds: Prefrontal brain regions differentially orchestrate presence experience in adults and children. *Frontiers in Human Neuroscience, 2*, 8.

Bechara, A., Damasio, H., Tranel, D., & Damasio, A. R. (1997). Deciding advantageously before knowing the advantageous strategy. *Science, 275*, 1293–1295.

Beyer, F., Munte, T. F., & Kramer, U. M. (2014). Increased neural reactivity to socio-emotional stimuli links social exclusion and aggression. *Biological Psychology, 96*, 102–110.

Birbaumer, N., Veit, R., Lotze, M., Erb, M., Hermann, C., Grodd, W., & Flor, H. (2005). Deficient fear conditioning in psychopathy: A functional magnetic resonance imaging study. *Archives of General Psychiatry, 62*, 799–805.

Blair, R. J. (2016). The neurobiology of impulsive aggression. *Journal of Child and Adolescent Psychopharmacology, 26*, 4–9.

Buckholtz, J. W., & Meyer-Lindenberg, A. (2008). MAOA and the neurogenetic architecture of human aggression. *Trends in Neurosciences, 31*, 120–129.

Bushman, B. J. (1995). Moderating role of trait aggressiveness in the effects of violent media on aggression. *Journal of Personality and Social Psychology, 69*, 950–960.

Caspi, A., McClay, J., Moffitt, T. E., Mill, J., Martin, J., Craig, I. W., … Poulton, R. (2002). Role of genotype in the cycle of violence in maltreated children. *Science, 297*, 851–854.

Cheetham, M., Pedroni, A. F., Antley, A., Slater, M., & Jancke, L. (2009). Virtual Milgram: Empathic concern or personal distress? Evidence from functional MRI and dispositional measures. *Frontiers in Human Neuroscience, 3*, 29.

Cherek, D. R. (1981). Effects of smoking different doses of nicotine on human aggressive behavior. *Psychopharmacology, 75*, 339–345.

Cima, M., Tonnaer, F., & Hauser, M. D. (2010). Psychopaths know right from wrong but don't care. *Social Cognitive and Affective Neuroscience, 5*, 59–67.

Coccaro, E. F., McCloskey, M. S., Fitzgerald, D. A., & Phan, K. L. (2007). Amygdala and orbitofrontal reactivity to social threat in individuals with impulsive aggression. *Biological Psychiatry, 62*, 168–178.

Comai, S., Tau, M., Pavlovic, Z., & Gobbi, G. (2012). The psychopharmacology of aggressive behavior: A translational approach: Part 2: Clinical studies using atypical antipsychotics, anticonvulsants, and lithium. *Journal of Clinical Psychopharmacology, 32*, 237–260.

Csikszentmihalyi, M. (1988). The flow experience and its significance for human psychology. In M. Csikszentmihalyi & I. Csikszentmihalyi (Eds.), *Optimal experience: Psychological studies of flow in consciousness* (pp. 15–35). Cambridge, England: Cambridge University Press.

da Cunha-Bang, S., Fisher, P. M., Hjordt, L. V., Perfalk, E., Persson Skibsted, A., Bock, C., … Knudsen, G. M. (2017). Violent offenders respond to provocations with high amygdala and striatal reactivity. *Social Cognitive and Affective Neuroscience, 12*, 802–810.

Decety, J., Chen, C., Harenski, C., & Kiehl, K. A. (2013). An fMRI study of affective perspective taking in individuals with psychopathy: Imagining another in pain does not evoke empathy. *Frontiers in Human Neuroscience, 7*, 489.

Denson, T. F., Pedersen, W. C., Ronquillo, J., & Nandy, A. S. (2009). The angry brain: Neural correlates of anger, angry rumination, and aggressive personality. *Journal of Cognitive Neuroscience, 21*, 734–744.

Ermer, E., Cope, L. M., Nyalakanti, P. K., Calhoun, V. D., & Kiehl, K. A. (2012). Aberrant paralimbic gray matter in criminal psychopathy. *Journal of Abnormal Psychology, 121*, 649–658.

Ferguson, C. J. (2007). The good, the bad and the ugly: A meta-analytic review of positive and negative effects of violent video games. *Psychiatric Quarterly, 78*, 309–316.

Gan, G., Preston-Campbell, R. N., Moeller, S. J., Steinberg, J. L., Lane, S. D., Maloney, T., … Alia-Klein, N. (2016). Reward vs. retaliation: The role of the mesocorticolimbic salience network in human reactive aggression. *Frontiers in Behavioral Neuroscience, 10*, 179.

Geniole, S. N., MacDonell, E. T., & McCormick, C. M. (2017). The Point Subtraction Aggression Paradigm as a laboratory tool for investigating the neuroendocrinology of aggression and competition. *Hormones and Behavior, 92*, 103–116.

Giancola, P. R., & Parrott, D. J. (2008). Further evidence for the validity of the Taylor Aggression Paradigm. *Aggressive Behavior, 34*, 214–229.

Glenn, A. L., & Raine, A. (2009). Psychopathy and instrumental aggression: Evolutionary, neurobiological, and legal perspectives. *International Journal of Law and Psychiatry, 32*, 253–258.

Gregory, S., ffytche, D., Simmons, A., Kumari, V., Howard, M., Hodgins, S., & Blackwood, N. (2012). The antisocial brain: Psychopathy matters. *Archives of General Psychiatry, 69*, 962–972.

Guo, X., Zheng, L., Wang, H., Zhu, L., Li, J., Wang, Q., … Yang, Z. (2013). Exposure to violence reduces empathetic responses to other's pain. *Brain and Cognition, 82*, 187–191.

Hare, R. D., & Neumann, C. S. (2008). Psychopathy as a clinical and empirical construct. *Annual Review of Clinical Psychology, 4*, 217–246.

Harlow, J. M. (1868). Recovery from the passage of an iron bar through the head. *Publications of the Massachusetts Medical Society, 2*, 327–347.

Hummer, T. A., Kronenberger, W. G., Wang, Y., Anderson, C. C., & Mathews, V. P. (2014). Association of television violence exposure with executive functioning and white matter volume in young adult males. *Brain and Cognition, 88*, 26–34.

Kalnin, A. J., Edwards, C. R., Wang, Y., Kronenberger, W. G., Hummer, T. A., Mosier, K. M., … Mathews, V. P. (2011). The interacting role of media violence exposure and aggressive-disruptive behavior in adolescent brain activation during an emotional Stroop task. *Psychiatry Research, 192*, 12–19.

Kätsyri, J., Hari, R., Ravaja, N., & Nummenmaa, L. (2013a). Just watching the game ain't enough: Striatal fMRI reward responses to successes and failures in a video game during active and vicarious playing. *Frontiers in Human Neuroscience, 7*, 278.

Kätsyri, J., Hari, R., Ravaja, N., & Nummenmaa, L. (2013b). The opponent matters: Elevated fMRI reward responses to winning against a human versus a computer opponent during interactive video game playing. *Cerebral Cortex, 23*, 2829–2839.

Kelly, C. R., Grinband, J., & Hirsch, J. (2007). Repeated exposure to media violence is associated with diminished response in an inhibitory frontolimbic network. *PLoS One, 2*, e1268.
Kiehl, K. A. (2006). A cognitive neuroscience perspective on psychopathy: Evidence for paralimbic system dysfunction. *Psychiatry Research, 142*, 107–128.
Kiehl, K. A., Smith, A. M., Hare, R. D., Mendrek, A., Forster, B. B., Brink, J., et al. (2001). Limbic abnormalities in affective processing by criminal psychopaths as revealed by functional magnetic resonance imaging. *Biological Psychiatry, 50*, 677–684.
Klasen, M., Weber, R., Kircher, T. T., Mathiak, K. A., & Mathiak, K. (2012). Neural contributions to flow experience during video game playing. *Social Cognitive and Affective Neuroscience, 7*, 485–495.
Klasen, M., Wolf, D., Eisner, P. D., Habel, U., Repple, J., Vernaleken, I., … Mathiak, K. (2018). Neural networks underlying trait aggression depend on MAOA gene alleles. *Brain Structure and Function, 223*, 873–881.
Klasen, M., Zvyagintsev, M., Schwenzer, M., Mathiak, K. A., Sarkheil, P., Weber, R., & Mathiak, K. (2013). Quetiapine modulates functional connectivity in brain aggression networks. *Neuroimage, 75*, 20–26.
Ko, C. H., Liu, G. C., Hsiao, S., Yen, J. Y., Yang, M. J., Lin, W. C., … Chen, C. S. (2009). Brain activities associated with gaming urge of online gaming addiction. *Journal of Psychiatric Research, 43*, 739–747.
Koepp, M. J., Gunn, R. N., Lawrence, A. D., Cunningham, V. J., Dagher, A., Jones, T., et al. (1998). Evidence for striatal dopamine release during a video game. *Nature, 393*, 266–268.
Krämer, U. M., Jansma, H., Tempelmann, C., & Munte, T. F. (2007). Tit-for-tat: The neural basis of reactive aggression. *Neuroimage, 38*, 203–211.
Krämer, U. M., Riba, J., Richter, S., & Munte, T. F. (2011). An fMRI study on the role of serotonin in reactive aggression. *PLoS One, 6*, e27668.
Kühn, S., Kugler, D., Schmalen, K., Weichenberger, M., Witt, C., & Gallinat, J. (2018). The myth of blunted gamers: No evidence for desensitization in empathy for pain after a violent video game intervention in a longitudinal fMRI study on non-gamers. *Neurosignals, 26*, 22–30.
Lam, B. Y. H., Yang, Y., Schug, R. A., Han, C., Liu, J., & Lee, T. M. C. (2017). Psychopathy moderates the relationship between orbitofrontal and striatal alterations and violence: The investigation of individuals accused of homicide. *Frontiers in Human Neuroscience, 11*, 579.
Lamsma, J., Mackay, C., & Fazel, S. (2017). Structural brain correlates of interpersonal violence: Systematic review and voxel-based meta-analysis of neuroimaging studies. *Psychiatry Research, 267*, 69–73.
Lotze, M., Veit, R., Anders, S., & Birbaumer, N. (2007). Evidence for a different role of the ventral and dorsal medial prefrontal cortex for social reactive aggression: An interactive fMRI study. *NeuroImage, 34*, 470–478.
Malloy, P., Bihrle, A., Duffy, J., & Cimino, C. (1993). The orbitomedial frontal syndrome. *Archives of Clinical Neuropsychology, 8*, 185–201.
Mathiak, K. A., Klasen, M., Weber, R., Ackermann, H., Shergill, S. S., & Mathiak, K. (2011). Reward system and temporal pole contributions to affective evaluation during a first-person shooter video game. *BMC Neuroscience, 12*, 66.
Mathiak, K. A., Klasen, M., Zvyagintsev, M., Weber, R., & Mathiak, K. (2013). Neural networks underlying affective states in a multimodal virtual environment: Contributions to boredom. *Frontiers in Human Neuroscience, 7*, 820.
Mathiak, K., & Weber, R. (2006). Toward brain correlates of natural behavior: fMRI during violent video games. *Human Brain Mapping, 27*, 948–956.
Menon, V. (2015). Salience network. In A. W. Toga (Ed.), *Brain mapping: An encyclopedic reference* (Vol. 2, pp. 597–611). Amsterdam, The Netherlands: Elsevier.
Miller, N. E., Sears, R. R., Rosenzweig, S., Bateson, G., Lewy, D. M., Hartmann, G. W., & Maslow, A. H. (1941). Symposium on the frustration-aggression hypothesis. *Psychological Review, 48*, 337–366.
Molenberghs, P., Ogilvie, C., Louis, W. R., Decety, J., Bagnall, J., & Bain, P. G. (2015). The neural correlates of justified and unjustified killing: An fMRI study. *Social Cognitive and Affective Neuroscience, 10*, 1397–1404.

Montag, C., Weber, B., Trautner, P., Newport, B., Markett, S., Walter, N. T., et al. (2012). Does excessive play of violent first-person-shooter-video-games dampen brain activity in response to emotional stimuli? *Biological Psychology, 89*, 107–111.

Nelson, R. J. (ed.). (2005). *Biology of aggression.* New York, NY: Oxford University Press.

Pan, W., Gao, X., Shi, S., Liu, F., & Li, C. (2017). Spontaneous brain activity did not show the effect of violent video games on aggression: A resting-state fMRI study. *Frontiers in Psychology, 8*, 2219.

Pavlov, K. A., Chistiakov, D. A., & Chekhonin, V. P. (2012). Genetic determinants of aggression and impulsivity in humans. *Journal of Applied Genetics, 53*, 61–82.

Pietrini, P., Guazzelli, M., Basso, G., Jaffe, K., & Grafman, J. (2000). Neural correlates of imaginal aggressive behavior assessed by positron emission tomography in healthy subjects. *American Journal of Psychiatry, 157*, 1772–1781.

Regenbogen, C., Herrmann, M., & Fehr, T. (2010). The neural processing of voluntary completed, real and virtual violent and nonviolent computer game scenarios displaying predefined actions in gamers and nongamers. *Social Neuroscience, 5*, 221–240.

Seeley, W. W., Menon, V., Schatzberg, A. F., Keller, J., Glover, G. H., Kenna, H., ... Greicius, M. D. (2007). Dissociable intrinsic connectivity networks for salience processing and executive control. *Journal of Neuroscience, 27*, 2349–2356.

Skibsted, A. P., Cunha-Bang, S. D., Carre, J. M., Hansen, A. E., Beliveau, V., Knudsen, G. M., & Fisher, P. M (2017). Aggression-related brain function assessed with the Point Subtraction Aggression Paradigm in fMRI. *Aggressive Behavior, 43*, 601–610.

Strenziok, M., Krueger, F., Deshpande, G., Lenroot, R. K., van der Meer, E., & Grafman, J. (2011). Fronto-parietal regulation of media violence exposure in adolescents: A multi-method study. *Social Cognitive and Affective Neuroscience, 6*, 537–547.

Szycik, G. R., Mohammadi, B., Munte, T. F., & Te Wildt, B. T. (2017). Lack of evidence that neural empathic responses are blunted in excessive users of violent video games: An fMRI study. *Frontiers in Psychology, 8*, 174.

Taylor, S. P. (1967). Aggressive behavior and physiological arousal as a function of provocation and the tendency to inhibit aggression. *Journal of Personality, 35*, 297–310.

Weber, R., Behr, K.-M., Tamborini, R., Ritterfeld, U., & Mathiak, K. (2009). What do we really know about first-person shooter games? An event-related, high-resolution content analysis. *Journal of Computer-Mediated Communication, 14*, 1016–1037.

Weber, R., Ritterfeld, U., & Mathiak, K. (2006). Does playing violent video games induce aggression? Empirical evidence of a functional magnetic resonance imaging study. *Media Psychology, 8*, 39–60.

Yu, R., Mobbs, D., Seymour, B., Rowe, J. B., & Calder, A. J. (2014). The neural signature of escalating frustration in humans. *Cortex, 54*, 165–178.

Zvyagintsev, M., Klasen, M., Weber, R., Sarkheil, P., Esposito, F., Mathiak, K. A., ... Mathiak, K. (2016). Violence-related content in video game may lead to functional connectivity changes in brain networks as revealed by fMRI-ICA in young men. *Neuroscience, 320*, 247–258.

16

Virtual Reality for Communication Neuroscience

Bradly Alicea

Virtual Reality (VR) is an emerging and useful tool for behavioral scientists, in general, and communication neuroscience, more specifically. The point of VR is to create simulations that correspond either to reality as it has been experienced or to the world of sensory experience as it might be. Given the right mix of technology and representation, we can simulate the feeling of being in a race car or riding a rollercoaster. We can transport the viewer to the surface of distant planets or other alternative realities with ease. Most importantly, we can control the stimulus and guide the experience to something that is quantifiable using contemporary experimental methods.

This chapter will discuss why VR is interesting to communication neuroscientists. One of these reasons involves the varieties of virtual experience available to the experimental participant. We will also engage in a discussion of the pros and cons related to using VR in the study of cognitive processes. There are also a number of ways VR is useful to communication neuroscience in terms of both stimulus design and psychological effects. Finally, we will discuss naturalistic investigations, the development of open cognitive models, and potential future research directions.

BASIC QUESTIONS

This section will discuss a host of questions and issues related to introducing readers to the world of using VR to study communication neuroscience. The basic questions include why VR is interesting to neuroscientists and how VR works in terms of technology and media content. Two main issues are also discussed: (1) the technical components of a VR system that is used for research; and (2) the varieties of virtual experience that result from exposure to a virtual environment.

Why Is VR Interesting?

VR has been part of the behavioral research landscape for decades, although its widespread adoption has been much more recent. We can use VR to study a wide range of cognitive functions, including spatial cognition (Hardiess, Meilinger, & Mallot, 2015), skill learning and expertise (Thompson, Blair, Chen, & Henrey, 2013), social connectivity (Tarr, Slater, & Cohen, 2018), cognitive performance and training (Bohil, Alicea, & Biocca, 2011), and attention (Weber, Alicea, Huskey, & Mathiak, 2018). It is easy to see the advantages of using VR in an experimental context, as VR also provides a high-resolution, multimodal stimulus with good ecological validity (Bohil et al., 2011; Reggente et al., 2018). This enables new types of experimental design, such as the free-viewing condition (Bartels & Zeki, 2004) and other naturalistic approaches to the study of cognition and media (Hasson & Honey, 2012).

Yet there is a more fundamental advantage of using VR systems. According to the media equation hypothesis (Reeves & Nass, 1996), high-resolution and interactive media are a surrogate of the natural world. For example, one can use a virtual avatar to explore a virtual model of a city park. Whereas the avatar's movements are indirectly initiated using a controller, the human can nevertheless be mentally invested in the behavior of the avatar. In addition, there is an equivalence in terms of complex skill learning between the real and simulated environments. Such an outcome has been observed for a rowing simulation that includes physical cues (Rauter et al., 2013). Moreover, people react to a familiar-looking virtual simulation without conscious effort (Reeves & Nass, 1996), much as they would respond to a well-known natural scene (Martens et al., 2012). This type of immersive telepresence (Slater & Sanchez-Vives, 2016) allows us to study both seamless cognitive experiences and controlled dissection of cognitive processes.

How Does VR Work?

VR takes advantage of our ability to observe action, plan for the future, and synthesize sensorimotor inputs. Although virtual worlds are often immersive, they are also highly interactive. According to McLuhan (1964), VR can generally be categorized as both "hot" and "cool" media. Hot media engage the senses briefly and intensely, often with one sensory input (vision) dominating all of the others. Yet VR also has components of "cool" media, notably in the case of narrative-based video games that provide opportunity for extended engagement. Truly immersive VR experiences can stimulate multiple senses simultaneously, allowing for a greater degree of virtual-to-real world equivalence. In terms of the medium's inherent semantic content (McLuhan, 1964), VR is similar to movies in that it serves as a transition from static images and linear action to configurations of images and the potential for non-linear action. The specifics of how this occurs depends upon the virtual stimulus itself and how it is presented to the nervous system.

Technical Components

The technical components of VR are also flexible and involve interfaces specific to each sense that is engaged. A wide range of visual and non-visual stimuli can be delivered to the nervous system, from vibratory sensations along the arms and legs to simulated hands in the field of vision. These stimuli can in turn be delivered using a number of perspectives, from first-person shooter video games to race car simulations with motion cues.

Visual information can be displayed in a Cave-Automatic Virtual Environment (CAVE) screen, a Head Up Display (HUD), or a Head Mounted Display (HMD). Auditory information can be delivered using a pair of earphones. Haptic information (touch) can be delivered using pinch gloves or a vibrating video game controller.

Application of Technology

Technical choices will also depend on what the experimenter wants to simulate and the physical constraints of the experimental setting. In a functional resonance magnetic imaging (fMRI) scanner, specialized equipment will be necessary. Due to the requirements of motion capture and potential for artifact generation, the normal operation of the VR interface may constrain mobility. In general, coupling VR and neural measurement provides a number of challenges. Neural measurement techniques such as functional near-infrared spectroscopy (fNIR) or electroencephalography (EEG) do not require highly specialized VR configurations. However, they can be affected by motion artifact, requiring users to limit their limb and trunk movements. VR systems can also be affected by motion artifact, particularly during the registration of body movements with positions of objects in the virtual environment. Fortunately, the limitations of neuroimaging technique and VR system configuration can be dealt with in tandem. Related to this, a major challenge of integrating neuroimaging and VR technologies is the synchronization of action in the simulation with continuous measurements of neural activity. Although such synchronization is difficult, it does provide a rich and highly informative dataset.

Varieties of Experience

Virtuality encompasses more than simply a collection of interfaces and content. According to Steuer (1992), VR is defined as any environment in which an individual experiences telepresence. Thus, experience is driven not by the act of simulation itself, but by some undefined cognitive function. The experiential response is elicited by stimulus form: virtual stimuli must be not only realistic, but realistic in context. For our purposes, VR encompasses three major types of simulation: immersive VR, augmented reality (AR), and mixed reality (MR). VR, AR, and MR are three variations on modifying a world of physical objects in a physical world. VR can be described as virtual objects in a virtual world. By contrast, AR involves virtual objects in a physical world. As the inverse of AR, MR presents viewers with physical objects in a virtual world (see Table 16.1).

Simulation and Experience Taxonomy

I propose that the broad variety of experiences generated by a VR simulation can be distilled into two categories: augmentative and narrative (see Table 16.1). Augmentative experience is the presentation of objects in locational context or depth perspective. A rollercoaster simulation is augmentative in that it provides a human in a non-moving room with the feeling of being on a rollercoaster. In contrast, narrative experience involves engagement with a story or a series of

Table 16.1 Major Types of Simulation and Two Types of Experience in Virtual Reality

	Virtual Reality (VR)	*Augmented Reality (AR)*	*Mixed Reality (MR)*
Type of virtuality	Virtual objects, virtual world	Virtual objects, physical world	Physical objects, virtual world
Augmentative	Pilot training	Annotated city map	Outer space simulation
Narrative	Fantasy games	Adventure games (Pokemon)	List of instructions

events linked by meaning. This includes both action sequences in video games and movies and stories that are based on a sequence of events. Augmentative and narrative experience can also merge to produce a coherent experience. In VR, we can combine an interactive narrative with augmentative features such as simulated motion or an increased flow of time.

Augmentative and narrative experiences combine into something often referred to as telepresence. Telepresence involves both the feeling of being transported to the actual environment by the virtual world (Gerrig, 1993) and how a certain experience feels in general (Rheingold, 1991). The feeling of realness comes from the features of the medium, or media form (Kim & Biocca, 1997). Telepresence can also involve competition between real-world and virtual stimuli to create a perceptual illusion (Biocca & Delaney, 1995). Variables such as the visual angle of the display or illumination level of the room can lead to effects such as sensory saturation and sensory suppression, respectively (Kim & Biocca, 1997). These can enhance the level of augmentative experience directly and the level of narrative experience indirectly. This sets the stage for numerous advantages and disadvantages to the use of VR technology as a real-world analogue.

ADVANTAGES AND LIMITATIONS OF THE VR MEDIUM

In this section, the advantages and limitations of using VR in communication neuroscience will be discussed. The advantages include ease of use and experimental implementation, the ability to take advantage of perceptual affordances, and the ability to trigger a neuroplastic response with neuroimaging-friendly stimuli. Limiting factors include the presence of perceptual gaps and the need to achieve coherent perceptual states for maximum experimental effect.

Advantages of the VR Medium

There are several advantages to using VR in an experimental setting. These advantages include ease of use, controllability, and customization. Unlike a sequential movie or static set of experimental images, the stimulus can be explored in a combinatorial manner (McLuhan, 1964). This allows not only for participants to interact with the stimulus more naturally, but also for the stimulus to be reconfigured in an experiment-specific manner. Hasson et al. (2008) discuss the use of neuroimaging to study how people respond to films, in particular, the continuous, sequential nature of the medium. Using a specialized type of analysis (inter-subject correlation), it becomes easy to examine the effectiveness of a media stimulus on a common cognitive process without the bias of subjective experience (Hasson et al., 2008).

Ease of Use

VR environments are easily constructed and become highly customizable at relatively low cost. Widely available software tools and programming languages provide an experimenter with many possible designs for environments, such as a maze or a racetrack. Not only are these environments highly reconfigurable, they also exhibit a high degree of replicability across participants, conditions, and experiments. As object-based simulations of real-world phenomena, VR worlds can be controlled simply by changing the programming. This enables a high degree of experimental control. The programmer can determine how the environment is represented across conditions and trials, in addition to reducing experimental noise by guiding the participant toward specific pathways and outcomes.

Experimental Implementation

VR systems can also be easy to implement in an experimental setting. For example, the ability to reconfigure virtual scenes in an experimentally reproducible manner (for an example using simple visual tasks, see Saleh et al., 2013) ultimately results in increased ecological validity, thus making the application of naturalistic methods more reliable as well. There are also advantages that enable unique models of experimental investigation. This allows us to ask compelling questions not typically possible using more traditional methods. For example, a free-viewing condition can be implemented without the potential confounds of itinerant behavior and uncharacterized environmental interactions. We can also explore phenomena such as proprioceptive and vestibular inputs during spatial cognition (for an example using path integration, see Chance, Gaunet, Beall, & Loomis, 1998).

Affordances

Another advantage to increased experimental control involves the ability to use the principle of perceptual affordances (Good, 2007). Affordances are features of the object and/or environment that have significance or even meaning to humans. The handle of a door or coffee mug is one such example: handles are interpreted as a place to manipulate an object. The ubiquity of perceptual affordances in VR environments allows an experimenter to guide perception and action in a desired way, or even uncover behaviors not easily observed in the real world. We will discuss the critical role affordances play in determining perception and action as a unique mode of interaction. Given the flexible control of virtual environments, affordances can be enhanced, deemphasized, or associated with unexpected places to elicit behaviors such as confusion or exploration.

Neuroplasticity and Longer-Term Changes

Yet a fourth advantage of using VR as an experimental stimulus is the ability to find people with different levels of skill in terms of their experience with the environment. Furthermore, participants can be trained on a specific VR system (e.g., a video game) to acquire expertise. Although controversial (see Simons et al., 2016), the acquisition of expertise provides the brain with enhanced or protective function that improves cognitive performance. A number of studies have been done that assess differences between video game players (those with expertise) versus non-video game players (those without expertise). Such effects are heavily dependent on the type of video game, the degree of expertise held by the experimental group, and the experimental design, and can be subject to interpretation (Hambrick et al., 2014). Nevertheless, it has been demonstrated that performance improvements in visual attention and spatial learning (Green & Bavelier, 2007, 2012) as well as related anatomical differences (West et al., 2017) are possible long-term effects of repeated engagement with VR environments.

Neuroimaging-Friendly Stimuli

For brain scientists, all of these positives are underscored by the potential for neuroimaging-friendly implementation. Given specialized interfaces, VR can be easily interacted with an fMRI scanner or while wearing an EEG cap. This allows us to study a number of cognitive functions and effects in the brain that would not be possible without VR. VR can also be combined with neuromodulation techniques to produce closed-loop adaptive feedback. In the fields of neurorehabilitation (Teo et al., 2016) and human-computer interface design (Huster, Mokom, Enriquez-Geppert, & Herrmann, 2014), such systems have been used to improve cognitive and physical performance. Using other combinations of measurement, virtuality, and direct neural input, we might also be able to understand VR-specific behaviors and outcomes that have not previously been characterized.

Limitations of the VR Medium

There are a number of challenges and limitations to VR technologies that constrain their potential, at least until further research and technological development advance to a suitable point. The first such limitation is that performance in a virtual environment is not always equivalent to real-world performance. One rather contentious issue in the Human Factors literature is the equivalence of virtual and real-world training for tasks such as piloting an airplane (Myers, Starr, & Mullins, 2018). Virtual stimuli can also be exaggerated either in the timing of their delivery or in providing ambiguity in terms of their actual size. Although we often take the juxtaposition of perceptual inputs for granted in real-world contexts, their specificity is a key design feature of virtual worlds. Successful immersion in a virtual environment involves three properties: inclusive, extensive, and incongruence (Bystrom, Barfield, & Hendrix, 1999). These properties can enhance (or limit) perceptual fluidity, which in turn is affected by the absence of perceptual gaps and emergence of coherent percepts.

Perceptual Gaps

Perceptual gaps are both uncontrollable and poorly understood. Whereas unintended perceptual gaps can result from latency in the delivery of the stimulus (e.g., video frame rate lag), they can also arise from cognitive processing itself. One example of a perceptual gap has to do with the phenomenon of presence. Presence involves a number of cognitive functions and deeply influences how we judge the "realness" of a given stimulus (see the plausibility illusion of Slater, 2009). Although not the same as immersion, presence involves subconscious reality judgments with respect to the stimulus. Rather, presence arises as a consequence of whether or not the current experience is credible enough to engage with (Baños et al., 2000).

Perceptual gaps can also be found in and characterized by formal psychological effects related to the use of VR. One such example is the uncanny valley effect (Cheetham, 2017). The uncanny valley involves affective ambiguity toward a human-like stimulus (Urgen, Kutas, & Saygin, 2018). A very different example involves psychophysical effects such as the visually based prism adaptation (Chapman et al., 2010), which may result in significant misalignment between acquired neural representations of the action and the current state of the environment. One candidate cause of perceptual gaps is a lack of explicit 3-D feedback between the novelty of a virtual environment and the nervous system (Fulvio & Rokers, 2017). As visual, auditory, and haptic perception are all part of a complex stimulus, perceptual gaps can also occur in a way that allows not only for interference but also for the construction of an alternate reality. This can be seen in the cutaneous rabbit illusion (Geldard, 1982), which involves rapid and successive stimulation of closely spaced points along the skin. This leads to the illusion of a rabbit crawling up the arm. Although the multimodal aspects of a realistic simulation are difficult to synchronize, understanding how they are experienced in different contexts is of primary importance.

Coherent Percepts and Novel Behaviors

Another limitation (or potential design principle) involves the relationship between mental coherence and presentation of the virtual stimulus. Perception of objects in natural scenes often involves multisensory processing, even when the input to non-visual sensory systems is subtle. Coherent perception of events and awareness of experience rely on spatiotemporal synchronization between sensory modalities. This is known from both experimental decoupling of complex stimuli and multimodal fusion by specific neurons in the brain. When the different components of a complex stimulus are decoupled (e.g., the visual and auditory components are presented at different spatial locations), this has a direct effect on multimodal fusion. This in turn has a larger effect on the experience of sync (Huskey, Craighead, Miller, & Weber, 2018) and flow (Csikszentmihalyi, 1990), namely, in the disruption of such continuous states.

FUNDAMENTAL ISSUES

Perception-Action Coupling

One fundamental issue for implementation of VR in experimental contexts is the role of perception-action coupling, namely, can sensory stimulation be action-independent? We can seek insight from the action selection literature to understand the links between sensory experience and decision-making (Barron et al., 2015), motor learning/control, and agency (Sidarus, Vuorre, & Haggard, 2017). We can also use computational models to understand these relationships better. Using an artificial life model, Seth (2007) defines action selection as a set of coupled sensorimotor processes. Overall, action selection allows us to prioritize our actions given a continuous flow of stimuli. Viewing the flow of perception from an ecological standpoint (Gibson, 2015) is essential to understanding the nature and dynamic aspects of virtual experience.

Another way to understand the link between cognition and virtual reality is through more formal cognitive modeling (Figure 16.1). Whereas cognitive modeling has been used to predict and explain multiple interacting components of human performance in a wide range of task environments, the role of higher-order phenomena such as awareness and engagement has largely been avoided. One way to view a VR-specific cognitive model is to focus on the components of the hypothetical virtuality network.

Toward a "Virtuality Network"

It is unclear whether cognition in virtual environments is mediated by a combination of well-characterized cognitive functions or whether it requires a novel set of cognitive functions heretofore unknown. Much as there are attentional, emotional, and default activity networks in the brain, there might also be a "virtuality" network that mediates environmental features unique to VR (Figure 16.2).

The functional components of a virtuality network might involve interactions between premotor and motor cortex, spatial cognition centers, multimodal integration centers, and brain regions involved in object and face processing. Yet other types of functionality, such as spatial navigation in high-dimensional artwork, might also involve a mechanism related to abstract reasoning or consciousness. Alternatively, we might also learn something about the evolutionary cognitive substrate of virtuality by surveying the literature on the use of virtual reality in animal models (Alicea, 2015).

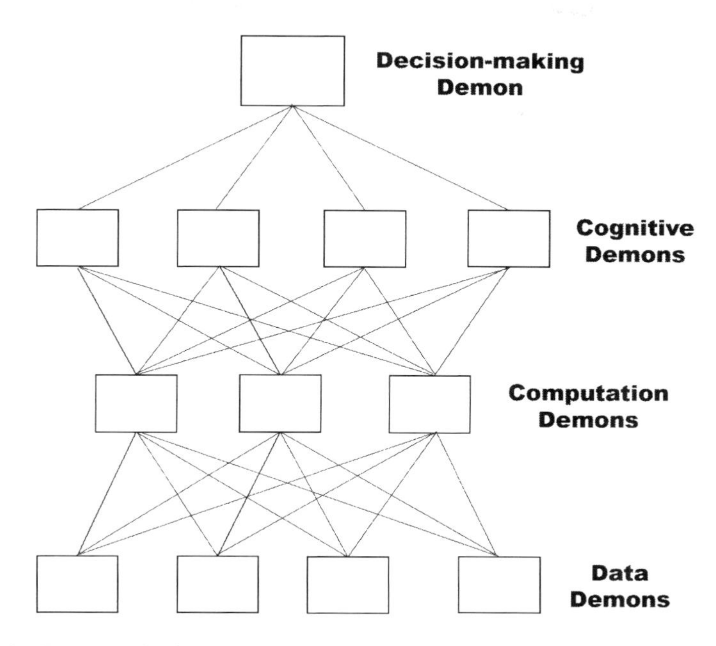

Figure 16.1 An Example of a Cognitive Mode Based on the Pandemonium Model of Selfridge (1959)

Note: Layers include decision-making (top), cognitive demons (second from top), computation demons (second from bottom), and data demons (bottom).

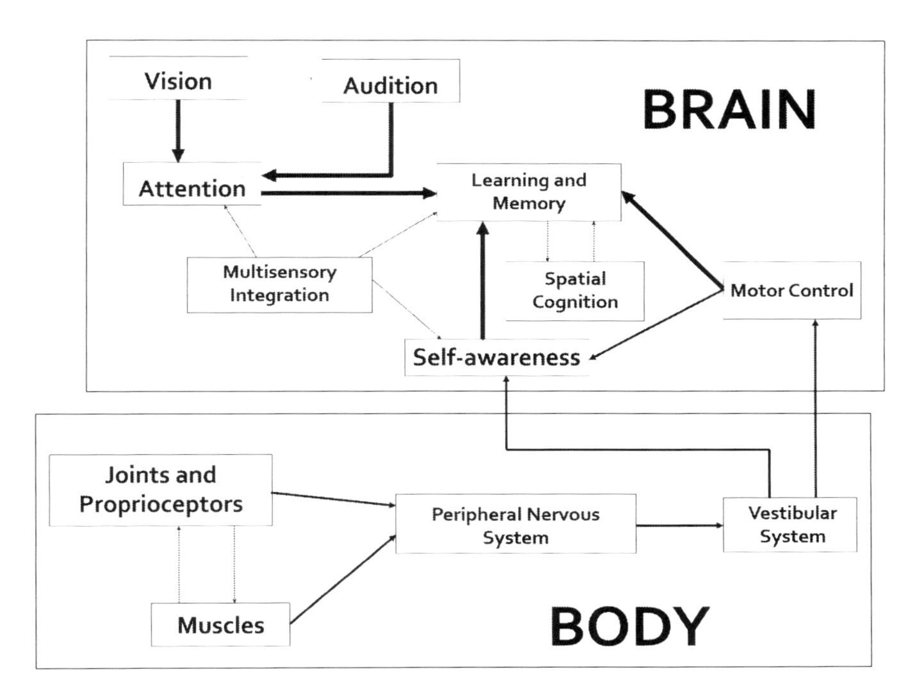

Figure 16.2 Model of a Virtuality Network as a Putative Cognitive Model

WHEN VIRTUAL REALITY BECOMES USEFUL

To discuss how virtual reality becomes useful, several topical areas will be highlighted. These include but are not limited to the necessary creation of new methods and phenomenology, the inclusion of sustained behavioral measurement and unique modes of interaction, and the enabling of unique behavioral modes.

New Methods and Phenomenology

Aside from considering how VR can be advantageous (or deleterious) in measuring neurobehavior, it is quite useful for enabling both new methods and phenomenology. These include, but are not limited to, sustained behavioral measurement, unique modes of interaction, and unique modes of behavior. As we will see, a combination of ingenious methods and the unique features of virtual environments can provide an exciting new perspective on cognition and brain function.

Sustained Behavioral Measurement

From a quantitative standpoint, permitting participants to engage in behaviors such as navigation, aggression, and attentional focusing over long periods of time (more than a single trial or set of trials) has a number of advantages. Whereas specific behaviors can be segmented and classified for hypothesis testing, continuous media stimuli can reveal longer-term cumulative and emergent cognitive effects. Continuous measurement allows us to acquire time-series data, which requires tools such as frequency domain and dynamic causality analysis (Ozaki, 2012). Continuous data may also provide a window into a whole class of dynamical phenomena such as self-organized criticality, phase transitions, and metastability to be observed (Beggs, 2008; Kozma & Freeman, 2017). More specifically, it allows for time-series data to be collected, which in turn allows for new types of analyses and contains information content not measurable as a set of discrete, randomized observations.

Unique Modes of Interaction

Aside from effects relevant to individual cognition, VR also enables people to interact in ways that are impossible in the physical world. When this is scaled up so that many people have access to the same content, it provides a unique window into social neuroscience by allowing us to understand how groups of people behave in a dynamic fashion. In particular, a method called hyperscanning (Balconi & Vanutelli, 2017) can provide a link between human social interactions and the simultaneous imaging of multiple brains. One example comes from the practice of neuroeconomics, wherein hyperscanning is used to enable the dynamic monitoring of continuous and virtually-mediated economic transactions (Rangel, Camerer, & Montague, 2008). Overall, hyperscanning and other social measurement techniques hold great promise for communication neuroscience.

The use of affordances in VR object design also provides a means to introduce new semantic information into the medium. In particular, affordances provide signifiers within the VR environment (Norman, 2008) in a way that provides meaning and constructivist potential (Olson, 2017). Rather than providing opportunities for perceptual illusion, signifiers provide the symbolic glue for perception and action by clarifying the specific role of affordances in the environment. Affordances also provide physical cues to the individual from the virtual environment. According to Regia-Corte et al. (2013), affordances (in this case, a slanted surface) allow an individual to infer physical properties that support judgments that influence perception and action.

Unique Modes of Behavior

Ultimately, exposure to VR may lead to new behaviors. At a basic level, perceptual adaptations, such as the prism effect (Chapman et al., 2010) are well-known after-effects of exposure to VR environments. According to Gonzalez-Franco and Lanier (2017), there are three types of illusion associated with VR: place, embodiment, and plausibility. The place illusion occurs when the virtual environment transports the user to a new spatial location. It could be a familiar location or an exotic one. The embodiment illusion involves experiencing the virtual environment either in a disembodied form or as an avatar. As the intersection of both the place and embodiment illusion, the plausibility illusion involves self-examination as to whether or not an experience is actually happening. In this way, the plausibility illusion overlaps with some of the theoretical issues faced by the presence community.

There are other perceptual illusions that are enabled as a consequence of the embodiment illusion. One of these is called body transfer and is exemplified by the rubber (sometimes called marble) hand illusion (Senna, Maravita, Bolognini, & Parise, 2014). In this example, body transfer occurs when a participant begins to treat a virtual hand as part of his or her own body. While this illusion becomes behaviorally manifest by the artificial hand becoming equivalent to an additional appendage, the phenomenon is represented in the brain as a complex set of multimodal interactions (Ehrsson, Holmes, & Passingham, 2005). Returning to the virtuality network shown in Figure 16.1, there are a number of other possible illusions involving the interactions between the peripheral nervous system and the central nervous system. This is particularly true in cases where the body is not in the same contextual frame as the visual and auditory systems.

ADDITIONAL CONSIDERATIONS

This section will highlight additional considerations, including emerging challenges, opportunities for open science, potential future directions for research, and the increasing omnipresence of virtual media.

Emerging Challenges

There are also several technical challenges when considering the future of VR as a research tool. These challenges can be summarized as a series of trends: portability, ubiquity, and usability. Whereas these issues have been partially overcome as computing power and visual display capacity have increased, they still serve as barriers depending on how VR systems might be used.

One major challenge to using VR and neuroimaging, particularly when taking advantage of naturalistic experimental designs, is to allow for the freedom of natural behavior without the movement artifact associated with many of these behaviors. Behaviors such as human mobility (e.g., walking, driving) require portable, unobtrusive interfaces that do not distract from the cognitive task being performed. There are differences between heads-up displays and monocles in terms of how information is displayed in the visual field as well as how aware people are of the hardware. This not only can degrade the ecological validity of experimental investigations but can also interfere with the feeling of immersion that the VR stimulus is intended to provide.

Opportunities for Open Science

The concurrence of Virtual Neuroscience and the Open Science movement provides a number of opportunities to produce better science. I will focus on two of these: open-source content and open cognitive models. One way to ensure reproducibility and stimulus effectiveness is to contribute to and explore the open-source VR community (see Additional Information section). Developing a set of best practices for content creation within and among research groups will bolster the experimental validity of the method. Another issue mentioned previously is to develop VR-specific cognitive models. While many detailed and even executable cognitive models exist, they are often either part of proprietary software packages or published in opaque form (e.g., non-accessible journals and programming languages). One sociology of science issue involves viewing the media-specific effects of virtual experience as epiphenomena. A focus on reproducible research and formal cognitive models may serve to mitigate skepticism from the wider community of cognitive neuroscientists.

Future Directions

One trend to follow is the explosion of data at the intersection of VR systems and neuroimaging. As VR-related equipment becomes increasingly more affordable in terms of fidelity per dollar, there will be an incentive to use such systems as experimental stimuli. This will lead to an explosion in the amount and variety of data centering on the cognitive neuroscience of virtual interactions. With this increase in data generation comes a need to adopt open science practices (Gilmore, Diaz, Wyble, & Yarkoni, 2017). For example, there is an opportunity for data to be systematically shared and reanalyzed. While this provides a virtuous resource that transcends the media and communication neuroscience community, these data require investments in organization and maintenance (Poldrack & Gorgolewski, 2014). Ultimately, look for a big data ecosystem (Khan et al., 2014) to emerge and help to improve our understanding of the major scientific issues. The use of big data in VR will be unique in that VR can serve as both the stimulus that enables data collection and an environment to visualize, evaluate, and meta-analyze the archived data (Matthews, 2018).

Virtual Media Omnipresence

Another trend to consider is the omnipresence of media content via VR and AR interfaces. As virtual and augmented reality equipment become more affordable and easy to wear, they will become integrated into our daily lives and social interactions. While we may not notice the distinction between real and virtual objects in such a world, there will be new sets of effects and behaviors nonetheless. Perhaps we will one day speak of a new field-specific term: "ubiquitous virtuality", or the neuroscience of ever-present media.

A third trend will involve exploring longer-term and global media effects. This is particularly the case as the effects of virtuality are experienced over longer durations of time by a more diverse group of people (e.g., non-WEIRD, i.e., Western educated, industrialized, rich, and democratic populations; see Henrich, Heine, & Norenzayan, 2010). As virtual technologies diffuse across the world and into various cultural/demographic groups, new effects will emerge that interface not only with cognitive processing, but also the cultural and biological components of global human variation.

Historical Serendipity

It is an opportune time to study the effects of VR on human nervous systems, as relatively affordable, easy-to-use systems are now becoming mainstream consumer goods and work tools. Using historical trends for Google Scholar search terms, Hamacher et al. (2016) have shown that the terms "virtual reality" and "mixed reality" started to grow exponentially in the early 1990s. This growth peaked around 2010 and resulted in a trough centered on the year 2014. While these trends are currently on the increase, the search term "augmented reality" shows a slightly different trend: exponential increase from 2000 to 2014, with a more recent slight decline. In both cases, it is clear that Virtual Neuroscience demonstrates ample opportunity for growth supported by a growing bibliography of prior studies. The Gartner report on technology hype cycles (Gartner Consulting, 2016) paints an even more encouraging picture: VR resides along the slope of enlightenment and AR along the trough of disillusionment. Both are expected to arrive on the plateau of productivity in roughly 5–10 years. With the introduction of easy-to-use consumer hardware and open-source software, using VR for research has never been more accessible.

ADDITIONAL INFORMATION

Here are a few examples of open source resources to get users started in creating and implementing virtual stimuli, as well as taking advantage of open datasets:

- Open Source Virtual Stimulus #1: Asteroid Impact: https://github.com/richardhuskey/asteroid_impact
- Open Source Virtual Stimulus #2: Space Fortress 5: https://github.com/CogWorks/SpaceFortress
- Open Source Content Generator #1: Mozilla A-frame: https://aframe.io/
- Open Source Content Generator #2: Open Space 3D: www.openspace3d.com
- Open Data Source #1: Open Neuro: https://openneuro.org/
- Open Data Source #2: Open fMRI: https://openfmri.org/

REFERENCES

Alicea, B. (2015). Animal-oriented virtual environments: Illusion, dilation, and discovery. *F1000 Research, 3*, 202.

Balconi, M., & Vanutelli, M. E. (2017). Cooperation and competition with hyperscanning methods: Review and future application to emotion domain. *Frontiers in Computational Neuroscience, 11*, article 86.

Baños, R. M., Botella, C., Garcia-Palacios, A., Villa, H., Perpina, C., & Alcaniz, M. (2000). Presence and reality judgment in virtual environments: A unitary construct? *CyberPsychology and Behavior, 3*, 327–335.

Barron, A. B., Gurney, K. N., Meah, L. F. S., Vasilaki, E., & Marshall, J. A. R. (2015). Decision-making and action selection in insects: Inspiration from vertebrate-based theories. *Frontiers in Behavioral Neuroscience, 9*, article 216.

Bartels, A., & Zeki, S. (2004). The chronoarchitecture of the human brain: Natural viewing conditions reveal a time-based anatomy of the brain. *NeuroImage, 22*, 419–433.

Beggs, J. M. (2008). The criticality hypothesis: How local cortical networks might optimize information processing. *Philosophical Transactions of the Royal Society A, 366*, 329–343.

Biocca, F., & Delaney, B. (1995). Immersive virtual reality technology. In F. Biocca & M. Levy (Eds.), *Communication in the age of virtual reality* (pp. 57–124). Hillsdale, NJ: Lawrence Erlbaum Associates.
Bohil, C., Alicea, B., & Biocca, F. (2011). Virtual reality in neuroscience research and therapy. *Nature Reviews Neuroscience, 12*, 752–762.
Bystrom, K.-E., Barfield, W., & Hendrix, C. (1999). A conceptual model of the sense of presence in virtual environments. *Presence, 89*, 241–244.
Chance, S. S., Gaunet, F., Beall, A. C., & Loomis, J. M. (1998). Locomotion mode affects the updating of objects encountered during travel: The contribution of vestibular and proprioceptive inputs to path integration. *Presence, 7*, 168–178.
Chapman, H. L., Eramudugolla, R., Gavrilescu, M., Strudwick, M. W., Loftus, A., Cunnington, R., & Mattingley, J. B. (2010). Neural mechanisms underlying spatial realignment during adaptation to optical wedge prisms. *Neuropsychologia, 48*, 2595–2601.
Cheetham, M. (2017). The Uncanny Valley Hypothesis and beyond. *Frontiers in Psychology, 8*, article 1738.
Csikszentmihalyi, M. (1990). *Flow: The psychology of optimal experience*. New York, NY: Harper & Row.
Ehrsson, H. H., Holmes, N. P., & Passingham, R. E. (2005). Touching a rubber hand: Feeling of body ownership is associated with activity in multisensory brain areas. *Journal of Neuroscience, 25*, 10564–10573.
Fulvio, J. M., & Rokers, B. (2017). Use of cues in virtual reality depends on visual feedback. *Scientific Reports, 7*, 16009.
Gartner Consulting. (2016). Gartner hype cycle for emerging technologies. Retrieved from www.gartner.com/smarterwithgartner/3-trends-appear-in-the-gartner-hype-cycle-for-emerging-technologies-2016/
Geldard, F. A. (1982). Saltation in somesthesis. *Psychological Bulletin, 92*, 136–175.
Gerrig, R. J. (1993). *Experiencing narrative worlds.* New Haven, CT: Yale University Press.
Gibson, J. J. (2015). *The ecological approach to visual perception.* New York, NY: Psychology Press.
Gilmore, R. O., Diaz, M. T., Wyble, B. A., & Yarkoni, T. (2017). Progress toward openness, transparency, and reproducibility in cognitive neuroscience. *Annals of the New York Academy of Sciences, 1396*, 5–18.
Gonzalez-Franco, M., & Lanier, J. (2017). Model of illusions and virtual reality. *Frontiers in Psychology, 8*, article 1125.
Good, J. M. M. (2007). The affordances for social psychology of the ecological approach to social knowing. *Theory and Psychology, 17*, 265–295.
Green, C. S., & Bavelier, D. (2007). Action-video-game experience alters the spatial resolution of vision. *Psychological Science, 18*(1), 88–94.
Green, C. S., & Bavelier, D. (2012). Learning, attentional control, & action video games. *Current Biology, 22*, R197–R206.
Hamacher, A., Kim, S.-J., Cho, S.-T., Pardeshi, S., Lee, S.-H., Eun, S.-J., & Whangbo, T.-K. (2016). Application of virtual, augmented, and mixed reality to urology. *International Neurourology Journal, 20*, 172–181.
Hambrick, D. Z., Oswald, F. L., Altmann, E. M., Meinz, E. J., Gobet, F., & Campitelli, G. (2014). Deliberate practice: Is that all it takes to become an expert? *Intelligence, 45*, 34–45.
Hardiess, G., Meilinger, T., & Mallot, H. A. (2015). Virtual reality and spatial cognition. In J. D. Wright (Ed.), *International encyclopedia of social and behavioral sciences* (pp. 133–137). Amsterdam, The Netherlands: Elsevier.
Hasson, U., & Honey, C. J. (2012). Future trends in neuroimaging: Neural processes as expressed within real-life contexts. *Neuroimage, 62*, 1272–1278.
Hasson, U., Landesman, O., Knappmeyer, B., Vallines, I., Rubin, N., & Heeger, D. J. (2008). Neurocinematics: The neuroscience of film. *Projections, 2*, 1–26.
Henrich, J., Heine, S. J., & Norenzayan, A. (2010). The weirdest people in the world? *Behavioral and Brain Sciences, 33*, 61–83.

Huskey, R., Craighead, B., Miller, M. B., & Weber, R. (2018). Does intrinsic reward motivate cognitive control? A naturalistic-fMRI study based on the synchronization theory of flow. *Cognitive, Affective, and Behavioral Neuroscience*. Advance online publication.

Huster, R. J., Mokom, Z. N., Enriquez-Geppert, S., & Herrmann, C. S. (2014). Brain-computer interfaces for EEG neurofeedback: Peculiarities and solutions. *International Journal of Psychophysiology, 91*, 36–45.

Khan, N., Yaqoob, I., Hashem, I. A. T., Inayat, Z., Ali, W. K. M., Alam, M., ... Gani, A. (2014). Big data: Survey, technologies, opportunities, and challenges. *Scientific World Journal, 14*, 712826.

Kim, T., & Biocca, F. (1997). Telepresence via television: Two dimensions of telepresence may have different connections to memory and persuasion. *Journal of Computer-Mediated Communication, 3*, article 325.

Kozma, R., & Freeman, W. J. (2017). Cinematic operation of the cerebral cortex interpreted via critical transitions in self-organized dynamic systems. *Frontiers in Systems Neuroscience, 11*, article 10.

Martens, U., Wahl, P., Hassler, U., Friese, U., & Gruber, T. (2012). Implicit and explicit contributions to object recognition: Evidence from rapid perceptual learning. *PLoS One, 7*, e47009.

Matthews, D. (2018). Virtual-reality applications give science a new dimension. *Nature, 557*, 127–128.

McLuhan, M. (1964). *Understanding media*. New York, NY: McGraw-Hill.

Myers, P. L., Starr, A. W., & Mullins, K. (2018). Flight simulator fidelity, training transfer, and the role of instructors in optimizing learning. *International Journal of Aviation, Aeronautics, and Aerospace, 5*, article 6.

Norman, D. (2008). Signifiers, not affordances. *ACM Interactions, 15*, 18–19.

Olson, B. (2017, February 3). It's not you, it's the interface. Retrieved from https://medium.com/@Ben.Olson/affordances-and-signifiers-in-mobile-interface-design-cf584696cda8.

Ozaki, T. (2012). *Time series modeling of neuroscience data*. New York, NY: CRC Press.

Poldrack, R. A., & Gorgolewski, K. J. (2014). Making big data open: Data sharing in neuroimaging. *Nature Neuroscience, 17*, 1510–1517.

Rangel, A., Camerer, C., & Montague, P. R. (2008). A framework for studying the neurobiology of value-based decision making. *Nature Reviews Neuroscience, 9*, 545–556.

Rauter, G., Sigrist, R., Koch, C., Crivelli, F., van Raai, M., Riener, R., & Wolf, P. (2013). Transfer of complex skill learning from virtual to real rowing. *PLoS One, 8*, e82145.

Reeves, B., & Nass, C. (1996). *The media equation: How people treat computers, television, and new media like real people and places*. Cambridge, England: Cambridge University Press.

Reggente, N., Essoe, J. K.-Y., Aghajan, Z. M., Tavakoli, A. V., McGuire, J. F., Suthana, N. A., & Rissman, J. (2018). Enhancing the ecological validity of fMRI memory research using virtual reality. *Frontiers in Neuroscience, 12*, article 408.

Regia-Corte, T., Marchal, M., Cirio, G., & Lecuyer, A. (2013). Perceiving affordances in virtual reality: Influence of person and environmental properties in perception of standing on virtual grounds. *Virtual Reality, 17*, 17–28.

Rheingold, H. (1991). *Virtual reality: Exploring the brave new technologies*. New York, NY: Simon & Schuster.

Saleh, G. M., Theodoraki, K., Gillan, S., Sullivan, P., O'Sullivan, F., Hussain, B., ... Athanasiadis, I. (2013). The development of a virtual reality training programme for ophthalmology: Repeatability and reproducibility. *Eye, 27*, 1269–1274.

Selfridge, O. G. (1959). Pandemonium: A paradigm for learning. *Proceedings of the Symposium on Mechanisation of Thought Processes*, 511–529.

Senna, I., Maravita, A., Bolognini, N., & Parise, C. V. (2014). The marble-hand illusion. *PLoS One, 9*, e91688.

Seth, A. K. (2007). The ecology of action selection: Insights from artificial life. *Philosophical Transactions of the Royal Society B, 362*, 1545–1558.

Sidarus, N., Vuorre, M., & Haggard, P. (2017). How action selection influences the sense of agency: An ERP study. *NeuroImage, 150*, 1–13.

Simons, D. J., Boot, W. R., Charness, N., Gathercole, S. E., Chabris, C. F., Hambrick, D. Z., & Stine-Morrow, E. A. L. (2016). Do "brain-training" programs work? *Psychological Science in the Public Interest, 17*, 103–186.

Slater, M. (2009). Place illusion and plausibility can lead to realistic behaviour in immersive virtual environments. *Philosophical Transactions of the Royal Society B, 364*, 3549–3557.

Slater, M., & Sanchez-Vives, M. V. (2016). Enhancing our lives with immersive virtual reality. *Frontiers in Robotics and AI: Virtual Environments, 3*, article 74.

Steuer, J. (1992). Defining virtual reality: Dimensions determining telepresence. *Journal of Communication, 42*, 73–93.

Tarr, B., Slater, M., & Cohen, E. (2018). Synchrony and social connection in immersive virtual reality. *Scientific Reports, 8*, article 3693.

Teo, W.-P., Muthalib, M., Yamin, S., Hendy, A. M., Bramstedt, K., Kotsopoulos, E., … Ayaz, H. (2016). Does a combination of virtual reality, neuromodulation and neuroimaging provide a comprehensive platform for neurorehabilitation? A narrative review of the literature. *Frontiers in Human Neuroscience, 10*, article 284.

Thompson, J. J., Blair, M. R., Chen, L., & Henrey, A. J. (2013). Video game telemetry as a critical tool in the study of complex skill learning. *PLoS One, 8*, e75129.

Urgen, B. A., Kutas, M., & Saygin, A. P. (2018). Uncanny Valley as a window into predictive processing in the social brain. *Neuropsychologia, 114*, 181–185.

Weber, R., Alicea, B., Huskey, R., & Mathiak, K. (2018). Network dynamics of attention during a naturalistic behavioral paradigm. *Frontiers in Human Neuroscience, 12*, 182.

West, G. L., Konishi, K., Diarra, M., Benady-Chorney, J., Drisdelle, B. L., Dahmani, L., … Bohbot, V. D. (2017). Impact of video games on plasticity of the hippocampus. *Molecular Psychiatry, 23*, 1566–1574.

17

Is There a Cultural Brain?

Analyzing Individual Differences in Processing Media Messages

Benjamin O. Turner

What is the proper role of the individual in communication science? Should phenomena be studied purely on an individual, case-by-case basis? Should individuals be treated as effectively interchangeable, with effects of interest requiring averaging across groups of individuals to be observable? Or are individuals the wrong level to think about altogether, with effects manifesting only at the level of organizations, institutions, or cultures? This chapter does not set out to provide a single correct answer to any of these questions, because no such answer exists. What it does instead is provide readers with a framework within which to think about these questions and to understand the conditions under which some answers might be more or less useful in their particular programs of research. To this end, this chapter introduces and defines the concept of individual differences. In doing so, it furthers previous attempts to highlight the idea of individual differences in communication (see, e.g., Oliver & Krakowiak, 2009) by drawing on some of the latest research in cognitive neuroscience, which demonstrates the degree to which individuals differ at the level of how their brains represent and process information.

Before turning to a description and discussion of individual differences, it is important to highlight that it is not a concept that will be useful to all communication scholars. For researchers from some intellectual traditions, or researchers studying certain phenomena, there is no role for the individual (Craig, 1999). For example, a researcher investigating how corporations communicate may have no use in considering the individual. This is because although corporations are made of up collections of individuals, their behavior can nonetheless be conceptualized as emergent and definable only at a level that abstracts away from individuals. Even here, there may be gray areas—for instance, a corporation's behavior may change drastically when its CEO or president is replaced—but the aim of this chapter is not to claim that the concept of individual differences is universally necessary, nor even universally helpful or possible. However, the hope is that many communication scientists will nonetheless find useful ideas herein.

Moreover, many phenomena that are typically conceptualized on a level abstracted away from individuals may benefit from a more explicit consideration of the individual. For example, theories such as agenda setting or cultivation are traditionally defined in terms of a "collective consciousness" or "the public," but of course studies based on these theories comprise groups of individuals and their effects depend to some degree on underlying processes happening within individuals. Therefore, whether they want to or not, scholars studying these sorts of phenomena must confront the question of individual differences. To be clear, this is hardly a new idea; for example, cultivation theory has been construed at the individual level for nearly two decades (Shrum, 2001). The point is that some theories or phenomena that were originally more sociological in nature may nonetheless be amenable to investigation at the individual level—in some cases, possibly including individual differences.

This chapter begins with an overview of individual differences, including definitions of some key terms, the various approaches researchers can take to dealing with individual differences, and what the latest research has to say about how individual differences can be used. Next, the chapter provides concrete examples of what cognitive neuroscience has demonstrated about the variability between individuals and how individual differences have already been studied in the nascent field of communication neuroscience. Finally, the chapter provides a brief discussion of the concept of culture through the lens of how the concepts of culture and individual differences can reciprocally inform one another, and it looks ahead to how future research might make use of both concepts to address outstanding questions in communication science.

INDIVIDUAL DIFFERENCES

Before this chapter can begin to review the state of the art with respect to individual differences, it is necessary to first define a few key terms.

- *Individual differences*: trivially, the differences that exist between people. This can be with respect to a particular variable or phenomenon; with respect to multiple variables or phenomena; or just a general observation of the fact that people differ.
- *Individual differences analysis*: an analysis that aims to leverage individual differences in some way, whether the empirical focus is on a particular individual difference factor per se, or whether the research merely uses such a factor to shed light on a theory or phenomenon that does not have anything intrinsically to do with individual differences. Most often, individual differences analyses will use correlational methods, but this includes many methods beyond simple Pearson correlation.
- *Individual difference factor*: a variable that might explain the differences between individuals, for example, personality traits, educational background, age, or gender.
- *Individual variability* (or *inter-individual variability*): essentially synonymous with individual differences, although this tends to be framed more explicitly with reference to a larger sample or population.

Here, the focus will be on individual differences in terms of relatively steady aspects of a person; in general, these will hew closely to the concept of a *trait* from psychology. This is in contrast to a *state*, which is a relatively more short-term aspect. However, the line between these two is not always clear, and the terms as used in psychology may not always apply. For example, if Brett is in a bad mood because he got too little sleep last night, and he is otherwise a positive person, most psychologists would define his bad mood as a state—but if Brett has a tendency to get too little sleep most nights, the question of what aspects are traits (only his sleep habits, or his frequent bad moods as well) versus states becomes trickier. For the purposes of this chapter, it is enough to use a common-sense definition and talk about individual differences as relating to aspects that are allowed to be somewhat probabilistic, without worrying about these semantic distinctions.

Individual differences are inherent to research across the entire field of behavioral science (a broad term that includes some areas of communication science). It is hardly controversial to observe that individuals differ from one another. However, individual differences have not always been treated in a uniform way by researchers in these domains. Roughly speaking, there are four main approaches researchers can take when confronting individual differences: (1) ignore them; (2) beg the question; (3) treat them as a *manageable* nuisance; or (4) treat them as a *signal* rather than noise. Each of these approaches will be addressed in greater detail below. Note that this chapter does not presume to name one approach as uniformly superior to any other, but instead, aims to point out the reasoning that may lead to each approach and the advantages and disadvantages of each.

First Approach to Individual Differences: Ignore Them

Simply ignoring individual differences is surely the most common approach across most fields, including communication science. In some cases, this is not only perfectly justifiable, but is the only logical option available. However, the validity of ignoring them depends on several factors, some of which researchers may not be aware of.

Before considering those factors, it is worth asking *why* scholars might choose to ignore individual differences, and to refute the motivations that are misguided. This chapter will focus only on those situations in which reasonable observers would agree that individual differences are in fact possible. Note also that this chapter will not address the utility of any particular individual difference factor, but instead, how individual differences in general are acknowledged. First, to some scholars, the idea that individuals differ may seem so trivially obvious as to be unworthy of investigation. This is primarily an argument against investigating individual differences per se, and less against using them to further our understanding of other topics. Nonetheless, this line of reasoning can be dismissed out of hand. After all, the fact that everyone could see that objects fall when dropped did not make Isaac Newton's observations on gravity any less useful to subsequent scientists. Second, and relatedly, individual differences may be seen as nothing more than noise that masks true population-level effects. In contrast to the first position, this is, in many cases, a defensible and correct viewpoint. Many differences between individuals cannot be linked to stable or recurring individual difference factors; moreover, some phenomena are robust to the observed degree of variability between individuals. However, as alluded to above and laid out in more detail below, there are a number of factors that determine whether treating individual differences as ignorable noise—in other words, treating individuals as interchangeable—is valid. Finally, the third argument for ignoring individual differences goes in the opposite direction, and views differences between individuals as being so idiosyncratic as to be beyond our ability to study (or perhaps, to offer too little benefit relative to the cost that would be incurred from studying them properly). This position may also be correct in some instances; much of the remainder of this chapter, however, is devoted to highlighting the degree to which individual differences are more quantifiable, and their study more beneficial, than adherents to this argument assume.

To return to the idea that individual differences reflect nothing more than noise, whether or not this is strictly accurate—a question for philosophy or chaos theory, perhaps—it is certainly practically true in many instances. The analysis approach that follows from this position is to treat individuals as interchangeable, and to consider effects at the level of some sort of group aggregate. This may be an average, such as the arithmetic mean rating across a group of participants who viewed an advertisement; a regression coefficient relating some experimental variable of interest such as emotional frame to ratings of ad liking; or any of a number of other approaches to aggregating data across a number of individuals.

This approach rests on a few assumptions, some of which can be taken as a given but others of which depend upon verification on the part of the researcher. The first will be familiar to all scientists, and does not need proof: under the assumption that the sampling of individuals from the population is random, or a sufficiently reasonable facsimile of random, then the law of large numbers says that an experiment's results will be an arbitrarily accurate estimate of the true (population) effect, in proportion to the sample size. In practice, the question of how close an experiment's results come to the true value depends on the effect size, the sample size, and myriad design, experimental, and institutional factors. In addition, the available evidence suggests that scientists at least occasionally fail to do a good job at capturing the true effect (Camerer et al., 2018; Open Science Collaboration, 2015).

The second assumption relates to the association between results obtained at the group level, on the one hand, and individuals, on the other hand. This may seem paradoxical in the context of the discussion about arguments in support of ignoring individual differences, but it is perfectly consistent, because this sort of group-to-individual projection does not necessarily refer to any individual in particular, but rather, individuals in general. There are two challenges to drawing this sort of link. The first has to do with the question of whether a group-level result accurately reflects any individual-level result or is instead an epiphenomenon or artifact of the aggregation procedure itself, whereas the second has to do with the issue of ergodicity.

For an example of an epiphenomenal result, consider the following simple scenario: a group of 200 individuals, comprising equal numbers of Republicans and Democrats, watches a campaign ad for a Republican politician. Each individual is asked to rate on a Likert scale how much he or she likes the ad, and the researcher aggregates across individuals by taking the arithmetic mean, which (in this contrived example) is precisely at the midpoint of the scale. Setting aside measurement theory's contention that the researcher's chosen aggregate statistic involves an invalid operation for ordinal data, the group-level result does a poor job of reflecting any of the individuals. In fact, even starker examples are common in psychology, where, for example, aggregating data that come from an all-or-nothing process (such as in insight problem-solving) can give the impression of a continuous process, which is fundamentally different psychologically. In both examples, the problem is the same if the researcher's empirical question is tied in any way to what actually happens at an individual level. Likewise, relating group-level results to individuals requires that the process(es) underlying the phenomenon being studied be ergodic—that is, homogeneous and stationary. As it happens, recent work in psychology suggests that many processes are in fact not homogeneous (i.e., across individuals) or stationary (i.e., stable over time within an individual), and therefore results related to the study of these processes at the group level cannot be translated one-to-one to individuals (Fisher, Medaglia, & Jeronimus, 2018; Gomes, de Araujo, do Nascimento, & Jelihovschi, 2018). Both of these are challenges to which this chapter will return later.

Second Approach to Individual Differences: Beg the Question

The second broad approach to dealing with individual differences is to beg the question. In other words, accept that there are individual differences, and get around them by doing all analyses on the individual level. In some cases, this might stem from the belief outlined above that individual differences are so widespread and complex that they are beyond study, but in this approach, the solution is not to ignore them, but to accommodate them. In the extreme, this approach is synonymous with the case study research method. The medical sciences have made good use of the case study approach for over a century, as has the field of psychology. This approach can certainly be employed usefully, but fewer types of generalizations can be drawn from a single case study than from multi-participant approaches. The same limitation of inferential scope applies to hybrid approaches that essentially bundle together multiple case studies. For example, Gordon et al. (2017) present extensive individualized results from ten individuals who were subjected to over 12 hours of testing (including brain imaging, behavioral task performance, and various psychological batteries). As much as anything, this study was presented as a proof of concept that individual-level testing can yield reliable results; however, with a sample size of ten, there is no possibility of testing for systematic differences between individuals, and the variability observed between them highlights the futility of combining across them using existing aggregation techniques.

Similarly, in a series of more communication-relevant studies, Gallant and colleagues have investigated the representation of images and concepts using small samples, with as few as two participants in the seminal study in this line (Kay, Naselaris, Prenger, & Gallant, 2008). This sort of quasi-case study research can yield important and wide-reaching results. For example, Kay et al. (2008) demonstrated a proof of concept that it is possible to identify with fairly high accuracy the images an individual was seeing based entirely on the patterns of brain activity that were measured while they were seeing those images; whether this is possible in *all* individuals is beside the point of this study. However, researchers who wish to understand neural representations, and in particular, what is universal or "typical," will generally be ill-served by this sort of approach.

Third Approach to Individual Differences: Treat Them as a Nuisance

The third broad approach to confronting individual differences is not only to acknowledge them, but indeed to use them—but only as potentially confounding variables. Researchers generally realize this approach in one of two ways: the individual difference factor is either included as a covariate in a statistical analysis, or else the analysis is carried out in a subset defined on the basis of an individual difference factor. For example, imagine a scenario in which a researcher is examining the effect of message framing on a measure of persuasion. If there is a sex effect, such that men have a higher mean than women, then it is possible that the framing effect will be rendered insignificant due to the additional variance introduced by the sex effect. If a researcher includes sex as a covariate, then this additional variance can be accounted for, rendering the framing effect significant. More extreme would be if the difference between men and women were not one of degree, but of kind: if women showed an effect but men did not, a researcher might carry out the analysis using data from women only.

Although controlling for confounds is reasonable in principle, in practice it is somewhat more complicated. The most immediate problem is the so-called "garden of forking paths," which is a well-known questionable research practice wherein researchers adapt their statistical tests based on properties of the observed data. There are a number of ways to deal with this problem (see, e.g., Rubin, 2017), but in general, decisions such as the cohort on which the analysis will be conducted should be made a priori on theoretical grounds. Even more principled approaches to controlling for confounds are not necessarily guaranteed to be free from problems, however. For example, researchers often use null-hypothesis significance tests in claiming that certain variables do *not* constitute a confound that must be corrected, but this is not an appropriate method for ascertaining a potential nuisance variable's influence (Sassenhagen & Alday, 2016). Likewise, the fact that some individual difference factors that may be treated as nuisance variables rely on measurement, and therefore are subject to measurement error, can complicate the use of these factors as nuisance variables (Westfall & Yarkoni, 2016). This is not to say that individual difference factors cannot be used as confound variables, just that doing so requires care. More important, however, if an individual difference factor is related to a construct of interest reliably enough that it can be used in this way, then by definition, it reflects a signal, not noise, and therefore an opportunity for theoretical advance. This perspective shift brings us to the final approach to dealing with individual differences.

Fourth Approach to Individual Differences: Leverage Them

The fourth approach to confronting individual differences is to conduct an individual differences analysis. The key idea of such an analysis is to use the patterns of variation in the phenomenon of interest across individuals to better understand the phenomenon or to advance the theory that explains the phenomenon. There are many benefits of this approach, but, first, it is important to point out the necessary conditions. First, the individual difference factor or factors being considered must reflect signal, not noise. This is difficult to ascertain in any given analysis, but depends on the validity and reliability of the measurements used to measure the factors, and if applicable, that the underlying processes are ergodic. Furthermore, because most individual differences analyses are correlational (whether or not the analysis itself actually relies on correlation), the inferential scope and attributions of causality are limited: it is invalid to ascribe explanatory power to any measured variable over any unmeasured variable. Finally, individual differences analyses, being correlational (i.e., based on covariances), typically require larger sample sizes than analyses focused on simple mean effects. Although the formulas for standard errors of the mean and (Pearson) correlation prevent direct comparison, a rule of thumb suggests that if an effect requires N participants to observe a mean effect, it will require $N^{1.34}$ to observe an equally large correlational effect (although note that any given effect will be well suited to a focus either on the mean or on variability; Hedge, Powell, & Sumner, 2018). Moreover, as will be discussed below, the best individual differences analyses will include multiple variables, which increases the required sample size even further (Maxwell, Kelley, & Rausch, 2008).

Despite these caveats, this approach is often the most informative and the most logically coherent of the four. That is, conditional on there being individual differences in the phenomenon of interest, it is often best to use those differences, rather than to ignore them or treat them as nothing more than noise to be removed. There are two major ways in which researchers can reap benefits when using individual differences analyses, namely, improving either prediction or theory. The first is straightforward: in applied contexts, if researchers can discover individual difference factors that influence a phenomenon of interest, then this can allow individualized prediction. For example, advertisers hope that by targeting ads based on demographic or other factors of their (presumed) audience, they can increase ad effectiveness, even up to the point of tailoring ads to specific viewers (Noar, Benac, & Harris, 2007). There is no doubt that, particularly as technological advances increasingly enable advertisers and other content creators or platforms to adapt to individual audience member/user characteristics, this sort of prediction can be practically useful (and in some contexts, potentially moral; Gabrieli, Ghosh, & Whitfield-Gabrieli, 2015). However, the bedrock of communication science is the testing and refining of theories, and it is on this front that individual differences analyses offer their greatest potential.

The cornerstone of individual differences analysis is the idea that understanding individual variability represents information that can be used to shed light on theory. For example, Vogel and Awh (2008) describe three types of theoretical contribution that individual differences analyses enable: validating cognitive constructs, demonstrating associations among (putatively separable) cognitive constructs, and demonstrating dissociations between similar constructs. How does this work? It is worth considering each contribution in turn using hypothetical examples from communication. First, imagine that a researcher has discovered what appears to be a hard limit in how individuals process persuasive messages, with arguments after the third (on average) failing to have an effect; moreover, suppose that the researcher has identified a psychophysiological measure that tracks monotonically with incoming arguments up to the third, and then asymptotes as with the behavioral measures. This psychophysiological measure may appear to be a biomarker of persuasion, but the causality is hard to establish. However, the evidence is much stronger if there are reliable differences between individuals in how many arguments they are able to accommodate, and if this psychophysiological measure likewise tracks across individuals.

As for demonstrating an association between constructs, suppose that in the above example, individuals' capacity for incorporating additional arguments were strongly correlated with a measure of their emotional regulation. Assuming this relationship is replicable, this suggests that emotional regulation processes are involved in argument processing. Not only does this inform theories of argument processing per se, but it has implications for any theory that is affected by argument processing. Finally, suppose that individual argument processing capacity were shown to be uncorrelated with another process that might be presumed to be related, such as inferential reasoning skill.[1] Again, beyond informing understanding of the underlying processes, this result would have implications for theories based on each.

There are many other examples from psychology and cognitive neuroscience from the past several years, as researchers increasingly recognize that the variability between individuals is often greater than the magnitude of the effects of interest (Huth, de Heer, Griffiths, Theunissen, & Gallant, 2016; Miller et al., 2009). In fact, in the last few years, there have been a number of high-profile calls for those fields to move more rapidly toward individual differences analyses (Dubois & Adolphs, 2016; Seghier & Price, 2018). One important idea articulated by these articles is that individual differences can often be studied with little cost alongside other, more central theoretical questions. Of course, every unplanned analysis must be treated as exploratory, and either corrected for multiple comparisons or, better yet, followed up with further studies examining promising trends. However, although individual differences analyses are still a relatively small niche in those fields, it is easy to see signs that they may soon compete with traditional group-average approaches, particularly in cases where those approaches have limitations of their own (e.g., Turner, Paul, Miller, & Barbey, 2018).

Leveraging Neural Variability to Understand Communication Phenomena

Now that this chapter has defined individual differences, discussed the various approaches to confronting them, and demonstrated that in at least some domains (especially psychology), individual differences can be capitalized upon to advance science, the question remains—how can this approach be employed in the field of communication? Previous work has partially addressed this question from a slightly different vantage point, by, for example, highlighting some of the ways in which particular individual difference factors can influence media effects (Oliver & Krakowiak, 2009). However, perhaps the most striking findings related to individual differences come from cognitive neuroscience, because of the high dimensional representation of an individual that cognitive neuroscience allows and because neuroimaging allows access to previously inaccessible information (as exemplified by the brain-as-predictor approach pioneered by Falk, Berkman, & Lieberman, 2012). Therefore, the focus here will be specifically on how individual differences can be leveraged using neuroscientific methods, especially functional magnetic resonance imaging (fMRI). For readers who are unfamiliar with fMRI, other chapters in this volume provide an introduction (see Hopp & Weber, Chapter 20).

In general, there are two forms of individual differences analyses with fMRI data, which will be described briefly here before turning to examples of individual difference analyses from communication neuroscience (for a more in-depth treatment, see Turner, Huskey, & Weber, 2019). The first form is to test for relationships between individual difference factors and outcome measures in their original space (or with some monotonic transformation applied). This can be coarse—for example, binarizing an individual difference factor such as sensation-seeking and doing a group comparison of activity in a particular brain region, although such binarization is generally inadvisable (see, e.g., Lemon, 2009)—or fine, as when using continuous variables. This approach is not limited in terms of the modality from which the neural data come: example analyses include correlating normalized volume or cortical thickness in a segmented brain region of interest (ROI) against a measure of task performance (structural); running a structural equation model relating personality factors and task-evoked brain activity to a measure of subsequent behavior change (activation); or running a regression analysis linking the degree centrality (a property of networks, obtained from a graph theoretic analysis) within a set of networks to measures of pre/post attitude change.

As should be evident from this small collection of hypothetical examples, this form of individual difference analysis allows for a considerable variety of different types of questions to be addressed. In general, the measures derived from fMRI serve as independent variables (when applicable), potentially alongside other individual difference factors, while the behavioral measures serve as dependent variables. Despite the flexibility this approach offers, however, it suffers from two major drawbacks: first, given that many measures from fMRI may exhibit various nonlinearities,[2] standard analytic methods such as ordinary least squares regression may fail, particularly if measures from multiple modalities are included simultaneously. Second, there is no simple way for this approach to accommodate multivariate data, particularly high-dimensional multivariate data. It may be feasible to include measures of activity or connectivity on the order of ten ROIs, given a large enough sample size, for instance—but trying to capture whole-brain phenomena or true multivariate patterns is unfeasible.

The second form of individual difference analysis overcomes both of these limitations. In this form of analysis, all measures of interest—whether neural, individual difference factors, behavioral, or so forth—are converted to similarities (or without loss of generality, differences) prior to analysis. This is generally inappropriate if the measures are all scalars, but if one or more measure is higher-dimensional, then this step has the effect of converting those measures into easy-to-analyze scalars, while at the same time potentially tempering nonlinearities in the original measures. Once all measures are converted to similarities, the typical next step would be to run a hierarchical regression model. The major drawback of this approach is the dependence that is introduced between observations (because each individual now contributes to N-1 similarities), but recent approaches for correcting for this sort of dependency structure may offer solutions for deriving valid inferential statistics (Chen, Taylor, Shin, Reynolds, & Cox, 2017).

The first approach has been used several times already in the young field of communication neuroscience. Examining some of these examples can help illustrate the benefits that can be gained from considering individual differences in fMRI, for instance, Pegors and colleagues used a sophisticated approach of comparing how the relationship between neural responses and message features itself related to behavior change, simultaneously demonstrating the importance of the selected message feature, and implicating one of the processes believed to be supported by the chosen brain region (Pegors, Tompson, O'Donnell, & Falk, 2017). Likewise, Cooper and colleagues related a set of scalar graph theoretic measures across individuals to various measures of behavior change, and demonstrated that a certain property of the so-called cognitive control network (see, e.g., Huskey, Craighead, Miller, & Weber, 2018), while individuals processed anti-smoking advertising, predicted individuals' intention to quit smoking (Cooper, Bassett, & Falk, 2017).

The second approach does not have any published examples from communication neuroscience. An example from the field of memory and decision-making in cognitive neuroscience comes from Miller and colleagues, who demonstrated that differences between individuals in individual difference factors such as cognitive style or task strategy were related to how similar individuals were in their patterns of brain activity (Miller, Donovan, Bennett, Aminoff, & Mayer, 2012). An unpublished example of this approach from communication involves comparing brain activity across a subset of regions—including a priori regions from the literature, *confirmatory* regions identified in earlier analyses of the same dataset, and *exploratory* regions selected on the basis of their high degree of inter-individual variance—in the activity maps from individuals who were exposed to anti-marijuana PSAs (see Weber, Huskey, Mangus, Westcott-Baker, & Turner, 2015, for more details on the broader project from which these analyses stem). This analysis has demonstrated the potential to fractionate the phenomenon along multiple dimensions: in terms of brain regions (and putative associated processes), stimulus features, and individual difference factors, which promises to enrich the theoretical understanding of the underlying processes.

INCORPORATING CULTURE AND LOOKING TO THE FUTURE

To complete the discussion of how individual differences can enrich understanding in communication science, this chapter turns briefly to a concept that, perhaps paradoxically, shares some features with individual differences, and that has been more widely employed in communication science, namely, culture. After briefly defining culture and presenting an overview of how it has been used in the field, this chapter will consider how individual differences and culture can be related and how existing examples demonstrate the potential of the individual differences approach.

Culture has been defined in innumerable ways, outside of the field of communication as well as within. These definitions depend on the intellectual tradition from which they come, as well as the approach being taken to studying culture. The two main approaches are the *emic* approach, which examines communication from a vantage point within a given culture (not unlike the case-study approach to individual differences), and the *etic* approach, which takes a vantage point outside any given culture (similar to the final approach to individual differences discussed above) and compares them across cultures (Gudykunst, 1997). The latter approach is a more natural analog to individual differences analysis, and therefore provides the most useful definitions of culture for the present purposes. For example, Geert Hofstede (2001) offers the compact definition of culture as "the collective programming of the mind that distinguishes the members of one group or category of people from another" (p. 9). More pointedly, Davis and Resnicow (2012) argue that "[w]hether a particular attribute is shared by multiple cultures is irrelevant; it is the unique combination of cultural attributes that makes a culture distinct" (p. 125). And from the standpoint of cross-cultural psychology, Lonner and Adamopoulos (1997) point out that culture can generally be conceptualized as acting as a moderating or mediating variable in relating individual difference factors to behavior—in the words of Triandis (1995), "people in each culture behave in ways that reflect unique configurations of attributes" (p. 16).

As these definitions should make clear, although culture is a phenomenon that on one hand involves large groups of people, it is nonetheless possible to view its manifestation through the lens of what impact it has on individuals. Moreover, it should be clear that culture is not monolithic or perfectly deterministic. Rather, it is better to think of culture as imposing some parameters on the distribution of a particular individual difference factor, which can in fact show greater variation within a culture than there is between the means of different cultures.[3] Of course, there are some individual difference factors which are inherently cultural, such as the well-known dimension of individualism versus collectivism, so this role of establishing something like priors is not the only way in which individual differences and culture can intersect—but it is the point of intersection that gives us the opportunity to use culture as an analog to individual differences in how the latter can be employed in communication science.

For instance, there are some lines of research that explicitly examine cultural effects at the individual level or that study how individual difference factors vary across cultures in terms of their effects on communication phenomena. Although it is not sophisticated, the general framework laid out by Singelis and Brown (1995), in which "culture affects the development of an individual's psychological makeup, which, in turn, affects communication behavior" (p. 355), serves as a reasonable guide to thinking about this type of research. One such integration of the two approaches comes from research on advertising and tailoring/customization, which typically finds cultural influences on the effectiveness of such strategies (Li, Kalyanaraman, & Du, 2011; Maslowska, Smit, & van den Putte, 2013). Another approach is to consider the sort of structural factors described above, but to assess how they interact with individual difference factors. Recent research in this mold has examined the relative influence of such system or individual factors with respect to political informedness (Fortin-Rittberger, 2016; Iyengar et al., 2010) and trust in the media (Tsfati & Ariely, 2014). Whether it is from a framework such as Singelis and Brown's or takes more of a structural angle, this sort of research serves as a stepping-stone to individual differences research for scholars who are more accustomed to thinking in cultural terms.

Despite these promising directions, the use of culture in communication serves as an analog of the use of individual differences in a less positive way as well. Often, the chosen cultures (or individual difference factors) seem to be a matter of convenience, leading to a haphazard pattern across the field. As described in the section on individual differences, individual differences analyses *can* be used to take advantage of patterns that exist in data collected for other purposes (when proper multiple comparison corrections are employed). However, a better path forward, and part of the call to arms made in this chapter, is to consider individual difference factors during the design of experiments, in exactly the same way as other variables are considered. The benefit of this approach is that, aside from potentially increasing sample size marginally, the design generally does not need to be changed to accommodate individual differences analyses. Moreover, as individual differences analyses become more common, not only will theories be refined on the basis of the new results yielded from this type of analysis, but a more systematic picture will emerge of which factors play a role in which phenomena. At some point, there will be a critical mass of relationships, and such analyses will be standard. Until then, researchers at the leading edge of communication science have an opportunity to establish key findings in what will surely be one of the new frontiers of the field.

NOTES

1. Note that drawing a statistical inference from this result requires more than a non-significant correlation between the two.
2. For instance, the saturation of the BOLD response (Friston, Mechelli, Turner, & Price, 2000), or the dramatic ways that the deletion of a single node in graph theoretic analyses can change network measures (Schlesinger, Turner, Grafton, Miller, & Carlson, 2017), to name just two examples.
3. Actually, this chapter remains agnostic regarding the direction of the causal arrow between culture and individual traits. It may be possible to construct an argument that geographic, genetic, or other forces lead some people to exhibit certain trait tendencies, which in turn shape their culture, but this question is beside the point of this chapter, and it is simpler to talk about culture in this context using the first formulation.

REFERENCES

Camerer, C. F., Dreber, A., Holzmeister, F., Ho, T.-H., Huber, J., Johannesson, M., ... Wu, H. (2018). Evaluating the replicability of social science experiences in *Nature* and *Science* between 2010 and 2015. *Nature Human Behavior, 2*, 637–644.

Chen, G., Taylor, P. A., Shin, Y.-W., Reynolds, R. C., & Cox, R. W. (2017). Untangling the relatedness among correlations, part II: Inter-subject correlation group analysis through linear mixed-effects modeling. *NeuroImage, 147*, 825–840.

Cooper, N., Bassett, D. S., & Falk, E. B. (2017). Coherent activity between brain regions that code for value is linked to the malleability of human behavior. *Scientific Reports, 7*, 43250.

Craig, R. T. (1999). Communication theory as a field. *Communication Theory, 9*, 119–161.

Davis, R. E., & Resnicow, K. (2012). The cultural variance framework for tailoring health messages. In H. Cho (Ed.), *Health communication message design: Theory and practice* (pp. 115–136). Thousand Oaks, CA: Sage.

Dubois, J., & Adolphs, R. (2016). Building a science of individual differences from fMRI. *Trends in Cognitive Sciences, 20*, 425–443.

Falk, E. B., Berkman, E. T., & Lieberman, M. D. (2012). From neural responses to population behavior: Neural focus group predicts population-level media effects. *Psychological Science, 23*, 439–445.

Fisher, A. J., Medaglia, J. D., & Jeronimus, B. F. (2018). Lack of group-to-individual generalizability is a threat to human subjects research. *Proceedings of the National Academy of Sciences, 115*, E6106–E6115.

Fortin-Rittberger, J. (2016). Cross-national gender gaps in political knowledge. *Political Research Quarterly, 69*, 391–402.

Friston, K. J., Mechelli, A., Turner, R., & Price, C. J. (2000). Nonlinear responses in fMRI: The Balloon model, Volterra kernels, and other hemodynamics. *NeuroImage, 12*, 466–477.

Gabrieli, J. D. E., Ghosh, S. S., & Whitfield-Gabrieli, S. (2015). Prediction as a humanitarian and pragmatic contribution from human cognitive neuroscience. *Neuron, 85*, 11–26.

Gomes, C. M. A., de Araujo, J., do Nascimento, E., & Jelihovschi, E. G. (2018). Routine psychological testing of the individual is not valid. *Psychological Reports, 122*(4), 1576–1593.

Gordon, E. M., Laumann, T. O., Gilmore, A. W., Newbold, D. J., Greene, D. J., Berg, J. J., ... Dosenbach, N. U. F. (2017). Precision functional mapping of individual human brains. *Neuron, 95*, 791–807.

Gudykunst, W. B. (1997). Cultural variability in communication. *Communication Research, 24*, 327–348.

Hedge, C., Powell, G., & Sumner, P. (2018). The reliability paradox: Why robust cognitive tasks do not produce reliable individual differences. *Behavior Research Methods, 50*, 1166–1186.

Hofstede, G. (2001). *Culture's consequences: Comparing values, behaviors, institutions and organizations across nations*. Thousand Oaks, CA: Sage.

Huskey, R., Craighead, B., Miller, M. B., & Weber, R. (2018). Does intrinsic reward motivate cognitive control? A naturalistic-fMRI study based on the synchronization theory of flow. *Cognitive, Affective, & Behavioral Neuroscience, 18*, 902–924.

Huth, A. G., de Heer, W. A., Griffiths, T. L., Theunissen, F. E., & Gallant, J. L. (2016). Natural speech reveals the semantic maps that tile human cerebral cortex. *Nature, 532*, 453–458.

Iyengar, S., Curran, J., Lund, A. B., Salovaara-Moring, I., Hahn, K. S., & Coen, S. (2010). Cross-national versus individual–level differences in political information: A media systems perspective. *Journal of Elections, Public Opinion & Parties, 20*, 291–309.

Kay, K. N., Naselaris, T., Prenger, R. J., & Gallant, J. L. (2008). Identifying natural images from human brain activity. *Nature, 452*, 352–355.

Lemon, J. (2009). On the perils of categorizing responses. *Tutorials in Quantitative Methods for Psychology, 5*, 35–39.

Li, C., Kalyanaraman, S., & Du, Y. R. (2011). Moderating effects of collectivism on customized communication: A test with tailored and targeted messages. *Asian Journal of Communication, 21*, 575–594.

Lonner, W. J., & Adamopoulos, J. (1997). Culture as antecedent to behavior. In J. W. Berry, Y. H. Poortinga, & J. Pandey (Eds.), *Handbook of cross-cultural psychology* (2nd ed., pp. 43–83). Boston, MA: Allyn & Bacon.

Maslowska, E., Smit, E. G., & van den Putte, B. (2013). Assessing the cross-cultural applicability of tailored advertising. *International Journal of Advertising, 32*, 487–511.

Maxwell, S. E., Kelley, K., & Rausch, J. R. (2008). Sample size planning for statistical power and accuracy in parameter estimation. *Annual Review of Psychology, 59*, 537–563.

Miller, M. B., Donovan, C., Bennett, C. M., Aminoff, E. M., & Mayer, R. E. (2012). Individual differences in cognitive style and strategy predict similarities in the patterns of brain activity between individuals. *NeuroImage, 59*, 83–93.

Miller, M. B., Donovan, C.-L., Van Horn, J. D., German, E., Sokol-Hessner, P., & Wolford, G. L. (2009). Unique and persistent individual patterns of brain activity across different memory retrieval tasks. *NeuroImage, 48*, 625–635.

Noar, S. M., Benac, C. N., & Harris, M. S. (2007). Does tailoring matter? Meta-analytic review of tailored print health behavior change interventions. *Psychological Bulletin, 133*, 673–693.

Oliver, M. B., & Krakowiak, M. (2009). Individual differences in media effects. In J. Bryant & M. B. Oliver (Eds.), *Media effects: Advances in theory and research* (pp. 517–531). New York, NY: Routledge.

Open Science Collaboration (2015). Estimating the reproducibility of psychological science. *Science, 349*, aac4716.

Pegors, T. K., Tompson, S., O'Donnell, M. B., & Falk, E. B. (2017). Predicting behavior change from persuasive messages using neural representational similarity and social network analyses. *NeuroImage, 157*, 118–128.

Rubin, M. (2017). An evaluation of four solutions to the forking paths problem: Adjusted alpha, preregistration, sensitivity analyses, and abandoning the Neyman-Pearson approach. *Review of General Psychology, 21*, 321–329.

Sassenhagen, J., & Alday, P. M. (2016). A common misapplication of statistical inference: Nuisance control with null-hypothesis significance tests. *Brain and Language, 162*, 42–45.

Schlesinger, K. J., Turner, B. O., Grafton, S. T., Miller, M. B., & Carlson, J. M. (2017). Improving resolution of dynamic communities in human brain networks through targeted node removal. *PloS One, 12*, e0187715.

Seghier, M. L., & Price, C. J. (2018). Interpreting and utilising intersubject variability in brain function. *Trends in Cognitive Sciences, 22*, 517–530.

Shrum, L. J. (2001). Processing strategy moderates the cultivation effect. *Human Communication Research, 27*, 94–120.

Singelis, T. M., & Brown, W. J. (1995). Culture, self, and collectivist communication: Linking culture to individual behavior. *Human Communication Research, 21*, 354–389.

Triandis, H. C. (1995). *Individualism and collectivism*. London, England: Routledge.

Tsfati, Y., & Ariely, G. (2014). Individual and contextual correlates of trust in media across 44 countries. *Communication Research, 41*, 760–782.

Turner, B. O., Huskey, R., & Weber, R. (2019). Charting a future for fMRI in communication science. *Communication Methods and Measures, 13*, 1–18.

Turner, B. O., Paul, E. J., Miller, M. B., & Barbey, A. K. (2018). Small sample sizes reduce the replicability of task-based fMRI studies. *Communications Biology, 1*(1), 1–10.

Vogel, E. K., & Awh, E. (2008). How to exploit diversity for scientific gain. *Current Directions in Psychological Science, 17*, 171–176.

Weber, R., Huskey, R., Mangus, J. M., Westcott-Baker, A., & Turner, B. O. (2015). Neural predictors of message effectiveness during counterarguing in antidrug campaigns. *Communication Monographs, 82*, 4–30.

Westfall, J., & Yarkoni, T. (2016). Statistically controlling for confounding constructs is harder than you think. *PLoS One, 11*, e0152719.

18

Neuromarketing

How to Choose the Right Measures

Steven Bellman and Duane Varan

The neuromarketing industry has expanded over the last decade. More buyers are using neuromeasures, and cheaper equipment has allowed new suppliers to enter the market. One of the main commercial uses of neuromeasures is for evaluating and pretesting TV advertising and video advertising. Neuromeasures, which tap rapid and unconscious processes that viewers cannot self-report, offer a potential advantage over traditional methods of pretesting. This chapter reviews evidence for the utility of neuromarketing measures for pretesting. The studies reviewed include the Advertising Research Foundation's (ARF) two neurostandards trials. These trials highlighted the need for vendors to offer greater transparency and continued dialog about the validity of their measures and recommendations. This chapter surveys how the industry has responded to these two challenges and provides the latest guidance on how to choose the right neuromarketing measures for pretesting TV/video ads.

REVIEW OF PRETESTING RESEARCH

Marketers should choose the neuromeasures that provide the most reliable evidence on which to base their decisions, such as whether to launch a new TV advertisement or keep the existing one. In this section, we review the results of TV advertising pretesting research using neuromeasures. We review this research through the lens of a framework, based on decades of pretesting research using traditional self-report measures, that allows us to identify whether a study is providing reliable evidence likely to generalize beyond the study's sample. But first, we briefly review the potential benefits of using neuromeasures for pretesting, rather than traditional measures.

THE PROMISE OF NEUROMARKETING MEASURES

Neuromeasures advance scientific knowledge in marketing by revealing the unconscious processes underlying responses to advertising and other aspects of consumer behavior (Shaw & Bagozzi, 2018). When pretesting TV and video advertising, neuromeasures potentially provide unique information beyond what traditional behavioral approaches can measure or observe (Bell et al., 2018; Hsu, 2017; Plassmann, Venkatraman, Huettel, & Yoon, 2015; Solnais, Andreu-Perez, Sánchez-Fernández, & Andréu-Abela, 2013). Neuromeasures are less likely than self-report measures to be affected by memory and social desirability biases (Crowne & Marlowe, 1960; Nisbett & Wilson, 1977). For example, in a functional magnetic resonance imaging (fMRI) study of antidrug message effectiveness, high-risk individuals rated all the messages as ineffective (Weber, Huskey, Mangus, Westcott-Baker, & Turner, 2015). This was most likely a social-desirability response (the messages contrasted with high-risk individuals' drug-friendly persona), which prompted counterarguments, resulting in negative evaluations of the messages. But in contrast to these stated negative evaluations, high-risk individuals' revealed evaluations (using fMRI) were very similar to those of low-risk individuals, and also to the perceived message effectiveness (PME) evaluations made by a separate large-scale sample. Since PME is predictive of actual behavior (Dillard, Weber, & Vail, 2007), these fMRI results suggested that these antidrug messages would be effective for both high- and low-risk individuals despite what high-risk individuals might claim in self-reported evaluations.

Harris and colleagues (2018) provide a comprehensive review of neuromarketing research methods. Some, such as fMRI, have been favored by academic researchers, whereas other less expensive techniques, such as eye-tracking, have been used more widely by consumer product companies. Some techniques have not been used much, such as positron emission topography (PET), which uses radioactivity, and functional transcranial Doppler sonography (fTCS), which cannot reveal cognitive processing. But another little-used technique, transcranial magnetic stimulation (TMS), has the potential to improve academic research by temporarily turning selected brain areas on and off (Klucharev, Munneke, Smidts, & Fernández, 2011). The next sections review research using neuromeasures for pretesting TV and video ads. We begin with a review of pretesting research using traditional measures. This research has identified these key factors to bear in mind when evaluating pretesting research:

- *Out-of-sample prediction.* The dependent variable should be actual behavior, observed in a different sample, and preferably the results should be replicated in a hold-out sample (Hakim & Levy, 2018). If neuromeasures are better than traditional measures, the test measures should not be "self-reported metrics—the very metrics they often aim to replace" (Kennedy & Northover, 2016, p. 188).
- *Controlling for brand.* Bigger brands get more sales, but the effect of advertising on sales is potentially greater for new brands (Lodish et al., 1995). Differences between brands should be controlled by testing strong and weak ads from each brand. Without controlling for brand and strong versus weak ads within brands, researchers might identify responses common to all ads as predictors of in-market success.

Pretesting Advertising with Traditional Measures

Post-launch testing of advertising (e.g., comparing differences in sales) dates to the beginnings of mail-order advertising in the nineteenth century (Caples, 1931/1974). Pretesting of TV commercials, prior to launch, was pioneered in the 1950s, using a variety of traditional measures including continuous-self-report (i.e., dial-turning; Mayer, 1958). In 1991, the ARF's Copy Research Validity Project (CRVP) designed a comprehensive test of traditional measures for pretesting TV ads (Haley & Baldinger, 1991). The design used six months of sales data for five brands as the out-of-sample behavioral dependent variable. These sales data came from split-cable studies, in which the test ad is shown only to one of two matched households, whose purchasing is tracked. The design also controlled for brand by comparing a strong and a weak ad from each brand. Two traditional measures, ad liking and category-cued brand recall, were identified as useful pretest measures, because they picked the successful ad more often than would be expected by chance. The ARF project did not test the persuasion-shift pretest measure, wherein participants put items into shopping baskets before and after exposure to the advertisement (Buzzell, 1964). The persuasion-shift test's vendor has reported very favorable results (e.g., Blair & Rosenberg, 1994). But a meta-analysis of 389 split-cable tests could not replicate these findings, and instead concluded that pretesting using traditional methods is unable to predict the sales-effectiveness of a TV advertisement (Lodish et al., 1995). For this reason, advertisers are currently advised *not* to pretest, but instead to monitor the sales performance of new ads after launch and revert to the old ads if the new ones are not working (Kennedy, Sharp, & Hartnett, 2017).

Functional Magnetic Resonance Imaging (fMRI) Studies

One of the first neuromarketing studies was a twist on the traditional "Pepsi Challenge" study (McClure et al., 2004). The study used fMRI to understand the results of blind-tasting tests, in which Pepsi is generally preferred even though Coca-Cola has a larger market share. The fMRI results revealed different brain activity for blind-tasting versus labeled-tasting. When blind-tasting, greater brain activity was seen in the ventromedial prefrontal cortex (VMPFC), which corresponded with behavior in a different blind-tasting sample (i.e., the out-of-sample dependent variable), in which most participants chose Pepsi as their preferred drink. Activity in the VMPFC is thought to reflect subjective value (Falk & Scholz, 2018). But when the drinks were labeled, subjective value (taste) was overridden by social conformity value (Falk & Scholz, 2018), measured by greater brain activity in areas outside the VMPFC (bilateral hippocampus, parahippocampus, midbrain, dorsolateral prefrontal cortex [DLPFC], thalamus, and left visual cortex). Again, this brain activity corresponded with out-of-sample behavior in a different labeled-tasting sample, in which most chose Coke. These results suggested that small-sample fMRI studies (the largest group had 18 participants) could potentially predict actual behavior.

However, to avoid the problem of reverse causality (Poldrack, 2006), fMRI researchers using the "brain-as-predictor" paradigm (Berkman & Falk, 2013) need to hypothesize in advance which areas of the brain will be associated with in-market success (Falk, Cascio, & Coronel, 2015). In-market success requires a large number of people to share the same favorable evaluation, which is more likely if the product or service appeals to a biologically relevant ("evolutionarily conserved") subjective-value response (Knutson & Genevsky, 2018), such as the taste of Pepsi (McClure et al., 2004). For this reason, most fMRI pretests use activity in the subjective-value areas of the brain (the ventral striatum [VS] as well as the VMPFC) as a predictor of success. But a pretest for an ad appealing to socially valued cultural knowledge (e.g., an ad for Coke) might use a "toolbox" approach, specifying different target areas of the brain (e.g., the hippocampus) to test whether an ad designed to activate social value processing does in fact achieve this objective (Cooper, Tompson, O'Donnell, & Falk, 2015).

An early use of fMRI for pretesting did not use an out-of-sample behavior as the dependent variable, or any controls for brand (Falk, Berkman, Mann, Harrison, & Lieberman, 2010). The study tested a single message to show that brain activity in the subjective-value area predicted behavior change in the same sample. A later study using three anti-smoking ads (low, medium, and high effectiveness) showed that subjective value predicted out-of-sample behavior as measured by the number of quit-calls (Falk, Berkman, & Lieberman, 2012). In both these studies, traditional measures (attitude, intention, ad ratings) did not predict behavior as well as subjective value measured by fMRI.

A similar result was found in the ARF's Neuro 2 study (Venkatraman et al., 2015). Like the ARF's earlier CRVP study, this study was designed to compare the relative effectiveness for pretesting of a variety of neuromeasures, including fMRI, plus implicit and explicit traditional measures. The out-of-sample behavioral dependent variable was sales-effectiveness, measured by advertising elasticity modeled from ratings and sales data. Brand was controlled for by using 26 strong and weak ads for ten brands from five companies. Four brain areas, associated with four different advertising constructs, were hypothesized in advance to predict sales-effectiveness: (1) *subjective value* ([desirability] measured by activity in the VMPFC or the VS); (2) *attention* (activity in the dorsolateral PFC [DLPFC]); (3) *emotion* (the amygdala); and (4) *memory* (the hippocampus). The results showed that the best measures for predicting sales were two traditional measures: brand recognition and persuasion-shift. A previous meta-analysis found no relationship between traditional measures and sales (Lodish et al., 1995), so this result may have been driven by the specific sample of ads. The only neuromeasure that significantly increased the explained variance beyond that explained by traditional measures was one fMRI measure of subjective value (activity in the VS). Interestingly, activity in the VS was not correlated with the traditional measures (so it added unique variance), and was negatively correlated with the number of fixations, an eye-tracking attention measure.

Despite these successful studies of the use of fMRI, Harris and colleagues (2018) argue that metabolic neuromeasures such as fMRI are not suitable for TV and video advertising research because of their low temporal resolution and blindness to certain cognitive processes (e.g., those that do not affect blood flow). Because of fMRI's ability to identify hypothesized activity in specific brain regions, Harris and colleagues suggest it might be suitable for measuring responses to audio and print stimuli (including static packaging), and for brand strategy and decision-making research. However, fMRI research is expensive, which limits its use in pretesting and prohibitively limits sample size, which suggests some published fMRI findings may have capitalized on chance (Button et al., 2013).

Electroencephalogram (EEG) Studies

Like fMRI studies, electroencephalogram (EEG) studies also reveal images of brain activity, but only of activity close to the skull's surface. EEG was used for pretesting TV ads in the 1970s (e.g., Appel, Weinstein, & Weinstein, 1979). The equipment used in those early EEG studies had low sensitivity and so measured the strongest EEG signal, which originates from the back of the head (where visual processing takes place). These studies compared the relative dominance in the brainwave spectrum of low-frequency alpha waves versus higher frequency beta waves. Alpha waves dominate when the eyes are closed and (in these studies) with watching television, whereas beta waves dominate when the eyes are open and (in these studies) when reading print advertising (Krugman 1971; although see Daugherty, Hoffman, Kennedy, & Nolan, 2018). EEG technology now allows the measurement of very-high-frequency gamma waves from the front of the head. In one study, EEG activity during action movie trailers was predictive of out-of-sample (different country) box-office sales (Boksem & Smidts, 2015). Like many fMRI studies, traditional measures (willingness-to-pay) did not predict box-office sales.

Early EEG pretesting studies, like early fMRI studies, also tended to use same-sample behavior as the dependent variable, without controlling for brand. For example, Silberstein and Nield (2008) compared EEG responses to a single TV ad with persuasion-shift in the same sample (the 18 [9%] of the 198 participants who shifted pre/post to choosing the advertised brand). Participants who shifted exhibited more EEG activity (measured by steady-state topography [SST]) in the left prefrontal area, compared with participants who stayed loyal to the competitor. This difference in brain activity was interpreted as reflecting long-term-memory storage of the brand. Later EEG studies used out-of-sample behavior in a large (population) sample as the dependent variable. Dmochowski and colleagues (2014) compared minute-by-minute inter-subject correlation (ISC) in EEG while watching a single TV program (*The Walking Dead*) with out-of-sample ratings data. Popular movies use framing and editing to direct viewers' attention (Bordwell & Thompson, 1990), and when ISC was low, indicating that synchronized attention was low, ratings were also low (especially during the ad breaks). It is difficult to associate an increase in ISC with an increase in ratings, as the viewers joining the audience to increase the rating would have been viewing different content.

The ARF's Neuro 2 study (Venkatraman et al., 2015) tested two EEG measures: (1) *(visual) attention* (occipital [back-of-the-head] alpha); and (2) *emotion* (frontal asymmetry in alpha). Neither EEG measure added unique variance to the prediction of sales-effectiveness by traditional measures. The two EEG measures were not correlated with traditional measures, but frontal asymmetry was positively correlated with skin conductance response (SCR), a measure of attention and emotion. Harris and colleagues (2018) suggest that electrical neural activity tools such as EEG are well suited for measuring responses to TV and video ads because of their high temporal resolution. However, the evidence so far suggests that these tools are good diagnostic measures of attention, but other measures may be needed for predicting in-market success (Bellman et al., 2017).

Physiological Measures

Neural activity often includes messages to organs of the body, such as the eyes, the heart, and the skin. The spinal column is connected to the brain and to the nerves traveling to the organs and muscles of the body, and for this reason neural activity outside of the brain can be an indicator of mental processes (Potter & Bolls, 2012). Messages to these peripheral areas are sent after the brain has filtered incoming information and decided which bits are relevant and worth acting on, so physiological measures can be simpler to interpret than the complexities of message processing in the brain. Physiological measures are also cheaper to gather, allowing the collection of larger samples that more reliably predict population behavior. For these reasons, physiological responses can be useful indicators of activity in the brain and the body elicited by viewing TV and video ads (see Chapter 19 by Potter & Bolls, in this volume).

Eye-tracking has high face validity as a measure of visual attention. Lohse (1997) showed that attention to larger Yellow Pages ads, measured by eye-tracking, justified their out-of-sample difference in cost, compared with simple text listings. The use of eye-tracking with dynamic TV stimuli is difficult, as potential areas of interest appear and disappear, and often move across the screen. One solution is to measure fixation dispersion, which, like ISC, measures synchronized attention. Teixeira and colleagues (2010) show that high dispersion, indicating low synchronized attention, is predictive of a participant's subsequent avoidance of a TV ad by skipping. The ARF's Neuro 2 study tested two eye-tracking measures, both measures of *attention*: (1) number of fixations; and (2) dwell time. Neither eye-tracking measure added to the variance explained by traditional measures in out-of-sample sales data. Eye-tracking was not significantly correlated with other measures of attention (e.g., EEG or SCR), but the number of fixations was negatively correlated with an fMRI measure of subjective value (activity in the ventral striatum).

Activity in the sympathetic nervous system (SNS) is associated with increases in sweating in preparation for fight or flight (Potter & Bolls, 2012). Sweating increases the conductivity of a small direct current between two electrodes, and so skin conductance (SC) is a good measure of SNS activity. However, SC cannot indicate the valence (positive or negative) of this emotional arousal. LaBarbera and Tucciarone (1995) provide evidence of SC predicting out-of-sample behavior. In one case study of a packaged-goods manufacturer, every TV ad that achieved its sales objective produced above-average SC scores. The ARF's Neuro 2 study (Venkatraman et al., 2015) measured SCR amplitude, which is a measure of attention as well as emotional arousal (Potter & Bolls, 2012). As mentioned above, SCR amplitude was correlated with the EEG measure of emotion, frontal asymmetry, but did not add to traditional measures' explained variance in sales.

Heart rate is increased by activity in the SNS but decreased by activity in the parasympathetic nervous system (PNS), to increase attention to external information (Potter & Bolls, 2012). In a study of responses to several brand logos (Maxian, Bradley, Wise, & Toulouse, 2013), in which effectiveness was measured out-of-sample by sales, bigger brands were associated with more attention, as measured by heart rate deceleration (HRD). In the ARF's Neuro 2 study, HRD measured attention whereas heart rate acceleration (HRA) measured emotion (Venkatraman et al., 2015). Neither measure improved the prediction of out-of-sample sales beyond that explained by traditional measures. Intriguingly, the heart-rate measure of attention (HRD) was positively correlated with the fMRI measure of emotion (activity in the amygdala).

Facial expression conveys emotion socially (Ekman, 2006), although electrical measurement of facial muscle activity (facial electromyography [EMG]) can detect implicit emotional responses that do not show on the face (Cacioppo, Petty, Losch, & Kim, 1986). Teixeira and colleagues (2012) showed that changes in two emotional responses, joy and surprise measured by facial coding, can help concentrate attention on TV and video advertising. In one study (McDuff, el Kaliouby, Cohn, & Picard, 2015), facial expressions (measured by webcams) in response to 170 ads, known to differ within brand in out-of-sample sales effectiveness, were compared with ad liking and purchase intent measures collected from the same large online sample ("crowdsourcing"). Because smiling is the most reliably measured facial expression, amusing ads (classified by MTurkers) had the highest rates of correct classification (76%) as high and liking (equal to the median or above) versus low. Purchase intention (high vs. low) was also predicted most accurately for amusing ads (75%). The ARF's Neuro 2 study (Venkatraman et al., 2015) did not test facial expression as a potential neuromeasure for pretesting.

A recent study improved on the design used in the ARF's Neuro 2 study by using a better out-of-sample behavioral dependent variable: sales-effectiveness measured by single-source data (like split cable data) rather than modeling of advertising elasticity (Bellman et al., 2017). As with the ARF's designs, brand was controlled for by testing strong and weak ads for each brand, but a much larger sample of ads was tested. The 109 ads with single-source data were seen in a lab, eight at a time, by a sample of over 1,000 consumers. Physiological measures (eye-tracking, SC, HRD, and facial expression) were collected, as well as traditional measures (brand recall, ad liking). The large number of test ads allowed a doubly out-of-sample test in which effectiveness signatures calibrated on two-thirds of the ads were validated in a holdout sample of the remaining ads. The use of a larger sample of ads than the 26 tested in the Neuro 2 study produced results that replicated meta-analysis of traditional pretesting measures (Lodish et al., 1995). Traditional measures did not perform better than chance when predicting sales-effectiveness in the whole sample (58% accuracy). In contrast, a combination of neuromeasures did significantly better than chance (78% in the whole sample, 69% in the holdout sample). This research showed the importance of using a "toolbox" approach to neuromeasures to select the right tool for testing the specific objectives of an advertising campaign. In the sample of ads tested in the study, the appropriate objective is to end with a positive-emotion (they were positive-ending "transformational" ads as opposed to negative-originating "informational" ads, as defined by Rossiter & Percy, 2017). A neuromeasure tool for measuring positive emotion at scale is facial expression. In this sample, ads with a rising trend in smiling were more likely to be sales-successful. Heart rate was another useful tool for detecting very weak performing ads, which failed to gain attention, as measured by HRD. But in this study, attention was a necessary but not sufficient condition for sales success.

Summary

This review suggests that the most useful neuromeasures for pretesting are physiological measures. These measures predict sales success out-of-sample, including to a holdout sample, after controlling for differences between brands. However, more work needs to be done to confirm that these measures forecast sales success. Future studies need to compare sales-effectiveness measured after pretesting. This review also suggests that metabolic neuromeasures (e.g., fMRI) are more useful for academic research than commercial pretesting. Electrical neuromeasures (e.g., EEG) are good diagnostic measures of attention but may not be related to sales effectiveness (Venkatraman et al., 2015).

Another advantage of physiological measures is their scalability. Eye-tracking (Papoutsaki et al., 2016), heart rate (Poh, McDuff, & Picard, 2011), and facial expression (McDuff et al., 2015) could potentially be census-measured from every member of the audience, using webcams. More work is needed to validate these webcam measures against high-quality lab-based measures. For example, the current evidence suggests that 20% of the audience will have missing data when webcam heart rate measures are used (Nahler et al., 2018).

For marketers wanting to gain insights from these new measures, the next section provides updated guidelines for choosing the right neuromeasure vendor.

CHOOSING THE RIGHT VENDOR

Prior to the ARF's Neuro 2 study, Neuro 1 asked eight commercial vendors to test eight commercials from eight different brands (Stipp & Woodard, 2011). The Neuro 1 report was careful not to disclose any ad-scores or even pictures of the vendors' results, because "the advertisers asked that specific findings about the commercials not be shared with other sponsors" (Stipp & Woodard, 2011, p. 20). Instead, an expert panel confidentially reviewed each vendor's results. This panel recommended steps that prospective buyers of neuromeasures needed to take (Varan, Lang, Barwise, Weber, & Bellman, 2015). First, given the complexity of the issues, buyers were encouraged to engage in independent assistance when evaluating the reliability and validity of these measures. Second, buyers were advised to gain experience with several vendors to understand which measures addressed their specific objectives. Third, buyers were admonished to be wary of vendors that provide only neuromeasures and no supporting traditional measures. Neuromeasures need cross-validation with well-understood traditional measures and constructs every time they are used. On the other hand, buyers were warned that sometimes vendors passed off combinations of traditional and neuroscience measures as neuromeasures. Finally, the expert panel urged vendors to have greater transparency and standardization around the constructs they were measuring, such as "engagement," and to provide more evidence about how these constructs related to objectives advertisers care about, such as sales.

Although the Neuro 2 trial did a lot to justify the credibility of neuromeasures, the problems identified by Neuro 1 remain. Most vendors still use "proprietary" measures, not the "standard" ones tested in Neuro 2. These standard measures generally had high construct validity when compared with traditional measures of the same construct (Venkatraman et al., 2015). But, the construct validity of propriety measures can be difficult to evaluate.

A Framework for Choosing Vendors

Kennedy and Northover (2016) provide a framework for the types of questions prospective buyers should ask of neuromeasure vendors. To properly answer these questions, vendors need robust answers underpinned by empirical validations. We summarize the framework, drawing attention to how vendors are responding to the needs of buyers at every stage.

Foundations

The best neuromeasures vendors have a firm foundation of understanding, not only of their technologies but also of marketing. Buyers need help understanding how neuromeasures relate to potential changes in consumers' behavior. But vendors lack the buyer's experience with the specific brand, so the buyer needs to translate the vendor's advice into relevant actions for the brand and its advertising (Stipp & Woodard, 2011; Varan et al., 2015). Perhaps the most notable change since Neuro 1 has been the addition to vendors' websites of pages devoted to "validation" and "due diligence" (Dooley, 2012).

Theory

Vendors need to use theoretically robust tools. Neuro 1 and Neuro 2 helped the industry by providing independent evidence of how different neuromeasures relate to different constructs of interest in theoretically defensible ways. The best vendors provide evidence for their measures by collaborating with academic researchers or publishing their own academic papers. The number of academic publications is now a key number advertised by vendors. The rigor of peer review endorses the validity and reliability of published vendors' methods. However, the relevant research for buyers is not basic research, but evidence of how the vendor's methods improve advertising.

Data

Vendors need to use quality data collection on an appropriate sample. Buyers should ask about the quality of both the people who will be carrying out the measurement and the equipment they will use. Facial EMG, for example, requires skilled placement of electrodes, by operators with PhDs if possible. The same goes for the quality of the hardware and software used. There are big differences in quality and sensitivity between medical-grade electrodes and amplifiers and cheaper equipment designed for use with video games.

Neuromeasures have not changed the laws of statistics. Neuromeasures need sample sizes with adequate power to find replicable results (Button et al., 2013). Sample location can also be important. Cross-cultural neuroscience shows that it is wrong to assume that "a brain is a brain is a brain" and location doesn't matter for neuromeasures (Han et al., 2013). In fact, because these measures tap unconscious responses, location may matter more than for traditional measures. Academics tend to use WEIRD (Western, educated, industrialized, rich, and democratic) samples of undergraduate students (Henrich, Heine, & Norenzayan, 2010; but see Burns et al., 2019). But to make predictions about sales, vendors should use samples of the advertised product's category users (Kennedy & Northover, 2016).

Analysis

Buyers should find out who interpreted the data and whether those interpretations are based on evidence or hypothesis. Vendors tend to be very secretive and proprietary about their analysis techniques. In Neuro 1, the eight vendors measured "attention," "positive emotion," and "engagement" using their own, usually proprietary, combination of measures and data reduction methods. Varan and colleagues (2015) recommended "reality checks" to understand a measure's construct validity. For example, convergence with a traditional measure such as recall increases confidence that a measure is tapping "attention" and does not necessarily mean there is no need for the neuromeasure, as it may predict other outcomes, such as low-involvement persuasion without recall (Heath, 2009). In Neuro 1, there was a striking lack of consistency between vendors' measures of the same constructs using supposedly error-free scientific measures (Varan et al., 2015).

Predictions

The number of studies showing a relationship between neuromeasures and in-market performance has grown. Kennedy and Northover (2016) recommended the use of single-source sales data as the out-of-sample dependent variable (e.g., as used by Bellman et al., 2017), rather than the modeling measure used in Neuro 2. We still need large-scale neuromeasures research using holdout samples and forecasting sales before those sales take place. In the meantime, neuromeasures seem more useful than traditional measures for diagnosing consumers' subtle and unconscious responses to different aspects of a commercial (with a view to improving it).

CONCLUSION

Advertisers want measures of advertising processing that go beyond mere exposure, and neuromeasures can provide continuous measures of viewers' responses, free of the memory and social desirability biases associated with self-reporting. For this reason, neuromeasures are potentially more predictive of ad effectiveness than traditional measures. Neuromeasures can also play a role in developing and refining the creative design elements used in the ad. However, even after decades of studying the reliability and validity of neuromeasures, the best advice for prospective buyers of these measures is that they exercise caution and a healthy skepticism about vendors' claims.

Chief among the recommendations of the Neuro 1 report (Stipp & Woodard, 2011) was that prospective buyers should discuss issues of reliability and validity with vendors. Later articles by Varan et al. (2015) and Kennedy and Northover (2016) provided tools and frameworks to help with these discussions. In this chapter, we have noted recent developments in published research and vendors' practices that help prospective buyers identify a short list of vendors to talk to.

Long-term theoretical research is still needed to understand why advertising "works" on these measures and how advertisers can optimize ad effectiveness using these measures. In the short term, buyers need more research that tests the validity of the non-standard, proprietary approaches offered by vendors. Vendors will have to show that they have enough confidence in their measures that they are willing to let others test them independently. Just because vendors use scientific tools doesn't inherently mean their work is scientific. In short, the main task for future research in this area should be to increase the transparency and credibility of commercial versions of neuromeasures so that these measures seem just as reliable and valid as traditional measures.

REFERENCES

Appel, V., Weinstein, S., & Weinstein, C. (1979). Brain activity and recall of TV advertising. *Journal of Advertising Research, 19*(4), 7–15.

Bell, L., Vogt, J., Willemse, C., Routledge, T., Butler, L. T., & Sakaki, M. (2018). Beyond self-report: A review of physiological and neuroscientific methods to investigate consumer behavior. *Frontiers in Psychology*. Advance online publication.

Bellman, S., Nenycz-Thiel, M., Kennedy, R., Larguinat, L., McColl, B., & Varan, D. (2017). What makes a television commercial sell? Using biometrics to identify successful ads. *Journal of Advertising Research, 57*(1), 53–66.

Berkman, E. T., & Falk, E. B. (2013). Beyond brain mapping. *Current Directions in Psychological Science, 22*(1), 45–50.

Blair, M. H., & Rosenberg, K. E. (1994). Convergent findings increase our understanding of how advertising works. *Journal of Advertising Research, 34*(3), 35–45.

Boksem, M. A. S., & Smidts, A. (2015). Brain responses to movie trailers predict individual preferences for movies and their population-wide commercial success. *Journal of Marketing Research, 52*(4), 482–492.

Bordwell, D., & Thompson, K. (1990). *Film art: An introduction* (3rd ed.). New York, NY: McGraw-Hill.

Burns, S. M., Barnes, L. N., McCulloh, I. A., Dagher, M. M., Falk, E. B., Storey, J. D., & Lieberman, M. D. (2019). Making social neuroscience less WEIRD: Using fNIRS to measure neural signatures of persuasive influence in a Middle East participant sample. *Journal of Personality and Social Psychology*. Advance online publication.

Button, K. S., Ioannidis, J. P. A., Mokrysz, C., Nosek, B. A., Flint, J., Robinson, E. S. J., & Munafò, M. R. (2013). Power failure: Why small sample size undermines the reliability of neuroscience. *Nature Reviews Neuroscience, 14*(5), 365–376.

Buzzell, R. D. (1964). Predicting short-term changes in market share as a function of advertising strategy. *Journal of Marketing Research, 1*(3), 27–31.

Cacioppo, J. T., Petty, R. E., Losch, M. E., & Kim, H. S. (1986). Electromyographic activity over facial muscle regions can differentiate the valence and intensity of affective reactions. *Journal of Personality and Social Psychology, 50*(2), 260–268.

Caples, J. (1931/1974). *Tested advertising methods* (4th ed.). Englewood Cliffs, NJ: Prentice-Hall.

Cooper, N., Tompson, S., O'Donnell, M. B., & Falk, E. B. (2015). Brain activity in self- and value-related regions in response to online antismoking messages predicts behavior change. *Journal of Media Psychology, 27*(3), 93–108.

Crowne, D. P., & Marlowe, D. (1960). A new scale of social desirability independent of psychopathology. *Journal of Consulting Psychology, 24*(4), 349–354.

Daugherty, T., Hoffman, E., Kennedy, K., & Nolan, M. (2018). Measuring consumer neural activation to differentiate cognitive processing of advertising. *European Journal of Marketing, 52*(1/2), 182–198.

Dillard, J. P., Weber, K. M., & Vail, R. G. (2007). The relationship between the perceived and actual effectiveness of persuasive messages: A meta-analysis with implications for formative campaign research. *Journal of Communication, 57*, 613–631.

Dmochowski, J. P., Bezdek, M. A., Abelson, B. P., Johnson, J. S., Schumacher, E. H., & Parra, L. C. (2014). Audience preferences are predicted by temporal reliability of neural processing. *Nature Communications, 5*, 4567.

Dooley, R. (2012, August 30). The neuromarketing challenge: First response. *Neuromarketing*. Retrieved from www.neurosciencemarketing.com/blog/articles/challenge-innerscope.htm.

Ekman, P. (Ed.). (2006). *Darwin and facial expression: A century of research in review*. Cambridge, MA: Malor Books.

Falk, E. B., Berkman, E. T., & Lieberman, M. D. (2012). From neural responses to population behavior: Neural focus group predicts population-level media effects. *Psychological Science, 23*(5), 439–445.

Falk, E. B., Berkman, E. T., Mann, T., Harrison, B., & Lieberman, M. D. (2010). Predicting persuasion-induced behavior change from the brain. *Journal of Neuroscience, 30*(25), 8421–8424.

Falk, E. B., Cascio, C. N., & Coronel, J. C. (2015). Neural prediction of communication-relevant outcomes. *Communication Methods and Measures, 9*(1–2), 30–54.

Falk, E., & Scholz, C. (2018). Persuasion, influence, and value: Perspectives from communication and neuroscience. *Annual Review of Psychology, 69*, 329–356.

Hakim, A., & Levy, D. J. (2018). A gateway to consumers' minds: Achievements, caveats, and prospects of electroencephalography-based prediction in neuromarketing. *Wiley Interdisciplinary Reviews: Cognitive Science, 10*(2), e1485.

Haley, R. I., & Baldinger, A. L. (1991). The ARF copy research validity project. *Journal of Advertising Research, 31*(2), 11–32.

Han, S., Northoff, G., Vogeley, K., Wexler, B. E., Kitayama, S., & Varnum, M. E. W. (2013). A cultural neuroscience approach to the biosocial nature of the human brain. *Annual Review of Psychology, 64*, 335–359.

Harris, J. M., Ciorciari, J., & Gountas, J. (2018). Consumer neuroscience for marketing researchers. *Journal of Consumer Behavior, 17*, 239–252.

Heath, R. (2009). Emotional engagement: How television builds big brands at low attention. *Journal of Advertising Research, 49*(1), 62–73.

Henrich, J., Heine, S. J., & Norenzayan, A. (2010). The weirdest people in the world? *Behavioral and Brain Sciences, 33*(2–3), 61–83.

Hsu, M. (2017). Neuromarketing: Inside the mind of the consumer. *California Management Review, 59*(4), 5–22.

Kennedy, R., & Northover, H. (2016). How to use neuromeasures to make better advertising decisions: Questions practitioners should ask vendors and research priorities for scholars. *Journal of Advertising Research, 56*(2), 183–192.

Kennedy, R., Sharp, B., & Hartnett, N. (2017). Advertising. In B. Sharp (Ed.), *Marketing: Theory, evidence, practice* (2nd ed., pp. 438–493). South Melbourne, Australia: Oxford University Press.

Klucharev, V., Munneke, M. A. M., Smidts, A., & Fernández, G. (2011). Downregulation of the posterior medial frontal cortex prevents social conformity. *Journal of Neuroscience, 31*(33), 11934–11940.

Knutson, B., & Genevsky, A. (2018). Neuroforecasting aggregate choice. *Current Directions in Psychological Science, 27*(2), 110–115.

Krugman, H. E. (1971). Brain wave measures of media involvement. *Journal of Advertising Research, 11*(1), 3–9.

LaBarbera, P. A., & Tucciarone, J. D. (1995). GSR reconsidered: A behavior-based approach to evaluating and improving the sales potency of advertising. *Journal of Advertising Research, 35*(5), 33–53.

Lodish, L. M., Abraham, M., Kalmenson, S., Livelsberger, J., Lubetkin, B., Richardson, B., & Stevens, M. E. (1995). How T.V. advertising works: A meta-analysis of 389 real world split cable T.V. advertising experiments. *Journal of Marketing Research, 32*(2), 125–139.

Lohse, G. L. (1997). Consumer eye movement patterns on Yellow Pages advertising. *Journal of Advertising, 26*(1), 61–73.

Maxian, W., Bradley, S. D., Wise, W., & Toulouse, E. N. (2013). Brand love is in the heart: Physiological responding to advertised brands. *Psychology & Marketing, 30*(6), 469–478.

Mayer, M. (1958). *Madison Avenue, U.S.A.: The inside story of American advertising*. London, England: Bodley Head.

McClure, S. M., Li, J., Tomlin, D., Cypert, K. S., Montague, L. M., & Montague, P. R. (2004). Neural correlates of behavioral preference for culturally familiar drinks. *Neuron, 44*, 379–387.

McDuff, D., El Kaliouby, R., Cohn, J. F., & Picard, R. (2015). Predicting ad liking and purchase intent: Large-scale analysis of facial responses to ads. *IEEE Transactions on Affective Computing, 6*(3), 223–235.

Nahler, C., Feldhofer, B., Ruether, M., Holweg, G., & Druml, N. (2018). Exploring the usage of time-of-flight cameras for contact and remote photoplethysmography. *Proceedings of the 21st Euromicro Conference on Digital System Design (DSD)*, 433–441.

Nisbett, R. E., & Wilson, T. D. (1977). Telling more than we can know: Verbal reports on mental processes. *Psychological Review, 84*(3), 231–259.

Papoutsaki, A., Daskalova, N., Sangkloy, P., Huang, J., Laskey, J., & James Hays, J. (2016). WebGazer: Scalable webcam eye tracking using user interactions. *Proceedings of the Twenty-Fifth International Joint Conference on Artificial Intelligence (IJCAI 2016)*, 3839–3845.

Plassmann, H., Venkatraman, V., Huettel, S., & Yoon, C. (2015). Consumer neuroscience: Applications, challenges, and possible solutions. *Journal of Marketing, 52*(4), 427–435.

Poh, M.-Z., McDuff, D. J., & Picard, R. W. (2011). Advancements in noncontact, multiparameter physiological measurements using a webcam. *IEEE Transactions on Biomedical Engineering, 58*(1), 7–11.

Poldrack, R. A. (2006). Can cognitive processes be inferred from neuroimaging data? *Trends in Cognitive Sciences, 10*(2), 59–63.

Potter, R. F., & Bolls, P. (2012). *Psychophysiological measurement and meaning.* New York, NY: Routledge.

Rossiter, J. R., & Percy, L. (2017). Methodological guidelines for advertising research. *Journal of Advertising, 46*(1), 71–82.

Shaw, S. D., & Bagozzi, R. P. (2018). The neuropsychology of consumer behavior and marketing. *Consumer Psychology Review, 1*(1), 22–40.

Silberstein, R. B., & Nield, G. E. (2008). Brain activity correlates of consumer brand choice shift associated with television advertising. *International Journal of Advertising, 27*(3), 359–380.

Solnais, C., Andreu-Perez, J., Sánchez-Fernández, J., & Andréu-Abela, J. (2013). The contribution of neuroscience to consumer research: A conceptual framework and empirical review. *Journal of Economic Psychology, 36*, 68–81.

Stipp, H., & Woodard, R. P. (2011). *Uncovering emotion: Using neuromarketing to increase ad effectiveness.* New York, NY: Advertising Research Foundation.

Teixeira, T. S., Wedel, M., & Pieters, R. (2010). Moment-to-moment optimal branding in TV commercials: Preventing avoidance by pulsing. *Marketing Science, 29*(5), 783–804.

Teixeira, T. S., Wedel, M., & Pieters, R. (2012). Emotion-induced engagement in Internet video advertisements. *Journal of Marketing Research, 49*(2), 144–159.

Varan, D., Lang, A., Barwise, P., Weber, R., & Bellman, S. (2015). How reliable are neuromarketers' measures of advertising effectiveness? Data from ongoing research holds no common truth among vendors. *Journal of Advertising Research, 55*(2), 176–191.

Venkatraman, V., Dimoka, A., Pavlou, P. A., Vo, K., Hampton, W., Bollinger, B., ... Winer, R. S. (2015). Predicting advertising success beyond traditional measures: New insights from neurophysiological methods and market response modeling. *Journal of Marketing Research, 52*(4), 436–452.

Weber, R., Huskey, R., Mangus, J. M., Westcott-Baker, A., & Turner, B. (2015). Neural predictors of message effectiveness during counterarguing in antidrug campaigns. *Communication Monographs, 82*(1), 4–30.

19

Investigating Communication Using Peripheral Nervous System Measurement

Robert F. Potter and Paul D. Bolls

A key assumption for the biological approach to studying human communication is that the mind is a part of a human body. In other words, signals not only travel from the brain to the periphery, but vital information is also received by the brain as efferent signals from peripheral systems. Communication researchers have measured signals in the peripheral nervous system (PNS) for decades. These measures are most useful for advancing communication theory when interpreted in light of how they correlate with psychological constructs such as attention, motivation, and emotion. When this happens, the enterprise is called psychophysiological research. Earlier volumes (Cacioppo, Tassinary, & Berntson, 2017; Stern, Ray, & Quigley, 2001) provide longer discussions of correct psychophysiological measurement. There are even sources (Lang, 1994; Potter & Bolls, 2012) written specifically to address how *communication* researchers can properly collect and interpret heart rate, skin conductance, and facial muscle activation data. This chapter allows only summaries of these longer treatments; interested readers are encouraged to consult them for additional detail.

We begin with a brief introduction to the PNS and its place within the structure of the larger human nervous system. This is followed by four sections focusing on different PNS measures. Each section provides: (1) a description of the physiology behind the particular measure; (2) a brief discussion of early communication research employing the measure; and (3) recent studies using the measure in interesting or/and novel ways. Table 19.1 provides a longer list of recent communication research employing peripheral measures. The number of studies included is striking; more communication scholars are using PNS measures than ever before. Furthermore, the breadth of technologies and communication scenarios being explored is impressive. Of course, traditional media such as radio and television continue to be studied, but they are joined by newer technologies such as virtual reality and video games. Also notable are the issues being addressed through psychophysiological measures, including parental co-viewing, effects of food and alcohol cues on attitudes and behaviors, and political decision-making.

One reason for this growth is the lower costs and increased portability associated with most PNS data collection hardware. Combine this affordability with greater user-friendliness in software, and more researchers can consider adding PNS measures to their methodological

Table 19.1 Recent Communication Research Employing Peripheral Measures

Study	*Media*	*Independent Variables*	*ECG*	*EDA*	*EMG*	*ET*
Bailey (2016)	TV food ads	Energy density of food, presence of food	–	SCL	CS	–
Bailey, Liu, & Wang (2018)	TV food ads	Energy density of food, sex appeal	HR	SCL	CS, OO	–
Bailey, Wang, & Kaiser (2018)	Obesity-prevention PSAs	Content arousal; content valence	HR	–	–	–
Banjo et al. (2017)	Sitcoms	Racial language, race of viewer	RSA	SCR	–	–
Bellman, Wooley, & Varan (2016)	TV programs	Content valence	–	SCL	–	–
Chang et al. (2015)	Online games	Problematic internet use	HRV	–	–	–
Chen & Tsai (2015)	Video games	Eye/hand coordination patterns	–	–	–	GD
Clayton, Lang, Leshner, & Quick (2018)	Anti-smoking PSAs	Threat frame; Smoking cue	HR	–	–	–
Clayton, Leshner, & Almond (2015)	Mobile phones	Separation from phone	HR	–	–	–
Clayton, Leshner, Bolls, & Thorson (2017)	Anti-smoking PSAs	Disgusting video; smoking cue	HR	–	–	–
Coronel & Federmeier (2016)	Computer images of candidates	Political sophistication; gender of image	–	–	–	FP
Coyne et al. (2015)	Video games	Problematic video game use	HRV	SCL	–	–
Cummins et al. (2014)	TV sports	Viewer fanship	–	–	–	FF, GD, OF
Cummins et al. (2017)	TV programming advisory	Advisory format	–	–	–	GD
Gao (2016)	YouTube clips	Topic interest; situational interest	–	–	CS, OO,	
LF, MS	–					
Grizzard et al. (2015)	Video games	Repeated exposure to violent game content	HR	–	–	–
Hart et al. (2018)	Moral narratives	Character morality; content valence	–	–	CS, ZM	–
Howell et al. (2018)	Safe driving TV PSAs	Aversive tone	HR	SCL	CS	–
Jabon et al. (2011)	Computer tasks	(In)correct performance	–	–	–	–
Jang et al. (2018)	Retail Storefronts	Visual complexity	–	SCL	ZM	–
Koruth et al. (2015)	Video clips	Production pacing; content arousal	HR, HRV	SCL	–	–
Kruikemeir et al. (2018)	News content	Modality of delivery (print v. web)	–	–	–	GD
Lang, Schwartz, & Mayell (2015)	TV clips	Production pacing; content arousal	HR	SCR	–	–
Lang & Yegiyan (2014)	Computer images	Alcohol images	–	–	CS, OO	–
Lee & Potter (2018)	Radio ads	Emotional words	HR	SCL	CS, ZM	–
Lee (2017)	Video game	Virtual & physical environment	–	–	EBS	–
Leshner et al. (2018)	Anti-smoking TV PSAs	Deception frame; disgusting video	HR	–	CS	–
Li, Ju, & Reeves (2017)	Robot body parts	Interaction type; body part accessibility rating	–	SCL	–	–
McDuff & el Kaliouby (2017)	Video ads	Product category	–	–	–	–
Missaglia et al. (2014)	Social media ads	Violent content	HR	SCL	CS	GD
Potter et al. (2016)	Radio ads	Announcer pitch	HR	–	–	–
Potter, Lynch, & Kraus (2015)	Radio jingles and effects	Repeated exposure	HR	–	–	–
Potter et al. (2018)	Radio jingles and effects	Cognitive load; repeated exposure	HR	–	–	–

continued

Table 19.1 Continued

Study	*Media*	*Independent Variables*	*ECG*	*EDA*	*EMG*	*ET*
Rassmussen et al. (2017)	TV segments	Child age; parent co-viewing; content arousal	HR	SCL	–	–
Read, van Driel, & Potter (2018)	TV ads	Couple sexuality in ad; viewer attitudes	HR	SCL	CS	–
Rodero, Potter, & Prieto (2017)	Radio ads	Announcer pitch	HR	SCL	–	–
Rubenking & Lang (2014)	TV and movie clips	Core and sociomoral disgust	HR	SCL	CS, LL	
Russell et al. (2017)	TV ads for fictional drug	Actor facial expressions; Text graphics	–	–	–	FF, PD
Shalom et al. (2015)	Computers	Face-to-face; computer communication	HR	SCL	–	–
Sparks & Lang (2015)	TV ads	Content sexiness & tone	HR	–	OO, PAR	–
Sukalla et al. (2016)	Narrative TV clips	Narrative cohesion; negative content	HR	SCL	CS	–
Vraga et al. (2016)	Facebook posts	Content category	–	–	–	GD
Wang, Morey, & Srivastava (2012)	TV political ads	Content arousal; content valence; candidate	HR	SCL	CS, ZM	–
Yegiyan (2015)	Movie clips	Content arousal; content valence	HR	SCL	CS, OO	–
Yegiyan & Bailey (2016)	Computer images of food	Healthiness; eating behaviors & food knowledge	–	–	CS, OO	–
Yeykelis, Cummings, & Reeves (2014)	Computer tasks	Task switching	–	SCL	–	–

Notes
CS = corrugator supercilii; EBS = eyeblink startle; FF = fixation frequency; HR = heart rate; LL = levator labii; GD = gaze duration; HRV = heart rate variability; MS = masseter; OF = observation frequency; OO = orbicularis oculi; RSA = respiratory sinus arrhythmia; SCL = skin conductance level; SCR = skin conductance response; PAR = post auricular response; PD = pupil dilation; ZM = zygomaticus major.

inventory. Plus, more researchers using PNS measures understand the assumptions underlying psychophysiology (Lang, Potter, & Bolls, 2009; Potter & Bolls, 2012) and apply them to their data, rather than viewing results through a more historic linear/causal conceptualization of how bodily signals reflect human communication. The assumptions of the psychophysiologist compel awareness of the interconnectedness of different subsystems of the PNS as well as the ways context perturbs them. The result is a nimbler approach to interpreting data and a more fruitful application of findings to understanding. Like any research method, gaining such dexterity takes practice. To begin, it requires a basic understanding of the PNS and where it fits in the complexity of the human body.

A BRIEF DESCRIPTION OF THE PERIPHERAL NERVOUS SYSTEM

The PNS is the portion of the body connecting the central nervous system (CNS; i.e., the brain and the spinal cord) to the rest of the organism. Some connections are in the form of sensory nerve cells. Cochlear nerves, for example, evolved to do the very specific task of transferring sound waves into bioelectric signals that get passed to the central nervous system. Other PNS nerve cells are called motor neurons, which are generally classified into two branches of the PNS. The somatic nervous system is the branch of motor neurons that control overt and conscious bodily movements. Motor neurons that regulate glands and organs, and operate largely below conscious awareness, make up the autonomic nervous system. It is in the autonomic nervous system where response patterns correlate with psychological variables.

The autonomic nervous system is further divided into the sympathetic nervous system and the parasympathetic nervous system. The sympathetic system activates when rapid and intense bodily response is called for: fighting, fleeing, procreating, etc. The parasympathetic system regulates the body during times of rest, information intake, and mild enjoyment. Research in human physiology shows that although some organs and glands receive input from only one of these systems, others are dually-innervated—controlled by both sympathetic and parasympathetic inputs. Therefore, psychophysiological data often represent a combination of different commands of varying intensities arriving at the organ. This is why a unitary view of arousal suggesting that all peripheral measures will increase as excitation rises (Duffy, 1962) is untenable. Instead, the human body exhibits *directional fractionation* (Lacey, 1967) in response to many environmental contexts. For example, suppose a person is watching a scene from a horror movie that initially evokes suspense as the fate of the protagonist is alluded to by music, lighting, and shot composition. Suddenly, the poor movie victim is eaten alive by the monster from Uranus in a particularly intense bit of cinematic gore. We would predict that measures of the viewer's skin conductance would increase throughout the scene because electrodermal activity (EDA) responds to singular inputs from the sympathetic nervous system. Predictions of cardiac response, on the other hand, are less easy to make because the heart is innervated by both the parasympathetic and sympathetic systems. So, at the beginning of the suspenseful scene, even though the viewer is aroused, he or she is also intrigued. Both systems send signals to the heart: the sympathetic says "speed up, get us out of here" whereas the parasympathetic says "hold on, slow down and take this in." Multiple studies indicate that the parasympathetic system overrides the sympathetic influence on the heart during most suspenseful (and even violent) media messages, resulting in slower heart rates. However, at a certain point of intensity the sympathetic inputs may outweigh the parasympathetic and the directionality of activation will flip, leading to cardiac acceleration. Further adding to the complexity of directional fractionation is the fact that there appear to be individual differences in patterns of physiological responses. Recent work demonstrates that individual differences of trait motivational system activation play a role in when this happens (Clayton, Lang, Leshner, & Quick, 2018).

Given this brief description of how the PNS fits into the knit of human bodily systems, and the challenge of interpretation that can arise when collecting peripheral data, we move to a discussion of specific measures. Before doing so, however, we refer again to Table 19.1 to note the number of studies that utilize multiple measures in their research. The mental processes that PNS measures index do not occur in isolation, nor does any single process ultimately drive message effects that emerge from media exposure. It is critical to combine multiples measures (including self-report measures) believed to index mental processes engaged by media exposure. The use of multiple PNS measures in psychophysiological research may also make interpretation of physiological responses reflecting mental processes a bit less daunting for the aspiring communication scientist.

MEASURING CARDIAC ACTIVITY

Cardiac activity is most commonly measured using the electrocardiogram (ECG). The process involves placing electrodes on the skin on both sides of the body in relation to the sinoatrial node of the heart, located just above the right atrium. Firing of muscle neurons here causes the heart to beat and form the electrical signal collected via ECG.

The structure of the ECG consists of three primary waves representing increases in bioelectrical voltage. The largest wave, the QRS complex, occurs with depolarization of the ventricles resulting in their contraction to pump blood through the body (Potter & Bolls, 2012). Because of its size, the QRS complex provides an easily visible amplitude in the ECG. Therefore, the time between subsequent QRS spikes is often used initially to quantify cardiac activity. This inter-beat interval (IBI) is measured in milliseconds and is usually converted into a weighted average of heart beats per minute (BPM) for each second of real time.

Heart rate is most often used as a measure of attention in communication research because of the strong influence of parasympathetic innervation in most communicative situations (Lang, 1994). Increased parasympathetic influence slows the heart, so heart rate deceleration has been used to indicate increases in attention. Lang (1990) provided early evidence for this by showing orienting responses (ORs)—brief decelerations in heart rate following the onset of signals or novelty—to aspects of television programs such as commercial onsets, camera changes within scenes, and character motion.

Although the cardiac OR can be used to identify evoked but fleeting increases in attention, deceleration in heart rate over the course of entire messages correlates with non-evoked increases in attention. For example, Lang, Bolls, Potter, and Kawahara (1999) collected cardiac data from subjects as they watched 30-second video clips. In what is called a tonic analysis—over the entirety of the media presentation—they showed that average heart rates were significantly slower during videos with arousing content compared to calm.

Current studies continue to use ECG to identify differences in tonic attention. Clayton, Leshner, Bolls, and Thorson (2017), for example, played a series of anti-smoking PSAs to participants who were regular smokers. They hypothesized that the presence of an actor smoking in the first 15 seconds of the message would result in greater attention compared to messages that did not have smoking cues. Cardiac data collected during the 30-second PSAs show that smoking cues led to increased attention (i.e., slower heart rates) early in all the messages. In an interesting demonstration of using cardiac data to explore tonic and phasic effects, half of the messages with smoking cues in Clayton et al. (2017) also contained disgusting images (e.g., tar-covered lungs) that appeared late in the message. With the onset of these images, heart rate increased as attention was gated to limit intake of disgusting information about the effect of smoking, as might be expected for smokers, who are likely to want to defensively avoid negatively intense images displaying the health consequences of tobacco use.

The work done by Potter and colleagues has tended to focus more on the OR, particularly looking at aspects of automatic attention capture by auditory structural features. One such feature is the voice change, when one speaker is replaced by another in a dialog. In one study (Potter, Jamison-Koenig, Lynch, & Sites, 2016) the speakers comprising a voice change were paired so that their vocal pitch was either similar, moderately different, or very different from each other. Time-locking the cardiac response to the onset of the voice change demonstrated significantly different ORs, with deeper decelerations in heart rate as the dissimilarity in vocal pitch increased. In another study (Potter, Sites, Jamison-Koenig, & Zheng, 2018), participants listened to a music radio station over headphones while working on a central task (studying, texting, playing a puzzle game, etc.). Orienting responses to the repeated onset of radio station identification jingles played between songs were shown to be significantly influenced by adding a cognitive load associated with a central task.

Heart rate is not the only index of cardiac activity. Two others are heart rate variability (HRV) and its subset measure, respiratory sinus arrhythmia (RSA). Ravaja (2004) expressed concern with using BPM as a measure of attention because of dual-innervation. He proposed using HRV, which is a general term for a series of analysis techniques that parse the separate influences of parasympathetic and sympathetic systems (see Allen, Chambers, & Towers, 2007). For example, computing an RSA value from the vector of IBIs over a set period of time is one way of quantifying the parasympathetic activation of the heart, independent of sympathetic influence. In a recent publication, the efficacy of HRV indices was compared to traditional BPM measures of attention when assessing attention hypotheses (Koruth, Lang, Potter, & Bailey, 2015). Results suggest that collapsing dynamic data to a single value, as is common in the HRV approach, was not as predictive as tonic BPM results. When IBI vectors across messages were parsed into smaller sequential segments, however, and multiple HRV values per message were calculated, this approach fared better.

MEASURING VISUAL ATTENTION

Because the optic nerve is connected directly to the brain stem, calling eye-tracking a PNS measure is a slight misnomer. Nevertheless, because most CNS researchers tend to limit the domain of their work to direct measures of brain activity—and because eye-tracking has been appearing more frequently in the communication literature—it is briefly covered here. Eye-tracking allows for precise determination of the point of visual attention. So, whereas tonic cardiac data can indicate that more attention is being paid to one type of message over another, they provide little detail about what drives the increased attention. Precise measurement of the eye's motion across the screen, and the time spent fixating on specific areas of interest, can be useful in this regard.

A common way of collecting eye motion and fixation data uses infrared light reflected off the participant's cornea to map eye movement during message processing. Common measures include the total number of fixations on a specific piece of content, the time to the first fixation on that content, and the total time spent fixated on that content during an entire message. Krugman, for example, used these measures to demonstrate that a very small proportion of time that adolescents spent looking at print cigarette advertising was focused on government-mandated health warnings (Fischer, Richards, Berman, & Krugman, 1989; Krugman, Fox, Fletcher, Fischer, Rojas, 1994). Recent examples use eye-tracking to quantify the effect of individual-difference variables on total time spent looking at certain screen elements (Cummins, Gong, & Kim, 2014; Cummins, Stone, Gong, Cui, 2017) or as an implicit measure of attitudes that is superior to self-report (Coronel & Federmeier, 2015; Vraga, Bode, & Troller-Renfree, 2016).

MEASURING ELECTRODERMAL ACTIVITY

Electrodermal activity (EDA) is a general term for the various measurements of the ease with which electricity moves across the surface of the skin. Sweat gland activity under the surface of the skin increases EDA, because both water and salt are electrical conductors. Unlike the dually innervated sinoatrial node, eccrine sweat glands are influenced only by the sympathetic nervous system. Increased sympathetic activation causes greater secretion from the glands into sweat ducts that open on the surface of the skin. Electrical conductance increases as more sweat fills the ducts, even if it does not reach the surface (Stern, Ray, & Quigley, 2000).

The highest density of eccrine glands is on the palmar surface of the hand and plantar surface of the foot. EDA is typically measured by placing two AG/AGCL electrodes on the non-dominant palm. Occasionally, experimental tasks require the use of the hands (e.g., keyboarding responses, manipulating a game controller) and then placement on the arch of the non-dominant foot is recommended to limit noise artifact and EDA resulting from somatic muscle activity. A small level of electrical voltage (0.5v) is sent to the electrodes and the level of conductance on the skin surface is sampled by a recording computer. Many conceptualize EDA measures as activation of the appetitive/approach motivational system, the aversive/avoidance system, or a combination of both (Potter & Bolls, 2012).

Just like the cardiac signal, EDA is operationalized as a phasic or tonic response. The phasic response occurs roughly 3 seconds after the onset of an eliciting source and is indicated by a steep-sloped increase in conductance known as a skin conductance response (SCR). SCRs often occur in response to known variables. However, nonspecific SCRs (nSCRs) also occur, due to unidentified environmental stimuli or endogenous sources of sympathetic activation. Given the likelihood of nSCRs in most communication-related data, they are sometimes left undifferentiated in analyses. Instead, SCRs are quantified without being time-locked to specific instances in the stimulus (Lang, Zhou, Schwartz, Bolls, & Potter, 2000; Schneider, Lang, Shin, & Bradley, 2004). Common measures used are SCR frequency, the number of nSCRs occurring during a communication event, average SCR amplitude, and largest SCR amplitude.

Tonic dynamic analysis of EDA aggregates the sampled data—often into 1-second increments—and then computes change scores from a baseline level just prior to stimulus onset. For example, if skin conductance is sampled at 1,000 Hz over the course of a 30-second television commercial, it is common for the average of each 1,000 samples to be exported as the SCL value for the corresponding second. Subtracting these values from a baseline measurement—1-second before the onset of the message is common—provides a sequence of 30 change scores representing the tonic change in sympathetic nervous system activation over the course of the commercial.

The two foundational HR pieces by Lang (1990; Lang et al., 1999) mentioned above also provide examples of using SCL to measure sympathetic nervous system activation. Detenber, Simons, and Bennet (1998) provide one of the earliest examples of using SCRs in the communication literature. Across two experimental blocks, participants watched a randomly presented set of images varying in content arousal. In one block, the images presented an object in motion for six seconds. In the other block, a still image version (a representative frame of the moving image) was presented for six seconds. Blocks were randomized between subjects and peripheral measures were collected. Results show significantly larger SCRs when the images were presented in motion compared to still. This effect was greater for arousing content compared to calm.

A search of recent research suggests that SCRs are uncommon measures, but not completely absent. Banjo et al. (2017) tested how the racial makeup of a trio of co-viewers affected cognitive and emotional responses to sitcoms containing racial slurs. Two of the co-viewers were confederates of the same race, selected to be either the same race as the participant (in-group condition) or a different race (out-group). Physiological data were collected while each trio watched episodes of the animated comedies *South Park* and *Boondocks* that contained racial slurs. Results show that as the number of slurs in the program increased, so did SCR frequency, regardless of the race of the participant. However, the effect of outgroup condition led to significantly more SCRs for Black viewers but not for White viewers.

As Table 19.1 suggests, there are more examples of tonic EDA analyses in the recent literature. Bailey (2016), for example, selected 12 television food advertisements in which half were for products high in energy density (kcal/gram) and half were for foods low in energy density. Participants watched the ads in a room where food was available for consumption. Results show a significant interaction between energy density and food presence.

When there was no food in the room, SCLs were significantly higher during energy-dense food ads than any other condition.

Yeykelis, Cummings, and Reeves (2014) used SCL during a field experiment to determine the relationship between arousal and task switching on personal laptops. Student participants had screen capture software installed that recorded their laptop activity every five seconds for most of an entire day. They also wore an EDA sensor that sampled SCL at 2 Hz. EDA data were aggregated in 30-second increments for between six and ten hours. The screen capture data were analyzed to determine whether the participant was at the computer and whether his or her computer task switched from one screen sample to another. Results show that SCL increased during the 12 seconds before switches from a work task to entertainment but decreased when switching from a source of entertainment back to work. Yeykelis et al. (2014) explained that the motivational relevance of the two types of tasks caused the pattern, with entertainment activating the appetitive system more than mundane work.

MEASURING ELECTROMYOGRAPHIC ACTIVITY

The positivity or negativity of a communication situation can have a substantial effect on the outcome. The most common way of harnessing the PNS to identify valence is by measuring the electrical activity in the muscles associated with emotional expression found below the skin of the face (Potter & Bolls, 2012). Through a process called electromyography (EMG), two small electrodes are placed close together directly over the muscle group of interest and the bioelectrical signals from the motor action units are sampled. There are three muscles frequently used in communication research to measure valence. The first is the corrugator supercilii (CS), located on the forehead just above the eyebrow and toward the nose. Increased impulses by this muscle group are associated with negative/aversive activation, and decreased activity corresponds with positivity. The other two muscle groups are primarily associated with appetitive activation: the zygomaticus major (ZM) and the orbicularis oculi (OO). ZM is located on the cheek in line with the base of the nose and down from the outer edge of the eye. Firing of ZM pulls the edge of the lips up in a smile. OO is located on the lower eyelid just below the pupil and has been described as bringing a sparkle to the eye during positive experience.

Hazlett and Hazlett (1999) were the first to publish work using EMG as a "way to get beyond the limitations and biases that often plague self-report" (p. 19–20). Bolls, Lang, and Potter (2001) provided a true test of validity for EMG in communication research, however, demonstrating that radio advertisements pretested via self-report to be highly positive or highly negative activated the ZM or CS respectively. Both of these early studies analyzed EMG data as a tonic response over the course of entire messages.

Recent literature also presents tonic EMG data. Sukalla, Bilandzic, Bolls, and Busselle (2016), for example, took five-minute segments from TV shows such as *Grey's Anatomy* and *Private Practice* and edited them to create two versions of each: the original and another in which scenes were presented in nonsequential order and therefore had lower narrative cohesion. Participants saw an equal number of high- and low-cohesion clips in the laboratory while peripheral data were collected. EMG results show greater CS activation during the low cohesion clips due to the negative affect elicited by narrative confusion.

Other scholars have recently presented phasic EMG results. Read, van Driel, and Potter (2018) collected PNS data while participants watched television advertisements featuring a moment when the couple featured in the ad was revealed as either homosexual or heterosexual. This "reveal moment" provided the evoked response in CS. After watching all the ads, participants completed self-report measures of implicit and explicit attitudes toward homosexuality. Results show that the more negative the participant's implicit attitude toward homosexuality, the greater the CS activation in response to the reveal of the same-sex couple compared to that of the other-sex couple.

A specific type of evoked measure involving EMG is the startle response (SR). Startle is an automatic response to a rapidly occurring, potentially harmful environmental event. SR is characterized by full-body muscle contraction, including eye-blinks originating in the OO and in the post-auricular (PA) muscle behind the ear. In the research lab, the startle response is often evoked by a 90 db/0 ms rise-time burst of white noise through headphones. Magnitudes of OO response to such a probe are well-established correlates of aversive activation caused by the pre-probe environment. Less recognized is the conclusion that larger PA responses to probes indicate a more appetitive state. Bradley (2007) was the first to use OO startle in the communication literature and it has had little use since, perhaps because of the irritating nature of the white noise probe. Two exceptions are Sparks and Lang (2015)—wherein PA startle response was used to differentiate processing of ads using humor and/or sex appeal—and Lee (2017), wherein OO startle was found to increase in players of the computer game *Skyrim* when the virtual environment was designed to be dark compared to light.

CONCLUSION

Peripheral nervous system measures are a valuable addition to the methodological catalog of the communication scientist. PNS measures index mental processes related to attention and emotion that underlie how individuals perceive and respond to communication content and technologies. Communication scientists can use PNS measures to observe moment-by-moment variation in mental processes across an entire communication episode or during specific meaningful moments that occur within an episode. This review of previous research illustrates how the use of PNS measures in communication science has a rich history and, we believe, an exciting future. Future research is likely to push communication science into new areas. New technology for collecting data—such as wireless mobile labs featuring PNS measures—can be used to significantly expand the range of contexts in which experiments can be conducted. These advances in technology have also made heart rate, skin conductance, and facial EMG measures common in neuromarketing, an applied form of communication science focused on testing and optimizing marketing communication. This newer area of application for research using PNS measures allows academic researchers to work with industry partners in ways that can expand the practical as well as theoretical value of this research.

We remind communication scientists who wish to apply PNS measures after reading this chapter that they should become familiar with the psychophysiological paradigm (Potter & Bolls, 2012). Further, this chapter has focused on measures, but mental processes unfolding across time as people experience communication episodes are the proper focus of the communication scientist. All measures, including PNS measures, are useful only to the degree that they provide valid data to advance understanding of these dynamic processes.

REFERENCES

Allen, J. J. B., Chambers, A. S., & Towers, D. N. (2007). The many metrics of cardiac chronotropy: A pragmatic primer and a brief comparison of metrics. *Biological Psychology, 74*(2), 243–262.

Bailey, R. L. (2016). Modern foraging: Presence of food and energy density influence motivational processing of food advertisements. *Appetite, 107*, 568–574.
Bailey, R. L., Liu, J., & Wang, T. (2018). Primary biological motivators in food advertisements: Energy density and sexual appeals compete for appetitive motivational activation. *Communication Research*. Advance online publication.
Bailey, R. L., Wang, T., & Kaiser, C. K. (2018). Clash of the primary motivations: Motivated processing of emotionally experienced content in fear appeals about obesity prevention. *Health Communication, 33*(2), 111–121.
Banjo, O. O., Wang, Z., Appiah, O., Brown, C., Walther-Martin, W., Tchernev, J., ... Irwin, M. (2017). Experiencing racial humor with outgroups: A psychophysiological examination of co-viewing effects. *Media Psychology, 20*(4), 607–631.
Bellman, S., Wooley, B., & Varan, D. (2016). Program–ad matching and television ad effectiveness: A reinquiry using facial tracking software. *Journal of Advertising, 45*(1), 72–77.
Bolls, P. D., Lang, A., & Potter, R. F. (2001). The effects of message valence and listener arousal on attention, memory, and facial muscular responses to radio advertisements. *Communication Research, 28*(5), 627–651.
Bradley, S. D. (2007). Examining the eyeblink startle reflex as a measure of emotion and motivation to television programming. *Communication Methods & Measures, 1*(1), 7–30.
Cacioppo, J. T., Tassinary, L. G., & Berntson, G. G. (2017). *Handbook of psychophysiology* (4th ed.). Cambridge, England: Cambridge University Press.
Chang, J. S., Kim, E. Y., Jung, D., Jeong, S. H., Kim, Y., Roh, M.-S., ... Hahm, B.-J. (2015). Altered cardiorespiratory coupling in young male adults with excessive online gaming. *Biological Psychology, 110*, 159–166.
Chen, Y., & Tsai, M.-J. (2015). Eye-hand coordination strategies during active video game playing: An eye-tracking study. *Computers in Human Behavior, 51*, 8–14.
Clayton, R. B., Lang, A., Leshner, G., & Quick, B. L. (2018). Who fights, who flees? An integration of the LC4MP and psychological reactance theory. *Media Psychology*. Advance online publication.
Clayton, R. B., Leshner, G., & Almond, A. (2015). The extended iself: The impact of iPhone separation on cognition, emotion, and physiology. *Journal of Computer-Mediated Communication, 20*(2), 119–135.
Clayton, R. B., Leshner, G., Bolls, P. D., & Thorson, E. (2017). Discard the smoking cues—keep the disgust: An investigation of tobacco smokers' motivated processing of anti-tobacco commercials. *Health Communication, 32*(11), 1319–1330.
Coronel, J. C., & Federmeier, K. D. (2015). The effects of gender cues and political sophistication on candidate evaluation: A comparison of self-report and eye movement measures of stereotyping. *Communication Research, 43*(7), 922–944.
Coyne, S. M., Dyer, W. J., Densley, R., Money, N. M., Day, R. D., & Harper, J. M. (2015). Physiological indicators of pathologic video game use in adolescence. *Journal of Adolescent Health, 56*(3), 307–313.
Cummins, R. G., Gong, Z., & Kim, H.-S. (2014). Individual differences in selective attention to information graphics in televised sports. *Communication & Sport, 4*(1), 102–120.
Cummins, R. G., Stone, C. H., Gong, Z., & Cui, B. (2017). Visual attention to and understanding of graphic program advisories: An eye-tracking study. *Journal of Broadcasting & Electronic Media, 61*(4), 703–722.
Detenber, B. H., Simons, R. F., & Bennett, G. G. (1998). Roll 'em!: The effects of picture motion on emotional responses. *Journal of Broadcasting & Electronic Media, 42*(1), 113–127.
Duffy, E. (1962). *Activation and behavior*. New York, NY: Wiley.
Fischer, P. M., Richards, J. W., Jr, Berman, E. J., & Krugman, D. M. (1989). Recall and eye tracking study of adolescents viewing tobacco advertisements. *Journal of the American Medical Association, 261*(1), 84–89.
Gao, Y. (2016). Interested processing of educational audiovisual messages: Attention, memory, and psychophysiological responses. Unpublished doctoral dissertation, Indiana University, Bloomington, IN.

Grizzard, M., Tamborini, R., Sherry, J. L., Weber, R., Prabhu, S., Hahn, L., & Idzik, P. (2015). The thrill is gone, but you might not know: Habituation and generalization of biophysiological and self-reported arousal responses to video games. *Communication Monographs, 82*(1), 64–87.

Hart, B. T., Struiksma, M. E., van Boxtel, A., & van Berkum, J. J. A. (2018). Emotion in stories: Facial EMG evidence for both mental simulation and moral evaluation. *Frontiers in Psychology, 9*, 613.

Hazlett, R. L., & Hazlett, S. Y. (1999). Emotional response to television commercials: Facial EMG vs. self-report. *Journal of Advertising Research, 39*(2), 7–23.

Howell, M., Ekman, D. S., Almond, A., & Bolls, P. (2018). Switched on: How the timing of aversive content in traffic safety videos impact psychophysiological indicators of message processing. *Health Communication*. Advance online publication.

Jabon, M. E., Ahn, S. J., & Bailenson, J. N. (2011). Automatically analyzing facial-feature movements to identify human errors. *IEEE Intelligent Systems, 26*(2), 54–63.

Jang, J. Y., Baek, E., Yoon, S.-Y., & Choo, H. J. (2018). Store design: Visual complexity and consumer responses. *International Journal of Design, 12*(2), 105–118.

Koruth, J., Lang, A., Potter, R. F., & Bailey, R. (2015). A comparative analysis of dynamic and static indicators of parasympathetic and sympathetic nervous system activation during TV viewing. *Communication Methods and Measures, 9*, 78–100.

Krugman, D. M., Fox, R. J., Fletcher, J. E., Fischer, P. M., & Rojas, T. H. (1994). Do adolescents attend to warnings in cigarette advertising? An eye-tracking approach. *Journal of Advertising Research, 34*(6), 39–52.

Kruikemeier, S., Lecheler, S., & Boyer, M. M. (2018). Learning from news on different media platforms: An eye-tracking experiment. *Political Communication, 35*(1), 75–96.

Lacey, J. L. (1967). Somatic response patterning and stress: Some revisions of activation theory. In M. H. Appley & R. Trumbull (Eds.), *Psychological stress: Issues in research* (pp. 14–38). New York, NY: Appleton-Century Crofts.

Lang, A. (1990). Involuntary attention and physiological arousal evoked by structural features and emotional content in TV commercials. *Communication Research, 17*, 275–299.

Lang, A. (Ed.). (1994). *Measuring psychological responses to media*. Hillsdale, NJ: Lawrence Erlbaum Associates, Inc.

Lang, A., Bolls, P., Potter, R. F., & Kawahara, K. (1999). The effects of production pacing and arousing content on the information processing of television messages. *Journal of Broadcasting & Electronic Media, 43*(4), 451–475.

Lang, A., Potter, R. F., & Bolls, P. D. (2009). Where psychophysiology meets the media: Taking the effects out of media research. In J. Bryant & M. B. Oliver (Eds.), *Media effects: Advances in theory and research* (pp. 185–206). New York, NY: Routledge.

Lang, A., Schwartz, N., & Mayell, S. (2015). Slow down you're moving too fast: Age, production pacing, arousing content, and memory for television messages. *Journal of Media Psychology: Theories, Methods, and Applications, 27*(2), 53–63.

Lang, A., & Yegiyan, N. (2014). Mediated substance cues: Motivational reactivity and use influence responses to pictures of alcohol. *Journal of Health Communication, 19*(11), 1216–1231.

Lang, A., Zhou, S. H., Schwartz, N., Bolls, P. D., & Potter, R. F. (2000). The effects of edits on arousal, attention, and memory for television messages: When an edit is an edit can an edit be too much? *Journal of Broadcasting & Electronic Media, 44*(1), 94–109.

Lee, J. (2017). Scared of the dark: Examining aversive activation during a virtual navigation task. Unpublished master's thesis, Indiana University, Bloomington, IN.

Lee, S., & Potter, R. F. (2018). The impact of emotional words on listeners' emotional and cognitive responses in the context of advertisements. *Communication Research*. Advance online publication.

Leshner, G., Clayton, R. B., Bolls, P. D., & Bhandari, M. (2018). Deceived, disgusted, and defensive: Motivated processing of anti-tobacco advertisements. *Health Communication, 33*(10), 1223–1232.

Li, J., Ju, W., & Reeves, B. (2017). Touching a mechanical body: Tactile contact with body parts of a humanoid robot is physiologically arousing. *Journal of Human-Robot Interaction, 6*(3), 118–130.

McDuff, D., & el Kaliouoby, R. (2017). Applications of automated facial coding in media measurement. *IEEE Transactions on Affective Computing, 8*(2), 148–160.

Missaglia, A. L., Oppo, A., Mauri, M., Ghiringhelli, B., Ciceri, A., & Russo, V. (2017). The impact of emotions on recall: An empirical study on social ads. *Journal of Consumer Behaviour, 16*(5), 424–433.

Potter, R. F., & Bolls, P. D. (2012). *Psychophysiological measurement and meaning: Cognitive and emotional processing of media.* New York, NY: Routledge.

Potter, R. F., Jamison-Koenig, E., Lynch, T., & Sites, J. (2016). Effect of vocal tonal difference on automatic attention to voice changes in audio messages. *Communication Research.* Advance online publication.

Potter, R. F., Lynch, T., & Kraus, A. (2015). I've heard that before: Habituation of the orienting response follows repeated presentation of auditory structural features in radio. *Communication Monographs, 82*(3), 359–378.

Potter, R. F., Sites, J., Jamison-Koenig, E., & Zheng, X. (2018). The impact of cognitive load on the cardiac orienting response to auditory structural features during natural radio listening situations. *Journal of Cognition.* Advance online publication.

Rasmussen, E. E., Keene, J. R., Berke, C. K., Densley, R. L., & Loof, T. (2017). Explaining parental coviewing: The role of social facilitation and arousal. *Communication Monographs, 84*(3), 365–384.

Ravaja, N. (2004). Contributions of psychophysiology to media research: Review and recommendations. *Media Psychology, 6*(2), 193–235.

Read, G. L., van Driel, I. I., & Potter, R. F. (2018). Same-sex couples in advertisements: An investigation of the role of implicit attitudes on cognitive processing and evaluation. *Journal of Advertising, 47*(2), 182–197.

Rodero, E., Potter, R. F., & Prieto, P. (2017). Pitch range variations improve cognitive processing of audio messages. *Human Communication Research, 43*(3), 397–413.

Rubenking, B., & Lang, A. (2014). Captivated and grossed out: An examination of processing core and sociomoral disgusts in entertainment media. *Journal of Communication, 64*(3), 543–565.

Russell, C. A., Swasy, J. L., Russell, D. W., & Engel, L. (2017). Eye-tracking evidence that happy faces impair verbal message comprehension: The case of health warnings in direct-to-consumer pharmaceutical television commercials. *International Journal of Advertising, 36*(1), 82–106.

Schneider, E. F., Lang, A., Shin, M., & Bradley, S. D. (2004). Death with a story: How story impacts emotional, motivational, and physiological responses to first-person shooter video games. *Human Communication Research, 30*(3), 361–375.

Shalom, J. G., Israeli, H., Markovitzky, O., & Lipsitz, J. D. (2015). Social anxiety and physiological arousal during computer mediated vs. face to face communication. *Computers in Human Behavior, 44*, 202–208.

Sparks, J. V., & Lang, A. (2015). Mechanisms underlying the effects of sexy and humorous content in advertisements. *Communication Monographs, 82*(1), 134–162.

Stern, R. M., Ray, W. J., & Quigley, K. S. (2000). *Psychophysiological recording* (2nd ed.). New York, NY: Oxford University Press.

Sukalla, F., Bilandzic, H., Bolls, P. D., & Busselle, R. W. (2016). Embodiment of narrative engagement: Connecting self-reported narrative engagement to psychophysiological measures. *Journal of Media Psychology, 28*(4), 175–186.

Vraga, E., Bode, L., & Troller-Renfree, S. (2016). Beyond self-reports: Using eye tracking to measure topic and style differences in attention to social media content. *Communication Methods & Measures, 10*(2–3), 149–164.

Wang, Z., Morey, A. C., & Srivastava, J. (2012). Motivated selective attention during political ad processing: The dynamic interplay between emotional ad content and candidate evaluation. *Communication Research, 41*(1), 119–156.

Yegiyan, N. S. (2015). Explicating the emotion spillover effect: At the intersection of motivational activation, resource allocation, and consolidation. *Journal of Media Psychology: Theories, Methods, and Applications, 27*(3), 134–145.

Yegiyan, N. S., & Bailey, R. L. (2016). Food as risk: How eating habits and food knowledge affect reactivity to pictures of junk and healthy foods. *Health Communication, 31*(5), 635–642.
Yeykelis, L., Cummings, J. J., & Reeves, B. (2014). Multitasking on a single device: Arousal and the frequency, anticipation, and prediction of switching between media content on a computer. *Journal of Communication, 64*(1), 167–192.

20

The State of the Art and the Future of Functional Magnetic Resonance Imaging in Communication Research

Frederic R. Hopp and René Weber

Functional magnetic resonance imaging (fMRI) has become an important methodological supplement for communication scholarship, spanning a wide array of communication sub-areas, including interpersonal communication (Silbert, Honey, Simony, Poeppel, & Hasson, 2014; Stephens, Silbert, & Hasson, 2010), health communication (Berkman & Falk, 2013; Falk, Berkman, Mann, Harrison, & Lieberman, 2010; Falk, Berkman, Whalen, & Lieberman, 2011; Weber, Huskey, Mangus, Westcott-Baker, & Turner, 2015), social media (Scholz et al., 2017), and media enjoyment (Anderson, Fite, Petrovitch, & Hirsch, 2006; Hasson, Nir, Levy, Fuhrmann, & Malach, 2004; Huskey, Craighead, Miller, & Weber, 2018; Schmälzle, Imhof, Grall, Flaisch, & Schupp, 2017). Special issues in flagship communication journals such as *Media Psychology* (Anderson et al., 2006), *Communication Monographs* (Afifi & Floyd, 2014), and *Communication Methods and Measures* (Weber, 2015) have recognized the potential of fMRI for the observation of neural activity associated with message processing at high temporal and spatial resolution.

A great proportion of fMRI's initial popularity in communication research has been attributed to its capability to provide moment-by-moment measures of individuals' responses to media stimuli that are difficult to measure reliably in self-reports (e.g., perceived message effectiveness; Weber, Huskey, Mangus, Westcott-Baker, & Turner, 2015) or are prone to social desirability biases (e.g., aggressiveness; Weber, Ritterfeld, & Mathiak, 2006). In addition, in the early 2000s, an increasing number of communication scholars were interested in understanding the roots and underlying cognitive mechanisms of communication phenomena, as a deeper understanding of brain function and the integration of modern brain science into communication research was recognized as an important additional source of scientific investigations in the field (Weber, Sherry, & Mathiak, 2008). More recently, increased access to scanning facilities,[1] as well as methodological advances within the broader neurosciences, in machine learning, and in computational approaches to content analysis, have significantly broadened the scope of questions communication researchers can address with fMRI (Turner, Huskey, & Weber, 2018). These developments culminated in novel, previously unconceived theoretical advances, and provided examples of method-theory synergy (Greenwald, 2012;

Weber, Fisher, Hopp, & Lonergan, 2018). With this background in mind, the aim of this chapter is threefold. First, we provide the reader with a basic overview of fMRI. Second, we introduce currently employed, state-of-the-art fMRI data analyses that are "pushing the envelope" of media neuroscience. Third, we illustrate potential future examples of method-theory synergy that may arise from the integration of natural language processing techniques into fMRI and from using fMRI simultaneously, in real time, at connected locations in order to account for the inherently interactive nature of human communication.

A WHIRLWIND TOUR OF FMRI

For readers unfamiliar with fMRI, this section provides a brief introduction to its basic methodology. Readers looking for more comprehensive introductions of fMRI methodology that are tailored to communication research are encouraged to consult recent reviews by Weber and colleagues (2015, 2018).

In a typical communication study using fMRI, participants are placed into a narrow bore of an MRI scanner that produces a strong magnetic field of usually three Tesla, which is about 60,000 times as strong as Earth's magnetic field. This strong magnetic field interacts with the magnetic properties of the hydrogen atom (Edelman & Warach, 1993), which exists in various bodily tissues. In simplified terms, the magnetic energy within the nuclei of these hydrogen atoms is sampled by a receiver inside the MRI scanner, which enables the construction of spatially encoded, high-resolution images that amplify contrasts in hydrogen atom concentration, for instance, to distinguish bodily tissue such as water, bone, or fat. In turn, two types of images are usually produced with MRI, depending on the specific use case. *Functional* magnetic resonance imaging (fMRI) samples magnetic resonance continuously at high temporal resolution (up to 500 milliseconds in modern scanners), but with lower spatial resolution to capture ephemeral changes in brain *activity*. In contrast, *structural* imaging (MRI) samples static magnetic resonance at a high spatial, but stationary timely resolution and thus is more commonly applied to study brain *structure* or to detect anatomical abnormalities.

To examine neural activity during some communication task, fMRI exploits the magnetic properties of hemoglobin, a protein that is responsible for binding oxygen in blood. Depending on its state of oxygenation (i.e., the saturation level of O_2), hemoglobin exhibits slightly different magnetic properties (Ogawa et al., 1992). Research has demonstrated that differences in oxygenated to deoxygenated hemoglobin can serve as a close *proxy* to neural activity (Logothetis, 2008; Logothetis, Pauls, Augath, Trinath, & Oeltermann, 2001). As neurons fire, they extract oxygen and glucose from neighboring blood vessels to sustain functioning, resulting in a short preponderance of deoxygenated to oxygenated hemoglobin. To counter this deficit, the metabolic system responds by channeling a significant amount (i.e., more than is consumed) of oxygenated blood to the active neurons, a biological pattern known as the hemodynamic response function (HRF: for a more detailed discussion, see Logothetis et al., 2001). As a result of the HRF, an overabundance of oxygenated to deoxygenated hemoglobin can be observed in brain areas of neural activity, which leads to disturbances of magnetic fields generated by the MRI scanner. This phenomenon is termed the blood-oxygen-level dependent (or BOLD) effect. State-of-the-art fMRI scanners can detect BOLD effects at a spatial resolution of up to 1 mm^3 (called a 1 mm^3 volume-element, or voxel, the 3D equivalent of a pixel) and at temporal intervals of 500 ms.

To use fMRI to gain insights into brain function (i.e., where and when neural activation occurred and how neural activation in one area is connected to activation in other areas), the classical fMRI task-based activation paradigm draws on a variety of visual and auditory stimuli, ranging from highly controlled and non-interactive stimuli (e.g., pictures, audio recordings) to more naturalistic and interactive (e.g., video games) stimuli. Typically, one or several experimental conditions are contrasted with a baseline condition to examine how brain activity changes in response to a given task or stimulus. A large proportion of these baseline conditions consists of resting tasks in which it is assumed that brains return to an "inactive" state, for instance, by instructing participants simply to look at a centered cross placed in the middle of the screen in order to keep head motion at a minimum. This assumption of inactivity has been challenged, however, and thus, *active* baselines or control conditions are usually considered superior. Active baselines resemble the stimulus as closely as possible in terms of potential confounds such as brightness, volume, color intensity, etc., and *only* remove the actual experimental condition of interest from the baseline stimulus.

Once the experimental paradigm has been designed and thoroughly pretested, data collection can begin. Each experimental session usually begins by collecting a series of structural (anatomical) scans to generate a static, high-resolution, anatomical image of the participant's brain. This image will be used later to locate neural activity in a participant's brain and to standardize locations across multiple participants' brains. Next, participants engage in the actual experimental task(s) and functional scans are taken that reveal participants' neural activity during task engagement and baselines. Each three-dimensional functional image within a time interval (TR), including all locations (i.e., all voxels) in a participant's brain, is referred to as a volume.

Following data collection, various important preprocessing steps need to be applied to fMRI data. In Weber, Mangus, and Huskey (2015), communication researchers interested in conducting or interpreting fMRI research can find more detailed and easy-to-comprehend information on why preprocessing is important, what the various preprocessing steps entail and accomplish, and, perhaps most important, how common mistakes can be avoided.

Subsequent to successful preprocessing, a standard statistical analysis of the data can begin (see Ashby, 2011, for a comprehensive introduction). Consistent with the previously described, standard task-based fMRI paradigm, researchers usually apply simple *subtraction logic* in their data analysis. This means that differences in neural activation between experimental and baseline conditions are tested for statistical significance. To do so, researchers fit a multi-level General Linear Model (GLM) to their fMRI data. This can be done for all voxel time-series in the entire brain (*whole brain analysis*) or for a more specific, theory-driven selection of a brain *region of interest* (ROI). Whole brain analysis aims to explore clusters of voxels that were activated in response to a given task (in contrast to some baseline). In comparison, ROI analyses focus on activation patterns within *a priori* selected brain regions that a body of previous research has identified to be involved in a given cognitive process. To account for individual differences in brain structure and connectivity (Davison et al., 2016), researchers increasingly employ *functional localization* tasks that specifically aim to better locate task-relevant ROIs for further analysis on a per-subject basis (Poldrack, 2006). For instance, a gambling task unrelated to the actual experimental task could be used to localize the reward network in a participant's brain.

Some Words of Caution

fMRI offers no panacea to investigations of human communication phenomena. It requires carefully designed experimental paradigms, preprocessing, and statistical analysis procedures aimed at addressing a quite restricted set of research questions (Weber, Mangus, & Huskey,

2015). Regardless of novel and exciting methodological and statistical advances discussed below, a healthy skepticism about conclusions drawn from fMRI data should be maintained. For instance, studies claiming to have discovered a specific, unique ROI that is responsible for encoding a multifaceted communication phenomenon are often prone to reverse inference fallacies (Poldrack, 2006). Whereas a given ROI might be activated when being exposed to a communication stimulus (e.g., a message designed to persuade receivers), the same ROI will likely also become active during exposure to many other stimuli. The critical point here is that the relationship between cognitive processes and brain activity is not bijective; that is, cognitive processes do not map one-to-one onto certain brain regions but follow a many-to-many mapping in which brain region X may be activated by cognitive process P, but also processes Q, R, and S. Likewise, cognitive process P might activate region X, but also regions W, Y, and Z. Furthermore, establishing that an a priori defined ROI was activated in response to a hypothesized task does not necessarily support the validity of the underlying theory for this a priori localization. Avoiding such consistency fallacies (Mole & Klein, 2010) often requires fMRI researchers to demonstrate that their study could have produced some specified alternative brain activation patterns that are inconsistent with the theoretical predictions and, despite all methodologically sound effort, these brain activity patterns were not among the observed results (Coltheart, 2013). Finally, novel statistical methods for brain imaging analysis increasingly resemble complex, black-box input-output models (e.g., deep neural networks; Silver et al., 2016) that are often coupled with sophisticated visualization tools to provide fascinating simulations of brain networks with seemingly convincing activation patterns (e.g., *Brain Viewer* developed by Jack Gallant's lab; http://gallantlab.org/huth2016/). Although we understand the excitement and persuasion that accompany these new approaches and tools, we advise our fellow communication researchers not to get too impressed and to critically assess the utility of these advances for the understanding, explanation, and real-world prediction of communication phenomena (Lin, 2015).

CURRENT ADVANCES IN COMMUNICATION NEUROSCIENCE

In recent years, great gains have been made in extending the methodological repertoire to collect and analyze fMRI data. Some of these advances have been inspired by new theorizing about how brains may work at a fundamental level and create complex, higher-order experiences such as empathy, morality, and consciousness. For instance, the theoretical conceptualization of the brain as a complex, emergent, dynamic system resulted in graph and network theoretical analyses of structural and functional brain imaging data (e.g., Bullmore & Sporns, 2009; Sporns, 2010). Similarly, advances in fMRI data collection and analyses such as the free viewing (Bartels & Zeki, 2004) and natural behavior (Mathiak & Weber, 2006) paradigm, in combination with intersubject correlation analysis (ISC: Hasson et al., 2004) led to previously inconceivable data and findings that allowed for important extensions of existing theory (e.g., the discovery of biologically encoded, shared memory structures in humans; Chen et al., 2017). Mounting evidence suggests that this method-theory synergy is becoming increasingly salient within media and communication neuroscience (Weber et al., 2018), pushing the application of state-of-the-art brain imaging methods to tackle traditional and novel research questions. Next, we briefly review a selection of current advances in fMRI research that, in our opinion, have most saliently innovated communication scholarship (see also Turner et al., 2018).

The Brain-as-Predictor Approach (BAP): From Neural Activity to Population Outcomes

The brain-as-predictor approach (Berkman & Falk, 2013; Falk et al., 2010) was inspired by a set of seemingly simple questions:

- Is it possible to predict humans' real-world behavior change from their neural responses to persuasive messages?
- Do these predictions generalize from typically small samples in fMRI studies to entire populations?
- Does neural activity predict behaviors better than common self-report measures of message persuasiveness?

Falk and colleagues have pioneered this approach and demonstrated that the simple answer to these questions is *yes*. Given the almost 100-year-long tradition of persuasion research in communication (O'Keefe, 2002), which evidences modest success in predicting behavior change, the brain-as-predictor approach was and still is a "game changer." Traditionally, persuasion scholars assessed the effectiveness of persuasive messages, conceptualized as modified behavioral intentions or actual behavior change, by relying on self-report measures following message exposure. Yet, individuals might not be able to reliably recall and verbalize whether a given message was crucial for engaging in or refraining from certain behaviors. Sometimes, individuals prefer to provide socially desirable responses to sensitive issues (e.g., issues related to racism or sexual behaviors), and changes in intentions and behaviors may proceed at a nonconscious level. In contrast to self-reports, fMRI—with its ability to provide a moment-by-moment measure of specific cognitive responses to persuasive messages—can circumvent these challenges and provide a more nuanced evaluation of message processing. In fact, Falk and colleagues have shown that adding neural activation into predictive models of behavior change improved the accuracy of these models by over 20% above and beyond self-reports (Falk et al., 2010). Furthermore, brain activation observed in small groups of participants (even $n < 30$) predicted population-level message effects when self-report data did not (Falk, Berkman, & Lieberman, 2012; Falk et al., 2016). Falk and colleagues also identified the specific networks that play a key role in these predictions, which are primarily networks belonging to the brain's mentalizing system (see Chapter 9 by Baek, Scholz, & Falk, in this volume). Since the appearance of the first brain-as-predictor studies in communication research almost ten years ago, the approach has introduced or refined theories on message and information sharing (O'Donnell & Falk, 2015a; Scholz et al., 2017), counterarguing during anti-drug messages (Weber, Huskey, Mangus, Westcott-Baker, & Turner, 2015), the design of antismoking health campaigns (Falk, Berkman, Whalen, & Lieberman, 2011; Falk et al., 2016), and many more (for an overview, see Falk, Cascio, & Coronel, 2015). Given further statistical refinements (Turner, Huskey, & Weber, 2018), we believe that the brain-as-predictor approach continues to demonstrate its central role in advancing communication theory and research.

Intersubject Correlation Analysis (ISC): Synchrony Across Brains

Another methodological innovation within neuroimaging has been focusing on recording how *similar* or *synchronized*—rather than different—patterns in brain activation are across participants in response to a stimulus. This shift in thinking has been driven partially by increased interest in more naturalistic, complex, and low-controlled stimuli that are common in communication research, such as movies and video games. Early research convincingly revealed that similarity in neural activation can reliably predict narrative engagement (Cohen, Henin, & Parra, 2017; Weber, Eden, & Mathiak, 2011) and interpretations of narratives (Yeshurun et al., 2017), perceptions of moral violations in political liberals and conservatives (Amir et al., 2017), convincingness of oral rhetoric (Schmälzle et al., 2015), interpersonal communication success (Hasson & Frith, 2016), and shared risk perception (Schmälzle, Häcker, Renner, Honey, & Schupp, 2013).

The commonly applied analytical procedure for assessing similarity in neural activation (also called synchrony or temporal reliability) is intersubject correlation analysis (ISC: Hasson et al., 2004), which typically calculates the pairwise average Pearson correlations of individuals' brain responses. In other words, ISC computes the average correlation of pairwise correlated brains (with n participants there are $\frac{n(n-1)}{2}$ pairs of participants). Whereas ISC studies in communication have convincingly demonstrated their vast utility in addressing various questions relevant to the field of communication (for an overview, see Chapter 8 by Schmälzle & Grall, in this volume), we suggest that further advances in fine-grained, computer-assisted, content-analytical procedures (discussed in more detail below), in combination with solid communication theory, will further help to interpret ISCs by demonstrating *why* specific brain regions fall in and out of synchrony with one another as a function of meaningful content features.

Multivoxel Pattern Analysis (MVPA): Decoding Brain Activity

Most fMRI research has been driven by *encoding* models that aim to establish how external stimulus information is represented internally across brain systems. Recent literature increasingly incorporates *decoding* models in fMRI research that examine how internal neural representations can be used to predict or recreate external stimuli (Haxby et al., 2001; Mitchel et al., 2004). In layman's terms, decoding models are known as "mind-reading." The statistical methodology that gave rise to the recent popularity of decoding models is called multi-voxel pattern analysis (MVPA: Norman, Polyn, Detre, & Haxby, 2006). MVPA distinguishes itself from individual-voxel-based analytical procedures such as the General Linear Model (see above) by applying multivariate pattern-classification algorithms. These algorithms can be rather simple and supervised, such as linear discriminant analysis (LDA) or support vector machines (SVM: Suykens & Vandewalle, 1999) that "learn" to discriminate which patterns of neural activity represent external stimuli with high probability. Decoding models can serve as complementary validation for encoding models by testing how accurate activation in a given ROI can discriminate between stimuli presentations. At the same time, mounting evidence suggests that decoding models can be successfully applied to generate predictions about an individual's *behavior*, for instance, by predicting task performance based on neural activity alone (Williams, Dang, & Kanwisher, 2007).

Although the application of decoding models to explain communication phenomena is still in its infancy, several researchers have started to incorporate MVPA in their analyses. For example, Amir and colleagues (2017) used MVPA to analyze whether exposure to moral adherences and violations elicits dissociable cortical activation. Findings suggest that activation in areas previously linked to moral judgments (e.g., Greene, Sommerville, Nystrom, Darley, & Cohen, 2001), such as the precuneus, the medial prefrontal cortex (MPFC), and the temporal pole (TP), can be used to predict which moral violations participants were exposed to. Another important example for the utility of decoding models in communication research is deception research (for an overview, see Langleben, 2008). Although there are theoretical, methodological, and ethical controversies surrounding deception research that need resolutions (e.g.,

Levine & McCornack, 2014), some communication neuroscience scholars have already begun studying deception using fMRI and a decoding perspective. Yang and colleagues (2014), for example, attempted to decode the veracity of participants' thoughts independent of their intent to conceal. By applying a decoding perspective, Yang and colleagues first demonstrated that mental representations preceding simple "yes/no" questions can be classified with above-chance accuracy. In addition, when participants were instructed to deceive during their response, the decoding classifiers that were trained on "honest" trails were still able to decode the neural responses to deception trails with above-chance accuracy. These findings provide initial evidence that concealed responses are, to some degree, independent of individuals' intent to tell the truth or to deceive. Furthermore, Cui and colleagues (2013) applied discriminant analysis to investigate whether participants' neural activity can reveal deceptive or truthful responses in a mock murder paradigm. Their findings suggest that activation patterns in the right ventral lateral prefrontal (VLPF) areas were most informative for discerning whether a participant was lying or telling the truth. These early examples demonstrate that communication scholars can learn much from attempts to decode brain activity patterns.

FUTURE AVENUES FOR FMRI IN COMMUNICATION RESEARCH

In this final section of the chapter, we discuss currently emerging methodologies that may have great potential to meaningfully shape the future of fMRI in communication research. Specifically, we believe that unprecedented theoretical advances will be gained by drawing on three fMRI research paradigms: (1) the integration of natural language processing (NLP) techniques and fine-grained, computational content-analyses into fMRI; (2) the interactive, simultaneous neuroimaging across multiple participants (i.e., hyperscanning, or HS); and (3) the manipulation of stimuli in real time and in response to neural activity (i.e., neurofeedback or real-time fMRI, or rtfMRI).

Combining fMRI and Computational Content Analysis

An often overlooked premise of fMRI assumes that understanding brain function is fundamentally constrained by the understanding of the stimulus features that give rise to the observed neural responses (Weber, 2008). This premise is especially important in low-controlled, naturalistic stimuli (e.g., a movie or an interactive environment), which are increasingly popular in media neuroscience studies. Yet, to date, most advances in fMRI research have focused on the analysis of neural activity (the dependent variable), while largely neglecting a fine-grained and meaningful analysis of *content* (the independent variable), which can separate important explanatory variables on theoretical grounds from confounding factors. This imbalance of focus might be due to neuroscientists' lack of training in content analytic procedures and the tremendous efforts associated with manual, theory-driven content labeling. Yet, we believe that media and communication neuroscientists are exceptionally well prepared to correct this imbalanced focus by combining their knowledge of communication theory and their expertise in rigorous content-analytic procedures with the advanced analyses of fMRI data introduced above.

A first step in this direction may aim to assess how well natural language processing algorithms (for an overview, see Grimmer & Stewart, 2013) are capable of automatically extracting meaningful content features. For example, popular automated content analysis packages such as the Linguistic Inquiry and Word Count (LIWC: Pennebaker, Francis, & Booth, 2001) apply dictionaries that simply count the occurrence of keywords in a text that are assumed to reflect psychological constructs, such as positive and negative sentiment. Next, these word counts can be related to neural activity during message processing. For instance, brain-as-predictor approaches may focus on assessing how message characteristics represented by word counts are related to neural activity in specified networks and predict behavioral outcomes, such as joining a social movement, voting, or message virality (O'Donnell & Falk, 2015a, 2015b). Given the increasingly unrestricted access to large-scale archives of automatically content-analyzed media content (see, e.g., the GDELT interface for communication research; iCoRe; Hopp, Schaffer, Fisher, & Weber, in press) and the recent popularity of computational approaches among communication researchers,[2] we predict intensified research activities in this area in the near future. In addition to automated *stimulus* content analysis, media neuroscientists have begun probing the feasibility of applying NLP to various types of participant *responses* following message exposure (e.g., think-aloud interviews, free-recall of perceived content) and linking these responses to neural activity during message processing. For instance, Falk, O'Donnell, and Lieberman (2012) examined how neural activity—while being exposed to descriptions of pilot TV shows—relates to subsequent sentiment in participants' evaluation of these episodes. By applying an automated sentiment analysis (SA) on the transcribed evaluations, Falk and colleagues (2012) found that more positive sentiment in these evaluations was associated with increased activity in areas linked to self-related processing and mentalizing (e.g., the medial prefrontal cortex; MPFC), areas that have also been shown to be predictive of spreading information (Scholz et al., 2017).

Moving beyond brain-as-predictor approaches, ISC analyses might aim to establish which specific narrative content features reliably predict increased neural synchrony (i.e., engagement with a narrative) across participants. For example, by relying on novel, crowd-sourced Moral Foundation Dictionaries (Hopp et al., 2018) that evaluate the occurrence of moral adherences, violations, and conflicts among actors in textual narratives, researchers could examine how the degree of neural synchrony fluctuates when individuals with similar or different moral sensibilities are exposed to these narratives.

In addition to computational analyses of textual data, great gains have recently been made in automatically extracting features from audio-visual stimuli such as movies or video games. For instance, the Python package *pliers* (McNamara, De La Vega, & Yarkoni, 2017) automatically extracts various features from audio-visual stimuli such as the occurrence of specific people, their faces, or specific objects. However, computational feature extraction algorithms for audio-visual stimuli are still experimental at the time of this writing and should be amended by manual annotations informed by theory (e.g., Ryan & Lenos, 2012). For instance, to study suspense during film reception (Grall & Schmälzle, 2018), researchers could automatically extract cut frequency, point of view, brightness, and sound amplitudes variations, and amend this information with the manual coding of information incongruency (e.g., the audience knows what is going to happen to a character, but the character in the narrative does not), which is one of many content indicators of suspense. Likewise, open-source, experimental video games in media neuroscience studies such as *Asteroid Impact*[3] provide a high-resolution content analysis of all events that are happening in the game. In turn, these fine-grained recordings of content features can significantly increase the accuracy and statistical power of fMRI analyses by increasing the frequency of content events and accurately recording their temporal alignment with the induced neural responses.

Hyperscanning

Hyperscanning entails *simultaneous* fMRI scanning of multiple participants who typically engage in some form of interaction or communication with each other. To enable simultaneous imaging, fMRI scanners are linked across geographical locations via high-speed data connections, although some special, latest generation, MRI scanners even allow imaging of two participants within the same scanner. Notwithstanding the many challenges that come along with the set-up and execution of a hyperscanning study, studying how brains encode human interaction in real time is intriguing and offers exciting directions for communication scholarship.

For instance, hyperscanning allows communication researchers to study how different forms (e.g., interpersonal, intercultural, group) and modalities (e.g., text-only, auditory-only, audio-visual) of communication may lead to different patterns of neural activation, in turn explaining variations in effective communication and cooperation (Schmälzle et al., 2015), interpersonal attraction and self-disclosure (Anders, de Jong, Beck, Haynes, & Ethofer, 2016), and social exclusion (Eisenberger, Lieberman, & Williams, 2003). In addition, hyperscanning might also contribute to our understanding of social anxiety disorders by exposing the neural correlates that precede and potentially undermine healthy and successful social interactions.

Real-time fMRI (Neurofeedback)

Real-time fMRI (rtfMRI; Cox, Jesmanowicz, & Hyde, 1995), commonly referred to as neurofeedback, captures "any process that uses functional information from an MRI scanner where the analysis and display of the fMRI keep pace with data acquisition" (Sulzer et al., 2013, p. 2). Traditional rtfMRI paradigms have aimed at inducing self-regulation of brain activity by presenting participants undergoing fMRI with a close to real-time measurement of their current brain activity, presenting this usually in visual form as a "thermometer display" or a color-coded image of participants' own brains. In these neurofeedback studies, participants are instructed to control (i.e., modify) the visualized brain activity, either with some motor movement such as finger tapping or by trying to change the cognitive states they are in. The visualized brain activity patterns may resemble the extracted BOLD time series of a specific ROI but can also be extended to reflect functional connectivity pattern in a participant's brain (i.e., correlated neural activity across multiple ROIs). Thus far, neurofeedback studies have focused mostly on training participants to "learn" how to control unwanted behavioral and neurological states arising from chronic pain, autism, or attention deficits. Applied to communication research, neurofeedback paradigms could use media stimuli for the treatment of disorders. For instance, the controls of a desirable video game could be linked to participants' neural activity pattern in brain regions known to regulate sustained attention (see Chapter 12 by Weber & Fisher, in this volume). The better participants learn to control their brain activity in these regions by sustaining attention, the more responsive will the video game be, leading to more successful game play.

Moving beyond self-regulation, researchers have started to modify the presented sensory stimulation in an attempt to exogenously manipulate real-time neural activity. Such "closed loop" paradigms are largely unexplored within rtfMRI but could open up interesting avenues within communication research. For example, researchers interested in media-induced flow experiences (Huskey et al., 2018) may aim to develop experimental stimuli that adapt in real time to maintain players' neural activation that is pleasurable and yields optimal performance. By understanding which content features and behaviors precede nonoptimal cognitive states, communication researchers could inform content designers to create interventions in real time that "push" players back into a more rewarding, optimally challenging state.

Moreover, future research on brain-computer interfaces (BCIs) might lead to applications relevant to the field of communication neuroscience. Although BCIs have mostly been studied with regard to controlling robots and prosthetic devices, considering their application to manipulate media content for furthering knowledge on human-computer interaction (HCI) is an intriguing perspective for media neuroscientists. The creation of novel BCIs in gaming and chat environments that can be entirely controlled via neural activity can provide unprecedented insights into how brain activity gives rise to communication outcomes.

Finally, the combination of neurofeedback with hyperscanning might reveal previously unconceived research directions. fMRI paradigms in which participants receive real-time feedback of other participants' brain states during social interactions or in response to interactive stimuli manipulations may provide interesting pathways to examine game-theoretical scenarios, deception and persuasion, or group coordination.

NOTES

1. In 2013, there were an average of 14 MRI machines per million people in the 35 developed countries that comprise the Organization for Economic Cooperation and Development (OECD: Health at a Glance, 2015, p. 27).
2. Since 2017, the computational methods division has been the fastest-growing division of the International Communication Association (httpa://icahdq.org).
3. https://github.com/medianeuroscience/asteroid_impact.

REFERENCES

Afifi, T., & Floyd, K. (2014). Biology and physiology in communication [Special Issue]. *Communication Monographs, 82*(1).

Amir, O., Huskey, R., Mangus, J. M., Swanson, R., Gordon, A., Khooshabeh, P., & Weber, R. (2017, March). Trump vs. Clinton: The role of moral intuition networks in processing political attack advertisements. Paper presented at the annual meeting of the Social and Affective Neuroscience Society, Los Angeles, CA.

Anders, S., de Jong, R., Beck, C., Haynes, J. D., & Ethofer, T. (2016). A neural link between affective understanding and interpersonal attraction. *Proceedings of the National Academy of Sciences, 113*(16), E2248–E2257.

Anderson, D. R., Bryant, J., Murray, J. P., Rich, M., Rivkin, M. J., & Zillmann, D. (2006). Brain imaging: An introduction to a new approach to studying media processes and effects. *Media Psychology, 8*(1), 1–6.

Anderson, D. R., Fite, K. V., Petrovich, N., & Hirsch, J. (2006). Cortical activation while watching video montage: An fMRI study. *Media Psychology, 8*, 7–24.

Ashby, F. G. (2011). *Statistical analysis of fMRI data*. Boston, MA: MIT Press.

Bartels, A., & Zeki, S. (2004). Functional brain mapping during free viewing of natural scenes. *Human Brain Mapping, 21*, 75–85.

Berkman, E. T., & Falk, E. B. (2013). Beyond brain mapping: Using neural measures to predict real-world outcomes. *Current Directions in Psychological Science, 22*(1), 45–50.

Bullmore, E., & Sporns, O. (2009). Complex brain networks: Graph theoretical analysis of structural and functional systems. *Nature Reviews Neuroscience, 10*(3), 186–198.

Chen, J., Leong, Y. C., Honey, C. J., Yong, C. H., Norman, K. A., & Hasson, U. (2017). Shared memories reveal shared structure in neural activity across individuals. *Nature Neuroscience, 20*(1), 115–125.

Cohen, S. S., Henin, S., & Parra, L. C. (2017). Engaging narratives evoke similar neural activity and lead to similar time perception. *Scientific Reports, 7*, 4578.

Coltheart, M. (2013). How can functional neuroimaging inform cognitive theories? *Perspectives on Psychological Science, 8*(1), 98–103.

Cox, R. W., Jesmanowicz, A., & Hyde, J. S. (1995). Real-time functional magnetic resonance imaging. *Magnetic Resonance in Medicine, 33*(2), 230–236.

Cui, Q., Vanman, E. J., Wei, D., Yang, W., Jia, L., & Zhang, Q. (2013). Detection of deception based on fMRI activation patterns underlying the production of a deceptive response and receiving feedback about the success of the deception after a mock murder crime. *Social Cognitive and Affective Neuroscience, 9*(10), 1472–1480.

Davison, E. N., Turner, B. O., Schlesinger, K. J., Miller, M. B., Grafton, S. T., Bassett, D. S., & Carlson, J. M. (2016). Individual differences in dynamic functional brain connectivity across the human lifespan. *PLoS Computational Biology, 12*(11), e1005178.

Edelman, R. R., & Warach, S. (1993). Magnetic resonance imaging. *New England Journal of Medicine, 328,* 785–791.

Eisenberger, N. I., Lieberman, M. D., & Williams, K. D. (2003). Does rejection hurt? An fMRI study of social exclusion. *Science, 302*(5643), 290–292.

Falk, E. B., Berkman, E. T., & Lieberman, M. D. (2012). From neural responses to population behavior: Neural focus group predicts population-level media effects. *Psychological Science, 23*(5), 439–445.

Falk, E. B., Berkman, E. T., Mann, T., Harrison, B., & Lieberman, M. D. (2010). Predicting persuasion-induced behavior change from the brain. *Journal of Neuroscience, 30*(25), 8421–8424.

Falk, E. B., Berkman, E. T., Whalen, D., & Lieberman, M. D. (2011). Neural activity during health messaging predicts reductions in smoking above and beyond self-report. *Health Psychology, 30*(2), 177–185.

Falk, E. B., Cascio, C. N., & Coronel, J. (2015). Neural prediction of communication-relevant outcomes. *Communication Methods and Measures*, 9(1–2), 30–54.

Falk, E., O'Donnell, M. B., & Lieberman, M. D. (2012). Getting the word out: Neural correlates of enthusiastic message propagation. *Frontiers in Human Neuroscience, 6*, 313.

Falk, E. B., O'Donnell, M. B., Tompson, S., Gonzalez, R., Dal Cin, S., Strecher, V., ... An, L. (2016). Functional brain imaging predicts public health campaign success. *Social Cognitive and Affective Neuroscience, 11*(2), 204–214.

Grall, C., & Schmälzle, R. (2018, May). From pixels to media effects: A study of 600 brains watching a suspenseful movie. Paper presented at the annual meeting of the International Communication Association, Prague, the Czech Republic.

Greene, J. D., Sommerville, R. B., Nystrom, L. E., Darley, J. M., & Cohen, J. D. (2001). An fMRI investigation of emotional engagement in moral judgment. *Science, 293*(5537), 2105–2108.

Greenwald, A. G. (2012). There is nothing so theoretical as a good method. *Perspectives on Psychological Science, 7*(2), 99–108.

Grimmer, J., & Stewart, B. M. (2013). Text as data: The promise and pitfalls of automatic content analysis methods for political texts. *Political Analysis, 21*(3), 267–297.

Hasson, U., & Frith, C. D. (2016). Mirroring and beyond: Coupled dynamics as a generalized framework for modelling social interactions. *Philosophical Transactions of the Royal Society B: Biological Sciences, 371*(1693), 20150366.

Hasson, U., Nir, Y., Levy, I., Fuhrmann, G., & Malach, R. (2004). Intersubject synchronization of cortical activity during natural vision. *Science, 303*(5664), 1634–1640.

Haxby, J. V., Gobbini, M. I., Furey, M. L., Ishai, A., Schouten, J. L., & Pietrini, P. (2001). Distributed and overlapping representations of faces and objects in ventral temporal cortex. *Science, 293*(5539), 2425–2430.

Hopp, F. R., Mangus, J. M., Swanson, R., Gordon, A., Khooshabeh, P., & Weber, R. (2018, May). Developing and validating the moral foundations dictionary for news narratives: A crowd-sourced approach. Paper presented at the annual meeting of the International Communication Association, Prague, the Czech Republic.

Hopp, F. R., Schaffer, J., Fisher, J., & Weber, R. (in press). iCoRe: The GDELT interface for the advancement of communication research. *Computational Communication Research.*

Huskey, R., Craighead, B., Miller, M. B., & Weber, R. (2018). Does intrinsic reward motivate cognitive control? A naturalistic-fMRI study based on the synchronization theory of flow. *Cognitive, Affective, & Behavioral Neuroscience, 18*(5), 902–924.

Langleben, D. D. (2008). Detection of deception with fMRI: Are we there yet? *Legal and Criminological Psychology, 13*(1), 1–9.

Levine, T. R., & McCornack, S. A. (2014). Theorizing about deception. *Journal of Language and Social Psychology, 33,* 431–440.

Lin, J. (2015). On building better mousetraps and understanding the human condition: Reflections on big data in the social sciences. *Annals of the American Academy of Political and Social Science, 659*(1), 33–47.

Logothetis, N. K. (2008). What we can do and what we cannot do with fMRI. *Nature, 453*(7197), 869–878.

Logothetis, N. K., Pauls, J., Augath, M., Trinath, T., & Oeltermann, A. (2001). Neurophysiological investigation of the basis of the fMRI signal. *Nature, 412*(6843), 150–157.

Mathiak, K., & Weber, R. (2006). Toward brain correlates of natural behavior: fMRI during violent video games. *Human Brain Mapping*, 27(12), 948–956.

McNamara, Q., De La Vega, A., & Yarkoni, T. (2017, August). Developing a comprehensive framework for multimodal feature extraction. In *Proceedings of the 23rd ACM SIGKDD International Conference on Knowledge Discovery and Data Mining*, Halifax, Canada.

Mitchell, T. M., Hutchinson, R., Niculescu, R. S., Pereira, F., Wang, X., Just, M., & Newman, S. (2004). Learning to decode cognitive states from brain images. *Machine Learning, 57*(1–2), 145–175.

Mole, C., & Klein, C. (2010). Confirmation, refutation and the evidence of fMRI. In S. J. Hanson & M. Bunzl (Eds.), *Foundational issues of human brain mapping* (pp. 99–112). Cambridge, MA: MIT Press.

Norman, K. A., Polyn, S. M., Detre, G. J., & Haxby, J. V. (2006). Beyond mind-reading: Multi-voxel pattern analysis of fMRI data. *Trends in Cognitive Sciences, 10*(9), 424–430.

O'Donnell, M. B., & Falk, E. B. (2015a). Big data under the microscope and brains in social context: Integrating methods from computational social science and neuroscience. *Annals of the American Academy of Political and Social Science, 659*(1), 274–289.

O'Donnell, M. B., & Falk, E. B. (2015b). Linking neuroimaging with functional linguistic analysis to understand processes of successful communication. *Communication Methods and Measures, 9*(1–2), 55–77.

Ogawa, S., Tank, D. W., Menon, R., Ellermann, J. M., Kim, S. G., Merkle, H., & Ugurbil, K. (1992). Intrinsic signal changes accompanying sensory stimulation: Functional brain mapping with magnetic resonance imaging. *Proceedings of the National Academy of Sciences, 89*(13), 5951–5955.

O'Keefe, D. J. (2002). *Persuasion: Theory and research.* Thousand Oaks, CA: Sage.

Pennebaker, J. W., Francis, M. E., & Booth, R. J. (2001). *Linguistic inquiry and word count: LIWC 2001.* Mahwah, NJ: Lawrence Erlbaum Associates.

Poldrack, R. A. (2006). Can cognitive processes be inferred from neuroimaging data? *Trends in Cognitive Sciences, 10*(2), 59–63.

Ryan, M., & Lenos, M. (2012). *An introduction to film analysis: Technique and meaning in narrative film.* New York, NY: Continuum.

Schmälzle, R., Häcker, F., Honey, C. J., & Hasson, U. (2015). Engaged listeners: Shared neural processing of powerful political speeches. *Social Cognitive and Affective Neuroscience, 10*(8), 1137–1143.

Schmälzle, R., Häcker, F., Renner, B., Honey, C. J., & Schupp, H. T. (2013). Neural correlates of risk perception during real-life risk communication. *Journal of Neuroscience, 33*(25), 10340–10347.

Schmälzle, R., Imhof, M. A., Grall, C., Flaisch, T., & Schupp, H. T. (2017). Reliability of fMRI time series: Similarity of neural processing during movie viewing. *Institutional Repository of the University of Konstanz*. doi: 10.1101/158188.

Scholz, C., Baek, E. C., O'Donnell, M. B., Kim, H. S., Cappella, J. N., & Falk, E. B. (2017). A neural model of valuation and information virality. *Proceedings of the National Academy of Sciences, 114*(11), 2881–2886.

Silbert, L. J., Honey, C. J., Simony, E., Poeppel, D., & Hasson, U. (2014). Coupled neural systems underlie the production and comprehension of naturalistic narrative speech. *Proceedings of the National Academy of Sciences, 111*(43), E4687–E4696.

Silver, D., Huang, A., Maddison, C. J., Guez, A., Sifre, L., Van Den Driessche, G., … Dieleman, S. (2016). Mastering the game of Go with deep neural networks and tree search. *Nature, 529*(7587), 484–489.

Sporns, O. (2010). *Networks of the brain*. Boston, MA: MIT Press.

Stephens, G. J., Silbert, L. J., & Hasson, U. (2010). Speaker-listener neural coupling underlies successful communication. *Proceedings of the National Academy of Sciences, 107*(32), 14425–14430.

Sulzer, J., Haller, S., Scharnowski, F., Weiskopf, N., Birbaumer, N., Blefari, M. L., … Herwig, U. (2013). Real-time fMRI neurofeedback: Progress and challenges. *NeuroImage, 76*, 386–399.

Suykens, J. A., & Vandewalle, J. (1999). Least squares support vector machine classifiers. *Neural Processing Letters, 9*, 293–300.

Turner, B., Huskey, R., & Weber, R. (2018). Charting a future for fMRI in communication science. *Communication Methods and Measures*, 13(1), 1–18.

Weber, R. (2008). *Connectivity of brain regions during social interactions: Theory-based, event-related content analysis of continuous, semi-natural stimuli as paradigm in functional magnetic resonance imaging*. Aachen, Germany: RWTH Library.

Weber, R. (2015). Biology and brains—methodological innovations in communication science: Introduction to the special issue. *Communication Methods and Measures, 9*(1–2), 1–4.

Weber, R., Eden, A., & Mathiak, K. (2011, June). Seeing bad people punished makes us think alike: Social norm violations in television drama elicit cortical synchronization in viewers. Paper presented at the annual meeting of the International Communication Association, Boston, MA.

Weber, R., Fisher, J. T., Hopp, F. R., & Lonergan, C. (2018). Taking messages into the magnet: Method-theory synergy in communication neuroscience. *Communication Monographs, 85*(1), 81–102.

Weber, R., Huskey, R., Mangus, J. M., Westcott-Baker, A., & Turner, B. O. (2015). Neural predictors of message effectiveness during counterarguing in anti-drug campaigns. *Communication Monographs, 82*(1), 4–30.

Weber, R., Mangus, J. M., & Huskey, R. (2015). Brain imaging in communication research: A practical guide to understanding and evaluating fMRI studies. *Communication Methods and Measures, 9*(1–2), 5–29.

Weber, R., Ritterfeld, U., & Mathiak, K. (2006). Does playing violent video games induce aggression? Empirical evidence of a functional magnetic resonance imaging study. *Media Psychology, 8*(1), 39–60.

Weber, R., Sherry, J., & Mathiak, K. (2008). The neurophysiological perspective in mass communication research. Theoretical rationale, methods, and applications. In M. J. Beatty, J. C. McCroskey, & K. Floyd (Eds.), *Biological dimensions of communication: Perspectives, methods, and research* (pp. 41–71). Cresskill, NJ: Hampton Press.

Williams, M. A., Dang, S., & Kanwisher, N. G. (2007). Only some spatial patterns of fMRI response are read out in task performance. *Nature Neuroscience, 10*(6), 685–686.

Yang, Z., Huang, Z., Gonzalez-Castillo, J., Dai, R., Northoff, G., & Bandettini, P. (2014). Using fMRI to decode true thoughts independent of intention to conceal. *NeuroImage, 99*, 80–92.

Yeshurun, Y., Swanson, S., Simony, E., Chen, J., Lazaridi, C., Honey, C. J., & Hasson, U. (2017). Same story, different story: The neural representation of interpretive frameworks. *Psychological Science, 57*, 307–319.

Part IV

INTERPERSONAL COMMUNICATION

21

Emotion and Emotional Communication

Colin Hesse and Alan C. Mikkelson

The field of communication has long maintained that emotions are central to any communication encounter, influencing virtually every facet of the interaction process. Several communication theories have placed both the experience and communication of emotions in key explanatory roles, including Affection Exchange Theory (AET: Floyd, 2006) and the Theory of Motivated Information Management (TMIM: Afifi & Weiner, 2004). Overall, emotional encoding and decoding skills are considered essential to communication competence in most any context (Spitzberg & Cupach, 2011).

However, these theories presuppose a general capability to engage in competent emotional communication, including facility at encoding one's own emotions, decoding the emotions of others, and even communicating *about* emotion. Differences in those abilities would, in turn, imply differences in the individual's communication competence, with possible implications for an individual's biopsychosocial wellbeing. This chapter explores three traits that inhibit emotional communication abilities: (1) alexithymia; (2) social anhedonia; and (3) autism. In each section, we will define the trait, discuss its psychological and relational outcomes and correlates, and then examine its physiological outcomes and correlates. The chapter concludes with a brief discussion of some implications and directions for future research.

ALEXITHYMIA

The term *alexithymia*, which translates to "a lack of words for emotions," was coined by Sifneos (1973) to help explain psychiatric patients who would appear distant and cold while recounting emotional events. Scholars later delineated four components of alexithymia: (1) difficulty identifying feelings and/or emotional arousal in the body; (2) difficulty describing feelings to others; (3) reduced ability to engage in fantasy; and (4) an externally oriented cognitive style (e.g., Taylor, Bagby, & Parker, 1997). In general, alexithymia represents an emotional processing deficit impairing the ability to translate emotional schema, including problems with understanding both nonverbal and verbal patterns of emotional arousal and meaning (Taylor et al., 1997).

Scholars currently believe that alexithymia emerges from gene x environment interactions. In two studies of monozygotic twins, genetic factors accounted for roughly 33% of the variability on alexithymic scores (Jørgensen et al., 2007; Picardi et al., 2011). Other studies have found correlations between alexithymia and adult insecure attachment styles (e.g., Lyvers, Edwards, & Thorberg, 2017) and being an unwanted child with poor family relationships (Joukamaa et al., 2003). These factors combine to create a stable trait whose prevalence in the population ranges from 10% (Joukamaa et al., 2007) to 25% (Hamaideh, 2017).

Psychological and Relational Outcomes

In recent decades, scholars have examined correlations between alexithymia and a host of outcomes dealing with both psychological and relational wellness. Alexithymia is comorbid with multiple psychological problems, several times more prevalent than in non-clinical samples (McGillivray, Becerra, & Harms, 2017). These include diagnosable mental illnesses such as depression (e.g., Hesse & Floyd, 2008; Yürümez, Akça, Uğur, Uslu, & Kılıç, 2014), anxiety disorders (Leweke, Leichsenring, Kruse, & Hermes, 2012), and borderline personality disorder (New et al., 2012), as well as conditions such as stress, loneliness, and even mobile phone addiction (e.g., Gao et al., 2018; Hesse & Floyd, 2008).

Alexithymia also is related to problems communicating with others and forming relationships. Specifically, alexithymic individuals struggle to find the proper words for their emotions (Suslow & Junghanns, 2002) and to experience dyadic coping (Gabriel, Untas, Lavner, Koleck, & Luminet, 2016). Relationally, alexithymia is inversely correlated with multiple relational outcomes, including closeness, satisfaction, quality, and the number of perceived close relationships (e.g., Hesse & Floyd, 2011; Hesse, Pauley, & Frye-Cox, 2015). In general, alexithymia appears to relate directly to relational and mental health problems.

Physiological Outcomes

Multiple studies also link alexithymia with a host of physiological outcomes. At the genetic level, alexithymia correlates with variants linked to immune function (TMEM88B), schizophrenia (ARHGAP32), and the production of dopamine (DRD2/ANKK1 and BDNF) (Mezzavilla et al., 2015; Walter, Montag, Markett, & Reuter, 2011). Another study found that variation in *OXTR* polymorphism rs53576 did not relate directly to alexithymia but interacted with childhood attachment security to affect alexithymia. Specifically, GG-homozygotes reported lower alexithymia when they had a secure versus insecure childhood attachment style (Schneider-Hassloff et al., 2016). As detailed below, other research has linked alexithymia with nervous system, hormonal, and neurological outcomes.

Autonomic Nervous System

Researchers have articulated two primary predictions regarding alexithymia and the autonomic nervous system (ANS). The hypoarousal prediction is that alexithymic individuals experience lower cardiovascular activation in response to a stressor due to their emotional processing deficits. For example, Linden, Lenz, and Stossel (1996) found that alexithymic individuals had a lower heart rate (HR) response to a discussion of an anger-provoking event than did non-alexithymic individuals. Other studies reported a similarly dampened HR response for alexithymic individuals while viewing disgusting film scenes (Friedlander, Lumley, Farchione, & Doyal, 1997) and giving a personal speech (Newton & Contrada, 1994). The hypoarousal prediction has also been supported with studies measuring skin conductance (e.g., Pollatos et al., 2011) and blood pressure (BP: Neumann, Sollers, Thayer, & Waldstein, 2004).

The second prediction is of a direct relationship between alexithymia and baseline cardiovascular indices. Scholars have found positive correlations between alexithymia and 24-hour systolic BP (Berra et al., 2017), BP taken over a three-week period (Jula, Salminen, & Saarijärvi, 1999), essential and secondary hypertension (e.g., Grabe et al., 2010), and several metabolic syndrome components, including waist circumference, triglycerides, and hypertensive BP (Karukivi, Jula, Hutri-Kähönen, Juonala, & Raitakari, 2016). Overall, alexithymia appears to be related both to the lack of physiological arousal to emotional stimuli as well as heightened baseline levels of cardiovascular function.

Endocrine System

As with the ANS, researchers have examined two principal predictions regarding alexithymia and the endocrine system. This includes the relationship of alexithymia and the activation of the stress response—particularly the hypothalamic-pituitary-adrenal (HPA) axis—and with baseline hormone levels. Research on the first prediction has been inconclusive, with one study finding increased cortisol for alexithymic individuals following a consult with a physician regarding fibromyalgia and another finding an inverse correlation with cortisol following a stressful situation for men at a fertility clinic (Conrad et al., 2002; Finset, Graugaard, & Holgersen, 2006).

More work on the second prediction has established associations between alexithymia and baseline levels of several hormones. This includes inverse correlations between alexithymia and baseline prolactin, oxytocin, cortisol, and ACTH, hormones that are implicated in relational bonding and the stress response (Conrad et al., 2002; Henry et al., 1992; Schmelkin et al., 2017). Another study found that high-alexithymic individuals experienced a lower cortisol awakening response (the change in cortisol levels the first hour after waking) than did low-alexithymic individuals (Härtwig, Aust, & Heuser, 2013). Finally, alexithymia is directly correlated for men with the norepinephrine:cortisol ratio (i.e., higher norepinephrine and lower cortisol), a marker that is implicated in the literature on post-traumatic stress disorder (Spitzer, Brandl, Rose, Nauck, & Freyberger, 2005). Taken as a whole, alexithymic individuals seem to possess fewer baseline physiological resources to deal with stressors.

Neurological Outcomes

Several lines of research connect neurological processes to alexithymia; the most important lies in socio-affective processing, meaning the ability to process emotional stimuli and reactions (Kano & Fukudo, 2013). This is examined largely through several particular ROIs, including the fusiform gyrus, the amygdala, and the anterior cingulate cortex (ACC), although other scholars look at general right-hemisphere activation. For example, one study found that alexithymic individuals had attenuated brain responses to facial emotion (sad, happy) in right frontal and right caudate nuclei, supporting the premise of a general deficit in the right hemisphere (Suslow et al., 2016). Another study also found lower right-hemisphere efficiency in response to a non-emotion-inducing task (Gavazzi et al., 2017).

This hypothesis regarding a socio-affective processing deficit for alexithymic individuals has continued to be addressed in numerous studies. Berthoz et al. (2002) found decreased activation for alexithymic individuals compared to nonalexithymic individuals in the left mediofrontal-paracingulate cortex in response to highly negative stimuli. Three subsequent studies looked at participant activation while viewing emotional faces. Comparison findings of alexithymic versus non-alexithymic individuals included lower blood flow in the right hemisphere (Kano et al., 2003), lower activation in the insula and amygdala (Reker et al., 2010), and lower activation in the fusiform gyrus, the amygdala, the hippocampus, and the insular cortex (Hesse et al., 2013). This weakness in socio-affective processing extends to the ability to accurately decode emotions. In one study, participants viewed a face and then selected the emotion they believed was portrayed. Alexithymic individuals experienced both decreased neurological activity in several ROIs (such as the amygdala and the ACC) and lower accuracy at decoding the viewed emotions (Ihme et al., 2014). Finally, van der Velde et al. (2013) concluded in a meta-analysis that alexithymic individuals have lower levels of activation in several areas of the brain than do non-alexithymic individuals, including the fusiform gyrus, the amygdala, the insula, and the prefrontal cortex. Compared to those with lower levels of alexithymia, highly alexithymic individuals appear to have a limited ability to process and understand emotional stimuli, largely supporting the socio-affective processing deficit hypothesis.

SOCIAL ANHEDONIA

Social anhedonia is typically understood as the reduced experience of positive emotions (Horan, Green, Kring, & Nuechterlein, 2006) or a diminished ability to experience pleasure (Silvia & Kwapil, 2011) in response to rewarding social stimuli. Common indicators of social anhedonia include social withdrawal and indifference to others (Blanchard, Gangestad, Brown, & Horan, 2000), decreased positive affect (Fung, Moore, Karcher, Kerns, & Martin, 2017), and greater perceived stress (Horan, Brown, & Blanchard, 2007). Social anhedonia is distinct from both introversion and social anxiety, representing genuine social disinterest and a reduced need to belong (Martin, Cicero, Bailey, Karcher, & Kerns, 2016; Silvia & Kwapil, 2011).

Psychological and Relational Outcomes

Social anhedonia is related to several psychological traits and outcomes. Cross-sectional studies have demonstrated negative associations with extraversion, openness to experience, and agreeableness (Kwapil et al., 2009). Social anhedonia is also negatively related to the extraversion dimensions of gregariousness, warmth, positive emotions, and excitement seeking (Kwapil, Barrantes-Vidal, & Silvia, 2008), and predicts poor evaluative processing (Martin & Kerns, 2010) and poor affective control (Martin, Cicero, & Kerns, 2012). Unsurprisingly, social anhedonia is also inversely related with pleasure in social situations (Lucas, Diener, Grob, Suh, & Shao, 2000). Social anhedonia often results in a lack of close relationships with others, including decreased rates of dating and marriage (Kwapil, 1998).

Finally, social anhedonia results in emotional communication impairment and generally poor psychosocial functioning. Social impairment is likely due to decreased attention to emotions, in general, and positive emotions, in particular (e.g., Martin, Becker, Cicero, Docherty, & Kerns, 2011). Social anhedonia can also influence communicative expressions. Collins, Blanchard, and Biondo (2005) reported that individuals high in social anhedonia exhibited constricted facial emotion displays, odd speech patterns, and reduced verbal and nonverbal expressions in comparison to control participants.

Physiological Outcomes

Physiologically, social anhedonia is connected with cardiovascular activity and is related to reduced activity in parts of the brain that experience pleasure. In this section, we will detail research on anhedonia and its relationship to the autonomic nervous system and neurological outcomes.

Autonomic Nervous System

Those with social anhedonia often experience reward insensitivity, or an inability to fully experience rewards, that may blunt or reduce cardiovascular reactivity to rewarding stimuli. Indeed, researchers have found that individuals with anhedonia demonstrated lower systolic BP reactivity to rewards than did control participants (Brinkmann, Schüpbach, Joye, & Gendolla, 2009). Similarly, reactivity of the cardiac pre-ejection period (a measure of sympathetic activity) and HR were also reduced for individuals with anhedonia as compared to control participants (Franzen & Brinkmann, 2015; Silvia, Nusbaum, Eddington, Beaty, & Kwapil, 2014). Further, in response to positive and neutral stimuli, individuals with anhedonia failed to demonstrate the typical association between HR and the emotional content of the stimuli (Fitzgibbons & Simons, 1992).

Anhedonia is the depression symptom with the strongest relationship to cardiac events and mortality after myocardial infarction (Davidson et al., 2010; Leroy, Loas, & Perez-Diaz, 2010). Specifically, Davidson et al. (2010) found that anhedonia was a significant predictor of all-cause mortality and major adverse cardiac events in patients with acute coronary syndrome (ACS). In a three-year study of ACS, only anhedonia was a significant predictor of severe cardiac events (Leroy et al., 2010).

Neurological Outcomes

Anhedonia is related to reduced activity in the ventral striatum and orbitofrontal cortex, both of which contribute to the experience of pleasure (Harvey, Armony, Malla, & Lepage, 2010; Keedwell, Andrew, Williams, Brammer, & Phillips, 2005). In addition, problems in parts of the prefrontal cortex (PFC) have also been associated with anhedonia (for review, see Der-Avakian & Markou, 2012). Specifically, both ventromedial activity and dorsolateral PFC activity were negatively related to anhedonia in schizophrenia patients (Park et al., 2009). Trait anhedonia was also negatively correlated with the activation of the anterior cingulate cortex (Wacker, Dillon, & Pizzagalli, 2009) and the dorsolateral PFC (Park et al., 2009) in a healthy population.

AUTISM SPECTRUM DISORDER

Autism spectrum disorder (ASD) constitutes a collection of neurodevelopmental abnormalities that typically appear early in childhood and are characterized by impairments in social communication and social interaction across numerous contexts, accompanied by restricted, repetitive behaviors and interests (American Psychiatric Association, 2013). The characteristics of ASD can vary widely and the symptoms of ASD can be mild to severe, making the study of ASD challenging (see Levy & Perry, 2011). According to a Centers for Disease Control and Prevention report (2017), ASD has been identified in approximately 1 in 68 children, with the male/female ratio of ASD being approximately 4/1. Not well understood, this sex difference may be due to genetic and hormonal differences that respond to various environmental factors (see Schaafsma & Pfaff, 2014).

According to the American Psychiatric Association (2013), ASD-related deficits in social communication include difficulties with social-emotional reciprocity, which can include problems establishing a back-and-forth conversation, sharing emotions, and failures to initiate or respond during social situations. Deficits in nonverbal communication are also observed, including atypical eye contact and body language, difficulty understanding nonverbal expressions, lack of facial expressions, and poorly synchronized verbal and nonverbal communication. Finally, deficits in developing and maintaining relationships include problems adjusting social behavior to various contexts, a lack of interest in relationships, and challenges forming and sustaining friendships.

Psychological and Relational Outcomes

Rates of psychiatric comorbidity for individuals with ASD are approximately 25–30% (Underwood, McCarthy, & Tsakanikos, 2010). Two common comorbidities are major depressive disorders and anxiety disorders (Lainhart, 1999; Volkmar & Klin, 2005). Rates of depression for youth with ASD are as high as 54%, whereas the rate of depression for typically developing (TD) youths is around 4–5% (Mayes, Calhoun, Murray, & Zahid, 2011; Solomon, Miller, Taylor, Hinshaw, & Carter, 2012). Not surprisingly, youths with ASD demonstrate higher levels of loneliness in comparison to their TD peers (Bauminger & Kasari, 2000). In adults and adolescents with ASD, only 8% interacted with a same-age peer outside an organized activity on a weekly basis and 46% reported having no same-age friends (Orsmond, Krauss, & Seltzer, 2004).

Researchers have found a higher than normal rate of anxiety in individuals with ASD (Tantam, 2000) and anxiety levels that are significantly higher than the general population (Bellini, 2004). Social skill deficiencies are a central characteristic in ASD and are one likely explanation for the development of anxiety. Individuals with ASD may be socially awkward and experience difficulties in nonverbal communication, which can interfere with the development of meaningful relationships (Tantam, 2000). Further, these social skill deficits increase the likelihood of negative peer interactions and peer rejection, both of which are related to social anxiety (La Greca & Lopez, 1998). Due to the presence of social anxiety, many individuals with ASD live without the support of close friends (Koning & Magill-Evans, 2001).

Physiological Outcomes

ASD is thought to be a result of abnormalities in the reward circuitry in the brain (Dawson, Webb, & McPartland, 2005) and under-connectivity between brain regions (Just, Keller, Malave, Kana, & Varma, 2012). Dysfunctions with the neurotransmitter dopamine have also been connected to ASD (Zeeland et al., 2010). In this section, we will detail research on ASD and its ANS, hormonal, and neurological outcomes.

Autonomic Nervous System

The literature connecting ASD to both cardiovascular and hormone levels operates under the theory that ASD can expose an individual to higher degrees of baseline stress (Baron, Groden, Groden, & Lipsitt, 2006). Individuals with ASD experience a number of autonomic dysfunctions linked to this claim (Ming, Patel, Kang, Chokroverty, & Julu, 2016). Specifically, higher resting HR, lower resting parasympathetic nervous system activity, and higher sympathetic nervous system activity exist in children with ASD compared to TD children (Ming, Julu, Brimacombe, Connor, & Daniels, 2005). Respiratory dysrhythmia has also been detected in children with ASD, along with greater variability in respiratory rhythm as compared to TD children (Ming et al., 2016).

The relationship between ASD and stress extends to research examining stress reactivity. This research finds a heightened response to stressors for individuals with ASD compared to TD, such as systolic BP reactivity (Bishop-Fitzpatrick, Minshew, Mazefsky, & Eack, 2017). Further, Croen et al. (2015) found that individuals with ASD were at higher risk of a range of cardiovascular diseases than were matched controls.

Endocrine System and Neurological Outcomes

The theories linking ASD to heightened stress and reduced social motivation extend to research measuring cortisol and key neurological mechanisms. Specifically, children diagnosed with ASD often exhibiting higher baseline cortisol levels than do TD children (Ogawa, Lee, Yamaguchi, Shibata, & Goto, 2017).

A key behavioral feature of individuals with ASD is reduced social motivation, which may be attributed to dysfunction in the mesocorticolimbic reward circuitry (Dawson et al., 2005). Neuroimaging studies of ASD individuals demonstrate abnormalities in the frontal cortex, cerebellum, hippocampus, and the amygdaloid nucleus (Santangelo & Tsatsanis, 2005). Zeeland et al. (2010) found that during an implicit learning task, autistic children had reduced nucleus accumbens activation in response to both monetary and social rewards. Further, ASD is linked with reward-processing deficits, specifically in the reduced activation in the amygdala and ventral anterior cingulate in response to social rewards (Kohls et al., 2013). Yet, the reward-processing deficits do not occur in response to all rewards. For example, in individuals with ASD, giving tangible items helped sustain a continuous performance task better than social attention and praise (Garretson et al., 1990). Finally, MRI studies have demonstrated lower connectivity between the frontal lobe and posterior cortical regions in ASD participants compared with TD participants (Just, Cherkassky, Keller, & Minshew, 2004).

CONCLUSION

Emotional competence, including skills involving encoding and decoding emotional messages, is key to several communication theories and concepts (e.g., Floyd, 2006). However, several psychological traits limit individual emotional competence, and those traits are linked to multiple biopsychosocial deficits. We conclude this chapter by identifying some conclusions of this research and offering calls for future study.

First, emotional competence is associated with stress, in terms of both physiological homeostasis and reactivity. With respect to homeostasis, the research on all three emotional communication deficits (alexithymia, social anhedonia, and autism spectrum disorders) supports the general claim that emotional incompetence is directly related with cardiovascular indices such as heart rate and blood pressure (e.g., Berra et al., 2017; Leroy et al., 2010; Ming et al., 2005). The relationship between emotional communication deficits and stress reactivity is more complex, with research pointing to both hypoarousal (e.g., Friedlander et al., 1997) and hyperarousal (e.g., Bishop-Fitzpatrick et al., 2017). This appears to be due to different interactions between the traits and the specific stimuli—specifically, the degree to which emotional decoding skills are necessary in order to experience the stress response. For example, individuals high in alexithymia or anhedonia experience hypoarousal to stimuli such as seeing disgusting images (Friedlander et al., 1997) or receiving an award (Brinkmann et al., 2009). In all cases, the stimuli required the participants to encode the emotional experience (whether disgust, happiness, or anger) in order to respond to the stimuli. On the other hand, a trait such as ASD can lead to hyperarousal, wherein the participant feels ill-prepared to handle stimuli focused on interpersonal interactions (e.g., Bishop-Fitzpatrick et al., 2017). This conclusion has mixed support, as highly alexithymic individuals still experience a blunted stress response in interpersonal situations such as the experience of social rejection through the game of Cyberball (Chester, Pond Jr., & DeWall, 2014). More research is needed to assess how, why, and when these traits are associated with either a blunted or exaggerated stress response.

Second, a burgeoning literature adds evidence that emotional competence is linked to decreased levels of neurological activity due to social stimuli. For alexithymia, this highlights areas that involve emotional processing, such as the fusiform gyrus and ACC (e.g., Ihme et al., 2014). Both ASD and social anhedonia involve decreased activity in pleasure centers of the brain, such as the ventral striatum (e.g., Kohls et al., 2013). Considered as a whole, this research supports the empirical claim that emotional competence is, to some degree, rooted in physiological/evolutionary factors. Future research should continue to explore the neurological substrates and correlates of these emotional communication deficits.

Finally, communication scholars should begin to reassess how best to assist individuals with a neurodevelopmental disorder. Current communication theories, textbooks, and training techniques presuppose a level of competence that fails to address physiological impairments essential to the communication encounter. Interventions tend to focus on changing individual behavior so they better fit into society (e.g., Levant, Halter, Hayden, & Williams, 2009). Recently, however, scholars have asked whether interventions should shift their focus to those without ASD or alexithymia, instead embracing the concept of "neurodiversity" (e.g., Baron-Cohen, 2017). In this worldview, people with ASD should not automatically be "normalized," but instead should be recognized as having a different type of brain and should be treated accordingly, and with respect (Baron-Cohen, 2017). Communication scholars, according to this call, should build theories and interventions designed to help neurotypical individuals to embrace and respect neurodiversity, including individuals with the emotional processing deficits discussed in the current chapter.

REFERENCES

Afifi, W. A., & Weiner, J. L. (2004). Toward a theory of motivated information management. *Communication Theory, 14*, 167–190.

American Psychiatric Association. (2013). *Diagnostic and statistical manual of mental disorders* (5th ed.). Washington, DC: Author.

Baron, M. G., Groden, J., Groden, G., & Lipsitt, L. P. (2006). *Stress and coping in autism*. New York, NY: Oxford University Press.

Baron-Cohen, S. (2017). Editorial perspective: Neurodiversity–a revolutionary concept for autism and psychiatry. *Journal of Child Psychology and Psychiatry, 58*, 744–747.

Bauminger, N., & Kasari, C. (2000). Loneliness and friendship in high-functioning children with autism. *Child Development, 71*, 447–456.

Bellini, S. (2004). Social skill deficits and anxiety in high-functioning adolescents with autism spectrum disorders. *Focus on Autism and Other Developmental Disabilities, 19*, 78–86.

Berra, E., Petit, G., George, C., Capron, A., Wallemacq, P., de Timary, P., & Persu, A. (2017). Twenty-four hour ambulatory blood pressure level is correlated with altered psychological profiles in patients with apparently treatment-resistant hypertension. *Journal of Hypertension, 35*, e129.

Berthoz, S., Artiges, E., Van de Moortele, P., Poline, J., Rouquette, S., Consoli, S. M., & Marinot, J. (2002). Effect of impaired recognition and expression of emotions on frontocingulate cortices: An fMRI study of men with alexithymia. *American Journal of Psychiatry, 159*, 961–967.

Bishop-Fitzpatrick, L., Minshew, N. J., Mazefsky, C. A., & Eack, S. M. (2017). Perception of life as stressful, not biological response to stress, is associated with greater social disability in adults with autism spectrum disorder. *Journal of Autism and Developmental Disorders, 47*, 1–16.

Blanchard, J. J., Gangestad, S. W., Brown, S. A., & Horan, W. P. (2000). Hedonic capacity and schizotypy revisited: A taxometric analysis of social anhedonia. *Journal of Abnormal Psychology, 109*, 87–95.

Brinkmann, K., Schüpbach, L., Joye, I. A., & Gendolla, G. H. (2009). Anhedonia and effort mobilization in dysphoria: Reduced cardiovascular response to reward and punishment. *International Journal of Psychophysiology, 74*, 250–258.

Centers for Disease Control and Prevention. (2017). *Autism spectrum disorder, 2017*. Atlanta, GA: Centers for Disease Control and Prevention, U.S. Department of Health and Human Services.

Chester, D. S., Pond Jr., R. S., & DeWall, C. N. (2014). Alexithymia is associated with blunted anterior cingulate response to social rejection: Implications for daily rejection. *Social Cognitive and Affective Neuroscience, 10*, 517–522.

Collins, L. M., Blanchard, J. J., & Biondo, K. M. (2005). Behavioral signs of schizoidia and schizotypy in social anhedonics. *Schizophrenia Research, 78*, 309–322.

Conrad, R., Schilling, G., Haidl, G., Geiser, F., Imbierowicz, K., & Liedtke, R. (2002). Relationships between personality traits, seminal parameters and hormones in male infertility. *Andrologia, 34*, 317–324.

Croen, L. A., Zerbo, O., Qian, Y., Massolo, M. L., Rich, S., Sidney, S., & Kripke, C. (2015). The health status of adults on the autism spectrum. *Autism, 19*, 814–823.

Davidson, K. W., Burg, M. M., Kronish, I. M., Shimbo, D., Dettenborn, L., Mehran, R., … Rieckmann, N. (2010). Association of anhedonia with recurrent major adverse cardiac events and mortality 1 year after acute coronary syndrome. *Archives of General Psychiatry, 67*, 480–488.

Dawson, G., Webb, S. J., & McPartland, J. (2005). Understanding the nature of face processing impairment in autism: Insights from behavioral and electrophysiological studies. *Developmental Neuropsychology, 27*, 403–424.

Der-Avakian, A., & Markou, A. (2012). The neurobiology of anhedonia and other reward-related deficits. *Trends in Neurosciences, 35*, 68–77.

Finset, A., Graugaard, P. K., & Holgersen, K. (2006). Salivary cortisol response after a medical interview: The impact of physician communication behavior, depressed affect and alexithymia. *Patient Education and Counseling, 60*, 115–124.

Fitzgibbons, L., & Simons, R. F. (1992). Affective response to color-slide stimuli in subjects with physical anhedonia: A three-systems analysis. *Psychophysiology, 29*, 613–620.

Floyd, K. (2006). *Communicating affection: Interpersonal behavior and social context*. Cambridge, England: Cambridge University Press.

Franzen, J., & Brinkmann, K. (2015). Blunted cardiovascular reactivity in dysphoria during reward and punishment anticipation. *International Journal of Psychophysiology, 95*, 270–277.

Friedlander, L., Lumley, M. A., Farchione, T., & Doyal, G. (1997). Testing the alexithymia hypothesis: Physiological and subjective responses during relaxation and stress. *The Journal of Nervous and Mental Disease, 185*, 233–239.

Fung, C. K., Moore, M. M., Karcher, N. R., Kerns, J. G., & Martin, E. A. (2017). Emotional word usage in groups at risk for schizophrenia-spectrum disorders: An objective investigation of attention to emotion. *Psychiatry Research, 252*, 29–37.

Gabriel, B., Untas, A., Lavner, J. A., Koleck, M., & Luminet, O. (2016). Gender typical patterns and the link between alexithymia, dyadic coping and psychological symptoms. *Personality and Individual Differences, 96*, 266–271.

Gao, T., Li, J., Zhang, H., Gao, J., Kong, Y., Hu, Y., & Mei, S. (2018). The influence of alexithymia on mobile phone addiction: The role of depression, anxiety and stress. *Journal of Affective Disorders, 225*, 761–766.

Garretson, H. B., Fein, D., & Waterhouse, L. (1990). Sustained attention in children with autism. *Journal of Autism and Developmental Disorders, 20*, 101–114.

Gavazzi, G., Orsolini, S., Rossi, A., Bianchi, A., Bartolini, E., Nicolai, E., … Mascalchi, M. (2017). Alexithymic trait is associated with right IFG and pre-SMA activation in non-emotional response inhibition in healthy subjects. *Neuroscience Letters, 658*, 150–154.

Grabe, H. J., Schwahn, C., Barnow, S., Spitzer, C., John, U., Freyberger, H. J., … Völzke, H. (2010). Alexithymia, hypertension, and subclinical atherosclerosis in the general population. *Journal of Psychosomatic Research, 68*, 139–147.

Hamaideh, S. H. (2017). Alexithymia among Jordanian university students: Its prevalence and correlates with depression, anxiety, stress, and demographics. *Perspectives in Psychiatric Care, 54*, 274–280.

Härtwig, E. A., Aust, S., & Heuser, I. (2013). HPA system activity in alexithymia: A cortisol awakening response study. *Psychoneuroendocrinology, 38*, 2121–2126.

Harvey, P. O., Armony, J., Malla, A., & Lepage, M. (2010). Functional neural substrates of self-reported physical anhedonia in non-clinical individuals and in patients with schizophrenia. *Journal of Psychiatric Research, 44*, 707–716.

Henry, J. P., Haviland, M. G., Cummings, M. A., Anderson, D. L., Nelson, J. C., MacMurray, J. P., … Hubbard, R. W. (1992). Shared neuroendocrine patterns of post-traumatic stress disorder and alexithymia. *Psychosomatic Medicine, 54*, 407–415.

Hesse, C., & Floyd, K. (2008). Affectionate experience mediates the effects of alexithymia on mental health and interpersonal relationships. *Journal of Social and Personal Relationships, 25*, 793–810.

Hesse, C., & Floyd, K. (2011). Affection mediates the impact of alexithymia on relationships. *Personality and Individual Differences, 50*, 451–456.

Hesse, C., Floyd, K., Rauscher, E. A., Frye-Cox, N. E., Hegarty, J. P., & Peng, H. (2013). Alexithymia and impairment of decoding positive affect: An fMRI study. *Journal of Communication, 63*, 786–806.

Hesse, C., Pauley, P. M., & Frye-Cox, N. E. (2015). Alexithymia and marital quality: The mediating role of relationship maintenance behaviors. *Western Journal of Communication, 79*, 45–72.

Horan, W. P., Brown, S. A., & Blanchard, J. J. (2007). Social anhedonia and schizotypy: The contribution of individual differences in affective traits, stress, and coping. *Psychiatry Research, 149*, 147–156.

Horan, W. P., Green, M. F., Kring, A. M., & Nuechterlein, K. H. (2006). Does anhedonia in schizophrenia reflect faulty memory for subjectively experienced emotions? *Journal of Abnormal Psychology, 115*, 496–508.

Ihme, K., Sacher, J., Lichev, V., Rosenberg, N., Kugel, H., Rufer, M., … Villringer, A. (2014). Alexithymic features and the labeling of brief emotional facial expressions: An fMRI study. *Neuropsychologia, 64*, 289–299.

Jørgensen, M. M., Zachariae, R., Skytthe, A., & Kyvik, K. (2007). Genetic and environmental factors in alexithymia: A population-based study of 8,785 Danish twin pairs. *Psychotherapy and Psychosomatics, 76*, 369–375.

Joukamaa, M., Kokkonen, P., Veijola, J., Läksy, K., Karvonen, J., Jokelainen, J., & Järvelin, M. (2003). Social situation of expectant mothers and alexithymia 31 years later in their offspring: A prospective study. *Psychosomatic Medicine, 65*, 307–312.

Joukamaa, M., Taanila, A., Miettunen, J., Karvonenm, J. T., Koskinen, M., & Veijola, J. (2007). Epidemiology of alexithymia among adolescents. *Journal of Psychosomatic Research, 63*, 373–376.

Jula, A., Salminen, J. K., & Saarijärvi, S. (1999). Alexithymia: A facet of essential hypertension. *Hypertension, 33*, 1057–1061.

Just, M. A., Cherkassky, V. L., Keller, T. A., & Minshew, N. J. (2004). Cortical activation and synchronization during sentence comprehension in high-functioning autism: Evidence of underconnectivity. *Brain, 127*, 1811–1821.

Just, M. A., Keller, T. A., Malave, V. L., Kana, R. K., & Varma, S. (2012). Autism as a neural systems disorder: A theory of frontal-posterior underconnectivity. *Neuroscience & Biobehavioral Reviews, 36*, 1292–1313.

Kano, M., & Fukudo, S. (2013). The alexithymic brain: The neural pathways linking alexithymia to physical disorders. *BioPsychoSocial Medicine, 7*, 1.

Kano, M., Fukudo, S., Gyoba, J., Kamachi, M., Tagawa, M., Mochizuki, H., ... Yanai, K. (2003). Specific brain processing of facial expressions in people with alexithymia: An H215O-PET study. *Brain, 126*, 1474–1484.

Karukivi, M., Jula, A., Hutri-Kähönen, N., Juonala, M., & Raitakari, O. (2016). Is alexithymia associated with metabolic syndrome? A study in a healthy adult population. *Psychiatry Research, 236*, 58–63.

Keedwell, P. A., Andrew, C., Williams, S. C., Brammer, M. J., & Phillips, M. L. (2005). The neural correlates of anhedonia in major depressive disorder. *Biological Psychiatry, 58*, 843–853.

Kohls, G., Perino, M. T., Taylor, J. M., Madva, E. N., Cayless, S. J., Troiani, V., ... Schultz, R. T. (2013). The nucleus accumbens is involved in both the pursuit of social reward and the avoidance of social punishment. *Neuropsychologia, 51*, 2062–2069.

Koning, C., & Magill-Evans, J. (2001). Social and language skills in adolescent boys with Asperger syndrome. *Autism, 5*, 23–36.

Kwapil, T. R. (1998). Social anhedonia as a predictor of the development of schizophrenia-spectrum disorders. *Journal of Abnormal Psychology, 107*, 558–565.

Kwapil, T. R., Barrantes-Vidal, N., & Silvia, P. J. (2008). The dimensional structure of the Wisconsin schizotypy scales: Factor identification and construct validity. *Schizophrenia Bulletin, 34*, 444–457.

Kwapil, T. R., Silvia, P. J., Myin-Germeys, I., Anderson, A. J., Coates, S. A., & Brown, L. B. (2009). The social world of the socially anhedonic: Exploring the daily ecology of asociality. *Journal of Research in Personality, 43*, 103–106.

La Greca, A. M., & Lopez, N. (1998). Social anxiety among adolescents: Linkages with peer relations and friendships. *Journal of Abnormal Child Psychology, 26*, 83–94.

Lainhart, J. E. (1999). Psychiatric problems in individuals with autism, their parents and siblings. *International Review of Psychiatry, 11*, 278–298.

Leroy, M., Loas, G., & Perez-Diaz, F. (2010). Anhedonia as predictor of clinical events after acute coronary syndromes: A 3-year prospective study. *Comprehensive Psychiatry, 51*, 8–14.

Levant, R. F., Halter, M. J., Hayden, E. W., & Williams, C. M. (2009). The efficacy of alexithymia reduction treatment: A pilot study. *The Journal of Men's Studies, 17*, 75–84.

Levy, A., & Perry, A. (2011). Outcomes in adolescents and adults with autism: A review of the literature. *Research in Autism Spectrum Disorders, 5*, 1271–1282.

Leweke, F., Leichsenring, F., Kruse, J., & Hermes, S. (2012). Is alexithymia associated with specific mental disorders? *Psychopathology, 45*, 22–28.

Linden, W., Lenz, J. W., & Stossel, C. (1996). Alexithymia, defensiveness and cardiovascular reactivity to stress. *Journal of Psychosomatic Research, 41*, 575–583.

Lucas, R. E., Diener, E., Grob, A., Suh, E. M., & Shao, L. (2000). Cross-cultural evidence for the fundamental features of extraversion. *Journal of Personality and Social Psychology, 79*, 452–468.

Lyvers, M., Edwards, M., & Thorberg, F. (2017). Alexithymia, attachment and fear of intimacy in young adults. *IAFOR Journal of Psychology & the Behavioral Sciences, 3*, 1–11.

Martin, E. A., Becker, T. M., Cicero, D. C., Docherty, A. R., & Kerns, J. G. (2011). Differential associations between schizotypy facets and emotion traits. *Psychiatry Research, 187*, 94–99.

Martin, E. A., Cicero, D. C., Bailey, D. H., Karcher, N. R., & Kerns, J. G. (2016). Social anhedonia is not just extreme introversion: Empirical evidence of distinct constructs. *Journal of Personality Disorders, 30*, 451–468.

Martin, E. A., Cicero, D. C., & Kerns, J. G. (2012). Social anhedonia, but not positive schizotypy, is associated with poor affective control. *Personality Disorders: Theory, Research, and Treatment, 3*, 263–272.

Martin, E. A., & Kerns, J. G. (2010). Social anhedonia associated with poor evaluative processing but not with poor cognitive control. *Psychiatry Research, 178*, 419–424.

Mayes, S. D., Calhoun, S. L., Murray, M. J., & Zahid, J. (2011). Variables associated with anxiety and depression in children with autism. *Journal of Developmental and Physical Disabilities, 23*, 325–337.

McGillivray, L., Becerra, R., & Harms, C. (2017). Prevalence and demographic correlates of alexithymia: A comparison between Australian psychiatric and community samples. *Journal of Clinical Psychology, 73*, 76–87.

Mezzavilla, M., Ulivi, S., La Bianca, M., Carlino, D., Gasparini, P., & Robino, A. (2015). Analysis of functional variants reveals new candidate genes associated with alexithymia. *Psychiatry Research, 227*, 363–365.

Ming, X., Julu, P. O., Brimacombe, M., Connor, S., & Daniels, M. L. (2005). Reduced cardiac parasympathetic activity in children with autism. *Brain and Development, 27*, 509–516.

Ming, X., Patel, R., Kang, V., Chokroverty, S., & Julu, P. O. (2016). Respiratory and autonomic dysfunction in children with autism spectrum disorders. *Brain and Development, 38*, 225–232.

Neumann, S. A., Sollers, J. J., Thayer, J. F., & Waldstein, S. R. (2004). Alexithymia predicts attenuated autonomic reactivity, but prolonged recovery to anger recall in young women. *International Journal of Psychophysiology, 53*, 183–195.

New, A. S., Aan Het Rot, M., Ripoll, L. H., Perez-Rodriguez, M. M., Lazarus, S., Zipursky, E., … Siever, L. J. (2012). Empathy and alexithymia in borderline personality disorder: Clinical and laboratory measures. *Journal of Personality Disorders, 26*, 660–675.

Newton, T. L., & Contrada, R. J. (1994). Alexithymia and repression: Contrasting emotion-focused coping styles. *Psychosomatic Medicine, 56*, 457–462.

Ogawa, S., Lee, Y., Yamaguchi, Y., Shibata, Y., & Goto, Y. (2017). Associations of acute and chronic stress hormones with cognitive functions in autism spectrum disorder. *Neuroscience, 343*, 229–239.

Orsmond, G. I., Krauss, M. W., & Seltzer, M. M. (2004). Peer relationships and social and recreational activities among adolescents and adults with autism. *Journal of Autism and Developmental Disorders, 34*, 245–256.

Park, I. H., Kim, J. J., Chun, J., Jung, Y. C., Seok, J. H., Park, H. J., & Lee, J. D. (2009). Medial prefrontal default-mode hypoactivity affecting trait physical anhedonia in schizophrenia. *Psychiatry Research: Neuroimaging, 171*, 155–165.

Picardi, A., Fagnani, C., Gigantesco, A., Toccaceli, V., Lega, I., & Stazi, M. A. (2011). Genetic influences on alexithymia and their relationship with depressive symptoms. *Journal of Psychosomatic Research, 71*, 256–263.

Pollatos, O., Werner, N. S., Duschek, S., Schandry, R., Matthias, E., Traut-Mattausch, E., & Herbert, B. M. (2011). Differential effects of alexithymia subscales on autonomic reactivity and anxiety during social stress. *Journal of Psychosomatic Research, 70*, 525–533.

Reker, M., Ohrmann, P., Rauch, A. V., Kugel, H., Bauer, J., Dannlowski, U., … Suslow, T. (2010). Individual differences in alexithymia and brain response to masked emotion faces. *Cortex, 46*, 658–667.

Santangelo, S. L., & Tsatsanis, K. (2005). What is known about autism. *American Journal of Pharmacogenomics, 5*, 71–92.

Schaafsma, S. M., & Pfaff, D. W. (2014). Etiologies underlying sex differences in autism spectrum disorders. *Frontiers in Neuroendocrinology, 35*, 255–271.

Schmelkin, C. B., Plessow, F., Thomas, J. J., Gray, E. K., Marangi, D. A., Pulumo, R., … Lawson, E. A. (2017). Low oxytocin levels are related to alexithymia in anorexia nervosa. *International Journal of Eating Disorders, 50*, 1332–1338.

Schneider-Hassloff, H., Straube, B., Jansen, A., Nuscheler, B., Wemken, G., Witt, S. H., … Kircher, T. (2016). Oxytocin receptor polymorphism and childhood social experiences shape adult personality, brain structure and neural correlates of mentalizing. *NeuroImage, 134*, 671–684.

Sifneos, P. E. (1973). The prevalence of ‘alexithymic’ characteristics in psychosomatic patients. *Psychotherapy and Psychosomatics, 22*, 255–262.

Silvia, P. J., & Kwapil, T. R. (2011). Aberrant asociality: How individual differences in social anhedonia illuminate the need to belong. *Journal of Personality, 79*, 1315–1332.

Silvia, P. J., Nusbaum, E. C., Eddington, K. M., Beaty, R. E., & Kwapil, T. R. (2014). Effort deficits and depression: The influence of anhedonic depressive symptoms on cardiac autonomic activity during a mental challenge. *Motivation and Emotion, 38*, 779–789.

Solomon, M., Miller, M., Taylor, S. L., Hinshaw, S. P., & Carter, C. S. (2012). Autism symptoms and internalizing psychopathology in girls and boys with autism spectrum disorders. *Journal of Autism and Developmental Disorders, 42*, 48–59.

Spitzberg, B. H., & Cupach, W. R. (2011). Interpersonal skills. In M. L. Knapp & J. A. Daly (Eds.), *The Sage handbook of interpersonal communication* (4th ed., pp. 481–524). Thousand Oaks, CA: Sage.

Spitzer, C., Brandl, S., Rose, H. J., Nauck, M., & Freyberger, H. J. (2005). Gender-specific association of alexithymia and norepinephrine/cortisol ratios: A preliminary report. *Journal of Psychosomatic Research, 59*, 73–76.

Suslow, T., & Junghanns, K. (2002). Impairments of emotion situation priming in alexithymia. *Personality and Individual Differences, 32*, 541–550.

Suslow, T., Kugel, H., Rufer, M., Redlich, R., Dohm, K., Grotegerd, D., ... Dannlowski, U. (2016). Alexithymia is associated with attenuated automatic brain response to facial emotion in clinical depression. *Progress in Neuro-Psychopharmacology and Biological Psychiatry, 65*, 194–200.

Tantam, D. (2000). Psychological disorder in adolescents and adults with Asperger syndrome. *Autism, 4*, 47–62.

Taylor, G. J., Bagby, R. M., & Parker, J. D. A. (1997). *Disorders of affect regulation: Alexithymia in medical and psychiatric illness.* New York, NY: Cambridge University Press.

Underwood, L., McCarthy, J., & Tsakanikos, E. (2010). Mental health of adults with autism spectrum disorders and intellectual disability. *Current Opinion in Psychiatry, 23*, 421–426.

van der Velde, J., Servaas, M. N., Goerlich, K. S., Bruggeman, R., Horton, P., Costafreda, S. G., & Aleman, A. (2013). Neural correlates of alexithymia: A meta-analysis of emotion processing studies. *Neuroscience & Biobehavioral Reviews, 37*, 1774–1785.

Volkmar, F. R., & Klin, A. (2005). Issues in the classification of autism and related conditions. In F. R. Volkmar, R. Paul, A. Klin, & D. Cohen (Eds.), *Handbook of autism and pervasive developmental disorders: Diagnosis, development, neurobiology, and behavior* (pp. 5–41). Hoboken, NJ: John Wiley & Sons.

Wacker, J., Dillon, D. G., & Pizzagalli, D. A. (2009). The role of the nucleus accumbens and rostral anterior cingulate cortex in anhedonia: Integration of resting EEG, fMRI, and volumetric techniques. *Neuroimage, 46*, 327–337.

Walter, N. T., Montag, C., Markett, S. A., & Reuter, M. (2011). Interaction effect of functional variants of the BDNF and DRD2/ANKK1 gene is associated with alexithymia in healthy human subjects. *Psychosomatic Medicine, 73*, 23–28.

Yürümez, E., Akça, Ö. F., Uğur, Ç., Uslu, R. I., & Kılıç, B. G. (2014). Mothers' alexithymia, depression and anxiety levels and their association with the quality of mother-infant relationship: A preliminary study. *International Journal of Psychiatry in Clinical Practice, 18*, 190–196.

Zeeland, S. V., Ashley, A., Dapretto, M., Ghahremani, D. G., Poldrack, R. A., & Bookheimer, S. Y. (2010). Reward processing in autism. *Autism Research, 3*, 53–67.

22

The Biology of Affectionate Communication

Kory Floyd, Nathan T. Woo, and Benjamin E. Custer

Few communicative behaviors are more consequential for the development, maintenance, and satisfaction of personal relationships than the exchange of affection (Floyd, 2006a). In addition to its social and cultural roots, affectionate communication also has distinct biological antecedents, consequences, and correlates. This chapter situates affectionate communication as an evolutionary adaptive behavioral tendency. It then describes the biological foundations of affectionate behavior; details its role in stress buffering, regulation, and recovery; identifies its associations with immunocompetence, relaxation, and rest; and notes the health detriments related to affection deprivation.

THE EVOLUTION OF AFFECTIONATE COMMUNICATION

Humans are a supremely social species. Baumeister and Leary (1995) assert that an innate need to belong motivates individuals to create, preserve, and safeguard close, social relationships that involve both frequent and meaningful contact. Such relationships are highly predictive of life satisfaction and happiness (e.g., Demir & Weitekamp, 2007) as well as physical wellness (House, Landis, & Umberson, 1988).

Unsurprisingly, deficits in social relationships are deleterious to welfare. Experiences such as bullying (Hansen et al., 2006), ostracism (Oaten, Williams, Jones, & Zadro, 2008), persistent loneliness (Cacioppo & Patrick, 2008), social rejection (Baumeister, Brewer, Tice, & Twenge, 2007), and stigmatization (Smart Richman & Leary, 2009) are related to multiple mental and physical health problems, including depression (Leary, 1990); suicide rumination (Stravynski & Boyer, 2001); drug abuse (Orzeck & Rokach, 2004); alcohol abuse (Åkerlind & Hörnquist, 1992); obesity (Hawkley & Caçioppo, 2010); cardiovascular problems (Sorkin, Rook, & Lu, 2002); problematic gambling behaviors (Trevorrow & Moore, 1998); and smoking (House et al., 1988).

Most humans are therefore highly motivated to establish and maintain meaningful interpersonal relationships, and the communication of affection is instrumental (see, e.g., Denes, 2012; Horan, 2012; Mansson & Booth-Butterfield, 2011). Affectionate communication comprises behaviors that express appreciation, love, and commitment (Floyd, Hesse, & Generous, 2015).

Verbal and nonverbal expressions of affection are used to establish close relationships and catalyze their progress. The tendency to express affection is strongly associated with multiple indices of relational quality, including satisfaction (Punyanunt-Carter, 2004); relational maintenance (Pauley, Hesse, & Mikkelson, 2014); commitment (Bell, Buerkel-Rothfuss, & Gore, 1987); sexual satisfaction (Muise, Giang, & Impett, 2014); and conflict management (Gulledge, Gulledge, & Stahmann, 2003; for review, see Floyd, 2019). Certain affectionate acts (e.g., a couple's first kiss) constitute significant relational turning points (Owen, 1987). Conversely, a deficiency of affectionate behaviors corresponds with emotional distress that may hasten relational turmoil (Coyne, Thompson, & Palmer, 2002; Gottman & Levenson, 1992; Gottman et al., 2003).

Floyd (2006a) asserts that the tendency to express affection is evolutionarily advantageous. His affection exchange theory (AET) proposes that individuals with a stronger inclination to express affection are advantaged with respect to both viability and fertility by contributing to the formation of close relationships (Floyd, in press a). Such an evolutionarily advantageous predisposition suggests genetic inheritance of some degree via natural and sexual selection. Some research supports this claim (see Floyd, in press b; Floyd & Denes, 2015), as detailed below.

If the tendency to communicate affection is evolutionarily adaptive, as AET suggests, then it likely has biological antecedents, correlates, and/or consequences. A growing literature has documented the biological and physiological components of affectionate behavior, many of which are addressed in the subsequent section.

AFFECTIONATE COMMUNICATION AND BIOLOGY

Recent years have seen significant advances in the articulation of biological and physiological dimensions of affectionate behavior. This section summarizes the biological underpinnings of affectionate communication and emphasizes how affectionate individuals are advantaged with respect to the management of stress, immunocompetence, and relaxation. Additionally, this section underscores adverse outcomes that characterize affection deprivation.

Affectionate Communication Has Biological Foundations

Humans differ markedly in their tendency to communicate affection, which has prompted researchers to investigate the factors accounting for such variation. A considerable proportion of variance may be attributed to enculturation and upbringing, insofar as cultures and families differ systematically in their propensities for affectionate behavior (e.g., Floyd & Morman, 2000). However, research has also begun to identify genetic and neurological factors that account for variation in the tendency to behave affectionately. When exploring genetic factors, Floyd and Denes (2015) focused their attention on how genotypic variation on the oxytocin receptor gene polymorphism rs53576—which past research has connected to other forms of prosocial behavior (e.g., Rodrigues, Saslow, Garcia, John, & Keltner, 2009; Tost et al., 2010)—interacts with attachment security to predict a predisposition for affectionate communication. As expected, individuals exhibiting two G alleles on rs53576 reported higher levels of trait-expressed affection than did those exhibiting two A alleles or an AG combination. Importantly, however, genotype interacted with attachment security, such that individuals with low attachment security showed more substantial differences in affectionate behavior between GG and AA/AG genotypes than did individuals with high attachment security. These findings were the first to indicate that genetic factors contribute at least partially to an individual's predisposition to communicate affection, especially if the individual holds an insecure attachment style. Other researchers who have studied rs53576 and related oxytocin receptor gene polymorphisms have similarly connected genotypic variation to empathic communication (Floyd, Generous, Clark, McLeod, & Simon, 2017) and nonverbal affiliation behavior (Floyd, Generous, & Clark, 2019), both of which share conceptual space with affectionate communication.

Varying degrees of affectionate communication are also accompanied by differences in neurological activity. Lewis, Heisel, Reinhart, and Tian (2011) speculated that the neurological mechanisms mediating the tendency to communicate affectionately may overlap with those responsible for personality. Gray (1994) argued that the principal personality dimensions emerge from variation in the approach system and the avoidance system. The former underlies reward, positive affect, and goal-oriented behavior, whereas the latter underlies anxiety, punishment, and sensitivity to negativity. Davidson (1995, 1998) observed that the left and right prefrontal cortices (PFCs) of the brain largely mediate approach and avoidance, respectively, although both hemispheres are implicated in both outcomes (see also Berkman & Lieberman, 2010). On that basis, Lewis et al. (2011) hypothesized that the tendency to convey affection—which can be characterized as an approach behavior—is associated with left PFC dominance, amplifying the motivational response to the potential reward of exchanging affectionate messages. Using EEG, Lewis and colleagues found that asymmetry evidencing greater dominance in the left PFC (which mediates the approach system) is positively associated with trait affectionate communication.

Affectionate Communication Helps to Manage the Stress Response

The manner and magnitude of the body's reaction to stressors play a pivotal role in wellness. Among the most common indices of the stress response is activity of the steroid cortisol. Cortisol is an adrenal hormone elevated by the hypothalamic-pituitary-adrenal (HPA) axis in response to the perception of a threat (Dickerson & Kemeny, 2004). The magnitude of cortisol elevation reflects the reaction to an acute stressor (Starcke, Wolf, Markowitsch, & Brand, 2008). In the absence of acute stress, however, cortisol observes a diurnal (24-hr) rhythm typically characterized by peak values within 30 minutes of awakening, a sharp decline throughout the morning and early afternoon, and nadir around midnight (Giese-Davis, Sephton, Abercrombie, Durán, & Spiegel, 2004). Chronic stress decreases diurnal cortisol variation, making the magnitude of variation a useful marker of chronic stress load (Giese-Davis et al., 2004). In addition to cortisol, blood glucose (in the form of glycosylated hemoglobin; Floyd, Hesse, & Haynes, 2007), blood pressure (Grewen, Anderson, Girdler, & Light, 2003), heart rate (HR: Ditzen et al., 2007), secretory immunoglobulin A (sIgA; Evans, Bristow, Hucklebridge, Clow, & Pang, 1994), and dehydroepiandrosterone-sulfate (DHEA-S; Cruess et al., 1999) have all been used as stress markers.

Research exploring the associations between affectionate communication and stress has adjudicated three principal relationships, each of which is described below: (1) affectionate communication buffers individuals from the effects of stressors; (2) affectionate communication aids regulation of the stress response; and (3) affectionate communication accelerates recovery from acute stress.

Affectionate Communication Acts as a Stress Buffer

Avoiding exaggerated stress responses is adaptive insofar as it protects the body from unnecessary damage. Various facets of the social environment, such as the presence of positive, supportive relationships, can protect individuals from overreacting to stressful situations and enhance well-being (Cohen & Wills, 1985), and there is evidence that the presence of affectionate communication has a similar buffering effect.

Grewen et al. (2003) tested the effects of affectionate touch on stress. Participants in the treatment group hugged and held hands with a romantic partner for ten minutes before participating in a speech stressor. Compared to controls who engaged in no affectionate contact, those in the treatment group evidenced significantly lower systolic and diastolic blood pressure reactivity to the stressor. In a later experiment, Ditzen et al. (2007) found that women who engaged in affectionate touch (in the form of a neck and shoulder massage) with their romantic partners demonstrated lower heart rate and cortisol reactivity to a subsequent public speaking stressor.

Similarly, after sharing affection with either a romantic partner or a platonic friend, participants in a study by Pauley, Floyd, and Hesse (2015) had significantly lower heart rate and systolic and diastolic blood pressure reactivity to standard laboratory stressors, compared to controls. Floyd, Mikkelson et al. (2007b) had documented similarly decreased cortisol and heart rate stress reactivity among participants who self-reported more supportive affection in their close relationships.

Although the stress-buffering effect of supportive and affectionate interaction is well documented, its mechanisms are not well understood. One plausible mechanism implicates the pituitary hormone oxytocin. Floyd, Pauley, and Hesse (2010) exposed 100 participants to standard laboratory stressors while assessing changes in cortisol and oxytocin. Prior to the laboratory procedure, participants kept a diary of the affection they received and the affection they gave to others. Although participants' stress was elevated by the laboratory stressors, those with high levels of affectionate interaction (both as a trait and in the previous week) also experienced highly elevated oxytocin in response to the stressors, which Floyd and colleagues proposed may at least partially account for the stress-buffering advantages of affectionate communication.

Affectionate Communication Aids Stress Regulation

Affectionate communication also aids the process of stress regulation. Two investigations, in particular, have identified associations between affectionate communication and HPA axis regulation. Floyd (2006b) collected four saliva samples from healthy young adults during a normal workday: upon awakening, at noon, in the late afternoon, and at bedtime. Floyd hypothesized a significant direct association between trait-level expressed affection (i.e., the amount of affection one typically communicates to others) and the magnitude of diurnal cortisol change, and the predicted correlation was significant ($r = .56$). Importantly, the association controlled for the effects of trait-level received affection, to account for the possibility that the observed benefits of expressing affection are simply those of receiving affection in return. A follow-up study by Floyd and Riforgiate (2008) measured affection communicated in the marital relationship, specifically. Following Floyd and Morman (1998), the researchers assessed affection expressed verbally, through direct nonverbal gestures, and through socially supportive behaviors. Saliva samples collected from one spouse in each couple were assayed for cortisol and DHEA-S. Results again demonstrated a significant direct correlation between affectionate communication and diurnal cortisol variation. Moreover, spouses' reports of affectionate behavior predicted participants' cortisol:DHEA-S ratio, a reliable indicator of stress (Cruess et al., 1999).

Ditzen, Hoppmann, and Klumb (2008) documented a different association between affectionate behavior and cortisol. The researchers had romantic partners from 51 couples complete relationship questionnaires and collect saliva samples six times per day for one week. At each time point, participants reported their total duration of time spent engaged in "physical affection, such as holding hands, touching, hugging, kissing, or having sexual intercourse" (p. 884), and took a saliva sample that was later analyzed for cortisol. Ditzen and colleagues hypothesized that participants display lower levels of cortisol on days with longer durations of physical affection, and that was the pattern that emerged. These beneficial effects have been shown to prevail the day after an affectionate communication experience (Burleson, Trevathan, & Todd, 2007).

Other biological markers of stress have been associated with being the recipient of affection, as well as having an affectionate disposition. Floyd, Hesse et al. (2007) found trait affection to be negatively associated with blood glucose levels, and other scholars have documented elevated oxytocin and lowered resting blood pressure when receiving affectionate communication in the form of hugging (Light, Grewen, & Amico, 2005).

Affectionate Communication Accelerates Recovery from Stress

Affectionate communication not only contributes to stress regulation and buffers against exaggerated responses, but also helps people recover more quickly from stress-inducing events. Recovery rate is clinically significant because a protracted stress response can damage muscle tissue and bone density, among other effects (Sapolsky, 2002). In one study, Floyd, Mikkelson et al. (2007a) induced stress with standard laboratory stressors and then randomly assigned adults to one of three groups. Experimental participants spent 20 minutes writing an affectionate letter to someone they cared for. Those in a comparison group thought about someone they cared for, and those in the control group sat quietly, both for 20 minutes. Stress recovery was assessed by participants' cortisol levels (relative to their baseline levels) after the 20-minute period. Floyd, Mikkelson et al. (2007a) found that participants who wrote affectionate letters recovered from the stressors more efficiently than those in the comparison and control groups, who did not differ significantly from each other.

Affectionate Communication Is Related to Immunocompetence

Immunocompetence indexes the immune system's ability to protect the body from infection and disease. No single, standard marker represents the potency of the immune response (Farnè, Boni, Corallo, Gnugnoli, & Sacco, 1994), but common markers in social science research include immunoglobulin levels, B and T cell counts, antibodies to Epstein-Barr virus, cytotoxicity of natural killer cells, and antibody response to specific antigens.

Three studies have directly examined the relationship between immunocompetence and affectionate communication. Floyd, Pauley et al. (2014) investigated trait-expressed affection in healthy adults and found that it was linearly related to circulating levels of immunoglobulin M (IgM), an antibody that provides immediate response to infection and leads other immune cells to destroy invading substances. Trait-expressed affection was also directly related to natural killer cell cytotoxicity, a measure of the efficacy of natural killer cells in killing target cells. Whereas Floyd, Pauley et al. (2014) measured trait-level affectionate communication, Floyd et al. (2018) specifically examined socially supportive affection among healthy adults. They again reported a linear association with IgM, as well as with immunoglobulin G (IgG), an antibody that binds to antigens and aids in antibody synthesis. Socially supportive affection was also directly associated with levels of CD3+, CD4+, CD8+, and CD19+ cells.

Both sets of findings suggest that communicating affection reflects (although does not necessarily cause) greater immunocompetence, at least as indexed by these specific markers. Kimata (2003, 2006) reached a similar conclusion in two experiments investigating the effects of romantic kissing on the allergic response. In the first study, Kimata had adult patients who were allergic to house dust mites and Japanese cedar pollen kiss their romantic partner for 30 minutes. He found that kissing significantly reduced skin wheal responses to dust mites and cedar pollen and significantly reduced plasma levels of nerve growth factor (NGF), brain-derived neurotrophic factor (BDNF), neurotrophin-3 (NT-3), and neurotrophin-4 (NT-4), whereas the same effects of kissing were not observed in non-patient controls. Kimata (2006) later found that kissing inhibited allergen-specific production of immunoglobulin E (IgE) among patients with mild atopic eczema and allergic rhinitis. Kimata concluded from these studies that, at least for allergic patients, kissing can alleviate some allergic symptoms, a finding that is also supportive of a positive association between affectionate behavior and immunocompetence.

Despite evidence linking affectionate behavior to immunocompetence, however, Floyd, Hesse, Boren, and Veksler (2014) documented that trait-expressed affection levels in healthy adults are positively related to antibody levels to latent Epstein-Barr virus (EBV). Higher antibody levels to EBV indicate a compromised ability of the immune system to hold the virus in a latent state. In contrast to other research, therefore, this finding suggests a negative association between affectionate communication and immunocompetence (again, as indexed by this specific marker). In light of conflicting results, additional research is certainly warranted to adjudicate the associations between immunocompetence and affectionate behavior.

Affectionate Communication Promotes Relaxation and Calm

Finally, affectionate communication corresponds with indices of relaxation and calm—such as decreased heart rate and increased calm-inducing hormones—when the body is in a restful state. Relaxation and calm relate to health and wellness to the extent they promote cellular restoration and may buffer individuals against the effects of ensuing stressors.

Two studies have identified links between affectionate communication and lower resting heart rate. Floyd, Pauley et al. (2014) discovered a moderately strong negative relationship between resting heart rate and trait-expressed affection. In other words, individuals with a tendency to communicate affectionately enjoyed lower resting heart rates. When focusing on affectionate communication within primary relationships, Floyd, Mikkelson et al. (2007b) discovered resting heart rate was negatively tied to both affection expressed verbally and through socially supportive behaviors, although no association emerged for nonverbal affectionate expressions. Researchers have also found affectionate communication to be tied to lower resting blood pressure. After controlling for the affection participants received—thereby isolating the influence of affection expressed to others—Floyd, Hesse, and Haynes (2007) discovered trait-expressed affection shared a strong, negative connection with both resting systolic and diastolic BP.

During a laboratory training session, Holt-Lunstad, Birmingham, and Light (2008) guided married couples in a “warm touch enhancement” procedure—composed of massage to the neck, shoulders, and hands—in advance of a four-week intervention in which the participants practiced the technique at home. After the intervention, couples who practiced warm touch experienced increased oxytocin and decreased alpha-amylase (a protein enzyme indicative of sympathetic nervous system arousal), relative to couples in an attention comparison group. Moreover, husbands in the warm touch group evidenced significantly lower systolic blood pressure than those in the comparison group. In an earlier study using the same warm contact intervention with married couples while accounting for effects of spouses’ relative emotional and social support, Grewen, Girdler, Amico, and Light (2005) discovered that participants who experienced greater support exhibited higher levels of oxytocin both before and after warm touch. Additionally, greater support predicted lower systolic BP in the wake of warm touch, albeit for women only.

Deprivation of Affection Is Associated with Physical Health Detriments

Affection deprivation occurs when individuals perceive themselves to be the recipients of insufficient amounts of affection. Whereas affection is positively related to enhanced immunocompetence, being affection deprived is negatively associated with multiple indices of mental, physical, and social wellness (Floyd, 2014b). In particular, Floyd (2014b) found that affection deprivation is negatively related to happiness, general mental well-being, social support, and attachment security, whereas it is positively related to loneliness, alexithymia, depression, stress, the number of diagnosed mood/anxiety disorders, and the number of diagnosed secondary immune disorders.

In a set of three studies, Floyd (2016) reported that affection deprivation evidenced small but consistent relationships with physical pain and impaired sleep quality. These associations are suggested by neuroimaging research demonstrating that social pain (such as would be expected to accompany states of social deprivation, including affection deprivation and loneliness) activates neurological responses similar to those activated by physical pain, particularly in the anterior cingulate cortex and its dorsal subdivision (e.g., Bush, Luu, & Posner, 2000; Eisenberger & Lieberman, 2004; but see Eisenberger, 2015).

CONCLUSION

Floyd (2006a, 2014a) has argued that the tendency to interact affectionately has deep evolutionary roots, insofar as affectionate communication helps humans form and maintain close, personal relationships that foster survival and reproductive success. If true, then affectionate communication should be reflected in beneficial physiological processes, such as those related to immunocompetence, stress regulation and recovery, and sleep quality, among others. Such processes cannot be fully adjudicated via the self-report and behavioral observation methods common in communication research; instead, psychophysiological research identifying the neurological, genetic, immune, endocrine, and nervous system processes related to affectionate behavior is necessary.

Considered collectively, this research suggests that an individual's predisposition for affectionate communication is not entirely environmentally determined, but also has both genetic and neurological substrates. Affectionate communicators are also advantaged in terms of stress regulation and recovery, immunocompetence, and ability to benefit from relaxation and calm. Unsurprisingly, being deprived of affection is linked to a host of adverse outcomes, many of which have specific biological roots.

Despite growing evidence regarding the biology of affectionate communication, much remains to be learned. One nascent line of inquiry is the association between affectionate communication and the pain response. Although studies have found ties between affectionate communication and chronic pain, continuing research is evaluating how received affectionate touch influences the response to painful stimuli. Such inquiry closely relates to research on touch tempering the stress response (e.g., Coan, Schaefer, & Davidson, 2006). Insofar as affectionate behavior mitigates reaction to pain, individuals can employ affection to assuage the experience of pain for those in their close relationships. A recent study found that cold pressor pain was tempered by receiving familiar but nonsexual touch from a romantic partner, but not from a friend or stranger (Floyd, Ray, van Raalte, Stein, & Generous, 2018). Much remains to be discovered about the mechanisms and conditions characterizing the association between affection and pain.

REFERENCES

Åkerlind, I., & Hörnquist, J. O. (1992). Loneliness and alcohol abuse: A review of evidences of an interplay. *Social Science & Medicine, 34*, 405–414.

Baumeister, R. F., Brewer, L. E., Tice, D. M., & Twenge, J. M. (2007). Thwarting the need to belong: Understanding the interpersonal and inner effects of social exclusion. *Social and Personality Psychology Compass, 1*, 506–520.

Baumeister, R. F., & Leary, M. R. (1995). The need to belong: Desire for interpersonal attachments as a fundamental human motivation. *Psychological Bulletin, 117*, 497–529.

Bell, R. A., Buerkel-Rothfuss, N. L., & Gore, K. E. (1987). "Did you bring the yarmulke for the Cabbage Patch Kid?" Idiomatic communication of young lovers. *Human Communication Research, 14*, 47–67.

Berkman, E. T., & Lieberman, M. D. (2010). Approaching the bad and avoiding the good: Lateral prefrontal cortical asymmetry distinguishes between action and valence. *Journal of Cognitive Neuroscience, 22*, 1970–1979.

Burleson, M. H., Trevathan, W. R., & Todd, M. (2007). In the mood for love or vice versa? Exploring the relations among sexual activity, physical affection, affect, and stress in the daily lives of mid-aged women. *Archives of Sexual Behavior, 36*, 357–368.

Bush, G., Luu, P., & Posner, M. I. (2000). Cognitive and emotional influence in anterior cingulate cortex. *Trends in Cognitive Sciences, 4*, 215–222.

Cacioppo, J. T., & Patrick, W. (2008). *Loneliness: Human nature and the need for social connection.* New York, NY: W. W. Norton.

Coan, J. A., Schaefer, H. S., & Davidson, R. J. (2006). Lending a hand: Social regulation of the neural response to threat. *Psychological Science, 17*, 1032–1039.

Cohen, S., & Wills, T. A. (1985). Stress, social support, and the buffering hypothesis. *Psychological Bulletin, 98*, 310–357.

Coyne, J. C., Thompson, R., & Palmer, S. C. (2002). Marital quality, coping with conflict, marital complaints, and affection in couples with a depressed wife. *Journal of Family Psychology, 16*, 26–37.

Cruess, D. G., Antoni, M. H., Kumar, M., Ironson, G., McCabe, P., Fernandez, J. B., ... Schneiderman, N. (1999). Cognitive-behavioral stress management buffers decreases in dehydroepiandrosterone sulfate (DHEA-S) and increases in the cortisol/DHEA-S ratio and reduces mood disturbance and perceived stress among HIV-seropositive men. *Psychoneuroendocrinology, 24*, 537–549.

Davidson, R. J. (1995). Cerebral asymmetry, emotion, and affective style. In R. J. Davidson & K. Hugdahl (Eds.), *Brain asymmetry* (pp. 361–387). Cambridge, MA: MIT Press.

Davidson, R. J. (1998). Affective style and affective disorders: Perspectives from affective neuroscience. *Cognition and Emotion, 12*, 307–330.

Demir, M., & Weitekamp, L. A. (2007). I am so happy 'cause today I found my friend: Friendship and personality as predictors of happiness. *Journal of Happiness Studies, 8*, 181–211.

Denes, A. (2012). Pillow talk: Exploring disclosures after sexual activity. *Western Journal of Communication, 76*, 91–108.

Dickerson, S. S., & Kemeny, M. E. (2004). Acute stressors and cortisol responses: A theoretical integration and synthesis of laboratory research. *Psychological Bulletin, 130*, 355–391.

Ditzen, B., Hoppmann, C., & Klumb, P. (2008). Positive couple interactions and daily cortisol: On the stress-protecting role of intimacy. *Psychosomatic Medicine, 70*, 883–889.

Ditzen, B., Neumann, I. D., Bodenmann, G., von Dawans, B., Turner, R. A., Ehlert, U., & Heinrichs, M. (2007). Effects of different kinds of couple interaction on cortisol and heart rate responses to stress in women. *Psychoneuroendocrinology, 32*, 565–574.

Eisenberger, N. I. (2015). Social pain and the brain: Controversies, questions, and where to go from here. *Annual Review of Psychology, 66*, 601–629.

Eisenberger, N. I., & Lieberman, M. D. (2004). Why rejection hurts: A common neural alarm system for physical and social pain. *Trends in Cognitive Sciences, 8*, 294–300.

Evans, P., Bristow, M., Hucklebridge, F., Clow, A., & Pang, F.-Y. (1994). Stress, arousal, cortisol and secretory immunoglobulin A in students undergoing assessment. *British Journal of Clinical Psychology, 33*, 575–576.

Farnè, M. A., Boni, P., Corallo, A., Gnugnoli, D., & Sacco, F. L. (1994). Personality variables as moderators between hassles and objective indications of distress (S-IgA). *Stress & Health, 10*, 15–20.

Floyd, K. (2006a). *Communicating affection: Interpersonal behavior and social context.* Cambridge, England: Cambridge University Press.

Floyd, K. (2006b). Human affection exchange XII. Affectionate communication is associated with diurnal variation in salivary free cortisol. *Western Journal of Communication, 70*, 47–63.

Floyd, K. (2014a). Humans are people, too: Nurturing an appreciation for nature in communication research. *Review of Communication Research, 2*, 1–29.

Floyd, K. (2014b). Relational and health correlates of affection deprivation. *Western Journal of Communication, 78*, 383–403.

Floyd, K. (2016). Affection deprivation is associated with physical pain and poor sleep quality. *Communication Studies, 67*, 379–398.

Floyd, K. (2019). *Affectionate communication in close relationships.* Cambridge, England: Cambridge University Press.

Floyd, K. (in press a). Affection exchange theory. In C. R. Berger & M. E. Roloff (Eds.), *International encyclopedia of interpersonal communication.* New York, NY: Wiley.

Floyd, K. (in press b). Evolutionary perspectives on affectionate communication. In C. R. Berger & M. E. Roloff (Eds.), *International encyclopedia of interpersonal communication.* New York, NY: Wiley.

Floyd, K., & Denes, A. (2015). Attachment security and oxytocin receptor gene polymorphism interact to influence affectionate communication. *Communication Quarterly, 63*, 272–285.

Floyd, K., Generous, M. A., & Clark, L. (2019). Nonverbal affiliation by physician assistant students during simulated clinical examinations: Genotypic effects. *Western Journal of Communication.* Advance online publication.

Floyd, K., Generous, M. A., Clark, L., McLeod, I., & Simon, A. (2017). Cumulative risk on the oxytocin receptor gene (*OXTR*) predicts empathic communication by physician assistant students. *Health Communication, 32*, 1210–1216.

Floyd, K., Hesse, C., Boren, J. P., & Veksler, A. E. (2014). Affectionate communication can suppress immunity: Trait affection predicts antibodies to latent Epstein-Barr virus. *Southern Communication Journal, 79*, 2–13.

Floyd, K., Hesse, C., & Generous, M. A. (2015). Affection exchange theory: A bio-evolutionary look at affectionate communication. In D. O. Braithwaite & P. Schrodt (Eds.), *Engaging theories in interpersonal communication: Multiple perspectives* (2nd ed., pp. 303–314). Thousand Oaks, CA: Sage.

Floyd, K., Hesse, C., & Haynes, M. T. (2007). Human affection exchange: XV. Metabolic and cardiovascular correlates of trait expressed affection. *Communication Quarterly, 55*, 79–94.

Floyd, K., Mikkelson, A. C., Tafoya, M. A., Farinelli, L., La Valley, A. G., Judd, J., … Wilson, J. (2007a). Human affection exchange: XIII. Affectionate communication accelerates neuroendocrine stress recovery. *Health Communication, 22*, 123–132.

Floyd, K., Mikkelson, A. C., Tafoya, M. A., Farinelli, L., La Valley, A. G., Judd, J., … Wilson, J. (2007b). Human affection exchange: XIV. Relational affection predicts resting heart rate and free cortisol secretion during acute stress. *Behavioral Medicine, 32*, 151–156.

Floyd, K., & Morman, M. T. (1998). The measurement of affectionate communication. *Communication Quarterly, 46*, 144–162.

Floyd, K., & Morman, M. T. (2000). Affection received from fathers as a predictor of men's affection with their own sons: Tests of the modeling and compensation hypotheses. *Communication Monographs, 67*, 347–361.

Floyd, K., Pauley, P. M., & Hesse, C. (2010). State and trait affectionate communication buffer adults' stress reactions. *Communication Monographs, 77*, 618–636.

Floyd, K., Pauley, P. M., Hesse, C., Eden, J., Veksler, A. E., & Woo, N. T. (2018). Supportive communication is associated with markers of immunocompetence. *Southern Communication Journal, 83*, 229–244.

Floyd, K., Pauley, P. M., Hesse, C., Veksler, A. E., Eden, J., & Mikkelson, A. C. (2014). Affectionate communication is associated with markers of immune and cardiovascular system competence. In J. M. Honeycutt, C. Sawyer, & S. Keaton (Eds.), *The influence of communication on physiology and health status* (pp. 115–130). New York, NY: Peter Lang Publishing.

Floyd, K., Ray, C. D., van Raalte, L. J., Stein, J. B., & Generous, M. A. (2018). Interpersonal touch buffers pain sensitivity in romantic relationships but heightens sensitivity between strangers and friends. *Research in Psychology and Behavioral Sciences, 6*, 27–34.

Floyd, K., & Riforgiate, S. (2008). Affectionate communication received from spouses predicts stress hormone levels in healthy adults. *Communication Monographs, 75*, 351–368.

Giese-Davis, J., Sephton, S. E., Abercrombie, H. C., Durán, R. E. F., & Spiegel, D. (2004). Repression and high anxiety are associated with aberrant diurnal cortisol rhythms in women with metastatic breast cancer. *Health Psychology, 23*, 645–650.

Gottman, J. M., & Levenson, R. W. (1992). Marital processes predictive of later dissolution: Behavior, physiology, and health. *Journal of Personality and Social Psychology, 63*, 221–233.

Gottman, J. M., Levenson, R. W., Gross, J., Fredrickson, B. L., McCoy, K., Rosenthal, L., ... Yoshimoto, D. (2003). Correlates of gay and lesbian couples' relationship satisfaction and relationship dissolution. *Journal of Homosexuality, 45*, 23–43.

Gray, J. A. (1994). Three fundamental emotion systems. In P. Ekman & R. J. Davidson (Eds.), *The nature of emotion* (pp. 243–247). New York, NY: Oxford University Press.

Grewen, K. M., Anderson, B. J., Girdler, S. S., & Light, K. C. (2003). Warm partner contact is related to lower cardiovascular reactivity. *Behavioral Medicine, 29*, 123–130.

Grewen, K. M., Girdler, S. S., Amico, J., & Light, K. C. (2005). Effects of partner support on resting oxytocin, cortisol, norepinephrine, and blood pressure before and after warm partner contact. *Psychosomatic Medicine, 67*, 531–538.

Gulledge, A. K., Gulledge, M. H., & Stahmann, R. F. (2003). Romantic physical affection types and relationship satisfaction. *American Journal of Family Therapy, 31*, 233–242.

Hansen, Å. M., Hogh, A., Persson, R., Karlson, B., Garde, A. H., & Ørbæk, P. (2006). Bullying at work, health outcomes, and physiological stress response. *Journal of Psychosomatic Research, 60*, 63–72.

Hawkley, L. C., & Cacioppo, J. T. (2010). Loneliness matters: A theoretical and empirical review of consequences and mechanisms. *Annals of Behavioral Medicine, 40*, 218–227.

Holt-Lundstad, J., Birmingham, W. A., & Light, K. C. (2008). Influence of a "warm touch" support enhancement intervention among married couples on ambulatory blood pressure, oxytocin, alpha amylase, and cortisol. *Psychosomatic Medicine, 70*, 976–985.

Horan, S. M. (2012). Affection exchange theory and perceptions of relational transgressions. *Western Journal of Communication, 76*, 109–126.

House, J. S., Landis, K. R., & Umberson, D. (1988). Social relationships and health. *Science, 241*, 540–545.

Kimata, H. (2003). Kissing reduces allergic skin wheal responses and plasma neutrophin levels. *Physiology & Behavior, 80*, 395–398.

Kimata, H. (2006). Kissing selectively decreases allergen-specific IgE production in atopic patients. *Journal of Psychosomatic Research, 60*, 545–547.

Leary, M. R. (1990). Responses to social exclusion: Social anxiety, jealousy, loneliness, depression, and low self-esteem. *Journal of Social & Clinical Psychology, 9*, 221–229.

Lewis, R. J., Heisel, A. D., Reinhart, A. M., & Tian, Y. (2011). Trait affection and asymmetry in the anterior brain. *Communication Research Reports, 28*, 347–355.

Light, K. C., Grewen, K. M., & Amico, J. A. (2005). More frequent partner hugs and higher oxytocin levels are linked to lower blood pressure and heart rate in premenopausal women. *Biological Psychology, 69*, 5–21.

Mansson, D. H., & Booth-Butterfield, M. (2011). Grandparents' expressions of affection for their grandchildren: Examining grandchildren's relational attitudes and behaviors. *Southern Communication Journal, 76*, 424–442.

Muise, A., Giang, E., & Impett, E. A. (2014). Post-sex affectionate exchanges promote sexual and relationship satisfaction. *Archives of Sexual Behavior, 43*, 1391–1402.

Oaten, M., Williams, K. D., Jones, A., & Zadro, L. (2008). The effects of ostracism on self-regulation in the socially anxious. *Journal of Social & Clinical Psychology, 27*, 471–504.
Orzeck, T., & Rokach, A. (2004). Men who abuse drugs and their experience of loneliness. *European Psychologist, 9*, 163–169.
Owen, W. F. (1987). The verbal expression of love by women and men as a critical communication event in personal relationships. *Women's Studies in Communication, 10*, 15–24.
Pauley, P. M., Floyd, K., & Hesse, C. (2015). The stress-buffering effects of a brief dyadic interaction before an acute stressor. *Health Communication, 30*, 646–659.
Pauley, P. M., Hesse, C., & Mikkelson, A. C. (2014). Trait affection predicts married couples' use of relational maintenance behaviors. *Journal of Family Communication, 14*, 167–187.
Punyanunt-Carter, N. M. (2004). Reported affectionate communication and satisfaction in marital and dating relationships. *Psychological Reports, 95*, 1154–1160.
Rodrigues, S. M., Saslow, L. R., Garcia, N., John, O. P., & Keltner, D. (2009). Oxytocin receptor genetic variation relates to empathy and stress reactivity in humans. *Proceedings of the National Academy of Sciences, 106*, 21437–21441.
Sapolsky, R. M. (2002). Endocrinology of the stress response. In J. B. Becker, S. M. Breedlove, D. Crews, & M. M. McCarthy (Eds.), *Behavioral endocrinology* (2nd ed., pp. 409–450). Cambridge, MA: Massachusetts Institute of Technology Press.
Smart Richman, L., & Leary, M. R. (2009). Reactions to discrimination, stigmatization, ostracism, and other forms of interpersonal rejection: A multimotive model. *Psychological Review, 116*, 365–383.
Sorkin, D., Rook, K. S., & Lu, J. L. (2002). Loneliness, lack of emotional support, lack of companionship, and the likelihood of having a heart condition in an elderly sample. *Annals of Behavioral Medicine, 24*, 290–298.
Starcke, K., Wolf, O. T., Markowitsch, H. J., & Brand, M. (2008). Anticipatory stress influences decision making under explicit risk conditions. *Behavioral Neuroscience, 122*, 1352–1360.
Stravynski, A., & Boyer, R. (2001). Loneliness in relation to suicide ideation and parasuicide: A population-wide study. *Suicide and Life-Threatening Behavior, 31*, 32–40.
Tost, H., Kolachana, B., Hakimi, S., Lemaitre, H., Verchinski, B. A., Mattay, V. S., … Meyer-Lindenberg, A. (2010). A common allele in the oxytocin receptor gene (OXTR) impacts prosocial temperament and human hypothalamic-limbic structure and function. *Proceedings of the National Academy of Sciences, 103*, 13936–13941.
Trevorrow, K., & Moore, S. (1998). The association between loneliness, social isolation, and women's electronic gaming machine gambling. *Journal of Gambling Studies, 14*, 263–284.

23

Social Support and Physiology

Current Trends and Future Directions

Alice E. Veksler, Justin P. Boren, and Jennifer Priem

Social support, comprising the provision of care, resources, assistance, or information before, during, or after times of burden or stress, has received decades of attention in the social sciences (Uchino, Bowen, Carlisle, & Birmingham, 2012). The concept of social support is multifaceted, such that it could include individuals' perceptions of the support that is available to them and/or it can be viewed as an interactive process in which a support provider engages in direct or indirect behaviors in the hopes of improving the situation of the support seeker who is attempting to garner support. Since the foundational work of Emile Durkheim, who discovered during the late 19th Century that social integration reduced the prevalence of suicide, researchers have attempted to connect social support to various health-related outcomes (Boren, 2017).

A large body of work has shown that social support is reliably associated with multiple physiological outcomes (Uchino, 2009). Results of meta-analyses suggest that the effects of social support on health and mortality are robust, with effect sizes comparable to or even exceeding those of exercise, obesity, or smoking cessation (Ditzen & Heinrichs, 2014; Uchino et al., 2012). Many questions remain, however, about the pathways and mechanisms by which social support affects human physiology and vice versa (Thoits, 2011; Uchino et al., 2012). Therefore, the aim of this chapter is to summarize the current body of knowledge on social support as it pertains to human communication and physiology and suggest future directions for communication researchers seeking to advance this area of study.

THEORETICAL AND OPERATIONAL DISTINCTIONS

In this section, we begin by identifying the primary theoretical underpinnings of social support and physiology research. Next, we explore how support can be studied from a perceived, received, or enacted perspective. Finally, we provide some specific examples of studies using each of these approaches.

Models of Support

Social support researchers have traditionally differentiated two major models by which support exerts effects on health and physiology: (1) the direct effects (or main effect) model; and (2) the buffering model. The direct effects model argues that support directly affects physical outcomes, not through mediated routes. For instance, one reason that supportive marital relationships foster more beneficial health outcomes is that health-modulating messages serve as pressure from a supportive spouse to correct bad behavior (such as smoking, eating poorly, etc.) thereby directly affecting health outcomes. Furthermore, in some cases, the experience of support is linked to automatic processes unaffected by psychological mechanisms (Uchino et al., 2012). For example, the neural structures involved in the processing of support are linked directly to the regulation of autonomic processes, which translate support into beneficial physiological outcomes such as lower blood pressure and cardiovascular reactivity (Ditzen & Heinrichs, 2014). The main effects model argues that social support can exert positive health effects independently of the presence or absence of a stressor.

On the other hand, the buffering model of support (Cohen & Wills, 1985) argues that support attenuates the experience of stress. In turn, support that is provided when individuals are stressed has the ability to reduce the deleterious effects of stress. In this model, support builds psychological or physical resources that aid in processing of, and coping with, distress (Ditzen & Heinrichs, 2014). For instance, marriage can serve as a significant source of perceived support when spouses feel they can turn to one another in times of need. In the event that a stressful event occurs, the knowledge that support is in fact available, functions to buffer the extent to which the stressor is negatively experienced. Because the model explains how support during stressful events builds coping resources, it has been the primary model by which interpersonal communication scholars have studied social support.

Perceived, Received, and Enacted Support

In addition to theories of how support influences individuals, previous research differs in how it operationalizes support. Perceived support, which is measured as either the extent of one's social integration or one's self-reported perceptions of availability and/or accessibility of support in terms of both quantity and quality (Baek, Tannenbaum, & Gonzalez, 2014), has been linked to the direct effects model. For instance, Roy, Steptoe, and Kirschbaum (1998) operationalized support availability as the number of people participants believed they would be able to turn to in each of 12 stressful scenarios. Their results indicated that in response to a stress induction, those high on perceived social support had greater heart rate reactivity but also faster recovery on subsequent cardiovascular measures.

Received support is typically conceptualized as the actual support given by members of a person's support network (Boren, 2017). Operationally, received support is typically evaluated based on measuring whether or not support was provided, or the number of supportive transactions a person has had over a set period of time. Research on received support, especially when contrasted with perceived support, has yielded mixed results with respect to effects on health-related and physiological outcomes (Holt-Lunstad, Smith, & Layton, 2010; Uchino, 2009). In some cases, received support has been associated with negative rather than positive effects. For example, Helgeson (1993) found that perceived support had a beneficial effect on heart attack patient adjustment, whereas received support from significant others was associated with maladaptive adjustment. Consequently, scholars posited that the divergent effects of perceived and received support may be due to costs of receiving support (Bolger & Amarel, 2007).

Some research has shown that received support may threaten face or independence and decrease esteem in ways that lead to negative outcomes (Bolger & Amarel, 2007; Floyd & Ray, 2017; Maisel & Gable, 2009). Furthermore, poor quality support may also lead to suboptimal outcomes, which highlights the potential limitation of studying the occurrence of support and not the actual behaviors that occurred during the support attempt.

Work on enacted support is based on examining the actual messages and behaviors individuals receive in supportive encounters. Much of this research reflects the foundational work on verbal person-centeredness (VPC). Messages high in VPC explicitly validate and legitimize the feelings of the distressed individual (see Burleson, 2003) and have been linked with greater perceptions of message sensitivity, effectiveness, and helpfulness, as well as emotional improvement and positive reappraisal of a stressor (Jones & Wirtz, 2006). Research has shown that particular message features facilitate faster physiological stress recovery. For example, Priem and Solomon (2018) had dating couples engage in a supportive interaction after one individual completed a series of stressor tasks that increased physiological stress, as indexed by salivary cortisol. Results showed that distressed individuals experienced significantly faster cortisol recovery when their partner used messages that explicitly validated the support seekers' emotions and reinforced their positive traits.

Previous researchers have also shown that enacted support may engender positive physiological outcomes when it is perceived as adequate and when it is well matched to the needs of the recipient (Faw, 2016; Priem & Solomon, 2015, 2018). For example, Priem and Solomon (2015) examined the effects of emotional support on physiological stress recovery after a stressful event under three different conditions from three different theoretical perspectives: (1) when the type of support provided matched the distressed individual's preferred type of support; (2) when the support provided was deemed as adequate (i.e., the discrepancy between the level of support provided and the level of support desired was small); and (3) when support was invisible (i.e., third-party observers rated the support provider as being supportive, but the support receiver rated the supportiveness of the interaction as low). Results showed that although invisibility of support did not have a significant dampening effect on cortisol recovery, support matching and support adequacy both increased the rate of cortisol recovery over time following a supportive interaction with a dating partner.

CURRENT RESEARCH TRENDS IN SOCIAL SUPPORT AND PHYSIOLOGY RESEARCH

In order to better understand the current research landscape, we briefly review distinctions made about sources of support and the circumstances under which support becomes ineffective or harmful. We conclude this section by overviewing common methodological approaches used to investigate support and physiology.

Primary vs. Secondary Sources of Support

Research on support generally focuses on either primary or secondary support providers. Primary support research examines the support exchanged among romantic partners, family members, and between friends. For instance, Heinrichs, Baumgartner, Kirschbaum, and Ehlert (2003) asked friends of participants to provide them instrumental and emotional support prior to participants undergoing the Trier Social Stress Test (TSST), a standardized public speaking and serial subtraction stress-induction task. They also introduced oxytocin intranasally and manipulated social support versus no support (control) conditions. Results indicated that salivary free cortisol increased in response to the TSST, however, social support attenuated this increase. Furthermore, the combination of support and oxytocin had an even greater effect than did support alone. Heinrichs et al. posited a central nervous system mechanism by which social support and oxytocin function together to ameliorate the negative psychological and physiological effects of stressors.

Others have looked at naturally occurring interactions among primary support providers. For example, Faw (2016) asked participants to discuss challenges they experienced while caring for a disabled child and instructed their conversational partner (a self-identified support provider) to respond in helpful and supportive ways. Results indicated that individuals receiving high person-centered support had the greatest cortisol reductions in response to support, relative to those receiving moderate or low person-centered messages.

Support research has also examined the effects of secondary support providers. These loose ties include support provided by strangers, community members, teachers, classmates, and co-workers or other organizational members. For instance, Kirsch and Lehman (2015) had participants interact with a confederate who provided support during a speech task and evaluated cardiovascular reactivity, whereas Taylor et al. (2010) evaluated the effects of a supportive vs. non-supportive audience during the TSST, and Hilmert, Christenfeld, and Kulik (2002) evaluated heart rate and blood pressure in the presence of positive or negative feedback during a speech task. A common model for these investigations involves introducing an artificial stressor, such as a speech task, and training confederates to respond in a scripted manner in order to precisely manipulate the nature of the supportive (or unsupportive) messages.

Other designs attempt to examine more naturally occurring forms of social support among secondary support providers. For instance, Priem and Solomon (2009) had a researcher provide various supportive messages to students in a public speaking class prior to an in-class speech activity and examined effects on cortisol. Results suggested using distraction from an immediately pressing stressor may be an effective form of support, as measured by cortisol reactivity. Overall, then, the existing research on support has shown that support has a robust and beneficial effect on physiological outcomes which has led to the development of hundreds of interventions that may help people deal with difficult circumstances (Taylor et al., 2010).

Ineffective or Harmful Support

As discussed earlier, social support can have substantial positive effects on individuals; however, it can also exacerbate stress in certain cases. In fact, a significant portion of recent research on support has focused on how and why support may be ineffective or even harmful. When the wrong type of support is provided, when support leads to decreased self-esteem, and when support leads to feelings of guilt or indebtedness (Bolger & Amarel, 2007), it can reduce or eliminate the expected benefits or in some cases even lead to harm. For instance, it appears that messages that are lower in quality than expected have no physiological stress-recovery benefit (Faw, 2016; Priem & Solomon, 2015) and Taylor et al. (2010) found that in some cases support does not ameliorate stress at all. For example, in their study of a supportive versus unsupportive audience during the TSST, Taylor and colleagues found that elevated cortisol levels, heart rate, and blood pressure were present for both groups. They hypothesized that this may be due to the fact that the source of support was also performing an evaluative function (serving as an audience for the stressful task). Therefore, in cases when people perceive support and judgment from the same source, it is possible that the effects of support on physiological stress are outweighed by the effects of the feeling evaluated.

One study indicated that cardiovascular reactivity was not reduced in response to social support when the support may have led to negative social evaluation (Kirsch & Lehman, 2015).

For instance, when the support provider is also serving as an evaluative audience, his or her support can exacerbate stress (Taylor et al., 2010). Priem and Solomon (2009) have also suggested that messages may differ in effectiveness when provided by a stranger versus a close relational partner, due to variations in person-centeredness.

Recent attention has also been given to the valence of the content of supportive messages. For instance, co-rumination occurs when two (or more) individuals decide in a socially supportive interaction to dwell on the negative elements of the problem (Rose, 2002). The interactants believe they are supporting each other, but they tend to focus on the problem, encourage each other to dwell on how bad the problem is, and generally fail to evaluate solutions. Although this behavior is linked to improvements in friendship quality (Rose, Carlson, & Waller, 2007), it can reduce the beneficial effects of social support (Boren, 2013, 2014), and increase physiological stress markers (Byrd-Craven, Geary, Rose, & Ponzi, 2008; Byrd-Craven, Granger, & Auer, 2011). Co-rumination has also been associated with elevated C-reactive protein and reduced interleukin-6 (Boren & Veksler, 2018). Recently, Rankin, Swearingen-Stanborough, Granger, and Byrd-Craven (2018) found that levels of adrenocortical attunement predicted the likelihood of engaging in co-rumination, indicating an interaction between the physiology of stress and the way that friends communicate in a perceived supportive interaction.

Methodological Approaches

Methodologically, researchers generally examine support either through self-report methods or through experimental interventions wherein a stressor is introduced, support is assessed or manipulated, and outcomes are evaluated. Some innovative work has tracked support over time. For instance, Taylor et al. (2010) gave participants small hand-held computers to track daily experiences of social support in real time. Overall perceived support from multiple sources has also been examined longitudinally in a sample of older adults, wherein it was associated with lower C-reactive protein (indicating lower chronic inflammation) and better mental health (Kim & Thomas, 2017).

Some researchers have linked support connections with long-term health outcomes using existing longitudinal data. For instance, framed through the Job Demand Control-Support Model (see Bakker & Demerouti, 2016), Shirom, Toker, Alkaly, Jacobson, and Balicer (2011) conducted a 20-year follow-up study of 820 workers using medical records. They discovered that risk of mortality among these workers was significantly lower for those who reported high levels of peer social support at work. In another study of 5,843 workers over a nearly two-year timeframe, Toker, Shirom, Melamed, and Armon (2012) found that work social support was a significant protective factor against Type 2 diabetes.

Although not an exhaustive review, this discussion offers a broad picture of where scholars have focused their attention in the extant social support literature. This work has given rise to new questions and revealed exciting nuances and complexities pertaining to the interplay of support and physiology that have yet to be fully explored. Accordingly, in the subsequent sections, we discuss what we feel are some promising avenues for future research.

FUTURE DIRECTIONS FOR STUDYING SOCIAL SUPPORT

In the remainder of this chapter, we outline what we believe to be some of the more exciting areas for future directions in studying social support, especially for scholars of communication and physiology.

Using Multiple Physiological Markers of Various Systems

As we have discussed throughout this chapter, the physiological outcomes of social support and supportive transactions are typically researched with the framework of the human stress response. This framework is important, given that social support is often studied in the context of some stressful situation. In fact, many social support researchers use physiological and health-linked outcomes of the human stress process to measure the efficacy of social support interventions or to explore the impact of perceived and received support. Therefore, for those researchers who explore the stress-ameliorating effects of social support, we propose taking a more comprehensive look at the various physiological systems at play.

In response to stressors, the hypothalamic-pituitary-adrenal (HPA) axis releases a cascade of hormones including corticotropin-releasing hormone (CRH), adrenocorticotropic hormone (ACTH), and glucocorticoids, including cortisol (Boren & Veksler, 2011; Floyd, 2014). These compounds act on various parts of the body, including the cardiovascular and immune systems. Because the HPA axis is a major component of the stress response and the "glucocorticoid cascade hypothesis" (see Sapolsky, Krey, & McEwen, 1986), and endocrine changes, including cortisol, are the first indicator of allostatic load (or wear and tear on the body associated with stress, see Levine, Zagoory-Sharon, Feldman, Lewis, & Weller, 2007), scholarship has extensively used cortisol as the standard endocrine marker of stress (Lee, Kim, & Choi, 2015). Although there are practical and financial reasons for using cortisol, focusing on a single marker of stress limits understanding of the nuances in the interplay between physiology and communication. Furthermore, as research advances, scholars should understand both the mechanisms that underlie disease along with those that protect health. Therefore, using multiple physiological markers of the human stress response is an important step in gaining a more complete understanding of support's role in health. Therefore, in the following sections, we explore various endocrine and immune markers that could be incorporated into social support literature to reach the goal of using multiple markers to understand the process of social support. We focus our attention on endocrine and immune markers of stress; however, we should note that other physiological markers (e.g., cardiovascular, musculature, neurological, and genetic), could also provide useful information in understanding the effects of social support on stress.

Markers of Endocrine Function

As scholars extend research on the physiological effects of social support, two markers that may be of interest given the characteristics of the support context are dehydroepiandrosterone-sulfate (DHEA-S) and oxytocin. DHEA-S is a steroid hormone released by the adrenal cortex and is typically elevated in times of stress, as a result of ACTH (Kroboth, Salek, Pittenger, Fabian, & Frye, 1999). In contrast to elevated cortisol, which, when prolonged, is associated with immunosuppression, hypertension, and other adverse effects, DHEA-S may dampen HPA activity, including cortisol, and protect against neurodegeneration (antiglucocorticoid and antiglutamatergic effects). In fact, researchers suggest that DHEA-S balances cortisol in ways that provide the motivation and performance-enhancing effects of acute stress, while limiting the negative effects. For example, Morgan, Rasmusson, Pietrzak, Coric, and Southwick (2009) found the DHEA-S/cortisol ratio significantly predicted greater performance of Special Forces soldiers during high-stress training. Furthermore, DHEA-S has been linked to positive mental (e.g., fewer depressive symptoms) and physical health outcomes (e.g., increase in bone density; Labrie et al., 2005). Therefore, DHEA-S may be a useful marker in studying how support enhances health during stressful situations.

Another hormone that may provide unique insight into effective support processes is oxytocin (Taylor, 2006). Oxytocin is a peptide hormone produced by the posterior pituitary gland (Heinrichs et al., 2003). Along with the biological functions, including initiating uterine contractions and stimulating the milk let-down reflex after childbirth, oxytocin is known for its mother-infant bonding effects during birth and breastfeeding (Feldman, Weller, Zagoory-Sharon, & Levine, 2007) and has been connected to human affection and love (Ditzen et al., 2009; Fisher, Aron, & Brown, 2006). In the context of social support interactions, increased oxytocin may motivate support-seeking behaviors because, according to Kemp and Guastella (2011), oxytocin is related to approach motivations. As reviewed by Ditzen et al. (2009), oxytocin has been linked to increased trust and positive communication. When taken in the context of support seeking (and consistent with tend and befriend theory, see Taylor, 2006), this could provide the impetus for distressed individuals to seek support in more direct and, therefore, more effective ways. Although more speculative, providers may be able to assist hesitant support seekers to speak up by engaging in behaviors, such as touch, that would increase oxytocin.

Markers of Immune System Health

Research involving stress and the immune system has typically taken one of two paths—an evaluation of inflammation and overall immune health or of immune reactivity. As part of the body's natural immune response, the immune system activates inflammation partly with the release of pro-inflammatory cytokines (Boren & Veksler, 2011). HPA activity from the endocrine system can impact the way that the body responds to antigens, by inhibiting the production of certain cytokines (such as interleukin-12 and tumor necrosis factor-alpha) and can increase the amounts of other chemicals (e.g., interleukins 4 and 10, and C-reactive protein) (Segerstrom & Miller, 2004). Therefore, researchers can analyze the underlying pathways by which social support processes interact with HPA activity linking to inflammatory function. As an example, Boren and Veksler (2018) found that trait co-rumination was positively associated with C-reactive protein and negatively associated with interleukin-6. From an immune health perspective, researchers typically gauge various functional metrics of the immune system. Working within this framework does require using multiple immune system markers beyond just cytokines, including natural killer cells, granulocytes, antibodies, monocytes, and C- and T-cells. Recently, Floyd and colleagues (2018) correlated social support with eight different immune markers. They found that perceived social support was associated with IgM, CD3+, and CD4+ and expressed support was associated with IgG, IgM, CD3+, CD4+, CD8+, and CD19+.

Other research programs have explored immune system health by looking at immune reactivity. From this perspective, scholars typically determine antibodies associated with a latent virus (instead of exposing someone to viruses). For instance, both the Epstein-Barr virus (EBV) and Cytomegalovirus (CMV) are present in latent forms in up to 80% of adults over the age of 40 (Boren & Veksler, 2011); therefore, evaluating how well the immune system suppresses these viruses can serve as a metric to immune health. For instance, Floyd, Hesse, Boren, and Veksler (2014) found that affectionate communication can be linked with increasing levels of latent-EBV titers, thereby indicating a reduction in the immune system's ability to suppress the virus. Taken together, we suggest that researchers explore the immune system by evaluating various metrics of immune health, including metrics of inflammation and reactive measures of immunosuppression as they relate to social support.

Interaction Outcomes for the Support Provider

Consistent with taking a dyadic perspective on the support process, the use of multiple markers could also provide insight into the effects of providing support on the support provider, an area of research currently lacking in communication work. One exception is a recent observational study in which participants observed a peer in distress, while their responses to the situation, both behavioral (i.e., whether they provided support or not), physiological, and emotional, were measured (Priem, Giles, & Rigau, 2016). Compared to controls who did not observe a distressed peer, potential support providers reported that the interaction was more stressful and showed increased markers of physiological stress. Although this extremely preliminary research supports the notion that providing support is stressful, it does not provide insight into the effects of that stress on the provider. Furthermore, research has argued that although providing support can be difficult, resulting in issues such as caregiver burnout, provision of support can also be salubrious (Faw, 2016).

Expanding the examination to include multiple physiological markers suggests ways in which the stress associated with providing support may have positive outcomes for the support provider. A study of the Changing Lives in Older Couples dataset (a decades long multi-wave study of over 1,500 married men and women) found that, controlling for the impact of receiving support, giving emotional support to spouses is associated with improved mortality over a 5-year period and receiving support has no effect on mortality after adjusting for giving support (Brown, Nesse, Vinokur, & Smith, 2003). One explanation for this is that providing support creates eustress or gives one positive feelings and fulfilment (Loving & Wright, 2013). To the extent that individuals can perceive the act of providing support to a loved one as a challenge, they may benefit from the adaptive physiological responses to stress. For example, feelings of fulfillment and happiness may be the result of elevated oxytocin.

Looking at immune system health may also help explain this phenomenon. For instance, a recent study of healthy adults found that expressed support within the context of non-distressed relationships was positively associated with multiple markers of immunocompetence. Furthermore, the relationship between immune health and support expression remained significant even when controlling for received support (Floyd et al., 2018). However, the mechanism through which providing support has positive or negative effects on providers is not fully understood. Given that the effects of various biomarkers are known, one objective way of assessing this may be to examine a variety of physiological outcomes for support providers and/or the dyad as a whole.

Examining the Nuances of Appraisals and Physiology in the Support Process

The transactional model of stress and coping (Lazarus & Folkman, 1984) highlights the central role of appraisals in physiological responses to events. Primary appraisals of event valence, relevance, and severity influence the extent of physiological responses and whether arousal is adaptive or maladaptive. Appraisals of threat create distress, in which physiological stress responses have detrimental effects, whereas appraisals of an event as a challenge are associated with eustress, which is more adaptive. The theory of conversationally induced reappraisal (Burleson & Goldsmith, 1996) posits that effective emotional support can facilitate emotional and physiological improvement. Although the role of appraisals has been theorized to be central, little research has explicitly examined its role.

Ways in which support may reduce the negative effects of stress have been documented; however, greater research into eustress provides an avenue for advanced inquiry into how emotional support may influence appraisals in a manner that facilitates positive outcomes. Support theory states that emotional support may function to facilitate stress recovery by prompting reappraisals of the problem or stressor (Burleson & Goldsmith, 1998). Recent research on appraisals of stress and arousal itself suggests that emotional support that helps distressed individuals reframe the stress response itself may facilitate more adaptive outcomes. For example, Keller et al. (2012) found that individuals who experience high levels of stress but do not appraise the stress as harmful had no ill effects on health. Jamieson et al. (2012) showed that reappraising arousal itself makes physiological and cognitive responses to stress more adaptive. Accordingly, a potentially fruitful venue for future research is to explore how emotional support messages can create eustress and facilitate reappraisal of the stress response itself to influence physiological responses and improve health outcomes.

Another potentially interesting avenue for research is to examine emotional support interactions, coping, and appraisals as a dyadic process. For example, according to the systemic transactional model of dyadic coping (Bodenmann, 1995), couples engage in dyadic appraisals of stressors. In a relationship, the origin of a stressor can be individual, but the stressor also becomes part of the partner's experience because the two individuals comprise an interdependent relationship (i.e., indirect dyadic stress). Alternatively, the stressor can be an event that directly affects both people in the relationship (i.e., direct dyadic stress); therefore, individuals engage in intrapersonal and dyadic appraisals of a situation.

The dyadic nature of appraisals has the potential to influence multiple parts of the support process. For example, Peters, Priem, and High (2017) found that congruence in partners' appraisal of the severity of a stressor predicted a support provider's motivation to produce a higher quality supportive message. Because we know that the message matters for physiological stress recovery, the appraisals of both individuals—not just the distressed individual—become important for understanding effective emotional support interactions. Furthermore, dyadic appraisals may influence emotional and physiological linking or contagion in ways that may have implications for message production and message effects in supportive interactions.

Research Directions with Older Populations and Longitudinal Data

Other areas for research should be engaged as well. For instance, scholars have proposed an increased focus on longitudinal research (Taylor et al., 2010), especially with specific demographics. For example, given the stresses of older age and the increased prevalence of illnesses such as cancer or dementia, social support may be of particular importance to the elderly population (Kim & Thomas, 2017). Furthermore, to the extent that long-established relationships may be lost over time as best friends and spouses pass away, the landscape of support in later years of life may be significantly different from that of more commonly studied (i.e., younger) populations. Examining the health effects of ongoing support and what the long-term effects of ongoing supportive relationships may be on both parties is also important (Faw, 2016).

Although some of this work has been done, examining the effects of various types of support, and looking at aging adults of differing socioeconomic strata, would be important extensions of the extant literature (Kim & Thomas, 2017). Furthermore, communication scholars can extend existing scholarship by utilizing more sophisticated measures of support than have been traditionally used in longitudinal multi-wave research. There may also be particularly interesting intersections with research on loneliness (Hawkley & Cacioppo, 2010) that would be particularly salient for older populations. Therefore, there are numerous new avenues for communication researchers to traverse with respect to support and physiology.

CONCLUSION

Research on social support and health is prolific within the social scientific literature. Although results have largely indicated that most forms of support improve physical health, more recent research has also found that certain types of support can be ineffectual or even detrimental. Additionally, much remains to be understood about the means by which support exerts its effects. Therefore, we have proposed a series of areas for future research within the field of communication that we believe can extend this program of work. Specifically, we believe that examining support at the dyadic level, looking at multiple physiological markers, and studying the effects of appraisals can further explicate the nuanced ways in which social support affects human health. Using longitudinal designs and stratifying across varying demographics including age and socioeconomic status can also provide greater insight into how support affects health across the lifespan.

REFERENCES

Baek, R. N., Tannenbaum, M. L., & Gonzalez, J. S. (2014). Diabetes burden and diabetes distress: The buffering effect of social support. *Annals of Behavioral Medicine, 48*(2), 145–155.

Bakker, A. B., & Demerouti, E. (2016). Job demands-resources theory: Taking stock and looking forward. *Journal of Occupational Health Psychology, 22*(3), 273–285.

Bodenmann, G. (1995). A system-transactional conceptualization of stress and coping in couples. *Swiss Journal of Psychology, 54*, 34–49.

Bolger, N., & Amarel, D. (2007). Effects of social support visibility on adjustment to stress: Experimental evidence. *Journal of Personality and Social Psychology, 92*(3), 458–475.

Boren, J. P. (2013). Co-rumination partially mediates the relationship between social support and emotional exhaustion among graduate students. *Communication Quarterly, 61*(3), 253–267.

Boren, J. P. (2014). The relationships between co-rumination, social support, stress, and burnout among working adults. *Management Communication Quarterly, 28*(1), 3–25.

Boren, J. P. (2017). Social support. In C. Scott & L. Lewis (Eds.), *The international encyclopedia of organizational communication*. New York, NY: John Wiley & Sons.

Boren, J. P., & Veksler, A. E. (2011). A decade of research exploring biology and communication: The brain, nervous, endocrine, cardiovascular, and immune systems. *Communication Research Trends, 30*(4), 1–31.

Boren, J. P., & Veksler, A. E. (2018). Co-rumination and immune inflammatory response in healthy young adults: Associations with interleukin-6 and C-reactive protein. *Communication Research Reports, 35*, 152–161.

Brown, S. L., Nesse, R. M., Vinokur, A. D., & Smith, D. M. (2003). Providing social support may be more beneficial than receiving it. *Psychological Science, 14*(4), 320–327.

Burleson, B. R. (2003). The experience and effects of emotional support: What the study of cultural and gender differences can tell us about close relationships, emotion, and interpersonal communication. *Personal Relationships, 10*, 1–23.

Burleson, B. R., & Goldsmith, D. J. (1996). How the comforting process works: Alleviating emotional distress through conversationally induced reappraisals. In P. A. Anderson & L. K. Guerrero (Eds.), *Handbook of communication and emotion* (pp. 246–280). Orlando, FL: Academic Press.

Byrd-Craven, J., Geary, D. C., Rose, A. J., & Ponzi, D. (2008). Co-ruminating increases stress hormone levels in women. *Hormones and Behavior, 53*, 489–492.

Byrd-Craven, J., Granger, D. A., & Auer, B. J. (2011). Stress reactivity to co-rumination in young women's friendships: Cortisol, alpha-amylase, and negative affect focus. *Journal of Social and Personal Relationships, 28*, 469–487.

Cohen, S., & Wills, T. A. (1985). Stress, social support, and the buffering hypothesis. *Psychological Bulletin, 98*, 310–357.

Ditzen, B., & Heinrichs, M. (2014). Psychobiology of social support: The social dimension of stress buffering. *Restorative Neurology and Neuroscience, 32*(1), 149–162.

Ditzen, B., Schaer, M., Gabriel, B., Bodenmann, G., Ehlert, U., & Heinrichs, M. (2009). Intranasal oxytocin increases positive communication and reduces cortisol levels during couple conflict. *Biological Psychiatry, 65*(9), 728–731.

Faw, M. H. (2016). Supporting the supporter. *Journal of Social and Personal Relationships, 35*(2), 202–223.

Feldman, R., Weller, A., Zagoory-Sharon, O., & Levine, A. (2007). Evidence for a neuroendocrinological foundation of human affiliation. *Psychological Science, 18*, 965–970.

Fisher, H. E., Aron, A., & Brown, L. L. (2006). Romantic love: A mammalian brain system for mate choice. *Philosophical Transactions of the Royal Society of London. Series B, Biological Sciences, 361*, 2173–2186.

Floyd, K. (2014). Humans are people, too: Nurturing an appreciation for nature in communication research. *Review of Communication Research, 2*, 1–29.

Floyd, K., Hesse, C., Boren, J. P., & Veksler, A. E. (2014). Affectionate communication can suppress immunity: Trait affection predicts antibody titers to latent Epstein-Barr Virus. *Southern Communication Journal, 79*, 2–13.

Floyd, K., Pauley, P. M., Hesse, C., Eden, J., Veksler, A. E., & Woo, N. T. (2018). Supportive communication is associated with markers of immunocompetence. *Southern Communication Journal, 83*, 229–244.

Floyd, K., & Ray, C. D. (2017). Thanks, but no thanks: Negotiating face threats when rejecting offers of unwanted social support. *Journal of Social and Personal Relationships, 34*, 1260–1276.

Hawkley, L. C., & Cacioppo, J. T. (2010). Loneliness matters: A theoretical and empirical review of consequences and mechanisms. *Annals of Behavioral Medicine, 40*(2), 218–227.

Heinrichs, M., Baumgartner, T., Kirschbaum, C., & Ehlert, U. (2003). Social support and oxytocin interact to suppress cortisol and subjective responses to psychosocial stress. *Biological Psychiatry, 54*(12), 1389–1398.

Helgeson, V. S. (1993). The onset of chronic illness: Its effect on the patient-spouse relationship. *Journal of Social and Clinical Psychology, 12*, 406–428.

Hilmert, C. J., Christenfeld, N., & Kulik, J. A. (2002). Audience status moderates the effects of social support and self efficacy on cardiovascular reactivity during public speaking. *Annals of Behavioral Medicine, 24*(2), 122–131.

Holt-Lunstad, J., Smith, T. B., & Layton, J. B. (2010). Social relationships and mortality risk: A meta-analytic review. *PLoS Medicine, 7*, e1000316.

Jamieson, J. P., Nock, M. K., & Mendes, W. B. (2012). Mind over matter: Reappraising arousal improves cardiovascular and cognitive responses to stress. *Journal of Experimental Psychology: General, 141*, 417–422.

Jones, S. M., & Wirtz, J. G. (2006). How does the comforting process work? An empirical test of an appraisal-based model of comforting. *Human Communication Research, 32*, 217–243.

Keller, A., Litzelman, K., Wisk, L. E., Maddox, T., Cheng, E. R., Creswell, P. D., & Witt, W. P. (2012). Does the perception that stress affects health matter? The association with health and mortality. *Health Psychology, 31*, 677–684.

Kemp, A. H., & Guastella, A. J. (2011). The role of oxytocin in human affect: A novel hypothesis. *Current Directions in Psychological Science, 20*, 222–231.

Kim, S., & Thomas, P. A. (2017). Direct and indirect pathways from social support to health? *The Journals of Gerontology: Series B*, 1–9.

Kirsch, J. A., & Lehman, B. J. (2015). Comparing visible and invisible social support: Non-evaluative support buffers cardiovascular responses to stress. *Stress and Health, 31*(5), 351–364.

Kroboth, P. D., Salek, F. S., Pittenger, A. L., Fabian, T. J., & Frye, R. F. (1999). DHEA and DHEA-S: A review. *Journal of Clinical Pharmacology, 39*(4), 327–348.

Labrie, F., Belanger, A., Lin, S. X., Simard, J., Pelletier, G., & Labrie, C. (2005). Is dehydroepiandrosterone a hormone? *Journal of Endocrinology, 187*(2), 169–196.

Lazarus, R. S., & Folkman, S. (1984). *Stress, appraisal, and coping*. New York, NY: Springer.

Lee, D. Y., Kim, E., & Choi, M. H. (2015). Technical and clinical aspects of cortisol as a biochemical marker of chronic stress. *BMB Reports, 48*, 209–216.

Levine, A., Zagoory-Sharon, O., Feldman, R., Lewis, J. G., & Weller, A. (2007). Measuring cortisol in human psychobiological studies. *Physiological Behavior, 90*, 43–53.

Loving, T. J., & Wright, B. L. (2013). Eustress in romantic relationships. In L. Campbell, J. La Guardia, J. M. Olson, & M. P. Zanna (Eds.), *The science of the couple* (12th ed., pp. 181–196). Philadelphia, PA: Psychology Press.

Maisel, N. C., & Gable, S. L. (2009). The paradox of received social support. *Psychological Science, 20*, 928–932.

Morgan, C. A., Rasmusson, A., Pietrzak, R. H., Coric, V., & Southwick, S. M. (2009). Relationships among plasma dehydroepiandrosterone and dehydroepiandrosterone sulfate, cortisol, symptoms of dissociation, and objective performance in humans exposed to underwater navigation stress. *Biological Psychiatry, 66*, 334–340.

Peters, L., Priem, J. S., & High, A. C. (2017, November). A test of dyadic appraisals: Whose perception matters in motivation to provide quality emotional support? Paper presented at the annual meeting of the National Communication Association, Dallas, TX.

Priem, J. S., Giles, S., & Rigau, E. (2016, May). A mixed-method approach to understanding supportive interactions: Support seekers' problem disclosures and support provider reactions. Paper presented at the annual meeting of the International Communication Association, Fukuoka, Japan.

Priem, J. S., & Solomon, D. H. (2009). Comforting apprehensive communicators: The effects of reappraisal and distraction on cortisol levels among students in a public speaking class. *Communication Quarterly, 57*(3), 259–281.

Priem, J. S., & Solomon, D. H. (2015). Emotional support and physiological stress recovery: The role of support matching, adequacy, and invisibility. *Communication Monographs, 82*(1), 88–112.

Priem, J. S., & Solomon, D. H. (2018). What is supportive about supportive conversation? Qualities of interaction that predict emotional and physiological outcomes. *Communication Research, 45*(3), 443–473.

Rankin, A., Swearingen-Stanborough, C., Granger, D. A., & Byrd-Craven, J. (2018). The role of co-rumination and adrenocortical attunement in young women's close friendships. *Psychoneuroendocrinology, 98*, 61–66.

Rose, A. J. (2002). Co-rumination in the friendships of girls and boys. *Child Development, 73*, 1830–1843.

Rose, A. J., Carlson, W., & Waller, E. M. (2007). Prospective associations of co-rumination with friendship and emotional adjustment: Considering the socioemotional trade-offs of co-rumination. *Developmental Psychology, 43*, 1019–1031.

Roy, M. P., Steptoe, A., & Kirschbaum, C. (1998). Life events and social support as moderators of individual differences in cardiovascular and cortisol reactivity. *Journal of Personality and Social Psychology, 75*(5), 1273–1281.

Sapolsky, R. M., Krey, L. C., & McEwen, B. S. (1986). The neuroendocrinology of stress and aging: The glucocorticoid cascade hypothesis. *Endocrine Review, 7*, 284–301.

Segerstrom, S. C., & Miller, G. E. (2004). Psychological stress and the human immune system: A meta-analytic study of 30 years of inquiry. *Psychological Bulletin, 130*, 601–630.

Shirom, A., Toker, S., Alkaly, Y., Jacobson, O., & Balicer, R. (2011). Work-based predictors of mortality: A 20-year follow-up of healthy employees. *Health Psychology, 30*, 268–275.

Taylor, S. E. (2006). Tend and befriend: Biobehavioral bases of affiliation under stress. *Current Directions in Psychological Science, 15*, 273–277.

Taylor, S. E., Seeman, T. E., Eisenberger, N. I., Kozanian, T. A., Moore, A. N., & Moons, W. G. (2010). Effects of a supportive or an unsupportive audience on biological and psychological responses to stress. *Journal of Personality and Social Psychology, 98*(1), 47–56.

Thoits, P. A. (2011). Mechanisms linking social ties and support to physical and mental health. *Journal of Health and Social Behavior, 52*(2), 145–161.

Toker, S., Shirom, A., Melamed, S., & Armon, G. (2012). Work characteristics as predictors of diabetes incidence among apparently healthy employees. *Journal of Occupational Health Psychology, 17*, 259–267.

Uchino, B. N. (2009). Understanding the links between social support and physical health: A life-span perspective with emphasis on the separability of perceived and received support. *Perspectives on Psychological Science, 4*(3), 236–255.

Uchino, B. N., Bowen, K., Carlisle, M., & Birmingham, W. (2012). Psychological pathways linking social support to health outcomes: A visit with the "ghosts" of research past, present and future. *Social Science and Medicine, 74*(7), 949–957.

24

Developmental Psychophysiology and the Human Stress Response during Communication

Chris R. Sawyer

Recent advances in developmental psychophysiology reveal that the interaction of environmental sensitivity and the quality of early life experiences moderates physiological reactions to social stressors (Ellis, Boyce, Belsky, Bakerman-Kranenburg, & van IJzendoorn, 2011). These new facts, which are based on evolutionary biology, help to explain a common, although counterintuitive, finding. Specifically, far fewer highly reactive individuals are disabled by stress than researchers previously believed. Moreover, individuals with higher levels of reactivity to stress often outperform their less reactive counterparts.

This chapter examines the phylogeny and ontogeny of the human stress response (HSR) in communication. Consistent with its evolutionary history, the HSR comprises the defensive reactions adapted to the ancestral environment of humans that are governed or regulated by specialized neural circuits called substrates. The operation of these circuits is most often accompanied by markedly increased arousal compared to an individual's typical or baseline level. Properly employed, current technologies used in psychophysiology provide researchers with unbiased measures of the HSR in communication. Unlike biological traits that are inherited directly, individual differences in the HSR arise from the interaction of early life experiences and a person's susceptibility to environmental influence. Throughout this chapter, recent scientific studies of communication will be examined in relation to this emerging paradigm.

SPECIES-SPECIFIC CHARACTERISTICS OF HUMAN STRESS

Phylogeny refers to the evolutionary history of organisms, their behaviors, anatomical forms, and related physiology. Accordingly, stress responses emerged very early among *Homo sapiens* and were useful in helping them manage imminent dangers as well as other threats to survival. Among modern humans, stress is usually associated with deleterious effects to mental and physical health. Examining the phylogeny of the HSR, however, reveals the contours of these underlying mechanisms. Specifically, the HSR constitutes a species-specific survival mechanism shaped by natural selection, regulated by neural anatomy, and observable in human physiology. Further, chronic stress permeates industrialized societies and negatively affects physical health, mental well-being, and social functioning for millions in developed countries.

Shaped by Natural Selection

Biologists and physical anthropologists maintain that the human brain and nervous system evolved during the Pleistocene epoch, which lasted from approximately 2.58 million years ago to around 12,000 years ago (Ash & Gallup, 2007; Cosmides & Tooby, 2002). Throughout much of this era, *Homo sapiens* roamed the African Savanna in bands of hunter-gatherers numbering 50 or so related kin. This physical setting and the social structure of early humans are known as the *environment of evolutionary adaptedness* (EEA: Bowlby, 1969). Predation and resource scarcity were recurring challenges for early humans. Consequently, inhibition, vigilance, escape, and aggression became indispensable to the success of *Homo sapiens* (Tovote et al., 2016). Each of these defensive responses was accompanied by its own pattern of physiological arousal and behaviors (Duke, Bègue, Bell, & Eisenlohr-Moul, 2013; Panayiotou, Karelka, Georgiou, Constanitou, & Paraskeva-Siamata, 2017; Sege, Bradley, & Lang, 2018). In the aggregate, these arrangements permit the immediate availability of responses to physical danger or other imminent threats (Nesse, Bhatnagar, & Young 2010). This capacity, in turn, provided a selective advantage for humans. Therefore, the HSR is a species-specific adaptation by *Homo sapiens* to their ancestral environment aimed at survival.

Regulated by Neural Anatomy

Broadly defined, stress is the body's response to a loss of equilibrium in the processes that maintain proper functioning, such as body temperature, fluid balance, and nourishment. Stressors, therefore, are stimuli or conditions that disrupt homeostasis. Stressors can include psychological distress and challenging social situations, such as public speaking. Irrespective of the conditions from which they arise, most stressors begin as sensory experiences that are processed within the cortical and subcortical regions of the nervous system. Neural substrates regulate these reactions, often autonomously. From a neurobiology perspective, the HSR is composed of a network of these circuits, operating at both subcortical and cortical levels, that enables humans to adapt appropriately to stressors (Friedman, 2015). Subcortical reactions are comparatively fast and often take place without conscious awareness. In contrast, cortical-level responses are frequently the product of conscious, effortful processing. Although considerably slower than their subcortical counterparts, cortical processes can result in more accurate assessments of potential threats. In keeping with recent advances in social cognition, the HSR can be alternatively an intuitive and a deliberative process. Although scholarly writings are replete with slightly differing accounts of how neural substrates operate in the HSR, most stress researchers describe three ongoing operations, namely, sensory processing, threat detection, and physiological adaptation. Figure 24.1 is a schematic diagram of the neurophysiology involved in the HSR. The following explanation emphasizes how communication situations, such as public speaking, serve as a trigger for the HSR.

The HSR begins when events or other stimuli are registered by the sensory thalamus, a mass of gray matter located near the center of the midbrain and just above the brainstem. The thalamus relays impulses to the sensory cortex and the amygdala, also known as the brain's threat detector (LeDoux, 2007, 2012). Threat detection, the second stage of the HSR, occurs when the amygdala receives threat-relevant impulses directly from the sensory thalamus or indirectly from either the sensory cortex or the prefrontal cortex (PFC). Once it receives these

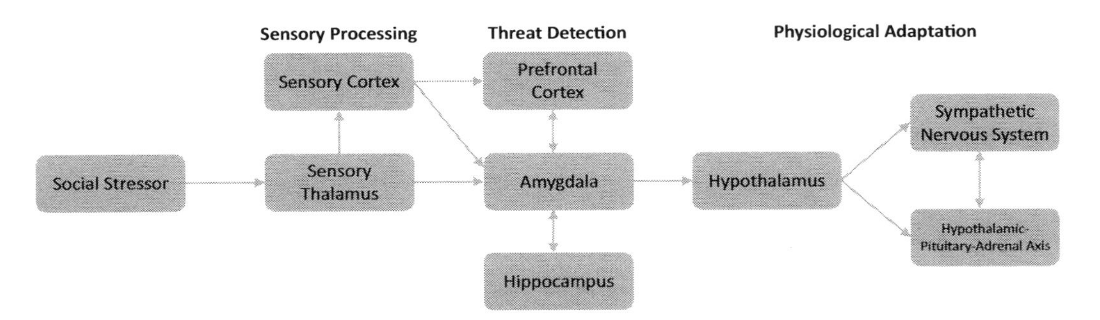

Figure 24.1 Schematic Diagram of Neurophysiology for the Human Stress Response during Communication

impulses, the amygdala then recruits memory resources from the hippocampus. In turn, the hippocampus alerts the amygdala as to whether previous experiences with a stimulus were negative. From this point, the amygdala sends a distress signal to the hypothalamus, which in turn triggers the sympathetic nervous system and hypothalamic-pituitary-adrenal axis. Among the physiological reactions that accompany the HSR are accelerated heart rate, changes in skin resistance, increased muscular tension, and the profusion of stress hormones such as cortisol and alpha-amylase. By marshalling energy resources throughout the body, the HSR ensures the levels of arousal required for defensive responses, such as vigilance, avoidance, escape, or active engagement with environmental stressors. Operationally, these neurological, physiological, and behavioral concomitants of survival constitute the HSR (Everly & Lating, 2013).

Previous research provides evidence for the operation of these mechanisms during challenging communication situations such as public speaking. First, the thalamus is a powerful integrator of sensory experiences related to social threats. In previous studies, sensory-rich audience characteristics, such as size, composition, and nonverbal cues, have been linked to speaker anxiety (Seta, Wang, Crisson, & Seta, 1989; Taylor et al., 2010). Speakers often gauge the potential for social embarrassment by attending to the facial and vocal reactions of their audiences. Likewise, activity in the PFC has been associated with excessive worrying (Hofmann et al., 2005) and contributes to the level of speaker anxiety *before* giving a presentation (Davidson, Marshall, Tomarken, & Henriques, 2000; Glassman et al., 2014). Anxious arousal or panic *during* a speech performance is associated with over-stimulation in the amygdala combined with catastrophic thinking produced by the PFC (Brinkmann et al., 2017; Finn, Sawyer, & Behnke, 2009; Gellatly & Beck, 2016; Henderson et al., 2016). Numerous physiological indicators of the HSR also serve as reliable indexes of speaker state anxiety. As a result, public speaking situations have provided the testbed for programs of research on the HSR in communication (Bodie, 2010; Sawyer & Behnke, 2009) and neuroscience, such as the Trier Social Stress Test (Allen, Kennedy, Cryan, Dinan, & Clarke, 2014; Kudielka, Hellhammer, & Kirschbaum, 2007).

Although modern civilization mitigates most physical dangers for humans, the social interactions upon which societies depend are frequently unpleasant and threatening. During these encounters, the ancient physiological reactions associated with human defensive reactions are mobilized. Social interaction thereby contributes to the prevalence of chronic stress in the general population (Booij, Snippe, Jeronimus, Wichers, & Wigman, 2018). Consequently, the psychophysiology of speakers and listeners often reveals the effects of the HSR during other forms of communication. Table 24.1 contains a summary of recent research in which psychophysiological methods were used to study the HSR in various communication contexts in addition to public speaking.

Table 24.1 Recent Psychophysiological Studies of Human Stress Response (HSR) Markers in Various Communication Contexts

	Endocrine System Studies	
HSR Marker	*Communication Context*	*Study*
Salivary cortisol	Public speaking	Kothgassner et al. (2016)
Salivary cortisol	Public speaking	Van Den Bos & Westenberg (2015)
Salivary α-amylase Cortisol Interleukin-1β	Public speaking	Auer et al. (2018)
Salivary α-amylase	Group & public speaking	Katz & Peckins (2017)
Salivary α-amylase Salivary cortisol	Marital conflict	Denes et al. (2014)
Salivary cortisol	Public speaking interviewing	Abakov, Voyt, & Shaboltas (2015)
Salivary cortisol Immunoglobulin-A	Job interviews	Campisi et al. (2012)
Immunoglobulin-M	Affection	Floyd et al. (2014)
Salivary cortisol	Dating partner conflict	Aloia & Solomon (2015)
Salivary cortisol	Dating partner emotional support	Priem & Solomon (2015)
	Hemodynamic Studies	
HSR Marker	*Communication context*	*Study*
Cardiovascular reactivity	Everyday conversations	Tardy (2014)
Heart rate	Family storytelling	Honeycutt et al. (2014a)
Heart rate	Public speaking	Panayiotou et al. (2017)
Heart rate variability	Public speaking	Kothgassner et al. (2016)
Heart rate variability	Observing marital conflict	Honeycutt et al. (2014b)
Heart rate	Public speaking	Van Den Bos & Westenberg (2015)
Systolic blood pressure Diastolic blood pressure Heart rate	Public speaking	Elfering & Grebner (2012)
Heart rate	Group communication	von Dawans et al. (2011)
Heart rate	Public speaking	Chadwick et al. (2016)
	Electrodermal Activity	
HSR Marker	*Communication Context*	*Study*
Skin conductance	Public speaking	Panayiotou et al. (2017)
Skin conductance	Public speaking	Van Den Bos & Westenberg (2015)
Galvanic skin response	Dyadic touch	Lougheed et al. (2016)
Skin conductance Skin temperature	Listening	Mackersie & Cones (2011)
	Neuromuscular Reactions	
HSR Marker	*Communication Context*	*Study*
Laryngeal muscle response	Public speaking	Helou et al. (2018)
Electromyographic activity	Listening	Mackersie & Cones (2011)
Laryngeal tightness (lump in throat)	Conversation	Holmqvist et al. (2013)
Electromyogram	Facial recognition	Künecke et al. (2017)
Eye blink	Face-to-face conversations	Hömke et al. (2017)

EXPLANATIONS FOR INDIVIDUAL DIFFERENCES IN THE HSR

Unlike phylogeny, which systematically pieces together the evolutionary history of a species, *ontogenesis* describes how the distinguishing characteristics or traits of individuals develop across the lifespan. Many scholars once presumed that the HSR was bound to species-level effects. That is, stress was viewed as a mismatch of the responses adapted to the EEA and the social demands of modern life. Despite the pervasiveness of stress in human societies, some individuals were more resistant to its effects than were others. Researchers then turned their attention to discover whether certain developmental influences, such as adversity during childhood, contributed to susceptible to stress-borne disease. Recently, this diathesis-stress model has been superseded by a nuanced perspective that also takes into account individual differences in environmental sensitivity. Each of these perspectives informs the psychophysiology of human communication.

The Diathesis-Stress Model

Medical and clinical studies frequently report that stress is detrimental to the well-being and functioning of humans (Bergeman & Deboeck, 2014). Individuals who experience chronic or repeated stressors frequently develop disability and disease. These include an array of physical illnesses (Shields & Slavich, 2017), psychiatric conditions (van Nierop et al., 2018; Zorn et al., 2017), and behavioral disorders (Reijman et al., 2016), including social anxiety (Fehm, Beesdom, Jacobi, & Fiedler, 2008). One explanation for these effects is that by inducing stress, various environmental conditions reveal individual vulnerabilities to disease or psychological disorders. According to this position, positive conditions reveal little if any vulnerability, whereas negative conditions markedly increase a person's susceptibility to stress-related ailments. Furthermore, adversity during early life appears to trigger vulnerability to stress-related disease or emotional disorders that can persist throughout life. That is, children who are maltreated by parents (Rothbart & Bales, 2006) or suffer privation during early life (Burmeister, McInnis, & Zollner, 2008) are more likely to develop negative conditions, but show little or no susceptibility when reared in supportive environments. This idea is called diathesis-stress (Monroe & Simons, 1991) or the transactional/dual risk model (Sameroff, 1983). From this point of view, human stress represents a throwback or *atavistic* response to imminent danger, triggered by the challenges of modern life.

Many scholars presume that stress is inherently unhealthy. This recurrent theme has formed the basis for studies of organizational (Keashly & Neuman, 2009), family (Theiss, 2018), and health communication (Moreland & Apker, 2016). In addition, the cognitive and physiological concomitants of stress often impair communication competence (LeFebvre, LeFebvre, & Allen, 2018; Sawyer, 2016; Sawyer & Richmond, 2015), such as impeding vocal fluency during assigned public speaking (Choi, Honeycutt, & Bodie, 2015). Increasingly, researchers have used psychophysiology as a lens through which to understand how stress affects communication (Honeycutt, Sawyer, & Keaton, 2014). These include studies of marital conflict (Weusthoff, Baucom, & Hahlweg, 2013), caregiving and comfort (Priem & Solomon, 2015), affectionate communication (Floyd et al., 2009), family systems (Afifi, Granger, Benes, Joseph, & Aldeis, 2011), and public speaking (Bodie, 2012; Sawyer & Behnke, 2014).

Despite an extensive corpus of literature supporting the diathesis-stress model (DSM), scholars have recently pointed out several of its shortcomings. First, although chronic stress is often associated with increased health risk, some individuals are more resistant to stress and recover more quickly from its effects than do others. These individuals function well under adverse conditions, rarely succumb to physical illnesses, and report greater emotional stability than their less-stress-resistant counterparts (Bergeman & Deboeck, 2014; Luthar, 2006). Consequently, stress does not always reveal physical or psychological susceptibility due to individual differences in resistance. Next, the DSM does not explain how supportive or resource-enriched environments affect human development. According to Ellis et al. (2011), many DSM studies look only at negative behavioral responses (e.g., depressed vs. not depressed) to a few environmental conditions (adversity vs. no adversity). As a result, diathesis-stress research focuses disproportionately on the negative effects of adversity, without considering how positive conditions influence human functioning and health (Belsky & Pluess, 2009). Third, natural selection favors a range of phenotypic responses, including those who adapt better to either positive or negative environments (Ellis & Boyce, 2008). Greater plasticity, or the ability to adapt favorably to changing conditions, improves the reproductive fitness of a species throughout its evolutionary history (Del Giudice, 2016). Although life on the Savanna provided early humans with times of comparative safety and abundance as well as times of peril and want, the DSM does not address adaptation to alternating environmental conditions. Finally, in many studies, participants who were most susceptible to environmental influence were helped as much by positive conditions as they were hurt by negative ones. For example, Blair (2010) observed children who were unable to regulate their negative emotions under low-quality environments but were much better behaved in high-quality ones. This finding and others like it directly contradict the DSM. In response to these criticisms, scholars have formulated alternative explanations for individual differences in human stress response based on developmental neurobiology.

Differential Sensitivity Theories

Ellis and Boyce (Boyce & Ellis, 2005; Ellis & Boyce, 2008) advance the idea that both highly stressful and highly protective environments yield disproportionate numbers of highly reactive children. According to this account, natural selection tends to favor a range of phenotypes in response to recurring environmental conditions (Ellis & Boyce, 2008). This capacity is known as adaptive phenotypic plasticity. Accordingly, up-regulated stress responses in children interact with protective environments to produce high levels of cognitive and social competence. Conversely, up-regulated responses to negative conditions, such as harsh upbringing, lead to greater sensitivity to threat cues and lower stress thresholds. Accordingly, these alternative responses reflect a biological sensitivity to context (BSC).

In addition to Ellis and Boyce's BSC, Belsky's differential sensitivity theory (DST) (Belsky 1997, 2005; Belsky & Pluess, 2013) predicts that the interaction between highly reactive phenotypes and their environments increases plasticity to both positive and negative conditions. That is, adaptiveness reflects individual differences in the neurobiology of the HSR. Thus, children who are more physiologically responsive to stressors will have greater plasticity than those with dampened stress reactions. According to Del Giudice (2016), studies of twins and adoptees are consistent with this prediction. Specifically, individuals with greater plasticity seem to have less of a trait in negative environments and more of it in positive ones. Adaptive phenotypic plasticity is a developmental response to one's early environment, such as parental warmth or family conflict. Studies in communication have used sensory processing sensitivity (SPS; Aron & Aron, 1997) as a measure of general biological reactivity to the environment (Loersch & Arbuckle, 2013). Previous researchers have reported that SPS is a strong predictor of communication adaptability (Glonek et al., 2007), that it increases distraction during listening (Gearhart, 2011, 2014), and that it is positively correlated with both communication apprehension and stress among college students (Gearhart & Bodie, 2012). Consequently, the relationship between environment and responsiveness profoundly affects individuals "for better and for worse" (Belsky, Bakermans-Kranenburg, & van IJzendoorn, 2007).

CONCLUSION

According to Albert Ax (1964), there appears to be a confusing interweaving of psyche and soma among *Homo sapiens*, due to their evolutionary history and ontogenetic growth. Consequently, the general goal of psychophysiology is the discovery of the translator mechanism that permits the software (or mind) and the hardware of the body to work together seamlessly. More recently, scholars have begun to use dual-process models when explaining communication that distinguish faster cognitive responses (System 1) from slower, more deliberative ones (System 2; Bodie, Keaton, & Jones, 2018; Carpenter, 2015; Kahneman, 2011). In line with Ax's (1964) data processing analogy, the limbic system enfolds a kind of emotional computer that operates on the cortical and subcortical levels of the nervous system. The amygdala, which is connected to subcortical as well as cortical structures, allows for both System 1 and System 2 processing during stress. The direct pathway between the sensory thalamus and the amygdala provides greater speed when dealing with imminent threats. This System 1 response allows for the detection of threats without a person's conscious awareness or control. In contrast, the effortful processing of social stressors on the cortical level can result in greater emotional control and more accurate threat assessment, but at the expense of longer processing times.

Combined, these systems explain a wide range of emotional processes, including emotional resilience (Salehinejad, Nejati, & Derakhshan, 2017) and emotion regulation (Dixon, Thiruchselvam, Todd, & Christoff, 2017). For example, despite the efficacy of certain coping strategies, the strength of unconscious bottom-up processes can frequently override them. The neural architecture of the HSR represents the adaptation by our species to life on the Savanna. However, humans are not born with a particular degree of reactivity to stressors. Rather, individual differences in responsiveness begin in early life and continue to develop throughout adolescence and into early adulthood. Fundamentally, the alignment of the stress response system with these formative social experiences constitutes a type of translator mechanism for psychophysiological responses during communication.

REFERENCES

Abakov, V. A., Voyt, T., & Shaboltas, A. (2015). Psychological and cortical human reactions in experimental stress. *Archive of Psychiatry and Psychotherapy, 17*, 31–40.

Afifi, T. D., Granger, D. A., Denes, A., Joseph, A., & Aldeis, D. (2011). Parents' communication skills and adolescents' salivary α-amylase and cortisol response patterns. *Communication Monographs, 78*, 273–295.

Allen, A. P., Kennedy, P. J., Cryan, J. F., Dinan, T. G., & Clarke, G. (2014). Biological and psychological markers of stress in humans: Focus on the Trier Social Stress Test. *Neuroscience and Biobehavioral Reviews, 38*, 94–124.

Aloia, L. S., & Solomon, D. H. (2015). Conflict intensity, family history, and physiological stress reactivity to conflict within romantic relationships. *Human Communication Research, 41*, 367–389.

Aron, E. N., & Aron, A. (1997). Sensory-processing sensitivity and its relation to introversion and emotionality. *Journal of Personality and Social Psychology, 73*, 345–368.

Ash, J., & Gallup, G. (2007). Paleoclimatic variation and brain expansion during human evolution. *Human Nature, 18*, 109–124.

Auer, B. J., Calvi, J. L., Jordan, N. M., Schrader, D., & Byrd-Craven, J. (2018). Communication and social interaction anxiety enhance interleukin-1 beta and cortisol reactivity during high-stakes public speaking. *Psychoneuroendocrinology, 94*, 83–90.

Ax, A. F. (1964). Goals and methods of psychophysiology. *Psychophysiology, 1*, 10–25.

Belsky, J. (1997). Variation in susceptibility to rearing influences: An evolutionary argument. *Psychological Inquiry, 8*, 182–186.

Belsky, J. (2005). Differential susceptibility to rearing influences: An evolutionary hypothesis and some evidence. In B. Ellis & D. Bjorklund (Eds.), *Origins of the social mind: Evolutionary psychology and child development* (pp. 139–163). New York, NY: Guilford Press.

Belsky, J., Bakermans-Kranenburg, M., & van IJzendoorn, M. H. (2007). For better *and* for worse: Differential susceptibility to environmental influences. *Current Directions in Psychological Science, 16*, 300–304.

Belsky, J., & Pluess, M. (2009). Beyond diathesis stress: Differential susceptibility to environmental influences. *Psychological Bulletin, 135*, 885–908.

Belsky, J., & Pluess, M. (2013). Beyond risk, resilience, and dysregulation: Phenotypic plasticity and human development. *Development and Psychopathology, 25*, 1243–1261.

Bergeman, C. S., & Deboeck, P. R. (2014). Trait stress resistance and dynamic stress dissipation on health and well-being: The reservoir model. *Research in Human Development, 11*, 108–125.

Blair, C. (2010). Stress and the development of self-regulation in context. *Child Development Perspectives, 4*, 181–188.

Bodie, G. D. (2010). A racing heart, rattling knees, and ruminative thoughts: Defining, explaining, and treating public speaking anxiety. *Communication Education, 59*, 70–105.

Bodie, G. D. (2012). Task stressfulness moderates the effects of verbal person centeredness on cardiovascular reactivity: A dual-process account of the reactivity hypothesis. *Health Communication, 27*, 569–580.

Bodie, G. D., Keaton, S. A., & Jones, S. M. (2018). Individual listening values moderate the impact of verbal person centeredness on helper evaluation: A test of the dual-process theory of supportive message outcomes. *International Journal of Listening, 32*, 127–139.

Booij, S. H., Snippe, E., Jeronimus, B. F., Wichers, M., & Wigman, J. T. W. (2018). Affective reactivity to daily life stress: Relationship to positive psychotic and depressive symptoms in a general population sample. *Journal of Affective Disorders, 225*, 474–481.

Bowlby, J. (1969). *Attachment and loss* (Vol. 1). New York, NY: Basic Books.

Boyce, W. T., & Ellis, B. J. (2005). Biological sensitivity to context: I. An evolutionary-developmental theory of the origins and functions of stress reactivity. *Development and Psychopathology, 17*, 271–401.

Brinkmann, L., Buff, C., Feldker, K., Tupak, S. V., Becker, M. P. I., Hermann, M. J., & Straube, T. (2017). Direct phasic and sustained brain responses and connectivity of amygdala and bed nucleus of the stria terminalis during threat anticipation in panic disorder. *Psychological Medicine, 47*, 2675–2688.

Burmeister, M., McInnis, M. G., & Zollner, S. (2008). Psychiatric genetics: Progress and controversy. *Nature Reviews Genetics, 9*, 527–540.

Campisi, J., Bravo, Y., Cole, J., & Gobeil, K. (2012). Acute psychological stress differentially influences salivary endocrine and immune measures in undergraduate students. *Physiology and Behavior, 107*, 317–321.

Carpenter, C. J. (2015). A meta-analysis of the ELM's argument quality x type predictions. *Human Communication Research, 41*, 501–534.

Chadwick, A. E., Zoccola, P. M., Figueroa, W. S., & Rabindeau, E. M. (2016). Communication and stress: Effects of hope evocation and rumination messages on heart rate, anxiety, and emotions after a stressor. *Health Communication, 31*, 1447–1459.

Choi, C. W., Honeycutt, J. M., & Bodie, G. D. (2015). Effects of imagined interactions and rehearsal of speaking performance. *Communication Education, 64*, 25–44.

Cosmides, L., & Tooby, J. (2002). Unraveling the enigma of human intelligence: Evolutionary psychology and the multi-modular mind. In R. Sternberg & J. Kaufman (Eds.), *The evolution of intelligence* (pp. 145–198). Mahwah, NJ: Lawrence Erlbaum Associates.

Davidson, R. J., Marshall, J. R., Tomarken, A. J., & Henriques, J. B. (2000). While a phobic waits: Regional brain activity and autonomic activity in social phobics during anticipation of public speaking. *Biological Psychiatry, 47*, 85–95.

Del Giudice, M. D. (2016). Differential susceptibility to the environment: Are developmental models compatible with evidence from twins studies? *Developmental Psychology, 52*, 1330–1339.

Del Giudice, M., Hinnant, J. B., Ellis, B. J., & El Sheikh, M. (2012). Adaptive patterns of stress responsivity: A preliminary investigation. *Developmental Psychology, 48*, 775–790.

Denes, A., Afifi, T. D., Granger, D. A., Joseph, A., & Aldeis, D. (2014). Interparental conflict and parents' inappropriate disclosures: Relations to parents' and children's salivary α-amylase and cortisol. In J. M. Honeycutt, C. R. Sawyer, & S. A. Keaton (Eds.), *The influence of communication on physiology and health* (pp. 131–152). New York, NY: Peter Lang.

Dixon, M. L., Thiruchselvam, R., Todd, R., & Christoff, K. (2017). Emotion and the prefrontal cortex: An integrative review. *Psychological Bulletin, 143*, 1033–1081.

Duke, A. A., Bègue, L., Bell, R., & Eisenlohr-Moul, T. (2013). Revisiting the serotonin-aggression relation in humans: A meta-analysis. *Psychological Bulletin, 139*, 1148–1172.

Elfering, A., & Grebner, S. (2012). Getting used to academic public speaking: Global self-esteem predicts habituation in blood pressure to repeated thesis presentations. *Applied Psychophysiology and Biofeedback, 37*, 109–120.

Ellis, B. J., & Boyce, W. T. (2008). Biological sensitivity to context. *Current Directions in Psychological Science, 17*, 183–187.

Ellis, B. J., Boyce, W. T., Belsky, J., Bakerman-Kranenburg, M. J., & van IJzendoorn, M. H. (2011). Differential susceptibility to the environment: An evolutionary-neurodevelopmental theory. *Development and Psychopathology, 23*, 7–28.

Everly, G. S., & Lating, J. M. (2013). *A clinical guide to the treatment of the human stress response* (3rd ed.). New York, NY: Springer.

Fehm, L., Beesdom, K., Jacobi, F., & Fiedler, A. (2008). Social anxiety disorder above and below the diagnostic threshold: Prevalence, comorbidity, and impairment in the general population. *Social Psychiatry and Psychiatric Epidemiology, 43*, 257–265.

Finn, A. N., Sawyer, C. R., & Behnke, R. R. (2009). A model of anxious arousal for public speaking. *Communication Education, 58*, 417–432.

Floyd, K., Boren, J. P., Hannawa, A. F., Hesse, C., McEwan, B., & Veksler, A. E. (2009). Kissing in marital and cohabitating relationships: Effects on blood lipids, stress, and relational satisfaction. *Western Journal of Communication, 73*, 113–133.

Floyd, K., Pauley, P. M., Hesse, C., Veksler, A. E., Eden, J., & Mikkelson, A. C. (2014). Affectionate communication is associated with markers of immune and cardiovascular system competence. In J. M. Honeycutt, C. R. Sawyer, & S. A. Keaton (Eds.), *The influence of communication on physiology and health* (pp. 115–130). New York, NY: Peter Lang.

Friedman, M. J. (2015). The human stress response. In N. C. Bernardy & M. J. Friedman (Eds.), *A practical guide to PTSD treatment: Pharmacological and psychotherapeutic approaches* (pp. 9–19). Washington, DC: American Psychological Association.

Gearhart, C. C. (2011, November). The impact of sensory-processing sensitivity on communication: A preliminary self-report study. Paper presented at the annual meeting of the National Communication Association, Chicago, IL.

Gearhart, C. C. (2014). Sensory-processing sensitivity and nonverbal decoding: The effect of listening ability and accuracy. *The International Journal of Listening, 28*, 88–111.

Gearhart, C. C., & Bodie, G. D. (2012). Sensory processing sensitivity and communication apprehension: Dual influences on self-reported stress in a college student sample. *Communication Reports, 25*, 27–39.

Gellatly, R., & Beck, A. T. (2016). Catastrophic thinking: A transdiagnostic process across psychiatric disorders. *Cognitive Therapy and Research, 40*, 441–452.

Glassman, L. H., Herbert, J. D., Forman, E. M., Bradley, L. E., Izzetoglu, M., Ruocco, A. C., & Goldstein, S. P. (2014). Near-infrared spectroscopic assessment of in vivo prefrontal activation in public speaking anxiety: A preliminary study. *Psychology of Consciousness: Theory, Research, and Practice, 1*, 271–283.

Glonek, K., Nash, E., Shields, V., Sawyer, C. R., & Behnke, R. R. (2007, November). Communication adaptability as a function of nervous system mobility and sensory processing sensitivity. Paper presented at the meeting of the National Communication Association, Chicago, IL.

Helou, L. B., Rosen, C. A., Wang, W., & Verdolini Abbot, K. (2018). Intrinsic laryngeal muscle response to a public speech preparation stressor. *Journal of Speech, Language, and Hearing Research, 61*, 1525–1543.

Henderson, L. A., Akhter, R., Youseef, A. M., Reeves, J. M., Peck, C. C., Murray, G. M., & Svensson, P. (2016). The effects of catastrophizing on central motor activity. *European Journal of Pain, 20*, 639–651.

Hofmann, S. G., Moscovitch, D. A., Litz, B. T., Kim, H. J., Davis, L. L., & Pizzagalli, D. A. (2005). The worried mind: Automatic and prefrontal activation during worrying. *Emotion, 5*, 464–475.

Holmqvist, S., Santtila, P., Lindstrom, E., Sala, E., & Simberg, S. (2013). The association between possible stress markers and vocal symptoms. *Journal of Voice, 27*, 787-e1.

Hömke, P., Holler, J., & Levinson, S. C. (2017). Eyeblink as addressee feedback in face-to-face conversation. *Research on Language and Social Interaction, 50*, 54–70.

Honeycutt, J. M., Bannon, B. D., & Hatcher, L. C. (2014). Effects of positive family conflict-renewal stories on heart rate. In J. M. Honeycutt, C. R. Sawyer, & S. A. Keaton (Eds.), *The influence of communication on physiology and health* (pp. 11–31). New York, NY: Peter Lang.

Honeycutt, J. M., Keaton, S. A., Hatcher, L. C., & Hample, D. (2014). Effects of rumination and observing marital conflict on observers' heart rates as they advise and predict the use of conflict tactics. In J. M. Honeycutt, C. R. Sawyer, & S. A. Keaton (Eds.), *The influence of communication on physiology and health* (pp. 73–92). New York, NY: Peter Lang.

Honeycutt, J. M., Sawyer, C. R., & Keaton, S. A. (Eds.). (2014). *The influence of communication on physiology and health*. New York, NY: Peter Lang.

Kahneman, D. (2011). *Thinking, fast and slow*. New York, NY: Farrar, Straus, & Giroux.

Kanazawa, S. (2004). The Savanna principle. *Managerial and Decision Economics, 25*, 41–54.

Katz, D. A., & Peckins, M. K. (2017). Cortisol and salivary alpha-amylase trajectories following a group social-evaluative stressor with adolescents. *Psychoneuroendocrinology, 86*, 8–16.

Keashly, L., & Neuman, J. H. (2009). Building a constructive communication climate: The Workplace Stress and Aggression project. In P. Lutgen-Sandvik & B. D. Sypher (Eds.), *Destructive organizational communication: Processes, consequences, and constructive ways of organizing* (pp. 339–362). New York, NY: Routledge.

Kothgassner, O. D., Feinhofer, A., Hiavais, H., Beutl, L., Palme, R., Kryspin-Exner, I., & Glenk, L. M. (2016). Salivary cortisol and cardiovascular reactivity to a public speaking task in a virtual and real-life environment. *Computers in Human Behavior, 62*, 124–135.

Kudielka, B. M., Hellhammer, D. H., & Kirschbaum, C. (2007). Ten years of research with the Trier Social Stress Test–revisited. In E. Harmon-Jones & P. Winkielman (Eds.), *Social neuroscience: Integrating biological and psychological explanation of social behavior* (pp. 58–83). New York, NY: Guilford Press.

Künecke, J., Wilhelm, O., & Sommer, W. (2017). Emotion recognition in nonverbal face-to-face communication. *Journal of Nonverbal Communication, 41*, 221–238.

LeDoux, J. E. (2007). The amygdala. *Current Biology, 17*, R868–R874.

LeDoux, J. E. (2012). Rethinking the emotional brain. In J. M. Honeycutt, C. R. Sawyer, & S. A. Keaton (Eds.), *The influence of communication on physiology and health* (pp. 207–227). New York, NY: Peter Lang.

LeFebvre, L., LeFebvre, L. E., & Allen, M. (2018). Training the butterflies to fly in formation: Cataloguing student fears about public speaking. *Communication Education, 67*, 348–362.

Loersch, C., & Arbuckle, N. L. (2013). Unraveling the mystery of music: Music as an evolved group process. *Journal of Personality and Social Psychology, 105*, 777–798.

Lougheed, J. P., Koval, P., & Hollenstein, T. (2016). Sharing the burden: The interpersonal regulation of emotion in mother-daughter dyads. *Emotion, 16*, 83–93.
Luthar, S. S. (2006). Resilience in development: A synthesis of research across five decades. In D. Cicchetti & D. J. Cohen (Eds.), *Developmental psychopathology: Risk, disorder, and adaptation* (Vol. 3, 2nd ed., pp. 739–795). New York, NY: John Wiley & Sons.
Mackersie, C. L., & Cones, H. (2011). Subjective and psychophysiological indexes of listening effort in a competing-taker task. *Journal of the American Academy of Audiology, 22*, 113–122.
Monroe, S. M., & Simons, A. D. (1991). Diathesis-stress theories in the context of life stress research: Implications for depressive disorders. *Psychological Bulletin, 110*, 406–425.
Moreland, J. J., & Apker, J. (2016). Conflict and stress in hospital nursing: Improving communication responses to enduring professional challenges. *Health Communication, 31*, 815–823.
Nesse, R. M., Bhatnagar, S., & Young, E. A. (2010). Evolutionary origins and functions of the stress response. In G. Fink (Ed.), *Stress of war, conflict, and disaster* (pp. 11–16). San Diego, CA: Elsevier Academic Press.
Panayiotou, G., Karelka, M., Georgiou, D., Constanitou, E., & Paraskeva-Siamata, M. (2017). Psychophysiological and self-reported reactivity associated with social anxiety and public speaking fear symptoms: Effects of fear versus distress. *Psychiatry Research, 255*, 278–286.
Priem, J. S., & Solomon, D. H. (2015). Emotional support and physiological stress recovery: The role of support matching, adequacy, and invisibility. *Communication Monographs, 82*, 88–112.
Reijman, S., Bakermans-Kranenburg, M. J., Hiraoka, R., Crouch, J. L., Milner, J. S., Alink, L. R. A., & van IJzendoorn, M. H. (2016). Baseline functioning and stress reactivity in maltreating parents and at-risk adults: Review and meta-analysis of autonomic nervous system studies. *Child Maltreatment, 21*, 327–342.
Rothbart, M. K., & Bates, J. E. (2006). Temperament. In E. Eisenberg, W. Damon, & R. M. Lerner (Eds.), *Handbook of child psychology*: Vol. 3. *Social, emotional, and personality development* (6th ed., pp. 99–166). Hoboken, NJ: John Wiley & Sons.
Salehinejad, M. A., Nejati, V., & Derakhshan, M. (2017). Neural correlates of trait resiliency: Evidence from electrical stimulation of the dorsolateral prefrontal cortex (dlPFC) and the orbitofrontal cortex (OFC). *Personality and Individual Differences, 106*, 209–216.
Sameroff, A. J. (1983). Developmental systems: Contexts and evolution. In P. Mussen (Ed.), *Handbook of child psychology* (Vol. 1, pp. 237–294). New York, NY: John Wiley & Sons.
Sawyer, C. R. (2016). Communication apprehension and public speaking instruction. In P. L. Witt (Ed.), *Handbooks of communication science*: Vol. 16. *Communication and learning* (pp. 397–426). Berlin, Germany: De Gruyter-Mouton.
Sawyer, C. R., & Behnke, R. R. (2009). Psychophysiological patterns of arousal in communication. In M. J. Beatty, J. C. McCroskey, & K. Floyd (Eds.), *Biological dimensions of communication: Perspectives, methods, and research* (pp. 197–210). Cresskill, NJ: Hampton Press.
Sawyer, C. R., & Behnke, R. R. (2014). Profiles of response stereotypy and specificity for public speaking state anxiety. In J. M. Honeycutt, C. R. Sawyer, & S. A. Keaton (Eds.), *The influence of communication on physiology and health* (pp. 55–72). New York, NY: Peter Lang.
Sawyer, C. R., & Richmond, V. P. (2015). Motivational factors and communication competence. In A. F. Hannawa & B. H. Spitzberg (Eds.), *Handbooks of communication science*: Vol. 22. *Communication competence* (pp. 193–212). Berlin, Germany: De Gruyter Mouton.
Sege, C. T., Bradley, M. M., & Lang, P. J. (2018). Avoidance and escape: Defensive reactivity and trait anxiety. *Behaviour Research and Therapy, 104*, 62–68.
Seta, J. J., Wang, M. A., Crisson, J. E., & Seta, C. E. (1989). Audience composition and felt anxiety: Impact averaging and summation. *Basic and Applied Social Psychology, 10*, 57–72.
Shields, G. S., & Slavich, G. M. (2017). Lifetime stress exposure and health: A review of contemporary assessment methods and biological mechanisms. *Social and Personality Compass, 11*, 1–17.
Tardy, C. H. (2014). Cardiovascular reactivity of social interaction: Predictors and consequences of physiological changes while speaking in everyday life. In J. M. Honeycutt, C. R. Sawyer, & S. A. Keaton (Eds.), *The influence of communication on physiology and health* (pp. 33–54). New York, NY: Peter Lang.

Taylor, S. E., Seeman, T. E., Eisenberger, N. I., Kozanian, T. A., Moore, A. N., & Moons, W. G. (2010). Effects of a supportive and unsupportive audience on biological and psychological responses to stress. *Journal of Personality and Social Psychology, 98*, 47–56.

Theiss, J. A. (2018). Family communication and resilience. *Journal of Applied Communication Research, 46*, 10–13.

Tovote, P., Esposito, M. S., Botta, P., Chaudun, F., Fadok, J. P., Markovic, M., … Lüchi, A. (2016). Midbrain circuits for defensive behavior. *Nature, 534*, 206–212.

van Den Bos, E., & Westenberg, P. M. (2015). Two-year stability of individual differences in (para)sympathetic and HPA-axis responses to public speaking in childhood and adolescence. *Psychophysiology, 52*, 316–324.

van Nierop, M., Lecei, A., Myin-germeys, I., Collip, D., Viechtnauer, W., Jacobs, N., … van Winkel, R. (2018). Stress reactivity links childhood trauma exposure to an admixture of depressive, anxiety, and psychosis symptoms. *Psychiatry Research, 260*, 451–457.

von Dawans, B., Kirschbaum, C., & Heinrichs, M. (2011). The Trier Social Stress Test for Groups (TSST-G): A new research tool for controlled simultaneous social stress exposure in a group format. *Psychoneuroendocrinology, 36*, 514–522.

Weusthoff, S., Baucom, B. R., & Hahlweg, K. (2013). Fundamental frequency during couple conflict: An analysis of physiological, behavioral, and sex-linked information in vocal expression. *Journal of Family Psychology, 27*, 212–220.

Zorn, J. V., Schür, R. R., Boks, M. P., Kahn, R. S., Joëls, M., & Vinkers, C. H. (2017). Cortisol stress reactivity across psychiatric disorders: A systematic review and meta-analysis. *Psychoneuroendocrinology, 77*, 25–36.

25

Physiological Arousal while Ruminating about Conflict with a Quantum Application to Relational Observation

James M. Honeycutt and Ryan D. Rasner

We endure conflict every day in varying degrees. As Lennon and McCartney sang in their classic song, "A Day in the Life," people experience and are exposed to news about others dealing with sadness, war, and accidents while reacting with approach, avoidance, and freezing. These reactions reflect the instincts of fight, flight, and fright in the sympathetic nervous system in response to threats. Physiological arousal occurs within the autonomic nervous system as adrenaline (epinephrine) is released, and is indexed by cardiovascular activity (heart rate, blood pressure), and electrodermal activity (EDA). This chapter reviews physiological research on imagined interactions, a type of social cognition in which people imagine conversations with others. Cardiovascular research is reviewed in the areas of road rage, marital conflict, positive storytelling, sports fandom, and signal detection. The chapter concludes by commenting on current and future research in quantum relational observation.

Early in the 20th Century, Walter Cannon's (1929) research in psychobiology led him to describe the "fight-or-flight" response of the sympathetic nervous system (SNS) to threats. A third option is freezing or fright in which an organism is startled due to fear and freezes for possible survival (e.g., deer frozen in headlights, surviving a grizzly attack by playing dead; Honeycutt, Keaton, Hatcher, & Hample, 2014; LeDoux, 2014). The SNS is one of two parts of the autonomic nervous system concerned with managing threats. Once activated, the SNS releases epinephrine, norepinephrine, and cortisol into the bloodstream. Correspondingly, there is increased respiration, blood pressure (BP), heart rate (HR), and muscle tension. Sweating is controlled by the SNS and skin conductance indicates physiological arousal. If the SNS is highly activated, then sweat gland activity also increases, which in turn increases skin conductance. In this way, EDA can be a measure of emotional and sympathetic responses (Carlson, 2013).

The second part of the autonomic nervous system is the parasympathetic nervous system (PNS). The PNS does essentially the opposite of the SNS: it decreases HR, increases digestion, and so on. Hence, the relaxation response turns off SNS arousal by activating the PNS. So, in essence, one doesn't really control the relaxation response; instead, one does the things that result in the PNS taking control (Honeycutt, Sawyer, & Keaton, 2014).

The PNS and SNS can be activated internally through introspective thoughts about pleasing and displeasing encounters in the form of imagined interactions (IIs) in which individuals use cognition and mental imagery to envision encounters in the past (retroactivity), the future (proactivity) and in the present (Honeycutt, 2003). Physiological arousal involves the autonomic nervous system in a state of adrenaline release that includes pulse (heart rate beats per minute [bpm]) or interbeat intervals (IBI). IBI is a measure of time in milliseconds between adjacent heartbeats. High IBI rates are related to increased levels of adrenaline, anxiety, and arousal (Porges, 2011). The lower the IBI value, the shorter the cardiac cycle, which reflects a faster HR.

Under normal conditions, the heart's rate is under control of the PNS. Generally, healthy resting heart rates are 70 BPM for men and 80 BPM for women, according to the American Heart Association. Heart rates above 105 are high and above the effects of exercise (Rowell, 1986). Women tend to have slightly higher rate averages because the left ventricle of males is usually larger, which allows more blood to pump with each beat while distributing oxygen and nutrients to tissues. People's resting HR typically decreases with age. HR is also affected by environmental factors; for example, it increases with extremes in temperature and altitude (Honeycutt, 2012).

A wide array of research in intrapersonal communication reveals that physiological arousal is associated with imagined interactions, as the subsequent section details.

IMAGINED INTERACTIONS

Imagined interactions (IIs) are a type of daydreaming and social cognition involving mental imagery that is theoretically grounded in symbolic interactionism and script theory, in which individuals imagine conversations with significant others for a variety of purposes (Honeycutt, 2003, 2015). They are associated with changes in the SNS depending on whether the imagined conversations are positive, negative, or deal with conflict. Honeycutt and Sheldon (2018) discussed how scripts are a type of "automatic pilot," providing guidelines on how to act when one encounters new situations. Scripts are activated mindlessly and created through IIs as people envision contingency plans for actions. Following is a review of findings from the research on the physiology of IIs.

Physiology of IIs and Road Rage

We have measured change in adrenaline and anxiety levels in terms of imagined interaction and emotion as automobile drivers "vent" at offending drivers in heavy traffic conditions while late for an important meeting (Honeycutt, 2006). We have employed Applied Simulation Technologies that are used to train state police officers in driving skills. Aggressive driving reflects lack of dialogue with the offending driver; this represents the I-It metaphor in which the self is stereotyped as being "good" while the offending driver is "evil." However, even in some road rage incidents, there are anecdotal reports of dialogue with passengers who may encourage the driver to be aggressive. Indeed, social comparison theory could be operating in the situation in which there is a diffusion of responsibility, as car occupants may feel uninhibited in expressing coercive actions. Comparisons were made of driving alone with a condition in which the driver has an intimate partner as a passenger.

There are individual differences in having imagined interactions while driving. Some of the personality differences in driving profiles involve being an angry, impatient, competing, or punishing driver. We have measured the use of IIs for venting at offending drivers reflecting the catharsis and conflict-linkage functions. The presence of passengers, music, weather conditions, drinking patterns, and driving personality profiles are correlated with aggressive driving tendencies, including acceleration, braking, swerving, and tailgating.

In such instances, we have found consistent associations between IIs and physiological outcomes (for review, see Honeycutt, 2015). For example, controlling for the baseline HR and HR variability, using IIs to vent at offending drivers is negatively associated with HR (partial $r = -.36$).

Cardiovascular Findings in Marital Conflict

A program of research has revealed that thinking about conflict and actually arguing are associated with the activation of SNS (Honeycutt, 2014). Studies in physiology and partner interaction have determined that couples do not need to engage in conflict, but merely imagine a conflict scenario with their partner, to increase their physiological arousal (Honeycutt, 2010; Honeycutt, Sawyer, & Keaton, 2014). Additional data examining the use of proactive IIs revealed that imagining dialogue with relational partners increases IBI (i.e., reduces HR). More specifically, ruminating about conflict prior to actual conversations results in increased heart rate while "there may be a latent reward when it comes to the actual discussion as revealed in the lowering of blood pressure" (Honeycutt, 2010; p. 55).

Research has found that when people take conflict personally, they experience increased diastolic (but not systolic) BP (Honeycutt, Hample, & Hatcher, 2016). One of the major causes of elevated diastolic pressure is stress or anxiety. Thus, as stress or anxiety at the conflict escalation increased, diastolic pressure also increased.

Ruminating and Observing Conflict

Rumination in terms of IIs involves dwelling on past grievances. Individuals relive conflict while simultaneously planning for anticipated arguments (Honeycutt, 2004). Unfortunately, people often manage conflict through dysfunctional means due to an inability to effectively argue their points of view and emotions (Infante & Rancer, 1995). We have conducted research on how watching escalating conflict affects cardiovascular arousal. Arguments may linger in which there is no reconciliation. Conciliation such as apologizing may not be reciprocated, leaving feelings of inequity. *Imagined interaction conflict-linkage theory* explains the persistence of everyday conflict within the mind through a series of axioms and theorems that have been supported in numerous studies (e.g., Allen & Berkos, 2010; Hample, Richards, & Na, 2012; Honeycutt, 2003, 2010; Wallenfelsz & Hample, 2010). The basic idea in conflict-linkage theory is that IIs occur in between enacted episodes of the continuing conflict, keeping the conflict "alive" even when it is not behaviorally active. Correspondingly, there is an increase in physiological arousal.

We have examined II rumination in terms of people predicting the escalation of conflict in a series of studies examining conflict escalation in a domestic argument that ends with physical conflict (Honeycutt, Hample, & Hatcher, 2016). In these studies, taking conflict personally did not predict conciliatory or aggressive reactions or conflict linkage rumination, but it did predict changes in diastolic blood pressure as conflict increased. Victims predicted more conciliation as conflict escalated. This finding was especially interesting because it can represent a state of learned helplessness and survival.

Our studies further revealed that individuals could become aroused when watching others argue. Indeed, this is a primary assumption behind Bandura's (1977) classical social learning theory with behavioral and physiological outcomes. Honeycutt and associates (2014b) had viewers watch a five-scene escalating argument in which conflict was initiated either by the husband or wife. We asked viewers what they thought the people in the videos were going to do. Separate groups of viewers watched a clip that was identical except for the sex of the conflict initiator. Viewers were presented with a list of nine possible reactions, representing three conciliatory actions (apologize, ask for forgiveness, discuss the issue calmly) and five aggressive reactions (insult, swear, or curse; push, grab, shove partner; slap partner; throw, smash, hit, or kick something; hit or try to hit partner with something). Additionally, II rumination items were based on the II proactivity and retroactivity attributes taken from the abbreviated SII Scale (Honeycutt, 2010). A sample item from the wife-initiated conflict version for proactivity is, "What is the probability that the wife (husband) will think about what to say next?" A sample item for retroactivity is, "What is the probability that the wife (husband) is replaying the previous scene in her (his) mind?"

When watching an escalating fight between a husband and wife, latent growth curve modeling revealed that there is a moderate level of using retroactive IIs to replay conflict as well as rehearsing what to say next when husbands initiate the conflict. However, II rumination dramatically increases for viewers as the conflict rapidly escalates (e.g., yelling and shoving). HR is stable for viewers as the conflict evolves, except it decreases when the conflict initiator throws objects. As Honeycutt et al. (2014b) stated:

> The bottom line in this analysis is that the sympathetic nervous arousal is activated in terms of HR when individuals watch an escalating conflict. Part of this arousal may be due to II rumination, in which individuals are ruminating about the argument while it is occurring (replaying of prior scenes and forecasting of ensuing scenes).
>
> (p. 85)

Excitation-transfer theory explains how we may become physiologically aroused while watching videos or being exposed to stimuli that consciously or subconsciously remind us of earlier events. We saw a slight elevation in HR while observing the climax of the aggression. Hence, there is a psychological transference of past events to the current stimuli. The classic research of Zillmann (1971) revealed that levels of excitation do not drop abruptly with termination of exposure to an excitation-producing communication but linger for some time. When excitation occurs in the form of physiological arousal, individuals examine their immediate environment for cues as to what their arousal is due to, and attribute emotion to it.

Effects of Heart Rate on Positive, Conflict-Resolution Storytelling

Gottman (1994) explicated the idea of physiological linkage in distinguishing happy from unhappy couples by finding a reciprocity of negative affect (misery loves company) as well as temporal predictability and "reciprocity in physiology" (p. 72). We have found that couples experience elevated HR as a result of discussing positive, conflictual topics (Honeycutt, Bannon, & Hatcher, 2014). Additionally, time-series analyses reveal that husbands' HR increased slightly when telling positive, conflict-resolution stories, whereas wives' continued to decrease after conflictual discussion. This result supports Kiecolt-Glaser et al.'s (1996) pioneering position that negativity in marital interaction can affect women's immune function just as much as men's. Often, marital conflict is a chronic stressor, which impairs the immune system's ability to prevent illness (Gottman, 2011; Honeycutt, Bannon, & Hatcher, 2014; Kiecolt-Glaser et al., 1993).

Positive stories can provide hope that people can survive even the most difficult times. Indeed, optimism is associated with cardiovascular health even when controlling for variables such as age, sex, ethnicity, marital status, and education (Hernandez et al., 2015). Seligman, Parks, and Steen (2006) also supported this stance by showing that optimists and pessimists differ markedly in how long they will live with optimists. The brain and body are interconnected so it comes as no surprise that a negative mind will impart a negative influence on the functioning of the body over time (Honeycutt et al., 2014a).

EFFECTS OF HEART RATE ON SPORTS FANDOM

While watching vested sporting events, fans often have imagined interactions with opposing players, referees, and coaches. Sports team identification is "the extent to which a fan feels a psychological connection to a team and the team's performances are viewed as self-relevant" (Wann, 2006, p. 332). Passionate fans tend to experience physiological arousal involving HR, BP, and levels of serotonin, dopamine, and cortisol (for a review, see Lövheim, 2012). It is important to understand why people may be inclined to act in this manner.

The functions of IIs, including catharsis and rehearsal, mediated the link between identifying with a team, social behaviors, and negative mental health outcomes, such as being verbally aggressive toward others at a sporting event, and positive mental health outcomes such as self-esteem (Keaton, Gearhart, & Honeycutt, 2014). Passionate fans maintain relationships with their favorite team's players through IIs and rehearse what they want to say or vent to players and coaches. While behaving in these manners, fans experience catharsis as emotions are released, and conflict continues depending on the team's performance. Sometimes, passionate fans even imagine calling plays and vocally communicate which plays should be called. Using latent growth curve modeling, Keaton and Honeycutt (2014) found that football fans experienced increases in HR while viewing highlights. Conversely, HR increased and decreased, showing an inverted parabola function when watching lowlights. During the lowlights, fans were yelling at the opposing coach when lifting the championship trophy after the game ended, which represents the conflict linkage and catharsis functions of imagined interactions.

SIGNAL DETECTION THEORY AND PHYSIOLOGICAL AROUSAL

Signal detection theory is relevant for communication research, as the theory predicts under what circumstances individuals communicate accurately about evolutionarily relevant information and accurately read the signs, where communication effectiveness is defined as congruence between the sender's intent behind a message (verbal or nonverbal) and the impact of the message on the receiver (Honeycutt & Sheldon, 2018). Signal detection theory was originally designed by Peterson, Birdsall, and Fox (1954) as a method for understanding and addressing problems with radar signals. Communication scholars have applied signaling theory to explain social influence in face-to-face communication (Pentland, 2010). The theory has been used to address a number of issues, including detection of health issues (Abbey, Eckstein, & Boone, 2009; Giral, Kurt, Yeğin, & Yeğin 2014), impending domestic violence (Honeycutt & Eldredge, 2015; Honeycutt, Sheldon, Pence, & Hatcher, 2015), social threats (Lynn & Barrett, 2014), memory (Wixted & Stretch, 2004), and intent (Reed, 2012). The theory claims that our ancestors who were able to detect aggressive signals were survivors in protecting offspring, whereas those who lacked that ability were extinguished (Buss, 2016).

We have examined how physical abuse victims detect signs of escalating conflict (Honeycutt, Frost, & Krawietz, 2017). Men and women watched the conflict escalation videos described earlier. Participants were informed that the objective of the study was to determine whether untrained individuals can detect escalation cues similarly to people with victimization experience. To take part, participants had to have been involved in arguments within the past year that escalated to involve name-calling, yelling, throwing objects, and/or physical contact such slapping or shoving. Perpetration and victimization were measured using the 20-item short form of the Conflict Tactics Scale, which has long history of usage with good external validity and consistent reliability (Straus & Douglas, 2004). Exploratory factor analysis identified two factors: perpetrator force and victim force.

Physiological arousal (in the form of EDA) interacted with perpetrator force and victim force. As conflict escalated beyond the baseline measure, perpetrator force was negatively associated with EDA arousal and declined over time. Conversely, the inverse pattern was revealed for being a victim from force, which was positively associated with EDA arousal.

QUANTUM-INSPIRED RELATIONAL OBSERVATION: IMPLICATIONS FOR FUTURE PHYSIOLOGICAL RESEARCH

Not only has imagined interactions theory helped spur the use of physiological data collection for investigations within the communication discipline, it has also sparked the imagination (pun intended) of various scholars across the social sciences. Here, we present a preliminary investigation incorporating physiological data (in the form of HR) with a newly emerging conceptual framework. The current section provides a precursory explication of the conceptual framework behind relational observation. Rasner has initiated the novel and abstract idea that incorporates bio-evolutionary and quantum-inspired approaches to advance the understanding of relationship development through interpersonal communication.

The current conceptual framework outlines relational observation as a mechanism through which relationships simultaneously evolve and devolve through processes that are driven by observation and are recognized as chaotic, similar to human behavior. We attempt to ameliorate some of the unexplained mysteries surrounding human communicative behavior and subsequent relational processes within this conceptual framework. Scholars in the physical sciences (e.g., biology, chemistry, physics) have long understood that without increasing the range of observation, many nuances to the phenomenon under investigation may be overlooked. Social scientists have followed suit and begun to realize the importance of observational breadth.

Researchers are aware of the implications that observation has on participants during data collections (i.e., the Hawthorne effect). The mere presence of a researcher during social investigations affects participant behavior (Wickstrom & Bendix, 2000) and subsequent communication. It is therefore little wonder that many, if not most, social scientific theories allude to the notion of observation, yet they never fully illuminate the construct as a way to better understand interpersonal relationships. Individuals make observations about their relationships, and those observations are revisited in the form of imagined interactions (Honeycutt, 2003), but why do people observe in the first place? What communicative effects does observation have on each person? More important, how does observation affect the development of the relationship? These are the fundamental questions, among others, that relational observation seeks to answer while continuing to bridge the gap between intrapersonal and interpersonal communication with respect to human relationships.

The phenomenon is understood through two distinct fields of study: evolutionary psychology and quantum theory. First, evolutionary psychology provides an understanding of why humans observe in interpersonal relationships. Individuals face adaptive problems, and solutions to those problems are considered adaptations through natural selection, which aided the organism in survival and reproduction (Al-Shawaf, Conroy-Beam, Asao, & Buss, 2015). Had our ancestors been unable to detect certain verbal and nonverbal cues to relational threats, the human species surely would not have survived and prospered. Internally and externally observing relationships has addressed these adaptive problems and continues to do so.

Second, aspects of quantum theory allow researchers to understand the interpersonal implications of relational observation on communication within the context of relationship development. Quantum theory remains among the most successful theoretical frameworks known to scholars (Busemeyer & Bruza, 2012; Yukalov & Sornette, 2017), guiding researchers in understanding complexities surrounding microscopic and macroscopic phenomena, including the evolution of the universe. Two decades ago, a group of psychologists proposed quantum probability as a way to investigate human cognition mathematically (Bruza, Wang, & Busemeyer, 2015). Quantum theory allows social scientists to draw from its abstract structures; however, cognitive psychologists do not assume that cognition happens at the quantum level (Gabora & Kitto, 2016). Describing relationships through the lens of quantum mechanics provides researchers novel ways through which to understand observational effects on communication within relationships. One way to understand how observation operates in the context of human relationships is to look at the classic double-slit experiment (see Davisson, 1928). In this experiment, researchers found that a particle behaved like a wave of potential when it was shot at a screen. Between the screen and the particle-shooting device, the researchers placed a metal plate with two slits. Because they were shooting individual particles at the screen, they believed they would see two patterns emerge on the screen, as there were only two possible places the particles could go through. What they instead found was an interference pattern: the particles were behaving as waves of potential. When they attempted to observe (measure) the particles before they went through the slits, the particles went back to behaving like particles and not waves. The very act of observation had a profound effect on the particles' behavior. Although humans are not particles, their behavior may be better understood through a similar conceptualization.

Honeycutt (2003) articulated that as individuals make new observations about the relationship, the observations are revisited in imagined interactions, as earlier noted. Moreover, Kelly's (1955) classic theory of personality stated that humans are novice scientists who observe, hypothesize, and test their theories regularly; therefore, how we perceive of the world directly affects how we observe the world. "Specific relationship experiences are shaped as much by participants' views of those experiences as they are by anything they do, extrinsic events, and culturally constructed meanings" (Burnett, 1987, p. 77). In the words of quantum mechanics, *how* we measure affects and changes *what* we measure.

Relational observation is observation of a communicative behavior that elicits a verbal or nonverbal response, either consciously or unconsciously. Relational observation consists of three parts: internal observation, external observation, and observational bias. Relational observation provides an antecedent causal explanation for these ideas as well as a theoretical lens through which to understand interpersonal relationships. Observation is a mindful and present nonverbal communicative action—not a reflection on the past—driving cognitive processes that alter an individual's future communicative behaviors. Observation happens intrapersonally, driving the proceeding interpersonal communicative behaviors. Berkos and Denham (2017) stated that intimacy within a relationship could be obstructed by individuals observing external issues, such as their significant other speaking to a sexual rival or the happiness of another couple not found in their own relationship eliciting, among other things, uncertainty. Observation could drive the increase in uncertainty or uncertainty could drive observation.

CURRENT AND FUTURE DIRECTIONS

Busemeyer and Bruza (2012) discuss the notion that quantum probability allows for more chaos than does classic probability. In classic probability models, Markov chains are found. This idea allows researchers to understand human behavior in a linear fashion, wherein the relationship is either good or bad, on or off, and people are either coming together or coming apart. Quantum probability instead claims that people are simultaneously coming together and growing apart (not either/or), similar to the staircase model of Knapp, Vangelisti, and Caughlin (2014). This understanding is more in line with how humans actually behave: unpredictably, just like particles in a physicist's laboratory. With the following probability models, we hope to illuminate the interference found in relationships in terms of relational observation. We can measure cardiovascular reactivity as people view themselves during a reconstructed history. Because people, from this point of view, are simultaneously getting together and breaking up, we can find where the interference is happening (i.e., observation).

The notation derived from Busemeyer, Wang, Khrennikov, and Basieva (2014) helps explain the quantum probability model that we discuss. The decision system is represented by ψ, which is a unit length state vector that lies within a four-dimensional Hilbert space spanned by four basis vectors. Each basis vector represents one of the four combinations of categories and actions (e.g., *IT* is a basis vector corresponding to category *I*—internal observation—and action *T*—coming together). Thus, the state of a developing relationship would be roughly denoted as follows:

$$|\neg_{s}\rangle = \neg_{IT}\,|\,IT\rangle + \neg_{IA}\,|\,IA\rangle + \neg_{ET}\,|\,ET\rangle + \neg_{EA}\,|\,EA.$$

Theoretically, this formula represents the simultaneous nature of coming together and growing apart as a function of both internal and external observation. Individuals would be internally observing and coming together, internally observing and coming apart, externally observing and coming together, and externally growing apart. The probability, if explored empirically, should present the interaction of observation upon relational outcomes. This model and theoretical explication still require much work, but a basic understanding is currently underway. Future research seeks to create and validate scales aimed at measuring the latent construct of relational observation.

A naturally occurring observation point is to have couples view videos of their wedding and measure arousal before and after watching the video. Alternatively, we can conduct oral history interviews (Buehlman, Gottman, & Katz, 1992), asking about highlights and "lowlights" (periods of distance and coming apart) and use latent growth modeling to measure arousal during these reports. The interviews can be videotaped and couples can view themselves during the interview and explain what they are feeling (Honeycutt & Sheldon, 2018). Beyond that, we plan to experimentally test and validate the probabilistic model to provide a more reliable predictive model of relational outcomes as a result of observation. Recent physiological data has emerged from our laboratory indicating a relationship between relational observation and heart rate. A case study of two couples involving a video interview and a follow-up viewing of the video took place. Each couple was interviewed about how observation worked in their relationship over approximately 15 minutes. Two days later the couple returned to the laboratory where they were instrumented with ECG and watched the video. The researcher flagged each instance of the participants discussing observation in the relationship. Figure 25.1 reveals increased HR by the participant viewing his or her partner as they each describe an instance of relational observation. Similar to the double-slit experiment, the simple act of observation alters the state of individuals and their subsequent communication within the relationship. Experimental designs are currently being created to control for confounding variables and isolate the effects of observation on relationships through alterations in communication.

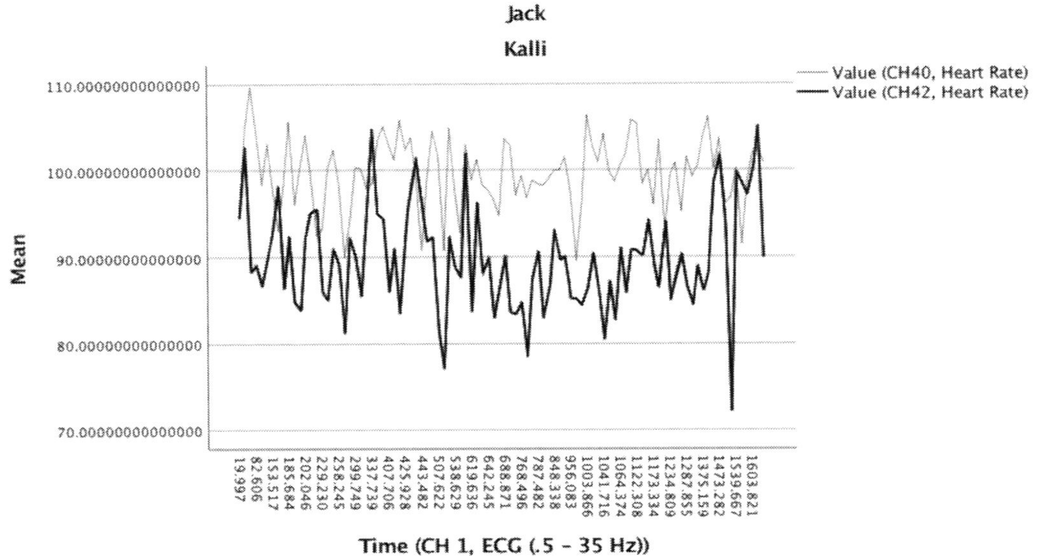

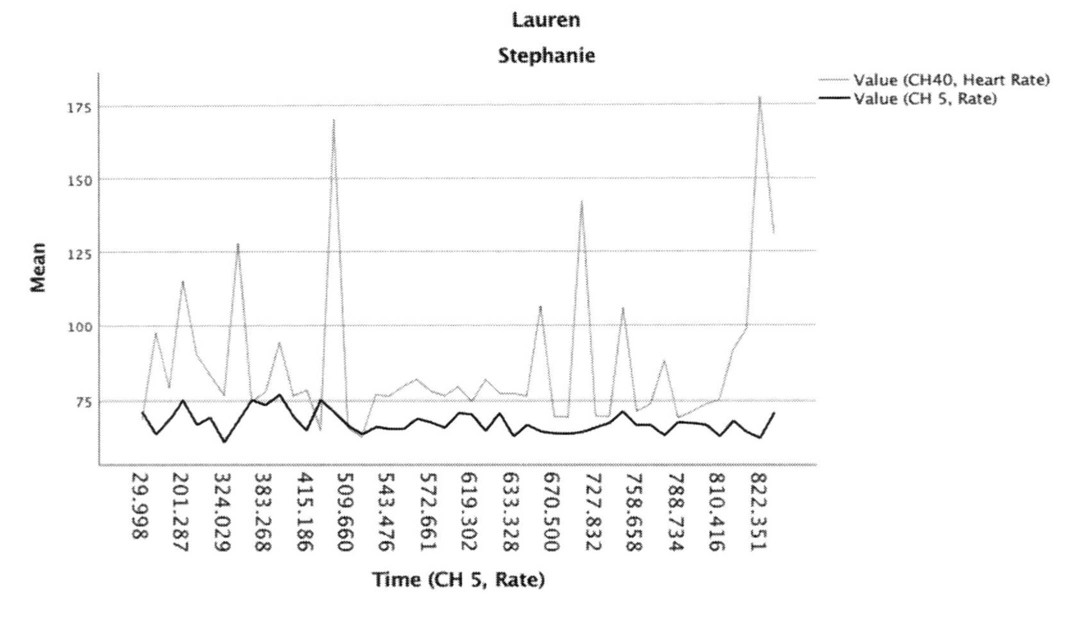

Figure 25.1 Heart Rate Effects by the Participants Viewing their Partners while Describing an Instance of Relational Observation

CONCLUSION

Physiological arousal can be beneficial or stressful as adrenalin is sent to the cardiovascular and endocrine systems. We have observed increased HR as people think about arguments. Unfortunately, society has a series of competing maxims that precisely reflect this arousal depending on how it is dealt with: I can forgive and forget (no arousal); I can forgive but not forget (long-term memory and some arousal); or I can neither forgive nor forget (most debilitating and destructive response). Our studies revealed increases in blood pressure and HR in terms of story-telling, sports fandom, road rage, and detecting signals of physical coercion (increase in EDA).

When we observe and think about our own relationships, evolutionary psychology and quantum observation offer a unique perspective into why relationships split apart. The simultaneous nature of coming together and growing apart is a function of both internal and external observation. What is perceived as "good" by one person" (e.g., doing someone a favor) may be labeled by the other partner as "bad" (e.g., ingratiation). Hence, we have IIs in which we think about our relationships, which can result in physiological arousal.

REFERENCES

Abbey, C. K., Eckstein, M. P., & Boone, J. M. (2009). An equivalent relative utility metric for evaluating screening mammography. *Medical Decision Making, 30*, 113–122.

Allen, T. H., & Berkos, K. M. (2010). Imagined interaction conflict-linkage theory: Examining accounts of recurring imagined interactions. In J. M. Honeycutt (Ed.), *Imagine that: Studies in imagined interaction* (pp. 43–64). Cresskill, NJ: Hampton.

Al-Shawaf, L., Conroy-Beam, D., Asao, K., & Buss, D. M. (2015). Human emotions: An evolutionary psychological perspective. *Emotion Review, 8*, 1–14.

Bandura, A. (1977). *Social learning theory*. Englewood Cliffs, NJ: Prentice Hall.

Berkos, K. M., & Denham, J. A. (2016). Intimacy, marital satisfaction, and third party imagined interactions. *Imagination, Cognition, and Personality, 36*, 312–330.

Bruza, P. D., Wang, Z., & Busemeyer, J. R. (2015). Quantum cognition: A new theoretical approach to psychology. *Trends in Cognitive Sciences, 19*, 383–393.

Buehlman, K. T., Gottman, J. M., & Katz, L. F. (1992). How a couple views their past predicts their future: Predicting divorce from an Oral History Interview. *Journal of Family Psychology, 5*, 295–318.

Burnett, R. (1987). Reflection in personal relationships. In R. Burnett, P. McGhee, & D. Clarke (Eds.), *Accounting for relationships: Explanation, representation and knowledge*. New York, NY: Methuen.

Busemeyer, J. R., & Bruza, P. D. (2012). *Quantum models of cognition and decision*. Cambridge, MA: Cambridge University Press.

Busemeyer, J. R., Wang, Z., Khrennikov, A., & Basieva, I. (2014). Applying quantum principles to psychology. *Physica Scripta, T163*, 1–10.

Buss, D. M. (2016). *Evolutionary psychology: The new science of the mind*. New York, NY: Routledge.

Cannon, W. B. (1929). *Bodily changes in pain, hunger, fear and rage: An account of recent research into the function of emotional excitement* (2nd ed.). New York, NY: Appleton Century-Crofts.

Carlson, N. (2013). *Physiology of behavior*. Essex, England: Pearson Education, Ltd.

Davisson, C. J. (1928). The diffraction of electrons by a crystal of nickel. *The Bell System Technical Journal, 1*, 90–105.

Gabora, L., & Kitto, K. (2016). Towards a quantum theory of humour. *Frontiers in Physics, 4*, 3–47.

Giral, A., Kurt, R., Yeğin, E. G., & Yeğin, K. (2014). Signal detection theory approach to gastro-esophageal reflux disease: A new method for symptom analysis of impedance-pH data. *Diseases of the Esophagus, 27*, 206–213.

Gottman, J. M. (1994). *What predicts divorce?* Hillsdale, NJ: Lawrence Erlbaum Associates.

Hample, D., Richards, A. S., & Na, L. (2012). A test of the conflict linkage model in the context of serial arguments. *Western Journal of Communication, 76*, 459–479.

Hernandez, R., Kershaw, K. N., Siddique, J., Boehm, J. K., Kubzansky, L. D., Diez-Roux, A., … Lloyd-Jones, D. M. (2015). Optimism and cardiovascular health: Multi-ethnic study of atherosclerosis (MESA). *Health Behavior & Policy Review, 2*, 62–73.

Honeycutt, J. M. (2003). *Imagined interactions: Daydreaming about communication.* Cresskill, NJ: Hampton.

Honeycutt, J. M. (2004). Imagined interaction conflict-linkage theory: Explaining the persistence and resolution of interpersonal conflict in everyday life. *Imagination, Cognition, and Personality, 23,* 3–25.

Honeycutt, J. M. (2006). Enhancing EI intervention through imagined interactions. *Issues and Recent Developments in Emotional Intelligence* [On-line serial], *1*(1), 1–4. Retrieved from www.eiconsortium.org.

Honeycutt, J. M. (2010). Physiology and imagined interactions. In J. M. Honeycutt (Ed.), *Imagine that: Studies in imagined interaction* (pp. 43–64). Cresskill, NJ: Hampton.

Honeycutt, J. M. (2012). Imagined interactions: On knowing what to say. In K. Lollar, M. Monsour, & J. Barwind (Eds.), *The talk within: Its central role in communication* (pp. 181–203). Dubuque, IA: Kendall Hunt.

Honeycutt, J. M. (2014). Imagined interactions. In W. Donsbach (Ed.), *International encyclopedia of communication* (2nd ed., pp. 249–271). Washington, DC: International Communication Association.

Honeycutt, J. M. (2015). Imagined interaction theory: Mental representations of interpersonal communication. In D. O. Braithwaite & P. Schrodt (Eds.), *Engaging theories in interpersonal communication* (2nd ed., pp. 75–87). Thousand Oaks, CA: Sage.

Honeycutt, J. M., Bannon, B., & Hatcher, L. C. (2014a). Effects of positive family conflict-renewal stories on heart rate. In J. M. Honeycutt, C. R. Sawyer, & S. A. Keaton (Eds.), *The influence of communication on physiology and health* (pp. 11–31). New York: Peter Lang.

Honeycutt, J. M., & Eldredge, J. H. (2015). Applying game theory and signal detection theory to conflict escalation: A case study of a police investigator viewing a domestic argument. In K. Chapman (Ed.), *Decision and game theory: Perspectives, applications and challenges* (pp. 17–30). New York: Nova Science.

Honeycutt, J. M., Frost, J. K., & Krawietz, C. E. (2017, November). Signal detection, victimization and domestic violence. Paper presented at the annual meeting of the National Communication Association, Dallas, TX.

Honeycutt, J. M., Hample, D. A., & Hatcher, L. C. (2016). A latent growth curve analysis of taking conflict personally as a consequence of sex, conflict initiation, victimization, conflict linkage, and cardiovascular reactivity. *Imagination, Cognition, and Personality, 35,* 325–350.

Honeycutt, J. M., & Keaton, S. A. (2015). A note on testing autoregressive time-series latent trajectory models for physiological behaviors that vary in stationarity. *Communication Research Reports, 32*, 353–359.

Honeycutt, J. M., Keaton, S. A., Hatcher, L. C., & Hample, D. (2014b). Effects of rumination and observing marital conflict on observers' heart rates as they advise and predict the use of conflict tactics. In J. M. Honeycutt, C. R. Sawyer, & S. A. Keaton (Eds.). *The influence of communication on physiology and health* (pp. 73–92). New York, NY: Peter Lang.

Honeycutt, J. M., Sawyer, C. R., & Keaton, S. A. (Eds.). (2014c). *The influence of communication on physiology and health.* New York, NY: Peter Lang.

Honeycutt, J. M., & Sheldon, P. A. (2018). *Scripts and communication for relationships* (2nd ed.). New York, NY: Peter Lang.

Honeycutt, J. M., Sheldon, P., Pence, M. E., & Hatcher, L. C. (2015). Predicting aggression, conciliation, and concurrent rumination in escalating conflict. *Journal of Interpersonal Violence, 30,* 133–151.

Infante, D. A., & Rancer, A. S. (1995). Argumentativeness and verbal aggressiveness: A review of recent theory and research. *Annals of the International Communication Association, 19*, 319–352.

Keaton, S. A., Gearhart, C. C., & Honeycutt, J. M. (2014). Sport fandom, intrapersonal communication, negative behaviors, and psychological enhancement: A mediated model of team identification, imagined interactions, and their effects. *Imagination, Cognition and Personality, 33*, 251–269.

Keaton, S. A., & Honeycutt, J. M. (2014). Sports fandom and physiological arousal while watching professional/college highlights and lowlights of "your" team. In J. M. Honeycutt, C. R. Sawyer, & S. A. Keaton (Eds.), *The influence of communication on physiology and health* (pp. 93–113). New York: Peter Lang.

Kelly, G. A. (1955). *A theory of personality: The psychology of personal constructs*. New York, NY: W. W. Norton & Company.

Kiecolt-Glaser, J., Malarkey, W. B., Chee, M., Newton, T., Cacioppo, J. T., Mao, H., & Glaser, R. (1993). Negative behavior during marital conflict is associated with immunological down-regulation. *Psychosomatic Medicine, 55*, 395–409.

Kiecolt-Glaser, J., Newton, T., Cacioppo, J., MacCallum, R., Glaser, R., & Malarkey, W. (1996). Marital conflict and endocrine function: Are men really more affected than women? *Journal of Consulting and Clinical Psychology, 64*, 324–332.

Knapp, M. L., Vangelisti, A. L., & Caughlin, J. P. (2014). *Interpersonal communication and human relationships* (7th ed.). New York, NY: Pearson.

LeDoux, J. E. (2014). Coming to terms with fear. *Proceedings of the National Academy of Sciences, 111*, 2871–2878.

Lövheim, H. (2012). A new three-dimensional model for emotions and monoamine neurotransmitters. *Medical Hypotheses, 78*, 341–348.

Lynn, S. K., & Barrett, L. F. (2014). "Utilizing" signal detection theory. *Psychological Science, 25*, 1663–1673.

Miller, G. R. (1976). *Explorations in interpersonal communication*. Beverly Hills, CA: Sage.

Pentland, A. S. (2010). To signal is human: Real-time data mining unmasks the power of imitation, kith and charisma in our face-to-face social networks. *American Scientist, 98*, 204–211.

Peterson, W. W., Birdsall, T. G., & Fox, W. C. (1954). The theory of signal detectability. *IRE Transactions: Information Theory*, PGIT-4, 171–212.

Porges, S. W. (2011). *The polyvagal theory: Neurophysiological foundations of emotions, attachment, communication, and self-regulation*. New York, NY: Norton.

Reed, L. I. (2012). Facial expressions as honest signals of cooperative intent in a one-shot anonymous prisoner's dilemma game. *Evolution and Human Behavior, 33*, 200–209.

Rowell, L. (1986). *Human circulation: Regulation during physical stress*. New York, NY: Oxford University Press.

Seligman, M. E. P., Parks, A. C., & Steen, T. (2006). A balanced psychology and a full life. In F. Huppert, B. Keverne, & N. Baylis (Eds.), *The science of well-being* (pp. 275–283). Oxford, England: Oxford University Press.

Straus, M. A., & Douglas, E. M. (2004). A short form of the revised conflict tactics scales, and typologies for severity and mutuality. *Violence and Victims, 19*, 507–520.

Wallenfelsz, K. P., & Hample, D. (2010). The role of taking conflict personally in imagined interactions about conflict. *Southern Communication Journal, 75*, 471–487.

Wann, D. L. (2006). The causes and consequences of sport team identification. In A. A. Raney & J. Bryant (Eds.), *Handbook of sports and media* (pp. 331–352). Mahwah, NJ: Lawrence Erlbaum Associates.

Wickstrom, G., & Bendix, T. (2000). The "Hawthorne effect"—What did the original Hawthorne studies actually show? *Scandinavian Journal of Work, Environment & Health, 26*, 363–367.

Wixted, J., & Stretch, V. (2004). In defense of the signal detection interpretation of remember/know judgments. *Psychonomic Bulletin & Review, 11*, 616–641.

Yukalov, V. I., & Sornette, D. (2017). Quantum probabilities as behavioral probabilities. *Entropy, 19*, 112–142.

Zillmann, D. (1971). Excitation transfer in communication-mediated aggressive behavior. *Journal of Experimental and Social Psychology, 7*, 419–434.

26

Communication, Stress, and Thriving in Close Relationships

Tamara D. Afifi, Kathryn Harrison, and Nicole Zamanzadeh

Stress and resilience are interdependent constructs that are studied extensively across disciplines. Nonetheless, they continue to be misunderstood, elusive, and examined in isolation from each other and relatively removed from the complex environment in which they are embedded. One reason why these constructs are misunderstood is because the vast majority of research focuses on the toxic nature of stress and implies that few people are resilient in the face of it. In reality, most people adapt positively to stressful situations, perhaps because it is evolutionarily advantageous to do so (Bonanno, 2004; Ellis & Del Giudice, 2014). Negative forms of stress can threaten the ability to cope, but individuals often adapt positively over time because they must; they have jobs they need to go to and/or children they need to parent. Stress is also natural and can be positive, propelling people to enhance their performance, grow, and experience a sense of discovery and growth from a challenge.

Even if most people are resilient, however, fewer people thrive in the face of adversity. Those who do thrive, by growing personally and relationally from stressful experiences, broadening their perspectives, and/or learning something new, are less understood than those who are merely resilient. Those who thrive amidst adversity, however, are the group from whom scholars and practitioners can learn the most. In particular, researchers do not know what thriving looks like physiologically when people are stressed or how it is developed and sustained in social relationships. People typically do not experience stress in isolation but manage it through their communication with others. In fact, social relationships are one of the most important factors in determining individuals' resilience (see Repetti, Robles, & Reynolds, 2011).

We argue in this chapter that to fully conceptualize, measure, and understand how people communicate when they are stressed and its impact on resilience and potential thriving, it is important to consider the sociocultural, psychological, biosocial, and social forces involved. In so doing, we discuss how thriving may manifest physiologically and psychologically, and how it can be fostered through social relationships. We also discuss the role of social relationships and communication as a part of this process. Finally, the theory of resilience and relational load (TRRL; Afifi, Merrill, & Davis, 2016) is outlined as a way to examine resilience and thriving in close relationships. We begin by defining stress, resilience, and thriving.

POSITIVE AND NEGATIVE FORMS OF STRESS

Stress is a tension that imperils an individual's homeostasis (Chrousos & Gold, 1992). The body's homeostasis, or equilibrium, is often threatened by outside (and inside) forces (Chrousos & Gold, 1992). Response from the mind and body to this tension is dependent on the appraisal of the stressor (Cohen, Kessler, & Gordon, 1995) and the coping resources of the individual (Lazarus & Folkman, 1984). It is well documented that the stress response process is composed of biological, psychological, and social factors, which produce a physiological, emotional, and behavioral response (Ice & James, 2006).

An important first step in understanding resilience in relation to stress is recognizing that stress is not inherently bad. In fact, stress is natural and can serve an important function during the performance of tasks. When stress is deemed beneficial to one's physical or mental health, performance, or motivation, it is called *eustress* (Seyle, 1987). The cerebral cortex has the ability to distinguish between hormones such as adrenocorticotropic hormone (ACTH) and glucocorticoids being released in response to something negative, such as harmful forms of conflict, or something positive, such as a first date. It is when the *stressor*, or the stimulus producing a demand on the individual, exceeds the individual's capacity to adapt that the individual experiences *distress*.

Personal benefits such as growing from challenging experiences, re-prioritizing goals, and increasing self-esteem are all potential outcomes of facing stressful life events (Cohen, Kessler, & Gordon, 1995). It is important for researchers to measure individuals' assessments of the events they feel they have control over and those they do not (Brulé & Morgan, 2018). All too often researchers inquire only about distress, ignoring the fact that the stress an individual is experiencing could actually be motivating, performance-enhancing, and positive. For example, caregiving for someone with Alzheimer's disease has been shown to be extremely stressful, but it can also be personally and relationally fulfilling, depending on how it is experienced and framed (see Beach, 1997). As Crawford, LePine, and Rich (2010) argue, eustress could be distinguished from distress when individuals view the stressor as a challenge or an opportunity instead of a demand. It is also important to acknowledge that eustress and distress may co-occur for an individual for the same stimulus (Kozusznik, Peiró, Lloret, & Rodriguez, 2015). This co-occurrence is important because it implies that individuals can be coached to reappraise a demand as a challenge or an opportunity.

CHRONIC AND ACUTE STRESS

It is also not uncommon for researchers to refer to acute and chronic stress without a clear delineation of when acute stress—resulting in demands or pressures in the recent past or anticipated in the near future—becomes chronic or longer-lasting stress. Generally, acute stress is time-limited, whereas chronic stress occurs continuously or periodically (Cohen et al., 1995) as a result of enduring problems, conflicts, or threats (Pearlin, 1989). These temporal characteristics can become difficult to articulate and measure, depending on the nature of the stressor and the individual appraisal. For example, the persistence of a stressful event becomes a chronic stressor if it is appraised as stressful and it elicits a stress response. Alternatively, a stressful event that has since ended but is still ruminated on by the individual may elicit a chronic stress response, even though it is no longer happening. This example highlights the artificiality of acute and chronic stress categorization (Ice & James, 2006). Perceptions of stress and the ensuing physiological responses to stressors are also often culturally based and context-specific.

For example, adolescents in a refugee camp might become accustomed to high levels of chronic stress produced by exposure to trauma, violence, and mistrust. Their levels of resilience and physiological/psychological responses to stress are likely very different than those of other adolescents put in that situation who are not exposed to the same types of stressors on a recurring basis (see Afifi, Afifi, Merrill, & Nimah, 2016).

Current research examining the physiological effects of stress has focused heavily on its lasting effects. To measure these long-term health consequences, many researchers have used salivary cortisol as a marker of the hypothalamic-pituitary-adrenal (HPA) axis response. This process helps the body combat the stressor. When the body faces chronic stress, however, it may begin releasing either too much or too little cortisol, which triggers a maladaptive recovery. When this happens, the body struggles to return to its natural homeostasis, which can adversely affect people's health. However, these health consequences can be minimized or prevented by prosocial and affiliative communication patterns such as social support, affection, and relationship maintenance behaviors (e.g., Afifi, Merrill, & Davis, 2016; Floyd, 2006).

Research on chronic stress and cortisol has been rather inconsistent. Some studies have found that chronic stress is associated with too much cortisol secretion—hypercortisolism—and others have found that it is associated with too little cortisol secretion—hypocortisolism. This contradiction can be difficult to understand at first glance. Miller, Chen, and Zhou (2007) suggested, however, that the culmination of cross-sectional studies on the subject has misled researchers to think that only one outcome or the other occurs when it is more likely that both outcomes occur at different times during response and recovery. In a review of research exploring HPA axis function, Miller et al. (2007) found inconsistency across studies regarding the point at which cortisol was measured in relation to stressors, with some studies measuring cortisol directly after the onset of stress and others measuring it after more time had passed. Moreover, measuring immediate responses to stress after a stress task (e.g., gathering saliva after a stress-inducing conversation task or after a stress-inducing artificial task like a difficult math problem or public speech) yields completely different results and is interpreted differently than measuring diurnal rhythm (e.g., gathering saliva at several points throughout the day, beginning immediately upon awakening to bedtime). They also found that there has not been consistent delineation between traumatic and non-traumatic stress and physical or social stress. The association between cortisol and various outcomes depends on how and when cortisol is measured and the perceptions and behaviors that are examined along with it.

People's stress responses are intrinsically connected to the social environment. Research has focused on how chronic stressors from the social environment influence people's physiological stress responses. Prolonged exposure to chronic stress from one's environment can cause "wear and tear" on the body's physiological stress response systems, compromising the immune system and providing a gateway to disease (McEwen & Stellar, 1993). This wear and tear, known as allostatic load (McEwen, 1998), is the cost to the body of the persistent requirement to adapt to change (McEwen, 2001). McEwen created an index of allostatic load (McEwen, 2001) as a way to gauge indicators of "system failure." Relevant biological markers often include high blood pressure, elevated urinary epinephrine, and cortisol. The index calculates the number of biological stress indicators and uses that measurement to make predictions about an individual's cognitive and physical functioning. What allostatic load does not account for, however, is why some people are resilient (mentally, physically, relationally) and even thrive in the face of prolonged adversity. Recent work has found that chronic exposure to stress is not always predictive of negative outcomes (Del Giudice, Ellis, & Shirtcliff, 2011). In fact, the adaptive calibration model states that, biologically, people can adapt to changing circumstances better that many stress researchers originally thought (Del Giudice et al., 2011). These revelations have led many scholars to define and measure the concept of resilience alongside stress as an explanatory factor of biological, psychological, and behavioral differences in individuals faced with the same stressors.

POSITIVE ADAPTATIONS TO STRESS: RESILIENCE AND THRIVING

Whereas some responses to challenges and threats produce distress, other responses can prevent or lessen distress and result in quicker recovery and potential thriving. Two trajectories of positive adaptation to stress are resilience and thriving. Ellis and Del Giudice (2014) argue that it is unlikely that human physiology would evolve such that adaptations to repeated stressors would be harmful. It is possible that biobehavioral systems over time are not always impaired by distress, but rather are calibrated to overcome distress. Resilience is the dynamic process by which people recover from distress and adapt positively to it (Bonanno, 2004; Bonanno & Mancini, 2008). People who are resilient adapt positively to distress so that they regain a sense of physiological and psychological well-being (Bonanno, 2004). Individuals who are more resilient demonstrate quicker recovery to baseline physiologically (e.g., recovery to a baseline cortisol level quicker after an acute stressor) and experience less distress over time as a result of a stressor (Bonanno & Mancini, 2008). This definition works well in terms of physiologically returning to baseline after a threat has passed but can be problematic if a person's baseline is unhealthy. Psychologically and relationally, we therefore prefer to consider resilience as adapting positively to a stressor rather than returning to a baseline per se. People also learn from experiencing stressors and never truly return to a prior state psychologically or relationally (Patterson, 2002).

Thriving builds on resilience and refers to the process and state of personal and/or relational growth, learning, discovery, and flourishing (Carver, 1998; Collins & Feeney, 2010). Thriving not only manifests in an absence of distress (e.g., lack of physical or mental illness) (Collins & Feeney, 2010), but also emerges via growth (e.g., gaining greater capacities) and optimal physiological and psychological well-being. This optimal state suggests that exceptional physical health for one's age is a part of thriving, as well as adaptive developmental plasticity and post-traumatic growth. Individuals who are categorized as thriving not only have recovered or become less reactive to stress but have gained personal and/or relational resources over time, improving their psychological, relational, and/or physiological well-being. Many investigations have examined resilience to answer questions about why some people are better able to adapt to adversity than others. However, the more interesting question is who is most likely to grow from distress and flourish. Moreover, what does thriving look like physiologically?

BIOLOGICAL MARKERS OF RESILIENCE AND THRIVING

Due to variation in conceptualizations of resilience, operational definitions of resilience also vary. Psychologically, it is often operationalized as self-report measures that assess protective factors of cognitive control (flexibility), self-efficacy, and positive versus negative affect states. Psychological resilience or positive adaptations are outcomes of protective factors, such as positive emotions, self-restorative/persevering thoughts, social support, and self-regulatory capacities (Walker, Pfingst, Carnevali, Sgoifo, & Nalivaiko, 2017). Yet, resilience is also conceptualized as the ability to withstand or adapt positively to distress. Therefore, it is often measured as the absence of self-reported mental or physical illness (Osório, Probert, Jones, Young,

& Robbins, 2017). Resilience research has been characterized by the interaction among cognitive, emotional, and dispositional (e.g., optimism and openness) attributes of individuals, attributes of one's social relationships and environment, and attributes of the genetic, epigenetic, neurological, hormonal (endocrine), and immunological systems within individuals. Validating operationalizations of resilience can be complex, as a result of the dynamic and reciprocal relationships among affect, thoughts, social environment, and physiology. However, some biological markers have demonstrated utility as resiliency measures.

Across the biological systems that serve as markers of resilience, two common approaches are used to measure resilience physiologically: (1) various biological markers are often correlated with self-report measures of psychological and physical health within a sample of individuals who have experienced adversity (especially over time); or (2) the biological markers are compared between clinical and non-clinical samples who have experienced an adverse life event or illness. Both approaches aim to identify individuals who are resilient in the face of adversity, which tends to be measured as an absence of distress. However, evolutionary psychology suggests that a single strategy for resilience, survival or well-being is unlikely (Ellis & Del Giudice, 2014). Research still often involves the study of severe stressors or significant life adversity rather than daily stressors. The limited scholarly understanding of resilience likely contributes to underestimating its prevalence (Bonanno, 2004). Similarly, although scholars are generally interested in understanding whether resilience is innate, evaluating individuals before and after a traumatic event is challenging. Nonetheless, using these existing methodologies, scholars have identified a range of potential physiological markers of resilience.

Sympathetic Nervous System and Cardiovascular Biomarkers

In the sympathetic nervous system (SNS), resilience can be measured by investigating the amplitude and duration of reactivity to stressors. Reactivity and recovery are often investigated by manipulating stimuli to imitate the severity or sudden onset of stressors. There are two common forms: short-term high-intensity stimuli known as acoustic start probes (e.g., sudden loud sound) and prolonged artificial distressing situations (e.g., Trier Social Stress Test). Low-resilient individuals identified as having symptoms of post-traumatic stress disorder (PTSD) or major depressive disorder (MDD) demonstrate higher heart-rate responses to bursts of threatening stimuli and slower recovery (or habituation) via skin conductance responses (Pole, 2007). Because much of this scholarship compares people who have had a traumatic experience to those who have not, it is unclear whether this reactivity is an antecedent or an outcome of traumatic events (Walker, Pfingst, Carnevali, Sgoifo, & Nalivaiko, 2017). Cardiovascular reactivity to the Trier Social Stress Test, involving a public speaking stressor and serial subtraction task, has predicted heart disease (Pfau & Russo, 2015; Walker et al., 2017). Both increased and decreased sensitivity or reactivity to these stimuli have been associated with poorer psychological health (e.g., PTSD, depressive symptoms). Of these measures of resilience, recovery time has demonstrated some of the most interpretable and consistent findings.

Neuroendocrine and Hormonal Biomarkers

Numerous biomarkers are used to identify resilience that are produced by the HPA axis, the hypothalamic-pituitary-gonadal (HPG) axis, and the posterior pituitary gland. Three of the most common hormones to emerge from these stress responses systems are discussed below.

Cortisol

Cortisol is a stress hormone that is both necessary for helping energize the body to combat daily challenges, but it can also be harmful if the body produces too much or too little of it. When there is the perception of an impending threat or stressor, the hypothalamus emits the corticotropin-releasing hormone (CRH), producing the release of the adrenocorticotropic hormone (ACTH) from the pituitary. This triggers the release of cortisol from the adrenal glands. Salivary cortisol levels have been shown to increase after acute social stress, usually peaking 10–15 minutes and recovering 30 minutes after the stressor ends (McEwen, 1998, 2001). The response and recovery patterns from an acute stressor are the body's way of returning it back to a state of homeostasis (Miller et al., 2007). This return to a natural homeostasis is what researchers often consider a form of resilience to an acute stress task. For short-term stressors, resilience can be examined via participants' cortical reactivity to stress-inducing situations via the Trier Social Stress Test or a natural conflict-inducing/stressful task with a partner (Pfau & Russo, 2015; Walker et al., 2017). As mentioned above, these stressors activate the HPA axis and elevate cortisol, which can be measured via saliva or blood samples. Resilience is demonstrated by both lower cortisol levels in reaction to the stressor and quicker recovery or return to baseline following the stressor.

In addition to short-term cortisol reactivity, the diurnal rhythm is used as a marker of resilience to chronic stress. Cortisol follows a natural diurnal rhythm in healthy individuals (McEwen, 2001; Saxbe & Repetti, 2010), wherein it peaks within 30 minutes of awakening (Adam, Schönfelder, Forneck, & Wessa, 2014), declines throughout the day, and reaches its lowest point around midnight (Saxbe & Repetti, 2010). To measure resilience, diurnal rhythms are compared between individuals who are experiencing similar traumatic experiences. In individuals who are coping with depression or anxiety, for instance, the diurnal rhythm approximates a flattened slope, with morning or evening cortisol levels that are abnormally low or high (Pendry & Adam, 2007). Additionally, if the diurnal rhythm demonstrates peaks throughout the day or fails to decline throughout the day, this could also be an indicator of less resilience. As opposed to short-term resilience, diurnal rhythms demonstrate resilience to chronic or prolonged stressors (Doane et al., 2006).

Reproductive Hormones

Reproductive hormones are also important to social interaction and are related to sex differences. Estrogens, progesterone, and testosterone are produced by the gonads and to a lesser extent by the adrenal glands. The HPA axis and gonadal axis are highly connected, so that hyperactivity in the HPA has been related to pubertal maturation (Ellis & Del Giudice, 2014; Pfau & Russo, 2015). Previous research has identified that estrogens and progesterone levels are associated with stress reactivity in women with mood disorders (Pfau & Russo, 2015). Even though testosterone is typically associated with negative characteristics, such as aggression and competition, it has also been associated with greater psychological resilience, such as increased feelings of social connection and positive mood via self-satisfaction and social dominance (Edwards, Wetzel, & Wyner, 2006).

Oxytocin

Oxytocin is a pituitary peptide that has been associated with human emotion and social cognition (Sippel et al., 2017). Previous research has related oxytocin to social connectedness and bonding as well as reward sensations (Feldman, 2009; Taylor et al., 2000). Similar to reproductive hormones, oxytocin has been associated with positive emotions and social resources that can serve as a protective factor from stress and is, therefore, believed to promote resilience (Ramsey & Gentzler, 2015). Oxytocin is theorized to be an essential stress response that leads to approach responses and developing social resources, especially in women (Taylor, 2006). For instance, studies applying tend-and-befriend theory have shown that oxytocin is positively associated with trust and altruistic behaviors (Sippel et al., 2017) and negatively associated with feelings of rejection and social threats (Zwolinski, 2008). As Kosfeld et al. (2005) argue, oxytocin encourages people to approach and bond with others in ways that might go against their natural instincts to be more reserved and protect themselves. These authors found that participants who were given intranasal doses of oxytocin were more likely to take social risks and initiate conversations with others compared to those who did not receive the oxytocin.

Oxytocin has also been examined in relation to attachment theory. At a very basic level, oxytocin helps muscle contractions throughout the labor process, produces lactation, and enhances mother-infant social bonds (Kosfeld, Heinrichs, Zak, Fischbacher, & Fehr, 2005). Because oxytocin is also often released in response to one's senses, parents often experience increases in oxytocin when they touch, smell, and feel the warmth of their baby's skin (Hiller, 2004). Even though the causal mechanisms are still being explored, children's oxytocin levels tend to be positively associated with secure attachments, which some researchers argue could help them form and maintain close relationships later in life (Masten & Cicchetti, 2016). Oxytocin has also been positively associated with attachments in romantic relationships. For example, Schneiderman, Zagoory-Sharon, Leckman, and Feldman (2012) examined plasma oxytocin levels in 163 young adults, 120 of whom were new lovers (60 couples) and 43 of whom were non-dating singles, at two time points, separated by six months. They found that oxytocin levels were higher in new lovers compared to single young adults and that these rates remained stable over the two time periods. The higher oxytocin levels were associated with greater positive affect, affectionate touch, synchronized dyadic communication, and worrying about one's partner. In general, oxytocin levels are believed to be positively correlated with secure attachments.

Research has also begun to demonstrate how oxytocin reacts in correspondence with other hormones in response to stress (Cardoso, Kingdon, & Ellenbogen, 2014; Rutigliano et al., 2016). Oxytocin may act as a moderator to buffer the negative impact of other hormones such as testosterone and cortisol on people's stress levels (Ellis, Bianchi, Griskevicius, & Frankenhuis, 2017). Oxytocin, for example, has been shown to interact with social support to buffer the negative effects of cortisol and subjective stress on health (Heinrichs, Baumgartner, Kirschbaum, & Ehlert, 2003).

Immunological Biomarkers

Within the immune system, stress is related to inflammatory responses (Boren & Veksler, 2011; Pfau & Russo, 2015). Specifically, cytokines such as interleukin-6 (IL-6) and interleukin 1-beta (IL-1B) interact with the HPA axis such that inflammation increases due to stress (Juster, McEwen, & Lupien, 2010). The presence of cytokines is theorized to be indicative of allostatic load (Danese & McEwen, 2012). Thus, resilience within the immune system is observable via diminished levels of cytokines and inflammatory responses. According to the conceptualization of resilience in the allostatic load theory, less wear and tear on the body's immune system is related to less illness and prolonged mortality (Juster et al., 2010). Under these circumstances, resilience is the absence of distress rather than the recovery from distress.

Potential Markers of Thriving

Whereas research has developed both psychological and physiological indicators of resilience, thriving is primarily measured as a psychological construct. Psychologically, flourishing and growth include experiences of greater hedonic pleasure, purpose, and self-confidence for both individuals and relational partners (Bonanno, 2004; Feeney & Collins, 2014). The parallel physiological indicators of thriving have yet to be operationalized. However, Ellis et al. (2017) noted that children may develop capacities when they have been sensitized and adapted to chronic stressful environments that might shed light on resilience. According to Ellis and Del Giudice (2014), the perspective of allostatic load biomarkers is incomplete, in that it merely distinguishes the maladaptiveness of biomarkers as opposed to the potential adaptive functions of these calibrations. Other researchers have found that greater cortisol reactivity and higher levels of salivary alpha-amylase have been associated with better performances or improved cognitive capacities during stressful tasks (e.g., Roos et al., 2018). These improved cognitive capacities could be physiological adaptations to stress that are context-dependent. In investigating the construct of thriving physiologically, scholars could investigate other forms of stress reactivity that can be learned, allowing individuals not only to meet the demands of stressful environments, but also to thrive in them. Context-specific sensitization and adaptation also may not only appear under acute or stressful situations but may increase the likelihood to anticipate potential stressors, engage in anticipatory coping, and thus successfully cope with more demands on a daily basis. Thus, it is valuable to identify the contexts in which markers of allostatic load are beneficial or at least adaptive.

THE ROLE OF COMMUNICATION AND SOCIAL RELATIONSHIPS IN STRESS RESPONSES

Finally, we focus on the effects of social environment on stress responses and resilience/thriving. When examining the research on both stress and resilience, it is clear that social relationships play a crucial role in the experience of stress and people's ability to cope with it. An abundance of research shows that soothing communicative behaviors such as social support, affection, and collaborative communication during conflict help manage stress. Research has also shown that larger social networks and higher quality social support, particularly emotional support, can buffer the effects of stress on emotional and physical health (Cobb, 1976). For instance, when participants are provided with social support after a stressful or hurtful interaction with a dating or marital partner, they experience a weaker cortisol response and faster cortisol recovery after the discussion task (e.g., Priem & Solomon, 2012). Similar benefits follow when people receive levels of support that they desire or expect (e.g., Priem & Solomon, 2015). Feelings of affiliation and receptivity by one's romantic partner have similarly soothing effects (Priem, McLaren, & Solomon, 2010). Research also shows that adolescents who perceive their parents as communicatively skilled (e.g., greater communication competence, socially supportive, collaborative conflict management, affiliative) tend to experience weaker cortisol responses after stressful conversations with their parents (Afifi, Granger, Denes, Joseph, & Aldeis, 2011). These communication patterns can also have long-term effects on children's social and personal development (see Repetti, Robles, & Reynolds, 2011). An underlying assumption that crosses these positive behaviors is that they tend to make people feel validated and secure. According to attachment theory, people long to experience security with close others, to give and receive love freely, and to feel as if they can approach others for support during times of distress (Bowlby, 1982).

One particularly prominent line of research has examined affectionate communication to understand the relationship between social relationships and stress management. Through the exploration of different hormonal reactions to environmental factors, expressing affection—and thus promoting the secretion of oxytocin—became a key explanatory variable in the reduction of unhealthy cortisol levels. Affection exchange theory (AET; Floyd, 2002, 2006) provides a bio-evolutionary explanation for the link between affectionate communication and positive health outcomes. AET assumes the evolutionary principle that humans have an innate need to pursue fertility and viability. Being affectionate signals to others that one is a viable mate and parent. Floyd hypothesized that only some communicative acts are conscious, whereas others have deep-rooted evolutionary motivations that influence hormonal reactivity. Floyd and his colleagues have found that sharing affectionate behavior increases oxytocin and reduces stress biomarkers, such as cortisol, blood pressure, and blood glucose (for review, see Floyd, 2019). AET declared that affection is so fundamental to survival that the giving and receiving (or lack thereof) directly covary with our physiological health. People differ, however, in how much and what kind of love and validation they need from their relationships (High & Steuber, 2014; Priem & Solomon, 2015).

Negative forms of conflict appear to have the opposite effect. For instance, research has shown that dysfunctional conflict, loneliness, and relationship dissatisfaction can produce chronic stress, which, in turn, can increase inflammation and promote illness (see Jaremka, Lindgren, & Kiecolt-Glaser, 2013). Violence and negative forms of conflict management (e.g., criticism, contempt, stonewalling, demand-withdrawal) over time can harm individuals' mental and relational well-being and also impair their physical health by weakening the immune system (see Jaremka et al., 2013).

We recently developed the theory of resilience and relational Load (TRRL: Afifi, Merrill, & Davis, 2016) to explain and predict stress and resilience in social relationships. The TRRL builds on the theory of emotional capital (Driver & Gottman, 2004; Feeney & Lemay, 2012), which assumes that when people validate their relational partners or family members over time, they accumulate positive emotional reserves or emotional capital upon which they can draw later when their relationships feel threatened. The TRRL views relationship maintenance behaviors and actions as forms of stress prevention and management. It argues that when people better maintain their relationships over time, they not only manage stress more effectively but also perceive events as less stressful in the first place. Giving maintenance behaviors and actions, and especially receiving them, fuels positive emotions toward one's partner or family member that helps buffer stress.

The TRRL also extends the theory of emotional capital by outlining who is likely to invest in their relationships in the first place. According to the TRRL, relational partners and family members who have more of a communal orientation or who are unified in combating a stressor and life in general are more likely to invest in their relationships and maintain them compared to people who have less of a communal orientation. When family members have a strong communal orientation, they feel as if they are a team, "in it together," and feel as if they are not alone in their ability to confront life's challenges. A sense of unity has been shown to be "integral to cooperation, companionship, negotiating differences, problem solving, feeling supported emotionally, and feeling motivated to support each other in being who the other is" (Reid, Dalton, Laderoute, Doell, & Nguyen, 2006, p. 248). Unlike communal coping, however, wherein dyads or groups of people perceive and act on a stressor together (Lyons, Mickelson, Sullivan, & Coyne, 1998), communal orientation is only the former and not the latter. That is, it is the perception of togetherness or unity against a stressor without necessarily acting on it. Relationship maintenance and communal orientation are reciprocal, such that having a stronger communal orientation makes people want to invest in their relationships and maintaining one's relationship makes people feel more communally oriented toward each other.

Together, relationship maintenance and communal orientation affect people's appraisals and communication about their stressors. Accrued relationship maintenance and communal orientation build emotional reserves, which allow people to appraise relationally stressful situations from a broader and more positive mindset (see also broaden and build theory; Fredrickson, 2001). The positive feelings and emotions about one's partner and the relationship also help people communicate in ways that uplift their partner and preserve their relationship. When people have not maintained their relationship, their tendency is to appraise stress and communicate about it in more threat-based ways by protecting themselves in times of stress/conflict rather than their partner or the relationship. For instance, research shows that partners in dissatisfying relationships engage in more attributional errors and blame each other for their stress during conflict-inducing conversations (Sillars, Roberts, Leonard, & Dun, 2000). Such threatening appraisals and communication patterns deplete people's emotional, behavioral, and psychological resources over time and exacerbate their degree of personal and relational stress. Using more secure appraisals and communication patterns during conflict creates resilience and allows for the possibility of thriving. The positive emotions provide opportunities for creative problem solving, listening, social support, and constructive conflict management when stressed. This should also help manage perceived and physiological stress and promote personal and relational health.

If relational partners/family members fail to maintain their relationships, they can become emotionally disconnected from each other (i.e., reduced communal orientation) and chronic conflict and stress can slowly wear away at the relationships, creating what we refer to as *relational load* (Afifi et al., 2016). Chronic conflict and stress can slowly deplete one's cognitive, emotional, and relational resources. As the dual process model of social support (Bodie et al., 2011) also indicates, when people become highly stressed or too emotional, their ability to process information and provide/receive adequate support wanes. Ongoing negative conflict (e.g., criticism, contempt, defensiveness, demand-withdraw patterns) and the stress produced by that conflict can create relational burnout. People become hardened to their relationships and less willing to invest in them. Relational load can also negatively affect their mental and physical health because of heightened physiological stress responses over extended periods of time.

CONCLUSION

Communication within social relationships is a fundamental way that resilience and thriving are created and sustained. This ultimately affects people's personal, relational, and physiological health. People's health also simultaneously affects how people communicate with their relational partners and how they feel in those relationships. According to the TRRL (Afifi et al., 2016), relational partners need to invest in their relationships by actively maintaining them to reduce the propensity for relational load and reinvigorate their relational and personal health. Resilience and thriving become a series of feedback loops wherein relational partners need to continually gather feedback about their stress, including how they are feeling toward each other and communicating about their stress, and to reinvest in their relationships through active maintenance efforts. Fortunately, because these behaviors are socially constructed, they can be learned and changed, providing ample opportunities for interventions for couples and families.

REFERENCES

Adam, R., Schönfelder, S., Forneck, J., & Wessa, M. (2014). Regulating the blink: Cognitive reappraisal modulates attention. *Frontiers in Psychology, 5*, 143.

Afifi, T. D., Afifi, W. A., Merrill, A., & Nimah, N. (2016). "Fractured communities": Uncertainty, stress, and (a lack of) communal coping in Palestinian refugee camps. *Journal of Applied Communication Research, 44*, 343–361.

Afifi, T. D., Granger, D., Denes, A., Joseph, A., & Aldeis, D. (2011). Parents' communication skills and adolescents' salivary α-amylase and cortisol response patterns. *Communication Monographs, 78*, 273–295.

Afifi, T. D., Merrill, A. F., & Davis, S. (2016). The theory of resilience and relational load. *Personal Relationships, 23*, 663–683.

Beach, D. (1997). Family caregiving: The positive impact on adolescent relationships. *Gerontologist, 37*, 233–238.

Bodie, G. D., Burleson, B. R., Holmstrom, A. J., McCullough, J. D., Rack, J. J., Hanosono, L. K., & Rosier, J. G. (2011). Effects of cognitive complexity and emotional upset on processing supportive messages: Two tests of a dual-process theory of supportive communication outcomes. *Human Communication Research, 37*, 350–376.

Bonanno, G. A. (2004). Loss, trauma, and human resilience: Have we underestimated the human capacity to thrive after extremely aversive events? *American Psychologist, 59*(1), 20–28.

Bonanno, G. A., & Mancini, A. D. (2008). The human capacity to thrive in the face of potential trauma. *Pediatrics, 121*, 369–375.

Boren, J. P., & Veksler, A. E. (2011). A decade of research exploring biology and communication. *Communication Research Trends, 30*(4), 4–31.

Bowlby, J. (1982). Attachment theory and its therapeutic implications. *Adolescent Psychiatry, 6*, 5–33.

Brulé, G., & Morgan, R. (2018). Working with stress: Can we turn distress into eustress? *Journal of Neuropsychology & Stress Management, 3*, 1–3.

Cardoso, C., Kingdon, D., & Ellenbogen, M. A. (2014). A meta-analytic review of the impact of intranasal oxytocin administration on cortisol concentrations during laboratory tasks: Moderation by method and mental health. *Psychoneuroendocrinology, 49*, 161–170.

Carver, C. S. (1998). Resilience and thriving: Issues, models, and linkages. *Journal of Social Issues, 54*, 245–266.

Chrousos, G. P., & Gold, P. W. (1992). The concepts of stress and stress system disorders: Overview of physical and behavioral homeostasis. *Journal of the American Medical Association, 267*, 1244–1252.

Cobb, S. (1976). Social support as a moderator of life stress. *Psychosomatic Medicine, 38*, 300–314.

Cohen, S., Kessler, R. C., & Gordon, L. U. (1995). Strategies for measuring stress in studies of psychiatric and physical disorders. In A. Cohen, R. C. Kessler, & L. Underwood Gordon (Eds.). *Measuring stress: A guide for health and social scientists* (pp. 3–26). New York, NY: Oxford University Press.

Collins, N. L., & Feeney, B. C. (2010). An attachment theoretical perspective on social support dynamics in couples: Normative processes and individual differences. In K. Sullivan & J. Davila (Eds.), *Support processes in intimate relationships* (pp. 89–120). New York, NY: Oxford University Press.

Crawford, E. R., LePine, J. A., & Rich, B. L. (2010). Linking job demands and resources to employee engagement and burnout: A theoretical extension and meta-analytic test. *Journal of Applied Psychology, 95*, 834–848.

Danese, A., & McEwen, B. S. (2012). Adverse childhood experiences, allostasis, allostatic load, and age-related disease. *Physiology & Behavior, 106*, 29–39.

Del Giudice, M., Ellis, B. J., & Shirtcliff, E. A. (2011). The adaptive calibration model of stress responsivity. *Neuroscience & Biobehavioral Reviews, 35*(7), 1562–1592.

Doane, A. S., Danso, M., Lal, P., Donaton, M., Zhang, L., Hudis, C., & Gerald, W. L. (2006). An estrogen receptor-negative breast cancer subset characterized by a hormonally regulated transcriptional program and response to androgen. *Oncogene, 25*, 3994–4008.

Driver, J. L., & Gottman, J. M. (2004). Daily marital interactions and positive affect during marital conflict among newlywed couples. *Family Process, 43*, 301–314.

Edwards, D., Wetzel, K., & Wyner, D. (2006). Intercollegiate soccer: Saliva cortisol and testosterone are elevated during competition, and testosterone is related to status and social connectedness with teammates. *Physiology & Behavior, 87*, 135–143.

Ellis, B. J., Bianchi, J., Griskevicius, V., & Frankenhuis, W. E. (2017). Beyond risk and protective factors: An adaptation-based approach to resilience. *Perspectives on Psychological Science, 12*, 561–587.

Ellis, B. J., & Del Giudice, M. (2014). Beyond allostatic load: Rethinking the role of stress in regulating human development. *Development and Psychopathology, 26*, 1–20.

Feeney, B. C., & Collins, N. L. (2014). A new look at social support: A theoretical perspective on thriving through relationships. *Personality and Social Psychology Review, 19*, 113–147.

Feeney, B. C., & Lemay, E. P. (2012). Surviving relationship threats: The role of emotional capital. *Personality and Social Psychology Bulletin, 38*, 1004–1017.

Feldman, R. (2009, September). Oxytocin and bonding: The neurobiological foundation of human affiliation. Paper presented at Frontiers in Behavioral Neuroscience Conference Abstract: 41st European Brain and Behaviour Society Meeting, Rhodes, Greece.

Floyd, K. (2002). Human affection exchange: V. Attributes of the highly affectionate. *Communication Quarterly, 50*, 135–152.

Floyd, K. (2006). *Communicating affection: Interpersonal behavior and social context.* Cambridge, England: Cambridge University Press.

Floyd, K. (2019). *Affectionate communication in close relationships.* Cambridge, England: Cambridge University Press.

Fredrickson, B. (2001). The role of positive emotions in positive psychology. *American Psychologist, 56*, 218–226.

Heinrichs, M., Baumgartner, T., Kirschbaum, C., & Ehlert, U. (2003). Social support and oxytocin interact to suppress cortisol and subjective responses to psychosocial stress. *Biological Psychiatry, 54*, 1389–1398.

High, A. C., & Steuber, K. R. (2014). An examination of support (in)adequacy: Types, sources, and consequences of social support among infertile women. *Communication Monographs, 81*, 157–178.

Hiller, J. (2004). Speculations on the links between feelings, emotions, and sexual behavior: Are vasopressin and oxytocin involved? *Sexual and Relationship Therapy, 19*, 1468–1479.

Ice, G. H., & James, G. D. (2006). *Measuring stress in humans: A practical guide for the field.* Cambridge, England: Cambridge University Press.

Jaremka, L. M., Lindgren, M. E., & Kiecolt-Glaser, J. K. (2013). Synergistic relationships among stress, depression, and troubled relationships: Insights from psychoneuroimmunology. *Depression and Anxiety, 30*, 288–296.

Juster, R. P., McEwen, B. S., & Lupien, S. J. (2010). Allostatic load biomarkers of chronic stress and impact on health and cognition. *Neuroscience and Biobehavioral Reviews, 35*, 2–16.

Kosfeld, M., Heinrichs, M., Zak, P. J., Fischbacher, U., & Fehr, E. (2005). Oxytocin increases trust in humans. *Nature, 435*, 637–676.

Kozusznik, M., Peiró, J. M., Lloret, S., & Rodriguez, I. (2016). Hierarchy of eustress and distress: Rasch calibration of the Valencia eustress-distress appraisal scale. *Central European Journal of Management, 2*, 67–79.

Lazarus, R. S., & Folkman, S. (1984). Coping and adaptation. In W. D. Gentry (Ed.), *The handbook of behavioral medicine* (pp. 282–325). New York, NY: Guilford.

Lyons, R. F., Mickelson, K. D., Sullivan, M. J., & Coyne, J. C. (1998). Coping as a communal process. *Journal of Social and Personal Relationships, 15*, 579–605.

Masten, A. S., & Cicchetti, D. (2016). Resilience in development: Progress and transformation. In D. Cicchetti (Ed.), *Developmental psychopathology*: Vol. 4. *Risk, resilience, and intervention* (3rd ed., pp. 271–333). Hoboken, NJ: John Wiley & Sons.

McEwen, B. S. (1998). Protective and damaging effects of stress mediators. *New England Journal of Medicine, 338*, 171–179.

McEwen, B. S. (2001). Plasticity of the hippocampus: Adaptation to chronic stress and allostatic load. *Annals of the New York Academy of Science, 933*, 265–277.
McEwen, B. S., & Stellar, E. (1993). Stress and the individual: Mechanisms leading to disease. *Archives of Internal Medicine, 153*, 2093–2101.
Miller, G. E., Chen, E., & Zhou, E. S. (2007). If it goes up, must it come down? Chronic stress and the hypothalamic-pituitary-adrenocortical axis in humans. *Psychological Bulletin, 133*(1), 25–45.
Osório, C., Probert, T., Jones, E., Young, A. H., & Robbins, I. (2017). Adapting to stress: Understanding the neurobiology of resilience. *Behavioral Medicine, 43*, 307–322.
Patterson, J. M. (2002). Understanding family resilience. *Journal of Clinical Psychology, 58*, 233–245.
Pearlin, L. (1989). The sociological study of stress. *Journal of Health and Social Behavior, 30*, 241–256.
Pendry, P., & Adam, E. K. (2007). Associations between parents' marital functioning, maternal parenting quality, maternal emotion and child cortisol levels. *International Journal of Behavioral Development, 31*, 218–231.
Pfau, M. L., & Russo, S. J. (2015). Peripheral and central mechanisms of stress resilience. *Neurobiology of Stress, 1*, 66–79.
Pole, N. (2007). The psychophysiology of posttraumatic stress disorder: A meta-analysis. *Psychological Bulletin, 133*, 725–746.
Priem, J. S., McLaren, R. M., & Solomon, D. H. (2010). Relational messages, perceptions of hurt, and biological stress reactions to a disconfirming interaction. *Communication Research, 37*, 48–72.
Priem, J. S., & Solomon, D. H. (2015). Emotional support and physiological stress recovery: The role of support matching, adequacy, and invisibility. *Communication Monographs, 82*, 88–112.
Ramsey, M. A., & Gentzler, A. L. (2015). An upward spiral: Bidirectional associations between positive affect and positive aspects of close relationships across the life span. *Developmental Review, 36*, 58–104.
Reid, D. W., Dalton, E. J., Laderoute, K., Doell, F. K., & Nguyen, T. (2006). Therapeutically induced changes in couple identity: The role of we-ness and interpersonal processing in relationship satisfaction. *Genetic, Social, and General Psychology Monographs, 132*, 241–284.
Repetti, R., Robles, R., & Reynolds, B. (2011). Allostatic processes in the family. *Development and Psychopathology, 23*, 921–938.
Roos, L. E., Beauchamp, K. G., Giuliano, R., Zalewski, M., Kim, H. K., & Fisher, P. A. (2018). Children's biological responsivity to acute stress predicts concurrent cognitive performance. *Stress*, 1–8.
Rutigliano, G., Rocchetti, M., Paloyelis, Y., Gilleen, J., Sardella, A., Cappucciati, M., … McGuire, P. (2016). Peripheral oxytocin and vasopressin: Biomarkers of psychiatric disorders? A comprehensive systematic review and preliminary meta-analysis. *Psychiatry Research, 241*, 207–220.
Saxbe, D., & Repetti, R. L. (2010). For better or worse? Coregulation of couples' cortisol levels and mood states. *Journal of Personality and Social Psychology, 98*, 92–103.
Schneiderman, I., Zagoory-Sharon, O., Leckman, J., & Feldman, R. (2012). Oxytocin during the initial stages of romantic attachment: Relations to couples' interactive reciprocity. *Psychoneuroendocrinology, 37*, 1277–1285.
Seeman, T. E., McEwen, B. S., Rowe, J. W., & Singer, B. H. (2001). Allostatic load as a marker of cumulative biological risk: MacArthur studies of successful aging. *Proceedings of the National Academy of Sciences, 98*, 4770–4775.
Selye, H. (1936). A syndrome produced by diverse nocuous agents. *Nature, 138*(3479), 32–33.
Selye, H. (1987). Stress without distress. In L. Levi (Ed.), *Society, stress, and disease* (Vol. 5, pp. 257–262). Oxford, England: Oxford University Press.
Sillars, A., Roberts, L. J., Leonard, K. E., & Dun, T. (2000). Cognition during marital conflict: The relationship of thought and talk. *Journal of Social and Personal Relationships, 17*, 479–502.
Sippel, L. M., Allington, C. E., Pietrzak, R. H., Harpaz-Rotem, I., Mayes, L. C., & Olff, M. (2017). Oxytocin and stress-related disorders: Neurobiological mechanisms and treatment opportunities. *Chronic Stress, 1*, 1–15.
Taylor, S. E. (2006). Tend and befriend: Biobehavioral bases of affiliation under stress. *Current Directions in Psychological Science, 15*(6), 273–277.

Taylor, S. E., Klein, L. C., Lewis, B. P., Gruenewald, T. L., Gurung, R. A. R., & Updegraff, J. A. (2000). Biobehavioral responses to stress in females: Tend-and-befriend, not fight-or-flight. *Psychological Review, 107*, 411–429.

Walker, F. R., Pfingst, K., Carnevali, L., Sgoifo, A., & Nalivaiko, E. (2017). In the search for integrative biomarker of resilience to psychological stress. *Neuroscience and Biobehavioral Reviews, 74*, 310–320.

Zwolinski, J. (2008). Biophysical responses to social rejection in targets of relational aggression. *Biological Psychology, 79*, 260–267.

27

Sexual Communication and Biology

Amanda Denes, Margaret Bennett, and Maria DelGreco

Research exploring the interconnections of sexual communication and biology has grown over the past decade as scholars turn their attention to the ways that hormones, genes, and other relevant biomarkers correlate with behavior in and outside the bedroom. Although still in its nascent stages, empirical work to date has revealed several associations between sexual communication and biology, and the field is ripe with future directions for scholars interested in understanding the links between sexual communication and physiological responses. This line of work has revealed a number of bidirectional relationships, indicating that various biomarkers both influence and are influenced by sexual communication. Research investigating the links between communication before, during, and after sexual activity and biology can be divided into two primary domains: (1) the influence of trait-like measures of biology on sexual communication; and (2) the bi-directional relationship between sexual communication and state changes in biology. A new line of research also considers the lasting effects of sexual communication on biology. This chapter overviews each of these domains of sexual communication scholarship and discusses past, present, and future research directions on sexual communication and its biological associations.

TRAIT-LIKE MEASURES OF BIOLOGY AND SEXUAL COMMUNICATION

One approach to understanding the connections between biology and sexual communication involves focusing on trait-like measures of biology and their correlations with communication behavior. Such studies often measure a hormone or gene of interest at one time point to capture a baseline measurement and then test whether that measure is correlated with some form of sexual communication. Within the communication literature specifically, the few studies of sexual communication using this approach have generally focused on individuals' testosterone levels and genetic variation in the oxytocin receptor gene, and how each of these indicators correlates with post-sex communication.

Testosterone is an androgen produced in both females and males, although in greater levels in the latter (Mazur & Booth, 1998). The hormone has been linked to a range of social behavior, including aggressive and dominant behavior (although there is mixed evidence as to whether higher levels of testosterone are associated with more aggression; see Archer, 2006; Archer, Birring, & Wu, 1998; Book, Starzyk, & Quinsey, 2001, for reviews). Testosterone is of particular interest to relationship researchers, given its associations with interpersonal behavior. Research suggests that high levels of testosterone may impede healthy relational functioning. For example, testosterone is negatively associated with relationship satisfaction, commitment, and spending time with one's partner (Edelstein, van Anders, Chopik, Goldey, & Wardecker, 2014; Gray et al., 2002; Julian & McKenry, 1989). Furthermore, being in a relationship is associated with lower testosterone levels (Burnham et al., 2003; Gray et al., 2002; van Anders, Hamilton, & Watson, 2007; van Anders & Watson, 2007), and individuals with higher testosterone levels are less likely to marry and more likely to experience marital problems if they do marry (Booth & Dabbs, 1993).

For sex researchers, specifically, testosterone is of interest given its connection to sexual behavior. The hormone has been found to increase prior to and the day following sexual activity (Dabbs & Mohammed, 1992; Hamilton & Meston, 2010) and in response to sexual stimuli (Hellhammer, Hubert, & Schürmeyer, 1985). Testosterone has also been positively linked to the frequency and likelihood of orgasm (Knussmann, Christiansen, & Couwenbergs, 1986; van Anders, Hamilton, Schmidt, & Watson, 2007), although the findings on this point are inconsistent (e.g., Kruger et al., 1998; Lee, Jaffe, & Midgley, 1974).

Given the associations between testosterone levels and social and sexual behavior, recent studies have examined the role of testosterone in sexual communication. For example, Denes, Afifi, and Granger (2017) investigated post-sex disclosures and found that the higher individuals' testosterone levels, the more risks and fewer benefits they saw to disclosing to their partners after sex, which in turn predicted less-positive disclosures after sex. They also found that orgasm moderated the association between testosterone and post-sex talk, such that testosterone predicted post-sex communication for those who did not experience orgasm but not for those who did. Denes, Speer, and Speer (2015) also found that testosterone was a better predictor of post-sex communication than biological sex, suggesting that scholars would benefit from investigating intrasexual variation in biomarkers rather than simply comparing female and male individuals.

Other research has indicated that sexual communication is associated with genotypic variation. Such studies have focused on the oxytocin receptor gene. As discussed below, the pituitary hormone oxytocin is associated with a range of prosocial behaviors. When the hormone is released, it binds with a receptor that activates the hormone, and that receptor is encoded by the oxytocin receptor gene (OXTR). Researchers have recently investigated the single nucleotide polymorphism (SNP) rs53576 on OXTR. For this SNP, individuals can have one of three genotypes: GG, AG, or AA, with the presence of an A allele (i.e., the AG or AA genotypes) generally being considered a risk factor for social impairments (Tost et al., 2010).

For communication scholars, rs53576 is of interest, given its association with social behavior. For example, research suggests that GG individuals are more empathic, better able to read emotional cues, and more likely to engage in affiliative nonverbal behavior than individuals with an A allele (Kogan et al., 2011; Rodrigues, Saslow, Garcia, John, & Keltner, 2009). Within the communication literature specifically, genetic variation in rs53576 has been found to interact with attachment security to predict communication outcomes. For example, Floyd and Denes (2015) found stronger genotypic effects on affectionate communication when individuals were low, rather than high, in attachment security, with GG individuals low in attachment security reporting more trait-expressed affection than individuals with an A allele. Denes (2015) similarly found that genotypic variation was more predictive of communicative and relational outcomes for insecurely attached individuals. Insecure individuals with the GG genotype reported more closeness with their romantic partners and perceived fewer risks to disclosing compared to individuals with an A allele, but such associations were not significant for securely attached individuals, perhaps suggesting that positive relational experiences may counteract some of the social "risks" of having an A allele.

Within the domain of sexual communication, specifically, research has revealed associations between variations in rs53576 and post-sex communication. Previous work has found that individuals who orgasm from sexual activity disclose more positive thoughts and feelings for their partners during pillow talk (Denes, 2012). Denes (in progress) similarly found that orgasm predicted more positive post-sex communication, which in turn predicted greater increases in relationship satisfaction, but only for individuals with the GG genotype. However, preliminary evidence from a recent study (Denes et al., 2018) suggests that individuals with an A allele may benefit more from interventions aimed at promoting sexual communication. For women with an A allele, increasing post-sex communication over the course of three weeks predicted greater closeness at the end of the intervention, compared to a control condition, but the same pattern did not emerge for GG women. Given previous research suggesting that AA and AG individuals have greater difficulty reading and displaying empathic behavior, interventions aimed at promoting bonding and connection may have a stronger effect on these individuals than on those with the GG genotype, who may already be predisposed to intimacy and closeness.

Although the research on sexual communication and its associations with trait-like biological measures is in its early stages, the line of work developed thus far suggests that individuals' biological makeup is associated with their communication in sexual contexts. The importance of understanding how individuals' biology influences everyday behavior is also reinforced in recent theorizing from other fields. For example, the steroid/peptide theory of social bonds (S/P theory; van Anders, Goldey, & Kuo, 2011) details the ways that three hormones—testosterone, oxytocin, and arginine vasopressin—link to nurturant and sexual behavior. Van Anders et al. (2011) offer a framework for understanding the associations among these hormones and interpersonally focused behavior, such as social bonding and competition. Whereas the theory offers a broader framework for understanding the role of biology in a range of social interactions, it also specifically notes the ways that hormones can change in response to certain stimuli and contexts, suggesting that sexual intimacy increases both oxytocin and testosterone. Research on sexual communication and its association with changes or state measures of biology is discussed next.

SEXUAL COMMUNICATION AND STATE MEASURES OF BIOLOGY

A second domain of sexual communication scholarship focuses on the ways that biomarkers react to sexual communication and how changes in hormones influence sexual communication. Whereas the first domain of research examines the ways that individuals' hormone levels and genes associate with communication, the second domain focuses on the ways that bodies change in response to sexual communication, or conversely, how state changes in hormones influence sexual communication. Two techniques for assessing bodily changes to sexual behavior or arousal are described next: (1) measuring changes in salivary or plasma oxytocin; and (2) measuring genital blood flow via vaginal photoplethysmograph, and penile plethysmograph.

Changes in Oxytocin During Sexual Activity

One hormone that responds to sexual behavior is oxytocin. Oxytocin is a neuropeptide produced in the hypothalamus (see Crespi, 2015, for a review) and is associated with a range of prosocial behaviors in interpersonal relationships, such as pair bonding and attachment

(Apter-Levi, Zagoory-Sharon, & Feldman, 2014), as well as relationship quality (Holt-Lunstad, Birmingham, & Light, 2014; see Denes, Dhillon, Ponivas, & Winkler, in press, for a review). Conversely, oxytocin has been connected to ingroup favoritism and negative behavior toward outgroups (e.g., De Dreu, Greer, Van Kleef, Shalvi, & Handgraaf, 2011; see Crespi, 2015, for a review).

Studies of oxytocin's role in sexual interactions outside of the field of communication find that the hormone is elevated in response to sexual climax (Blaicher et al., 1999; Carmichael et al., 1987; Murphy, Seckl, Burton, Checkley, & Lightman, 1987), intensifies orgasms (Behnia et al., 2014), and increases the likelihood of having an orgasm and experiencing post-orgasm feelings of contentment (Zhang et al., 2015). Within the field of communication, scholars have speculated about the role of oxytocin in communication processes, focusing on the post-sex context specifically. Denes (2012, 2018) suggests that climax may serve as a proxy for understanding fluctuations in oxytocin and its effect on sexual communication. More specifically, they suggested that because the surge in oxytocin that accompanies orgasm produces feelings of elation (which may promote connection and bonding), orgasm is also likely to predict positive disclosures to one's partner in the "afterglow" (Veenestra, 2007) of sexual activity. In several studies, they found support for the link between orgasm and sexual communication, as individuals who experienced orgasm reported disclosing more positive thoughts and feelings to their partners after sexual activity than those who did not orgasm (Denes, 2012, 2018).

Although such studies are rooted in understandings of the body's physiological response to sexual arousal, they do not directly capture changes in the hormones of interest. To address this gap, Denes and colleagues (in progress) collected saliva samples during an actual sexual episode. Prior to the saliva collection, participants (who constituted romantic couples) completed a survey within two hours of sexual activity asking about their communication during and after sexual activity. The researchers then collected a baseline sample during a lab visit as a measure of oxytocin outside of the sexual context. They then had participants take home saliva collection kits and collect saliva samples before the next time they engaged in sexual activity, as well as 5, 20, and 40 minutes after they completed sexual activity. These data are currently being analyzed and will help researchers better understand whether individuals' oxytocin response is associated with their sexual communication and sexual satisfaction. Such approaches to measuring hormonal changes as they occur will help substantiate the links between physiological changes and communication behavior proposed in earlier work.

State Measures of Sexual Arousal

Understanding bodily responses to sexual activity can be achieved not only through tracing hormonal trajectories but also by focusing on sexual arousal as measured through blood flow. Despite competing definitions of sexual arousal and its interchangeable use with sexual desire in both academic research (see Janssen, 2011, for a review) and colloquial language (Mitchell, Wellings, & Graham, 2014), many scholars agree that this complex concept is physiological, psychological, and behavioral (Janssen, 2011; Rosen & Beck, 1988). As such, sexual arousal is defined here as "an emotional/motivational state that can be triggered by internal and external stimuli and that can be inferred from central (including verbal), peripheral (including genital), and behavioral (including action tendencies and motor preparation) responses" (Janssen, 2011, p. 708).

Scholars disagree as to the most appropriate ways of measuring sexual arousal. Two subjective measurement strategies exist. The first approach involves a survey where participants review a list of specific behaviors and indicate how arousing they would find each one (e.g.,

Sakheim, Barlow, Beck, & Abrahamson, 1985). The second method employs a lever system in which individuals indicate their current sexual arousal by pulling a lever from 0 degrees (which indicates no sexual arousal) to 180 degrees (which indicates maximum sexual arousal, e.g., Chivers, Rieger, Latty, & Bailey, 2004). Both measurement strategies rely on individuals' self-awareness of their current or imagined sexual arousal, which is questionable for several reasons. First, some individuals may not be attuned to their own bodies or the extent of their physiological arousal. Second, laboratory settings or demand characteristics that leave participants feeling that a certain sexual response is expected or desired by the researcher may influence such measurements (Laan et al., 1993; Rellini, McCall, Randall, & Meston, 2005). Third, individuals may be influenced by societal norms that discourage sexual arousal in response to certain stimuli or in specific contexts (see Bhugra, Popelyuk, & McCullen, 2010, for a review), and may therefore not report sexual arousal even in situations in which they experience it. Self-report measures therefore present challenges for researchers interested in accurately assessing sexual arousal, especially its physiological component.

More objective measures of physiological sexual arousal may help address potential self-report bias in sexual communication research. Sexual arousal can be measured with such tools as a vaginal photoplethysmograph for women and a penile plethysmograph for men. Both devices measure vasocongestion in the genitals (Chivers, 2005). A vaginal photoplethysmograph is inserted into a female's vagina to measure vaginal pulse amplitude (VPA), blood flow to the vagina that coincides with the female's heartbeat (Pieterse et al., 2008). A penile plethysmograph is placed around the base of the penis with a strain gauge to measure penile blood flow through the expansion of the penis (e.g., Tong, 2007).

Researchers have also focused on how well individuals' evaluation of their subjective sexual arousal aligns with physiological sexual arousal measures. The term "concordance" or "subjective-genital agreement" was advanced by Chivers et al. (2010) to describe the agreement of subjective and objective (i.e., genital) sexual arousal. Studies investigating both self-report and physiological sexual arousal have found discrepancies between these measures, especially among women (see Chivers et al., 2010, for a review). Chivers et al. note that although some women have high sexual concordance, many women also display a weak or even negative association between their subjective and genital arousal. Sexual concordance also appears to be highly variable and does not display stability over time in sexually functional individuals (Suschinsky & Lalumière, 2011). Thus, measures of physiological sexual arousal may be especially beneficial in distinguishing between the mental and physical experiences of sexual interactions and cognitions, as well as in understanding the relationship between sexual arousal and sexual communication.

Despite a focus on sexual arousal in many studies investigating sexual behavior, few studies have examined the relationship between sexual communication and sexual arousal. The limited research that examines communication during sexual activity has focused on communication about sexual pleasure and has assessed sexual satisfaction (rather than sexual arousal) using self-report measures. Further research is needed to examine how sexual communication affects physiological sexual arousal, specifically in heterosexual women, given that previous research has noted issues with sexual concordance in this group (Chivers et al., 2010). Future research would benefit from investigating types and forms of phrases or utterances that may be more or less physiologically arousing to an individual. Being cognitively distracted during sexual activity is also negatively associated with women's subjective and physiological sexual arousal (Elliott & O'Donohue, 1997) Improving communication focused on connection during sexual activity may help keep the focus on the sexual activity itself and improve sexual and relational outcomes.

For example, communication scholars may focus on promoting "erotic talk." One form of erotic talk—mutualistic erotic talk—focuses on the partner and the relationship and has been positively associated with sexual satisfaction (Jonason, Betteridge, & Kneebone, 2016). Thus, an intervention aimed at individuals who may have trouble being open with their partners or feel uncomfortable talking during sexual activity, might help improve sexual arousal and satisfaction and offer important implications for both sexual communication researchers and therapists. Such studies would offer a more complex understanding of the ways that sexual communication is associated with both *perceptions* of biological experiences (i.e., self-reported sexual arousal) and actual measurements of bodily changes in sexual contexts (i.e., increased blood flow due to sexual arousal).

FUTURE DIRECTIONS

As noted throughout this chapter, studies of sexual communication and biology have largely focused on either trait or state measures of various biomarkers and their associations with sexual communication. However, two areas worthy of further investigation involve understanding the lasting effects of sexual communication on physiological responses and considering the links between biology and more socially problematic forms of sexual communication, such as sexual harassment and deceptive affection. These areas of current and future research are each discussed next.

Lasting Effects of Sexual Communication

One emerging area of sexual communication scholarship involves investigating the effects of sexual communication on communication in non-sexual contexts. Recently, Denes et al. (in press) studied the effects of post-sex communication on other facets of relational functioning. Some couples in the study were instructed to double their pillow talk over the course of three weeks, whereas control participants received no such directions. After three weeks, couples discussed a recent conflict issue in the laboratory and reported on their relational well-being. The study revealed that men (but not women) in the experimental condition reported greater relationship satisfaction (controlling for pre-intervention levels) and experienced decreases in the stress hormone cortisol from pre- to post-conversation, compared to men in the control condition. This finding may suggest that increasing certain forms of sexual communication might invite benefits that extend beyond the sexual episode itself. Although it is only speculation at this point, couples who increase their intimate post-sex communication may feel more connected, which may aid in relational maintenance and well-being by helping couples feel more unified and better equipped to manage stressors. This possibility aligns with recent theorizing suggesting that couples who build emotional reserves through the consistent use of relationship maintenance behaviors are better able to manage stress and face challenges as a team (Afifi, Merrill, & Davis, 2016). This line of research reinforces the need for longitudinal investigations that consider the links between communication, relational quality, and biological outcomes inside and outside the bedroom.

The Dark Side of Sexual Communication

Thus far, the research reviewed in this chapter has primarily focused on assuming prosocial forms of communication, such as sharing positive thoughts and feelings with one's partner after sexual activity, communication patterns linked to sexual climax, and communication during sexual activity. However, it is possible that more negative or problematic forms of sexual communication are also associated with various state and trait biological measures.

Sexual harassment, along with street harassment, is an under-researched form of sexual communication that may be associated with a perpetrator's biology and a victim's physiological response. Sexual harassment is defined as "unwelcome sexual advances, requests for sexual favors, or other verbal or physical conduct of a sexual nature" (U.S. Equal Employment Opportunity Commission, 2018, para. 1). Although both women and men can be perpetrators and victims of sexual harassment and street harassment, women are overwhelmingly the victims and men are overwhelmingly the perpetrators of these behaviors (Stop Street Harassment, 2014). Surveys have shown that "more than half of American women have experienced unwanted and inappropriate sexual advances from men" (*ABC News-Washington Post*, 2017, p. 1). Additionally, the U.S. Equal Employment Opportunity Commission (2016) estimated that 75% of all workplace harassment incidents go unreported, suggesting that the actual number of sexual harassment incidents is even higher.

Street harassment is considered a more specific type of sexual harassment. Street harassment (sometimes referred to as "stranger harassment") is defined as unwanted sexual attention, harassment, or objectification by a stranger in a public space, such as streets, parks, or public transportation (Bowman, 1993; Wesselmann & Kelly, 2010). Behaviors considered street harassment can include catcalling, stalking, gesturing, groping, exposing, or any other unwanted behaviors that attempt to sexualize the victim and make him or her feel unsafe or uncomfortable (Stop Street Harassment, 2018). In studies of street harassment conducted around the world, at least 65%, but sometimes up to 90%, of the people surveyed reported experiencing some form of street harassment throughout their lifetime (Stop Street Harassment, 2014).

Both sexual harassment and street harassment have been shown to have significant negative effects on victims. A meta-analysis of 41 studies of the consequences of workplace sexual harassment with a total sample size of nearly 70,000 respondents found that sexual harassment in the workplace is associated with less job satisfaction and organizational commitment, and more withdrawal behaviors (Willness, Steel, & Lee, 2007). Sexual harassment also has negative mental health effects for victims, such as increased anxiety, depression, and post-traumatic stress disorder (Willness et al., 2007). Studies have also shown that women who experience street harassment feel more unsafe, have a higher fear of sexual assault, are more likely to change their lives in some way, and experience higher self-objectification than women who do not (Fairchild & Rudman, 2008; Macmillan, Nierobisz, & Welsh, 2000; Stop Street Harassment, 2014). Additionally, street harassment is significantly related to several health outcomes, such as increased anxiety and depression and poorer sleep quality (DelGreco & Christensen, 2019). This body of research suggests that problematic forms of sexual communication such as sexual and street harassment may also be linked to biological responses. Threats to safety and feelings of fear may increase anxiety and activate physiological stress response systems. For example, when encountering street harassment, individuals may experience elevations in cortisol and alpha-amylase. Additionally, the negative health outcomes identified by DelGreco and Christensen (2019) may be rooted in biological responses to harassment experiences that have negative consequences for mental health and sleep patterns.

Research on perpetrators' experiences is sparse but may further reveal links between instigating harassing behavior and biological measures. For example, it is possible that individuals with higher levels of testosterone may engage in more harassing behavior as a means of asserting themselves and demonstrating dominance. Similar power dynamics were revealed in recent work, which found that men who perceived themselves as having less power than women were more likely to use street harassment in an attempt to control women's behavior (DelGreco, Ebesu Hubbard, & Denes, 2018). It may be the case that such pathways are moderated by testosterone levels (e.g., individuals with higher testosterone levels may be more likely to engage in street harassment with the goal of controlling women's behavior than those with lower testosterone levels), but such a possibility awaits future testing. Furthermore, given the association between oxytocin and ingroup favoritism, it is possible that individuals who engage in harassing behavior as a means of supporting some aspect of their social identity may also vary in their oxytocin levels. For example, research reveals that for some men, their street harassing behavior may be rooted in male bonding and asserting one's masculinity, and *not* focused on the female target of the communication (Quinn, 2002). Thus, intergroup dynamics may be at play and be associated with oxytocin levels, as increases in oxytocin have been linked to ingroup-promoting and outgroup-derogating behaviors (see Crespi, 2015, for a review).

Finally, deceptive affectionate messages (DAMs) during sexual activity may also have biological consequences. DAMs are defined "as overt expressions of affection that are not consistent with sources' internal feelings" (Horan & Booth-Butterfield, 2011, p. 79) and are fairly common in relationships (Horan & Booth-Butterfield, 2013). In a sexual context, Bennett and Denes (2018) found that engaging in deceptive affectionate communication during sexual activity (e.g., expressing affection one is not actually feeling, expressing more affection than one feels) was negatively associated with sexual satisfaction, which in turn predicted less post-sex communication. Given the potential physiological benefits of post-sex communication (e.g., Denes et al., in press) and the links between orgasm and oxytocin, and oxytocin and prosocial behavior, it is possible that using DAMs during sexual activity impedes the physiological and relational benefits of sexual interactions and pillow talk. A valuable future direction for this line of work might involve tracking patterns of deceptive affectionate communication inside and outside the bedroom, and discovering how the use of such messages coincides with hormones relevant to sexual interactions (e.g., testosterone, oxytocin) and their lasting effects (e.g., cortisol).

CONCLUSION

This chapter offered a review of the developing scholarship on sexual communication and biology. Research in this domain has primarily focused on exploring either trait or state-like measures of hormones and genes, and their associations with communication in sexual contexts. Together, this line of research suggests that individuals' biological makeup can both impede (e.g., high testosterone levels, "risky" genotypes) and promote (e.g., oxytocin increases) sexual communication and should be considered in concert with other relevant relational and environmental factors. Research on sexual communication and biology is still in the nascent stage and much work remains to be done to substantiate the links between sexually related biomarkers and communication behavior. As such, this growing line of research offers ample opportunities for both emerging and established scholars to explore the many ways that biology and sexual communication influence each other.

REFERENCES

ABC News-Washington Post. (2017). Unwanted sexual advances: Not just a Hollywood story. Retrieved from www.langerresearch.com/wp-content/uploads/1192a1SexualHarassment.pdf.

Afifi, T. D., Merrill, A. F., & Davis, S. (2016). The theory of resilience and relational load. *Personal Relationships, 23*, 663–683.

Apter-Levi, Y., Zagoory-Sharon, O., & Feldman, R. (2014). Oxytocin and vasopressin support distinct configurations of social synchrony. *Brain Research, 1580*, 124–132.

Archer, J. (2006). Testosterone and human aggression: An evaluation of the challenge hypothesis. *Neuroscience and Biobehavioral Reviews, 30*, 319–345.

Archer, J., Birring, S. S., & Wu, F. C. W. (1998). The association between testosterone and aggression among young men: Empirical findings and a meta-analysis. *Aggressive Behavior, 24*, 411–420.

Babin, E. A. (2013). An examination of predictors of nonverbal and verbal communication of pleasure during sex and sexual satisfaction. *Journal of Social and Personal Relationships, 30*, 270–292.

Behnia, B., Heinrichs, M., Bergmann, W., Jung, S., Germann, J., Schedlowski, M., ... Kruger, T. H. (2014). Differential effects of intranasal oxytocin on sexual experiences and partner interactions in couples. *Hormones and Behavior, 65*, 308–318.

Bennett, M., & Denes, A. (2018). Lying in bed: An analysis of deceptive affectionate messages in casual and committed sexual encounters. *Communication Quarterly, 67*, 140–157.

Bhugra, D., Popelyuk, D., & McMullen, I. (2010). Paraphilias across cultures: Contexts and controversies. *Journal of Sex Research, 47*, 242–256.

Blaicher, W., Gruber, D., Bieglmayer, C., Blaicher, A. M., Knogler, W., & Huber, J. (1999). The role of oxytocin in relation to female sexual arousal. *Gynecological and Obstetric Investigation, 47*, 125–126.

Book, A. S., Starzyk, K. B., & Quinsey, V. L. (2001). The relationship between testosterone and aggression: A meta-analysis. *Aggression and Violent Behavior, 6*, 579–599.

Booth, A., & Dabbs, J. N. (1993). Testosterone and men's marriages. *Social Forces, 72*, 463–477.

Bowman, C. G. (1993). Street harassment and the informal ghettoization of women. *Harvard Law Review, 106*, 517–580.

Burnham, T. C., Chapman, J. F., Gray, P. B., McIntyre, M. H., Lipson, S. F., & Ellison, P. T. (2003). Men in committed, romantic relationships have lower testosterone. *Hormones and Behavior, 44*, 119–122.

Carmichael, M. S., Humbert, R., Dixen, J., Palmisano, G., Greenleaf, W., & Davidson, J. M. (1987). Plasma oxytocin increases in the human sexual response. *Journal of Clinical Endocrinology and Metabolism, 64*, 27–31.

Chivers, M. L. (2005). Leading comment: A brief review and discussion of sex differences in the specificity of sexual arousal. *Sexual and Relationship Therapy, 4*, 377–390.

Chivers, M. L., Rieger, G., Latty, E., & Bailey, J. M. (2004). A sex difference in the specificity of sexual arousal. *Psychological Science, 15*, 736–744.

Chivers, M. L., Seto, M. C., Teresa, L., Lalumière, M. L., Laan, E., & Grimbos, T. (2010). Agreement of self-reported and genital measures of sexual arousal in men and women: A meta-analysis. *Archives of Sexual Behavior, 39*, 5–56.

Crespi, B. J. (2015). Oxytocin, testosterone, and human social cognition. *Biological Reviews, 91*, 390–408.

Dabbs Jr, J. M., & Mohammed, S. (1992). Male and female salivary testosterone concentrations before and after sexual activity. *Physiology & Behavior, 52*, 195–197.

De Dreu, C. K. W., Greer, L. L., Van Kleef, G. A., Shalvi, S., & Handgraaf, M. J. J. (2011). Oxytocin promotes human ethnocentrism. *Proceedings of the National Academy of the Sciences, 108*, 1262–1266.

DelGreco, M., & Christensen, J. (2019). Effects of street harassment on anxiety, depression, and sleep quality of college women. *Sex Roles*. Advance online publication.

DelGreco, M., Ebesu Hubbard, A., & Denes, A. (in press). Power dynamics and communicative motivations in street harassment. *Violence against Women.*

Denes, A. (2012). Pillow talk: Exploring disclosures after sexual activity. *Western Journal of Communication, 76*, 91–108.

Denes, A. (2015). Genetic and individual influences on predictors of disclosure: Exploring variation in the oxytocin receptor gene and attachment security. *Communication Monographs, 82*, 113–133.

Denes, A. (2018). Toward a post sex disclosures model (PSDM): Exploring the associations among orgasm, self-disclosure, and relationship satisfaction. *Communication Research, 45*, 297–318.

Denes, A. (in progress). Genetic correlates of pillow talk: Exploring the oxytocin receptor gene and its relationship to post-sex communication.

Denes, A., Afifi, T. A., & Granger, D. (2017). Physiology and pillow talk: Relations between testosterone and communication post sex. *Journal of Social and Personal Relationships, 34*, 281–308.

Denes, A., Crowley, J. P., Bennett, M., Dhillon, A., Ponivas, A., & Winkler, K. L. (2018, February). Do genetics interact with pillow talk to influence relational outcomes? Investigating the role of the oxytocin receptor gene in the post-sex communication process. Paper presented at the annual meeting of the Western States Communication Association, Santa Clara, CA.

Denes, A., Crowley, J. P., Dhillon, A., & Granger, D. (in progress). Testing the influence of changes in salivary oxytocin pre- and post-sexual activity on post sex communication.

Denes, A., Crowley, J. P., Winkler, K. L., Ponivas, A., Dhillon, A., & Bennett, M. (2018). Exploring the effect of pillow talk on physiological stress responses to couples' difficult conversations. Manuscript submitted for publication.

Denes, A., Dhillon, A., Ponivas, A., & Winkler, K. L. (in press). The hormonal underpinnings of sexual communication. In L. S. Aloia, A. Denes, & J. P. Crowley (Eds.), *Oxford handbook of the physiology of interpersonal communication*. New York, NY: Oxford University Press.

Denes, A., Speer, A. C., & Speer, R. B. (2015, May). Beyond sex differences: Using testosterone to investigate communication during the post-coital time interval. Poster presented at the annual meeting of the International Communication Association, San Juan, Puerto Rico.

Edelstein, R. S., van Anders, S. M., Chopik, W. J., Goldey, K. L., & Wardecker, B. M. (2014). Dyadic associations between testosterone and relationship quality in couples. *Hormones and Behavior, 65*, 401–407.

Elliott, A. N., & O'Donohue, W. T. (1997). The effects of anxiety and distraction on sexual arousal in a nonclinical sample of heterosexual women. *Archives of Sexual Behavior, 26*, 607–624.

Fairchild, K., & Rudman, L. A. (2008). Everyday stranger harassment and women's objectification. *Social Justice Research, 21*, 338–357.

Floyd, K., & Denes, A. (2015). Attachment security and oxytocin receptor gene polymorphism interact to influence affectionate communication. *Communication Quarterly, 63*, 272–285.

Gray, P. B., Kahlenberg, S. M., Barrett, E. S., Lipson, S. F., & Ellison, P. T. (2002). Marriage and fatherhood are associated with lower testosterone in males. *Evolution and Human Behavior, 23*, 193–201.

Hamilton, L. D., & Meston, C. M. (2010). The effects of partner togetherness on salivary testosterone in women in long distance relationships. *Hormones and Behavior, 57*, 198–202.

Hellhammer, D. H., Hubert, W., & Schürmeyer, T. (1985). Changes in saliva testosterone after psychological stimulation in men. *Psychoneuroendocrinology, 10*, 77–81.

Holt-Lunstad, J., Birmingham, W. C., & Light, K. C. (2014). Relationship quality and oxytocin: Influence of stable and modifiable aspects of relationships. *Journal of Social and Personal Relationships, 32*, 472–490.

Horan, S. M., & Booth-Butterfield, M. (2011). Is it worth lying for? Physiological and emotional implications of recalling deceptive affection. *Human Communication Research, 37*, 78–106.

Horan, S. M., & Booth-Butterfield, M. (2013). Understanding the routine expression of deceptive affection in romantic relationships. *Communication Quarterly, 61*, 195–216.

Janssen, E. (2011). Sexual arousal in men: A review and conceptual analysis. *Hormones and Behavior, 59*, 708–716.

Jonason, P. K., Betteridge, G. L., & Kneebone, I. I. (2016). An examination of the nature of erotic talk. *Archives of Sexual Behavior, 45*, 21–31.

Julian, T., & McKenry, P. C. (1989). Relationship of testosterone to men's family functioning at mid-life: A research note. *Aggressive Behavior, 15*, 281–289.

Knussmann, R., Christiansen, K., & Couwenbergs, C. (1986). Relations between sex hormone levels and sexual behavior in men. *Archives of Sexual Behavior, 15*, 429–445.

Kogan, A., Saslow, L. R., Impett, E. A., Oveis, C., Keltner, D., & Saturn, S. R. (2011). Thin-slicing study of the oxytocin receptor (OXTR) gene and the evaluation and expression of the prosocial disposition. *Proceedings of the National Academy of Sciences, 108*, 19189–19192.

Kruger, T., Exton, M. S., Pawlak, C., von zur Muhlen, A., Hartmann, U., & Shedlowski, M. (1998). Neuroendocrine and cardiovascular response to sexual arousal and orgasm in men. *Psychoneuroendocrinology, 23*, 401–411.

Laan, E., Everaerd, W., Van Aanhold, M. T., & Rebel, M. (1993). Performance demand and sexual arousal in women. *Behaviour Research and Therapy, 31*, 25–35.

Lee, R., Jaffe, R., & Midgley, A. (1974). Lack of alteration of serum gonadotropins in men and women following sexual intercourse. *American Journal of Obstetrics and Gynecology, 120*, 985–987.

Macmillan, R., Nierobisz, A., & Welsh, S. (2000). Experiencing the streets: Harassment and perceptions of safety among women. *Journal of Research in Crime and Delinquency, 37*, 306–322.

Mazur, A., & Booth, A. (1998). Testosterone and dominance in men. *Behavioral and Brain Sciences, 21*, 353–363.

Mitchell, K. R., Wellings, K. A., & Graham, C. (2014). How do men and women define sexual desire and sexual arousal? *Journal of Sex & Marital Therapy, 40*, 17–32.

Murphy, M. R., Seckl, J. R., Burton, S., Checkley, S. A., & Lightman, S. L. (1987). Changes in oxytocin and vasopressin secretion during sexual activity in men. *Journal of Clinical Endocrinology and Metabolism, 65*, 738–741.

Pieterse, Q. D., Ter Kuile, M. M., Deruiter, M. C., Trimbos, J. B. M. Z., Kenter, G. G., & Maas, C. P. (2008). Vaginal blood flow after radical hysterectomy with and without nerve sparing: A preliminary report. *International Journal of Gynecological Cancer, 18*, 576–583.

Quinn, B. A. (2002). Sexual harassment and masculinity: The power and meaning of "girl watching." *Gender and Society, 16*, 386–402.

Rellini, A. H., McCall, K. M., Randall, P. K., & Meston, C. M. (2005). The relationship between women's subjective and physiological sexual arousal. *Psychophysiology, 42*, 116–124.

Rodrigues, S. M., Saslow, L. R., Garcia, N., John, O. P., & Keltner, D. (2009). Oxytocin receptor genetic variation relates to empathy and stress reactivity in humans. *Proceedings of the National Academy of Sciences, 106*, 21437–21441.

Rosen, R. C., & Beck, J. G. (1988). *Patterns of sexual arousal: Psychophysiological processes and clinical applications*. New York, NY: Guilford Press.

Sakheim, D. K., Barlow, D. H., Beck, J. G., & Abrahamson, D. J. (1985). A comparison of male heterosexual and male homosexual patterns of sexual arousal. *Journal of Sex Research, 21*, 183–198.

Stop Street Harassment. (2014). Unsafe and harassed in public spaces: A national street harassment report. Retrieved from www.stopstreetharassment.org/ourwork/nationalstudy/.

Stop Street Harassment. (2018). About. What is street harassment? Retrieved from www.stopstreetharassment.org/about/what-is-street-harassment/.

Suschinsky, K. D., & Lalumière, M. L. (2011). Category-specificity and sexual concordance: The stability of sex differences in sexual arousal patterns. *The Canadian Journal of Human Sexuality, 20*(3), 93–109.

Tong, D. (2007). The penile plethysmograph, Abel Assessment for Sexual Interest, and MSI-II: Are they speaking the same language? *The American Journal of Family Therapy, 35*, 187–202.

Tost, H., Kolachana, B., Hakimi, S., Lemaitre, H., Verchinski, B. A., Mattay, V. S., … Meyer-Lindenberg, A. (2010). A common allele in the oxytocin receptor gene (OXTR) impacts prosocial temperament and human hypothalamic-limbic structure and function. *Proceedings of the National Academy of Sciences, 103*, 13936–13941.

U.S. Equal Employment Opportunity Commission. (2016). Select task force on the study of sexual harassment in the workplace. Retrieved from www.eeoc.gov/eeoc/task_force/harassment/upload/report.pdf.

U.S. Equal Employment Opportunity Commission. (2018). Sexual harassment. Retrieved from www.eeoc.gov/laws/types/sexual_harassment.cfm.

van Anders, S. M., Goldey, K. L., Conley, T. D., Snipes, D. J., & Patel, D. A. (2012). Safer sex as the bolder choice: Testosterone is positively correlated with safer sex behaviorally relevant attitudes in young men. *Journal of Sexual Medicine, 9*, 727–734.

van Anders, S. M., Goldey, K. L., & Kuo, P. X. (2011). The steroid/peptide theory of social bonds: Integrating testosterone and peptide responses for classifying social behavioral contexts. *Psychoneuroendocrinology, 36*, 1265–1275.

van Anders, S. M., Hamilton, L. D., Schmidt, N., & Watson, N. V. (2007). Associations between testosterone secretion and sexual activity in women. *Hormones and Behavior, 51*, 477–482.

van Anders, S. M., Hamilton, L. D., & Watson, N. V. (2007). Multiple partners are associated with higher testosterone in North American men and women. *Hormones and Behavior, 51*, 454–459.

van Anders, S. M., & Watson, N. V. (2007). Testosterone levels in women and men who are single, in long-distance relationships, or same-city relationships. *Hormones and Behavior, 51*, 286–291.

Veenestra, M. (2007). Afterglow. In F. Malti-Douglas (Ed.), *Encyclopedia of sex and gender* (Vol. 1, pp. 39–40). Detroit, MI: Macmillan Reference.

Wesselmann, E. D., & Kelly, J. R. (2010). Cat-calls and culpability: Investigating the frequency and functions of stranger harassment. *Sex Roles, 63*, 451–462.

Willness, C. R., Steel, P., & Lee, K. (2007). A meta-analysis of the antecedents and consequences of workplace sexual harassment. *Personnel Psychology, 60*, 127–162.

Zhang, Y., Deiter, F., Jung, S., Heinrichs, M., Schedlowski, M., & Krüger, H. C. (2015). Differential effects of intranasal oxytocin administration on sexual functions in healthy females: A laboratory setting. *European Psychiatry, 30*, 264.

28

The State of the Art and the Future of "Wet Psychophysiology" in Communication Research

Kory Floyd and Colin Hesse

"Wet psychophysiology" refers to the analysis of psychological and emotional processes and their connections to chemicals—such as hormones, immune markers, and metabolic markers—measured in body fluids. In recent years, the communication discipline has applied the methods of wet psychophysiology to the task of illuminating social behaviors such as conflict, aggression, attachment, pair bonding, and affection (see, e.g., Floyd & Cole, 2009; Floyd & Roberts, 2009). This chapter provides an overview of methods and social applications of wet psychophysiology by addressing endocrine markers, immune/inflammatory markers, and metabolic markers separately. The conclusion acknowledges additional outcomes adjudicated in wet psychophysiology studies and identifies important logistical and safety considerations for this category of research.

ENDOCRINE MARKERS

In the communication discipline, most wet psychophysiological research has focused on the endocrine system. The endocrine system comprises a network of glands that controls the production and secretion of hormones. Hormones bind with specific receptor cells, influencing their metabolic activity, and aiding the body in the accomplishment of defense mobilization, the maintenance of nutrients in the bloodstream, and the regulation of energy balance (Marieb & Hoehn, 2010). Some commonly measured hormones are amino acid-based, such as epinephrine and melatonin; others are steroids, such as testosterone and cortisol; still others are peptides, such as oxytocin and prolactin (Gardner & Shoback, 2017). Although the body produces more than 90 unique hormones, this section focuses specifically on adrenal/gonadal and pituitary hormones, as they are most commonly represented in research on communication and social behavior.

Adrenal and Gonadal Hormones

In communication research, the principal hormones of interest produced by the adrenal gland and gonads are cortisol, androgens, estrogens, and progesterone.

Cortisol

Cortisol, a glucocorticoid, is elevated in response to acute stressors as a component of the hypothalamic-pituitary-adrenal (HPA) axis. Cortisol mobilizes and regulates energy resources that can be used to respond to the stressor (Dickerson & Kemeny, 2004). Cortisol follows a diurnal rhythm in which it is highest shortly after awakening, drops sharply over the day, decreasing more slowly throughout the evening, and reaching nadir around midnight (Kirschbaum & Hellhammer, 1989). Whereas acute stress is indexed by acute elevations in cortisol, a lack of variation in the diurnal rhythm is indicative of chronic stress (Ockenfels et al., 1995). Cortisol is typically measured in saliva or blood, although it can also be assessed in urine and even in hair (Short et al., 2016).

Scholars have generally predicted that positive communication behaviors (such as affection) would reduce both chronic and acute stress. In terms of chronic stress, two studies have shown that diurnal variation in cortisol is positively correlated with affectionate communication (Floyd, 2006; Floyd & Riforgiate, 2008). In terms of acute stress, levels of affection in a romantic relationship were also inversely correlated with cortisol increase in response to stressors (Floyd, Mikkelson, Tafoya et al., 2007). On the other hand, negative marital conflict patterns, such as demand-withdrawal, are directly correlated with cortisol increase in response to the conflict (Heffner et al., 2006).

Androgens

Androgens are gonadocorticoids, such as testosterone, synthesized in the testes, ovaries, and adrenal glands. Testosterone contributes to the growth and maintenance of male sex organs, helps spur spermatogenesis, and helps control the male sex drive (Marieb & Hoehn, 2010). Previous studies have largely used saliva (Denes, Afifi, & Granger, 2017) or blood (Krüger et al., 1998) to examine questions about baseline testosterone or testosterone reactivity.

Multiple studies have demonstrated a positive association between testosterone and aggression for both men (Ehrenkranz, Bliss, & Sheard, 1974) and women (Harris, Rushton, Hampton, & Jackson, 1996), although some question the strength of that association (Eisenegger, Haushofer, & Fehr, 2011). Testosterone levels also predict criminal violence—again, for both men (Dabbs, Frady, Carr, & Besch, 1987) and women (Dabbs, Ruback, Frady, Hopper, & Sgoutas, 1988). In terms of mating behaviors, trait testosterone is inversely related to pair bonding for both men and women, with levels dropping when individuals marry and have children (e.g., Barrett et al., 2013; Gray et al., 2002). However, state testosterone increases while individuals are engaged in mating behaviors and are sexually aroused (e.g., Krüger et al., 1998). The production of testosterone is thus activated during mating behaviors, dropping upon completion of the task. Individuals with higher levels of testosterone are also less likely to perceive a benefit to self-disclosure following sexual activity (Denes et al., 2017). The authors hypothesized that this is due to the suppression effect of testosterone on the production of oxytocin (Denes et al., 2017).

Estrogens and Progesterone

Estrogens and progesterone are produced by the ovaries and, to a smaller extent, the adrenal glands and other tissues. As with androgens, these hormones are linked to the development and maintenance of both reproductive organs and secondary sexual characteristics, while also regulating menstruation (Marieb & Hoehn, 2010). Estrogens and progesterone vary systematically across phases of the menstrual cycle. These hormones are measured primarily in saliva (e.g., Roney & Simmons, 2013) and urine (e.g., Feinberg et al., 2006).

Most questions regarding estrogens, progesterone, and social behavior address changes over the reproductive cycle. For instance, women report higher sexual desire during peak fertility, with a direct relationship with the estrogen estradiol and an inverse relationship with progesterone (e.g., Roney & Simmons, 2013). Estrogens are also linked to different mating preferences, with women more attracted to more masculine voices during the peak-fertility phase (Feinberg et al., 2006). Finally, women who rate their partner lower in sexual attractiveness decrease in relational closeness at peak fertility, while women who rate their partner higher in sexual attractiveness actually increase in relational closeness at the same point (Larson et al., 2013).

Pituitary Hormones

Although the pituitary gland produces several hormones, we focus here on three: oxytocin, vasopressin, and prolactin.

Oxytocin and Vasopressin

Oxytocin and vasopressin are neuropeptides (Donaldson & Young, 2008). Oxytocin stimulates uterine contractions and the letdown reflex in women, whereas vasopressin regulates the tonicity of body fluids (Marieb & Hoehn, 2010). Both hormones can be measured in blood (e.g., Floyd, Pauley, & Hesse, 2010) and urine (e.g., Bick & Dozier, 2010).

Social scientists believe that both hormones are linked to social bonding and attachment, with previous studies showing associations between oxytocin and empathy, social support, and affection (for review, see Carter, 2017). One study found a decrease in the perceived friendliness of male faces and an increase in the perceived friendliness of female faces following a dosage of vasopressin (Thompson et al., 2006). Another administered intranasal vasopressin to participants, who then played the Prisoner's Dilemma Game (wherein participants must decide whether to cooperate with their partner in order to win various levels of payment). For women, vasopressin increased their own conciliatory behavior (cooperating even when their partner was not), while men increased overall levels of partner cooperation (Rilling et al., 2014).

For oxytocin, a recent meta-analysis concluded that a single intranasal dosage improved emotional competence and increased the communication of positive emotions while not influencing the communication of negative emotions (Leppanen, Ng, Tchanturia, & Treasure, 2017). Floyd et al. (2010) found that affectionate communication was directly related to oxytocinergic reactivity to a series of stressors and speculated that oxytocin may be partially responsible for the stress-buffering effect of affectionate behavior. Whereas most hypothesized relationships between oxytocin and social behavior are positive, De Dreu and colleagues (2011) argued that the production of social bonding might also make individuals more biased toward their own groups, with oxytocin thus playing a role in inter-group conflict.

Prolactin

Prolactin is a protein hormone primarily associated with the lactation process, including lactogenesis and galactopoiesis (Freeman et al., 2000). Prolactin helps regulate both metabolic and immune functions and is secreted in response to light, auditory, and olfactory stimuli (Freeman et al., 2000). Prolactin can be assayed in both blood serum and milk (Marieb & Hoehn, 2010).

As with oxytocin and vasopressin, prolactin is normally studied as it relates to social bonding, especially between parents and children. For example, Delahunty and colleagues (2007) found that women's prolactin levels increased prior to the birth of a child upon exposure to infant cries, whereas men's prolactin levels decreased. In another study, fathers with higher prolactin were more alert to infant cries, with experienced fathers showing an even greater increase in prolactin than first-time fathers (Fleming, Corter, Stallings, & Steiner, 2002). Other research with married couples has shown that prolactin levels are decreased when spouses engage in conflict (Malarkey, Kiecolt-Glaser, Pearl, & Glaser, 1994).

IMMUNE AND INFLAMMATORY MARKERS

The immune system comprises a system of structures and processes that protect the body from disease through the recognition and destruction of pathogens. Both innate and adaptive immune processes are operative. The innate immune response, which mobilizes markers such as immunoglobulins, inflammatory cytokines (such as IL-1 and IL-6), and natural killer cells, responds to pathogens in a generic, non-specific manner (Segerstrom & Miller, 2004). The adaptive immune response, which mobilizes a variety of lymphocytes, is antigen-specific, conferring immunity. Many immune markers can be measured in either blood or saliva. This section discusses immunoglobulins, inflammatory cytokines, and natural killer cells to denote their connections to social behavior.

Immunoglobulins

Immunoglobulins, also referred to as antibodies, are proteins secreted by various plasma cells in order to bind themselves to an antigen. Three commonly measured in social science are immunoglobulin A (IgA), immunoglobulin M (IgM), and immunoglobulin G (IgG). IgA is found in regions of the body that produce and secrete mucous, leading some to dub it the body's "first line of defense" against pathogenic invasion (Underdown, 1998). IgM is one of the first antibodies to respond to antigen invasion (Casali, 1998), and IgG is the most prevalent antibody in humans (Painter, 1998). These markers are normally studied in blood serum (e.g., Hui, Hua, Diandong, & Hong, 2007) or saliva (e.g., Matos-Gomes et al., 2010).

Several studies link immunoglobulins to the experience of stress (Tsujita & Morimoto, 1999; but see Mouton, Fillion, Tawadros, & Tessier, 1989). For instance, IgA, IgM, and IgG are all elevated by sleep deprivation (Hui et al., 2007) and final exam stress (Matos-Gomes et al., 2010). During stressful events, immunoglobulin levels are affected by the provision of social support. Jemmott and Magloire (1988) found that college students' IgA levels were positively associated with received social support, whereas Gallagher, Phillips, Ferraro, Drayson, and Carroll (2008) found that IgM proliferation after pneumococcal vaccination was positively associated with tangible, emotional, and affectionate support (see also Ohira, 2004). Pressman et al. (2005) also found that social network size and lack of loneliness predicted IgG response to an influenza vaccination.

Inflammatory Cytokines

Cytokines are proteins implicated in multiple immune functions, including inflammation. Cytokine networks manipulate neurotransmitter metabolism and HPA axis hormones, such as serotonin (Raison, Capuron, & Miller, 2006). Some of the main cytokines examined in social science are interleukin-1 (IL-1), interleukin-2 (IL-2), and interleukin-6 (IL-6). Both IL-1 and IL-6 in particular promote inflammation along with the production of IL-2, which in turn activates the production of natural killer (NK) cells. Cytokine production is generally measured in blood serum (e.g., Jaremka et al., 2013).

As with antibody production, scholars have hypothesized that cytokine levels indicate stress and impaired immunocompetence. Multiple studies have found direct correlations between proinflammatory cytokines and both depression and stress (see Raison et al., 2006). Studies also point to a relationship between cytokine production and social interaction in general. In two separate samples, lonelier individuals experienced more pronounced cytokine production (specifically IL-6) in response to stressors than did less lonely individuals (Jaremka et al., 2013). One comprehensive review of the literature claimed that cytokines might be related to several social systems, including social stressors such as separation, social support, and attachment (Hennessy, Deak, & Schiml, 2014).

Natural Killer Cells

Natural killer (NK) cells are one of the main innate chemical weapons of the immune system, activated in response to acute stressors (Dopp, Miller, Myers, & Fahey, 2000). NK cells release proteins that kill infected cells via the process of apoptosis (Smyth et al., 2005). Using blood serum, scholars measure both NK cell expression (the number of NK cells in the bloodstream) and cytotoxicity (the efficacy for killing target cells; Dopp et al., 2000).

Both NK cell measures are connected to social behavior. For instance, expression and cytotoxicity are both elevated during marital conflict (Dopp et al., 2000), and expression is increased by a mental math stressor (Naliboff et al., 1991; but see Glaser, Rise, Speicher, Stout, & Kiecolt-Glaser, 1986). Perceived social support from a loved one also predicts higher NK expression in breast cancer patients (Levy et al., 1990; for a meta-analysis of the connection between social support and NK cell expression, see Miyazaki et al., 2003), whereas higher levels of affectionate communication correspond to greater cytotoxicity (Floyd, Pauley et al., 2014).

METABOLIC MARKERS

Metabolism comprises chemical transformations at the cellular level that facilitate the maintenance of the organism. Two metabolic markers assessed in communication and behavioral research are blood lipids and blood glucose.

Blood Lipids

Lipids are fat-soluble organic molecules that are essential for the provision of energy to cells, the formation of cell membranes, cellular growth, maintenance, and repair, and the synthesis of steroid hormones and bile acids. Lipids occur in multiple forms, but the two most commonly measured in social science research are cholesterol and triglycerides. Cholesterol is a sterol (also called a steroid alcohol) that is an essential component of cell membranes. It performs multiple essential functions, including maintaining membrane fluidity, producing bile, and contributing to the metabolism of fat-soluble vitamins (Shier, Butler, & Lewis, 2015). Triglycerides are esters formed by combining glycerol with three fatty acid molecules. They are the principal lipids in the blood and the chief constituent of body fat and skin oils.

Cholesterol can be assessed in terms of individual lipoproteins (the most commonly measured being high-density lipoproteins, or HDL, and low-density lipoproteins, or LDL), or in the form of "total cholesterol," which is a sum of HDL, LDL, and very-low-density lipoproteins, or VLDL. Triglycerides are measured separately, and both cholesterol and triglycerides can be measured in serum or plasma collected through venipuncture or fingerstick.

In behavioral research, lipids are typically studied due to their connection to stress. Multiple studies have demonstrated that both cholesterol and triglycerides are elevated in response to chronic or acute stressors (Bacon, Ring, Lip, & Carroll, 2004; Muldoon et al., 1995; Stoney, Niaura, Bausserman, & Metacin, 1999). Agarwal, Gupta, Singhal, and Bajpai (1997) found, for instance, that mental stress associated with high-stakes examinations increased both total cholesterol and serum triglycerides for male and female medical students, whereas McCann et al. (1996) showed that examination stress for law students increased LDL for women and men and HDL for women only. Similarly, Muldoon and colleagues (1995) documented that a 20-minute cognitive stressor increased total cholesterol, HDL, and LDL levels. Larger stress-induced elevations in LDL are observed in men than in women (Stoney, Matthews, Mcdonald, & Johnson, 1988; but see Van Doornen & Van Blokland, 1987), and the temporal stability of stress-induced elevations is moderately high, at least among men (Stoney, Niaura, & Bausserman, 1997).

Social behaviors show some connections to lipid values, although results are somewhat inconsistent. In a study of 300 healthy Swedish women, for instance, Horsten, Wamala, Vingerhoets, and Orth-Gomer (1997) reported a positive association between total cholesterol and the amount of social support women reported receiving. This may at first appear inconsistent with the well-documented effect of stress on lipids, as one might naturally expect social support to be *negatively related* to total cholesterol. Lipids are a vital source of energy, however, and Horsten and colleagues suggested that substantially low levels can have adverse psychosocial correlates, which might include depression, social withdrawal, and a corresponding lack of social support from others. Indeed, other research has found that higher levels of depression are associated with lower levels of total cholesterol (Steegmans, Hoes, Bak, van der Does, & Grobbee, 2000) and triglycerides (Suarez, 1999), although some research suggests that association with total cholesterol may be accounted for primarily by low levels of HDL (or "good cholesterol"; Chen, Lu, Wu, & Chang, 2001).

Similarly, in the communication discipline, Floyd and colleagues have examined the connection between lipid levels and affectionate behavior. Using the literature linking stress, affection, and cholesterol, Floyd, Mikkelson, Hesse, and Pauley (2007) predicted and found in two five-week trials that expressing affection (in the form of writing) reduces total cholesterol. Later, Floyd et al. (2009) found that increasing romantic kissing reduced total cholesterol in a six-week trial. Floyd (2019) points out, however, that although both studies demonstrated significant effects of affectionate behavior on cholesterol, some attempts from his laboratory to replicate these patterns have produced null results. He therefore urges caution in the interpretation and application of these findings.

Blood Glucose

Blood glucose (or blood sugar concentration) references the amount of glucose present in the blood at a given time. Glucose is a critical source of energy for the human body, one that supplies most of the energy required for cerebral metabolism (Huang et al., 1980). Glucose is assessed in blood, via fingerstick or venipuncture. Two measurement strategies for blood glucose are most common in social science research. Fasting glucose is measured by assessing glucose levels in the blood from eight to 14 hours after an individual's most recent meal.

Glycated hemoglobin (also referred to as HbA_{1c}) is a non-fasting measure of an individual's average blood glucose concentration over the preceding three months.

Like lipids, blood glucose is also elevated in response to stressors (Schuck, 1998). During episodes of acute stress, cortisol and epinephrine increase glucose to fuel the body's fight-or-flight response. Because affectionate communication reduces stress, Floyd, Hesse, and Haynes (2007) hypothesized that expressed affection is inversely related to glycated hemoglobin. Net of the effects of received affection, expressed affection exhibited a strong negative association with glycated hemoglobin, meaning that the more affection individuals typically communicated, the lower their average glucose levels over the previous three-month period. Floyd, Veksler et al. (2017) also reported an inverse relationship between fasting glucose and perceived social inclusion.

Stanton, Campbell, and Loving (2014) surmised that the experience of romantic love can produce eustress, a positive or euphoric form of stress that triggers activation of both the HPA axis and the sympathetic-adrenal-medullary (SAM) pathway of the endocrine system. Their study demonstrated that visualizing a romantic partner and thinking about feelings of love for that person increased blood glucose if the visualization was accompanied by positive affect. On the contrary, visualizing a platonic friend or thinking of one's morning routine produced decreases in glucose.

As noted, the practice of wet psychophysiology adjudicates a number of endocrine, immune, and metabolic markers that are highly relevant for understanding communication and social behavior. To conclude this chapter, we identify some additional markers gaining traction in the literature and denote some of the important logistical issues relevant to the practice of wet psychophysiology.

CONCLUSION

Although the outcomes discussed above represent the majority of psychophysiological studies on communication and social behavior, additional outcomes are also relevant to many behavioral questions. For instance, salivary α-amylase (sAA) is an enzyme that indexes both stress and anxiety. In the field of communication, Afifi and colleagues (2011) found that adolescents' perceptions of their parents' communication skills predicted their physiological reaction to a stressful interaction for sAA but not for cortisol (see also Nater & Rohleder, 2009). Similarly, immunocompetence can be assessed as a function of the number of antibodies to latent Epstein–Barr virus (EBV), and Floyd, Hesse, Boren, and Veksler (2014) discovered that healthy adults' trait affection levels were positively associated with EBV antibodies (indicating immunosuppression). More recently, Boren and Veksler (2018) found that co-rumination was positively related to levels of C-reactive protein (CRP), an acute-phase protein that is elevated in response to inflammation. These and other markers have only recently gained traction in communication research, but they offer the potential to further illuminate the relevant physiological correlates of social behaviors.

By its very nature, wet psychophysiology involves the collection and handling of body fluids (most often blood or saliva), which requires attention to important logistical and safety considerations. Many outcomes of interest to communication researchers can be measured in saliva, which is advantageous for multiple reasons. First, collection is non-invasive and not stress-inducing for most participants. Second, the procedure for salivary collection is simple, allowing for samples to be collected *in situ* when warranted. Third, saliva is a chemically stable substance that can be frozen, if necessary, with minimal loss to hormone or immune marker levels. Finally, saliva is not considered to be a biohazardous substance.

Many other outcomes require the collection of blood, either through a fingerstick procedure or through a venous blood draw. If venipuncture is required, it should be performed only by a properly trained professional (such as a nurse, phlebotomist, or medical assistant). Those collecting and handling blood samples should observe standard precautions (including the use of medical gloves and face shields), and all research personnel coming in contact with samples should be trained in the avoidance of bloodborne pathogens. When collecting blood samples, the researcher must know whether the intended analyses require whole blood, serum, or plasma.

In addition to safety considerations, the measurement of hormones, immune markers, and metabolic markers in body fluids frequently also necessitates the enforcement of specific inclusion and exclusion criteria and the simultaneous measurement of potential control variables. As one example, the accuracy of lipid assessments can be influenced by multiple behavioral and environmental factors. These include the use of alcohol and tobacco products, level of hydration, history of heart attack, vitamin consumption, and recent infectious illness (see, e.g., De Oliveira e Silva et al., 2000). Even seasonal effects are observed, wherein average total cholesterol levels are higher in winter than in summer (Ockene et al., 2004). In research, some of these factors are typically measured for use as exclusion criteria, whereas others are measured as potential moderating factors. Floyd, Mikkelson, Hesse et al. (2007), for instance, excluded prospective participants with a history of diagnosis or treatment for hypercholesterolemia and/or who were currently using prescribed anticoagulants, whereas they measured alcohol consumption, tobacco use, and frequency of exercise to control for these potential influences.

When logistical and safety requirements are properly observed, the practice of wet psychophysiology—alone or in conjunction with neuroimaging techniques and nervous system measures—offers communication scientists the ability to explore the physiological antecedents, correlates, and consequences of social behavior in a way that observational and self-report techniques cannot. This approach has already greatly expanded our understanding of interpersonal communication (for a review, see Floyd & Afifi, 2012), and it has the potential to do the same for a variety of other subdisciplines, including organizational communication, media/mass communication, instructional communication, and health communication, among others.

REFERENCES

Afifi, T. D., Granger, D. A., Denes, A., Joseph, A., & Aldeis, D. (2011). Parents' communication skills and adolescents' salivary α-amylase and cortisol response patterns. *Communication Monographs, 78*, 273–295.

Agarwal, V., Gupta, B., Singhal, U., & Bajpai, S. K. (1997). Examination stress: Changes in serum cholesterol, triglycerides and total lipids. *Indian Journal of Physiology and Pharmacology, 41*, 404–408.

Bacon, S. L., Ring, C., Lip, G. Y., & Carroll, D. (2004). Increases in lipids and immune cells in response to exercise and mental stress in patients with suspected coronary artery disease: Effects of adjustments for shifts in plasma volume. *Biological Psychology, 65*, 237–250.

Barrett, E. S., Tran, V., Thurston, S., Jasienska, G., Furberg, A.-S., Ellison, P. T., & Thune, I. (2013). Marriage and motherhood are associated with lower testosterone concentrations in women. *Hormones and Behavior, 63*, 72–79.

Bick, J., & Dozier, M. (2010). Mothers' concentrations of oxytocin following close, physical interactions with biological and nonbiological children. *Developmental Psychobiology, 52*, 100–107.

Boren, J. P., & Veksler, A. E. (2018). Co-rumination and immune inflammatory response in healthy young adults: Associations with interleukin-6 and C-reactive protein. *Communication Research Reports, 35*, 152–161.

Carter, C. S. (2017). The role of oxytocin and vasopressin in attachment. *Psychodynamic Psychiatry, 45*, 499–517.
Casali, P. (1998). IgM. In P. J. Delves (Ed.), *Encyclopedia of immunology* (2nd ed., pp. 1212–1217). Oxford, England: Academic Press.
Chen, C. C., Lu, F.-H., Wu, J.-S., & Chang, C.-J. (2001). Correlation between serum lipid concentrations and psychological distress. *Psychiatry Research, 102*, 153–162.
Dabbs, J. M., Frady, R. L., Carr, T. S., & Besch, N. F. (1987). Saliva testosterone and criminal violence in young adult prison inmates. *Psychosomatic Medicine, 49*, 174–182.
Dabbs, J. M., Ruback, R. B., Frady, R. L., Hopper, C. H., & Sgoutas, D. S. (1988). Saliva testosterone and criminal violence among women. *Personality and Individual Differences, 9*, 269–275.
De Dreu, C. K., Greer, L. L., Van Kleef, G. A., Shalvi, S., & Handgraaf, M. J. (2011). Oxytocin promotes human ethnocentrism. *Proceedings of the National Academy of Sciences, 108*, 1262–1266.
Delahunty, K. M., McKay, D. W., Noseworthy, D. E., & Storey, A. E. (2007). Prolactin responses to infant cues in men and women: Effects of parental experience and recent infant contact. *Hormones and Behavior, 51*, 213–220.
Denes, A., Afifi, T. D., & Granger, D. A. (2017). Physiology and pillow talk: Relations between testosterone and communication post sex. *Journal of Social and Personal Relationships, 34*, 281–308.
De Oliveira e Silva, E. R., Foster, D., Harper, M. M., Seidman, C. E., Smith, J. D., Breslow, J. L., & Brinton, E. A. (2000). Alcohol consumption raises HDL cholesterol levels by increasing the transport rate of apolipoproteins A-I and A-II. *Circulation, 102*, 2347–2352.
Dickerson, S. S., & Kemeny, M. E. (2004). Acute stressors and cortisol responses: A theoretical integration and synthesis of laboratory research. *Psychological Bulletin, 130*, 355–391.
Donaldson, Z. R., & Young, L. J. (2008). Oxytocin, vasopressin, and the neurogenetics of sociality. *Science, 322*, 900–904.
Dopp, J. M., Miller, G. E., Myers, H. F., & Fahey, J. L. (2000). Increased natural killer-cell mobilization and cytotoxicity during marital conflict. *Brain, Behavior, and Immunity, 14*, 10–26.
Ehrenkranz, J., Bliss, E., & Sheard, M. H. (1974). Plasma testosterone: Correlation with aggressive behavior and social dominance in men. *Psychosomatic Medicine, 36*, 469–475.
Eisenegger, C., Haushofer, J., & Fehr, E. (2011). The role of testosterone in social interaction. *Trends in Cognitive Sciences, 15*, 263–271.
Feinberg, D. R., Jones, B. C., Smith, M. L., Moore, F. R., DeBruine, L. M., Cornwell, R. E., … Perrett, D. I. (2006). Menstrual cycle, trait estrogen level, and masculinity preferences in the human voice. *Hormones and Behavior, 49*, 215–222.
Fleming, A. S., Corter, C., Stallings, J., & Steiner, M. (2002). Testosterone and prolactin are associated with emotional responses to infant cries in new fathers. *Hormones and Behavior, 42*, 399–413.
Floyd, K. (2006). Human affection exchange: XII. Affectionate communication is associated with diurnal variation in salivary free cortisol. *Western Journal of Communication, 70*, 47–63.
Floyd, K. (2019). *Affectionate communication in close relationships*. Cambridge, England: Cambridge University Press.
Floyd, K., & Afifi, T. D. (2012). Biological and physiological perspectives on interpersonal communication. In M. L. Knapp & J. A. Daly (Eds.), *The handbook of interpersonal communication* (4th ed., pp. 87–127). Thousand Oaks, CA: Sage.
Floyd, K., Boren, J. P., Hannawa, A. F., Hesse, C., McEwan, B., & Veksler, A. E. (2009). Kissing in marital and cohabiting relationships: Effects on blood lipids, stress, and relationship satisfaction. *Western Journal of Communication, 73*, 113–133.
Floyd, K., & Cole, T. (2009). Communication and biology: The view from evolutionary psychology and psychophysiology. In M. J. Beatty, J. C. McCroskey, & K. Floyd (Eds.), *Biological dimensions of communication: Perspectives, methods, and research* (pp. 15–30). Cresskill, NJ: Hampton Press.
Floyd, K., Hesse, C., Boren, J. P., & Veksler, A. E. (2014). Affectionate communication can suppress immunity: Trait affection predicts antibodies to latent Epstein-Barr virus. *Southern Communication Journal, 79*, 2–13.

Floyd, K., Hesse, C., & Haynes, M. T. (2007). Human affection exchange: XV. Metabolic and cardiovascular correlates of trait expressed affection. *Communication Quarterly, 55*, 79–94.

Floyd, K., Mikkelson, A. C., Hesse, C., & Pauley, P. M. (2007). Affectionate writing reduces total cholesterol: Two randomized, controlled trials. *Human Communication Research, 33*, 119–142.

Floyd, K., Mikkelson, A. C., Tafoya, M. A., Farinelli, L., La Valley, A. G., Judd, J., ... Wilson, J. (2007). Human affection exchange: XIII. Affectionate communication accelerates neuroendocrine stress recovery. *Health Communication, 22*, 123–132.

Floyd, K., Pauley, P. M., & Hesse, C. (2010). State and trait affectionate communication buffer adults' stress reactions. *Communication Monographs, 77*, 618–636.

Floyd, K., Pauley, P. M., Hesse, C., Veksler, A. E., Eden, J., & Mikkelson, A. C. (2014). Affectionate communication is associated with markers of immune and cardiovascular system competence. In J. M. Honeycutt, C. Sawyer, & S. Keaton (Eds.), *The influence of communication on physiology and health status* (pp. 115–130). New York, NY: Peter Lang Publishing.

Floyd, K., & Riforgiate, S. (2008). Affectionate communication received from spouses predicts stress hormone levels in healthy adults. *Communication Monographs, 75*, 351–368.

Floyd, K., & Roberts, J. B. (2009). Principles of endocrine system measurement in communication research. In M. J. Beatty, J. C. McCroskey, & K. Floyd (Eds.), *Biological dimensions of communication: Perspectives, methods, and research* (pp. 249–264). Cresskill, NJ: Hampton Press.

Floyd, K., Veksler, A. E., McEwan, B., Hesse, C., Boren, J. P., Dinsmore, D. R., & Pavlich, C. A. (2017). Social inclusion predicts lower blood glucose and low-density lipoproteins in healthy adults. *Health Communication, 32*, 1039–1042.

Freeman, M. E., Kanyicska, B., Lerant, A., & Nagy, G. (2000). Prolactin: Structure, function, and regulation of secretion. *Physiological Reviews, 80*, 1523–1631.

Gallagher, S., Phillips, A. C., Ferraro, A. J., Drayson, M. T., & Carroll, D. (2008). Social support is positively associated with the immunoglobulin M response to vaccination with pneumococcal polysaccharides. *Biological Psychology, 78*, 211–215.

Gardner, D. G., & Shoback, D. M. (2017). *Greenspan's basic & clinical endocrinology* (10th ed.). New York, NY: McGraw-Hill.

Glaser, R., Rice, J., Speicher, C. E., Stout, J. C., & Kiecolt-Glaser, J. K. (1986). Stress depresses interferon production by leukocytes concomitant with a decrease in natural killer cell activity. *Behavioral Neuroscience, 100*, 675–678.

Gray, P. B., Kahlenberg, S. M., Barrett, E. S., Lipson, S. F., & Ellison, P. T. (2002). Marriage and fatherhood are associated with lower testosterone in males. *Evolution and Human Behavior, 23*, 193–201.

Harris, J. A., Rushton, J. P., Hampson, E., & Jackson, D. N. (1996). Salivary testosterone and self-report aggressive and pro-social personality characteristics in men and women. *Aggressive Behavior, 22*, 321–331.

Heffner, K. L., Loving, T. J., Kiecolt-Glaser, J. K., Himawan, L. K., Glaser, R., & Malarkey, W. B. (2006). Older spouses' cortisol responses to marital conflict: Associations with demand/withdraw communication patterns. *Journal of Behavioral Medicine, 29*, 317–325.

Hennessy, M. B., Deak, T., & Schiml, P. A. (2014). Sociality and sickness: Have cytokines evolved to serve social functions beyond times of pathogen exposure? *Brain, Behavior, and Immunity, 37*, 15–20.

Horsten, M., Wamala, S. P., Vingerhoets, A., & Orth-Gomer, K. (1997). Depressive symptoms, social support, and lipid profile in healthy middle-aged women. *Psychosomatic Medicine, 59*, 521–528.

Huang, S. C., Phelps, M. E., Hoffman, E. J., Sideris, K., Selin, C. J., & Kuhl, D. E. (1980). Noninvasive determination of local cerebral metabolic rate of glucose in man. *American Journal of Physiology, 238*, E69–E82.

Hui, L., Hua, F., Diandong, H., & Hong, Y. (2007). Effects of sleep and sleep deprivation on immunoglobulins and complement in humans. *Brain, Behavior, and Immunity, 21*, 308–310.

Jaremka, L. M., Fagundes, C. P., Peng, J., Bennett, J. M., Glaser, R., Malarkey, W. B., & Kiecolt-Glaser, J. K. (2013). Loneliness promotes inflammation during acute stress. *Psychological Science, 24*, 1089–1097.

Jemmott, J. B., & Magloire, K. (1988). Academic stress, social support, and secretory immunoglobulin A. *Journal of Personality and Social Psychology, 55*, 803–810.

Kirschbaum, C., & Hellhammer, D. H. (1989). Salivary cortisol in psychobiological research: An overview. *Neuropsychobiology, 22*, 150–169.

Krüger, T., Exton, M. S., Pawlak, C., von zur Mühlen, A., Hartmann, U., & Schedlowski, M. (1998). Neuroendocrine and cardiovascular response to sexual arousal and orgasm in men. *Psychoneuroendocrinology, 23*, 401–411.

Larson, C. M., Haselton, M. G., Gildersleeve, K. A., & Pillsworth, E. G. (2013). Changes in women's feelings about their romantic relationships across the ovulatory cycle. *Hormones and Behavior, 63*, 128–135.

Leppanen, J., Ng, K. W., Tchanturia, K., & Treasure, J. (2017). Meta-analysis of the effects of intranasal oxytocin on interpretation and expression of emotions. *Neuroscience & Biobehavioral Reviews, 78*, 125–144.

Levy, S. M., Herberman, R. B., Whiteside, T., Sanzo, K., Lee, J., & Kirkwood, J. (1990). Perceived social support and tumor estrogen/progesterone receptor status as predictors of natural killer cell activity in breast cancer patients. *Psychosomatic Medicine, 52*, 73–85.

Malarkey, W. B., Kiecolt-Glaser, J. K., Pearl, D., & Glaser, R. (1994). Hostile behavior during marital conflict alters pituitary and adrenal hormones. *Psychosomatic Medicine, 56*, 41–51.

Marieb, E. N., & Hoehn, K. (2010). *Human anatomy & physiology* (8th ed.). San Francisco, CA: Pearson.

Matos-Gomes, N., Katsurayama, M., Makimoto, F. H., Santana, L. L. O., Paredes-Garcia, E., Becker, M. A. D. Á., & Dos-Santos, M. C. (2010). Psychological stress and its influence on salivary flow rate, total protein concentration and IgA, IgG and IgM titers. *Neuroimmunomodulation, 17*, 396–404.

McCann, B. S., Benjamin, G. A., Wilkinson, C. W., Carter, J., Retzlaff, B. M., Russo, J., & Knopp, R. H. (1996). Variations in plasma lipid concentration during examination stress. *International Journal of Behavioral Medicine, 3*, 251–265.

Miyazaki, T., Ishikawa, T., Iimori, H., Miki, A., Wenner, M., Fukunishi, I., & Kawamura, N. (2003). Relationship between perceived social support and immune function. *Stress and Health, 19*, 3–7.

Mouton, C., Fillion, L., Tawadros, É., & Tessier, R. (1989). Salivary IgA is a weak stress marker. *Behavioral Medicine, 15*, 179–185.

Muldoon, M. F., Herbert, T. B., Patterson, S. M., Kameneva, M., Raible, R., & Manuck, S. B. (1995). Effects of acute psychological stress on serum lipid levels, hemoconcentration, and blood viscosity. *Archives of Internal Medicine, 155*, 615–620.

Naliboff, B. D., Benton, D., Solomon, G. F., Morley, J. E., Fahey, J. L., Bloom, E. T., ... Gilmore, S. L. (1991). Immunological changes in young and old adults during brief laboratory stress. *Psychosomatic Medicine, 53*, 121–132.

Nater, U. M., & Rohleder, N. (2009). Salivary alpha-amylase as a non-invasive biomarker for the sympathetic nervous system: Current state of research. *Psychoneuroendocrinology, 34*, 486–496.

Ockene, I. S., Chiriboga, D. E., Stanek, E. J., Harmatz, M. G., Nicolosi, R., Saperia, G., ... Hebert, J. R. (2004). Seasonal variation in serum cholesterol levels: Treatment implications and possible mechanisms. *Archives of Internal Medicine, 164*, 863–870.

Ockenfels, M. C., Porter, L., Smyth, J., Kirschbaum, C., Hellhammer, D. H., & Stone, A. A. (1995). Effects of chronic stress associated with unemployment on salivary cortisol: Overall cortisol levels, diurnal rhythm, and acute stress reactivity. *Psychosomatic Medicine, 57*, 460–467.

Ohira, H. (2004). Social support and salivary secretory immunoglobulin A response in women to stress of making a public speech. *Perceptual and Motor Skills, 98*, 1241–1250.

Painter, R. H. (1998). IgG. In P. J. Delves (Ed.), *Encyclopedia of immunology* (2nd ed., pp. 1208–1211). Oxford, England: Academic Press.

Pressman, S. D., Cohen, S., Miller, G. E., Barkin, A., Rabin, B. S., & Treanor, J. J. (2005). Loneliness, social network size, and immune response to influenza vaccination in college freshmen. *Health Psychology, 24*, 297–306.

Raison, C. L., Capuron, L., & Miller, A. H. (2006). Cytokines sing the blues: Inflammation and the pathogenesis of depression. *Trends in Immunology, 27*, 24–31.

Rilling, J. K., DeMarco, A. C., Hackett, P. D., Chen, X., Gautam, P., Stair, S., ... Pagnoni, G. (2014). Sex differences in the neural and behavioral response to intranasal oxytocin and vasopressin during human social interaction. *Psychoneuroendocrinology, 39*, 237–248.

Roney, J. R., & Simmons, Z. L. (2013). Hormonal predictors of sexual motivation in natural menstrual cycles. *Hormones and Behavior, 63*, 636–645.

Schuck, P. (1998). Glycated hemoglobin as a physiological measure of stress and its relations to some psychological stress indicators. *Behavioral Medicine, 24*, 89–94.

Segerstrom, S. C., & Miller, G. E. (2004). Psychological stress and the human immune system: A meta-analytic study of 30 years of inquiry. *Psychological Bulletin, 130*, 601–630.

Shier, D., Butler, J., & Lewis, R. (2015). *Hole's human anatomy & physiology* (14th ed.). New York, NY: McGraw-Hill.

Short, S. J., Stalder, T., Marceau, K., Entringer, S., Moog, N. K., Shirtcliff, E. A., ... Buss, C. (2016). Correspondence between hair cortisol concentrations and 30-day integrated daily salivary and weekly urinary cortisol measures. *Psychoneuroendocrinology, 71*, 12–18.

Smyth, M. J., Cretney, E., Kelly, J. M., Westwood, J. A., Street, S. E., Yagita, H., ... Hayakawa, Y. (2005). Activation of NK cell cytotoxicity. *Molecular Immunology, 42*, 501–510.

Stanton, S. C. E., Campbell, L., & Loving, T. J. (2014). Energized by love: Thinking about romantic relationships increases positive affect and blood glucose levels. *Psychophysiology, 51*, 990–995.

Steegmans, P. H. A., Hoes, A. W., Bak, A. A., van der Does, E., & Grobbee, D. E. (2000). Higher prevalence of depressive symptoms in middle-aged men with low serum cholesterol levels. *Psychosomatic Medicine, 62*, 205–211.

Stoney, C. M., Matthews, K. A., Mcdonald, R. H., & Johnson, C. A. (1988). Sex differences in lipid, lipoprotein, cardiovascular, and neuroendocrine responses to acute stress. *Psychophysiology, 25*, 645–656.

Stoney, C. M., Niaura, R., & Bausserman, L. (1997). Temporal stability of lipid responses to acute psychological stress in middle-aged men. *Psychophysiology, 34*, 285–291.

Stoney, C. M., Niaura, R., Bausserman, L., & Metacin, M. (1999). Lipid reactivity to stress: I. Comparison of chronic and acute stress responses in middle-aged airline pilots. *Health Psychology, 18*, 241–250.

Suarez, E. C. (1999). Relations of trait depression and anxiety to low lipid and lipoprotein concentrations in healthy young adult women. *Psychosomatic Medicine, 61*, 273–279.

Thompson, R. R., George, K., Walton, J. C., Orr, S. P., & Benson, J. (2006). Sex-specific influences of vasopressin on human social communication. *Proceedings of the National Academy of Sciences, 103*, 7889–7894.

Tsujita, S., & Morimoto, K. (1999). Secretory IgA in saliva can be a useful stress marker. *Environmental Health and Preventative Medicine, 4*, 1–8.

Underdown, B. J. (1998). IgA. In P. J. Delves (Ed.), *Encyclopedia of immunology* (2nd ed., pp. 1196–1199). Oxford, England: Academic Press.

Van Doornen, L. J. P., & Van Blokland, R. (1987). Serum-cholesterol: Sex specific psychological correlates during rest and stress. *Journal of Psychosomatic Research, 31*, 239–249.

Part V

INTEGRATED PERSPECTIVES

29

How the LC4MP Became the DHCCST

An Epistemological Fairy Tale

Annie Lang

The thesis that the goal of science is substantial, significant, explanatory truth, has two elements: (1) scientists seek true answers to the questions that concern them; and (2) they concern themselves with substantial, significant questions (Haack, 2011, p. 136).

Truth in science is not relative, but it is conditional. Our best, most-supported answers are called true. And the more they successfully predict outcomes, lead to new technologies, and solve problems, the more confident we are that they are true. But their truth remains conditional. The things you say are true must continue to match the observations you make of the world. If our theory posits the existence of something we cannot see with our unaided senses, then it is up to us to devise a way of observing what we believe exists. One does not develop a neutrino detector unless one believes there are neutrinos and can, at least theoretically, describe their properties sufficiently to detect them. Similarly, one does not devise a method for observing cognitive resources unless one believes that a thing like cognitive resources exists.

According to Kuhn (2012), normal science requires a paradigm. A paradigm provides a set of assumptions, which are things that you assume are true. Assumptions are the accepted truths upon which theories are built. Scientists do not test them because they accept them as true and move forward from there. If assumptions turn out to be untrue, then paradigms and theories fall and must be rebuilt. Parts of them will remain the same, other parts will no longer be considered true. New, perhaps unrecognizable, questions, methods, and predictions will arise from the new set of assumptions/assumed truths.

A SCIENTIFIC FAIRY TALE: THE LC4MP

Once upon a time I quit my job selling newspaper advertising and wandered into graduate school to study communication. At the time I knew nothing about communication as an academic discipline. Hell, I didn't know what an academic discipline was. My first day of graduate school blew my mind when, in my *Introduction to Theory* class, we spent the whole day discovering that we didn't even know how to define communication; but that was okay—we could study it anyway. The whole idea that you didn't know what you were studying but could study it anyway amazed me. But, like everyone else in the field, I just went with it.

Looking back, this is the first assumption I dumped when I started to think that the theory I had been developing for 20 years (LC4MP; Lang 2000, 2006) was probably wrong in some very fundamental ways. Because you really cannot study something that you cannot define. It just means that, unconsciously, you're working with an implicit definition. For the LC4MP, I think my implicit definition of communication was that communication is primarily concerned with the sending and receiving of messages (though this was never clearly stated as an assumption of the LC4MP). This assumed truth can be found in the LC4MP's stated assumptions: First, messages have two primary parts: (1) content/meaning, and (2) structure/form. Second, messages occur over time. Third, the content and form of messages interact with the evolved perceptual, emotional, motivational, and cognitive systems that make up a human recipient through automatic (defined as fast, and outside conscious control and awareness) processes. Fourth, humans have a limited and finite processing capacity, allocating resources to processing brings the world into conscious awareness, and resources are allocated through automatic and controlled processes.

I would note that as I compare things I say here to things I have written about the LC4MP, I encourage readers to never, ever assume that what I considered to be conditionally true in 1990, or 2000, or 2010, is what I consider to be conditionally true in the present. I encourage people using the LC4MP to guide their research to use the most up-to-date version of the theory, which you will find published in Lang (2017). Although I no longer consider the LC4MP to be a conditionally true explanation of how communication works, that version of the model is the one I consider to be the most conditionally true. I would caution readers against relying on the descriptions of the LC4MP contained in recently published articles about the future and past of the LC4MP (Fisher, Huskey, Keene, & Weber 2018; Fisher, Keene, Huskey, & Weber 2018) as they misstate many aspects of the most up-to-date model and draw incorrect global conclusions about what it predicts. I will, later in this chapter, at the request of the editors of this volume, attempt to identify the parts of the model that I think are still conditionally true, the parts I think work well as predictors of message outcomes (though I no longer think the explanation for how they work is true), and the parts that I now think are simply not true.

How did the LC4MP come about? This is a story about what it means to be a data-driven model. I did not set out to create a theory. Rather, I asked and answered a series of what I hoped were substantial and significant empirical questions and tried to find true answers to them. My first question was: can I confirm that structural and content features of television elicit orienting responses (ORs)? To answer this question, I had to find an indicator of an OR that was not also an indicator of some other cognitive event. It turned out that, although there are many indicators of orienting, only the occurrence of a 5- to 7-second inverted U-shaped deceleratory cardiac response curve immediately following the structural feature of interest provides definitive evidence of the existence of an OR. I tested a number of structural and content features in television messages that had previously been shown to elicit alpha blocking and found that they did indeed elicit the expected cardiac response curve (Lang, 1990). One question asked, one conditionally true answer found. Later work by me and others asked and discovered which structural and content features in other media elicited orienting responses (Lang, Borse et al., 2002, Potter, Lang, & Bolls, 2008).

The next two questions I asked were: First, do different kinds of structural features elicit different kinds of orienting responses? And second, how do orienting responses influence cognitive processing? Several studies were conducted to arrive at the following conditionally true statements: First, different types of structural features could elicit different-shaped ORs (Lang, Chung, et al., 2005), though to date no one knows why. Second, ORs did influence memory, but the type of stimulus that elicited the OR did not affect the size of the effect on memory

(Thorson & Lang, 1992). Third, the effect of an OR on secondary task reaction times (a measure of attention) and memory was dynamic, with attention and memory peaking around 2–3 seconds after the OR (Geiger & Reeves, 1993; Lang, Geiger, Strickwerda, & Sumner, 1993). Fourth, the content of the message at the time of the OR-eliciting stimulus influenced the direction of that response. When content was more cognitively demanding, ORs resulted in decreases in memory. When content was less cognitively demanding, ORs resulted in increases in memory (Lang & Lanfear, 1990).

Each study cited here produced a piece of the data that were eventually combined to produce the first published version of the Limited Capacity Model of Mediated Message Processing (LC3MP; Lang, 2000). The aim of the model was to predict the direction and size of change in message recipients' attention and memory as a function of the messages' structural and content features. More importantly, this model also came with a toolbox: a set of measures that had been developed and tested for use in the media laboratory in the studies cited above, and that could covertly track these automatic, unconscious changes in cognitive processing.

To very briefly review the LC3MP: Attention was conceptualized using a single pool-limited capacity model. Processing resources could be allocated through controlled or automatic allocation processes. The allocation of processing resources was measured using phasic heart rate (HR) for OR detection and either secondary task reaction times (STRT) or tonic HR to assess changes in overall level of resources allocated. Memory was conceptualized as a set of three dynamic processes (encoding, storage, and retrieval) competing for and fueled by the automatic allocation of processing resources. Encoding was assessed by recognition tests, storage by cued recall tests, and retrieval by free recall. ORs resulted in the automatic allocation of resources to encoding. The effect of an OR on memory was dependent on the cognitive load of the ongoing message content.

Following the publication of the LC3MP, a lot more research was done by many people trying to figure out how dynamic change in the content and structure of messages influenced change in attention and memory over time (Bolls, 2002; Potter, 2000). I was particularly interested in identifying the types of message features and content that altered the cognitive load of a message.

The next stage of my research was guided by two substantial and significant questions. First, I wanted to understand why research consistently showed that fast-paced messages (defined as messages that elicited lots of ORs) elicited the fastest STRTs (indicating extremely low attention) when other indicators suggested that people were paying a lot of attention to those messages. Second, I wanted to understand why there were such large differences in attention and memory for emotional compared to neutral messages and for negative compared to positive emotional messages. Working with many colleagues on these questions led me to reconceptualize emotion as a by-product of activation in the motivational systems (Lang, Bolls, Potter, & Kawahara, 1999; Lang, Dhillon, & Dong, 1995; Lang, Newhagen, & Reeves, 1996), to add motivational system activation to the model as a tonic automatic resource allocation mechanism (Lang, Zhou et al., 2000), and to redefine STRT as a measure of available, rather than allocated, resources (Lang & Basil, 1998). These studies led to the Limited Capacity Model of Motivated Mediated Message Processing (LC4MP; Lang, 2006).

Again, to very briefly review, this version incorporated a biological model of the behavioral (approach and avoid) motivational systems into the previous LC3MP model and argued that automatic activation in the motivational systems leads to automatic allocation of resources to encoding and storage. In addition, it suggested that the dynamics of resource allocation and memory as a function of emotion differed for negative compared to positive messages (Lang, Park, Sanders-Jackson, & Wilson, 2007). Further, memory performance was based not on resources allocated to a message, but rather on available resources, that is, the resources allocated minus the resources required by the message to be processed (Lang, Bradley et al., 2006). Over time methods to assess resources required (information introduced (ii), a 7-dimension scale designed to measure the resources required to process the specific information introduced by a camera change) and resources allocated (camera changes (cc), the number of camera changes in a message) were developed along with the redefinition of STRT as a measure of available resources (Lang, Gao, et al., 2015; Lang, Kurita, et al., 2013). This version of the model did a credible job predicting, for a given message, when encoding would increase or decrease over time as a function of message cognitive load (resources required) and the level of automatic allocation of resources (as a function of emotion and orienting eliciting structural features). It also introduced levels of motivational reactivity as an individual difference (Lang, Bradley et al., 2007; Lang, Kurita, et al., 2011; Lang, Shin, & Lee, 2005) that influenced media preferences and message processing (Lang & Lee, 2014; Lang & Yegiyan, 2011, 2014). Further, the model could be used to guide the creation of messages that could achieve desired combinations of attention and memory (Lang, Sanders-Jackson, Wang, & Rubenking, 2013). For example, a message designed to maintain a consistent low level of cognitive load and a consistent moderately high level of automatic resource allocation would be consistently attended to and well-remembered, never causing cognitive overload or poor memory (Lang, Potter, & Grabe, 2003).

The LC4MP also made some predictions about memory as a function of motivational activation that were not very well supported by data. Highly arousing positive messages, which according to LC4MP should be remembered well, often are not. Research examining people's ability to notice and remember change in motivationally relevant aspects of messages turned out to be inexplicably terrible (Sanders-Jackson, 2007). A number of unpublished studies taught us that even though LC4MP predicted that you would encode motivationally relevant stuff best and first, that was not always the case. Further, in one study we found that the longer things were on screen, the worse they were remembered. Thus, even though many of the LC4MP's hypotheses were supported, there were enough anomalous results that I became more and more certain, as time went on, that something was rotten in the state of LC4MP. Most of the problems had to do with motivationally relevant material, suggesting that at least some of the model's conditional truths about how motivational activation influenced memory were not true. Another problem was that while LC4MP worked well to understand dynamic media (e.g., audio only/radio), it was not that successful when used to understand computer-presented text and pictures. For example, the appearance of novel sentences, shapes, and pictures on computer screens does not lead to an orienting response unless the content is motivationally relevant (Lang, Borse, Wise, & David, 2002). Why should that be?

At the same time that my LC4MP research was coming up with these problems, I was being heavily influenced by research on embedded cognition and on complex dynamic systems. As a psychophysiologist, I have always embraced a dynamic, embodied model of cognition, investigating the automatic and involuntary dynamic, biological, and physiological responses of our bodies to the ongoing content and structures of mediated messages. But as an experimentalist, I believed that I could put people in a lab and study pieces of processing and later add them up to get a generalizable whole. The more I read about Clark's work on embedded cognition (Clark, 2008) and Gibson's action perception theory (Gibson, 2014), the more I began to believe that we perceive, feel, think, and act not just in an environment, but in cooperation with a given environment. Thus, how we perceive, think, feel, and act is likely to change as a function of the environment in which we find ourselves. As a result, I was beginning to worry about applying the standard scientific reductionist analytical approach and the processing machine metaphor that drives the LC4MP to the study of human beings and human communication.

A SECOND SCIENTIFIC FAIRY TALE: DHCCST

This story is the story of an unintended intellectual journey caused by observations of the world that eventually stopped agreeing with the data-driven scientific theory I thought I was building. It was fine up to a point. But as I tried to fit more pieces into my solution to the mediated message processing puzzle, the picture was getting fuzzy and the model was becoming less rather than more predictive. In addition, my assessment of my own research and the research of others was undergoing a complete transformation. As I read new research, I found myself thinking that as a field we were no longer asking substantial and significant questions and that our data collection and analysis methods seemed to be poor at best and at times actually invalid.

I found myself wondering whether memory was representational at all. Do we really create representations? Do we encode information that we store and retrieve? Are we processing machines? And my answers became a series of "nos," which left me and the LC4MP on opposite sides of a growing divide. The editors of this book suggested that I tell you, the readers, which parts of it are still good and which are not, but that is a hard thing to do. Epistemologically, we know that concepts are not real (Schlick, 2012). We develop concepts and define them, then we create ways to measure them, then we collect data. But the data are our observations of the world, they are NOT proof that the concepts exist. So, I don't think I believe in the concept of resource allocation anymore. Which kind of eviscerates the LC4MP. I do believe in the OR, which I think of as directly linked to our motivational and direct perceptual systems. I think of it as part of the complex system that allows us to navigate the world and respond to change in our environments. So, I believe findings about which structural and content features of media elicit ORs and/or motivational activation are still conditionally true. I believe that conclusions about how messages' cognitive load influences memory for messages are still conditionally true. I believe that motivational activation leads to emotional experience and influences attention and memory.

I no longer think the following things are conditionally true: (1) that there are cognitive resources; (2) that encoding, storage, and retrieval are always ongoing or are the sole causes of memory; and (3) any of the specific, level, and directional predictions about how motivational activation influences memory for messages.

The LC4MP was a data-driven model. Each study I designed and conducted asked a specific question and produced a conditionally true answer that was then added to the assemblage of previously uncovered conditional truths. It was a model because the data produced the truths, and as each new conditional truth arose, the explanations of how the truths related to one another were added to the model. A theory of mediated message processing thus arose from the fitting together of data-driven results. It was inherently reductionist, as pieces were investigated and then added together in hopes of producing the whole.

The DHCCST (Dynamic Human-centered Communication Systems Theory, Lang, 2014), on the other hand, began solely as a revised set of assumptions/assumed truths, which were then applied logically to fundamental problems of communication research, such as defining communication, media, symbols, representations, etc. It provides a new set of assumed truths and begins the work of turning this new foundation into new questions. Thus, it began as a theoretical construction to which I and others (e.g., Bailey, 2015, 2016) have slowly been adding some data supporting conditional truths about predictions that follow from the assumptions. Though the name of the theory is quite a mouthful, it is also extremely descriptive of the goal of the theory. It is first and foremost a theory that tries to explain how communication systems that contain a human work over time. It aims, as did LC4MP, to be a general theory—applicable to any dynamic human-centered communication system you can think of, past, present, and future.

The theory is human-centered because the only constant over time in such a system is the presence of a human. The media, machines of communication, encoding systems, locations, and content of communication are all free to vary across systems and across historical and real time. Only the presence of a human is constant. The theory is human-centered because an attempt is made to conceptualize concepts and propositions in terms of aspects of humans—not machines, not contents, not locations, not media.

So what are the assumed truths of the DHCCST? That is, things we are not going to test, but rather the foundations on which the theory is built. There are nine assumptions, all of which are assumptions about humans. Five of them combine to define what it means to be a human:

1. Humans exist in linear dynamic time.
2. Humans are embedded at a given moment in a given location.
3. Humans evolved in THIS world and are specifically adapted to it.
4. Humans are embodied, that is, we exist as biological organisms.
5. Human beings are made up of a set of nested, biological, dynamic systems—including perceptual, motivational, physiological, cognitive, and experiential—which the theory shortens to "brain."

These five assumptions result in the definition of a human as an evolved, embodied, embedded brain (called an EEEB). The final four assumptions are:

1. Humans interact continuously over time with their environments in voluntary and involuntary ways.
2. Our nested systems interact continuously over time with one another and with the environment in voluntary and involuntary ways.
3. We evolved to save energy (and everything, even thinking, takes energy).
4. Because we are biological organisms, the behavior of our systems can be directed and limited by evolved biological imperatives.

DHCCST then uses these assumptions to conceptualize communication as the interaction over time of at least one EEEB with one or more other EEEBs or with one or more media. Because all concepts must have human-centered conceptualizations, media are conceptualized as brain-like creatures (BLC) because research shows that we automatically respond to media as if they were real and as if they were social actors/animals (Reeves & Nass, 1996).

How do these assumptions and definitions compare to those of the LC4MP? The LC4MP defined humans as limited capacity processors allocating resources using controlled and automatic processes. In other words, as machines with two kinds of sub-routines, one effortless and the other effortful. The DHCCST's new human is an animal first and a processor of information sometimes. However, information processing is expensive and therefore something that our new energy-saving human tries not to do. Rather our new human, who also has voluntary and involuntary behaviors, is assumed to rely not on energy-intensive thinking but rather on energy-conserving evolved, embodied responses to the environment.

The LC4MP defined communication as an interaction between a message and a human. Messages were defined as having form and content. And communication was characterized as a dynamic interaction between the humans' voluntary and involuntary resource allocation mechanisms and the form and content of the message. There is much that is still good in this definition: (1) it is dynamic; (2) it allows for voluntary and involuntary behavior; and (3) it separates form and content. But, for DHCCST, messages need to be defined not in terms of themselves, but in terms of human-centered concepts. And the whole thing needs to be embedded because humans are embedded and embodied and evolved. Hence I have replaced "messages" with "brain-like creatures"—which are specifically defined based on Gibson's definition of an animal as objects having a closed body envelope and displaying apparently animate behavior. Media range along a continuum from not very brain-like (e.g., a book) to very brain-like (e.g., television).

Similarly, I have redefined medium from a human-centered perspective. Specifically, there are two dimensions of mediation, each conceptualized as ranging along a continuum from evolved to man-made. The *carrier* dimension of mediation refers to how a message is carried from one EEEB to another or to a BLC. Carrier media can carry messages through time, space, or both time and space. Evolved carrier media include air to carry sound and light to carry vision. Man-made carrier media include cave walls to carry paintings, clay tablets and paper to carry words and pictures, the Pony Express to carry letters, the electromagnetic spectrum to carry moving audio and visual messages, etc.

The second dimension of mediation is the *encoding system*. Evolved encoding systems include things like body language, facial expressions, smell, touch, cries, and screams. Man-made systems can be representational (where the encoding system shares perceptual information with the thing it represents) such as pictures, videos, and audio recordings; or symbolic (where the encoding system shares no perceptual information with the thing being encoded) such as smoke signals, drumming, Morse code, and language.

Using these definitions of mediation, DHCCST then reconceptualizes the three major types of communication (interpersonal, mass, and human media interaction). Interpersonal communication is defined as interaction between two EEEBs who may or may not be co-located in space in time. Which media is used to carry messages from one EEEB to another depends on the state of co-location in time and space. EEEBs co-located in time and space may use air and light to communicate, though they might also text. Those separated in time and space could use cave paintings, books, letters, or email. Those separated in space might use phones, texts, etc.

Human/media interaction is defined as co-located EEEBs and BLCs interacting over time using a variety of evolved and man-made carrier media and encoding systems. Mass communication is defined as a message encoded into some system and sent to multiple EEEBs who may or may not be co-located in space or time using a variety of carrier media.

So, doing the theoretical and conceptual work to write the DHCCST took me about 4 years. At the same time, I was trying to figure out how to move forward to start collecting data to test hypotheses derived from the theory's assumptions. I knew that my overall research question had changed from "how do people process mediated messages?" to "how do dynamic human communication systems function?" But, to be honest, I had no idea how to answer that question. I took two different approaches to this problem. One continued the work begun in the DHCCST of reconceptualizing important variables from a human-centered point of view—and then, after doing that, deriving hypotheses about system outcomes based on those new conceptualizations. The second was to begin to approach the problem using the assumptions and concepts of Dynamic Systems Theory (DST; Kelso, 1995).

An example of the first approach is work done with Rachel Bailey (Lang & Bailey, 2015) to reconceptualize encoding from a DHCCST perspective. Here, we needed to think about memory from the point of view of a human navigating an environment—not from the point of view of a message being processed by a machine. A combination of Clark's embedded processing work and Gibson's ecological perception theory was used to conceptualize encoding (that is the creation of mental representations of the world) as an energy-intensive cognitive activity to be avoided when possible. As a result, the research question was changed from "How well do people encode messages?" to "When and which parts of messages do people bother to encode?" To answer this question, we conceptualized the television message as an environment through which the viewer navigates. From this perspective, we theorized that information had four important properties with respect to a person: (1) imminence (how close is it to me?); (2) stability (how long is it going to remain in the environment?); (3) motivational relevance (is it a threat or an opportunity?); and (4) task relevance (will it help me with my current goal?). We argued that people do not need to encode things that are stable (i.e., likely to stick around) and things that are close by (especially if they are task or motivationally relevant), since they can simply stay in perceptual contact with them, but that they do encode things that were going to disappear and things that were far away (again especially if they are motivationally or task-relevant). This led to the weird and supported prediction that the worst encoded information would be for things that were stable (on screen a long time), nearby (in the center of the screen), and motivationally relevant. Another study using this approach argued that stimuli using representational encoding systems (which share perceptual information with what they represent) would elicit faster and larger responses compared to those using symbolic encoding systems, because only information presented using a representational system would be available to the direct perceptual system and therefore be able to trigger a biological imperative to respond (Lang & Bailey, 2015).

The second approach was to use DST as a mathematical, not just a conceptual, theory. Mathematically, a dynamic system is defined by a pair of differential equations. Of course, you don't actually know what those equations are, and even if you did, most pairs of differential equations are not solvable. Nonetheless, I refreshed my calculus and took a course in DST analysis. It turns out that the way we learn about dynamic systems, mathematically, is by running them forward through time and seeing what happens. As time progresses, qualitative behavioral states (represented by points, spirals, and chaotic attractors, etc.) exist, appear, and disappear, until the system settles into a final attractor state. Once you have a basic understanding of a system, you can write some equations, based on what you know, and see if the resulting system exhibits over time the behaviors that you expect. But, you cannot do that until you know what the system does and how it changes over times and places.

So, I next began to look at how innovators in other fields started applying dynamic systems theory to their research questions. Here I came across a fabulous book on human development by Esther Thelen and Linda Smith (1996). In this book, they talk about how they began to investigate the development of human babies as a dynamic system using an empirical, observational, analytical strategy. The strategy has five basic steps:

1. Define your system.
2. Observe the system in action in order to learn all of its possible behavioral states, which is called the *state space* of the system.
3. Determine the relative stability of each of the behavioral attractors within the system. If a behavioral attractor is very stable (often called deep), most or all of the individuals in the system will exhibit the behavior. The less stable (shallower) the attractor is, within the system, the fewer individuals will exhibit it.
4. Determine the relative stability of each of the behavioral attractors within each individual. This is important because we know that initial conditions (prior to observation) and an individual's unique history cause large differences both in which behavioral attractors they exhibit and in the stability of those behavioral attractors. Within a given individual, behaviors range from very stable (i.e., exhibited frequently for long periods of time) to very unstable (rarely exhibited for not for very long).

5. The final step is to begin to look for candidate variables that act as order or control parameters. Order parameters are variables within the system that reduce the degrees of freedom (possible behaviors) of the system. Control parameters are linear variables that shift individuals from one qualitatively different behavior to another.

My colleagues and I made our first attempt at this kind of analysis by doing a secondary analysis of data collected by individual participants playing 30 minutes of Grand Theft Auto IV on an Xbox system in a laboratory (Lang, Han et al., 2019). We had video of the game play of 34 people. Our first step was to watch all of the videos and identify all of the system behaviors people exhibited and which behaviors were exhibited by which people. The state space turned out to have six behaviors which were, in order of stability (high to low), walking, driving, dying, aggressing, being arrested, and going on missions. Next, to determine behavioral stability within the system, we counted the number of people who exhibited each behavior. Players exhibited a range of behaviors from two (walking and driving) to all six, as would be expected if we are looking at a complex dynamic system. We also looked at two possible control parameters that might move participants from one set of behaviors to another. One was how much experience players had with the game Grand Theft Auto, thinking that the higher a player's skill level, the more s/he would have learned the affordances (i.e., possible behaviors and their consequences) of the game. This allowed us to examine change in behavioral stability over developmental time. The second parameter was learning. Here we compared behavioral stability within participants over the 30 minutes of game play by comparing the stability of their observed behaviors in the first and second half of play. Here we found that behaviors that were rewarded by the game (e.g., dying) were more stable in the second half of play while behaviors that were not rewarded by the game (e.g., being arrested) became less stable. This was particularly true for lower-skilled players who were still learning the game.

A CAUTIONARY TALE

And that brings me to June 2019. My current work is all aimed at rethinking and reconceptualizing existing communication theories and questions from a human-centered point of view and seeing what sorts of new predictions come out of that exercise. A recent paper (Han & Lang, 2019) provides an example, using the television co-viewing literature, of how questions, research designs, methods, and analysis all change as you move from a static linear approach to communication to a dynamic non-linear approach. Most of the work done so far by my colleagues and I using this approach encounters stiff opposition during the peer review process, not because of issues with the studies themselves, but because it is fairly alien. It just sounds weird and a lot of people don't get it. And I completely understand that! It took me nine years to get from accepting that the LC4MP was not an accurate explanation of how communication works and deciding that communication probably needed to be studied as a dynamic system to writing the DHCCST. Along the way, many studies were conducted to try out the ideas or test some of the hypothesis. Most of that time I haven't had any idea what I was doing. I just kept reading and thinking and writing and trying stuff. Much of that time was spent undoing habitual linear causal reductionist ways of thinking. But, as Kuhn says, once you succeed in doing that, it's like you are looking at a different world. It has been in some ways a difficult journey, but in all ways a rewarding one. I wouldn't go back to the way I used to think for anything.

However, it is true that modern academia, as a system, is not very conducive to the production of science. Today's universities reward speed and amount of publishing in top tier journals.

They hope that if a person is publishing in high quality journals, then this assures the asking of substantial and significant questions and the slow building of conditional truths. But that probably is not really the case. The best way to get published is to add a small conditional truth accretion to a "theory" or "model" that everyone already understands and is comfortable with. Taking chances, suggesting that some of our theories are wrong or are answering non-substantial questions, or even suggesting that your own theory is wrong and needs to be retired are all ways to become an academic pariah. But despite modern academia's fascination with translational research and the need to produce more and more credit hours and grant dollars, it continues to include tenure and academic freedom as core values. The purpose of these values is to allow scientists to pursue unpopular ideas and to put forward new and alien candidates for conditional truth. It also allows the tenured faculty member the luxury of slowing down and taking the time to think, not just about the findings that came out as predicted and got published, but also the implications of those which contradicted our theories and expectations. And they allow us the time to realize that our new ideas may mean that we need to take the time to learn new things, develop new methods, and learn new ways to think about and analyze data. It is to be hoped that in the rush to publish the most, the fastest, we don't forget that our goal is not to build a great vita but rather to discover new (conditional) explanatory truths about how the world actually is.

REFERENCES

Bailey, R. L. (2015). Processing food advertisements: Initial biological responses matter. *Communication Monographs, 82*, 163–178.

Bailey, R. L. (2016). Modern foraging: Presence of food and energy density influence motivational processing of food advertisements. *Appetite, 107*, 568–574.

Bolls, P. D. (2002). I can hear you, but can I see you? The use of visual cognition during exposure to high-imagery radio advertisements. *Communication Research, 29*, 537–563.

Clark, A. (2008). *Supersizing the mind: Embodiment, action, and cognitive extension.* Oxford, England: Oxford University Press.

Fisher, J. T., Huskey, R., Keene, J. R., & Weber, R. (2018). The limited capacity model of motivated mediated message processing: Looking to the future. *Annals of the International Communication Association, 42*(4), 291–315.

Fisher, J. T., Keene, J. R., Huskey, R., & Weber, R. (2018). The limited capacity model of motivated mediated message processing: Taking stock of the past. *Annals of the International Communication Association, 42*(4), 270–290.

Geiger, S., & Reeves, B. (1993). The effects of scene changes and semantic relatedness on attention to television. *Communication Research, 20*(2), 155–175.

Gibson, J. J. (2014). *The ecological approach to visual perception.* New York, NY: Psychology Press.

Haack, S. (2011). *Defending science-within reason: Between scientism and cynicism.* Amherst, NY: Prometheus Books.

Han, J., & Lang, A. (2019). It's a journey: From media effects to dynamic systems. *Media Psychology.* Advance online publication.

Kelso, J. S. (1995). *Dynamic patterns: The self-organization of brain and behavior.* Cambridge, MA: MIT Press.

Kuhn, T. S. (2012). *The structure of scientific revolutions.* Chicago, IL: University of Chicago Press.

Lang, A. (1990). Involuntary attention and physiological arousal evoked by structural features and emotional content in TV commercials. *Communication Research, 17*, 275–299.

Lang, A. (2000). The information processing of mediated messages: A framework for communication research. *Journal of Communication, 50*, 46–70.

Lang, A. (2006). Using the limited capacity model of motivated mediated message processing (LC4MP) to design effective cancer communication messages. *Journal of Communication, 56*, 1–24.

Lang, A. (2014). Dynamic human-centered communication systems theory. *The Information Society, 30*(1), 60–70.

Lang, A. (2017). LC4MP: Limited capacity model of motivated mediated message processing model. In P. Rössler (Ed.), *International encyclopedia of media effects*. Hoboken, NJ: Wiley-Blackwell.

Lang, A., & Bailey, R. L. (2015). Understanding information selection and encoding from a dynamic, energy saving, evolved, embodied, embedded perspective. *Human Communication Research, 41*(1), 1–20.

Lang, A., & Basil, M. D. (1998). Attention, resource allocation, and communication research: What do secondary task reaction times measure, anyway? *Annals of the International Communication Association, 21*, 443–458.

Lang, A., Bolls, P., Potter, R., & Kawahara, K. (1999). The effects of production pacing and arousing content on the information processing of television messages. *Journal of Broadcasting and Electronic Media, 43*(4), 451–476.

Lang, A., Borse, J., Wise, K., & David, P. (2002). Captured by the World Wide Web: Orienting to structural and content features of computer presented information. *Communication Research, 29*(3), 215–245.

Lang, A., Bradley, S. D., Park, B., Shin, M., & Chung, Y. (2006). Parsing the resource pie: Using STRTs to measure attention to mediated messages. *Media Psychology, 8*, 369–394.

Lang, A., Bradley, S. D., Sparks, J. V., & Lee, S. (2007). Measuring individual differences in motivation activation: Predicting physiological and behavioral indicators of appetitive and aversive activation. *Communication Methods and Measures, 1*(2), 113–136.

Lang, A., Chung, Y., Lee, S., & Zhao, X. (2005). It's the product: Do risky products compel attention and elicit arousal in media users? *Health Communication, 17*(3), 283–300.

Lang, A., Dhillon, P., & Dong, Q. (1995). Arousal, emotion, and memory for television messages. *Journal of Broadcasting and Electronic Media, 38*, 1–15.

Lang, A., Gao, Y., Potter, R. F., Lee, S., Park, B., & Bailey, R. L. (2015). Conceptualizing audio message complexity as available processing resources. *Communication Research, 42*(6), 759–778.

Lang, A., Geiger, S., Strickwerda, M., & Sumner, J. (1993). The effects of related and unrelated cuts on viewers' memory for television: A limited capacity theory of television viewing. *Communication Research, 20*(1), 4–29.

Lang, A., Han, J., Zheng, X., Almond, A., Lynch, T., & Matthews, N. (2019). Learning to play: How virtual world affordances drive adaptation and learning in Grand Theft Auto. In B. Liebold, D. Pietschmann, & B. Lange (Eds.), *Evolutionary psychology and digital games: Digital hunter-gatherers* (pp. 179–192). New York, NY: Routledge.

Lang, A., Kurita, S., Gao, Y., & Rubenking, B. (2013). Measuring television message complexity as available processing resources: Dimensions of information and cognitive load. *Media Psychology, 16*(2), 129–153.

Lang, A., Kurita, S., Rubenking, B., & Potter, R. F. (2011). MiniMAM: Developing a short version of the Motivation Activation Measure. *Communication Methods and Measures, 5*, 146–117.

Lang, A., & Lanfear, P. (1990). The information processing of televised political advertising: Using theory to maximize recall. In J. Muncy & M. Goldberg (Eds.), *Advances in consumer research* (Vol. 17, pp. 149–158). Provo, UT: Association for Consumer Research.

Lang, A., & Lee, S. (2014). Individual differences in trait motivational reactivity influence children and adolescents' responses to pictures of taboo products. *Journal of Health Communication, 19*(9), 1030–1046.

Lang, A., & Newhagen, J., & Reeves, B. (1996). Negative video as structure: Emotion, attention, capacity, and memory. *Journal of Broadcasting and Electronic Media, 40*, 460–477.

Lang, A., Park, B., Sanders-Jackson, A., & Wilson, B. D. (2007). Separating emotional and cognitive load: How valence, arousing content, structural complexity and information density affect the availability of cognitive resources. *Media Psychology, 10*, 317–338.

Lang, A., Potter, D., & Grabe, E. (2003). Making news memorable: Applying theory to the production of local television news. *Journal of Broadcasting and Electronic Media, 47*(1), 113–123.
Lang, A., Sanders-Jackson, A., Wang, Z., & Rubenking, B. (2013). Motivated message processing: How motivational activation influences resource allocation, encoding, and storage of TV messages. *Motivation and Emotion, 37*(3), 508–517.
Lang, A., Shin, M., & Lee, S. (2005). Sensation seeking, motivation, and substance use: A dual system approach. *Media Psychology, 7*, 1–29.
Lang, A., & Yegiyan, N. (2011). Individual differences in motivational activation influence responses to pictures of taboo products. *Journal of Health Communication, 16*, 1072–1087.
Lang, A., & Yegiyan, N. (2014). Mediated substance cues: Motivational reactivity and use influence responses to pictures of alcohol. *Journal of Health Communication, 19*(11), 1216–31.
Lang, A., Zhou, S., Schwartz, N., Bolls, P. D., & Potter, R. F. (2000). The effects of edits on arousal, attention, and memory for television messages: When an edit is an edit can an edit be too much? *Journal of Broadcasting and Electronic Media, 44*, 94–109.
Potter, R. F. (2000). The effects of voice changes on orienting and immediate cognitive overload in radio listeners. *Media Psychology, 2*, 147–177.
Potter, R. F., Lang, A., & Bolls, P. D. (2008). Identifying structural features of audio: orienting responses during radio messages and their impact on recognition. *Journal of Media Psychology, 20*(4), 168–177.
Reeves, B., & Nass, C. I. (1996). *The media equation: How people treat computers, television, and new media like real people and places*. New York, NY: Cambridge University Press.
Sanders-Jackson, A. (2007). The person behind the door: How motivation and structural features effect how we process information. Unpublished master's thesis, Indiana University, Bloomington, IN.
Schlick, M. (2012). *General theory of knowledge*. Chicago, IL: Open Court Publishing.
Thelen, E., & Smith, L. B. (1996). *A dynamic systems approach to the development of cognition and action*. Cambridge, MA: MIT Press.
Thorson, E., & Lang, A. (1992). Effects of television videographics and lecture familiarity on adult cardiac orienting responses and memory. *Communication Research, 9*(3), 346–369.

30

The Life of a Model

Commentary on "How the LC4MP Became the DHCCST"

Jacob T. Fisher, Richard Huskey, Justin Robert Keene, and René Weber

Lang (in Chapter 29, in this volume) presents an insightful and engaging overview of the birth and history of the Limited Capacity Model of Motivated Mediated Message Processing (LCM4MP) (Lang, 2000, 2006, 2009, 2017), highlighting the breadth of published and unpublished work that has played a role in the model's[1] development over the last several decades. In Lang's view, the LC4MP is a model beset by a burden of ambiguous or countervailing evidence for its core predictions and built upon assumptions that are largely untenable with dynamic systems theory. Accordingly, Lang advocates a departure from the LC4MP in favor of the Dynamic Human-Centered Communication Systems Theory (DHCCST; Lang, 2014), which is based on an entirely new set of assumptions and predictions grounded in dynamic systems theory (Kelso, 1995). We find ourselves in agreement with Lang on many of her arguments. Indeed, we admire and support her mission of innovation, taking chances, and pursuing unpopular ideas. That said, we disagree with Lang that our recent updates to the LC4MP (Fisher, Huskey, Keene, & Weber, 2018; Fisher, Keene, Huskey, & Weber, 2018) misrepresent the model's assumptions and predictions. In addition, we challenge the idea that the LC4MP should be discarded in favor of the DHCCST. Instead, we advocate for an update of the model's assumptions, for the refinement of its predictions, and for its co-existence with the DHCCST.

INCORRECT INTERPRETATIONS?

In Chapter 29, Lang (p. 398) writes:

> [She] would caution readers against relying on the descriptions of the LC4MP contained in recently published articles about the future and past of the LC4MP … as they misstate many aspects of the most up-to-date model and draw incorrect global conclusions about what it predicts.

We appreciate caution in systematic scientific debates, but Lang does not point out which aspects of the LC4MP have been misstated or where exactly conclusions about its predictions are incorrect. It is true that the LC4MP has been revised over time, most notably with the incorporation of the motivational aspect of the model in the mid-2000s (Lang, 2006). We are certain, however, that our description of the assumptions and predictions of the LC4MP is consistent with Lang's latest version of the model (as it is described in Lang, 2017).

With that said, in our updated model, we have made a good faith effort to highlight ways in which the new model differs from the latest version of the LC4MP, and we welcome any debate as to how we may better represent the core aspects of the model and its development over the last two decades. We completely agree with Lang's notion of "conditional truth" in science—the idea that in science all truth is preliminary and will change through evidence and informed debate. Our updates to the LC4MP describe what is conditionally true in the model, given the latest evidence from communication and a wide range of cognate fields.

OUTDATED ASSUMPTIONS?

Lang highlights several core components in the model in which she no longer believes: (1) the construct of cognitive resources and resource allocation; (2) the conceptualization of memory encoding as representational; and (3) the idea that humans are information processors. These assumptions are by no means unique to the LC4MP. Each of these assumptions has been successfully used to guide research in communication as well as a wide array of cognate fields. At the same time, however, auxiliary hypotheses (specific statements that enable a theory to be rendered testable; see e.g., Popper, 1985) associated with these assumptions have progressed markedly in the last three decades, driven by huge advances in our understanding of how the brain enables human cognition and behavior. These updates have served to modernize the "nomological network" (Cronbach & Meehl, 1955) surrounding the LC4MP's assumptions and render them tenable within the modern scientific milieu.

This updating process can be seen perhaps most clearly in the area of cognitive resources and resource allocation. Although the exact nature of cognitive resources is still a point of active investigation (for a review, see Shenhav et al., 2013), given current evidence, it is clear that cognitive resource availability is related to spatial, temporal, and/or metabolic constraints on neural activation and connectivity (Buschman, Siegel, Roy, & Miller, 2011; Feng, Schwemmer, Gershman, & Cohen, 2014; Kurzban, Duckworth, Kable, & Myers, 2013; Marois & Ivanoff, 2005). In addition, it is clear that humans take resource availability into account when deciding whether or not to engage in a task (Kool & Botvinick, 2014; Kool, McGuire, Rosen, & Botvinick, 2010; Westbrook & Braver, 2015; Westbrook, Kester, & Braver, 2013), that humans' resource availability meaningfully influences cognitive processing performance and the neural substrates thereof (Finc et al., 2017; Kurzban, Duckworth, Kable, & Myers, 2013; Lavie, Hirst, de Fockert, & Viding, 2004; Sweller, 1988), and that resources are allocated to tasks based on motivational considerations (Botvinick & Braver, 2015; Huskey, Craighead, Miller, & Weber, 2018; Pessoa, 2009; Shenhav, Botvinick, & Cohen, 2013).

Space does not permit a thorough discussion of the representational nature of encoding in the brain, or of recent advances in our understanding of the brain as a predictive information processor. Suffice it to say that in each of these areas, the assumptions of the LC4MP have been *strengthened* rather than weakened as research has progressed over the last several decades. Interested readers are encouraged to consult Binder, Desai, Graves, and Conant (2009), Haxby et al. (2011), and Henke (2010) for more in-depth treatments of memory and Feldman-Hall and

Shenhav, (2019), Krakauer, Ghazanfar, Gomez-Marin, MacIver, and Poeppel (2017), and Zénon, Solopchuk, and Pezzulo (2019) for a discussion of information processing. With these observations in mind, we call into question Lang's assertion that the assumptions of the LC4MP are untrue. Instead, we posit that these assumptions are quite well supported and that with small updates to their auxiliary hypotheses, they are still quite useful for understanding human cognition and behavior.

PROBLEMATIC PREDICTIONS?

Lang claims that a proliferation of anomalous results in published and unpublished literature led her to believe that "something was rotten" in the core of the LC4MP. Although we cannot speak to the unpublished data mentioned in Lang's chapter, it is quite clear from the literature (see, e.g., Fisher, Keene et al., 2018; Huskey, Wilcox, Clayton, & Keene, 2020) that the LC4MP has overall been impressively accurate in its "risky predictions" (Meehl, 1990; Popper, 1985). Furthermore, many areas in which the predictions of the LC4MP have been shown to be incorrect are areas in which further knowledge has been acquired through iterative scientific progress.

A notable example of this process can be seen in recent work by Clayton and colleagues (2018) characterizing individual differences in the progression of the defensive cascade. Here, the LC4MP clearly contributes to scientific knowledge of how humans process fear-inducing, disgusting, or otherwise threatening messages, as well as knowledge of the motivated processing system in general. Thus, in the ways in which the model is wrong, it is wrong in the right ways—ways that serve as footholds to catalyze scientific progress rather than quell it. As such, a handful of ambiguous or anomalous findings does not necessitate the abandonment of the LC4MP. On the contrary, these findings allow for the model to be strengthened and refined, increasing its utility for scientific inquiry.

COMPLEXITY IN COMMUNICATION

It is true that the human brain is a constellation of complex systems, and that human behavior is contingent upon interactions between the brain and the environment (which is itself an amalgamation of complex systems; Weber, Mathiak, & Sherry, 2008). In addition, it has been known from the earliest days of communication research that communication behavior is complex (Schramm, 1955; Shannon, 1948). As such, we are in full agreement with Lang that communication researchers stand to benefit markedly from an approach rooted in complex (dynamic) systems theory (Sanbonmatsu & Johnston, 2019; Sherry, 2014, 2015). In fact, our own theories draw heavily from a complex systems perspective (see, e.g., Chapter 12 by Weber & Fisher, in this volume; Weber, Tamborini, Westcott-Baker, & Kantor, 2009).

It is, however, not the case—as Lang suggests—that re-thinking communication research from the perspective of complex systems theory requires abandoning linear, causal ways of thinking. Not all predictions in a model informed by complex systems theory need be of non-static, non-linear nature in order to observe and meaningfully interpret complexity within a system. In fact, systems can be thought of as a set of interacting subsystems and there is a surprising amount of linearity and simple cause-and-effect relationships at play at *some level* of any system (Strogatz, 2004, 2014). With this in mind, communication researchers should work to incorporate emerging methods for studying complex phenomena, such as computational and agent-based modelling (Madsen, Bailey, Carrella, & Koralus, 2019), network science (Barabási & Pósfai, 2016; Bassett & Gazzaniga, 2011; Newman, 2010), and "computational thinking" in general (Jolly & Chang, 2019), while retaining a focus on methods and models that can precisely characterize causes and effects within a system and its subsystems.

Furthermore, there is a trade-off between the intuitive generality of complex system theories and the precision required to make predictions that are useful for applied solutions to relevant problems (Sanbonmatsu & Johnston, 2019; Watts, 2017). Our communication theories should accomplish at least three goals: (1) they should be inclusive of a wide range of phenomena; (2) they should strive for explanatory and predictive power; and (3) they should strive for practical utility in real-world communication scenarios (Chaffee & Berger, 1987). The LC4MP has proven quite laudable in its practical utility for designing messages in a wide variety of contexts, including persuasive messaging, news, video games, advertisements, and many more (Fisher, Keene et al., 2018). It has also shown itself to be quite amenable to consideration within complex systems-based theorizing (see, e.g., Fisher, Lonergan, Hopp, & Weber, in press). In fact, Lang notes that the DHCCST contains many elements of the LC4MP. As such, we assert that the LC4MP need not be discarded to make way for a complex systems approach—rather, the refined LC4MP and the DHCCST can co-exist.

BELIEF VS. EVIDENCE

Finally, and perhaps most pressingly, we reject the notion that a scientific model should be supported or discarded based on anything but empirical evidence for or against its predictions. In Chapter 29, Lang states that she advocates for discarding the LC4MP in favor of the DHCCST in large part because she does not "believe" in its core concepts anymore. The incremental progress of scientific inquiry, however, has little to do with belief (Kuhn, 2012; Meehl, 1978; Popper, 1985). Cumulative scientific progress is contingent upon the idea that models are appraised, amended, or abandoned in a rigorous, systematic fashion, wherein precise predictions are tested in order to gather evidence for or against the model (see, e.g., Meehl, 1990). In our updated version of the LC4MP (Fisher, Huskey et al., 2018), we outline a clear set of falsifiable predictions that remain faithful to the core of the LC4MP (see, e.g., Lang, 2000, 2006, 2009, 2017), while providing grounds for systematically determining whether or not the model should be abandoned.

If the LC4MP is to be relegated to the dustbin of history, we argue that this should be done based on a process of rigorous falsification of its predictions. Given these standards, and our assertions outlined herein, the evidence does not warrant abandoning the LC4MP. We applaud Dr. Lang on her development of the DHCCST, and we hope that it proves to be a useful theoretical framework for the investigation of complex communication behavior. In large part, the business of science is the creation and testing of "risky predictions" (Popper, 1985), and we believe that Lang's journey as outlined in Chapter 29 is in many ways one to be emulated. We hope that our brief response sparks an engaged, good-natured, and constructive debate on the continuing value of the LC4MP for communication scholarship.

NOTE

1. In Chapter 29, in this volume, Lang refers to the LC4MP as a "theory" whereas in other works it is referred to as a "model." Although the LC4MP undeniably has characteristics of a theory (demarcated assumptions, description of scope, etc.), in order to maintain consistency with the larger body of LC4MP literature, it will herein be referred to as a model.

REFERENCES

Barabási, A.-L., & Pósfai, M. (2016). *Network science*. Cambridge, England: Cambridge University Press.

Bassett, D. S., & Gazzaniga, M. S. (2011). Understanding complexity in the human brain. *Trends in Cognitive Sciences, 15*(5), 200–209.

Binder, J. R., Desai, R. H., Graves, W. W., & Conant, L. L. (2009). Where is the semantic system? A critical review and meta-analysis of 120 functional neuroimaging studies. *Cerebral Cortex, 19*(12), 2767–2796.

Botvinick, M., & Braver, T. (2015). Motivation and cognitive control: From behavior to neural mechanism. *Annual Review of Psychology, 66*(1), 83–113.

Buschman, T. J., Siegel, M., Roy, J. E., & Miller, E. K. (2011). Neural substrates of cognitive capacity limitations. *Proceedings of the National Academy of Sciences, 108*(27), 11252–11255.

Chaffee, S. H., & Berger, C. R. (1987). What communication scientists do. In C. R. Berger & S. H. Chaffee (Eds.), *Handbook of communication science* (pp. 99–122). Newbury Park, CA: Sage.

Clayton, R. B., Lang, A., Leshner, G., & Quick, B. L. (2018). Who fights, who flees? An integration of the LC4MP and psychological reactance theory. *Media Psychology, 22*, 545–571.

Cronbach, L. J., & Meehl, P. E. (1955). Construct validity in psychological tests. *Psychological Bulletin, 52*(4), 281–302.

Feldman-Hall, O., & Shenhav, A. (2019). Resolving uncertainty in a social world. *Nature Human Behaviour, 3*, 426–435.

Feng, S. F., Schwemmer, M., Gershman, S. J., & Cohen, J. D. (2014). Multitasking versus multiplexing: Toward a normative account of limitations in the simultaneous execution of control-demanding behaviors. *Cognitive, Affective, & Behavioral Neuroscience, 14*(1), 129–146.

Finc, K., Bonna, K., Lewandowska, M., Wolak, T., Nikadon, J., Dreszer, J., Duch, W., & Kühn, S. (2017). Transition of the functional brain network related to increasing cognitive demands. *Human Brain Mapping, 38*, 3659–3674.

Fisher, J. T., Huskey, R., Keene, J. R., & Weber, R. (2018). The limited capacity model of motivated mediated message processing: Looking to the future. *Annals of the International Communication Association, 42*(4), 291–315.

Fisher, J. T., Keene, J. R., Huskey, R., & Weber, R. (2018). The limited capacity model of motivated mediated message processing: Taking stock of the past. *Annals of the International Communication Association, 42*(4), 270–290.

Fisher, J. T., Lonergan, C., Hopp, F. R., & Weber, R. (in press). Media entertainment, flow experiences, and the synchronization of audiences. In P. Vorderer & C. Klimmt (Eds.), *The Oxford handbook of media entertainment research.* Oxford, England: Oxford University Press.

Haxby, J. V., Guntupalli, J. S., Connolly, A. C., Halchenko, Y. O., Conroy, B. R., Gobbini, M. I., ... Ramadge, P. J. (2011). A common, high-dimensional model of the representational space in human ventral temporal cortex. *Neuron, 72*(2), 404–416.

Henke, K. (2010). A model for memory systems based on processing modes rather than consciousness. *Nature Reviews Neuroscience, 11*(7), 523–532.

Huskey, R., Craighead, B., Miller, M. B., & Weber, R. (2018). Does intrinsic reward motivate cognitive control? A naturalistic-fMRI study based on the synchronization theory of flow. *Cognitive, Affective, & Behavioral Neuroscience, 18*(5), 902–924.

Huskey, R., Wilcox, S., Clayton, R., & Keene, J. R. (2020). The limited capacity model of motivated mediated message processing: A meta-analytic summary of two decades of research. Manuscript submitted for publication.

Jolly, E., & Chang, L. J. (2019). The flatland fallacy: Moving beyond low-dimensional thinking. *Topics in Cognitive Science, 11*(2), 433–454.

Kelso, J. S. (1995). *Dynamic patterns: The self-organization of brain and behavior*. Cambridge, MA: MIT Press.

Kool, W., & Botvinick, M. (2014). A labor/leisure tradeoff in cognitive control. *Journal of Experimental Psychology: General, 143*, 131–141.

Kool, W., McGuire, J. T., Rosen, Z. B., & Botvinick, M. M. (2010). Decision making and the avoidance of cognitive demand. *Journal of Experimental Psychology: General, 139*, 665–682.

Krakauer, J. W., Ghazanfar, A. A., Gomez-Marin, A., MacIver, M. A., & Poeppel, D. (2017). Neuroscience needs behavior: Correcting a reductionist bias. *Neuron, 93*(3), 480–490.

Kuhn, T. S. (2012). *The structure of scientific revolutions* (4th ed.). Chicago, IL: University of Chicago Press.

Kurzban, R., Duckworth, A., Kable, J. W., & Myers, J. (2013). An opportunity cost model of subjective effort and task performance. *Behavioral and Brain Sciences, 36*(06), 661–679.

Lang, A. (2000). The limited capacity model of mediated message processing. *Journal of Communication, 50*(1), 46–70.

Lang, A. (2006). Using the limited capacity model of motivated mediated message processing to design effective cancer communication messages. *Journal of Communication, 56*, S57–S80.

Lang, A. (2009). The limited capacity model of motivated mediated message processing. In R. Nabi & M. B. Oliver (Eds.), *The Sage handbook of media processes and effects* (pp. 193–204). Thousand Oaks, CA: Sage.

Lang, A. (2014). Dynamic human-centered communication systems theory. *The Information Society, 30*(1), 60–70.

Lang, A. (2017). Limited capacity model of motivated mediated message processing (LC4MP). In P. Rössler (Ed.), *The international encyclopedia of media effects*. Hoboken, NJ: John Wiley & Sons.

Lavie, N., Hirst, A., de Fockert, J. W., & Viding, E. (2004). Load theory of selective attention and cognitive control. *Journal of Experimental Psychology: General, 133*, 339–354.

Madsen, J. K., Bailey, R., Carrella, E., & Koralus, P. (2019). Analytic versus computational cognitive models: Agent-based modeling as a tool in cognitive sciences. *Current Directions in Psychological Science, 28*(3), 299–305.

Marois, R., & Ivanoff, J. (2005). Capacity limits of information processing in the brain. *Trends in Cognitive Sciences, 9*(6), 296–305.

Meehl, P. E. (1978). Theoretical risks and tabular asterisks: Sir Karl, Sir Ronald, and the slow progress of soft psychology. *Journal of Consulting and Clinical Psychology, 46*, 806–834.

Meehl, P. E. (1990). Appraising and amending theories: The strategy of Lakatosian defense and two principles that warrant it. *Psychological Inquiry, 1*(2), 108–141.

Newman, M. (2010). *Networks: An introduction*. Oxford, England: Oxford University Press.

Pessoa, L. (2009). How do emotion and motivation direct executive control? *Trends in Cognitive Sciences, 13*(4), 160–166.

Popper, K. (1985). The problem of demarcation. In D. Miller (Ed.), *Popper selections* (pp. 118–130). Princeton, NJ: Princeton University Press.

Sanbonmatsu, D. M., & Johnston, W. A. (2019). Redefining science: The impact of complexity on theory development in social and behavioral research. *Perspectives on Psychological Science, 14*(4), 672–690.

Schramm, W. (1955). Information theory and mass communication. *Journalism Quarterly, 32*(2), 131–146.

Shannon, W. (1948). A mathematical theory of communication. *Bell System Technical Journal, 27*(3), 379–423.

Shenhav, A., Botvinick, M. M., & Cohen, J. D. (2013). The expected value of control: An integrative theory of anterior cingulate cortex function. *Neuron, 79*(2), 217–240.

Sherry, J. L. (2014). Media effects, communication, and complexity science insights on games for learning. In F. C. Blumberg (Ed.), *Learning by playing* (pp. 104–120). Oxford, England: Oxford University Press.

Sherry, J. L. (2015). The complexity paradigm for studying human communication: A summary and integration of two fields. *Review of Communication Research, 1*, 22–65.

Strogatz, S. (2004). *Sync: The emerging science of spontaneous order*. London, England: Penguin.

Strogatz, S. (2014). *Nonlinear dynamics and chaos*. Philadelphia, PA: Taylor & Francis.

Sweller, J. (1988). Cognitive load during problem solving: Effects on learning. *Cognitive Science, 12*, 257–285.

Watts, D. J. (2017). Should social science be more solution-oriented? *Nature Human Behaviour, 1*(1), 0015.

Weber, R., Mathiak, K., & Sherry, J. L. (2008). The neurophysiological perspective in mass communication research. In M. Beatty, J. McCroskey, & K. Floyd (Eds.), *Biological dimensions of communication: Perspectives, methods, and research* (pp. 43–73). New York, NY: Hampton Press.

Weber, R., Tamborini, R., Westcott-Baker, A., & Kantor, B. (2009). Theorizing flow and media enjoyment as cognitive synchronization of attentional and reward networks. *Communication Theory, 19*(4), 397–422.

Westbrook, A., & Braver, T. S. (2015). Cognitive effort: A neuroeconomic approach. *Cognitive, Affective, & Behavioral Neuroscience, 15*, 395–415.

Westbrook, A., Kester, D., & Braver, T. S. (2013). What is the subjective cost of cognitive effort? Load, trait, and aging effects revealed by economic preference. *PLoS One, 8*(7), e68210.

Zénon, A., Solopchuk, O., & Pezzulo, G. (2019). An information-theoretic perspective on the costs of cognition. *Neuropsychologia, 123*, 5–18.

31

Bodies and Minds in Sync

Forms and Functions of Interpersonal Synchrony in Human Interaction

Gary Bente and Eric Novotny

Humans are fundamentally social. We live in groups and our survival and life achievements very much depend on our ability to communicate and coordinate our actions with others. A very basic constituent of this ability is the evolutionary deep-rooted tendency to attune our movements to our conspecifics. Whenever two or more humans show some kind of repetitive motion, and they can perceive each other, they are prone to engage in a shared rhythm: they synchronize. Interpersonal synchrony (IPS) occurs for people walking together, sitting in rocking chairs, rowing boats, cheering their teams, and, more subtly, in our everyday conversations. Not only in a phylogenetic but also in an ontogenetic perspective, IPS appears to be foundational for bonding and social learning. IPS can be observed in early mother-infant interactions and has been posited as a mechanism through which infants develop language capabilities (Delaherche et al., 2012; Feldman, 2007) and self-control mechanisms (Feldman, Greenbou, & Yirmiya, 1999).

IPS is an emergent phenomenon and a widely automatic, uncontrolled and unconscious process comprising the mutual entrainment of at least two individuals. In this sense, we can conceive IPS as a core mechanism of group formation, action coordination, and collaboration. It has been shown that simple rhythmic alignment of body movements is closely tied to cooperativeness and perceived group entitativity (Lakens & Stel, 2011; Wiltermuth & Heath, 2009), relational quality and rapport (Ramseyer & Tschacher, 2011), and a variety of other social outcomes (cf. Rennung & Göritz, 2016). Cultural practices as present in religion, sports, military, arts, etc. make use of this seemingly hardwired connection between synchrony and the sense of communality to strengthen the bonds between the group members (Noy, Levit-Binun, & Golland, 2015; Reddish, Fischer, & Bulbulia, 2013; Tunçgenç & Cohen, 2016).

Despite the growing attention synchrony has received across various disciplines in the last decade, the concept as applied to human communication remains somehow elusive. Varied definitions in social science and metaphorical uses of the term have led to some confusion and a watering-down of a powerful concept. For instance, defining synchrony as "the temporal linkage of the nonverbal behavior of two or more interacting individuals" (Won, Bailenson, Stathatos, & Dai, 2014, p. 390) leads to confusions with other concepts referring to movement coordination and postural similarity, such as imitation, mirroring, or mimicry. In contrast to these phenomena, synchrony is an emergent interpersonal phenomenon that does not require a leader and a follower (Strogatz 2003; Strogatz & Stewart, 1993). As Noy, Dekel, and Alon (2011) showed, joint movement improvisation without designated leaders and followers generates even better results in temporal alignment and less jitter (i.e., corrective micro-movements) than leader-follower scenarios (see also Fairhurst, Janata, & Keller, 2014; Konvalinka et al., 2014). To preserve the potential explanatory power of the concept, we suggest staying close to the definitions of synchrony as formulated in the natural sciences, physics and biology, and to examine how far the concept can be stretched to match the particularities of human communication without losing its boundaries and becoming fuzzy and meaningless.

This chapter conceives synchrony as a ubiquitous "force of nature" and approaches the special case of interpersonal synchrony (IPS) as an essential component in human communication and collaboration. We will start with a line-up of the major constituents of synchrony as they appear in physics and biology and ask how these can be used to understand IPS in humans. This implies the questions of why higher-developed beings show behavioral synchrony and how this might be functional for individual and group survival. We then turn to the distinction of various types and instances of synchrony and scientific ways to look at these. The focus here is on those features relevant to communication research and not on completeness as a research review. Excellent overviews and meta-analyses can be found in Chetouani, Delaherche, Dumas, and Cohen (2017), as well as in Rennung and Göritz (2016). We follow this by discussing the effects of synchrony, including positive as well as potentially negative outcomes. Finally, our concluding sections examine methodological challenges and novel research perspectives that have emerged from new technologies.

PRINCIPLES: WHAT METRONOMES, FIREFLIES, AND HUMANS HAVE IN COMMON

Synchrony is a ubiquitous phenomenon, which can be observed in the inanimate and animated nature. It occurs when two or more oscillators are weakly coupled. An oscillator is any system that shows periodic behavior. Of particular interest in our case are oscillators with an inherent energy source, which controls frequency and amplitude of the periodic behavior. This is true for some manufactured devices, such as metronomes or clockwork pendulums, but most importantly for all biological systems with some kind of pacemaker that drives periodic behaviors, such as our heartbeat, breathing activity, hormone cycles, walking steps, and more (Strogatz, 2003; Strogatz & Stewart, 1993). It is important to note that synchrony is not identical with simultaneity. Oscillating systems can synchronize in-phase or anti-phase, or with a certain phase shift. This is true for metronomes or the flashing of fireflies, but also for intentional, goal-directed synchrony in human behavior. For instance, in-phase synchrony can enable the accumulation of forces in the service of maximal impact at a point in time (e.g., pulling a rope), whereas anti-phase synchrony can enable the iterative application of forces in service of a continuous impact (e.g., moving a two-handed saw back and forth). In both cases, the behaviors of the two agents are phase-locked, and both might be equally functional. In-phase synchrony appears to be the more stable state though, which is more difficult to influence by perturbation (Schmidt, Carello, & Turvey, 1990).

Weak coupling means that two or more oscillating systems are connected in a way that energy, respectively information, can travel between the systems. The term "weak coupling" distinguishes a synchronized system of independent oscillators from two components that are forced into mutual rhythm, such as the rotation of two wheels that share an axis (Pikovsky & Rosenblum, 2007). For instance, placing two metronomes on a moving board that can pick up and transmit the kinetic energy of the metronomes will lead to synchronized movements (see www.youtube.com/watch?v= yysnkY4WHyM). A prominent example of coupled oscillators in living organisms is the unison flashing of fireflies (see www.youtube.com/watch?v=ZGvtnE1Wy6U). This requires only a light sensor with limited range and a neural connection between the visual input and the pacemaker's output to synchronize the flashing of the neighboring individuals. Synchronization then can spread over a large population without each individual necessarily perceiving each other. As Strogatz and Stewart (1993) hold: "Synchronization emerges cooperatively. If a few oscillators happen to synchronize, their combined, coherent signal rises above the background din, exerting a stronger effect on the others" (p. 107). Importantly, in contrast to the metronomes, the coupling between biological systems, as well as between more complex machines, such as, for instance, robots, relies on sensing and some kind of processing. To understand synchrony in human interactions, we have to take into account that beyond mere exposure of the periodic behavior to a perceptual channel, some attention must be devoted to this channel and the transmitted signal (Koban, Ramamoorthy, & Konvalinka, 2019). Moreover, entrainment might depend on cognitive representations that relate behavioral synchrony to specific situational contexts, task demands, social norms, and values.

Whereas simple motor coordination exists in many social species, including humans (Nagasaka, Zenas, Hasegawa, & Fujii, 2013), we might expect synchrony at higher behavioral complexity levels to occur only in humans. Hence, it might be important to distinguish different types or instances of synchrony within the human species. Koban et al. (2019), for instance, differentiate two types of IPS, one being unintentional or spontaneous and the other being intentional or instructed/induced (see also Keller, Novembre, & Hove, 2014). A more specific example of spontaneous synchrony is the entrainment of body sway in conversations (Shockley, Richardson, & Dale, 2009). With a few exceptions though, behaviors that tend to show spontaneous synchrony, such as marching in stride, clapping in concert, bending down in prayers, or the unison movements of side-by-side rocking chairs, can also be induced by social rules or instructed by an experimenter. Whether these different instantiations are processed in our brains in similar or distinct ways, and whether they produce differential socio-emotional effects, are still open research questions. Another differentiation, which refers to different types of entrainment that entails interpersonal synchrony, has been suggested by Cacioppo et al. (2014). The first type, "unilateral entrainment," involves one person becoming entrained to the behavior of another in the sense of a leader-follower relationship. The second type, "orchestral entrainment," involves actors entraining their behavior to an external pacing driver, such as a ticking clock or metronome (e.g., Hove & Risen, 2009). The third type, "reciprocal entrainment," involves two actors mutually adapting to each other's behavior, such that the person acting as the referent continuously switches during the interaction (Noy et al., 2011; Oullier, De Guzman, Jantzen, Lagarde, & Scott Kelso, 2008). Referring to the basic definitions above, one might question whether the term synchrony in interpersonal contexts should be preserved for the emergent phenomenon described as type 3 while excluding types 1 and 2. In fact, we still know too little about the underlying neural processes that might group these phenomena together or afford a clear separation. It is important to note though, that these types of entrainment are not mutually exclusive. It might well be that IPS in a motor task, which might also occur spontaneously, is just supported by an orchestration. This is putatively the case in songs that once were sung during harvests. In this example, the visually perceived motor behavior is overlaid by a rhythm in the auditory channel that can help stabilize IPS as it makes it independent from visual attention demanded by the task. We might also find examples of leader-supported IPS in cultural practices where a central figure serves as pacemaker, as seen, for instance, in the role of the imam during prayers in a mosque. Most likely, the worshiping community would also synchronize in their bending movements spontaneously, but by facing the same direction and avoiding visual distraction, the imam makes it easier to stay in sync.

EXPLANATIONS: WHY WE SYNCHRONIZE

It is a plausible assumption that the urge to synchronize has evolved in humans because synchrony is adaptive, enabling coordination and collaboration in social groups and bringing about synergies in joint goal achievement (Sebanz, Bekkering, & Knoblich, 2006; Valdesolo, Ouyang, & DeSteno, 2010). In this sense, synchrony is seen as primarily motivated by external rewards, which putatively resonate in the activation of internal neural reward mechanisms (Kokal, Engel, Kirschner, & Keysers, 2011). But why then do we spontaneously attune our movements to the rhythms of others when no task at hand would benefit from synchrony and joint effort? This is the case in many cultural practices, such as in ritual dance or rhythmic praying movements, and we find it in many examples of mundane life. Consider two contrasting examples of types of synchrony that are evidently functional and "non"-functional. In one, two people are sitting in a rowboat, each holding one oar at opposite sides of the boat. Optimal speed, balance, straight movement and least effort will be achieved if they synchronize their strokes. Not only does such synchrony lead to higher efficiency on the dyad level, it also has rewarding effects on the individual level. As Cohen, Ejsmond-Frey, Knight, and Dunbar (2010) showed, synchronized rowing training, compared with a training regimen carried out alone, creates a heightened endorphin surge in the brain. Now imagine a second example of people sitting side-by-side in rocking chairs on a porch. They will also synchronize their rhythmic motion after a while, although there is no specific task that might benefit from synchrony (Richardson, Garcia, Frank, Gregor, & Marsh, 2012). Interestingly though, it has been shown that the social effects of synchronized rocking generalized to upcoming group tasks, as it "enhanced individuals' perceptual sensitivity to the motion of other entities and thereby increased their success in a subsequent joint-action, that required the ability to dynamically detect and respond appropriately to a partner's movements" (Valdesolo, Ouyang, & DeSteno, 2010, p. 693).

The reason why synchrony occurs in the absence of specific task demands can be conceptualized in various ways. On one hand, we might assume that, because of its adaptive function and the paramount importance of synchrony for collaboration, the mere perception of emergent synchrony has become intrinsically rewarding. We might further speculate that beyond the rewarding experience of being in sync, experienced synchrony might have diagnostic value. For instance, synchrony achieved in simple conversations might indicate the ease or difficulty in synchronizing with a respective partner later, when facing collaborative task demands. The feeling of getting along well, liking each other, or in other words experiencing rapport, might be the socio-emotional consequence of a successful behavioral "conversational synchrony game," which identifies the other as potential collaborator or even life partner when things become serious. Simply put, conversational synchrony might become a predictor for action coordination in future, sometimes already anticipated, collaborative tasks and consequently promote partner selection.

This assumption is supported by a series of studies directly addressing synchronizations of brain activities and physiological activation patterns. Neural synchronization has been found in various interaction settings including face-to-face communication, motor coordination, and joint decision-making tasks (Hasson, Ghazanfar, Galantucci, Garrod, & Keysers, 2012; King-Casas et al., 2005). Yun, Watanabe, and Shimojo (2012) showed "spontaneous bi-directional improvisation (implicit synchrony) increased motor synchrony" (p. 5) as evidenced by increased inter-brain connectivity of beta and theta bands. It has also been shown that IPS of facial expressions correlates with IPS in peripheral-physiological activation patterns in real-life communication as well as in virtual settings such as game interactions (Chanel, Kivikangas, & Ravaja, 2012; Sovijärvi-Spapé et al., 2013). Analyzing the coordinated activity of singers, Müller and Lindenberger (2011) concluded that "oscillatory coupling of cardiac and respiratory patterns provide a physiological basis for interpersonal action coordination" (p. 1). IPS thus can be conceived as a crucial mechanism to establish cognitive and motivational readiness states, reflected in the entrainment of brain activity and physiological arousal, which prepares the interactors for coordinated perception and action. This position has been recently articulated in the concept of bio-behavioral synchrony (Atzil & Gendron, 2017; Feldman, 2017), which, in our opinion, holds transformational potential for theories of communication and social behavior.

Overall, this view holds that IPS is primarily functional for social groups, because it contributes to the optimization of shared attention and motor coordination in group efforts, and is secondarily functional in conversations, as it possesses signaling functions to this end, which are subjectively perceived as group entitativity or rapport (Lakens, 2010; Lakens & Stel, 2011). Although this perspective provides an interesting framework for communication research in itself, it does not fully explain why spontaneous synchrony happens in the absence of specific motor tasks or communication efforts and which basic biological mechanisms drive its occurrence. Koban et al. (2019) suggest an alternative explanation, which emphasizes the principle of energy optimization on the individual brain level instead of the optimization of motor energy on the group level. The authors surmise "the phenomenon of interpersonal motor synchronization occurs due to the tendency of brains to conserve computational resources (Laughlin & Sejnowski, 2003), thereby resulting in nearness of self and other representations" (p. 13). In this view, the somehow "merged" cognitive representation of self and other is supposed to reduce complexity of the social environment and promote less effortful predictions of the conspecific's behavior. It is important to note that approaches focusing on the optimization of motor resources on the group level and those emphasizing the optimization of cognitive resources on the individual brain level are not mutually exclusive: both have the potential to spawn creative research in future communication research.

EFFECTS: HOW SYNCHRONY AFFECTS SOCIAL AND INDIVIDUAL OUTCOMES

IPS has been studied mostly as an independent variable, with questions examining the effects of synchrony on a variety of outcome variables (e.g., Hove & Risen, 2009; Wiltermuth & Heath, 2009). Less frequently, IPS has been conceptualized as a dependent variable, (e.g., Miles, Lumsden, Richardson, & Macrae, 2011; Paxton & Dale, 2013). Also, most of the research has focused on social effects, whereas outcomes of synchrony for the individual have received much less attention (cf. Vicaria & Dickens, 2016). For instance, individual outcomes of IPS have been documented regarding enhancement of self-esteem (Lumsden, Miles, & Macrae, 2014) or improved memory for social information (Macrae, Duffy, Miles, & Lawrence, 2008). These effects were related to rewarding experiences of successful communication and to cognitive optimization resulting from merged other-self representation, respectively. Although established on the individual level, these variables clearly tie in to social functioning in more general terms.

Social outcomes of IPS in fact have been framed in multiple ways referring to constructs such as group entitativity, rapport, affiliation, cooperation, trust, and others. There is insufficient space in this chapter to provide a comprehensive overview of the various research lines. Instead, we aim to sketch some of the major results to illustrate their common thread. As has been shown, the mere observation of people in sync causes the impression of group entitativity, or the perception that individuals constitute a social unit (Lakens, 2010). Using animated stick-figures or real people hand waving, this research found that only synchrony caused the impression of group entitativity, while other variables such as physical similarity did not. In follow-up studies, perceived entitativity led to the impression of relational quality in terms of rapport. These effects, however, were only evident for dyads that supposedly showed spontaneously emerging synchrony (mutual entrainment) and not for induced synchrony of the leader-follower type (Lakens & Stel, 2011). Direct associations between synchrony and perceived rapport have been found repeatedly (Bernieri, 1988; Vacharkulksemsuk & Fredrickson, 2012). Miles, Nind, and Macrae (2009) exposed participant observers to pairs of walking stick-figures that portrayed varying configurations of coordinated movement. These stimuli were manipulated such that the phase relationship between their strides varied from 0° to 360°. The authors discovered that the most stable phase relationship, or 0° difference (i.e., in-phase synchrony) led to the highest perceptions of rapport. This effect was duplicated in a second study where only the walkers' audio cues were provided to participants. Closely related to these results are findings regarding the influence of IPS on affiliation and liking. In a series of studies, Hove and Risen (2009) demonstrated that synchronized motor activity, in his case, tapping to a metronome alongside a confederate, increased affiliation and liking. Again, this effect only held when synchrony was interpersonal in nature and not orchestrated (attributed to the metronome) or led.

Other studies have more directly investigated the behavioral outcomes of IPS. For instance, Launay, Dean, and Bailes (2013) found that participants instructed to tap synchronously as compared to asynchronously with a beat given through headphones contributed more money to their partner in a subsequent economic game. In the same sense, Cirelli, Einarson, and Trainor (2014) found that infants who were bounced by an adult synchronously (versus asynchronously) to music were more likely to subsequently help an assistant when she dropped objects on the floor. This effect was shown even when the movements were in anti-phase synchrony, suggesting that even a less stable form of synchrony can have prosocial effects. A number of studies also established a connection between IPS and collaboration, including cooperativeness (Cohen, Mundry, & Kirschner, 2013; Reddish, Fischer, & Bulbulia, 2013; Wiltermuth & Heath, 2009). Wiltermuth and Heath (2009) conducted a series of studies in which participant triads synchronized their steps while walking, or their voices and hand movements while singing and swinging a cup. Across these studies, groups who synchronized (whether it was movements, voices, or both) were more likely to choose the "best for the group" option in economic games than asynchronous triads. Reddish et al. (2013) found that the effect of synchrony on cooperation is magnified when there is shared intentionality among interactants, as opposed to incidental synchrony or intentional *a*synchrony (another form of coordination). IPS not only increases the motivation to cooperate, but it also increases one's ability to cooperate (Valdesolo, Ouyang, & DeSteno, 2010), suggesting a preparatory mechanism for later collaboration as mentioned in our introduction.

Across these results covering a variety of psychological constructs, the common thread is that IPS promotes group formation and strengthens the social bonds within. Yet, where there is an ingroup there is also an outgroup, and it is an interesting question whether synchrony can be observed with outgroup members at all, and how ingroup synchrony relates to feelings and behaviors toward the respective outgroup. Using a minimal group paradigm, Miles, Lumsden, Richardson, and Macrae (2011) found, counterintuitively, that synchronization worked even better with designated outgroup members. They attributed this result to attentional factors as well as the motivation to bridge intergroup differences. In line with this, Tunçgenç and Cohen (2016) showed that biases toward outgroup members can be mitigated by behavioral synchrony. A more complex and potentially negative relation between group membership and social outcomes of IPS, however, emerged in one of our own studies (Tamborini et al. 2018). Using a dance videogame, we asked participants to synchronize dance moves with a black or white in-game character. We measured the influence of group membership and synchrony on trust toward other ingroup and outgroup members using a standard online trust game (cf. Bente, Baptist, & Leuschner, 2012). Results suggested that synchrony with an ingroup member can decrease general trust toward outgroup members. Further, it has been shown that synchrony with an authority figure can promote destructive obedience toward other living things, such as killing insects on command (Wiltermuth, 2012). Thus, it appears that synchrony can be socially beneficial as well as detrimental, depending on the salience of the respective reference group.

METHODS: HOW TO ASSESS IPS IN NATURAL INTERACTIONS

A major challenge in identifying signatures of IPS lies in the measurement and statistical modeling of the mutual interdependencies and temporal dynamics in free-running human interactions, such as conversations. Human interactions unfold as a continuous, dynamic, and multilevel phenomenon among dyads or groups (Poyatos, 1983). Such interactions are characterized by (1) high dimensional complexity, and (2) high processual complexity (Vogeley & Bente, 2010). Dimensional complexity relates to the fact that various channels, such as facial expressions, gaze, gestures, body postures, and movements show multiple interactions within and across the communicators. This includes idiosyncratic channel preferences and different levels of information density in the nonverbal signals (Chovil, 1991). Processual complexity implies that meaningful information is coded in dynamic aspects of the nonverbal behavior, as for instance, in response delays, recurrence patterns, and movement durations (Krumhuber, Manstead, & Kappas, 2007; Provost, Troje, & Quinsey, 2008). This is particularly important when looking at emergent dyadic interaction patterns such as IPS.

Previous approaches to understanding the dynamics of IPS in natural human interactions are limited in their explanatory value, as they all start with complexity reduction at the level of data collection, either taking a restrictive or a generic measurement perspective on the interaction behavior. Restrictive approaches reduce the complex interaction behavior to specific and easily measurable rhythmic movement phenomena with a characteristic form. In fact, the majority of studies on IPS have focused on simple repetitive movements such as marching in stride (Wiltermuth & Heath, 2009), waving (Lakens, 2010), swinging a pendulum (Schmidt & O'Brien, 1997), rocking chairs (Richardson, Marsh, Isenhower, Goodman, & Schmidt, 2007), body sway (Reynolds & Osler, 2014), or jointly moving sliders in the so-called mirror game (Noy et al., 2011). Schmidt, Morr, Fitzpatrick, and Richardson (2012) summarize:

> The disadvantage of these studies is that the interactions between participants were not very natural. All the tasks involved having the subjects produce stereotyped rhythmic movements, which are obviously not present in everyday interactions. Moreover, the social interactions prescribed in the tasks were artificial.
>
> (p. 268)

As such, the existing approaches are often a pale resemblance of the truly social phenomenon they were designed to study.

Whereas restrictive approaches select one behavioral aspect to study IPS, generic approaches are based on aggregates of communication behavior, for instance, representing the multichannel nonverbal interaction behavior as a single activity vector. An example of this approach is Motion Energy Analysis (MEA), which has been suggested as a possibility to track synchrony in free-running conversations (Ramseyer & Tschacher, 2011). In MEA, the movement activity of interaction partners is automatically extracted from video recordings by essentially quantifying pixel changes between pre-filtered sequential video frames. Although an updated approach includes a method to separate body and head movement within MEA (Ramseyer & Tschacher, 2014), nonverbal behavior remains constrained to global activity measures that miss out on postural dynamics, specific gestures, and variations in distance and orientation. Furthermore, MEA is based on 2D data and ignores depth information, a key reason why this approach has been limited to seating dyads.

As a powerful alternative to restrictive as well as generic approaches in communication research, we have suggested motion capture technology (for a comprehensive overview, see Bente, 2019), which generates detailed protocols of body movement consisting of rotation and translation information for all joints. Although the resulting data protocols can be aggregated to produce a one-dimensional activity time-series as does MEA, they also allow for a micro-analysis of the various nonverbal subsystems such as gestures, head and body movements, body orientation, locomotion, etc., and their interpersonal crosstalk. Accurate motion capture systems (e.g., OptiTrack) are still dependent on markers and/or special cameras, which restricts their application to experimental studies in the lab. Current progress in motion analysis based on video, using deep machine learning, can be expected to help overcome these constraints in the near future (Neverova, 2016).

It is important to note that capturing the processual and dimensional complexity of human communication is a necessary but not a sufficient precondition for the study of IPS. A second important requirement concerns the analytical methods to detect entrainment patterns and to quantify the level of synchrony, which is particularly challenging when studying natural interactions. Various solutions have been suggested, such as cross-lagged correlation analysis (Ramseyer & Tschacher, 2014), cross-recurrence analysis (Coco & Dale, 2014), cross-spectral analysis (Schmidt et al., 2012) and wavelets (Fujiwara & Daibo, 2016), to name a few. Cross-recurrence quantification (CRQ), for instance, has been used to study coupled postural sway patterns (Shockley, 2005). CRQ extends the general idea of recurrence analysis where signals are embedded with their own time-shifted versions, by embedding the motions of one person with those of another at some time shift. Because recurrence methods are comprehensive and flexible with regards to assumptions, Marwan, Romano, Thiel, and Kurths (2007) argue that these methods can be considered as a generalized form of the linear cross-correlation function. However, at least in part due to the higher mathematical complexity compared to MEA, for example, their application in practice remains rare. In addition to time-domain methods (MEA, cross-lagged correlation, CRQ), researchers have also adapted frequency-domain methods (e.g., Pikovsky et al., 2003). Frequency-domain methods have been applied mainly to simple motor trajectories such as swinging a pendulum and moving a rocking chair (e.g., Richardson et al., 2007; Schmidt & O'Brien, 1997). To quantify IPS effects, movement data are transformed using Fourier analysis, or using the more potent wavelet approach. Fujiwara and Daibo (2016) applied cross-wavelet coherence analysis to movement data, extracted from video recordings of dyadic conversations. Movement protocols of the real pairings were compared to random combinations of individual protocols in a so-called pseudo-synchrony paradigm. They found higher values of cross-wavelet coherence in the real than in the pseudo dyads, specifically in the frequency band under 0.5 Hz. Although the modeling approach appears promising, the input data is rather crude, using an aggregate of movement activity based on the spatial differences of extracted nose and hand positions and ignoring the potential crosstalk between different nonverbal subsystems. The different approaches show specific advantages and downsides. As such, it is important that future approaches provide comparative data on their predictive value regarding the occurrence and various social outcome of IPS.

OUTLOOK: WHERE TO GO FROM HERE

We aimed to show that IPS constitutes a crucial element in human interaction and development, and that it creates a basis for the synchronization of minds and joint action. Numerous studies point to the close relation between behavioral and neurophysiological synchrony and their effects on individual and social outcomes. Causality between social outcomes and IPS evidently goes both ways. Yet, investigations into IPS as a dependent variable are scarce, and it might be particularly interesting to develop paradigms to explore social and individual factors that promote or hinder IPS in human interactions. We made a case for studying IPS as a dynamic, emergent process, encouraging precise definition and measurement, based on the concept of coupled oscillators. This implies the assessment of repetitive patterns in overt motion or covert neurophysiological processes. Although there are similarities between metronomes, fireflies, and humans in this regard, our perspective notes key differences among these categories. Metronomes are not living creatures, whereas fireflies are—and though fireflies and humans are both living, humans possess the additional ability to attend to and perceive (rather than just sense) informational changes in the environment. These comparisons could be conceived as continuous, from purely mechanical (or “off-line”) synchrony to a combination of physical and perceptual aspects (“on-line” synchrony). We see this continuum as useful in defining the type of synchrony upon which researchers focus. For example, physicists and psychologists likely have different aims when examining properties of synchronized behaviors and should conceptualize their version of synchrony accordingly.

We also see a particular challenge in distinguishing more clearly synchrony from other similar coordination behaviors such as mimicry or imitation. Some work (i.e., Hove & Risen, 2009; Noy et al., 2011) already shows that synchrony has stronger effects than mimicry, though more work is needed to disentangle the mechanisms leading to these effects. If mimicry and synchrony both involve shared representations of self and other, and they both impact social outcomes, in what ways are they different? Approaches to answer this question could strongly benefit from a combination of behavioral and neurophysiological methods looking at interpersonal synchrony in EEG patterns as well as distinctive activation patterns by means of brain imaging. Most important, the study of IPS in natural human interactions will have to overcome different methodological challenges, which include the multidimensional assessment of communication behavior, the identification of interrelated channels, quantification in terms of frequencies, amplitudes and phase shifts, and the establishment of robust interrelations between IPS and relevant social outcomes.

Last but not least, we see a major challenge for communication research in identifying periodic phenomena and their interpersonal entrainment within ongoing social interactions. Conversations, discussions or negotiations are not steady processes. They are progressive, implying changes in topics and relational quality over time. Depending on these changes, one can expect dyads to repeatedly fall into sync and out of sync over the course of one interaction or at different stages of a relationship. Moreover, the frequency of the entrained behaviors can change, and phase transitions between in-phase and anti-phase synchrony, or changes in the phase shift, can occur. For instance, in an early stage of relationships, high levels of motor synchrony might be more important to facilitate bonding, promote trust, and foster self-disclosure. In later stages, IPS might be only necessary to overcome conflicts and re-adjust the relationship balance. In fact, we know very little about these temporal dynamics, and their study represents a particular theoretical and methodological challenge for future communication research.

REFERENCES

Atzil, S., & Gendron, M. (2017). Bio-behavioral synchrony promotes the development of conceptualized emotions. *Current Opinions in Psychology, 17*, 162–169.

Bente, G. (2019). New tools—new insights: Using emergent technologies in nonverbal communication research. In S. W. Wilson & S. W. Smith (Eds.), *Reflections on interpersonal communication* (pp. 161–188). San Diego, CA: Cognella.

Bente, G., Baptist, O., & Leuschner, H. (2012). To buy or not to buy: Influence of seller photos and reputation on buyer trust and purchase behavior. *International Journal of Human-Computer Studies, 70*(1), 1–13.

Bernieri, F. J. (1988). Coordinated movement and rapport in teacher-student interactions. *Journal of Nonverbal Behavior, 12*(2), 120–138.

Cacioppo, S., Zhou, H., Monteleone, G., Majka, E. A., Quinn, K. A., Ball, A. B., ... Cacioppo, J. T. (2014). You are in sync with me: Neural correlates of interpersonal synchrony with a partner. *Neuroscience, 277*, 842–858.

Chanel, G., Kivikangas, J. M., & Ravaja, N. (2012). Physiological compliance for social gaming analysis: Cooperative versus competitive play. *Interacting with Computers, 24*, 306–316.

Chetouani, M., Delaherche, E., Dumas, G., & Cohen, D. (2017). Interpersonal synchrony: From social perception to social interaction. In J. K. Burgoon, N. Magnenat-Thalmann, M. Pantic, & A. Vinciarelli (Eds.), *Social signal processing* (pp. 202–212). Cambridge, England: Cambridge University Press.

Chovil, N. (1991). Social determinants of facial displays. *Journal of Nonverbal Behavior, 15*(3), 141–154.

Cirelli, L. K., Einarson, K. M., & Trainor, L. J. (2014). Interpersonal synchrony increases prosocial behavior in infants. *Developmental Science, 17*(6), 1003–1011.

Coco, M. I., & Dale, R. (2014). Cross-recurrence quantification analysis of categorical and continuous time series: An R package. *Frontiers in Psychology, 5*, 510.

Cohen, E., Ejsmond-Frey, R., Knight, N., & Dunbar, R. (2010). Rowers' high: Elevated endorphin release under conditions of active behavioural synchrony. *Biology Letters, 6*(1), 106–108.

Cohen, E., Mundry, R., & Kirschner, S. (2014). Religion, synchrony, and cooperation. *Religion, Brain & Behavior, 4*(1), 20–30.

Delaherche, E., Chetouani, M., Mahdhaoui, A., Saint-Georges, C., Viaux, S., & Cohen, D. (2012). Interpersonal synchrony: A survey of evaluation methods across disciplines. *IEEE Transactions on Affective Computing, 3*(3), 349–365.

Fairhurst, M. T., Janata, P., & Keller, P. E. (2014). Leading the follower: An fMRI investigation of dynamic cooperativity and leader-follower strategies in synchronization with an adaptive virtual partner. *Neuroimage, 84*, 688–697.

Feldman, R. (2007). Parent-infant synchrony: Biological foundations and developmental outcomes. *Current Directions in Psychological Science, 16*(6), 340–345.

Feldman, R. (2017). The neurobiology of human attachments. *Trends in Cognitive Science, 21*(2), 80–99.

Feldman, R., Greenbaum, C. W., & Yirmiya, N. (1999). Mother-infant affect synchrony as an antecedent of the emergence of self-control. *Developmental Psychology, 35*(1), 223–231.

Fujiwara, K., & Daibo, I. (2016). Evaluating interpersonal synchrony: Wavelet transform toward an unstructured conversation. *Frontiers in Psychology, 7*, 516.

Hasson, U., Ghazanfar, A. A., Galantucci, B., Garrod, S., & Keysers, C. (2012). Brain-to-brain coupling: A mechanism for creating and sharing a social world. *Trends in Cognitive Sciences, 16*(2), 114–121.

Hove, M. J., & Risen, J. L. (2009). It's all in the timing: Interpersonal synchrony increases affiliation. *Social Cognition, 27*(6), 949–960.

Keller, P. E., Novembre, G., & Hove, M. J. (2014). Rhythm in joint action: Psychological and neurophysiological mechanisms for real-time interpersonal coordination. *Philosophical Transactions of the Royal Society B: Biological Sciences, 369*, 20130394.

King-Casas, B., Tomlin, D., Anen, C., Camerer, C. F., Quartz, S. R., & Montague, P. R. (2005). Getting to know you: Reputation and trust in a two-person economic exchange. *Science, 308*(5718), 78–83.

Koban, L., Ramamoorthy, A., & Konvalinka, I. (2019). Why do we fall into sync with others? Interpersonal synchronization and the brain's optimization principle. *Social Neuroscience, 14*(1), 1–9.

Kokal, I., Engel, A., Kirschner, S., & Keysers, C. (2011). Synchronized drumming enhances activity in the caudate and facilitates prosocial commitment-if the rhythm comes easily. *PLoS One, 6*, e27272.

Konvalinka, I., Bauer, M., Stahlhut, C., Hansen, L. K., Roepstorff, A., & Frith, C. D. (2014). Frontal alpha oscillations distinguish leaders from followers: Multivariate decoding of mutually interacting brains. *Neuroimage, 94*, 79–88.

Krumhuber, E., Manstead, A. S., & Kappas, A. (2007). Temporal aspects of facial displays in person and expression perception: The effects of smile dynamics, head-tilt, and gender. *Journal of Nonverbal Behavior, 31*(1), 39–56.

Lakens, D. (2010). Movement synchrony and perceived entitativity. *Journal of Experimental Social Psychology, 46*(5), 701–708.

Lakens, D., & Stel, M. (2011). If they move in sync, they must feel in sync: Movement synchrony leads to attributions of rapport and entitativity. *Social Cognition, 29*(1), 1–14.

Laughlin, S. B., & Sejnowski, T. J. (2003). Communication in neuronal networks. *Science, 301*(5641), 1870–1874.

Launay, J., Dean, R. T., & Bailes, F. (2013). Synchronization can influence trust following virtual interaction. *Experimental Psychology, 60*(1), 53–63.

Lumsden, J., Miles, L. K., & Macrae, C. N. (2014). Sync or sink? Interpersonal synchrony impacts self-esteem. *Frontiers in Psychology, 5*, 1064.

Macrae, C. N., Duffy, O. K., Miles, L. K., & Lawrence, J. (2008). A case of hand waving: Action synchrony and person perception. *Cognition, 109*(1), 152–156.

Marwan, N., Romano, M. C., Thiel, M., & Kurths, J. (2007). Recurrence plots for the analysis of complex systems. *Physics Reports, 438*(5–6), 237–329.

Miles, L. K., Lumsden, J., Richardson, M. J., & Macrae, C. N. (2011). Do birds of a feather move together? Group membership and behavioral synchrony. *Experimental Brain Research, 211*(3–4), 495–503.

Miles, L. K., Nind, L. K., & Macrae, C. N. (2009). The rhythm of rapport: Interpersonal synchrony and social perception. *Journal of Experimental Social Psychology, 45*(3), 585–589.

Müller, V., & Lindenberger, U. (2011). Cardiac and respiratory patterns synchronize between persons during choir singing. *PLoS One, 6*, e24893.

Nagasaka, Y., Chao, Z. C., Hasegawa, N., Notoya, T., & Fujii, N. (2013). Spontaneous synchronization of arm motion between Japanese macaques. *Scientific Reports, 3*, 1151.

Neverova. N. (2016). Deep learning for human motion analysis. Artificial Intelligence [cs.AI]. Université de Lyon. Retrieved from https://tel.archives-ouvertes.fr/tel-01470466/file/these.pdf

Noy, L., Dekel, E., & Alon, U. (2011). The mirror game as a paradigm for studying the dynamics of two people improvising motion together. *Proceedings of the National Academy of Sciences, 108*(52), 20947–20952.

Noy, L., Levit-Binun, N., & Golland, Y. (2015). Being in the zone: Physiological markers of togetherness in joint improvisation. *Frontiers in Human Neuroscience, 9*, 187.

Oullier, O., De Guzman, G. C., Jantzen, K. J., Lagarde, J., & Scott Kelso, J. A. (2008). Social coordination dynamics: Measuring human bonding. *Social Neuroscience, 3*(2), 178–192.

Paxton, A., & Dale, R. (2013). Argument disrupts interpersonal synchrony. *Quarterly Journal of Experimental Psychology, 66*(11), 2092–2102.

Pikovsky, A., & Rosenblum, M. (2007). *Synchronization*. Retrieved from www.scholarpedia.org/article/Synchronized.

Pikovsky, A., Rosenblum, M., & Kurths, J. (2003). *Synchronization: A universal concept in nonlinear sciences* (Vol. 12). Cambridge, England: Cambridge University Press.

Poyatos, F. (1983). Language and nonverbal systems in the structure of face-to-face interaction. *Language & Communication, 3*(2), 129–140.

Provost, M. P., Troje, N. F., & Quinsey, V. L. (2008). Short-term mating strategies and attraction to masculinity in point-light walkers. *Evolution and Human Behavior, 29*(1), 65–69.

Ramseyer, F., & Tschacher, W. (2011). Nonverbal synchrony in psychotherapy: Coordinated body movement reflects relationship quality and outcome. *Journal of Consulting and Clinical Psychology, 79*(3), 284–295.

Ramseyer, F., & Tschacher, W. (2014). Nonverbal synchrony of head- and body-movement in psychotherapy: Different signals have different associations with outcome. *Frontiers in Psychology, 5*, 979.

Reddish, P., Fischer, R., & Bulbulia, J. (2013). Let's dance together: Synchrony, shared intentionality and cooperation. *PLoS One, 8*(8), e71182.

Rennung, M., & Göritz, A. S. (2016). Prosocial consequences of interpersonal synchrony. *Zeitschrift für Psychologie, 224*(3), 168–189.

Reynolds, R. F., & Osler, C. J. (2014). Mechanisms of interpersonal sway synchrony and stability. *Journal of the Royal Society Interface, 11*(108), 20140751.

Richardson, M., Garcia, R. L., Frank, T. D., Gregor, M., & Marsh, K. L. (2012). Measuring group synchrony: A cluster-phase method for analyzing multivariate movement time-series. *Frontiers in Physiology, 3*, 405.

Richardson, M. J., Marsh, K. L., Isenhower, R. W., Goodman, J. R., & Schmidt, R. C. (2007). Rocking together: Dynamics of intentional and unintentional interpersonal coordination. *Human Movement Science, 26*, 867–891.

Schmidt, R. C., Carello, C., & Turvey, M. T. (1990). Phase transitions and critical fluctuations in the visual coordination of rhythmic movements between people. *Journal of Experimental Psychology: Human Perception and Performance, 16*(2), 227–247.

Schmidt, R. C., Morr, S., Fitzpatrick, P., & Richardson, M. J. (2012). Measuring the dynamics of interactional synchrony. *Journal of Nonverbal Behavior, 36*(4), 263–279.

Schmidt, R. C., & O'Brien, B. (1997). Evaluating the dynamics of unintended interpersonal coordination. *Ecological Psychology, 9*(3), 189–206.

Sebanz, N., Bekkering, H., & Knoblich, G. (2006). Joint action: Bodies and minds moving together. *Trends in Cognitive Sciences, 10*(2), 70–76.

Shockley, K. (2005). Cross recurrence quantification of interpersonal postural activity. In M. A. Riley & G. C. Van Orden (Eds.), *Tutorials in contemporary nonlinear methods for the behavioral sciences* (pp. 142–177). Retrieved from www.researchgate.net/profile/Joseph_Hamill/publication/51404288_Developmental_changes_in_the_dynamical_structure_of_postural_sway_during_a_precision_fitting_task/links/56eff9eb08ae52f8ad7f86e7.pdf#page=149.

Shockley, K., Richardson, D. C., & Dale, R. (2009). Conversation and coordinative structures. *Topics in Cognitive Science, 1*(2), 305–319.

Sovijärvi-Spapé, M. M., Kivikangas, J. M., Järvelä, S., Kosunen, J., Jacucci, G., & Ravaja, N. (2013). Keep your opponents close: Social context affects EEG and fEMG linkage in a turn-based computer game. *PLoS One, 8*(11), e78795.

Strogatz, S. (2003). *Sync*. New York, NY: Theia.

Strogatz, S. H., & Stewart, I. (1993). Coupled oscillators and biological synchronization. *Scientific American, 269*(6), 102–109.

Tamborini, R., Novotny, E., Prabhu, S., Hofer, M., Grall, C., Klebig, B., ... Bente, G. (2018). The effect of behavioral synchrony with black or white virtual agents on outgroup trust. *Computers in Human Behavior, 83*, 176–183.

Tunçgenç, B., & Cohen, E. (2016). Movement synchrony forges social bonds across group divides. *Frontiers in Psychology, 7*, 782.

Vacharkulksemsuk, T., & Fredrickson, B. L. (2012). Strangers in sync: Achieving embodied rapport through shared movements. *Journal of Experimental Social Psychology, 48*(1), 399–402.

Valdesolo, P., Ouyang, J., & DeSteno, D. (2010). The rhythm of joint action: Synchrony promotes cooperative ability. *Journal of Experimental Social Psychology, 46*(4), 693–695.

Vicaria, I. M., & Dickens, L. (2016). Meta-analyses of the intra- and interpersonal outcomes of interpersonal coordination. *Journal of Nonverbal Behavior, 40*(4), 335–361.

Vogeley, K., & Bente, G. (2010). "Artificial humans": Psychology and neuroscience perspectives on embodiment and nonverbal communication. *Neural Networks, 23*(8–9), 1077–1090.

Wiltermuth, S. (2012). Synchrony and destructive obedience. *Social Influence, 79*(2), 78–89.

Wiltermuth, S. S., & Heath, C. (2009). Synchrony and cooperation. *Psychological Science, 20*(1), 1–5.

Won, A. S., Bailenson, J. N., Stathatos, S. C., & Dai, W. (2014). Automatically detected nonverbal behavior predicts creativity in collaborating dyads. *Journal of Nonverbal Behavior, 38*(3), 389–408.

Yun, K., Watanabe, K., & Shimojo, S. (2012). Interpersonal body and neural synchronization as a marker of implicit social interaction. *Scientific Reports, 2*, 959.

32

Physically, Biologically, and Socially Constructed Notions of Sex and Gender in Communication Science

Chelsea A. Lonergan and Nicholas A. Palomares

Sex and gender are two of the most common, ubiquitous, and fundamental constructs of theory and research in the social sciences, including in the discipline of human communication. Much research includes sex and/or gender either as key demographic variables or as explanatory or moderating variables playing central roles in theoretical accounts for data. However, whereas sex and gender remain undeniably prevalent within research, they are shockingly untethered to detailed conceptualization, operationalization, and *a priori* theoretically-based hypotheses or research questions. In this chapter, we define sex and gender, including where they overlap and deviate. Then, we discuss and critique how sex and gender have been employed in research both broadly and more specifically within communication theory and research. Finally, we provide implications for future research and how researchers might adjust how they integrate the two constructs in their own research.

EXPLICATING SEX AND GENDER

One's sex generally refers to a biological attribute of a person, whereas one's gender refers to a social manifestation. Understanding the distinctions between sex and gender is a useful first step toward a better understanding of each construct. The distinction, however, is not always clear. Consistent with this conceptual murkiness, the terms "sex" and "gender" are often employed interchangeably in research (Canary & Hause, 1993). Moreover, a distinction between sex and biological sex is often blurred with the second term only clarifying the origin of the first one. To gain a more specific understanding of sex and gender, we first discriminate between *genotype* and *phenotype* (see, e.g., Tooby & Cosmides, 1992). Genotype refers to the set of genes an organism possesses, whereby genes are grouped in chromosomes or long chains of genes. The collection of all genotypes within an individual is referred to as the individual's genome. Distinct from genotype, which can be assessed only via specialized analytical tools and methods, phenotype refers to all characteristics of an individual that are visible with the "naked eye" (e.g., one's facial symmetry, level of introversion, or likelihood of buying a yacht). Having articulated some preliminary distinctions, we move on to explicating the concept of sex.

Sex

Biological sex encompasses all sex-relevant baseline genotypical variation that predicts phenotypical variation. Sex can therefore be understood as anatomically dictated, often binary distinctions into either male or female; this can be based off of genetic, physiological, anatomical, or reproductive characteristics (Ritz et al., 2014). Genetically, embryos inherit 23 pairs of chromosomes, or long chains of genes. One pair contains the "sex chromosomes" with either XX for female or XY for male. The Y chromosome, which is generally possessed only by males, is where the gene for 'maleness' exists (i.e., SRY gene; Griffin & Ellis, 2018). When an embryo possesses the Y chromosome, this gene is set off, beginning a complex and not entirely understood cascade of hormonal (e.g., testosterone and dihydrotestosterone; Manning, Fink, & Trivers, 2018) release that allows the fetus to develop male sex organs. Prenatal androgens (primarily testosterone) are secreted as part of this process and development (Halpern, 2013). When this gene is not turned on, usually in the case of females with XX sex chromosomes, the fetus phenotypically develops female sex organs at the 'absence' of the maleness hormonal switch and subsequent cascade. Often, biological sex is assessed by considering one's genitals and physical features such as body composition and facial structure, a process that is mostly automatic and doesn't generally require individuals' conscious cognitions and appraisals.

Although biological sex development can seem like a straightforward process—we automatically assess baseline phenotypical expressions to label someone as male or female—ambiguities exist both physically and genetically. As Ainsworth (2015) explains, "almost everyone is, to varying degrees, a patchwork of genetically distinct cells, some with a sex that might not match that of the rest of their body" (p. 288). Fetal sex development in utero is rather complex, and depending on how it is hormonally executed, one's genotype (biological sex) may predict ambiguous sex phenotypes (one's outward appearance). For example, if one possesses the Y chromosome (with the male SRY gene) but lacks testosterone receptors, then the initial process to develop as a male will be activated and male androgens will be secreted. Yet, such a process may be ineffective because the body will not respond to the secreted hormones and therefore go on to form female genitalia. This outcome would occur despite the presence of XY chromosomes. In this specific case, the lack of receptors for the released testosterone to latch onto would prevent the male hormones from communicating with the rest of the body, which therefore wouldn't 'know' to begin male development. Essentially the testosterone would 'fall on deaf ears', so to speak. To state in more technical terms, fetuses with a male (46, XY) karyotype but no functional androgen (i.e., testosterone) receptors may instead develop female sex organs; they may therefore develop as phenotypical women with complete androgen insensitivity syndrome (CAIS; Savic, Frisen, Manzouri, Nordenstrom, & Lindén Hirschberg, 2017). Strikingly, current research suggests that as many as 1 in 100 people have some form of these so-called disorders of sex development (DSDs; Ainsworth, 2015).

Gender

Conceptually distinct from sex, gender is a social construct of what it means to be a man, woman, or some other manifestation of the socially defined and germane set of attributes partly based on individuals' phenotypical expression and also on arbitrary attributes with little or no genotypical or phenotypical origins. That is, the conceptualization of gender focuses on the outward appearance and behavior of people, which varies for men and women in ways that reflect a system of symbols representing masculinity and femininity. Gender is connected to sex because it is predicted to an extent by genotype through outward appearance and/or behavior. The link between gender and genotype has been used to refer to the social aspects and processes associated with roles, relationships, behaviors, power, and anything else culturally and environmentally projected onto the definitions of what it means to be a woman or a man (Ritz et al., 2014). This often includes stereotypical appearances (e.g., women have soft feminine features and wider hips; men have harsher masculine features and broader shoulders) or social roles (e.g., women as nurturers; men as providers). Importantly, gender is often outwardly assessed based on surface-level features (phenotypes), although these phenotypes can be due to underlying genotype, environmental influences, or an interaction of the two. Of course, gender manifestation does not have to have biological origins (e.g., women have long hair, use more emotional language, and provide more effective social support; men have short hair, use less emotional language, and provide less effective social support).

Just as sex is not always contained entirely within the biological realm, neither is gender always contained within the social one. For example, testosterone levels can affect the pitch of one's voice, lowering it for males or increasing it for females, which would be a phenotype based somewhat on genotype (Gugatschka et al., 2010). However, environmental influences such as smoking or laryngeal damage can affect the pitch of one's voice, perhaps making one's gender somewhat more ambiguous (Re, O'Connor, Bennett, & Feinberg, 2012). Moreover, transgender women can actively change their pitch, intonation, volume, articulation and resonance in order to sound more female, with inconclusive findings regarding transgender men deepening their pitch to sound more male (Hancock, Krissinger, & Owen, 2011; Van Borsel, de Pot, & De Cuypere, 2009). In a similar example, even something as seemingly clear-cut and biologically based as bone density can be influenced by both biological and environmental inputs. Although perhaps initially determined via sex hormones and genetics, often-gendered social behaviors such as attire, occupation, and physical activity level can influence how much sun exposure (and subsequent Vitamin D synthesis) or weight-bearing levels one has—ultimately affecting one's bone density (Fausto-Sterling, 2005; Ritz et al., 2014).

Because gender and sex are often intertwined, with an overlap of the two almost 99% of the time (Ainsworth, 2015), they are conflated in a majority of conceptual, methodological, and colloquial contexts. Due to the growing body of evidence illuminating the complexities and non-linear nature of sex and gender, it is becoming increasingly clear that the "boxes" we have put them in may no longer contain them (Fausto-Sterling, 2005, 2012; Fine, 2017; Unger, 1979). Nonetheless, our aim is to explicate sex and gender in a way that highlights their unique features and origins, and shows how they are connected despite the need to keep them conceptually distinct.

RESEARCH ON SEX AND GENDER

A major challenge in reviewing the literature on sex, gender, and communication is that its trajectory has not been defined nor has it been linear. Whereas studies examining sex/gender in terms of communication exist, the programs of such researchers have not followed the theoretical and methodological steps often seen for research on other constructs. Most major concepts or constructs within communication research can be traced back to key players in the field, with a documented, almost family-tree-like pattern to the maturation of a concept as it is defined, measured, and branched into sub-constructs and other areas of study. For example, research within the area of media multitasking has grown from the initial arguments about human-computer interaction (i.e., Reeves & Nass, 1996), and has since branched off to also include topics such as children, depression, anxiety, cognitive load, and neural correlates (i.e., van Der Schuur, Baumgartner, Sumter, & Valkenburg, 2015). Notably, this area has been saturated with methodological inquiry, leading researchers to continually define and adapt new ways of measuring media multitasking (i.e., Baumgartner, Lemmens, Weeda, & Huizinga, 2017).

Research on sex/gender has not followed this pattern. No key players exist in terms of scholars devoting their careers to defining, measuring, adapting, and furthering the area of sex/gender research, which is not to say that scholars have ignored sex, gender, or related concepts as key variables in their research. However, the pieces of sex/gender and communication research remain scattered and untethered to an overall history of explicating and measuring the two constructs. That is, no single literature of gender/sex and communication exists because the assorted programs of research do not interact with a primary goal of understanding sex and gender. Rather, disconnected programs of research learn aspects of how gender and sex are involved in restricted domains of communication with little cross-fertilization across those domains because gender and sex are usually secondary foci.

In surveying how sex and gender have been employed as variables of interest within communication research, then, we acknowledge the inherent difficulty in interpreting which concept researchers were exploring because of how often scholars use the terms interchangeably. Indeed, scholars often report the proportions of males/men to females/women without explicitly specifying how sex and gender were conceptualized and how those theoretical definitions were responsible for their operationalizations of the constructs (Fausto-Sterling, 2012). Therefore, readers are often left to guess the relevance of sex and gender and how they are theoretically relevant to the research, as sex and gender are commonly measured through simplistic self-reported responses to a single item (Fine, 2017; Unger, 1979) or through raters' subjective assessment of the participant (i.e., Balachandra, Briggs, Eddleston, & Brush, 2017). Both of these have distinct theoretical and methodological drawbacks, as we elaborate below.

Treatment of Sex and Gender

Admittedly, categorizing research into the extent to which any given study is focused on sex versus gender is often difficult, if not impossible, because researchers infrequently make a clear and theoretically-based case for how and why their research draws on sex versus gender, which can mean the constructs are often intertwined, yielding a conceptually messy and unwieldy understanding. Rather than thoroughly defining and distinguishing sex and gender, scholars often employ them as primitive terms—constructs that "are accepted as commonly understood or given" (Chaffee, 1991). Treating sex and gender as primitive terms typically means that thoroughly defining them is unnecessary, which is acceptable at times, such as when sex and/or gender may serve as an extraneous variable (e.g., control). Sex and gender are often erroneously employed as primitive terms, however. In fact, much research employing sex and/or gender includes the constructs as part of the central explanatory mechanisms, which means attendant explications are necessary.

Indeed, sex and gender affect many areas of scholarly interest, separately and also together, as well as directly and indirectly. Baird's (1976) review of differences research in communication reveals the most consistent trends in the literature:

> Males, encouraged to be independent, aggressive, problem-oriented, and risk-taking generally are more task-oriented in their interactions, more active and aggressive verbally, more interested

> and capable in problem-solving, more willing to take risks, more resistant to social influence, more competitive when bargaining, and more likely to assume leadership in task-oriented situations. Females, taught to be noncompetitive, dependent, empathic, passive, and interpersonally oriented, typically are more willing to self-disclose, more expressive of emotions and perceptive of others' emotional states, more sensitive to nonverbal cues, less interested and able in problem-solving, relatively unwilling to assume risks, more yielding to social pressure, and less likely to assume leadership, although capable of providing leadership in certain situations.
>
> (p. 192)

Of the more recent work that has looked explicitly at biological sex, the majority of it has focused on topics such as hormonally-based differences in communication (Denes, Afifi, & Granger, 2017), differences in a wide array of communication styles (i.e., nonverbal behavior, eye contact, smiling, etc.; Adams, Miles, Dunbar, & Giles, 2018), or emotion recognition (Thompson & Voyer, 2014), and evolutionary theories of mating and sexual selection (see, e.g., Buss & Schmitt, 1993; Keblusek, 2018; Purvis, 2017). In all of these cases, the researchers specified sex as their variable of interest, but measurement of it was done either as a dichotomous self-report (male versus female) or was reported without explanation of measurement. Other research has employed the construct of gender without defining it. To make matters murkier, gender is often measured via a dichotomous nominal variable of man/woman, which runs the risk of either not capturing extant variation or leading to miscommunication if participants interpret the choices as biological rather than gendered, or vice versa. Although not as precise as is ideal in sex and gender research, it is important to note that researchers seem to be increasingly aware of the trials and tribulations of sex and gender research. Denes and colleagues (2017) make the point in their study that sex differences in communication are often overestimated, leading them to avoid assuming there would be sex differences to begin with. Adams and colleagues (2018) also state that the "nuanced discussion on the effect of gender identity and one's biologically assigned sex is beyond the scope of this research" (p. 479), helping the reader to at least be aware of the limitations of conflating the two. Whereas acknowledgments of complications with the use of sex and gender as primitive terms are steps in the right direction, admitting limitations is not as effective as directly addressing and mitigating those limitations. Some scholars are beginning to be more proactive.

A CHANGING LANDSCAPE

This is not the first study to address sex and gender as highly conflated, misunderstood, unexplicated, and/or treated without the care and precision it necessitates. Decades ago, Wright (1988) criticized the tendency of social scientists to focus on and report mostly differences. He argued that particularly in gender research, problems arise when: (1) researchers assume statistically significant results equate to very large and inherently important findings; (2) within-group variation is overlooked and often negligible group differences are overstated; and (3) the role of gender as a subject variable (i.e., subject to often overlapping correlates and lacking interpretive clarity similar to other personality or psychological variables) is ignored. In a similar vein, Canary and Hause (1993) touch on sex, writing in their seminal piece:

> The operationalization of sex is perhaps the easiest research task going. Researchers only have to record the sex of the participant, typically using a dichotomous code. Many researchers probably include sex as an independent variable because it is so easy to do and it presents another legitimate way to probe for significant differences.... Although the operationalization of sex is

> easy, measuring it in this way is not valid when researchers are more interested in testing effects due to gender.
>
> (p. 137)

Similarly, Floyd (2014) argues that "we claim to study gender differences, but often we do not" (p. 1) due to the propensity to define gender by cultural markers but measure it by comparing biological "males" and "females." In addition, Dainton and Zelley (2018) criticize the ubiquitous proclamation that "men inhabit one planet while women occupy another" despite little-to-no communication theory or research to support these vast and common claims of sex or gender differences (p. 56). This stereotype most likely exists, they argue, because sex/gender is perceived as the simplest set of categories into which humans can be classified—even easier than race, age, or other observable traits—a process that is often done immediately and subconsciously. In other words, sex and gender are seen as primitive terms immune from explication or differentiation in their operationalizations.

Indeed, the differences hypothesis—that males and females are vastly different psychologically and otherwise—has sustained widespread popularity among both scholars and the general public (Hyde, 2005), and is adopted by both biological sex and gender researchers. The traits and characteristics within which these differences exist are presumed to be vast, although the male advantage in spatial tasks and the math and sciences is among the more consistent (and "supported") developmental trends (Miller & Halpern, 2014). A potential explanation for these differences, which many researchers within this area support, is hormonal influence during fetal development, which ultimately affects brain organization and cognition (Lombardo et al., 2012). Females exposed to higher levels of prenatal androgens (e.g., via congenital adrenal hyperplasia) exhibit higher mental rotation skill along with other sex-differentiated traits such as sensation-seeking and tooth size (Miller & Halpern, 2014). Still, in a meta-analysis of sex differences across many cognitive domains, Linn and Petersen (1985) found inconsistent differences that were larger only for specific tasks and specific age ranges (e.g., over 18). They cite hormonal changes during puberty as a potential explanation for these inconsistencies, although this has not been validated.

With minimal support for consistent sex and/or gender differences over the last several decades, the similarities hypothesis—that males and females are in fact more alike than they are different—has gained traction among all areas of social science (Belle, 1985; Deaux, 1984; Eagly, 2013; Hyde, 1981). Strikingly, in a meta-meta-analysis of psychological sex differences, Hyde (2005) found that 78% of reported differences are small or close to zero. Similarly, Canary and Hause (1993) analyzed over 1,200 studies on differences in their own meta-meta-analysis, also using standard Cohen's (1977, 1988) interpretation of effect size *d*, whereby a *d* of .2 represents a "small" effect and accounts for 1% of variance. They found that among communication research on sex/gender differences, the average effect sizes were small (about .23 or less) and accounted for only about 1% of the variance.

In a more recent critique of researchers' propensity to maximize assumptions of sex differences, a team of neuroscientists analyzed over 1,400 human brains (via MRI scans) and assessed whether anatomical sex differences in region size and connectivity exist (Joel et al., 2015). From a total of 116 brain regions, they specifically assessed 10 regions that were documented as having the most consistent and/or robust sex differences (Cohen's $d > 0.70$); in doing so, they maximized their chances of finding significantly and internally-consistent "male" vs. "female" brains. They instead found that they could not distinguish between "male" or "female" brains, and that the vast majority of individuals had brains that possessed both male-typical and female-typical characteristics (i.e., having both a larger hypothalamus and smaller amygdala, both presumed to be larger in males than females). Further, they repeated their methodology using behavioral (i.e., more gendered) rather than anatomical data from over 5,500 participants, specifically assessing 10 highly gender-stereotyped activities taken from the literature, such as video game playing. Again, they found that very few individuals are consistently at one end of the behavioral spectrum, with the majority behaving in both male-typical and female-typical ways. With the suggestion that brains are most often simultaneously "male" and "female"-typical, alongside more extreme evidence that cells throughout one's body can be genetically distinct (Ainsworth, 2015), potentially leading to various degrees of chimerism (i.e., both 46,XX and 46,XY) or other genetic conditions, it seems that humans are to some degree mosaics of both biologically and gender-based "male" and "female" traits and characteristics.

In other social science fields, scholars and organizations are beginning to make these theoretical and methodological cries louder. Fine (2010) criticizes what she deems as rampant "neurosexism," whereby popularized neuroscientific claims about sex/gender differences are used to reinforce or justify stereotypes that are not actually scientifically supported. In 2016, the National Institutes of Health (NIH) issued new guidelines for the treatment of sex as a variable of study. In their published reviewer guide to evaluate Sex as a Biological Variable (SABV), they now require all researchers studying humans: (1) to know the difference between biological sex and gender; (2) to assess whether their study is intended to test for sex differences; and (3) to include both sexes in any studies of differences unless a strong justification for a single-sex study is given. Further, in January 2017, the *Journal of Neuroscience Research* (JNR) released a special issue devoted entirely to sex differences on all levels of the nervous system, simultaneously implementing a policy requiring all authors to consider biological sex as a variable within their studies (Lee, 2018). Authors must complete a questionnaire in which they (along the same lines as NIH) must (1) identify whether sex is considered as a biological variable; (2) report how many and what kinds of subjects are included; and (3) address any limitations including the reason why or why not potential sex differences were addressed (Prager, 2017).

RECOMMENDATIONS FOR FUTURE DIRECTIONS

The ways in which scholars think of and research sex/gender are changing rapidly. The contentious nature of these foundational variables has sparked drastic re-labeling and policy changes in major organizations and journals, such as NIH and JNR. Still, the issue is far from resolved and scholars are urged to reconsider what we know about sex and gender before reconsidering how we go about studying it. Rippon, Jordan-Young, Kaiser, Joel, and Fine (2017) bring up many clarifications needed by JNR's editorial policy announcement. First, they discuss the fact that many variables considered peripheral may in fact be influencing studies' dependent measures, and should be "identified, recorded and analysed, in order to ensure control of co- or potentially confounding variables" (p. 1357). Additionally, they suggest that researchers should report effect sizes and both similarities *and* differences found. In interpreting findings of differences, they argue that inferences should not be immediately attributed to context-independent, pre-programmed, or persistent factors, and that post-hoc exploratory analyses of potential differences may do more harm than good for studies that did not make a theoretical case for them *a priori*. With all this, they once again echo the sentiment that sex is inherently neither a "cleanly binary" nor "independent" variable and may be better thought of in terms of sex "influences" rather than "differences" as a way to avoid interpretive inflations (Rippon et al., 2017, p. 1358).

Consistent with and extending these suggestions, we have our own set of recommendations. We take a theoretically motivated perspective when offering suggestions on how to integrate sex and/or gender into research. First, we urge researchers interested in studying sex and/or gender to consider the extent to which sex and gender are primitive terms or constructs in need of thorough explication. We argue that if sex and/or gender are of theoretical interest, in that they represent at least a portion of the mechanistic process for the theoretical rationale, then they need conceptual definitions. Of course, the extent to which sex and gender are theoretically relevant will depend on the outcomes being assessed. For instance, if researchers are trying to understand how nonverbal behaviors are products of socialization, then gender, gender identity, and gender stereotypes are likely germane, all of which are in need of conceptual definitions, with sex likely being a primitive term potentially serving as a control. If, on the other hand, researchers aim to understand trait levels of aggressions between men and women, then sex and gender are likely equally relevant because both genetic and environmental influences likely play a role. Thus, taking theoretical stock of the roles of sex and gender is an effective first step at being proactive to mitigate common limitations involved in sex and gender research.

Second, we recommend that scholars specify the theoretical distinctions between sex and gender and how these variables may need to be employed in tandem or separately, such that only one of the two is of theoretical interest. In other words, it is useful to determine whether sex and gender play different explanatory roles in research. If gender and socialization processes are contingent on genotypical variations across males and females, then it is particularly meaningful and effective to distinguish between sex and gender. Perhaps only sex is relevant, whereas gender is not related to the dependent variable(s) of interest. It may benefit researchers who are unable to cleanly untangle (from a theoretical perspective) the relative influences of each to include both as part of the theoretical process, with perhaps one serving as a control given the theoretical ambiguities between the two constructs. Just as with any communication study, researchers should consult prior literature to develop a strong theoretical argument for why sex and/gender may predict their dependent variable(s).

Once sex and gender are conceptually defined, researchers should consider the best method to measure them. A consensus has not yet been reached on what the most valid and reliable ways of measuring sex and/or gender are. Consensus at this level may be unrealistic, however, because, as with all constructs, no sole means to assess a concept in research exists, which is to say constructs can be measured in numerous and diverse ways. That is, the best method depends on the construct and its particular conceptual definition. A "consensus" then would need to be reached not in terms of one singular way to measure sex/gender, but in terms of how researchers explicate sex/gender, which broader methods of measurement might be available, and how findings would be interpreted. A consensus may be reached when it comes to measuring each specific, conceptually defined construct within sex/gender research (i.e., gender roles or testosterone level), but to do so first requires a broader consensus in sex/gender theorization, and much more methodological research surrounding each construct. Thus, although we could offer several suggestions for measuring sex and gender, we opt to suggest strategies for determining measures. A synergistic relationship between theory and method exists and helps to propel both forward. Thus, the means by which sex and gender are measured should be derived directly from the theoretical definitions of the constructs. As methods develop over time, more micro-level theory may also develop. Again, considering several options for a measurement may help to narrow down which are most consistently and robustly predicting sex and/or gender influences.

Apart from traditional categorical self-report items for either biological sex (i.e., "male," "female," "intersex", etc.) and gender (i.e., "man," "woman," "trans," "non-binary," etc.), which might be appropriate if sex and gender are primitive terms, researchers should consider including measures that may better tap into the construct at hand. For example, scholars who wish to study biological sex as a predictor of aggression may really be referring to the hormonal effects of increased testosterone in males versus females. In this case, the researcher should measure hormones (state, trait, or both), rather than if someone indicates male or female in a single self-report item. As can be seen throughout this book's chapters, there are many methodological avenues with hormonal research. For example, levels of sex or stress hormones can be exogenously measured via blood or saliva, as well as administered in experimental work via intranasal spray.

Additionally, more phenotypical measures of hormones are sometimes used, such as 2D:4D finger ratio. Initially introduced in the field of anthropology, this ratio between the length of the second and fourth digits has been implicated as an indicator of testosterone levels in utero, with males generally having lower ratios than females (Manning, Scutt, Wilson, & Lewis-Jones, 1998). Research has used the 2D:4D ratio to predict performance in a wide variety of sexually dimorphic tasks and attributes, including communication-relevant variables such as verbal aggression (Putz, Gaulin, Sporter, & McBurney, 2004).

Scholars interested in gender research are urged to include more precise measures of what they hope to gain insight into. For example, are they interested in gendered self-identification, gender role identity, stereotypes of gender about *others*, or some other gender-based specific construct? Classical scales (i.e., Bem's Sex Role Inventory; Bem, 1981), better-validated scales (i.e., Beliefs About Women Scale; Belk & Snell, 1986), or even novel scales may be adopted or developed depending on the theory and hypotheses at hand (Luyt, 2015). Research on gender-based language use is concerned with gender and the extent to which people situationally identify with their gender as a means of understanding variation in language use typical of men and women (Palomares, 2012). All of these examples demonstrate the need to define constructs before selecting measures for sex and gender because the explication process will play a major role in which measures are employed and their inherent limitations which will provide meaningful caveats for how results are interpreted.

Finally, we suggest that scholars report and interpret all findings (or lack thereof) of sex and/or gender differences delicately. Effect sizes should be calculated and compared to Cohen's *d* interpretations (i.e., small, medium, or large) as well as compared to prior literature. Further, equivalence testing should be conducted to test for the *significance of equivalence between groups* (see e.g., Weber & Popova, 2012). This is calculated using effect sizes, which should again be drawn from previously established average effect sizes in the literature. Importantly, results should not be overstated, and special attention should be paid to what findings mean or do not mean when considering theories of sex and/or gender differences. How might these findings illustrate overlap, or not, between sex and gender? Do these findings have implications for the interplay between them? These are just a couple of the questions scholars should consider when interpreting findings.

CONCLUSION

As with any other variable of interest, proper operationalization stems from proper conceptualization. Depending on the hypothesis at hand, researchers may be tapping into biological sex more than gender, or vice versa. Measuring, for example, participants' gender when a biological and hormonally relevant dependent variable is the focus (i.e., stress response, arousal, or aspects of aggression) is not as precise as measuring sex or some other related variable (i.e., ingoing testosterone levels). Similarly, a researcher interested in predicting one's identification with media characters would do well to measure other socially based traits such as gender (identification, role, etc.) rather than biological sex. This is especially the case given our current climate of increasing line distortion between sex and gender, as self-identifying non-binary, trans, or LGBTQ+ members have an ever-growing voice and presence (Clark, Veale, Townsend, Frohard-Dourlent, & Saewyc, 2018). The first step in proper and precise measurement is therefore to develop a theoretically valid case for the inclusion of a sex- and/or gender-based measurement. Although the concerns we illuminate are not new, it is increasingly important that communication scholars lead the way for others adopting sex and gender into their social scientific repertoire.

REFERENCES

Adams, A., Miles, J., Dunbar, N. E., & Giles, H. (2018). Communication accommodation in text messages: Exploring liking, power, and sex as predictors of textisms. *The Journal of Social Psychology, 158*(4), 474–490.

Ainsworth, C. (2015). Sex redefined. *Nature, 518*(7539), 288–291.

Baird, J. E. (1976). Sex differences in group communication: A review of relevant research. *Quarterly Journal of Speech, 62*, 179–192.

Balachandra, L., Briggs, T., Eddleston, K., & Brush, C. (2017). Don't pitch like a girl! How gender stereotypes influence investor decisions. *Entrepreneurship Theory and Practice, 43*(1), 116–137.

Baumgartner, S. E., Lemmens, J. S., Weeda, W. D., & Huizinga, M. (2017). Measuring media multitasking: Development of a short measure of media multitasking for adolescents. *Journal of Media Psychology, 29*(4), 188–197.

Belk, S. S., & Snell Jr, W. E. (1986). Beliefs about women: Components and correlates. *Personality and Social Psychology Bulletin, 12*(4), 403–413.

Belle, D. (1985). Ironies in the contemporary study of gender. *Journal of Personality, 53*(2), 400–405.

Bem, S. L. (1981). *Bem sex-role inventory*. Palo Alto, CA: Mind Garden.

Buss, D. M., & Schmitt, D. P. (1993). Sexual strategies theory: An evolutionary perspective on human mating. *Psychological Review, 100*(2), 204–232.

Canary, D. J., & Hause, K. S. (1993). Is there any reason to research sex differences in communication? *Communication Quarterly, 41*(2), 129–144.

Chaffee, S. H. (1991). *Explication*. Newbury Park, CA: Sage.

Clark, B. A., Veale, J. F., Townsend, M., Frohard-Dourlent, H., & Saewyc, E. (2018). Non-binary youth: Access to gender-affirming primary health care. *International Journal of Transgenderism, 19*(2), 158–169.

Cohen, J. (1977). *Statistical power analysis for the behavioral sciences* (Rev. ed.). New York, NY: Academic Press.

Cohen, J. (1988). *Statistical power analysis for the behavioral sciences* (2nd ed.), Hillsdale, NJ: Lawrence Erlbaum Associates.

Dainton, M., & Zelley, E. D. (2018). *Applying communication theory for professional life: A practical introduction.* London, England: Sage.

Deaux, K. (1984). From individual differences to social categories: Analysis of a decade's research on gender. *American Psychologist, 39*(2), 105–116.

Denes, A., Afifi, T. D., & Granger, D. A. (2017). Physiology and pillow talk: Relations between testosterone and communication post sex. *Journal of Social and Personal Relationships, 34*(3), 281–308.

Eagly, A. H. (1987). *Sex differences in social behavior: A social-role interpretation*. Hillsdale, NJ: Lawrence Erlbaum Associates.

Fausto-Sterling, A. (2005). The bare bones of sex: Part 1—sex and gender. *Signs: Journal of Women in Culture and Society, 30*(2), 1491–1527.

Fausto-Sterling, A. (2012). *Sex/gender: Biology in a social world.* New York, NY: Routledge.

Fine, C. (2010). *Delusions of gender: How our minds, society, and neurosexism create difference*. New York, NY: W. W. Norton.
Fine, C. (2017). *Testosterone rex: Myths of sex, science, and society*. New York, NY: W. W. Norton & Company.
Floyd, K. (2014). Taking stock of research practices: A call for self-reflection. *Communication Monographs, 81*(1), 1–3.
Griffin, D. K., & Ellis, P. J. (2018). The human Y-chromosome: Evolutionary directions and implications for the future of "maleness". In G. D. Palermo & E. S. Sills (Eds.) *Intracytoplasmic sperm injection: Indications, techniques and applications* (pp. 183–192). Cham, Switzerland: Springer.
Gugatschka, M., Kiesler, K., Obermayer-Pietsch, B., Schoekler, B., Schmid, C., Groselj-Strele, A., & Friedrich, G. (2010). Sex hormones and the elderly male voice. *Journal of Voice, 24*(3), 369–373.
Halpern, D. F. (2013). *Sex differences in cognitive abilities* (4th ed.). New York, NY: Psychology Press.
Hancock, A. B., Krissinger, J., & Owen, K. (2011). Voice perceptions and quality of life of transgender people. *Journal of Voice, 25*(5), 553–558.
Hyde, J. S. (1981). How large are cognitive gender differences? A meta-analysis using ω^2 and *d*. *American Psychologist, 36*(8), 892–901.
Hyde, J. S. (2005). The gender similarities hypothesis. *American Psychologist, 60*(6), 581–592.
Joel, D., Berman, Z., Tavor, I., Wexler, N., Gaber, O., Stein, Y., … Assaf, Y. (2015). Sex beyond the genitalia: The human brain mosaic. *Proceedings of the National Academy of Sciences, 112*(50), 15468–15473.
Keblusek, L. M. (2018). Socio-sexuality, self-reported physical formidability, and the dark triad: Predictors of gossip behavior and cognitions. Doctoral dissertation, University of California, Santa Barbara.
Lee, S. K. (2018). Sex as an important biological variable in biomedical research. *BMB Reports, 51*(4), 167–173.
Linn, M. C., & Petersen, A. C. (1985). Emergence and characterization of sex differences in spatial ability: A meta-analysis. *Child Development, 56*(6), 1479–1498.
Lombardo, M. V., Ashwin, E., Auyeung, B., Chakrabarti, B., Taylor, K., Hackett, G., … Baron-Cohen, S. (2012). Fetal testosterone influences sexually dimorphic gray matter in the human brain. *Journal of Neuroscience, 32*(2), 674–680.
Luyt, R. (2015). Beyond traditional understanding of gender measurement: The gender (re)presentation approach. *Journal of Gender Studies, 24*(2), 207–226.
Manning, J. T., Fink, B., & Trivers, R. (2018). The biology of human gender. In S. K. Todd & W. S. A. Viviana (Eds.), *Encyclopedia of evolutionary psychological science* (pp. 1–6). Cham, Switzerland: Springer.
Manning, J. T., Scutt, D., Wilson, J., & Lewis-Jones, D. I. (1998). The ratio of 2nd to 4th digit length: A predictor of sperm numbers and concentrations of testosterone, luteinizing hormone and oestrogen. *Human Reproduction, 13*(11), 3000–3004.
Miller, D. I., & Halpern, D. F. (2014). The new science of cognitive sex differences. *Trends in Cognitive Sciences, 18*(1), 37–45.
National Institutes of Health (NIH). (2016). Reviewer guidance to evaluate sex as a biological variable (SABV). Retrieved from https://grants.nih.gov/grants/peer/guidelines_general/Sabv_decision_tree_for_reviewers.pdf.
Palomares, N. A. (2012). Gender and intergroup communication. In H. Giles (Ed.), *The handbook of intergroup communication* (pp. 197–210). New York, NY: Routledge.
Prager, E. M. (2017). Addressing sex as a biological variable. *Journal of Neuroscience Research, 95*(1–2), 11.
Purvis, J. L. (2017). Strategic interference and Tinder use: A mixed-method exploration of romantic interactions in contemporary contexts. Doctoral dissertation, University of Hawai'i at Manoa.
Putz, D. A., Gaulin, S. J., Sporter, R. J., & McBurney, D. H. (2004). Sex hormones and finger length: What does 2D:4D indicate? *Evolution and Human Behavior, 25*(3), 182–199.
Re, D. E., O'Connor, J. J. M., Bennett, P. J., & Feinberg, D. R. (2012). Preferences for very low and very high voice pitch in humans. *PLoS One, 7*(3), e32719.

Reeves, B., & Nass, C. I. (1996). *The media equation: How people treat computers, television, and new media like real people and places*. New York, NY: Cambridge University Press.

Rippon, G., Jordan-Young, R., Kaiser, A., Joel, D., & Fine, C. (2017). *Journal of Neuroscience Research* policy on addressing sex as a biological variable: Comments, clarifications, and elaborations. *Journal of Neuroscience Research, 95*(7), 1357–1359.

Ritz, S. A., Antle, D. M., Côté, J., Deroy, K., Fraleigh, N., Messing, K., … Mergler, D. (2014). First steps for integrating sex and gender considerations into basic experimental biomedical research. *The FASEB Journal, 28*(1), 4–13.

Savic, I., Frisen, L., Manzouri, A., Nordenstrom, A., & Lindén Hirschberg, A. (2017). Role of testosterone and Y chromosome genes for the masculinization of the human brain. *Human Brain Mapping, 38*(4), 1801–1814.

Thompson, A. E., & Voyer, D. (2014). Sex differences in the ability to recognise non-verbal displays of emotion: A meta-analysis. *Cognition and Emotion, 28*(7), 1164–1195.

Tooby, J., & Cosmides, L. (1992). The psychological foundations of culture. In J. H. Barkow, L. Cosmides, & J. Tooby (Eds.), *The adapted mind: Evolutionary psychology and the generation of culture* (pp. 19–132). New York, NY: Oxford University Press.

Unger, R. K. (1979). Toward a redefinition of sex and gender. *American Psychologist, 34*(11), 1085–1094.

van Borsel, J., de Pot, K., & De Cuypere, G. (2009). Voice and physical appearance in female-to-male transsexuals. *Journal of Voice, 23*(4), 494–497.

van Der Schuur, W. A., Baumgartner, S. E., Sumter, S. R., & Valkenburg, P. M. (2015). The consequences of media multitasking for youth: A review. *Computers in Human Behavior, 53*, 204–215.

Weber, R., & Popova, L. (2012). Testing equivalence in communication research: Theory and application. *Communication Methods and Measures, 6*(3), 190–213.

Wright, P. H. (1988). Interpreting research on gender differences in friendship: A case for moderation and a plea for caution. *Journal of Social and Personal Relationships, 5*, 367–373.

33

Communication and Quantum Cognition

Lorraine Borghetti, Zheng Wang, and Jerome Busemeyer

Quantum cognition (QC) is a new development in the field of cognitive science, which provides potentially important applications to human communication processes. QC takes the basic mathematical principles from quantum theory (without the physics) and applies them to theorize and model human judgment and decision-making, perception, attention, memory, and other cognitive behaviors (for introductions and reviews, see Bruza, Wang, & Busemeyer, 2015; Busemeyer & Bruza, 2012; Busemeyer, Pothos, Franco, & Trueblood, 2011; Busemeyer & Wang, 2015; Pothos & Busemeyer, 2013; Pothos, Busemeyer, & Trueblood, 2013; Wang, Busemeyer, Atmanspacher, & Pothos, 2013).[1] It may seem surprising to connect these two areas together—quantum theory and communication—but there are good reasons for doing so.

Communication is an inherently sequential process, and the order of information strongly influences its effects, such as its effectiveness for changing attitudes, opinions, and choices. Classic communication theory is based on classical information theory, which is derived from classical probability theory (Shannon, 1948). Classical probability theory relies on principles that include commutativity (order independence) of information, that is,

$$prob(A) \cdot prob(B) = prob(B) \cdot prob(A)$$

This commutative property makes it difficult to apply classical probability theory to account for sequential effects of information presentation on inference, reasoning, attitude, memory, and decision-making in a principled and *a priori* manner (certainly, post-hoc explanations and descriptive explanations are always possible and relatively easy). For example, according to classical probability theory, the order of prosecutor and defense information should not influence the final judgment of guilt; but on the contrary, it is well known and documented in research that these judgments are strongly influenced by order of presentation (Trueblood & Busemeyer, 2011). Different from classical probability theory, quantum probability theory is based on non-commutative (order dependent) principles, that is,

$$prob(A) \cdot prob(B) \neq prob(B) \cdot prob(A),$$

which are especially designed for sequential effects observed in the physics world. For this reason, quantum probability theory has been successfully used to predict the effects of presentation order on various kinds of opinions and judgments (Trueblood & Busemeyer, 2011; Wang & Busemeyer, 2013, 2016; Wang, Solloway, Shiffrin, & Busemeyer, 2014). Beyond non-commutativity, there are many other fundamental differences between the quantum probability theory and classical probability theory. QC, based on quantum probability theory, has been able to account for a large range of paradoxical findings in the decision and cognition literature using a small set of coherent, principled rules (for reviews, see Bruza et al., 2015; Busemeyer & Bruza, 2012; Busemeyer & Wang, 2015; Wang et al., 2013).

To introduce and demonstrate the applicability of QC to communication, this chapter (1) summarizes the basic QC concepts relevant to communication; and (2) provides a case study where QC concepts are used to guide hypotheses for psychophysiological research in communication and decision processes. This introduction to QC will be a conceptual summary of some general principles (requiring no background in quantum theory), rather than a presentation of the formal mathematics. For readers interested in the mathematical formalism, see many other publications in QC with formal mathematical and computational models (e.g., Busemeyer & Bruza, 2012; Busemeyer et al., 2009; Busemeyer & Wang, 2018, in press; Busemeyer, Wang, & Shiffrin, 2015; Trueblood, Yearsley, & Pothos, 2017; Wang & Busemeyer, 2015a, 2016).

COMMUNICATION RESEARCH AND QUANTUM COGNITION

To get a quick idea about how quantum probability theory works in a communication research setting, consider the following communication message evaluation experiment reported in Wang and Busemeyer (2015a). In the study, 131 participants were shown 12 anti-smoking public service announcements (PSAs), and then asked to rate how effective the PSAs were from a "self" and "other" person perspective, with the perspective taking conditions counterbalanced so that half rated "self" first, and the other half rated "other" first. The results indicated that the PSAs were rated more effective for "self," as predicted by the well-known first-person effect in communication research; however, this effect depended on order, because the effect only occurred when "self" was rated first. The study used a 9-point rating scale, which produced two (one for each order) 9 × 9 contingency tables with 81 joint frequencies, such as the frequency of "9 = certainly yes" to "self" and "1 = certainly no" to "other." However, for simplicity, here we consider a 3-point scale (Y = yes effective, U = uncertain, N = no effective), which would produce two 3 × 3 contingency tables, one for each order.

Classical probability could try to model the two 3 × 3 joint probability tables with a single 3 × 3 joint probability distribution with cell entries such as *prob(Self = N and Other = Y)*. This model is testable because it uses only 9 joint probabilities to predict 18 joint frequencies. This model assumes commutativity, however, and fails to account for the order effect. To account for order effects, the classical model would have to form two different 3 × 3 joint probability distributions, one for each order. This model, however, becomes untestable because it now requires using 18 joint probabilities to predict 18 joint frequencies, which is simply tautological.

In contrast, quantum probability theory provides a model of the judgments more efficiently by using a common three-dimensional vector space for both perspectives, which provides strong testable predictions about the order effects. Figure 33.1 illustrates a simple quantum model for "self" versus "other" ratings of the effectiveness of PSAs. The left panel represents the "self" perspective, and the right panel represents the "other" perspective. In this example, "self" is rated first. The three lines labeled Y, N, U within each panel form three basis vectors, one representing each answer. The basis vectors used for "other" are obtained by a rotation of the basis vectors for "self" (where the rotation is a critical component of the model and can be psychologically meaningfully parameterized through formal model estimation, and it indicates

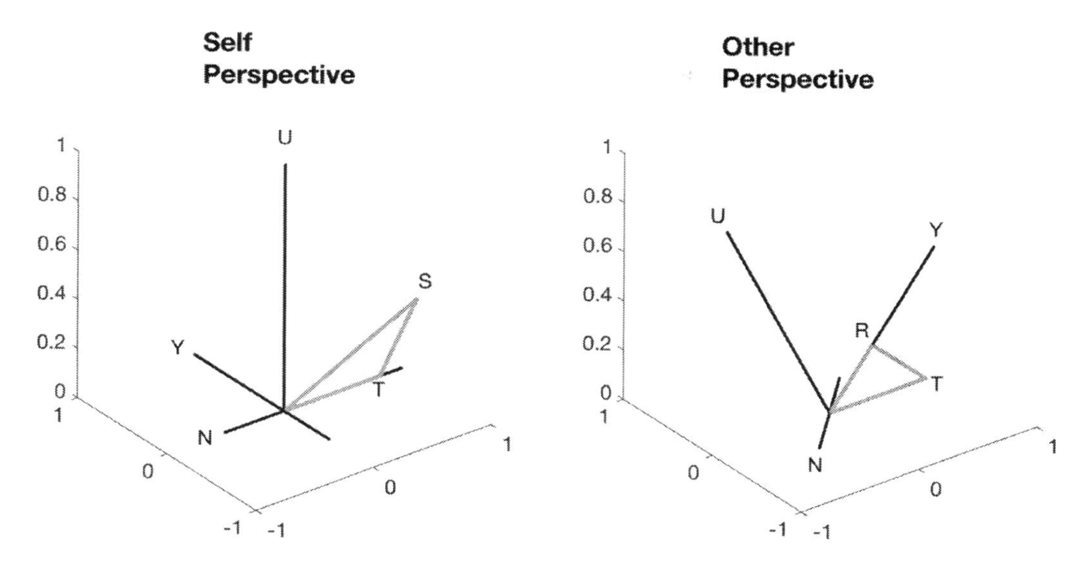

Figure 33.1 Illustration of a Quantum Model for "Self" Versus "Other" Ratings of the Effectiveness of PSAs

the relationship between the two measures/basis, here in the example, "self" and "other"; for formal modeling examples, see Busemeyer & Wang, 2018, in press; Trueblood et al., 2017; Wang & Busemeyer, 2015a).

The line labeled S in the left panel is a unit length vector representing a person's initial state before answering any questions. In this example, the state S is a superposition state because it is a linear combination of all three basis vectors. Psychologically, this means that in state S, each answer has some potential to be selected. The projection of S onto the basis vector N produces the projection T, and the squared length of this projection equals the probability of N to the "self" question (which happens to equal .67 in the example). If the person first answered N to "self," then the person's state would "collapse" and become aligned with the vector T.

To answer the question about "other," the person rotates the basis vectors for Y, N, U to a new basis. The vector T is the same vector in both figures, but now evaluated with respect to a different basis. The probability of first answering N to "self" and then Y to "other" is obtained by projecting T onto the basis vector for Y in the "other" basis to produce the final projection R. The squared length of R equals the sequential probability of first rating N to "self" and then rating Y to "other" (which happens to equal .17 in the example).

If the questions were answered in the opposite order, using the same bases and initial state S, then the probability of Y to "other" and N to "self" equals .20, producing a small order effect in the direction of the observed results. As illustrated in this simple example, the advantage of the quantum model is that it only needs two rotation parameters to make predictions for 18 joint frequencies obtained from both 3×3 contingency tables, making it strongly testable.

The left panel represents the "self" perspective, and the right panel represents the "other" perspective. The three lines labeled Y, N, U within each panel represent three basis vectors, but the basis used for "other" is obtained by a rotation of the basis used for "self." The rotation is parameterized in the model and quantifies the relationship between the two perspectives. The line labeled S in the left panel represents a person's disposition before answering any questions.

The line labeled T in both panels is the projection after answering N to "self." The line labeled R is the projection representing Y to "other" after saying N to "self."

Perhaps more important, the quantum model also provides strong *a priori*, parameter-free predictions about order effects. Wang and colleagues (2014) derived a powerful test of the quantum model for measurement order effects in typical social and behavioral scientific research, called the QQ equality. This test is obtained from pairs of 2×2 contingency tables for binary answers obtained to questions asked in both orders. This equality essentially states that a predetermined linear combination of the observed cell relative frequencies must equal zero. Rarely in social and behavioral sciences has such an exact *a priori* prediction been possible. Wang et al. (2014) tested this *a priori* prediction using 70 US national representative surveys examining order effects and found strong support for the surprising QQ equality predicted by the quantum model.

More generally, questions and measurements, such as the "self" versus "other" questions, that require changing perspectives (formally, changing the basis) are described in QC as incompatible. *Incompatibility* is a fundamental concept in QC and in quantum theory (Busemeyer & Bruza, 2012; Pothos & Busemeyer, 2013; Wang & Busemeyer, 2015b; Wang et al., 2013). According to QC, these questions cannot be answered simultaneously, because the answers require rotating to different bases, which cannot be evaluated at the same time—likely due to the limited experience and limited capacity of most people to perform such tasks (for a discussion, see Busemeyer & Wang, 2018; Wang & Busemeyer, 2015b). Also, the projection that occurs when answering one question (e.g., N to "self") changes the state (from S to T in Figure 33.1), and thus disturbs the mental state of thoughts prior to answering the other question (about the "other"). Order effects are the result of being asked about questions that have incompatible representations (Bruza et al., 2015; Busemeyer & Wang, 2018; Trueblood & Busemeyer, 2011; Wang & Busemeyer, 2013, 2015b; Wang et al., 2014). QC allows all different kinds of questions to be represented and answered simply by using rotation from one basis to another within the same dimensional space. This way, QC can provide an understanding of how a person can answer an infinite number of questions using relatively limited resources. Thus, from the QC perspective, although the disturbance produced by incompatible questions is a limitation of our mental machinery, the advantage is the representation of many questions within an efficient low-dimensional cognitive space and with limited cognitive resources (Busemeyer & Wang, 2018; Wang & Busemeyer, 2013, 2015b).

Incompatible representations are responsible for producing what is called a *superposition* state—mathematically, this means a state that is a linear combination of basis vectors representing possible answers to a question, and psychologically, this means the person feels ambiguous or uncertain about the answers and all the possible answers have more or less potential to be selected. The concept of superposition captures the deep meaning of uncertainty in human cognition. If a person is certain about an answer to one question, then the person must be uncertain about the answers to an incompatible question. (This is the famous Heisenberg uncertainty principle.) Suppose a person decides that N is true for "self," and therefore is directly aligned with the vector N in the "self" basis. Then the person has to be in a superposition state with respect to "other" because the vector N in the "self" basis is a linear combination of the three basis vectors representing the answers in the "other" basis. Superposition states are indefinite or uncertain states, which describe a potential to respond to each answer to a question at any moment.

A major way that communication enters into QC theory concerns how it modifies a superposed state of a person. Communication of facts can reduce an indefinite superposition state to a definite state, which is a process called *collapse* of the state vector in quantum theory.

For example, a person may initially be superposed over the possible effectiveness of a PSA for another person. However, if informed that in fact the PSA is effective for the person, then the uncertainty is resolved, and the state moves to be aligned with the basis vector representing the communicated fact. As another example, consider the case of the juror prior deciding about the defendant. When in a superposition state, the juror is superposed between guilt and innocence. However, if an incontrovertible fact of guilt is presented, then the juror collapses to a state of certain guilt.

According to QC, a superposition state might also be "collapsed" when a person makes a "self-measurement," that is, when a person is asked to report if a fact is true or not (Busemeyer et al., 2009; Wang & Busemeyer, 2016; White, Pothos, & Busemeyer, 2014). This self-expression is a form of implicit "self" communication. Therefore, either implicit or explicit communication, either expressed-by or received-by the decision-maker, acts as a measurement to disturb the superposition state (White et al., 2014). Conceptually, a self-measurement or observed fact initiates a transition from being uncertain to a state of certainty (Busemeyer & Bruza, 2012). This is the equivalent of a projection onto the subspace representing the fact that can be viewed as a belief-updating process. Conversely, when not disturbed by a measurement or fact, an individual remains in a superposition or uncertain state.

The uncertainty produced by superposition states, and the reduction produced by communication, provide a natural link to a prominent theory in neuroscience: cognitive control.

Cognitive Control and Conflict Processing

Cognitive control refers to the ability to respond flexibly to changes in the environment in the service of goal optimization (Botvinick et al., 2001; Cavanagh & Frank, 2014; Gratton et al., 2018; Ridderinkhof, Ullsperger, Crone, & Nieuwenhuis, 2004). Of particular interest here are cognitive control processes modulating conflict and uncertainty. Defined as the neural activity elicited when deciding between two or more valid alternatives (Cavanagh & Frank, 2014; Cohen, 2014b; Cohen & Ridderinkhof, 2013), conflict processing closely corresponds to the characterization of the process used to construct an answer from a superposition state. Further, conflict processing reflects sequential reasoning sensitivity (Cohen, 2014b).

Neural Mechanisms of Conflict

The study of neural oscillations, or frequency band signals, recorded by electroencephalography (EEG) has produced important insights about temporally sensitive conflict processing under uncertainty (Banich, 2018; Cavanagh & Frank, 2014; Cohen, 2014a). The conflict processing system is composed of three basic mechanisms: (1) monitoring and detection of uncertainty producing conflict; (2) transmission of a conflict signal to relevant brain regions; and (3) response to the signal to adapt behavior. This system manifests sequentially and consistently in two frequency bands: the frontal-midline theta (FMθ; 3–8 Hz) and the posterior alpha band (8–14 Hz; Cavanagh & Frank, 2014; Cohen, 2014b; Hanslmayr, Staresina, & Bowman, 2016; Khader & Rösler, 2011; Klimesch, 1999, 2012). Exposure to a conflict-eliciting event, e.g., deciding between potential alternatives, increases FMθ power; the FMθ signal is followed by the modulation of parietal alpha power. The theta-alpha coupling indexes cognitive functions important to adaptive conflict processing: increases in FMθ power signals the need to implement adaptive control processes to relevant network structures, particularly the medial frontal cortex (MFC; Cavanagh & Frank, 2014; Cohen, 2014b); subsequent suppression (decrease) of parietal alpha power mediates attention to and processing of information content needed for the response (Hanslmayr et al., 2016; Khader & Rösler, 2011; Klimesch, 1999, 2012).

Frontal-Midline Theta

FMθ is associated with the monitoring and detection of uncertainty and conflict and then recruits resources for an adaptive resolution (Botvinick et al., 2001; Cavanagh & Frank, 2014; Cohen, 2014b; Ridderinkhof et al., 2004). The signal's magnitude indicates the degree of control implemented; for instance, lower FMθ power indicates less conflict detection (Cohen & Donner, 2013). Uncertainty increases the instantiation of FMθ power (Cavanagh & Frank, 2014). For example, exposure to probabilistic (80% valid) stimulus cues, as opposed to deterministic (100% valid) cues, led to longer response times correlated with greater theta band power (van Driel et al., 2015).

Parietal Alpha Band

Parietal alpha power indexes attention and memory processes in response to a preceding FMθ conflict signal (Hanslmayr et al., 2016; Khader & Rösler, 2011; Klimesch, 1999, 2012). Increased alpha power represents the inhibition of irrelevant information—a key process of selective attention. By contrast, alpha suppression has been interpreted as a search of memory in the absence of inhibitory processes. Though raw power decreases during alpha suppression, the search process places more demands on cognitive resources, not less (Hanslmayr et al., 2016; Khader & Rösler, 2011; Klimesch, 1999, 2012).

Hypotheses

QC and cognitive control converge on some critical concepts and mechanisms: (1) deciding among multiple competing alternatives engages conflict control processes; (2) these conflict processes occur under uncertainty; (3) communicating information can reduce the uncertainty prior to the decision; and (4) resolution of the uncertainty facilitates forming an appropriate response. Such convergence suggests that it may be plausible to use a superposition state and its reduction from communication as key concepts for explicating cognitive control under uncertainty during decision-making. In particular, guided by the QC conception of incompatible measures, superposition, and "collapses" as reviewed earlier, the sequence of communication of information (i.e., measurements), or presence versus absence of communication, can make a fundamental difference in the final observed (i.e., measured) cognition and behavior. Although not developed in the current chapter, formal computational models based on quantum probability theory and QC theory can be developed to predict the EEG outcomes in communication and decision tasks to improve our understanding of the neural and cognitive mechanisms predicting choice behaviors. Such new model development and comparison with established traditional methods require significant effort, and in the current analysis, we start to test whether the observed EEG data are consistent with the basic concepts of QC, in particular, superposition and "collapse" of superposition (upon communication of information or measurement). Specifically, we test the following two hypotheses:

> *Hypothesis 1*: (a) when the absence of communication leaves uncertainty about potential alternatives unresolved, we expect no collapse of the superposition state and less uncertainty (conflict) reduction during decision-making information processing in comparison to conditions with

expressed or received information; and (b) we expect the expressing condition to result in less complete or certain collapse ("partial collapse") and uncertainty reduction in comparison to the received communication condition. FMθ power will serve as an indicator of the neural resources recruited for conflict processing and indicate the level of uncertainty or being superposed.

Hypothesis 2: (a) when the absence of communication leaves uncertainty about potential alternatives unresolved, we expect more persistent evaluation of potential alternatives in comparison to explicitly communicated information; and (b) we expect self-expressing to result in more persistent evaluation in comparison to received communication. Similar to the first hypothesis, if received information is believed, then additional searches in memory would be unnecessary. Parietal alpha suppression indexing searches in memory and attentional inhibitory processes will serve an indicator of the persistent evaluation of potential alternatives.

As explained earlier, in this chapter we provide an illustration of applying QC concepts to understand communication and conflict processing, and to guide hypotheses regarding EEG data during these processes. However, here the chapter shows no formal modeling of EEG data using the quantum probability theory and the QC theory yet, which should be developed in the future. Here, we simply test whether the EEG data are consistent with the hypotheses based on QC, and we test the hypotheses using traditional (classical) statistics. If the EEG data are consistent with the QC predictions for the experimental paradigm, this means at least the EEG data support that QC is plausible for predicting the differences that communication of information can make on decision making. Next, we describe the EEG experiment and data.

QUANTUM COGNITION, COGNITIVE CONTROL, AND SENDER-RECEIVER PROCESSES

While information processing research in communication science and psychology has primarily focused on a better understanding of information receiving (e.g., attentional and emotional responses to incoming information), a handful of studies have suggested that sender processes are also important as part of the communication (Namkoong et al., 2017; Pingree, 2007; Yoo, Kim, & Gil de Zúñiga, 2017). A few studies examine how expressing information shapes the cognitions of information senders.

Our study investigates sender-receiver information processing and the effect of communication on subsequent decision-making. We designed our experiment to afford a direct assessment of how communication disturbs pre-decisional conflict processing and resolves uncertainty. Our study design drew from earlier studies of QC inspired by the double-slit experiment in quantum physics which developed our views on the critical role of measurement in observed physical "reality." Our study adopted a categorization-decision experimental paradigm (Busemeyer et al., 2009; Townsend, Silva, Spencer-Smith, & Wenger, 2000; Wang & Busemeyer, 2015a). The basic procedure is to present a face, categorize the face as belonging to a "good" guy or "bad" guy category, and then decide to act "attack" or "withdraw." The main design consisted of a within-subjects factorial design: three types of communicating categorization information (no communication, expressing communication, receiving communication) were crossed with two types of faces (stereotype either wide or narrow). In total, 31 participants completed two blocks of 34 trials for the three communication conditions, for a total of 204 trials each. The experiment lasted about one hour.

The cover story described two novel humanoid-like species that were discovered in a fictitious NASA mission. They tended to have different types of faces (wide, narrow) and correlated with different types of personality (good, bad). The task of the participants was to follow the described face type-personality associations (i.e., wide tend to be good and narrow tend to be bad) to categorize which species the face belonged to, and then make a decision to act appropriately: to defend themselves (from the aggressive individuals) or act friendly (to friendly individuals). However, as in real life, only a portion (60%) of the faces were consistent with the stereotypes described in the cover story. The computer assigned a wide face to the good category 60% of the trials and assigned the narrow faces to the bad category 60% of the trials.

Given that a category was known to be present, the probability of getting rewarded or punished for each action was also probabilistic. If the face was categorized as a "bad" guy, then 70% of the trials resulted in a reward for "attacking," and otherwise punished. Likewise, if the face was categorized as a "bad" guy, then 70% of the trials resulted in a reward for "attacking," and otherwise punished. Because of the uncertainty about face and personality and action, both the category and the appropriate action remained uncertain after presenting the face.

The key manipulation varied communication about the face categorization after exposure and prior to deciding to act (see Figure 33.2). In the no-communication condition, participants were not provided with any categorization information; instead, the categorization process was presumed to occur implicitly (Busemeyer et al., 2009; Townsend et al., 2000; Wang & Busemeyer, 2016). In the received condition, the computer explicitly told the participants about the face's categorization. In the expressed condition, participants explicitly reported a categorization response to the face themselves. Communicating information about the categorization provided explicit evaluative information for deciding. Differences between the implicit process in the no-communication and the expressed and received conditions constituted a manipulation of communication information.

The window of analysis, outlined in bold frame, occurred upon exposure to a decision question ("Act friendly or defensive?") and prior to clicking a computer button to indicate the decision. Participants completed 68 trials for each communication condition across two blocks.

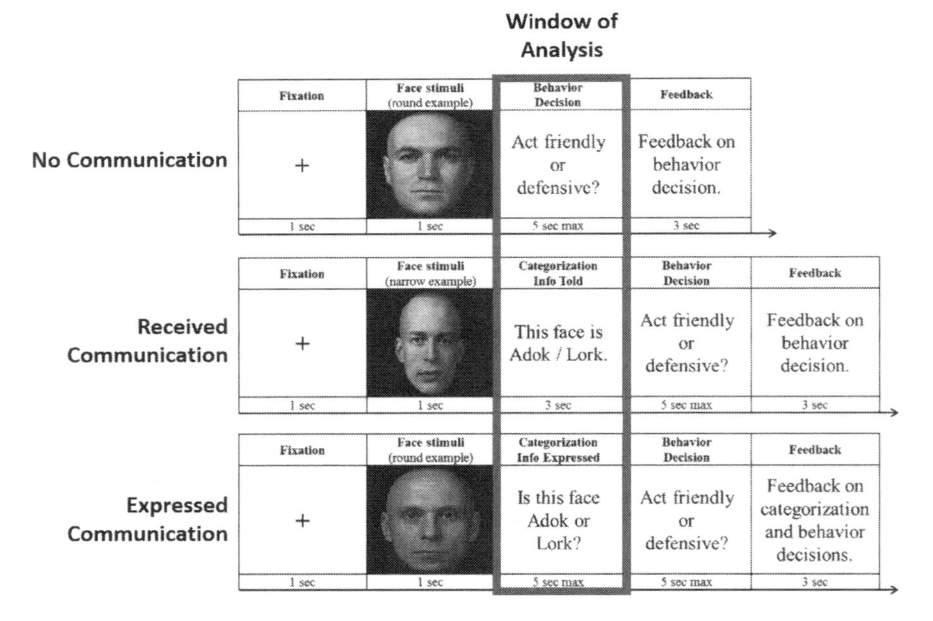

Figure 33.2 Experimental Design

THE EXPERIMENT AND FINDINGS

EEG Data Acquisition and Preprocessing

EEG data was recorded during the experiment described above using a 64-channel BioSemi ActiveII system at a sampling rate of 512 Hz. The 64-electrode scalp placement used an elastic electrode cap (Electro-cap International, Inc.) corresponding to the 10–20 International System. Offline preprocessing was conducted using EEGLAB Toolbox (Delorme & Makeig, 2004) within the MATLAB environment. Preprocessing included applying a 1.0 Hz high pass filter, using the *cleanline* method to reduce sinusoidal artifacts and line noise, and an independent components analysis using the *runica* algorithm (Delorme & Makeig, 2004). Based on the experimental design, data was epoched from -4,500 ms to 2,000 ms for each trial. To test the hypotheses, the window of analysis included only the portion of the epoch between the onset of the decision question and the behavioral response to the question (clicking a button on the screen to indicate the decision). Figure 33.2 illustrates the experimental design and the analysis time window.

Behavioral Data Analysis

Response times (RTs) for decisions to act were reciprocally transformed, 1/RT (Ratcliff, 1993), prior to analysis and computed using a repeated-measures ANOVA for each communication condition. As expected, RTs were significantly slower in the no-communication in comparison to both explicit communication conditions $F(2, 311.40)$, $p<0.001$. In addition, RTs in the expressed communication condition were significantly slower than those in the received communication condition, $t(4954.5)=3.04$, $p<0.005$. For ease of interpretation, raw RTs are presented in Table 33.1. Overall, the findings showed a robust effect of communication such that slower response times indicated more conflict processing occurred in the absence of communication, and to a lesser extent for expressed communication. Greater conflict processing in the no-communication is consistent with the superposition state predicted by the QC, and in support of our first hypothesis.

Time-Frequency Analysis of EEG Data

We used frequency band power to estimate the effect of communication on conflict processing/dissonance reduction (FMθ) and evaluative response processes (parietal alpha). Power was represented by event-related spectral perturbations (ERSP) and calculated using the EEGLAB toolbox v14.1 within the MATLAB environment (Delorme & Makeig, 2004). ERSPs are a measure of time-frequency power relative to the onset of an experimental event and represent

Table 33.1 Mean Response Times for Decisions to Act

Communication Condition	*Mean*	*SD*
No communication	821.97[a,b]	581.79
Expressing information	626.26[a,c]	547.78
Receiving information	530.95[b,c]	398.78

Notes
Mean RTs were reciprocally transformed for analysis but presented here in raw values for ease of interpretation. There was a significant main effect for communication, F (2, 311.4), $p<0.001$. Similarly, all pairwise comparisons were significant: (a) t (4,766.8) = 21.77, $p<0.001$; (b) t (4,854.5) = 3.04, $p=0.002$; (c) t (5,335.7) = 22.54, $p<0.001$.

the synchronous activity of neuron populations (Makeig, 1993). ERSPs are calculated as a moving average of frequency band activity across trial time points, normalized by dividing by mean baseline power, and converted to log power in decibels (dB) for analysis and plotting (for detailed information, see Delorme & Makeig, 2004).

We tested our hypotheses in two ways. First, parametric statistical analyses were performed on ERSP values using EEGLAB's study functions (Delorme & Makeig, 2004). One factor repeated-measures ANOVAs were carried out for the condition of communication (expressed, received, or none) on FMθ and parietal alpha power, respectively. The statistical threshold was set to $p < .05$ and the FDR method was used to correct for multiple comparisons. F and p-values were produced at every time/frequency point. Second, the ERSP values for the respective time, frequencies, and channel were averaged and exported to the statistical program R for repeated measures ANOVAs using the *lme4* package (Bates, Meachler, & Bolkner, 2012) and planned comparisons, corrected with the Holm method, using the *multcomp* package (Hothorn, Bretz, & Westfall, 2008).

Hypothesis 1

Results from the repeated-measures analyses supported the hypothesis that communication reduces uncertainty or conflicts during information processing prior to a decision. Consistent with previous research (Cavanagh & Frank, 2014; Cohen & Donner, 2013), we examined FMθ (3–8 Hz) from the canonical frontal midline FCz node from 200–600 ms after the onset of the decision question. In support of *H1a* and as shown in Figure 33.3, there was a main effect for our manipulated factor: communication, $F(2, 89) = 12.88$, $p < 0.001$. Preplanned Tukey contrasts showed that FMθ power in the absence of communication ($M = 1.09$ dB, $SD = 0.79$) was greater in comparison to expressed communication ($M = 0.52$ dB, $SD = 0.55$), $z = 3.36$, $p = 0.001$, as well as received communication ($M = 0.17$ dB, $SD = 0.76$), $z = 5.53$, $p < 0.001$. The results suggest that without definite information about the face category, there was uncertainty with respect to potential alternatives, signaling the need to respond to a conflict. Conversely, explicit communication reduced conflict processing by increasing certainty about potential alternatives and diminishing the need for a conflict response. In support of *H1b*, less FMθ power was observed for receiving information than for expressing information, $z = -2.17$, $p = 0.03$. The difference between the two conditions suggests received information is more helpful for resolving uncertainty during conflict processing than expressed information. That is, receiving information provided more certainty about the potential alternatives.

In Figure 33.3, on the left, the scalp map depicts the location of the FCz node used in our analysis. Four additional nodes (Fz, FC1, FC2, and Cz) comprising the frontal-midline region

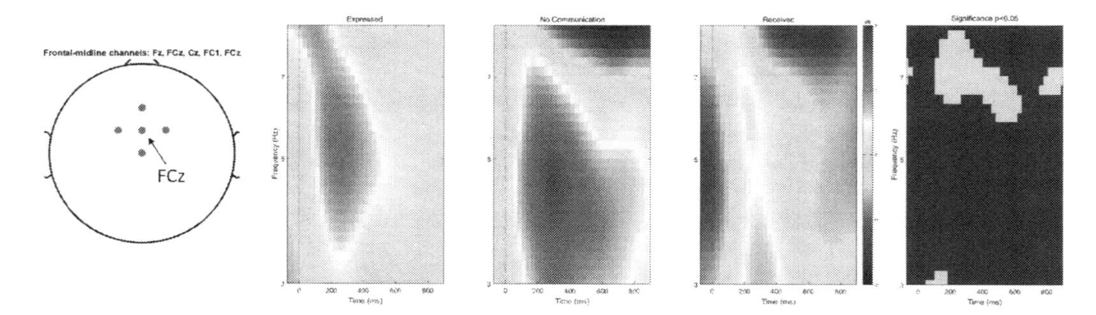

Figure 33.3 Communication, FMθ Power, and Conflict Processing/Dissonance Resolution

are also depicted for visualization purposes although their data were not included in the present analysis. In the center of Figure 33.3, ERSP values in dB representing FMθ (displayed in heat maps) are plotted for the three communication conditions: expressed, none, and received. Areas in dark and light shades indicate active conflict processing. The Y-axis represents the frequency range in our analysis, 3–8 Hz. The X-axis represents the window of analysis, where 0 indicates the onset of the decision question. On the right in Figure 33.3, significant p-values for the main effect of communication—for each frequency at each time point—are depicted in black.

Hypothesis 2

Results of parietal ERSP values provided mixed support for hypotheses about the extent to which communication disturbs the persistent evaluation of potential alternatives. Based on the existing research (Cohen & Ridderinkhof, 2013; Khader & Rösler, 2011), we examined alpha (8–14 Hz) power at canonical central parietal Pz node from 300–800 ms after the onset of the decision question. As shown in Figure 33.4, the main effect for communication was significant, $F(2, 89)=5.43$, $p=0.005$. However, the planned Tukey contrasts revealed only partial support of *H2a*: Alpha suppression—indicating memory search and absence of inhibitory processes in this context—was greater in the absence of communication ($M=-2.67$, $SD=2.19$) in comparison to self-expressing ($M=-1.15$, $SD=1.43$), $z=-4.27$, $p<0.001$, but no different from receiving information ($M=2.17$, $SD=1.83$), $p=0.16$. *H2b* was not supported: Alpha suppression was less for expressing than for receiving communication, $z=-2.86$, $p=0.009$. Though *H2a* was partially supported, the results were generally surprising since the weak conflict signal for received communication was expected to be followed by the absence of a memory search. Similarly, given the larger conflict signal for expressed communication, a more persistent search in memory was expected. The implications of these results are further addressed in the discussion section.

In Figure 33.4, on the left, the scalp map depicts the location of the Pz node used in our analysis. Four additional nodes (CPz, POz, P1, and P2) additionally representing the parietal region are displayed for visualization purposes although their data were not included in the present analysis. In the center of Figure 33.4, ERSP values in dB representing alpha power are plotted for the communication conditions (expressed, none, and received). Areas in dark shades suggest active memory processes and the release of attentional inhibition. The Y-axis represents the frequency range in our analysis, 8–14 Hz. The X-axis represents the analysis window. In Figure 33.4, on the right, significant p-values for the main effect of communication—for each frequency at each time point—are depicted in black.

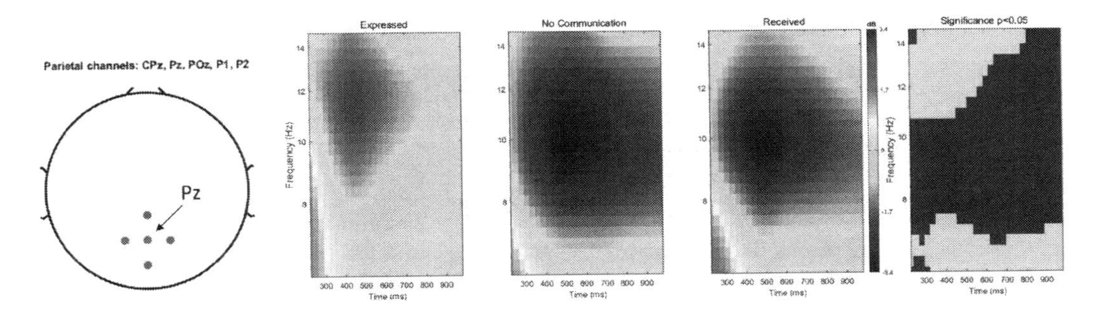

Figure 33.4 Communication, Parietal Alpha Power, and Evaluation of Potential Alternatives

DISCUSSION

Communication provides important information used in decision-making (Fiske & Taylor, 1991). QC views the role of communication of information as essential as the role of measurement in observed "reality" in quantum physics. QC models have been successfully theorizing and understanding human cognition and decision processes, and QC has accounted for a large range of paradoxical findings in the literature of human cognition and decision. This chapter proposes reasons to apply QC to communication research and start to test the plausibility of such application by examining whether the EEG data from a communication-decision experiment, inspired by the two-slit experiment in quantum physics, are consistent with the hypotheses predicted by QC.

QC often develops highly testable mathematical and computational models based on quantum probability theory and compares the new QC model with existing models to show the advantage of the QC models in predicting the empirical data (e.g., Busemeyer & Bruza, 2012; Busemeyer & Wang, 2018, in press; Trueblood et al., 2017; Wang et al., 2014; Wang & Busemeyer, 2016). The current work only starts to explore the plausibility of the QC theorizing of communication, and no model has yet been developed for the EEG data from the communication-decision experiment. Significant effort is needed in future research to develop QC models to theorize the neural mechanism under communication and decision, and to predict the EEG data along with behavioral data. Only in that way can QC models and other competing theories be quantitatively compared and tested. In the current study, we were not interested, and were unable, at such a general prediction level, to differentiate QC from other theories that could predict similar patterns. Instead, we were interested in whether predictions from QC, particularly based on the new concepts of superposition and "collapse"/reduction of the superposition state, would be consistent with neural activity in the brain known to respond to uncertainty and its reduction. In particular, inspired by the two-slit experiment in quantum physics, we additionally expected the way information was communicated—expressing, receiving, or none—would distinguish neural patterns not readily proposed by other theories. Overall, although with mixed findings, the EEG data suggest that the QC theoretical and computational approach might be a useful way to formalize communication of information and its fundamental impact on subsequent cognition and decision.

Our analysis of temporally sensitive frequency band power fully supported our first hypothesis: Communication resolves conflict processing. Specifically, communication conveying evaluative information reduced conflict as indexed by FMθ power, and that receiving information resulted in less conflict than expressing it. Response times conformed to FMθ activity across the three communication conditions.

The mixed findings for our second hypothesis—communication reduces the persistence evaluation of alternatives—are intriguing. Recall, evaluative processes were operationalized as attentional release and search in memory indexed by parietal alpha power. As expected, the absence of communication, in comparison to self-expressing, resulted in the persistent evaluation of possible alternatives. However, surprisingly, the evaluation of potential alternatives did not differ between the absence of and receiving communication. Also surprisingly, self-expressing resulted in less evaluative processing in comparison to receiving information.

Our surprise is based on assumptions about the relationship between conflict processing and related memory retrieval activity. Consistent with QC predictions and behavioral evidence (Busemeyer et al., 2009; Wang & Busemeyer, 2016) as well as neural activity indexing memory and attention operations (Khader & Rösler, 2011; Klimesch, 2012), we expected to observe a symmetrical relationship between conflict processing and subsequent evaluative processes.

However, this relationship only emerged in the no-communication condition where a large conflict signal was followed by prominent evaluative processes. For self-expressing, a moderate conflict signal was followed by very little evaluative processing. For receiving communication, a small conflict signal resulted in a moderate amount of evaluative processing. An explanation for such findings may be that people are more certain about their own estimations (Clark et al., 2013), an internal process not evident in the conflict signal but observable during the subsequent evaluation stage.

In sum, by applying cognitive control theory to neural oscillations in EEG data, we found support for QC predictions about the effect of communication on a superposition state and state reduction (Busemeyer et al., 2009; Wang & Busemeyer, 2016). This very initial exploration of the plausibility of applying QC to understanding communication processes and effects shows the good potential of the QC approach. The QC approach may be especially useful to communication research. First, it places communication of information at a central stage of human cognition and decision, and views it as playing an essential "measurement" role in human cognitive and behavioral systems. Second, the nature of communicating information sequentially makes the theory especially good at explaining sequential effects based on quantum probability theory (a non-commutative probability theory), as it is naturally suitable. Third, besides new concepts and principles, such as incompatibility, superposition, state reduction, and the uncertainty principle, QC provides a coherent set of mathematical and computational tools, principled on quantum probability theory, to model communication processes and effects. With the increasing complexity of communication phenomena (e.g., neural process in communication, communication networks), a mathematical and computational approach has become essential for understanding communication. The QC approach provides an innovative theorizing and modeling opportunity for communication researchers.

ACKNOWLEDGMENTS

This research has been supported by the U.S. National Science Foundation (SES-1560501 and SES-1560554) and the U.S. Air Force grant of Scientific Research (FA9550-15-1-0343) to Wang and Busemeyer.

NOTE

1. Notably, we do not claim that (part of) the brain is some type of quantum computer (cf., Hameroff & Penrose, 1996; Stapp, 1993; Vitiello, 1995) as our QC does not depend on the quantum brain hypothesis (Bruza et al., 2015; Wang et al., 2013). Instead, we simply use the mathematics of quantum theory, particularly quantum probability theory, to model and predict cognition and behavior.

REFERENCES

Banich, M. T. (2018). Emerging themes in cognitive control: Commentary on the special issue of Psychophysiology entitled "Dynamics of Cognitive Control: A View Across Methodologies." *Psychophysiology, 55*(3), e13060.

Bates, D., Maechler, M., & Bolker, B. (2012). lme4: Linear mixed-effects models using S4 classes. R package version 0.999999-0.

Bates, D., Maechler, M., Bolker, B., & Walker, S. (2015). Fitting linear mixed-effects models using lme4. *Journal of Statistical Software, 67*(1), 1–48.

Botvinick, M. M., Braver, T. S., Barch, D. M., Carter, C. S., & Cohen, J. D. (2001). Conflict monitoring and cognitive control. *Psychological Review, 108*(3), 624–652.

Bruza, P. D., Wang, Z., & Busemeyer, J. R. (2015). Quantum cognition: A new theoretical approach to psychology. *Trends in Cognitive Sciences, 19*(7), 383–393.

Busemeyer, J. R., & Bruza, P. (2012). *Quantum models of cognition and decision.* Cambridge, England: Cambridge University Press.

Busemeyer, J. R., Pothos, E. M., Franco, R., & Trueblood, J. (2011). A quantum theoretical explanation for probability judgment errors. *Psychological Review, 118*, 193–218.

Busemeyer, J. R., & Wang, Z. (2015). What is quantum cognition, and how is it applied to psychology? *Current Directions in Psychological Science, 24*, 163–169.

Busemeyer, J. R., & Wang, Z. (2018). Hilbert space multi-dimensional modeling. *Psychological Review, 125*, 572–591.

Busemeyer, J. R., & Wang, Z. (in press). Hilbert space multidimensional modeling of continuous variables. *Philosophical Transactions of the Royal Society A: Mathematical, Physical and Engineering Sciences.*

Busemeyer, J. R., Wang, Z., & Lambert-Mogiliansky, A. (2009). Empirical comparison of Markov and quantum models of decision making. *Journal of Mathematical Psychology, 53*(5), 423–433.

Busemeyer, J. R., Wang, Z., & Shiffrin, R. (2015). Bayesian comparison of a quantum versus a traditional model of human decision making. *Decision, 2*, 1–12.

Cavanagh, J. F., & Frank, M. J. (2014). Frontal theta as a mechanism for cognitive control. *Trends in Cognitive Sciences, 18*(8), 414–421.

Clark, A., Adams, F., Aizawa, K., Alais, D., Burr, D., Alink, A., ... Der, R. (2013). Whatever next? Predictive brains, situated agents, and the future of cognitive science. *Behavioral and Brain Sciences, 36*(3), 181–204.

Cohen, M. X. (2014a). *Analyzing time series data: Theory and practice.* Cambridge, MA: MIT Press.

Cohen, M. X. (2014b). A neural microcircuit for cognitive conflict detection and signaling. *Trends in Neurosciences, 37*(9), 480–490.

Cohen, M. X., & Donner, T. (2013). Midfrontal conflict-related theta-band power reflects neural oscillations that predict behavior. *Journal of Neurophysiology, 110*(12), 2752–2763.

Cohen, M. X., & Ridderinkhof, K. R. (2013). EEG source reconstruction reveals frontal-parietal dynamics of spatial conflict processing. *PLoS One, 8*(2), e57293.

Delorme, A., & Makeig, S. (2004). EEGLAB: An open source toolbox for analysis of single-trial EEG dynamics including independent component analysis. *Journal of Neuroscience Methods, 134*(1), 9–21.

Fisher, M. (2015). Quantum cognition: The possibility of processing with nuclear spins in the brain. *Annals of Physics, 362*, 593–602.

Fiske, S., & Taylor, S. (1991). *Social cognition.* New York, NY: McGraw-Hill.

Gratton, G., Cooper, P., Fabiani, M., Carter, C. S., & Karayanidis, F. (2018). Dynamics of cognitive control: Theoretical bases, paradigms, and a view for the future. *Psychophysiology, 55*(3), e13016.

Hameroff, S. R., & Penrose, R. (1996). Conscious events as orchestrated spacetime selections. *Journal of Consciousness Studies, 3*, 36–53.

Hanslmayr, S., Staresina, B. P., & Bowman, H. (2016). Oscillations and episodic memory: Addressing the synchronization/desynchronization conundrum. *Trends in Neurosciences, 39*(1), 16–25.

Hothorn, T., Bretz, F., & Westfall, P. (2008). Simultaneous inference in general parametric models. *Biometrical Journal, 50*, 346–363.

Khader, P. H., & Rösler, F. (2011). EEG power changes reflect distinct mechanisms during long-term memory retrieval. *Psychophysiology, 48*(3), 362–369.

Klimesch, W. (1999). EEG alpha and theta oscillations reflect cognitive and memory performance: A review and analysis. *Brain Research Reviews, 29*(2–3), 169–195.

Klimesch, W. (2012). Alpha-band oscillations, attention, and controlled access to stored information. *Trends in Cognitive Sciences, 16*(12), 606–617.

Makeig, S. (1993). Auditory event-related dynamics of the EEG spectrum and effects of exposure to tones. *Electroencephalography and Clinical Neurophysiology, 86*(4), 283–293.

Namkoong, K., Shah, D. V., McLaughlin, B., Chih, M.-Y., Moon, T. J., Hull, S., & Gustafson, D. H. (2017). Expression and reception: An analytic method for assessing message production and consumption in CMC. *Communication Methods and Measures, 11*(3), 153–172.

Pingree, R. J. (2007). How messages affect their senders: A more general model of message effects and implications for deliberation. *Communication Theory, 17*(4), 439–461.

Pothos, E. M., & Busemeyer, J. R. (2013). Can quantum probability provide a new direction for cognitive modeling? *Behavioral and Brain Sciences, 36*(3), 255–274.

Pothos, E. M., Busemeyer, J. R., & Trueblood, J. S. (2013). A quantum geometric model of similarity. *Psychological Review, 120*, 679–696.

Ratcliff, R. (1993). Method for dealing with reaction time outliers. *Psychological Bulletin, 114*(3), 510–532.

Ridderinkhof, K. R., Ullsperger, M., Crone, E. A., & Nieuwenhuis, S. (2004). The role of the medial frontal cortex in cognitive control. *Science, 306*(5695), 443–447.

Shannon, C. E. (1948). A mathematical theory of communication. *Bell System Technical Journal, 27*, 379–423.

Stapp, H. P. (1993). *Mind, matter, and quantum mechanics*. New York, NY: Springer.

Townsend, J. T., Silva, K. M., Spencer-Smith, J., & Wenger, M. J. (2000). Exploring the relations between categorization and decision making with regard to realistic face stimuli. *Pragmatics & Cognition, 8*(1), 83–105.

Trueblood, J. S., & Busemeyer, J. R. (2011). A quantum probability account of order effects in inference. *Cognitive Science, 35*(8), 1518–1552.

Trueblood, J. S., Yearsley, J. M., & Pothos, E. M. (2017). A quantum probability framework for human probabilistic inference. *Journal of Experimental Psychology: General, 146*, 1307–1341.

van Driel, J., Swart, J. C., Egner, T., Ridderinkhof, K. R., & Cohen, M. X. (2015). (No) time for control: Frontal theta dynamics reveal the cost of temporally guided conflict anticipation. *Cognitive, Affective, & Behavioral Neuroscience, 15*(4), 787–807.

Vitiello, G. (1995). Dissipation and memory capacity in the quantum brain model. *International Journal of Modern Physics B, 9*, 973–989.

Wang, Z., & Busemeyer, J. R. (2013). A quantum question order model supported by empirical tests of an a priori and precise prediction. *Topics in Cognitive Science, 5*(4), 689–710.

Wang, Z., & Busemeyer, J. R. (2015a). Comparing quantum versus Markov random walk models of judgements measured by rating scales. *Philosophical Transactions of the Royal Society A: Mathematical, Physical and Engineering Sciences, 374*(2058), 20150098.

Wang, Z., & Busemeyer, J. (2015b). Reintroducing the concept of complementarity into psychology. *Frontiers in Psychology, 6*, 1822.

Wang, Z., & Busemeyer, J. R. (2016). Interference effects of categorization on decision making. *Cognition, 150*, 133–149.

Wang, Z., Busemeyer, J. R., Atmanspacher, H., & Pothos, E. M. (2013). The potential of using quantum theory to build models of cognition. *Topics in Cognitive Science, 5*, 672–688.

Wang, Z., Solloway, T., Shiffrin, R. M., & Busemeyer, J. R. (2014). Context effects produced by question orders reveal quantum nature of human judgments. *Proceedings of the National Academy of Sciences, 111*(26), 9431–9436.

White, L. C., Pothos, E. M., & Busemeyer, J. R. (2014). Sometimes it does hurt to ask: The constructive role of articulating impressions. *Cognition, 133*(1), 48–64.

Yoo, S. W., Kim, J. W., & Gil de Zúñiga, H. (2017). Cognitive benefits for senders. *Journalism & Mass Communication Quarterly, 94*(1), 17–37.

34

Advancing the Model of Intuitive Morality and Exemplars

Ron Tamborini and René Weber

The Model of Intuitive Morality and Exemplars (MIME) is a comprehensive model developed to explicate processes governing the selection, valuation, and production of media content (Tamborini, 2011, 2013). The framework provided by the model combines logic from several well-known media theories within the social intuitionist perspective from moral psychology to describe how media content and moral intuitions interact to influence each other and subsequent media consumption. As such, the MIME is one of few models in communication science that combines ultimate with proximate causes of human behavior (Tinbergen, 1963) to explain and predict media selection and its consequences.

In terms of *ultimate* causes, the MIME builds on the social intuitionist approach (Haidt, 2001) and moral foundations theory (Haidt & Joseph, 2007), both of which posit a set of universally held intuitive moral motivations that cut across groups, cultures, civilizations, and—from the perspective of *phylogeny*—even across closely related species (e.g., Brosnan, 2013). Intuitive moral motivations (or, simply, moral intuitions) can be understood as biologically rooted sensitivities that produce positive or negative affect in response to actions that are alternatively beneficial or detrimental to small groups and society at large (Haidt & Joseph, 2007). The logic underlying the MIME suggests that each moral intuition is distinguished by the unique adaptive function it serves in providing humans an evolutionary advantage (Haidt & Joseph, 2007; Lazarus, 1991). Drawing on Tinbergen's influential terminology, innate moral intuitions possess *adaptive value*. Separately, each intuition is understood as a psychological system that, when activated, produces an instinctive, preconscious judgment of right or wrong in a corresponding domain of social behavior.

These moral drivers, referred to in MIME research as *altruistic* intuitions, compel humans to act in a manner that benefits the group or others rather than themselves. Although discussion continues regarding the number of unique moral intuitions thought to exist (see Iyer, Koleva, Graham, Ditto, & Haidt, 2012, for discussion), five seem largely agreed upon. They include:

1. care, which is characterized by a desire to help others in need and to mitigate harm;
2. fairness, the innate drive to see other individuals treated justly, with equality and equity;
3. ingroup loyalty, which highlights a bias favoring benefit toward one's ingroup and against outgroup members;

4. respect for authority, characterized by the desire to follow benevolent leaders and obey the traditions of legitimate hierarchies;
5. purity, or the urge to pursue cleanliness, godliness, and a noble lifestyle while rebuffing social contamination.

In terms of *proximate* causes of human behavior, the MIME is sensitive to the fact that whereas moral intuitions are assumed to be innate and biologically rooted, *ontogeny* suggests that they are flexible, and are amended, shaped, and contextualized by experiences over a human's lifetime. For instance, humans are said to possess a "preparedness" through which they are able to gain specific types of moral knowledge related to intuitions; and, as a result, people "can very easily be taught or made to care [more] about harm, fairness, ingroups, [or] authority" (Haidt & Bjorklund, 2008, p. 204). In this regard, several agents have been noted for their ability to make intuitions either temporarily or chronically salient (Haidt & Bjorklund, 2008). These include messages received through direct moral education, various cultural or societal influences and, particularly relevant here, exposure to different forms of narrative media.

Importantly, the MIME's explanation of humans' selection, valuation, and production of media content does not end with ultimate general and proximate developmental processes; rather, it posits specific neurological and behavioral *mechanisms* behind media selection and its consequences. Since its first publication in 2011 (Tamborini, 2011, 2013), the model has sparked extensive research examining these mechanisms and their affordances. By our count, to date, the MIME has been cited in close to 200 published research papers and tested in nearly 80 empirical studies. Although research has supported many of the model's initial predictions, some studies have produced unexpected results. Other studies have discovered potential shortcomings in measurement and suggested methodological improvements. Thus, we believe the time is ripe to summarize the accomplishments, shortcomings, and latest advances of the MIME. In the sections that follow, we review recent literature regarding the presence of moral exemplars in media content, and evidence of the model's short-term and long-term mechanisms. We follow this with a discussion of theoretical and methodological advances to the MIME, and end by examining its future as a comprehensive, biologically rooted model of media production, selection, and influence.

INTUITION EXEMPLARS IN MESSAGE CONTENT

The proliferation of media content that exemplifies adherence to or contravention of specific intuitions is at the root of MIME predictions. Using tools designed to identify both manifest and latent content, MIME research examining national and international media supports suggestions that moral exemplars are replete across a wide range of content (see Weber et al., 2018). For example, several studies on entertainment produced for U.S. child audiences found a predominance of content exemplifying adherence to altruistic intuitions, and noted that this tendency decreased in content for older children (Tamborini, Hahn, Prabhu, Klebig, & Grall, 2017) and adults (Hahn, Tamborini, Novotny, Grall, & Klebig, 2018). Other studies on general audience programming support predicted differences in cross-cultural entertainment. For instance, research shows that exemplars of care and ingroup loyalty were more common in English-language soap operas than in Spanish-language telenovelas (Mastro, Enriquez, Bowman, Prabhu, & Tamborini, 2013), and that respect for authority was exemplified more often in a Hindi-language serial than an English-language U.S. soap opera (Prabhu & Tamborini, 2018).

Although most MIME-based content analyses have focused on entertainment, the model's predictions cut across other narrative forms. In the political domain, research on U.S. newspaper reports of Osama bin Laden's death (Bowman, Lewis, & Tamborini, 2014) showed that headlines from conservative areas of the country framed the event as patriotic (highlighting loyalty and authority), whereas those from liberal regions framed it more in terms of justice restoration (highlighting fairness and reciprocity). Other research on the mission statements of terrorist organizations found that, much like content produced by conventional media organizations, terrorist group messaging demonstrates a hierarchy of intuitive values (Hahn, Tamborini, Novotny et al., 2018). Overall, findings from research across a broad spectrum of media content are consistent with the MIME's notion regarding the ubiquitous presence and pattern of exemplars emphasizing altruistic intuitions. Moreover, research combining content analysis with audience surveys (Prabhu & Tamborini, 2018) provides initial evidence in line with the model's predictions regarding the effects of exposure to these exemplars.

THE MIME'S SHORT-TERM AND LONG-TERM COMPONENTS

The reciprocal processes depicted in the MIME outline a dynamic relationship between individuals and their environments that can add to our understanding of issues involving moral judgments. The short-term and long-term components of the model describe how exposure to exemplars (in media or non-media environments) emphasizing the adherence to or contravention of distinct moral intuitions can influence the temporary and persisting salience of those intuitions. Patterns of intuition salience (i.e., the relative accessibility of different intuitions) and exemplar salience (i.e., the accessibility of entities or events that activate certain intuitions) at any given moment combine to shape instinctive and reflective processes that govern an individual's moral judgement, appraisal of media content, and subsequent patterns of media selection. Aggregate patterns of selective exposure to media upholding salient intuitions alter features of subsequently manufactured content.

The Model's Short-term Component

The MIME's short-term logic outlines dual-process mechanisms through which viewers engage in media appraisal (Tamborini, 2011, 2013). The model begins by describing how exposure to exemplars can increase the accessibility of specific moral intuitions in conscious and preconscious processing, and thus increase the weight of those intuitions in any subsequent appraisal. Appraisal outcomes are governed by two different thought processing systems: one that is quicker and instinctive and another that is slower and deliberative (Evans & Stanovich, 2013). Slower, controlled appraisals are produced when media content prompts conflict between intuitions in conscious or preconscious processing, whereas fast, intuitive appraisals occur when media content is void of noticeable conflict. In either case, positive appraisal results from observing adherence to more heavily weighted intuitions, and negative appraisal from their contravention.

Applied to narratives, the model's dual-process mechanisms are used to differentiate forms of positive media appraisal referred to as *enjoyment* and *appreciation*. Enjoyment is conceptualized as a positive reaction produced by fast, intuitive processing that occurs when narrative circumstances satisfy one (or more) of an audience member's salient intuitions without raising any thought of conflict with other instinctive drives. As such, enjoyment is understood as the immediate satisfaction of all salient intuitions. Correspondingly, appreciation is defined as a positive response produced by deliberative processing that arises when reflecting on circumstances in which two or more intuitions are in conflict. This type of positive response to conflict often occurs in narratives when characters in difficult situations have no choice but to violate a salient intuition in order to uphold another intuition weighed more heavily. MIME-based reasoning suggests that in order to appraise the character's actions positively, audience members must first accept the decision to sublimate the intuition violated to the one that was upheld. Both enjoyment and appreciation of media content predict positive evaluation of actions observed and the characters that performed them, a desire for similar content, and selective exposure to media that promises such action, characters, and content.

MIME scholars hold that skillful writers regularly craft narratives to vary the salience of intuitions that subsequently will be upheld or violated. As such, if a narrative makes loyalty dominantly salient (i.e., so accessible that any additional concerns go unconsidered) and pays little attention to other intuitions, audiences should greatly enjoy seeing a star athlete choose to play for the hometown team instead of playing for some other college. In comparison, if a narrative makes fairness overridingly salient (i.e., accessible enough to prevail over other conscious concerns, but not enough to preclude their consideration), audiences should appreciate it when a protagonist decides to hire a stranger over his own brother (violating ingroup loyalty), given the audience knows that the stranger is slightly more qualified (upholding fairness). Of course, had loyalty been made overridingly salient, audiences should appreciate seeing the protagonist bypass the stranger (violating fairness) to hire his devoted sibling (upholding loyalty).

Extant research is consistent with the MIME's predictions that intuitions made salient play a central role in shaping appraisal of narratives, subsequent moral judgments, and related behaviors, and selective exposure to similar media content. Initial evidence that intuitions emphasized in media can influence intuition accessibility has been found in research on general audience exposure to both narrative entertainment (Tamborini, Lewis et al., 2016; Tamborini, Prabhu, Lewis, Grizzard, & Eden, 2016) and news (Tamborini, Prabhu, Wang, & Grizzard, 2013). Similar support can be found for media's influence on intuition accessibility in children (Hahn, Tamborini, Weber, Bente, & Sherry, 2019).

The short-term component's predictions about how intuition accessibility can influence subsequent appraisal have found support in research directly testing the dual-process model of enjoyment and appreciation (Lewis, Tamborini, & Weber, 2014). This study showed that scenarios without intuition conflict were responded to faster and *enjoyed* more, whereas those containing conflict produced slower reactions and greater *appreciation*. Other research has produced similar findings in line with this dual-process logic (Janicke & Raney, 2018).

Research has also demonstrated that media-induced intuition salience can shape moral judgment and behavior. For example, exposure to narratives with a clear moral message were found to increase cooperation among group members (Lewis, Grizzard, Mangus, Rashidian, & Weber, 2017). Related research demonstrated not only that watching news of a terrorist attack increased the salience of the authority intuition, but this salience mediated decisions to donate money to outgroup members in need (Tamborini, Hofer et al., 2017). Finally, combining the full logic of the MIME's short-term processes, one study supported a model suggesting that (1) exposure to media can temporarily increase the salience of altruistic intuitions; (2) intuition salience affects the enjoyment (or appreciation) of content that highlights (or downplays conflict between) those intuitions; and (3) enjoyment of that content predicts selective exposure to media emphasizing related intuitions (Prabhu et al., 2014).

The Model's Long-term Component

The model's long-term component identifies macro-level and cross-level processes that affect media production and the resulting message environments within which audiences dwell. At the macro level, the model suggests that the representation of moral intuitions in media content is influenced by aggregate patterns of audience evaluation and choice. The preference of aggregate audiences for media that adhere to salient intuitions drives the production of content upholding those intuitions. As such, content produced will emphasize adherence to moral intuitions in a manner that conforms to the relative salience of those intuitions in target audiences. Notably, the relative salience of intuitions in different aggregate target audiences is expected to vary across media systems and cultures. Similarly, the entities and events that come to exemplify adherence to or contravention of specific intuitions are expected to vary from one target audience to another.

The model's cross-level processes describe how the production of content adhering to audience preferences (i.e., upholding aggregate patterns of intuition salience) cycles back to permeate the individual's media environment. Adopting logic from exemplification theory (Zillmann, 2002), the MIME suggests that recurrent exposure to intuition-adhering exemplars in media (or non-media) environments reinforces the shared patterns of intuition salience residing in audience members served by different media systems. These processes form the foundation for predictions regarding (1) the patterns of adherence to different moral intuitions found in media content produced for different audiences; (2) the selection and appraisal of that content by members of those audiences; and (3) the prolonged effect of repeated exposure to this content on shared patterns of intuition salience that shape the moral judgments of individuals and aggregate groups.

Several recent studies have supported the MIME's long-term predictions. For example, evidence consistent with the predictions that content produced for different target audiences will adhere to intuition salience patterns extant in those audiences can be seen in research showing that intuition emphasis in different television serials covaried with trait patterns of intuition salience in respective viewers (Prabhu & Tamborini, 2018; Prabhu, Tamborini, Klebig, Grall, & Hahn, 2015). Evidence also supports the contention that chronic intuition salience drives patterns of media appraisal and selective exposure. One study found that trait differences in intuition salience predicted preferences for different narratives that varied in adherence to those intuitions (Tamborini, Eden et al., 2013). Related research (Eden & Tamborini, 2017) demonstrated that trait salience could also predict appraisal of narrative characters (i.e., disposition formation). Finally, claims regarding repeated exposure's enduring effect on intuition salience were corroborated in an experiment that varied exposure to an online serial over eight weeks (Eden et al., 2014). The findings suggested that recurring exposure to the soap opera's care and purity intuition-laden content predicted greater salience for those intuitions a week after final viewing.

CONCEPTUAL ADVANCES TO THE MIME

Since the MIME's inception, researchers have attempted to advance understandings of the theoretical mechanisms that underlie the model as well as the methodological procedures needed to study them. Among the most recent efforts are those that use the MIME's understanding of moral intuition salience to define prosocial and antisocial media effects, and to examine the role of egoistic intuitions in the model's long-term and short-term processes.

Defining Prosocial and Antisocial Media Effects

MIME research in its infancy has attempted to reduce ambiguity associated with use of the terms *prosocial* and *antisocial* in media research. Despite being a major focus of study, no commonly shared conceptualization of prosocial and antisocial appears to exist among investigators examining media content or its effects. MIME scholars argue that the model's dual-process understanding of moral intuition salience can be used to address potential confusion in this area of research (Tamborini, Hahn, Novotny et al., 2018).

Equivocality in what constitutes a prosocial or antisocial act can lead researchers to label the same act differently or to use the same label for starkly different acts. For instance, if the town sheriff shoots a person robbing a bank in order to enjoy the thief's suffering, the act might be called brutality. But if the same act were performed out of a sense of duty to the town, the act might be called honorable. These types of motivations are central to classifying behavior as prosocial or antisocial, and the absence of a comprehensive scheme that explicates this process can confound research in this area. The potential for this problem is amplified in narratives that show conflict where the salience of one altruistic intuition overrides another. Because by definition the sublimated intuition in such cases is also salient, awareness of its violation could make it difficult to label the act as prosocial or antisocial.

To address this concern, MIME researchers have promoted a definition for prosocial and antisocial behavior based on salient motivations (Tamborini, Hahn, Novotny et al., 2018). Using the model's framework, they define prosocial and antisocial actions in line with the dual-process model's determination of positive and negative appraisal. When only one altruistic intuition is made salient, and an act upholds that dominantly salient intuition over other unnoticed intuitions, the act is easily classified as prosocial and its violation antisocial (as no other salient concern is present). In these cases, the audience's experience of enjoyment or repugnance (respectively) shows the actions' intuitive prosocial or antisocial natures. In comparison, when two intuitions are salient, violating a subordinated altruistic intuition in order to uphold an overridingly salient one would be classified as prosocial (and its violation antisocial). In these cases, the audience would experience appreciation or discontent (respectively), again showing the actions' intuitive prosocial or antisocial natures. However, if neither intuition is made overridingly salient, the viewer experiences a moral dilemma, regardless of which intuition is upheld, which results in affective ambivalence.

Pilot tests examining the appraisal of text depicting non-conflicted acts suggest that both mass communication researchers and members of the general public intuitively rely on altruistic intuitions to label acts that adhere to or contravene altruistic intuitions as prosocial and antisocial respectively. This line of thinking offers practical advantages to those attempting to conduct research in this area, as it helps establish boundary conditions to distinguish these concepts. The framework offers guidelines not only for identifying prosocial and antisocial content in media, but also for predicting the short-term and long-term influence of exposure to this content, and for resolving ambiguity in research by different scholars examining issues related to prosocial and antisocial media content or exposure.

The Model of Intuitive Motivations and Exemplars

In addition to emphasizing how altruistic intuitions shape media appraisal and related outcomes, MIME research has recently begun to examine egoistic intuitive drives. Whereas altruistic intuitions mostly compel benefit to others or an individual's group, egoistic intuitions compel benefit to self. Early MIME work identified egoistic motivations as central to the conflict found in many narrative plots (Tamborini, 2013) and promoted efforts to define and identify these motivations in content. While acknowledging that it is sometimes difficult to distinguish whether compulsions to benefit others may be rooted in benefit to self, recent MIME efforts adopted six egoistic motivations from two bodies of research in preliminary attempts to develop a comprehensive scheme (Prabhu et al., 2014). The first set of egoistic motivations came from self-determination theory (SDT; Deci & Ryan, 1985), which identifies three innate motivations as driving forces in a person's emotional well-being: (1) competence, the desire for recognition of one's skill and accomplishments; (2) autonomy, a penchant to feel in command of one's life; and (3) relatedness, the need to feel connected to others. The second set of egoistic motivations was adopted from research on universal human values (UHV; Schwartz, 1994). Schwartz (1994) determined "ten motivationally distinct types of values" (p. 21), seven of which overlap with either SDT's egoistic or MFT's altruistic intuitions. MIME researchers adopted the remaining three for inclusion in its scheme of egoistic intuitive motivations: (1) hedonism, the yearning for physical pleasure; (2) power, the desire for control over resources and other people; and (3) security, the desire for personal safety and stability. In total, the scheme of 11 intuitive motivations comprised the five altruistic motivations (care, fairness, ingroup loyalty, authority, and purity) and six egoistic motivations (competence, autonomy, relatedness, hedonism, power, and security).

This broader scheme has underpinned studies on the model of intuitive *motivations* and exemplars examining the presence of egoistic motivations in a wide range of media content (Hahn, Tamborini, Klebig et al., 2018; Lewis & Mitchell, 2014; Tamborini, Hahn, Klebig et al., 2018; Tamborini, Hahn, Prabhu et al., 2017). Recent MIME-based investigations have attempted to distinguish the unique roles of altruistic and egoistic intuitions in disposition theory's claim that audiences like to see good things happen to people they like. Although lengthy discussion is beyond the scope of the present chapter, this work focuses on the importance of egoistic intuitions in shaping narrative enjoyment. In line with most disposition theory reasoning, logic here contends that adherence to altruistic intuitions is central in audience appraisal of whether characters are "people they like." The more novel part of this research contends that the satisfaction of protagonist's egoistic needs are central in audience appraisal of whether the outcome for a character is seen as a "good thing." The MIME's discussion of dominant and overriding salience plays a central role in locating this appraisal along a continuum from repulsion through dissatisfaction, ambivalence and appreciation to joy.

METHODOLOGICAL ADVANCES OF THE MIME

Recent methodological advances of the MIME are primarily concerned with two major challenges. First, how can MIME researchers measure representations of moral intuitions in media content (exemplars) with high reliability and validity? If moral information in media content is context-dependent and assumed to trigger an intuitive response among individuals that differ in their sensibilities toward different types of moral information, then the application of traditional content analyses seems limited, as these approaches require a deliberate and consistent judgment from trained expert coders. Second, how can we measure intuition salience in individuals, which, again, is assumed to be inaccessible to conscious deliberations and as such difficult to measure with instruments that are based on self-reports?

Measuring the Representation of Moral Intuitions in Media Content

Early MIME researchers (Mastro et al., 2013) developed the MIME coding scheme, a codebook designed to describe the core essence of each altruistic intuition, which prompts human coders to mark the presence of a specific intuition when they see behavior *motivated* by that intuition. For instance, the essence of care deals with whether an act gives aid to or hinders the well-being of others in need. If Daniel Tiger picked up his friend who fell and wanted help, the act would *uphold care*. But if Daniel instead threw down the friend who was trying to arise, this would *violate care* (i.e., inflict harm). Over time, the MIME coding scheme has been expanded to identify the presence of both altruistic and egoistic intuitions, and it has been applied to areas of media content including newspaper headlines (Bowman et al., 2014), serial television (Mastro et al., 2013; Prabhu et al., 2015; Prabhu & Tamborini, 2018), children's television (Hahn et al., 2017; Lewis & Mitchell, 2014), children's books and movies (Tamborini, Hahn, Klebig et al., 2018), song lyrics (Hahn, Tamborini, Klebig et al., 2018), and experimental text and video stimuli (e.g., Tamborini, Prabhu et al., 2016).

More recent advances in MIME content analyses have taken a different track. In six studies, Weber et al. (2018) demonstrated that, under conditions of truly independent content coding procedures, the high reliabilities of early MIME coding schemes could not be replicated. Instead of a conventional approach using lengthy procedures that train a few people to carefully deliberate whether content fits predefined categories to identify the "essence" of each intuition, Weber et al. (2018) developed an online platform named the Moral Narrative Analyzer (MoNA; https://mnl.ucsb.edu/mona) and implemented a crowd-sourced procedure in which a relatively large number of coders are provided a simple overview of each of the intuitions, and then rate content presented in MoNA based on their intuitive responses to that content. Weber et al. show that this intuition-based, crowd-sourced approach to coding moral themes in media content produces acceptable inter-coder reliabilities among coders, and importantly, more valid codings of moral intuitions represented in media content.

Dictionary Approaches

Early moral foundations researchers attempting to extract moral intuitions from media content largely relied on the Moral Foundations Dictionary (MFD; Graham, Haidt, & Nosek, 2009). Notably, the use of the MFD to capture the representation of moral themes in media content is limited by the fact that the original MFD is simply a subjective lexicon of virtue and vice words across the five intuitions (compiled by Graham, Haidt, & Nosek without validation). By nature, the MFD is only capable of extracting *manifest* moral content, and even then, only when the specific words used appear in the MFD. This is problematic considering that moral content is often inherently *latent* and that many words in the MFD can have moral or non-moral meanings depending on context (Weber et al., 2018). The MFD is also limited by the fact that the list of words it offers to represent each intuition is far from comprehensive. To address these shortcomings, Hopp, Cornell, Fisher, Huskey, and Weber (2018) have used the MoNA platform and collected tens of thousands of content codings among a large crowd of coders. These researchers used modern natural language processing tools and created an extended, crowd-based Moral Foundation Dictionary (eMFD). Validations of this new MFD are currently underway. The eMFD is available to the research community at https://osf.io/vw85e/.

Measuring the Affective Component of Intuition Salience: The MF-AMP

Affect misattribution procedures (AMP) have been implemented in past research to assess implicit attitudes related to moral affect (Hofmann & Baumert, 2010). Because preconscious attitudes in these areas may not attain the level of consciousness required for awareness, AMPs assess attitudes implicitly through misattribution, or attributing the effect of one stimulus as the effect of another. AMPs rely on projection, which is one form of misattribution. In this procedure, individuals mistakenly attribute the affect associated with their preconscious attitudes toward a target stimulus for the affect associated with a neutral symbol.

Because moral intuitions operate preconsciously, previous attempts to measure their salience using self-report instruments have been marred by characteristically low reliability. The moral foundations-affect misattribution (MF-AMP) procedure's ability to assess intuition accessibility at a preconscious level was designed to address these concerns (Tamborini, Prabhu et al., 2016). The MF-AMP begins by briefly exposing people to a target word exemplifying the upholding or violation of an intuition (e.g., obedience or disrespect) before asking participants to appraise a neutral symbol (e.g., an unfamiliar Chinese logogram) as pleasant or unpleasant. The speed with which participants appraise the neutral symbol in concordance with the word prime's affective valence (e.g., pleasant for obedience and unpleasant for disrespect) compared to the response speed for control words is used to indicate the intuition's accessibility. This procedure has been used successfully to measure the temporary accessibility of intuitions in audiences after exposure to intuition-laden stimuli (Tamborini, Prabhu et al., 2016). Recently, the procedure has been updated to eliminate some of its early methodological shortcomings (Hopp et al., 2019), and include measures of the egoistic intuitions (Prabhu et al., 2014; Tamborini, Lewis et al., 2016).

Measuring the Semantic Component of Intuition Salience: The MF-LDT

The MIME conceives intuitions (whether altruistic or egoistic) to comprise both affective and semantic components. Initial efforts to measure accessibility devised the MF-AMP to measure the intuition's affective component. The decision to focus initially on this component was prompted by the belief that affect might play a bigger role in determining an intuition's motivating capacity, and thus might be more central to the appraisal processes stipulated in the MIME. More recent efforts to complement the MF-AMP have developed the moral foundations lexical decision-making task (MF-LDT; Tamborini et al., 2019) to measure the semantic component of moral intuitions (for previous applications of a moral words lexical decision task, see Gantman & Van Bavel, 2016). The MF-LDT is designed to measure the accessibility of intuition-specific information in pre-cognitive processing. As is traditional with most lexical decision-making tasks, the MF-LDT flashes letter combinations quickly (<500 ms) on a screen and asks participants to choose if the display is a word (e.g., butter) or a non-word (e.g., grinter). Response speed and accuracy indicate the word's accessibility. The MF-LDT consists of 200 words comprising 50 moral words (5 adhering to and 5 contravening each of 5 moral intuitions), 50 neutral words, and 100 non-words. To test the MF-LDT's ability to measure intuition accessibility, separate groups of subjects engaged in activities designed to prime one of the five different moral intuitions (care, fairness, ingroup loyalty, respect for authority, and purity) or a control. Following this activity, they completed the MF-LDT. In initial tests, after separating stimulus words into those that adhere to or contravene an intuition, confirmatory factor analyses showed good fit for the model representing the five moral intuitions.

Although still in early stages of development, the MF-LDT holds considerable potential. If validated, it could help researchers identify the semantic nature of a person's mental response to media experience. Moreover, efforts to combine the MF-LDT with the MF-AMP could offer a more comprehensive assessment and understanding of audience response to media than is available through other means. An improved MF-LDT procedure using the eMFD (see above; Hopp et al., 2018) as a basis for stimulus words is currently being validated (Hopp et al., 2019).

Neurological Indicators of Intuition Salience

Using functional magnetic resonance imaging (fMRI), one study (Eden, Tamborini, Wang, & Sarinopoulos, 2012) attempted to investigate differences in the neural processing of altruistic intuitions, specifically regarding perceptions of media characters' moral behaviors. After presenting participants with a person's face and describing a moral/neutral behavior relating to one of the five altruistic intuitions, Eden et al. (2012) asked participants to rate the perceived morality of the person's behavior. Study results suggested two conclusions. First, compared to morally irrelevant acts, moral acts were correlated with activation in a distinct moral neural network including the left supplementary motor cortex, superior temporal sulcus, left and right dorsolateral prefrontal cortex (DLPFC), left operculum, left amygdala, right cerebellum, left thalamus, left insula, and right TPJ. Second, separate neural regions were associated with distinct moral scenarios as they relate to a focus on the community or the self. Moral scenarios focused on the self activated the right precuneus whereas those associated with community activated the superior medial DLPFC.

A similar study (Amir et al., 2017) exposed participants to moral foundations vignettes developed by Clifford, Iyengar, Cabeza, and Sinnott-Armstrong (2015). Using multivoxel-pattern analysis (MVPA) and representational similarity analysis (RSA), the researchers found that different types of moral intuitions elicit dissociable cortical activation patterns within areas that commonly define the "moral brain." For instance, while the dorsomedial prefrontal cortex (DMPFC) seems to be involved in all moral evaluation tasks, evaluating the wrongness of violations in binding foundations (loyalty, authority, purity) elicited greater activation in the temporoparietal junction (TPJ), an area implicated in processing others' state of mind. In contrast, care and fairness evaluations showed specific activation pattern in the precuneus and insula. Overall, findings of studies analyzing the neurological indicators of individuals' intuition salience are surprisingly consistent. This consistency provides additional evidence for the innate character of humans' moral intuition salience.

Measures of Intuition Salience for Children

Recent MIME research has developed child-friendly instruments to measure moral intuition salience (see Hahn et al., 2019). For example, the moral measure of intuition accessibility (M-MIA) was designed to present adolescents with sets of words that represent separate intuitions. Across six different response items, adolescents are asked to select which word they think is *better* (for words that uphold the intuitions) or *worse* (for words that violate the intuitions) as something to be. For instance, one item might ask respondents to choose which of the following they think it is better to be: kind (for the care intuition), honest (for fairness), loyal (for ingroup loyalty), or respectful (for authority). Another item might ask which it is worse to be: harmful (for care), unfair (for fairness), a traitor (for ingroup loyalty), or disobedient (for authority). The M-MIA has been used successfully in three separate studies examining child audiences (Hahn et al., 2019).

Other efforts have attempted to develop a measure of behaviors in children linked to moral intuitions. The result is a procedure termed the *moral measure of intuitively motivated behavior* (M-MIMB; Hahn et al., 2019). Using a protocol similar to the dictator game, which is a popular economic game (Engel, 2011), the M-MIMB evaluates children's behavioral responses when confronted by a moral dilemma (see Hahn et al., 2019, for procedural details). The M-MIMB is an especially valuable addition to research on children's media effects as, to date, researchers investigating behavioral effects in young audiences have relied largely on self-report measures of children's behavioral intentions in hypothetical situations (e.g., Mares & Braun, 2013; Martins & Wilson, 2012). Reliance on behavioral intention is of concern given recent literature documenting large disparities across research examining individuals confronted with hypothetical versus real-life moral dilemmas (e.g., Bostyn, Sevenhant, & Roets, 2018).

CONCLUSION

Tinbergen's four questions (1963) provide a comprehensive approach to the study of human behavior from a biological perspective. In this framework, a deeper and integrative understanding of human behavior must include both an ultimate analysis (i.e., an analysis of the behaviors' adaptive value and evolution) and a proximate analysis (i.e., an analysis of the behaviors' development and underlying mechanisms). As this chapter has demonstrated, the MIME provides a comprehensive model of the processes governing the selection, valuation, and production of media narratives *across* all four levels of analysis. To our knowledge, only a few models that have been developed and refined by communication scholars are characterized by this level of in-depth analysis. We encourage continued efforts to advance the MIME conceptually and methodologically while keeping all four levels of Tinbergen's analysis in mind. We hope the advances outlined in this chapter will jump-start fellow MIME scholars interested in new avenues of exploration and provide a reference for the progress MIME researchers will make within the next five to ten years.

REFERENCES

Amir, O., Huskey, R., Mangus, J. M., Swanson, R., Gordon, A., Khooshabeh, P., & Weber, R. (2017, March). Moral intuitions elicit dissociable cortical activation. Paper presented at the annual meeting of the Social Cognitive and Affective Neuroscience Society (SANS), Los Angeles, CA.

Bostyn, D. H., Sevenhant, S., & Roets, A. (2018). Of mice, men, and trolleys: Hypothetical judgment versus real-life behavior in trolley-style moral dilemmas. *Psychological Science, 29*, 1084–1093.

Bowman, N., Lewis, R. J., & Tamborini, R. (2014). The morality of May 2, 2011: A content analysis of US headlines regarding the death of Osama bin Laden. *Mass Communication and Society, 17*, 639–664.

Brosnan, S. F. (2013). Justice- and fairness-related behaviors in nonhuman primates. *Proceedings of the National Academy of Sciences, 110*, 10416–10423.

Clifford, S., Iyengar, V., Cabeza, R., & Sinnott-Armstrong, W. (2015). Moral foundations vignettes: A standardized stimulus database of scenarios based on moral foundations theory. *Behavior Research Methods, 47*, 1178–1198.

Deci, E. L., & Ryan, R. M. (1985). *Intrinsic motivation and self-determination in human behavior*. New York, NY: Plenum.

Eden, A., & Tamborini, R. (2017). Moral intuitions: Morality subcultures in disposition formation. *Journal of Media Psychology, 29*, 198–207.

Eden, A., Tamborini, R., Grizzard, M., Lewis, R. J., Weber, R., & Prabhu, S. (2014). Repeated exposure to narrative entertainment and the salience of moral intuitions. *Journal of Communication, 64*, 501–520.

Eden, A., Tamborini, R., Wang, L., & Sarinopoulos, I. (2012, May). Morality and media: Neural indicators of moral processing within news stories. Paper presented at the annual meeting of the International Communication Association, Phoenix, AZ.

Engel, C. (2011). Dictator games: A meta study. *Experimental Economics, 14*, 583–610.

Evans, J. S. B., & Stanovich, K. E. (2013). Dual-process theories of higher cognition: Advancing the debate. *Perspectives on Psychological Science, 8*, 223–241.

Gantman, A. P., & Van Bavel, J. J. (2016). Exposure to justice diminishes moral word detection. *Journal of Experimental Psychology: General, 145*, 1728–1739.

Graham, J., Haidt, J., & Nosek, B. A. (2009). Liberals and conservatives rely on different sets of moral foundations. *Journal of Personality and Social Psychology, 96*, 1029–1046.

Hahn, L., Tamborini, R., Klebig, B., Novotny, E., Grall, C., Hofer, M., & Lee, H. (2018). The representation of altruistic and egoistic motivations in popular music over 60 years. *Communication Studies*. Advance online publication.

Hahn, L., Tamborini, R., Novotny, E., Grall, C., & Klebig, B. (2018). Applying moral foundations theory to identify terrorist group motivations. *Political Psychology*. Advance online publication.

Hahn, L., Tamborini, R., Prabhu, S., Klebig, B., Grall, C., & Pei, D. (2017). The importance of altruistic versus egoistic motivations: A content analysis of conflicted motivations in children's television programming. *Communication Reports, 30*, 67–79.

Hahn, L., Tamborini, R., Weber, R., Bente, G., & Sherry, J. (2019, May). Can moral narratives increase moral intuition accessibility and behavior in pre-teen children? Testing the model of intuitive morality and exemplars in young audiences. Paper presented at the annual meeting of the International Communication Association, Washington, DC.

Haidt, J. (2001). The emotional dog and its rational tail: A social intuitionist approach to moral judgment. *Psychological Review, 108*, 814–834.

Haidt, J., & Bjorklund, F. (2008). Social intuitionists answer six questions about morality. In W. Sinnott-Armstrong (Ed.), *Moral psychology*: Vol. 2. *The cognitive science of morality* (pp. 181–217). Cambridge, MA: MIT Press.

Haidt, J., & Joseph, C. (2007). The moral mind: How 5 sets of innate intuitions guide the development of many culture-specific virtues, and perhaps even modules. In P. Carruthers, S. Laurence, & S. Stich (Eds.), *The innate mind* (Vol. 3, pp. 367–391). New York, NY: Oxford University Press.

Hofmann, W., & Baumert, A. (2010). Immediate affect as a basis for intuitive moral judgement: An adaptation of the affect misattribution procedure. *Cognition and Emotion, 24*, 522–535.

Hopp, F. R., Cornell, D., Fisher, J. T., Huskey, R., & Weber, R. (2018, November). The Moral Foundations Dictionary for News (MFD-N): A crowd-sourced moral foundations dictionary for the automated analysis of news corpora. Paper presented at the annual meeting of the National Communication Association, Salt Lake City, UT.

Hopp, F. R., Fisher, J. T., Prabhu, S., Cornell, D., Tamborini, R., & Weber, R. (2019, May). Revisiting the moral foundations-affect misattribution procedure (MF-AMP): An extended, open-source tool for measuring the accessibility of moral intuitions. Paper presented at the annual meeting of the International Communication Association, Washington, DC.

Iyer, R., Koleva, S., Graham, J., Ditto, P., & Haidt, J. (2012). Understanding libertarian morality: The psychological dispositions of self-identified libertarians. *PLoS One, 7*, e42366.

Janicke, S. H., & Raney, A. A. (2018). Modeling the antihero narrative enjoyment process. *Psychology of Popular Media Culture, 7*, 533–546.

Lazarus, R. S. (1991). *Emotion and adaptation*. New York, NY: Oxford University Press.

Lewis, R. J., Grizzard, M., Mangus, J. M., Rashidian, P., & Weber, R. (2017). Moral clarity in narratives elicits greater cooperation than moral ambiguity. *Media Psychology, 20*, 533–556.

Lewis, R. J., & Mitchell, N. (2014). Egoism versus altruism in television content for young audiences. *Mass Communication and Society, 17*, 597–613.

Lewis, R. J., Tamborini, R., & Weber, R. (2014). Testing a dual-process model of media enjoyment and appreciation. *Journal of Communication, 64*, 397–416.
Mares, M. L., & Braun, M. T. (2013). Effects of conflict in tween sitcoms on US students' moral reasoning about social exclusion. *Journal of Children and Media, 7*, 428–445.
Martins, N., & Wilson, B. J. (2012). Social aggression on television and its relationship to children's aggression in the classroom. *Human Communication Research, 38*, 48–71.
Mastro, D., Enriquez, M., Bowman, N. D., Prabhu, S., & Tamborini, R. (2013). Morality subcultures and media production: How Hollywood minds the morals of its audience. In R. Tamborini (Ed.), *Media and the moral mind* (pp. 75–92). London, England: Routledge.
Prabhu, S., & Tamborini, R. (2018, May). Corresponding morals featured in media content to moral intuitions in media users: A test of the MIME in two cultures. Paper presented at the annual meeting of the International Communication Association, Prague, the Czech Republic.
Prabhu, S., Tamborini, R., Idzik, P., Hahn, L., Grizzard, M., & Wang, L. (2014, May). The role of intuition accessibility on the appraisal and selection of media content. Paper presented at the annual meeting of the International Communication Association, Seattle, WA.
Prabhu, S., Tamborini, R., Klebig, B., Grall, C., & Hahn, L. (2015, November). Correlating the salience of intuitive motivations detected in the content TV serials and viewers of those serials: A test of the MIME. Paper presented at the annual meeting of the National Communication Association, Las Vegas, NV.
Schwartz, S. H. (1994). Are there universal aspects in the structure and contents of human values? *Journal of Social Issues, 50*, 19–45.
Tamborini, R. (2011). Moral intuition and media entertainment. *Journal of Media Psychology, 23*, 39–45.
Tamborini, R. (2013). Model of intuitive morality and exemplars. In R. Tamborini (Ed.), *Media and the moral mind* (pp. 43–74). London, England: Routledge.
Tamborini, R., Baldwin, J., Goble, H., Hofer, M., Aley, M., Grady, S., & Kryston, K. (2019). Measuring the semantic component of moral intuitions' accessibility: Developing a moral foundations-lexical decision-making task (MF-LDT). Work in progress.
Tamborini, R., Eden, A., Bowman, N. D., Grizzard, M., Weber, R., & Lewis, R. J. (2013). Predicting media appeal from instinctive moral values. *Mass Communication and Society, 16*, 325–346.
Tamborini, R., Hahn, L., Klebig, B., Walling, B., Kryston, K., & Aley, M. (2018, May). The representation of altruism and egoism in children's books and movies. Paper presented at the annual meeting of the International Communication Association, Prague, the Czech Republic.
Tamborini, R., Hahn, L., Novotny, E., Klebig, B., Hofer, M., Prabhu, S., ... Baldwin, J. (2018, November). Defining prosocial and antisocial media content in terms of intuitive motivations. Paper presented at the annual meeting of the National Communication Association, Salt Lake City, UT.
Tamborini, R., Hahn, L., Prabhu, S., Klebig, B., & Grall, C. (2017). The representation of altruistic and egoistic motivations in children's television programming. *Communication Research Reports, 34*, 58–67.
Tamborini, R., Hofer, M., Prabhu, S., Grall, C., Novotny, E., Hahn, L., & Klebig, B. (2017). The impact of terror attack news on moral intuitions and moral behavior towards outgroups. *Mass Communication and Society, 20*, 800–824.
Tamborini, R., Lewis, R. J., Prabhu, S., Grizzard, M., Hahn, L., & Wang, L. (2016). Media's influence on the accessibility of altruistic and egoistic motivations. *Communication Research Reports, 33*, 177–187.
Tamborini, R., Prabhu, S., Lewis, R. L., Grizzard, M., & Eden, A. (2016). The influence of media exposure on the accessibility of moral intuitions. *Journal of Media Psychology, 30*, 79–90.
Tamborini, R., Prabhu, S., Wang, L., & Grizzard, M. (2013, June). Setting the moral agenda: News exposure's influence on the salience of moral intuitions. Paper presented at the annual meeting of the International Communication Association, London, England.
Tinbergen, N. (1963). On the aims and methods of ethology. *Zeitschrift für Tierpsychologie, 20*, 410–433.

Weber, R., Mangus, J. M., Huskey, R., Amir, O., Swanson, R., Gordon, A., … Tamborini, R. (2018). Extracting moral foundations from text narratives: Relevance, challenges, and solutions. *Communication Methods and Measures, 12*, 119–139.

Zillmann, D. (2002). Exemplification theory of media influence. In J. Bryant & D. Zillmann (Eds.), *Media effects: Advances in theory and research* (2nd ed., pp. 19–41). Mahwah, NJ: Lawrence Erlbaum Associates.

Index

Page numbers in **bold** denote tables, those in *italics* denote figures.

THE PSYCHOLOGY OF READING

The Psychology of Reading reviews what has been learned about skilled reading and dyslexia using research on one of the most important but often overlooked languages and writing systems – Chinese. It provides an overview of the Chinese language and writing systems, discusses what is known about the cognitive and neural processes that support the skilled reading of Chinese, as well as its development and impairment, and describes the computer models that have been developed to understand these topics. It is written in an accessible way to appeal to anyone with an interest in cognitive psychology, language, or education.

ERIK D. REICHLE is a professor of Cognitive Psychology at Macquarie University, Australia. His research interests include computer models of reading, the cognitive and neural systems that support skilled reading, and how those systems are affected by different writing systems. He is the author of *Computational Models of Reading: A Handbook* (2021).

LILI YU is a lecturer at Macquarie University, Australia. Her research uses eye tracking and other behavioral methods to understand the cognitive processes that support the skilled reading of Chinese, and how those processes might differ across languages and writing systems.